DATE DUE

VIRAL AND OTHER INFECTIONS OF THE HUMAN RESPIRATORY TRACT

A festschrift for Dr David Tyrrell CBE DSc MD FRCP FRCPath FRS

VIRAL AND OTHER INFECTIONS OF THE HUMAN RESPIRATORY TRACT

Edited by

S. Myint

Department of Microbiology and Immunology, University of Leicester, UK

and

D. Taylor-Robinson

St Mary's Hospital Medical School, London, UK

CHAPMAN & HALL

London · Glasgow · Weinheim · New York · Tokyo · Melbourne · Madras

Published by Chapman & Hall, 2–6 Boundary Row, London SE1 8HN, UK

Chapman & Hall, 2–6 Boundary Row, London SE1 8HN, UK

Blackie Academic & Professional, Wester Cleddens Road, Bishopbriggs, Glasgow G64 2NZ, UK

Chapman & Hall GmbH, Pappelallee 3, 69469 Weinheim, Germany

Chapman & Hall USA, 115 Fifth Avenue, New York, NY 10003, USA

Chapman & Hall Japan, ITP-Japan, Kyowa Building, 3F, 2-2-1 Hirakawacho, Chiyoda-ku, Tokyo 102, Japan

Chapman & Hall Australia, 102 Dodds Street, South Melbourne, Victoria 3205, Australia

Chapman & Hall India, R. Seshadri, 32 Second Main Road, CIT East, Madras 600 035, India

First edition 1996

© 1996 Chapman & Hall

Typeset in Great Britain by Saxon Graphics Ltd, Derby
Printed in Great Britain by The Alden Press, Oxford

ISBN 0 412 60070 6

A catalogue record for this book is available from the British Library

Library of Congress Catalog Card Number: 95-78835

CONTENTS

LIST OF CONTRIBUTORS

ANDERS AKERLUND
Consultant, Department of
Otorhinolaryngology, Central Hospital, S-50/82
Boras, Sweden

ALI AL-JABRI
Department of Medical Microbiology, London
Hospital Medical School, London E1 2AD, UK

WIDAD AL-NAKIB
Director, Advanced Pathology Services, 101
Harley Street, London W1N 1DF, UK

MATS BENDE
Consultant, Department of
Otorhinolaryngology, Central Hospital, S-541
85 Skovde, Sweden

HARRY CAMPBELL
Consultant in Public Health, Fife Health
Board, Sprigfield House, Cupar, Fife
KY15 5UP, UK

SHELDON COHEN
Professor of Psychology, Department of
Psychology, Carnegie Mellon, University,
Pittsburgh, USA

NIGEL DIMMOCK
Professor of Virology, Department of
Biological Sciences, University of Warwick,
Coventry CV4 7AL, UK

VINCENT EMERY
Senior Lecturer in Virology, Department of
Virology, Royal Free Hospital Medical School,
London NW3 2PF, UK

PAT FURR
Senior Scientific Officer, MRC Sexually
Transmitted Diseases Research Group, St
Marys Hospital Medical School, London
W2 1PG, UK

PETER HIGGINS
'Thornhedge', Poulton, Cirencester, UK

STEPHEN HOLGATE
MRC Professor of Immunopharmacology,
Faculty of Medicine, University of
Southampton, Southampton General
Hospital, Southampton SO16 6YD, UK

DAVID ISAACS
Head, Department of Immunology and
Infectious Diseases, The Children's Hospital,
Camperdown, Sydney, NSW 2050, Australia

SEBASTIAN JOHNSTON
Lecturer, Faculty of Medicine, University of
Southampton, Southampton General
Hospital, Southampton SO16 6YD, UK

ALBERT Z. KAPIKIAN
Professor and Head, Epidemiology Section,
Laboratory of Infectious Diseases, National
Institutes of Health, Bethesda, Maryland
20892, USA

PAUL LAMBDEN
Scientist, Department of Molecular
Microbiology, University of Southampton,
Southampton SO16 6YD, UK

ELLIOT LARSON
Consultant in Infectious Diseases, 126 Union
Street, Southborough, Massachussetts 01752,
USA

ETHNE MACMAHON
Lecturer, Department of Virology, Royal Free
Hospital Medical School, London NW3 2PF,
UK

JOYCE MCQUILLIN
Department of Virology, University of
Newcastle, Medical School, The Royal
Victoria Infirmary, Queen Victoria Road,
Newcastle-upon-Tyne NE1 4LP, UK

DICK MADELEY
Professor and Head of Virology, Department
of Virology, University of Newcastle Medical
School, The Royal Victoria Infirmary,
Queen Victoria Road, Newcastle-upon-Tyne
NE1 4LP, UK

MALCOLM MOLYNEAUX
Senior Lecturer/Consultant, Liverpool School
of Tropical Medicine, Pembroke Place,
Liverpool L3 5QA, UK

STEVEN MYINT
Professor of Clinical Microbiology,
Department of Microbiology, University of
Leicester, Leicester LE1 9HN, UK

KARL NICHOLSON
Senior Lecturer in Infectious Diseases,
Infectious Diseases Unit, Leicester Royal
Infirmary, Leicester LE1 5WW, UK

WALTER NIMMO
Chief Executive, Inveresk Clinical Research
Limited, Riccarton, Edinburgh, UK

JOHN OXFORD
Professor of Virology, Department of Medical
Microbiology, London Hospital Medical
College, Turner Street, London E1 2AD, UK

MALIK PEIRIS
Consultant Virologist, Department of
Virology, University of Newcastle Medical
School, The Royal Victoria Infirmary,
Queen Victoria Road, Newcastle-upon-Tyne
NE1 4LP, UK

YURI PERVIKOV
Virologist, World Health Organization,
CH-1211, Geneva-27, Switzerland

CRAIG PRINGLE
Professor of Virology, Department of
Biological Sciences, University of Warwick,
Coventry CV4 7AL, UK

SYLVIA REED
57, Watford Road, St Albans, Herts AL1 2AE,
UK

JIM ROBERTSON
National Institute for Biological Standards &
Control, Blanche Lane, South Mimms, Potters
Bar, Herts EN6 3QG, UK

GEOFFREY SCHILD
Head, National Institute for Biological
Standards & Control, Blanche Lane, South
Mimms, Potters Bar, Herts EN6 3QG, UK

GEOFF SCOTT
Consultant Microbiologist, Department of
Microbiology, University College Hospital,
London WC1E 6DB, UK

STUART SIDDELL
Professor of Virology, Institut fur Virologie
und Immunologie, Universitat Wuerzburg,
Versbacher Str. 7, 97078 Wuerzburg, Germany

ANDREW SMITH
Professor of Psychology, University of Bristol,
Woodland Road, Bristol BS8 1TN, UK

JIM STOTT
National Institute for Biological Standards &
Control, Blanche Lane, South Mimms, Potters
Bar, Herts EN6 3QG, UK

SIR CHARLES STUART-HARRIS
Professor Emeritus, Department of Medical
Microbiology, University of Sheffield, Beech
Hill Road, Sheffield S10 2RX, UK

GERALDINE TAYLOR
Principal Research Scientist, Institute for
Animal Health, Compton, Berks
RG16 0NN, UK

DAVID TAYLOR-ROBINSON
Professor and Head, MRC Sexually
Transmitted Diseases Research Group, The
Jefferiss Wing, St Marys Hospital, London
W2 1NY, UK

BRENDA THOMAS
Senior Scientific Officer, MRC Sexually
Transmitted Diseases Research Group, The
Jefferiss Wing, St Marys Hospital, London W2
1NY, UK

PETER WATT
Professor and Head, Department of Molecular
Microbiology, University of Southampton,
Southampton General Hospital, Southampton
SO16 6YD, UK

NORMA WATSON
Inveresk Clinical Research Limited, Riccarton,
Edinburgh, UK

DAVID WEBSTER
Senior Lecturer in Immunology and Head of
Immunodeficiency Unit, Royal Free Hospital,
Rowland Hill Street, London NW3 2PF,
England

JOHN WOOD
Scientist, National Institute for Biological
Standards & Control, Blanche Lane, South
Mimms, Potters Bar, Herts EN6 3QG, UK

Dr D. A. J. Tyrrell CBE DSc MD FRCP FRCPath FRS.

PREFACE

This book is a festschrift to mark the career of Dr David Tyrrell, eminent virologist and physician. Almost all of the contributors have been colleagues or students at some time during his career. Unlike most festschrifts, however, the essays have been integrated to produce a comprehensive book that covers the entire field of non-bacterial infections of the human respiratory tract. This is a measure of the breadth of interest of the research undertaken by David, spanning a period of over 40 years.

We hope that this book will be of interest to microbiologists, virologists and in particular, to physicians. All the major virus groups are covered, as well as chlamydial and mycoplasmal infections. Each of these chapters (or in the case of respiratory syncytial virus and influenza virus, pairs of chapters) comprises a review of the virology/microbiol-

ogy, immunology and clinical aspects of each group. There are also more general chapters overviewing the clinical manifestations, treatment and pathophysiology of respiratory virus infections. In addition, chapters in which psychological aspects and the ethical use of human volunteers are discussed will be of interest to all those involved with respiratory infectious agents.

We hope also that this book serves as a fitting tribute to a man who is a true polymath: a physician, researcher, teacher, mentor, linguist and ambassador for science, among other things. To those of us who have contributed to this book, he is even more: a friend.

Steven Myint
David Taylor-Robinson
London
August 1995

FOREWORD

When David Tyrrell was a young medical student in the Medical School in Sheffield in July 1946, he had already begun to be interested in the science of microbiology. Later that interest grew when he was a house-physician and became concerned with patients with pneumonia when admitted to hospital during a short influenza epidemic in 1949. For David, the experience in this epidemic and his subsequent work in the recovery of influenza A virus from garglings and sputa were seminal in their impact. From now on his training in the laboratory was directed against viruses as primary targets for attack.

There followed a couple of years at the Royal Hospital in Sheffield until one day an unexpected letter came from Frank Horsfall in New York enquiring whether we had a promising young postgraduate in Virology who would like to spend a year at the Rockefeller Institute for Medical Research. This proved an attractive prospect to David who immediately agreed to forward his name to New York as a possible candidate for the post. However, instead of one year the Americans subsequently made him serve two more years at the Institute before returning to Sheffield. Those years in New York City gave David a marvellous grounding in the basic virology of viruses, and virus neutralization in particular. He met men whose interests and work have become legendary in the techniques of handling viruses and have established their names for ever in the science of microbiology.

Back in Sheffield, in an MRC post held in a new laboratory established at the Lodge Moor Hospital, there were colleagues with whom to get acquainted and the shaky unit must have seemed a huge contrast to the experience in New York. There was work to be done in the isolation of new respiratory viruses including poliovirus in the laboratory. There then came an opportunity at Lodge Moor to study a community outbreak of an illness due to ECHO 9 virus, which caused fever in children with sore throat, a rash and sometimes aseptic meningitis. This provided an opportunity for epidemiological study as well as that of virology.

Soon, however, a call came to help out an old friend, Sir Christopher Andrewes, and this was to join the Common Cold Unit in Salisbury. This wing of the National Institute for Medical Research at Mill Hill had been running quite well with several virologists working closely with Andrewes at the helm [1]. Their work had become somewhat sterile for, after a promising start with a successful run of serial cultures of a common cold, Helio Pereira had found the ninth culture had shown a lack of growth. This was with cultures of human cells and this one promising result suddenly had become negative.

To David, this positive result was a challenge as well as an indication and he started in 1958 using cultures of human embryo kidney cultures. First he decided to culture cells at 33°C, knowing that lowering the temperature of incubation might increase the cells' sensitivity to viruses. Next, the protein content (p.c.) of the culture was reduced using a bicarbonate buffer at 0.09 p.c. instead of the usual strength of 0.16 p.c. Finally, bovine plasma was added to the medium and these three methods were each useful in producing sensitivity to the growth of the common cold virus, re-christened 'Rhinovirus'. Persistence with these cultural techniques was indeed instrumental in producing positive success in culturing the agents of the common cold [2]. Never mind that a complex antigenic structure was found indicating the existence of a very large number of serotypes of rhinovirus (100 at least) [3]. The apparent lack of immunity to the virus was thus explained and some strains were found with apparent sensitivity to antiviral action from convalescent patients.

In the meantime, the success of work with a virus now known as coronavirus was established that was distinct from rhinovirus [4]. This agent appears to be responsible for up to about 20% of all colds and its antigenic make-up has verified that it belongs to the group typified as 'avian bronchitis virus'. Human coronavirus has become recognized as another agent responsible, like rhinoviruses, for common colds.

Much of David's time throughout his career has been spent on antiviral substances, though negative results have been usual. Recently antivirals have been found with a switch to affinity for a pocket in the virus capsid. This line of enquiry was at first negative in therapeutic trials. However, Janssen in Belgium found that a compound R61837 given before exposure to virus or during the incubation period showed definite action against rhinovirus type 9. Unfortunately it requires to be given frequently by nasal spray so at present the need is to continue research on the compound. It can only be hoped that drug resistance does not appear to put an end to this promising development.

Professor Sir Charles Stuart-Harris

REFERENCES

1. Andrewes, C.H. (1962) Lecture III: The Common Cold, Epidemiology and Possibilities of Control. *J. R. Inst. Public Health Hyg.*, **25**, 79.
2. Tyrrell, D.A.J. (1965) *Common Colds and Related Diseases*. Edward Arnold Ltd., London.
3. Tyrrell, D.A.J. (1993) The Common Cold Unit 1946–1990; Farewell to a much loved British Institution. *Public Health Laboratory Service Microbiology Digest*, **6**, 74-76.
4. Tyrrell, D.A.J. and Bynoe, M.L. (1965) Cultivation of a novel type of common cold virus in organ cultures. *Br. Med. J.*, **1**, 1467–70.

EPIDEMIOLOGY OF VIRAL RESPIRATORY TRACT INFECTIONS

Sebastian Johnston and Stephen Holgate

Viral and Other Infections of the Human Respiratory Tract.
Edited by S. Myint and D. Taylor-Robinson. Published in
1996 by Chapman & Hall. ISBN 0 412 60070 6

1.1 INTRODUCTION

Episodes of respiratory viral infection were recognized long before their causal agents were discovered, and viral epidemiology was one of the first branches of virology to be developed. Studies of respiratory viral epidemiology have been limited by the capabilities of the diagnostic methods available at the time. These have improved steadily over the years and are continuing to do so, particularly with the recent development of molecular methods of viral detection. Early studies tended to concentrate on those viruses that could be most easily detected using traditional methods, such as influenza viruses, adenoviruses and respiratory syncytial (RS) virus. As time passed rhinoviruses and coronaviruses were discovered, and as methods for their detection have improved, so has come an appreciation of their importance in respiratory viral illness. These two viruses are now thought to account for between 50% and 75% of upper respiratory tract infections, and as such, due emphasis will be given to these virus types in this chapter.

As our knowledge of the pathogenesis and molecular basis of respiratory virology has increased, a better understanding of the epidemiology of respiratory viral infection has resulted. In this chapter we will consider some general principles of respiratory virology, the biology of the common respiratory viral types and its influence on respiratory viral epidemiology, and finally the epidemiology of the two most common respiratory viral syndromes: the common cold and acute exacerbations of wheezing illness.

The recent advance in molecular biology has led to the development of polymerase chain reaction (PCR)-based assays for the detection of most respiratory viral types, and use of these methods in carefully designed studies promises to increase even further our understanding of respiratory viral epidemiology. At the end of the chapter we discuss the results of some very recent studies using these new methodologies.

The epidemiology of bacterial infections of the respiratory tract, infections of the sinuses and ear, and respiratory infections associated with immunodeficiency are not discussed in this chapter.

1.2 EPIDEMIOLOGICAL DEFINITIONS

1.2.1 INCIDENCE AND PREVALENCE

Epidemiology quantifies disease occurrence within populations, and to achieve this, criteria need to be established to define rates of disease occurrence. The most commonly used measures are incidence and prevalence. Rates are fractions in which the numerator is the number of cases of disease and the denominator is a measure of the population.

Incidence (also called attack rate) is mainly used for acute diseases of short duration, or to define the number of new cases of a more chronic disease occurring in a certain period of time. A population and a time frame are defined, and the number of new cases occurring in that population during that interval of time is counted. The denominator includes both size of population and time frame. The incidence rate is then expressed as 'cases per million person-years' or a similar term. However, the time element is often omitted in expressing incidence, leaving the reader to determine the time frame from the context.

Prevalence is most frequently used for chronic diseases, particularly where onset is insidious and not readily dated. When point prevalence is calculated, a particular date is selected and the population recorded for that day constitutes the denominator; all new and previously existing cases identified on that date constitute the numerator. Prevalence is expressed simply as a rate with no time parameter. Frequently prevalence is expanded into a finite time, such as a month or year, in which case the rate is denoted as period prevalence.

1.2.2 STUDY DESIGNS

Prospective studies

Two groups are selected from a population under study, one with (index) and one without (control) a specified attribute. Both groups are followed prospectively for the incidence of the disease under study, and incidence rates for both groups are calculated. An important assumption governing the validity of such studies is that the two groups are at equal risk of the disease under study. For a variety of reasons, this assumption is often not fulfilled; differences between the two groups may then be minimized by dividing each according to parameters such as age, race, socioeconomic status, and then comparing rates for each subgroup.

Longitudinal studies

It is not always appropriate to a study design, or practically possible to select accurately matched index and control groups. Thus for some studies, a great deal of information about a particular disease can be obtained by studying a single group longitudinally. A great many studies of respiratory viral epidemiology have used such a study design, which has the important advantage that the outcome being studied is defined before the beginning of the study. In particular, many of the studies examining the epidemiology of the common cold, and attempting to demonstrate an association between respiratory viral infections and exacerbations of asthma, have used such a study design.

Retrospective studies

Prospective studies are expensive and time consuming because they require the enrolment of large numbers of subjects who must be followed for a period of months or years: the less frequent the expectation of outcome under study, the larger the population and the longer the follow-up that will be needed. An alternative approach is to carry out retrospective case-control studies. These necessarily have a great potential for the introduction of bias in the selection of the groups to be studied, as the outcome is already known. However, they are frequently used for initial studies to attempt to identify associations, with prospective studies then being justified to confirm or refute the findings of retrospective studies. It is of great importance that the two samples (cases and controls) are representative of the populations from which they are drawn, as the main flaw in retrospective studies is the choice of cases and particularly the selection of representative controls.

1.3 FACTORS INFLUENCING RESPIRATORY VIRAL EPIDEMIOLOGY

1.3.1 VIRAL DIAGNOSIS

Diagnosis of respiratory viral infections has classically been based on detection of the virus in nasal secretions, or on detection of an immune response in the blood (see Chapter 4). Nasal secretions (either washings or an aspirate) are the best specimens for viral detection, with maximal isolation rates being achieved within 1 or 2 days of the onset of symptoms. Specimens are inoculated onto a variety of cultured cell lines and viral presence detected by typical cytopathic effects, or by other testing such as haemagglutination. In addition nasal

epithelial cells can be cytospun onto microscope slides, and virus detected by fluorescence with specific antibodies. Detection of a humoral immune response initially used tests such as complement fixation, but in general these have now been superseded by more sensitive and specific tests such as enzyme-linked immunosorbent assays (ELISAs).

These classical methods are relatively successful for the diagnosis of respiratory viral infections such as influenza, parainfluenza, RS virus and adenovirus, though even with these, great care in the handling and processing of specimens is required. However, rhinoviruses and coronaviruses (which together account for between 50–75% of acute respiratory viral infections), do not grow on standard cell lines, and antibodies are not detected easily. These viruses are therefore frequently not sought in epidemiological studies, and even if they are, it is generally acknowledged that the methods used are suboptimal. Even with the most exhaustive testing including the use of organ cultures and passage in human volunteers, under-diagnosis is likely, with infectious agents being demonstrable in only 60% of common colds [146].

Recent advances in molecular biology have permitted the sequencing of entire genomes of representative serotypes of almost all of the common respiratory viruses. This knowledge has permitted the development of PCR assays to provide sensitive and specific diagnosis. Such assays are under development for all the respiratory viruses, and in some cases have been used in epidemiological studies. Particularly for rhinoviruses (where PCR has been shown to be between three and five times more sensitive than culture) and coronaviruses, this has allowed much greater insight into the true role played by these viruses in acute respiratory illnesses [125,126,179]. The results of previous epidemiological studies should therefore be interpreted in the light of this knowledge, and it should be particularly borne in mind that rhinoviruses and coronaviruses are likely to be underrepresented in figures quoted.

1.3.2 IMMUNITY

Most acute viral infections confer life-long immunity, mediated by a combination of local and systemic antibody responses, and cell-mediated immunity. Re-exposure at any interval after initial infection then results in (a) a re-infection with minimal virus replication, or (b) an anamnestic immune response. Such re-infections are frequently covert, and result in minimal shedding of infectious virus. For certain viruses, such as poliovirus or rhinovirus, immunity is type-specific and confers little protection against exposure to a different serotype. These facts have implications for respiratory viral epidemiology, as a population may therefore be divided into susceptible and immune subjects. Immune subjects are largely exempt from disease and are inefficient links in a transmission chain, whereas susceptible subjects can both spread the agent and experience disease. There are, however, well-proven exceptions to this pattern. Thus, most children will be infected with RS virus within the first year of life, but re-infection later in childhood and indeed in adulthood is more the rule than the exception. Similarly, re-infections occur with influenza virus, as a result of antigenic drift, and antigenic shift (see below), and it is likely that they also occur with the human strains of coronavirus, and several of the other respiratory viruses as well.

1.3.3 SEASONALITY

Many acute viral infections exhibit striking seasonal patterns in incidence which are very consistent. These patterns reflect seasonal differences in transmission of infection. Respiratory infections spread more readily in the winter, although they may peak at different times, and these times may vary from one year to another.

The underlying biological explanation for seasonal differences has remained elusive. However, it is clear that these variations correlate with changes in climate, including changes in humidity, barometric pressure, winds, precipitation and drought. These factors can affect infectious disease agents directly. Many viruses in the free state are vulnerable to heat, radiation and drying.

Overall, respiratory infections are more frequent in the colder months but within this period there are considerable variations in the relative prevalences of the individual viruses. Rhinoviruses, for example, peak in early and late winter, while influenza viruses are most active in midwinter, and at least in the UK, RS virus has a well-defined peak in mid–late winter. Both older and more recent studies have highlighted the strength of the relationship between viral respiratory infections and school term times, suggesting that for rhinoviruses and coronaviruses, the most important factor controlling the seasonal variability in incidence is the increased congregation of preschool and school age children indoors facilitating airborne transmission [27,125,180] (Figure 1.1). For other viruses such as influenza or RS virus, with more clearly demarcated mini-epidemics, it is likely that factors such as fluctuations in temperature and humidity are also important.

1.3.4 SOCIOECONOMIC FACTORS

The influence of socioeconomic factors depends to varying degrees on the density and distribution of populations, the level of social, political, cultural and scientific development and, most importantly, the inter-relations of people. Socioeconomic factors typically affect health by indirect means, and because they are often closely inter-related, the impact of individual factors is very difficult to assess. The relation of population density to the occurrence of infectious disease is substantial. Increasing density favours the spread of

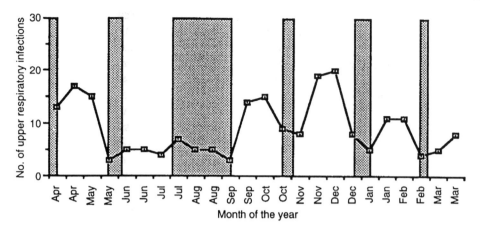

Figure 1.1 The seasonal variation in upper respiratory tract viral infections and its relationship to school attendance. Half-monthly incidence of viral infections in a cohort of 108, 9 to 11-year-old school children in Southampton, UK for the year April 1989–1990. School holiday periods are marked in shaded areas. (Adapted from [125].)

infectious agents, the occurrence of related disease and the development of immunity. In large and dense populations, agents typically infect early in childhood and persist because sufficient new susceptible subjects are added continuously by birth. In smaller populations the agents are unable to persist and are reintroduced at unpredictable intervals so that childhood diseases may be long delayed. Populations of urban and rural areas differ not only in relative density but also in other important ways, such as social structure, nutrition, air pollution, smoking, housing and family structure. The influence of many of these factors on respiratory viral infections has not been studied in detail, but several studies have addressed the important influence of family structure (see below).

The basic population unit is the household. Family members are genetically similar, share a common diet and economic status, are subject to the same cultural, religious and educational influences, and are exposed to common physical and biological environments. Most important for contact-transmitted disease, intra-familial contacts are prolonged and increase in intimacy with household crowding.

Family size, regardless of degree of crowding, is particularly important for acute respiratory infections, since it determines the number of potential introducers who bring home infections acquired elsewhere.

1.3.5 FAMILY STRUCTURE

Much of our current knowledge on the epidemiology of respiratory viral infections comes from studies involving the continuing observation of family units for episodes of infection. These studies, the Virus Watch programs in New York and Seattle, monitored family members for specific infections revealed by virus isolation and/or antibody response, whether related to illness or not [31,32,51,53,56–63,91,98,137,142,211]. The family studies began with one member's infection, acquired from outside the house. This member then exposed his or her fellow family members. The introductory infection and any infections in those exposed constituted a family episode that was described in terms of the time of onset of the related infections and the identities (age, sex, position in the family) of the introducer and both the infected and uninfected contacts. School-age and pre-school age children were identified as the most frequent introducers (hence important in community spread). This is consistent with the observed relationship between the crowding of children at school and the spread of respiratory viral infections, as these same children acquiring infections at school will then act as the introducers to the family unit.

Analysis also yielded estimates of cross-infection risks within the family, expressed in terms of secondary attack rates among specified members (for example, younger children) exposed to specified introducers (for example, a school child or a parent). In general, risk of contacts was related inversely by age overall, reflecting the influence of immunity, and of intimacy of within-family contact (ready exchange between spouses and between children nearest in age). Finally, the time relation between onsets of illness in the introducer and those exposed was used to define the range of incubation periods.

Studies of the Virus Watch type made it possible to identify and analyse family episodes caused by specific viruses (influenza A) or groups of viruses (adeno- or rhinoviruses) including both subclinical and overt infections. Analysis of the episodes also yielded information concerning the mode and duration of viral shedding; the spectrum of clinical response to infection, including the proportion that was subclinical; and the significance of prior immunity in the face of close exposure, as measured by the frequency and clinical consequences of re-infections that result.

1.4 SPECIFIC VIRUSES

There are in the region of 300 viruses or atypical bacteria that are known to be associated with respiratory tract infections. The more common of these will be considered individually, and aspects of their biology that relate to their epidemiology will be discussed. Details on those that have not been mentioned specifically in this chapter can be found within their own relevant chapters elsewhere in this text. The most common agents causing respiratory tract disease are listed in Table 1.1.

1.4.1 RHINOVIRUSES

Virus types and characteristics

Over 115 different serotypes of human rhinoviruses have been identified on the basis of serum neutralization studies [34,100,133,158]. It is likely, however, that new serotypes are constantly emerging under pressure from immune surveillance. Although most of the human rhinoviruses are antigenically distinct,

Table 1.1 The common infectious agents causing acute respiratory tract disease

Common viruses that cause acute respiratory illness	Rhinoviruses Coronaviruses 229E and OC43 Parainfluenza viruses types 1–4 Influenza viruses A, B and C Respiratory syncytial virus types A and B
Common infectious agents that may cause acute respiratory illness	Adenoviruses Enteroviruses Cytomegalovirus Herpes simplex virus *Mycoplasma pneumoniae* *Chlamydia pneumoniae* *Chlamydia psittaci*

cross-relationships between them have been detected [33]. Recent studies have revealed that 90% of rhinoviruses (major group) use the intercellular adhesion molecule-1 (ICAM-1) as their cellular receptor, while all but one of the remaining rhinoviruses (minor group) use a separate, and as yet only partially characterized receptor [223]. The minor group receptor has now been identified as the low density lipoprotein (LDL) receptor. All the rhinoviruses so far identified are capable of causing the common cold, and appear to cause a similar spectrum and severity of clinical illness.

Mode of transmission

Rhinovirus infections are spread from person to person by means of virus-contaminated respiratory secretions. Rhinoviruses are present in particularly high concentrations in nasal secretions rather than in pharyngeal secretions or saliva [86,109]. In studies of natural infections, higher rates of transmission were observed with longer exposures [41,57,86], and as discussed in previous sections, close contact (e.g. among family members) and crowding (e.g. school children) increase the transmission of rhinoviruses [42,115]. Early workers failed to demonstrate significant airborne transmission [37,89]. A more recent report, however, confirmed that rhinoviruses are transmitted by aerosol: in an experimental setting in which finger and fomite transmissions were prevented mechanically, transmission continued at the same rate [47], while the use of virucidal tissues in a family setting had no significant effect on rates of spread of colds [52]. These data, and the wealth of epidemiological data certainly support airborne transmission as the more important method. However, it is likely that both airborne and direct contact routes (contaminating hands, fingers and fomites with direct inoculation of virus through the conjunctiva and/or nasal epithelium) play a role in the transmission of rhinoviruses, as volunteer experiments show that saliva, lips and the external nares of infected volunteers contain small concentrations of virus [41], sufficient to initiate an infection in an individual provided that the virus successfully reaches the appropriate portal of entry [86,221].

Clinical course of rhinoviral infection

Clinical infection in the vast majority of cases is characterized by the typical symptoms of a cold which are indistinguishable from those caused by other viruses. The common cold is characterized by rhinorrhoea, nasal obstruction, pharyngitis and cough. Fever and systemic illness are infrequent though they may occur. The length of illness is normally around 7 days, with a peak of symptoms and signs between the second and third days. The acute phase of rhinitis usually corresponds with peak virus excretion [50]. Symptoms usually subside by the eighth day, although in some patients symptoms may persist for up to 1 month [39,86]. Lower respiratory tract symptoms are frequently present during common colds, especially cough which was present in 86% [54], 71% [175] and 40% [198] of normal subjects with naturally occurring infection.

There is also increasing evidence that rhinoviruses may be implicated in more severe lower respiratory tract illness, such as bronchiolitis, pneumonia and exacerbations of asthma, and indeed that they may directly infect the lower respiratory tract. A large number of epidemiological studies have strongly implicated rhinoviruses in exacerbations of asthma and as a cause of 'wheezy bronchitis' in children [126,191], and experimental rhinoviral infection studies have demonstrated the development of bronchial hyperreactivity during rhinoviral infection [8], as well as the development of late asthmatic responses in a group of atopic subjects [150].

Horn *et al.* have provided evidence for lower airway rhinoviral infection in that sputum cultures from 22 children with wheezy bronchitis were more often positive than either nose or throat swabs, suggesting that viral replication had occurred in the lower airways [118]. Cytopathic effects (CPE) developed more quickly in cultures from expectorated sputum as compared with upper airway samples implying that more virus was present in the lower airways.

Post-mortem studies have demonstrated that rhinoviruses are able to infect lung tissue in patients with compromised immunity. This has been seen in a patient with myelomatosis in whom RV13 was recovered from three specimens of lung tissue [39], and rhinovirus 47 was cultured post mortem from an 11-month-old infant with a history suggestive of bronchial asthma who had died suddenly at night [147]. Similar severe lower respiratory tract illnesses have been ascribed to rhinoviral infection in recent case reports [202].

Further evidence for a role for rhinoviruses in severe lower respiratory tract illnesses comes from three recent studies in New Zealand, Europe and the USA. In the first rhinoviruses were isolated from 20 children with bronchiolitis, and from 12 children with pneumonia over a 3-year period [155]. It was concluded that rhinoviral infections are less frequent than RS virus infections in infants (0.7% versus 8.2% of hospital admissions), but that the severity of illness and clinical presentations were similar. In the second study, rhinovirus was also identified as a cause of severe lower respiratory tract illness, including cough, wheeze, fever, respiratory distress and feeding difficulties; 70% of cases also had new chest X-ray abnormalities [136]. In the third study, rhinoviruses were detected in 12% of children requiring hospitalization for acute respiratory diseases including bronchiolitis and pneumonia; in comparison, RS virus was detected in 23%, but there were no differences between the clinical diseases caused by the two virus types [143]. It should be noted that these studies may well have underestimated the importance of the contributions of rhinoviruses as they employed cell culture only; the use of PCR would probably have increased rhinovirus detection considerably.

Prevalence of rhinovirus infections, age and sex distributions

The large number of different serotypes of rhinovirus together with the fact that infection with one serotype does not confer immunity to another, dictates that rhinovirus infections are very common. It has been estimated from studies using standard methods of detection in the USA that on average the incidence of rhinovirus infection is about 0.5 per person per year [86]. The mean rate for Seattle families was 0.68 infections per person-year [57], and rates in Virginia were higher [87]. Adjustment of the Seattle rates to reflect the Virginia rates indicates that infants and small children average more than one rhinovirus infection per year. Rates for young adult females in Virginia and in a Tecumseh, Michigan survey were higher than those for males of similar age [87,171]. A greater exposure to young children is thought to account for this difference.

All these studies relied upon viral culture to detect rhinovirus infection, and are therefore likely to have considerably underestimated the true prevalence of infection. A recent study using PCR in 9 to 11-year-old children in the UK found infection rates of 1.5 per child per year [126]. True rates of infection are probably even higher, since even with the use of PCR, the detection of rhinovirus infection is likely to be less than 100% as a result of incomplete reporting of symptoms, so a true estimate of the rate of infection is very difficult to obtain. The same UK study has revealed a great deal of unreported symptomatology, as well as an appreciable asymptomatic infection point prevalence of 12% [126]. Indeed, illness induced by rhinoviruses probably represents the most common acute infectious illness of humans.

Infection is most common in early life, and incidence declines with increase in age, probably because of a combination of environmental factors such as reduced inter-personal contact, and the higher frequency of immunity among adults who have experienced infection by a larger number of rhinoviruses. Indeed, studies on the distribution of neutralizing antibodies among different age groups in the USA indicate that sera from children aged 2 to 4 years neutralized about 10% of the rhinoviruses investigated (a total of 56 rhinoviruses; rhinovirus 1A-55) whereas sera from adults 30 to 40 years of age neutralized just over half of the rhinoviruses investigated [48,86]. It is likely that measures of antibody prevalence by more sensitive assays such as ELISA will reveal a much higher proportion of individuals with antibody to rhinoviruses [7].

Rhinovirus infection in families

The family unit is a major site for the spread of rhinoviruses. Young children and primarily those less than 2 years of age were identified as the major introducers in Seattle [57,58]. In Virginia, the school-aged child was a more common introducer and adults were also prominent. Family size is directly related to number of family episodes of rhinovirus infections [57]. Secondary infections appearing at 2 to 5-day intervals, are most common among young children and mothers but occur in all members of the household [57,58]. Secondary attack rates in the family have varied between 30% and 70%, but most studies have revealed rates of about 50%. When antibody-free (susceptible) members only are considered, secondary rates of 70% are common. Secondary infection rates in families fall with increasing age and are higher among mothers than fathers, presumably because of their close exposure to children in the family.

Characteristics of the introducer influence the secondary attack rates in families. Among Seattle families, the secondary attack rate was 71% if the introducer was ill but only 27% if he/she was well, a finding similar to that noted for spread between spouses [57].

Rhinovirus infection in schools

Efficient spread of rhinoviruses among children in a nursery school was demonstrated by Beem [13]. Rhinoviruses clearly spread among university students, medical students, boarding school populations, and probably other groups in dormitory-type situations where there is little or no contact with children [88,103,104,164]. The finding that illness peaks were simultaneous amongst adults with and without children, has been interpreted as suggesting that something other than the opening of school in September accounted for the early winter peak in prevalence [108]. However, attempts to identify a correlative meteorological event were not successful. It is in fact quite possible that the primary determinant of a community's prevalence is indeed the attendance of children at school, and that the peaks observed in adults without children are a result of secondary spread through the community as a whole, while the peaks observed in adults with children are more likely to be a result of direct spread from children within the family.

A recent study has identified very strong correlations between rhinoviral infections and school attendance, and suggests that although other factors may play a part, attendance of children at school is a very powerful determinant of the community load of infection in both adults and children [125] (Figures 1.1 and 1.2). The prime influence presumably is infection in children crowded together at schools, with secondary attacks within and then without family units accounting for the passage of infections throughout the community.

Geographical distribution and seasonal pattern of rhinovirus infection

Rhinovirus infection is distributed worldwide [20], having been detected in the USA in many

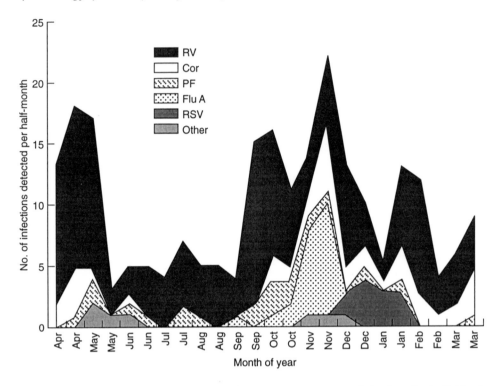

Figure 1.2   The seasonal variation in numbers of upper respiratory tract infections for individual organisms per half-month detected in the Southampton cohort of children in the year April 1989 to March 1990. Individual organisms are identified as shown: RV, rhinovirus; Cor, coronaviruses 229E and OC43; PF, parainfluenza viruses types 1–3; Flu A, influenza type A; RSV, respiratory syncytial virus. (Adapted from [125].)

studies [57,137]; in the UK [126]; in eastern Europe [101]; in a variety of South and Central American countries, including Panama with an epidemiological pattern similar to that reported for the USA [166,169], Amazon Indians [216] and Brazil [178], where almost continuous infection has been documented in very crowded impoverished urban surroundings [6]; and in isolated communities such as Antarctica [227], the bushmen of the Kalahari, Canadian Eskimos and natives of Tristan da Cunha [20,218].

Data from the USA suggest two main seasonal peaks – one major peak during late summer and early autumn (August to October), and a second peak in spring (April or May) [104,133]. In the tropics, the peak of rhinovirus

infection occurs during the rainy season [169]. Schools, nurseries and family groups are the major sites for the spread of rhinoviruses in any community, and in a recent study in the UK, school attendance appeared to be the main determinant of seasonal distribution [126].

Epidemiology of serotypes

Initial observations on serotyping of rhinovirus isolates indicated that a multiplicity of serotypes existed and that surveillance of a population for at least several months would result in isolation of several distinct types. Officially numbered serotypes now extend to 100 [100], while over 115 serologically distinct serotypes have been described. Prevalent serotypes vary from year to year, but at any one time, a small number of the serotypes tend to cause most of the illnesses [55,58,165,168]. The distribution of serotypes recovered over 7 years in Virginia changed, with an increase in higher-numbered serotypes and in non-typeable strains with time [22]. Analysis of serotypes occurring at New York, Seattle, and Michigan during this same period produced similar findings [58]. These findings suggested an emergence of new strains with time, probably as a result of selection pressure from immune surveillance.

Antigenic characterization of recent isolates, however, indicates that rhinoviruses identified earlier are still prevalent. Antisera to types 1–89 identified 15 of 16 (94%) isolates in Boston in 1982–1983, 752 of 790 (95%) isolates in Seattle in 1975–1979, and 194 of 209 (93%) isolates in Tecumseh, Michigan in 1976–1981 [100].

1.4.2 CORONAVIRUSES

Virus types

There are two major serotypes of human coronaviruses, designated 229E and OC43. Epidemiological studies of these viruses have been limited as they are very difficult to identify. Antigen for enzyme immunoassays has very limited availability, and virus culture is also difficult, as 229E will only grow in a specially adapted cell line called Clone 16, while OC43 can only be propagated reliably in suckling mouse brain.

It is probable that there are other similar but distinct coronaviruses, as the limited epidemiological studies that have been carried out have detected some viruses with atypical features; for example OC43-like viruses that have grown in cell culture, or antibodies to coronaviruses that react to both serotypes in enzyme immunoassays.

The recent development of PCR assays for the detection of both virus types should improve detection rates in future epidemiological studies [176].

Mode of transmission

There are few data available specifically on the transmission of coronaviruses, but such epidemiological data as do exist suggest that they follow the same general pattern as other respiratory viruses, in that droplet spread is the major route of transmission, and crowding of pre-school and school children the most important influence on the incidence of infection.

All the well-characterized human coronaviruses have been transmitted in human volunteers by nasal inoculation [15,16,197]. All have caused clinical illness manifest chiefly by nasal symptoms. The incubation period is around 3 days, and the duration of illness about 7 days.

Clinical features

Human coronaviruses are thought to cause around 15% of acute upper respiratory tract infections, and are also implicated in lower respiratory disease such as exacerbations of asthma [21,85,125,126,207]. However, a recent Southampton, UK survey found that of upper respiratory viral infections, those with coronaviruses were less likely to result in lower airway symptoms than upper respiratory infections with the other respiratory viruses as a group [126].

The main symptoms are rhinorrhoea, sore throat, cough, malaise and fever. Pneumonia and a pleural reaction were observed in 33% of military recruits with coronavirus infection [230].

Minor illness appears to occur throughout life, and mortality at any age is exceedingly rare.

Prevalence, age and sex distributions

Antibody to both the OC43 and 229E viruses appears in early childhood and increases in prevalence rapidly with age [154]. The incidence of coronavirus infection among subjects with upper respiratory illness varies markedly from one season to another. In a 6-year study of 229E infections among medical students, Hamre and Beem [102] found a low incidence of 1% in 1964–1965 and a high of 35% in 1966–1967. An infection rate of 34% was found in Tecumseh, Michigan during the same peak year [170], and a rate of 24% in Bethesda, Maryland [134].

Serosurveys of OC43 infection have shown quite similar incidence rates. During peak seasons, 25% [170] to 29% [154] of adult colds could be associated with OC43 infection; overall, 17% of individuals develop antibody rises each year. In a single study in children, rates were lower, with a peak infection rate of 19% in one winter outbreak of respiratory disease but an overall winter rate of only 5% [135].

Infections seem to occur throughout life with no particular predilection for either the young or the old. No significant difference in the incidence of infection between the sexes has been demonstrated.

A recent study has found that the use of PCR increased the detection rate for both serotypes by about two-fold over cell culture and ELISA combined. This study found coronaviruses accounted for 13% of proven viral respiratory infections in a year-long survey of 9 to 11-year-old children in Southampton, UK [126].

Seasonal distribution

Both OC43-like and 229E-like viruses are epidemic, with peak incidence in the winter or spring and well-defined outbreaks. Monto has noted that 229E-like strains have appeared to cause nationwide outbreaks in the USA at roughly 2-year intervals whereas OC43-like outbreaks have been more localized [167]. Serosurveys of 229E and OC43 infections correlated with respiratory disease have not been performed since the early 1970s, so recent information on this point is lacking.

The recent study in Southampton employing cell culture, ELISA and PCR to detect both serotypes revealed mini-peaks of infection in November and April, but there was a background of endemic infection throughout the winter months [125,126] (Figure 1.2).

Geographical distribution

Antibodies to human respiratory coronaviruses have been found in all areas of the world where they have been sought, including North and South America, and Europe. Of normal healthy adults in England in 1976, 100% and 94% had antibody by ELISA to the OC43 and 229E virus types, respectively [105]. This was similar to the prevalences of 87% and 86%, respectively, that were found in adult sera collected in the winter of 1981 in southern Iraq and tested at the same time by the same method [105].

1.4.3 RESPIRATORY SYNCYTIAL VIRUS

Types of virus

Analysis with monoclonal antibodies has allowed the separation of RS viruses into two subgroups, designated A and B. This antigenic polymorphism has been shown to be largely due to variability in the G surface glycoprotein, and has remained relatively stable throughout the last 30 or more years.

Clinical illness

Most individuals infected with RS virus have upper respiratory illness. Around 25–40% of infants between 6 weeks and 6 months of age infected with RS virus for the first time, will develop bronchiolitis or pneumonia, with the peak incidence at 2 months of age [70,185]. RS virus is the major cause of bronchiolitis during early infancy and one of the major causes of pneumonia during the first few years of life. Mortality is low in modern westernized society, occurring in less than 2% of hospitalized patients. Mortality rises in the presence of other cardiovascular or respiratory disease, and is higher in the less developed world. Re-infection is common, may occur at all ages, and is usually symptomatic. In adults RS virus normally causes simple upper respiratory infection, but cases of severe pneumonia (especially in the elderly) have been documented.

Mode of transmission and nosocomial infection

As with other respiratory viruses, RS virus is spread by infected respiratory secretions. However, the major mode of spread appears to be by large droplets or through fomite contamination rather than through droplet nuclei or small-particle aerosols [92,95]. Spread requires either close contact with infected individuals or contamination of the hands and subsequent contact between fingers and nasal or conjunctival mucosa ('self-inoculation').

RS virus is a major cause of nosocomial infection [71,94], and is a particular hazard for premature infants, infants with congenital heart disease or bronchopulmonary dysplasia, and infants and children who are immunodeficient. The rate of hospital-acquired infection for infants and children during an RS virus season ranges from 20% to 40%, and is associated with appreciable mortality in those with concurrent disease [96,152]. Hospital staff appear to play a major role in nosocomial spread through exposure to infected caregivers and transmission of fomites on hands. The likelihood of nosocomial RS virus infection increases with the duration of stay and the number of individuals housed in a patient's room [209].

Nosocomial spread in nurseries and paediatric wards can be reduced by a combination of measures that include (a) limitation of visitors, (b) active surveillance for RS virus infection among patients and new admissions, (c) cohorting of RS virus infected patients, and (d) institution of various barriers to virus spread [225], such as strict hand washing and the use of gloves and gowns [149,156].

Seasonality and geographical distribution

RS virus has a clear seasonality in temperate zones of the world. In urban centres, epidemics occur yearly in the winter months but not during the summer [139,174,231]. In the northern hemisphere, the virus is rarely isolated during August or September. Each RS virus epidemic lasts approximately 5 months, with 40% of infections occurring during the peak month in the temporal centre of the outbreak [17]. The peak may occur as early as December or as late as June, but most outbreaks peak in February or March.

The behaviour of RS virus in tropical or semi-tropical areas is somewhat different, with epidemics occurring during rainy seasons [210,215]. In Calcutta, epidemics have been found to occur during religious festivals when large groups of people gather together and overcrowding is common [113].

RS virus has a worldwide distribution and, throughout the world, is one of the most if not the most important paediatric viral respiratory tract pathogen.

Socioeconomic status, age and gender

Hospitalization for RS viral disease is more frequent in children from families with lower socioeconomic status and in heavily industrialized areas [76,205]. Hospitalization rates in lower socioeconomic groups are around 1 in 70 infections [79,80], while in middle- and high-income families, the rate is only one in 1000 infections [74].

Serological surveys indicate that infection with RS virus is almost universal during the first few years of life. Attack rates of around 50% have been noted in most epidemics [80,185,205], though in day-care centres the attack rate approaches 100% during epidemics [106]. The infection rate for RS virus is 69% during the first year of life and 83% during the second year; virtually all children are infected by 24 months of age and 50% are re-infected at least once, the risk of re-infection decreasing to 33% by 48 months of age [80]. In seronegative infants and young children, 98% were infected with RS virus when exposed during an outbreak, and the re-infection rates were 74% and 65% during two subsequent epidemics [106]. Re-infection of adults is also common, particularly when exposure to virus is heavy. In hospital staff on paediatric wards, infection rates of 25–50% have been observed during epidemics [93,94].

Serious lower respiratory tract disease occurs 30% more commonly in infant males than in females [185], perhaps as a consequence of males having smaller airways [29].

Epidemiology of serotypes

The two subgroups of RS viruses have remained stable, and have co-circulated during most yearly epidemics that have been studied, with subgroup A normally predominating [1,110]. The impact of subgroup antigenic variation on the epidemiology of RS virus appears to be limited. Infants undergoing primary RS virus infection develop a broad serum neutralizing antibody response. Thus, although antigenic dimorphism probably contributes to the high incidence of initial re-infection during early childhood, its effect appears to be modest, especially as many re-infections involve viruses of the same subgroup [173].

Lower respiratory tract disease and mortality

Lower respiratory tract disease, principally bronchiolitis and pneumonia, are common during RS virus infections, particularly in infants. Some 40% of institutionalized infants developed pneumonia during an outbreak of RS virus infection [130]. In family studies, 25% of infants undergoing their first infection with RS virus were considered to have lower respiratory tract disease [80].

Rates of hospitalization of infants for RS virus infections vary with the setting. In Washington, DC, one in 200 infants required hospital care for RS virus pneumonia or bronchiolitis during the first year of life [139]; in Norway, hospitalization was required for one of 100 infants under 1 year of age [182], and in Great Britain, one in 120 [28]. RS virus is detected in around 50% of hospital admissions with a diagnosis of bronchiolitis, in 25% of those with pneumonia, in 11% of those with bronchitis, in 10% of those with croup and in 5.4% of infants and children seen in the clinic or admitted to hospital for non-respiratory illness [139].

Death due to RS virus infection is not common in developed countries. There are no accurate estimations of the overall death rate, but several surveys of hospitalized children with RS virus infection report a fatality rate of 0.5–2.5% [28]. Death occurs most often in infants with underlying illnesses, with a mortality rate of 37% in children with congenital heart disease, the highest mortality being in those with pulmonary hypertension [152]. High mortality has also been observed in infants and children who are immunosuppressed for treatment of other diseases such as cancer [97]. Several cases of severe or fatal disease have been described in children with disorders of cell-mediated immunity [160]. Infants with bronchopulmonary dysplasia are also at high risk for severe or fatal illness [83].

1.4.4 INFLUENZA VIRUSES

Virus types

There are three serotypes of influenza viruses that cause disease in humans: types A, B and C. Type A is by far the most common, and causes the vast majority of clinical illness related to influenza viruses.

Influenza virus type A is further divided into subtypes depending on the antigenic composition of the major surface antigens, haemagglutinin and neuraminidase (see

Chapter 13). The subtypes are designated H and N numbers to classify them. For example the first human influenza virus isolated in 1933, and that which caused the devastating 1918 pandemic, was designated H1N1, while the most recent epidemic in 1989–1990 was related to type H3N2 [26].

The influenza viruses are unique among the respiratory tract viruses in that they undergo significant antigenic variation. Both of the surface antigens of the influenza A viruses undergo two types of antigenic variation: antigenic drift and antigenic shift. Antigenic drift involves minor antigenic changes in the haemagglutinin (H) and neuraminidase (N), while shift involves major antigenic changes in these molecules.

Clinical illness

Influenza infections can cause a broad spectrum of clinical illness, varying from mild upper respiratory tract disease to life-threatening pneumonia (see Chapter 14). The severity of illness will be determined by a variety of host, viral and environmental factors, but a powerful influence is the degree of immunity. This will be determined to a large extent by factors such as previous exposure and the extent of antigenic variation from previously experienced viruses.

Antigenic drift in the haemagglutinin and neuraminidase molecules

The haemagglutinin is the major surface antigen of influenza virus and is type specific. Antigenic drift occurs in types A, B and C virus haemagglutinins but is most pronounced in human influenza A strains. After the appearance of a new subtype, antigenic differences between isolates can be detected within a few years using antisera, and studies with monoclonal antibodies indicate that minor antigenic heterogeneity is detectable among different influenza virus isolates at any time [183,206,213]. Antigenic differences have been

detected among the haemagglutinins of H2N2 viruses isolated in 1957 and among the H3N2 viruses isolated in 1968. Analysis of influenza B viruses indicates that antigenically distinguishable viruses co-circulate during an epidemic [228]. Studies during an outbreak of influenza B in a boarding school in England revealed marked antigenic microheterogeneity, suggesting that antigenic drift occurs during an outbreak [183].

Antigenic drift also occurs in the neuraminidase of influenza viruses [40,184], and has been correlated with differences in amino acid sequences [30]. The frequency of isolation of N variants in *in vitro* systems is similar to that for the H, and sequence analysis shows single amino acid substitutions in the molecule at similar sites to those occurring in antigenic variants of the N that appear in the new epidemic strains arising in the human population [148].

Mechanism of antigenic drift

Antigenic drift clearly occurs by accumulation of a series of point mutations. However, single amino acid changes frequently have little effect on the antigenic properties of the surface protein, raising the question of how variants with epidemic potential arise in nature. Sequence analysis of naturally occurring drift strains reveals point mutations in two or more of the epitopes of the haemagglutinin, suggesting that in nature two or more mutations must be acquired before a new strain emerges that is able to escape neutralization by the antibody induced by preceding viruses. Since it is unlikely that two mutations simultaneously occur in one virus, it is likely that the mutations occur sequentially [226].

Nature of antigenic shift

Antigenic shifts in type A influenza viruses have occurred in 1957 when the H2N2 subtype (Asian influenza) replaced the H1N1 subtype, in 1968 when the Hong Kong (H3N2) virus appeared, and in 1977 when the H1N1 virus reappeared. All these major antigenic shifts in the virus occurred in China and anecdotal records suggest that previous epidemics also had their origin in China. Serological and virological evidence suggests that since 1890 there have been six instances of the introduction of a virus bearing a haemagglutinin subtype that had been absent from the human population for some time. For the haemagglutinin, there has been a cyclical appearance of the three human subtypes with the sequential emergence of H2 viruses in 1890, H3 in 1900, H1 in 1918, H2 again in 1957, H3 in 1968, H1 in 1977 and most recently H3 again in 1990.

The mechanisms through which these new human viruses emerge are uncertain, but there is evidence that the changes may be derived from genetic re-assortment in humans, or from genetically re-assorted animal or avian viruses [38,229]. Genetic and biochemical studies suggest that the 1957 and 1968 strains arose by a process of genetic re-assortment [72], with the new H3 in 1968 probably being donated by an avian influenza A virus [138]. However, the donating virus could have been one that contained H3 that had persisted unchanged since the 1900 epidemic. The lack of antigenic shift in influenza B and C viruses may be due to the absence of a significant gene pool in mammals or birds.

A second explanation for the origin of pandemic viruses is that the 'new' virus which may have caused an epidemic many years previously remains hidden and unchanged in some unknown place since that time. The strain of H1N1 which appeared in Anshan in northern China in May of 1977 and spread to the rest of the world, appears to be identical in all genes to the virus that caused an influenza epidemic in 1950 [177]. The most likely explanations for the prolonged absence of this virus include preservation in a frozen state or preservation in an animal reservoir. The animal reservoir option is a possibility for H3N2 viruses, since they have been found in pigs many years after they disappeared from

humans [203]. However, the RNAs of animal influenza viruses are very variable [114], making it unlikely that a human strain would be conserved in all genes for many years. The absence of antigenic drift over a number of years is also difficult to explain. The third way in which new viruses could appear in the human population would be if an animal or bird virus became infectious for humans: the transmission of swine influenza viruses to humans has been documented [114].

Transmission of influenza A virus in humans

The virus is maintained in humans by direct person-to-person spread during acute infection. Isolated communities become infected by the introduction of virus by an infected individual often resulting in an explosive, but discrete, epidemic. Subsequent epidemics are initiated by re-introduction of virus [19]. Influenza virus activity can be detected in a large population centre during each month of the year [62,140]. On a global scale, influenza viruses are isolated from humans during most months of the year somewhere in the world. Influenza virus epidemics tend to occur over the winter months, as with other respiratory viruses.

Pre-school and school-age children are major vectors in the transmission of influenza A viruses in the community [63,91]. After the introduction of the Asian influenza A virus (H2N2) in 1957 and the Hong Kong virus (H3N2) in 1968, epidemics occurred soon after schools opened in September and October [128]. As with other respiratory viruses, the crowding that occurs in schools favours the rapid spread of virus by aerosol transmission [27]. The incubation period for influenza viruses is about 3 days for influenza A virus and 4 days for influenza B viruses [64].

Factors influencing the size of an epidemic

Immunological factors clearly influence the size of an epidemic: in 1957 a pandemic was caused by an H2N2 virus to which the vast majority of the population was fully susceptible, as they had not had previous experience with influenza A viruses bearing a related H or N antigen. During 1968, the first year of spread of the Hong Kong (H3N2) virus, mortality was about one-half that caused by the H2N2 virus during its first year of prevalence. This dampening of the first H3N2 epidemic was a consequence of partial immunity induced by prior infection with H2N2 viruses that possessed a related N antigen. The size of epidemics occurring during interpandemic periods is always smaller than that of those occurring during the introduction of a new virus subtype. There is no regular periodicity in the occurrence of epidemics, indicating that the appearance of an epidemic in any given year represents a subtle interplay between the extent of antigenic drift of the virus and the waning immunity in the population. The time course of an epidemic is usually a sharp, single peak of virus activity, which then wanes as virus ultimately disappears from the population.

Morbidity

Influenza viruses have a major impact on morbidity leading to increases in hospitalization and medical consultations. High rates of hospitalization occur in the elderly, and in children less than 5 years of age with the highest rates in children less than 1 year of age [193]. Morbidity is also highest where immunity is lowest, such as in infants and young children undergoing first infection [193], and in all age groups following the introduction of new pandemic viruses. Age at the time of infection appears to be another factor that contributes independently to morbidity, as at each level of antibody to the H, infection in adults is more often asymptomatic than in children [63,75]. Infection rates and attack rates reflect a delicate interplay between the immunity in the population and the relative virulence of the virus.

Immunity to influenza viruses is incomplete, and morbidity following re-infection with the same strain or with a drift strain does occur, although it is usually less severe [44,45,62,65,66]. Significant morbidity and mortality with influenza A viruses can occur during nosocomial spread of infection [71,90].

Mortality

Appreciable mortality associated with influenza B virus has been seen in only two influenza seasons, 1961–1962 and 1979–1980, whereas influenza A has caused appreciable mortality in well over 20 epidemics since 1934 [180,181]. The highest mortality rates occur in people over 65 years of age [4,11,151]. However, mortality is seen at all ages; during the 1918 epidemic, the mortality rate was extremely high in young adults as well as in those over 65 years of age. The coexistence of cardiovascular, pulmonary, metabolic or neoplastic disease and pregnancy all increase the risk of mortality [9,10,82]. In epidemic years such as 1957–1958, the mortality totalled 70 000, and 1 in 300 of adults over 65 years of age died from influenza infection. Although excess mortality is highest during the first year of circulation of a new subtype, the cumulative interpandemic mortality significantly exceeds that caused during the first year [180].

Strain differences

The clinical illnesses caused by influenza A viruses belonging to different subtypes are remarkably similar. There are, however, some differences between virus strains in both the illnesses they cause, and in their epidemic behaviour [140,225,232]. H3N2 viruses are more frequently associated with croup, and the H1N1 virus that caused the 1918 pandemic was associated with very high mortality in young adults, an age group that was not at high risk for mortality in subsequent H1N1, H2N2, or H3N2 epidemics [225]. It is also likely that a variety of host, viral and environmental factors play a role in this variability. Encephalitis lethargica has been associated with the early H1N1 virus, but not with any others [196], while the H1N1 virus introduced in 1977 caused a relatively mild illness in susceptible children and young adults who would not have encountered it previously [69,180,232].

Immunity

Immunity to influenza viruses is primarily mediated by responses to the surface glycoproteins, is long-lived and is subtype-specific [63,68,73,180,194]. People infected previously with H1N1 viruses which circulated between 1918 and 1957 were resistant to infection or disease in 1977. The duration of immunity in the presence of antigenic drift has been estimated to be between 1 and 5 years depending on the extent of the drift [65,66].

Although immunity can be long-lived, reinfection with homologous influenza A viruses occurs, indicating that immunity induced by a single infection can be incomplete [65,66]. This incomplete immunity is likely to be a result of a gradual diminution in local and serum antibodies within the first year following first infection [123]. In addition, infection with a new influenza virus subtype leads to generation of antibodies that react with only a limited number of antigenic sites, whereas after second or subsequent infections, antibodies with a broad range of specificities are produced.

1.4.5 PARAINFLUENZA VIRUSES

Virus types

There are four serotypes of parainfluenza viruses causing disease in humans, designated types 1–4, with type 4 being subdivided into subtypes 4A and 4B. Each of the four parainfluenza virus types can cause acute respiratory tract disease in humans.

Mode of transmission

Transmission of parainfluenza viruses is by direct person-to-person contact or by droplet spread, but the viruses do not persist in the environment. All four serotypes have been transmitted in experimental infection studies, and cause upper respiratory tract symptoms [132,219,220]. The high rate of infection early in life, coupled with the frequency of re-infection, suggests that the viruses spread readily from person to person. The type 3 virus appears to be the most efficient in its ability to spread, and generally infects all susceptible individuals in an enclosed population in a relatively short time, while types 1 and 2 are less effective, infecting 40–69% of susceptible individuals [25,131]. Parainfluenza viruses are introduced into the family primarily by preschool children.

In experimental infection of adult volunteers, the interval between administration of type 1, 2 or 3 virus and onset of upper respiratory tract symptoms ranges from 3 to 6 days [132,208,219,220]. The interval between exposure to type 3 virus and subsequent viral shedding is 2–4 days [25]. Type 3 virus is shed from the oropharynx for 3–10 days (median 8) during initial infection, and for a shorter period during re-infection [25]. However, prolonged shedding in infants and young children has been observed for as long as 3–4 weeks [67]. Prolonged shedding of type 3 virus has also been observed occasionally in adults with underlying chronic lower respiratory tract disease [84].

Clinical course of parainfluenza virus infections

All four serotypes of parainfluenza viruses cause acute respiratory illness in humans, with the most common symptoms being rhinitis, pharyngitis, bronchitis and fever [186,187].

Parainfluenza virus type 1 is the principal cause of croup (laryngotracheobronchitis) in children, and parainfluenza virus type 3 is second only to RS virus as a cause of pneumonia and bronchiolitis in infants less than 6 months of age [24,25,46,186]. Parainfluenza virus type 2 resembles type 1 virus in the clinical manifestations it causes, but serious illness occurs less frequently; infections with type 4 parainfluenza virus are detected infrequently, and associated illnesses are usually mild.

The parainfluenza viruses are most important as respiratory tract pathogens during infancy and childhood, when types 1, 2 and 3 viruses can cause anything from asymptomatic infection to life-threatening lower respiratory illness. In addition to croup, types 1, 2 and 3 viruses are also responsible for a smaller but appreciable proportion of other acute respiratory tract diseases of infancy and early childhood.

Geographical distribution

Parainfluenza viruses have a wide geographical distribution. The first three types have been identified in most areas where appropriate diagnostic techniques have been used to investigate childhood respiratory tract disease. Type 4 viruses, which are more difficult to culture, have been isolated in fewer areas, but serological studies suggest that they are as prevalent as the other types.

Prevalence of parainfluenza virus infections

The parainfluenza viruses are exceeded only by RS virus as an important cause of lower respiratory tract disease in young children, and they commonly re-infect older children and adults to produce upper respiratory tract disease. Re-infection of adults as well as children has been recognized on a number of occasions, particularly with type 3 virus. Although the frequency of re-infection is not known, it is probable that most individuals have repeated infections with types 1, 2 and 3 viruses. In a study of three outbreaks of type 3 virus infection in a nursery population, 17% of the children infected during one outbreak were re-infected during a subsequent outbreak, although the interval between the first and last outbreaks was only 9 months [25]. Re-infection

of pre-school children living at home also occurs with high frequency [77]. Illness usually occurs less often (and is less severe) during re-infection than during primary infection.

Type-specific infection rates are difficult to estimate because there is considerable cross-reactivity between the serotypes. Family studies reveal rapid spread of virus within infected families, with 64% of family members developing a serum antibody response.

Age distribution

Parainfluenza infection generally occurs very early in life. Serological surveys indicate that at least 60% of children are infected with parainfluenza type 3 virus by 2 years of age and that over 80% are infected by 4 years of age [178,179]. Longitudinal studies suggest that serological surveys may underestimate the prevalence [77]. Infection with type 1 or type 2 virus generally occurs somewhat later in childhood, but by 5 years of age a majority of children have been infected with type 2 virus, and over 75% have been infected with type 1 virus [186,187].

Type 3 virus often causes illness during the first months of life at a time when infants still possess circulating neutralizing antibodies from their mothers. In contrast, in young infants, maternally derived antibodies appear to prevent both infection and severe disease with either type 1 or type 2 virus [24,25,76]. After 4 months of age, there is a rise in the number of cases of croup and other lower respiratory tract diseases caused by type 1 and type 2 viruses. This high incidence continues until approximately 6 years of age.

After school-age, there is a much lower incidence of severe lower respiratory tract disease caused by type 1 and type 2 parainfluenza viruses, and the occurrence of lower respiratory tract illness in individuals infected with either virus during adolescence or adult life is unusual, although this does occur on occasion. Parainfluenza viruses have been associated with exacerbations of asthma in several studies [126,191].

Seasonal pattern

Some investigators have shown parainfluenza virus infections to follow an endemic pattern, with infection occurring sporadically and without a definite seasonal pattern [24,25,125] (Figure 1.2), while others have shown sharp autumn outbreaks occurring in 2-yearly cycles [25,78].

1.4.6 ADENOVIRUSES

Virus types

There are over 50 serotypes of adenovirus described so far, and new strains continue to be identified, particularly in the immunocompromised patient.

Serologic surveys have shown that antibodies to types 1, 2 and 5 are most common and are present in 40% of children [18,129]. The incidence of antibodies to types 3, 4 and 7 is low at the same ages. These antibody results probably explain why adults are more commonly infected with types 3, 4, and 7. During the surveillance for the Virus Watch studies it was documented that only about 75% of the adenovirus isolates were accompanied by an antibody response, as measured by complement fixation [56]. The reason for some of the serological non-responders may have been the insensitivity of the assay or the failure of some children to produce such antibodies.

Clinical illness

Adenovirus infections can cause a wide variety of clinical illnesses including asymptomatic infection, upper respiratory tract diseases, pharyngo-conjunctival fever, severe, life-threatening pneumonia, and enteric and hepatic infections.

Approximately 5% of acute respiratory disease in children under the age of 5 years is thought to be associated with adenoviral infection [18]. The most frequent symptoms are nasal congestion, coryza and cough, though exudative pharyngitis may also occur. Systemic symptoms such as fever and myalgia are common.

Adenoviruses are thought to account for about 10% of cases of childhood pneumonia. Most children will recover, but mortality has been considerable in some epidemics, and it is thought that bronchiectasis may follow severe infection early in childhood [145,204].

Geographical distribution

Adenovirus infections occur worldwide in humans. The transmission of infection and disease varies from sporadic to epidemic. The pattern often correlates very well with the viral serotype and the age of the population. Adenoviruses probably account for 3% of respiratory infections in civilian populations and approximately 7% if only febrile illnesses are considered [56]. The corresponding figures for young children are approximately 5% and 10%, respectively. However, in the same studies a large number of completely asymptomatic adenovirus infections have also been documented.

Mode of transmission

Faecal–oral transmission accounts for many infections in young children. Initial spread may occur via the respiratory route, but the prolonged carriage in the intestine makes the faeces a more common source during both the acute illness and intermittent recurrences of shedding [56]. Controlled studies of routes of infectivity for adenoviruses that caused epidemic acute respiratory disease among military recruits have demonstrated that aerosolized virus inhaled into the lungs of volunteers produced the disease, whereas application to the mouth, nasal mucosa or intestine did not [36].

Several of the adenovirus serotypes have probably also been spread as nosocomial infections [14,224].

Epidemic adenovirus infections in military recruits

Epidemics of acute respiratory disease were well known during World War II, and this awareness preceded the isolation and the characterization of the first adenovirus by approximately one decade. However, the results of later studies suggested that adenovirus type 4 or 7 infection caused most outbreaks [37,111,112]. These epidemics occurred almost exclusively in recently assembled military recruits, and were most common in winter. They did not occur in seasoned personnel in close contact with the recruits, suggesting that seasoned personnel had some form of immunity. However, the disease did not occur in similarly congregated college students, suggesting that more crowded sleeping conditions or the fatigue associated with training were contributing factors. The importance of these cofactors was supported by the observation that the adenovirus infections did not spread to civilian personnel in contact with the military. Adenovirus-induced respiratory disease often affected 80% of the recruits, with 20–40% hospitalized.

1.4.7 ENTEROVIRUSES

Virus types

The major human enteroviruses include poliovirus types 1–3, echoviruses [31 serotypes), Coxsackie viruses, of which there are two major groups, A and B, with many different serotypes within each group and at least five other known enteroviruses. Many are capable of causing acute respiratory illness, as well as gastrointestinal illness for which they are named. They share many structural and biological similarities with human rhinoviruses.

Mechanisms and routes of transmission

Humans are the only known reservoir for members of the human enterovirus group, and close human contact appears to be the primary mechanism of spread. Almost all enteroviruses can be recovered from the oropharynx and intestine of individuals with symptomatic and asymptomatic infections. As with other respiratory viruses, droplets or aerosols from coughing or sneezing are the major source of transmission when the infection is upper respiratory (faecal–oral spread is important for enteric infections). Coxsackievirus A21 has been transmitted experimentally from infected volunteers by airborne aerosols produced by natural coughing [37].

Family studies

Outbreaks of aseptic meningitis and other related illness due to echovirus 30 began to spread along the Pacific Coast of the USA in 1968, and its arrival in Washington State coincided with the initiation of the Seattle Virus Watch Program, thus permitting close observation of infection and illness among the regularly studied Virus Watch families [99]. A total of 64 families containing 291 members were studied with continuing virus isolation and serology; infection was documented in 70 (79%) of 88 members of 18 families; in the total observed Virus Watch population, the rate was 24%. The affected families tended to be of larger size, included more children 5–9 years of age, and included only three persons who had antibody before the epidemic. Mild febrile illness was reported in only 47% of those shedding virus; few of these episodes were serious enough to require medical attention, and only one subject developed aseptic meningitis. Extending these observations to the general population suggests that there must have been many thousands of echovirus 30 infections in the area, more than half of which were without symptoms, during a period when 44 virologically confirmed cases of echovirus 30 aseptic meningitis occurred in the city [217].

Geographical distribution and age

Enteroviruses are found throughout the world. In tropical regions they are present throughout the year, but in temperate climates there is a late summer, early autumn peak [172]. It is well recognized that enteroviral infections are more common among lower social classes and in the presence of poor hygiene. Most reported cases of enteroviral infection occur in children, with 56% being under 10 years of age, and 26% under 1 year [172].

Clinical course of enterovirus infection

Respiratory disease associated with enteroviruses is commonly upper respiratory symptoms indistinguishable from those caused by other respiratory viruses. They have also been implicated in episodes of pneumonia and bronchiolitis [201].

A survey of enteroviral disease in the USA from 1976 to 1979 revealed that meningitis was present in 35% of patients, respiratory disease in 21%, encephalitis in 11% and non-specific febrile illness in 6%, with a variety of other syndromes making up the remainder [172].

Several Coxsackieviruses and echoviruses have been associated with mild upper respiratory illness, and also occasionally with mild or severe lower respiratory disease [43,157].

1.4.8 CYTOMEGALOVIRUS

Virus type

Cytomegalovirus (CMV) is a herpes virus that is ubiquitous, infecting many animals as well as humans. Humans are believed to be the only reservoir for human CMV infection. It is thought that there may be genetically distinct strains of CMV that are continuously circulating throughout the world [3].

Mechanism of spread

Transmission occurs by direct or indirect person-to-person contact. Virus has been detected in almost all bodily fluids. However, because of the lability of CMV to environmental factors, close or even intimate contact is believed to be required for horizontal spread. The community prevalence is enhanced by the fact that virus excretion persists for years after congenital, perinatal and early postnatal infections [2,144], and prolonged replication can follow primary infection in older children and adults. Recurrent infections are also associated with shedding of CMV from many sites in a significant proportion of seropositive young adults [141,199].

Oral and respiratory spread appear to be the dominant routes of transmission during childhood and probably adulthood as well. CMV is also spread through venereal routes: in many populations there is a burst of infection with the advent of puberty, and infection rates are much increased in promiscuous populations. Multiple blood transfusions or transfusions of large quantities of blood are associated with an increased risk of both primary and recurrent CMV infection.

Clinical course of CMV infection

Intrauterine infection can cause abortion and choroidoretinitis. The vast majority of peri- and postnatal CMV infections are likely to be asymptomatic, or to present with mild flu-like symptoms indistinguishable from those caused by other respiratory viruses. CMV may also cause more prolonged and severe illness similar to infectious mononucleosis, and in the immunocompromised CMV frequently produces life-threatening pneumonitis and retinitis; it may also be implicated in episodes of graft rejection in transplanted subjects.

Prevalence, age and geographical distribution

CMV infection is endemic rather than epidemic and is present throughout the year rather than being seasonal [81]. Climate does not appear to affect infection rates. Ill-defined socioeconomic factors do predispose to higher infection rates, both by vertical (intrauterine) and horizontal (extrauterine) transmission [81,212]. Poor hygiene alone cannot explain the higher infection rates: once again, the closeness of contacts within population groups is important. Very high rates of infection among children have been recorded in isolated locations, as well as crowded areas of Africa, the Orient and the Middle East, irrespective of hygienic practices [81]. Transmission among toddlers is exceptionally high in day care centres and boarding schools [120,188]. After infancy in most developed countries, infection rates increase slowly until the age of entry into school, at which time they rise more rapidly; 40–80% of children are infected before puberty [81]. In other areas of the world, 90–100% of the population may be infected during childhood, even as early as 6 years of age [81]. Young infants and children with subclinical infection appear to be the major source for primary infection in pregnant women [190]. Day care centres and similar settings where pregnant women are in daily contact with children, especially toddlers, pose a high-risk setting for primary infection with its increased risk for intrauterine transmission [120,189].

Recurrent and chronic infections with CMV

Recurrent infection is defined as intermittent excretion of virus from single or multiple sites for a number of years and is to be distinguished from 'chronic' or 'prolonged excretion' of virus, which characterizes certain forms of CMV infection, particularly in the immunocompromised. Recurrent infection can result from one or more of three mechanisms. First, after primary infection, a low-grade chronic infection may be established in which virus excretion reaches detectable levels only periodically. Second, re-infection may occur in immune people because of antigenic and

genetic disparity among CMV strains. Third, CMV may become latent in various organs during the primary infection, as with herpes simplex virus, and be repeatedly reactivated in later life in response to different stimuli.

1.5 THE COMMON COLD

1.5.1 HISTORY

The term 'common cold' describes the universally recognized short mild illness in which the main symptoms involve the upper respiratory tract and in which nasal symptoms usually predominate. The symptoms usually comprise some or all of the following: nasal stuffiness, sneezing, coryza, pharyngitis, throat irritation, and mild fever. The name 'common cold' probably arose from a combination of the fact that the onset of symptoms included the feeling of chilliness on exposure to cold, and the increased prevalence during the winter months, giving the impression that there may be a cause-and-effect relationship. This is still a commonly (and erroneously) held view, even though as long as 200 years ago Benjamin Franklin pointed out that colds were caught from other people rather than by exposure to cold. In 1914 the infectious nature of colds in humans was demonstrated by the instillation of filtered, cell-free nasal washings from ill people with colds into the nares of subjects who subsequently developed colds. These findings were confirmed in 1930 by Dochez *et al.*, who provided the first evidence that colds were caused by a virus and not bacteria by inoculating volunteers with bacteria-free filtrates from individuals with symptoms of colds [49]. In 1933, the isolation and cultivation of an influenza A virus from a human was reported, thus opening avenues for further detailed study of common cold viruses.

Rhinoviruses were first reported in 1953 to have been cultured in explants of human embryonic lungs [5]. However, it was not until some years later that this work could be repeated when the virus was shown to produce a cytopathic effect in human embryo kidney cells [221,222]. In the meantime, two groups in the USA reported the isolation in monkey kidney cell cultures of a virus involved in upper respiratory illness, and this was subsequently designated rhinovirus 1A [192,195]. Since then, other rhinovirus-sensitive cells have been described, and over 115 different serotypes of rhinoviruses have been isolated, all of which have been found to cause common colds.

A major contribution to our present understanding of the common cold has been the use of human volunteers under carefully controlled conditions. The Common Cold Research Unit at Salisbury, England, was established in 1946, and only closed recently on the retirement of its director, David Tyrrell in June 1990. Volunteer studies here as well as those performed in the USA during the last 40 years are to a large degree responsible for our present understanding of colds in adults. Although studies of respiratory illness in children have also been extensive (in particular those concerned with the spectrum of clinical manifestations by age group, and seasonal prevalence rates of the different respiratory viruses), controlled volunteer trials have for ethical as well as practical reasons, not been performed.

1.5.2 AETIOLOGICAL AGENTS

It has been known for some time that rhinoviruses are responsible for the majority of common colds, and that coronaviruses, RS viruses, influenza viruses, parainfluenza viruses, adenoviruses, enteroviruses, and a variety of less common viruses and atypical bacteria may also cause common colds and contribute to the total disease load (see Table 1.1). These agents are all covered in detail in other chapters; in this chapter an overview of the epidemiology of each virus type has been presented.

As a group, rhinoviruses are the most common cause of colds in children as well as adults.

Also of major importance in the aetiology of colds are re-infections with parainfluenza viruses and RS virus. Although relatively little data exist on the contribution of coronaviruses to colds, the data that have been collected in those studies in which they have been looked for suggest that they contribute around 15–20% of common colds [125,153,154,159,230].

1.5.3 EPIDEMIOLOGY

The common cold is a frequent illness of childhood, with prevalence decreasing with age as a result of increasing immunity and reduced opportunity for spread, itself a result of changes in behavioural patterns. Despite the fact that over 200 serologically different viral types are likely to cause this illness, there is a general predictability of incidence and seasonal occurrence.

Prevalence rates vary considerably depending on age, the time of year, year-to-year variation, and on the population being studied. An example of the seasonal pattern for individual viruses, in 9 to 11-year-old children in the UK for the year 1989–1990 is given in Figure 1.2. From this it can be seen that as a general principle, rhinoviral infections occur all-year round, but peak in early and late winter, coronaviral and parainfluenza viral infections occur all-year round, though again with peaks in winter, and that influenza and RS virus have discrete winter peaks of infection.

Although numerous epidemiological studies have been conducted on the occurrence of respiratory illnesses, it is difficult to calculate a precise incidence of the common cold because criteria of disease classification have been different in different studies, incidence varies greatly with population characteristics and age, and because viral detection methods have been only partially successful in detecting the causative agents. It has been estimated that common colds occur at rates of two to five per person per year [35,59,60], though recent epidemiological data have shown that in school-children, the figure is more likely to be in the region of 7–10 colds per year [126], with younger children being likely to have more, and adults less, than that figure (many previous studies have suggested that adults have about half the number of colds as do children). Common colds are estimated to cost billions of dollars each year in terms of lost working days, cold remedies and analgesics [35].

The conventional spread of colds has its initial focus in the school. School-age children become infected and introduce secondary infections in the home. In these homes the secondary attack rate is highest in other school-age children and pre-school-age children, with the secondary attack rate in adult family members being half that for the children. The introduction of infection in the family by adults is unusual. Modern-day trends toward day-care centres and pre-schooling are likely to have increased primary infections in these younger children and will have made them the source from which secondary family infections frequently occur. Among children, boys tend to have more colds than girls. On the other hand, in the conventional family setting mothers tend to have at least one more cold per year than their spouses.

Colds occur throughout the world [6,20,166, 169,178,216,218,227]. In non-isolated populations, colds are more frequent during the winter months than during the summer time. In the tropics, colds are more prevalent during the rainy season. In isolated populations in which the number of people is fixed (such as members of Antarctic exploration teams or isolated island communities), colds do not occur unless introduced by a visiting person [227].

Although colds can be produced regularly in volunteers, the method(s) of transmission of viruses, which result in colds under natural circumstances, are still debated. The greatest concentration of common cold virus is in the nasal secretions. Children tend to have greater concentrations of virus than adults and they tend to shed virus for a longer period.

Sneezing, nose blowing, and the general contamination of external surfaces (including the sufferer's hands) with nasal secretions are the main sources for viral transmission. The route of acquisition of virus is by the nose and possibly the conjunctiva. Susceptible individuals become infected by inhalation of virus in small-particle aerosols generated by sneezes, by direct nasal impaction of virus-containing large droplets from a sneeze, by nose blowing, or by the inoculation of virus (usually by the fingers) from nasal secretions that have been transferred directly or indirectly from infected subjects.

There is considerable folklore related to the catching of a cold. There is no evidence to date to indicate that cold weather *per se*, chilling, wet feet, or draughts play any role in the susceptibility of people to colds. Epidemiological data suggest that school attendance and other forms of crowding populations (particularly children) are the major factors influencing common cold virus transmission rates.

1.6 RESPIRATORY VIRAL INFECTIONS AND EXACERBATIONS OF ASTHMA

Over the last 30 years there has been a large number of epidemiological studies in which the association of respiratory viral infections with asthma attacks has been investigated. Although many of these studies had shortcomings in their design, and all acknowledged the likelihood of underestimating the contribution of viral infections as a result of difficulties in viral detection, they provide considerable evidence to support a positive association [191].

The identification rate for viruses during exacerbations of asthma (10–50%) is similar to that generally found during respiratory infections, and more specifically so in studies that have investigated episodes of respiratory infection with and without asthma or wheezing [116,117]. This rate is much higher than the viral identification rate generally found during asymptomatic periods in asthmatics and non-asthmatics, which is around 3–5% [117,119, 122,163]. A recent study in children, using much more sensitive modern detection techniques, has produced similar findings, with the detection rate in colds without asthma, and colds with asthma being 80–85%, while the detection rate in the same children when asymptomatic was 12% [126].

There is also a close temporal relationship between viral infections and asthma exacerbations, both within individuals and within populations [107,153]. The rate of virus identification decreases after the acute stage of respiratory illness [116,118], which makes chance coincidence of virus identification with wheezing episodes unlikely. In three separate studies in which monitoring was intensive and specimens were obtained on a regular basis between episodes, virus identifications in individuals clearly coincided in time with asthma exacerbations [153,161,200]. Wheezing associated with infection has a characteristic pattern: in a recent study of children aged 1–6 years, the wheezing started 2 days after the first symptoms of respiratory infection, and lasted for 4 days [159], while in a study in 9 to 11-year-old children (see below), lower respiratory symptoms followed 1 day after the onset of upper respiratory symptoms, and lasted for 7 days, and peak expiratory flow recordings took longer to recover, at least 14 days [126].

A study of paediatric hospital admission rates for asthma provided indirect evidence: there was a striking relationship to school holiday periods, which were associated with troughs in admission rates, followed by a sharp rise to a peak after the beginning of each school term [214]. This pattern is similar to that observed for respiratory virus infections, and the authors suggested that viruses were acquired from other localities during holiday travel and were then rapidly spread through a susceptible population, largely as a result of crowding of the children at school. This

hypothesis is supported by a recent study in which proven viral infections and asthma admissions were shown to be closely correlated in time in both adults and children [125] (see below).

The results of numerous studies have documented an association between the viral identification rate and the severity of the wheezing illness, an association that argues for a causal relationship between the two [12,118,161, 162,200].

The individual viruses most likely to be associated with wheezing attacks vary with age, but rhinoviruses, RS virus and parainfluenza viruses are the predominant organisms, although it should be noted that in only one study were methods used specifically for coronavirus [23].

The associations listed above argue for a causal link, but are not conclusive, as the viral identification rates (typically 20–30% of exacerbations) in many of the studies were low (particularly for rhinoviruses and coronaviruses), and many studies had weaknesses in design, such as lack of objective assessment of the episodes being studied, and retrospective or cross-sectional design.

Virus detection rates have been increased remarkably by the recent development of PCR-based assays for the major respiratory viruses, most particularly for rhinoviruses [121,127] and coronaviruses [176], which are difficult to detect using standard methods.

These methods have been used recently in a detailed, intensive prospective study of asthma exacerbations in 108 school-age children in Southampton, UK. In this study, detailed diary card recordings were made of both upper and lower respiratory symptoms, and there were peak expiratory flow recordings for a 13-month period from April 1989 to 1990. Using precise objective criteria to define exacerbations, it was found that viral infections were associated with between 80% and 85% of exacerbations depending upon the method of definition of the episode [126]. An example of the asthma exacerbations studied and the virus infections associated with these episodes is depicted in a chart drawn of the recordings made by three of these children in Figure 1.3. The close association of the viral infections with the asthma exacerbations is clearly seen.

Rhinoviruses accounted for almost two-thirds of the viruses associated with exacerbations, with coronaviruses being next most common at 13% of viruses detected [126].

The same authors also examined the association between the seasonal pattern of viral

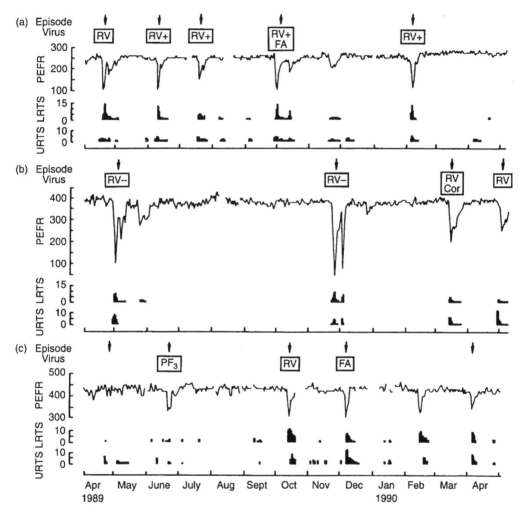

Figure 1.3 Example of charts drawn of the peak flow recordings and respiratory symptom scores for three of the children (A–C) taking part in the Southampton study of the relationship of upper respiratory viral infections to asthma exacerbations in school-age children. The horizontal axis represents the 13 months of the study. On the vertical axes, URTS and LRTS represent the upper and lower respiratory tract symptom scores respectively, PEFR represents the morning peak expiratory flow rate. Reported episodes are indicated by vertical arrows, and the viruses detected by the following symbols: RV, rhinovirus (+, major group; –, minor group); FA, influenza type A; Cor, coronavirus; PF_3, parainfluenza virus. Those reports with no such symbol had no virus detected. (Adapted from [126].)

respiratory infections in this sentinel cohort of children, and hospital admissions for asthma in both adults and children for the hospitals serving the areas from which the cohort was drawn. Strong correlations were found for both adult ($r = 0.53$; P <0.01) and paediatric ($r = 0.68$; P <0.0001) asthma admissions, though the relationship was clearly stronger in children [125]. Once again a profound influence of school attendance on infection rates (Figure 1.1), and therefore asthma exacerbation rates was observed, with 84% of combined adult and paediatric admissions occurring during school term times, compared with only 16% in school holiday periods [125] (P <0.0001).

A similar, but less intensive study of asthma exacerbations in adults has supported these findings, with viral infections being detected in 44% of adult asthma exacerbations, a rate much higher than any previous study in adults [179].

These recent studies confirm the findings of the previous studies, and in particular confirm the importance of rhinovirus infections, which were the major virus type detected. Further studies in different age groups, geographical locations and seasons, and using equivalent methods for all the respiratory viruses are now required before a fully informed picture will be gained.

1.7 CONCLUSIONS

The main causative agents of respiratory tract infections are now known to be a large number (well in excess of 200) of respiratory viruses. Rhinoviruses are the most important virus type in school-age children and adults, while RS virus is important in pre-school-age children and infants. Influenza virus type A causes periodic severe epidemics with appreciable morbidity and mortality.

The main vectors for transmission are preschool, and school-children, and the major factor influencing transmission rates is crowding together of children.

Detailed epidemiological studies of the Virus Watch type have provided a great deal of information with regard to the prevalence of infections, both symptomatic and asymptomatic, patterns of illness, and the influence of immunity, socioeconomic and other factors on disease prevalence. These studies were carried out with great thoroughness, but were hampered by the lack of availability of good detection methods. Many of the conclusions still hold true, particularly with regard to patterns of illness, rather than absolute numbers. The use of molecular methods of virus detection should allow a more precise picture of virus-related illness to be gained in the future.

Despite the many advances in medical science, infections caused by the majority of the agents discussed remain difficult to treat. Partial success has been achieved in the case of influenza virus type A infection, which is potentially preventable with vaccination, but there are great problems with delivery of the vaccine to those at greatest risk (the elderly, and those with chest and other serious disease), and with keeping abreast of antigenic variation.

Interest in common cold viruses has recently increased as a result of modern investigative techniques that have vastly improved our understanding of respiratory viruses, and which may in the future lead to the development of effective treatments [124].

1.8 REFERENCES

1. Akerlind, B. and Norrby, B. (1986) Occurrence of respiratory syncytial virus subtypes A and B strains in Sweden. *J. Med. Virol.*, **19**, 241–7.
2. Alford, C.A., Stagno, S. and Pass, R.F. (1980) Natural history of perinatal cytomegaloviral infection, in *Perinatal Infections*. Ciba Foundation Symposium. Amsterdam: Excerpta Medica, pp. 125–47.
3. Alford, C.A., Stagno, S., Pass, R.F. and Huang, E.S. (1981) Epidemiology of cytomegalovirus, in *The Human Herpesviruses: An Interdisciplinary Perspective*, (eds A. Nahmias *et al.*), Elsevier, New York, pp. 159–71.

4. Alling, D.W., Blackwelder, W.C. and Stuart-Harris, C.H. (1981) A study of excess mortality during influenza epidemics in the United States, 1968–1976. *Am. J. Epidemiol.*, **113**, 30–43.

5. Andrewes, C.H., Chapronieri, D.M., Gompels, A.E.H. *et al.* (1953) Propagation of common cold virus in tissue cultures. *Lancet*, **i**, 546–7.

6. de Arruda, E., Hayden, F.G., McAuliffe, J.F. *et al.* (1991) Acute respiratory infections in ambulatory children of urban northeast Brazil. *J. Infect. Dis.*, **164**, 252–8.

7. Barclay, W.S. and Al-Nakib, W. (1987) An ELISA for the detection of rhinovirus specific antibody in serum and nasal secretions. *J. Virol. Methods*, **15**, 53–64.

8. Bardin, P.G., Pattemore, P.K. and Johnston, S.L. (1992) Viruses as precipitants of asthma. II. Physiology and mechanisms. *Clin. Exp. Allergy*, **22**, 809–22.

9. Barker, W.H. and Mullooly, J.P. (1980) Impact of epidemic of type A influenza in a defined adult population. *Am. J. Epidemiol.*, **112**, 798–813.

10. Barker, W.H. and Mullooly, J.P. (1980) Influenza vaccinations of elderly persons. Reduction in pneumonia and influenza hospitalizations and deaths. *JAMA.*, **244**, 2547–9.

11. Barker, W.H. and Mullooly, J.P. (1982) Pneumonia and influenza deaths during epidemics. *Arch. Intern. Med.*, **142**, 85–9.

12. Beasley, R., Coleman, E.D., Hermon, Y. *et al.* (1988) Viral respiratory tract infection and exacerbations of asthma in adult patients. *Thorax*, **43**, 679–83.

13. Beem, M. (1969) Rhinovirus infections in nursery school children. *J. Pediatr.*, **74**, 818–20.

14. Bell, J.A., Rowe, W.P., Engler, J.I. et al. (1955) Pharyngoconjunctival fever. Epidemiological studies of a recently recognised disease entity. *JAMA*, **175**, 1083–92.

15. Bradburne, A.F., Bynoe, M.L. and Tyrrell, D.A.J. (1967) Effects of a 'new' human respiratory virus in volunteers. *Br. Med. J.*, **3**, 767–9.

16. Bradburne, A.F. and Somerset, B.A. (1972) Coronavirus antibody titers in sera of healthy adults and experimentally infected volunteers. *J. Hyg. (Lond)*, **70**, 235–44.

17. Brandt, C.D., Kim, H.W., Arrobio, J.O. *et al.* (1973) Epidemiology of respiratory syncytial virus in Washington D.C. III. Composite analysis of eleven yearly epidemics. *Am. J. Epidemiol.*, **98**, 355–64.

18. Brandt, C.D., Kim, H.W., Vargosdo, A.J. *et al.* (1969) Infections in 18,000 infants and children in a controlled study of respiratory tract disease. I. Adenovirus pathogenicity in relation to serologic type and illness syndrome. *Am. J. Epidemiol.*, **90**, 484–500.

19. Brown, P., Gajdusek, D.C. and Morris, J.A. (1966) Epidemic A2 influenza in isolated Pacific Island populations without pre-epidemic antibody to influenza virus types A and B, and discovery of other still unexposed populations. *Am. J. Epidemiol.*, **83**, 176–88.

20. Brown, P.K. and Taylor-Robinson, D. (1966) Respiratory virus antibodies in sera of persons living in isolated communities. *Bull. WHO.*, **34**, 895–900.

21. Buscho, R.O., Saxtan, D., Schultz, P.S., Finch, E. and Mufson, M.A. (1978) Infections with viruses and *Mycoplasma pneumoniae* during exacerbations of chronic bronchitis. *J. Infect. Dis.*, **137**, 377–83.

22. Calhoun, A.M., Jordan, W.S. Jr and Gwaltney, J.M. Jr (1974) Rhinovirus infections in an industrial population. V. Change in distribution of serotypes. *Am. J. Epidemiol.*, **99**, 58–64.

23. Carlsen, K.H., Orstavik, I., Leegaard, J. and Hoeg, H. (1984) Respiratory virus infections and aeroallergens and acute bronchial asthma. *Arch. Dis. Child.*, **59**, 310–15.

24. Chanock, R.M. and Parrott, R.H. (1965) Acute respiratory disease in infancy and childhood: present understanding and prospects for prevention. *Pediatrics*, **36**, 21–39.

25. Chanock, R.M., Parrott, R.H., Johnson, K.M. *et al.* (1963) Myxoviruses: parainfluenza. *Am. Rev. Respir. Dis.*, **88**, 152–66.

26. Chapman, L.E., Tipple, M.A., Schmeltz, L.M. *et al.* (1992) Influenza – United States, 1989–90 and 1990–91 seasons. *MMWR CDC Surveillance Summaries*, **41**, 35–46.

27. Chin, T.-D.Y., Mosley, W.H., Poland, J.D. *et al.* (1963) Epidemiologic studies of type B influenza in 1961–62. *Am. J. Public Health*, **53**, 1068–74.

28. Clarke, S.K.R., Gardner, P.S., Poole, P.M., Simpson, H. and Tobin, J.O.H. (1978) Respiratory syncytial virus infection: admissions to hospital in industrial urban, and rural areas: report to the Medical Research Council subcommittee on respiratory syncytial virus vaccines. *Br. Med. J.*, **2**, 796–8.

29. Clough, J.B. (1993) The effect of gender on the prevalence of atopy and asthma. *Clin. Exp. Allergy*, **23**, 883–5.
30. Colman, P.M. and Ward, C.W. (1985) Structure and diversity of influenza virus neuraminidase. *Curr. Top. Microbiol. Immunol.*, **11**, 177–255.
31. Cooney, M.K., Fox, J.P. and Hall, C.E. (1975) The Seattle Virus Watch. VI. Observations of infections with and illness due to parainfluenza, mumps and respiratory syncytial viruses and *Mycoplasma pneumoniae*. *Am. J. Epidemiol.*, **101**, 532–51.
32. Cooney, M.K., Hall, C.E. and Fox, J.P. (1972) The Seattle Virus Watch. III. Evaluation of isolation methods and summary of infections, detected by virus isolations. *Am. J. Epidemiol.*, **96**, 286–305.
33. Cooney, M.K., Wise, J.A., Kenny, G.E. and Fox, J.P. (1975) Broad antigenic relationships among rhinovirus serotypes revealed by cross-immunization of rabbits with different serotypes. *J. Immunol.*, **114**, 635–9.
34. Cooper, P.D., Agol, V.I., Bachrach, H.L. *et al.* (1978) Picornaviridae: second report. *Intervirology*, **10**, 165–80.
35. Couch, R.B. (1984) The common cold: control? *J. Infect. Dis.*, **150**, 167–73.
36. Couch, R.B., Cate, T.R., Douglas, R.G. Jr *et al.* (1966) Effect of route of inoculation on experimental volunteers and evidence for airborne transmission. *Bacteriol. Rev.*, **30**, 517–29.
37. Couch, R.B., Douglas, R.G. Jr, Lindgren, K.M., Gerone, P.J. and Knight, V. (1970) Airborne transmission of respiratory infection with coxsackievirus A type 21. *Am. J. Epidemiol.*, **91**, 78–86.
38. Cox, N.J., Bai, Z.S. and Kendall, A.P. (1983) Laboratory-based surveillance of influenza A (H1N1) and A (H3N2) viruses in 1980–81: antigenic and genomic analyses. *Bull. WHO*, **61**, 143–52.
39. Craighead, J.E., Meier, M. and Cooley, M.H. (1969) Pulmonary infection due to rhinovirus type 13. *N. Engl. J. Med.*, **281**, 1403–4.
40. Curry, R.L., Brown, J.D., Baker, F.A. and Hobson, D. (1974) Serological studies with purified neuraminidase antigens of influenza B viruses. *J. Hyg.*, **72**, 197–204.
41. D'Alessio, D.J., Meschievitz, C.K., Peterson, J.A. *et al.* (1984) Short-duration exposure and the transmission of rhinoviral colds. *J. Infect. Dis.*, **150**, 189–94.
42. D'Alessio, D.J., Peterson, J.A., Dick, C.R. and Dick, E.C. (1976) Transmission of experimental rhinovirus colds in volunteer married couples. *J. Infect. Dis.*, **133**, 28–36.
43. Dalldorf, G. and Melnick, J.L. (1965) Coxsackieviruses, in *Viral and Rickettsial Infections of Man*, 4th edn, (eds F.L. Hosfall Jr and I. Tamm), Lippincott, Philadelphia, pp. 474–512.
44. Davies, J.R., Grilli, E.A. and Smith, A.J. (1984) Influenza A: infection and reinfection. *J. Hyg. (Camb.)*, **92**, 125–7.
45. Davies, J.R., Grilli, E A. and Smith, A.J. (1986) 1. Infection with influenza A H1N1, 2. The effect of past experience on natural challenge. *J. Hyg. (Camb.)*, **96**, 345–52.
46. Denny, F.W., Murphy, T.F., Clyde, W.A. Jr *et al.* (1983) An 11 year study in a pediatric practice. *Pediatrics*, **71**, 871–6.
47. Dick, E.C., Jennings, L.C., Mink, K.A. *et al.* (1987) Aerosol transmission of rhinovirus colds. *J. Infect. Dis.*, **156**, 442–8.
48. Dick, E.C. and Chesney, P.J. (1981) Rhinoviruses, in *Textbook of Paediatric Infectious Diseases*, Vol. 11, (eds R.D. Feigin and J.D. Cherry), W.B. Saunders, Philadelphia, pp. 1167–86.
49. Dochez, A.R., Shibley, G.S. and Mills, K.C. (1930) Studies in the common cold. IV. Experimental transmission of the common cold to anthropoid apes and human beings by means of a filterable agent. *J. Exp. Med.*, **52**, 701–16.
50. Douglas, R.G. Jr (1970) Pathogenesis of rhinovirus common colds in human volunteers. *Ann. Otol. Rhinolaryngol.*, **79**, 563–71.
51. Elveback, L.R., Fox, J.P., Ketler, A. *et al.* (1966) The Virus Watch Program: a continuing surveillance of viral infections in metropolitan New York families. lll. Preliminary report on association of infections with disease. *Am. J. Epidemiol.*, **83**, 436–54.
52. Farr, B.M., Hendley, J.O., Kaiser, D.L. and Gwaltney, J.M. (1988) Two randomized controlled trials of virucidal nasal tissues in the prevention of natural upper respiratory tract infection. *Am. J. Epidemiol.*, **128**, 1162–72.
53. Fawzy, K.Y., Fox, J.P., Ketler, A. *et al.* (1967) The Virus Watch program: a continuing surveillance of viral infections in metropolitan New York families. V. Observations in employed adults on etiology of acute upper respiratory

disease and heterologous antibody response to rhinoviruses. *Am. J. Epidemiol.*, **86**, 653–72.

54. Forsyth, B.R., Bloom, H.H., Johnson, K.M. and Chanock, R.M. (1963) Patterns of illness in rhinovirus infections of military personnel. *N. Engl. J. Med.*, **269**, 602–6.

55. Fox, J.P. (1976) Reviews and commentary – Is a rhinovirus vaccine possible? *Am. J. Epidemiol.*, **103**, 345–54.

56. Fox, J.P., Brandt, C.D., Wassermann, F.E. *et al.* (1969) The Virus Watch program: a continuing surveillance of viral infections in metropolitan New York families. VI. Observations of adenovirus infections: virus excretion patterns, antibody response, efficiency of surveillance, patterns of infection, and relation to illness. *Am. J. Epidemiol.*, **89**, 25–50.

57. Fox, J.P., Cooney, M.K. and Hall, C.E. (1975) The Seattle Virus Watch. V. Epidemiologic observations of rhinovirus infections, 1965–1969, in families with young children. *Am. J. Epidemiol.*, **101**, 122–43.

58. Fox, J.P., Cooney, M.K., Hall, C.E. *et al.* (1985) Rhinoviruses in Seattle families, 1975–1979. *Am. J. Epidemiol.*, **122**, 830–46.

59. Fox, J.P., Elveback, L.R., Spigland, I. *et al.* (1966) The Virus Watch program: a continuing surveillance of viral infections in metropolitan New York families. l. Overall plan, methods of collecting and handling information, and a summary report of specimens collected and illnesses observed. *Am. J. Epidemiol.*, **83**, 389–412.

60. Fox, J.P., Hall, C.E., Cooney, M.K. *et al.* (1972) The Seattle Virus Watch. II. Objectives, study population and its observation, data processing and summary of illnesses. *Am. J. Epidemiol.*, **96**, 270–85.

61. Fox, J.P., Hall, C.E. and Cooney, M.K. (1977) The Seattle Virus Watch. Vll. Observations of adenovirus infections. *Am. J. Epidemiol.*, **105**, 362–86.

62. Fox, J.P., Hall, C.E., Cooney, M.K. *et al.* (1982) Influenza virus infections in Seattle families, 1975–1979. I. Study design, methods and the occurrence of infections by time and age. *Am. J. Epidemiol.*, **116**, 212–27.

63. Fox, J.P., Hall, C.E., Cooney, M.K. *et al.* (1982) Influenza virus infections in Seattle families, 1975–1979. II. Pattern of infection in invaded households and relation of age and prior antibody to occurrence of infection and related illness. *Am. J. Epidemiol.*, **116**, 228–42.

64. Foy, H.M., Cooney, M.K., Allan, I.D. and Albrecht, J.K. (1987) Influenza B in households: virus shedding without symptoms or antibody response. *Am. J. Epidemiol.*, **126**, 506–15.

65. Frank, A.L., Taber, L.H., Glezen, W.P., Paredes, A. and Couch, R.B. (1979) Reinfection with influenza A (H3N2) virus in young children and their families. *J. Infect. Dis.*, **140**, 829–36.

66. Frank, A.L., Taber, L.H. and Porter, C.M. (1987) Influenza B virus reinfection. *Am. J. Epidemiol.*, **125**, 576–86.

67. Frank, A.L., Taber, L.H., Wells, C.R. *et al.* (1981) Patterns of shedding of myxoviruses and paramyxoviruses in children. *J. Infect. Dis.*, **144**, 433–41.

68. Frank, A.L., Taber, L.H. and Wells, J.M. (1983) Individuals infected with two subtypes of influenza A virus in the same season. *J. Infect. Dis.*, **147**, 120–4.

69. Frank, A.L., Taber, L.H. and Wells, J.M. (1985) Comparison of infection rates and severity of illness for influenza A subtypes H1N1 and H3N2. *J. Infect. Dis.*, **151**, 73–80.

70. Gardner, P.S. (1973) Respiratory syncytial virus infections. *Postgrad. Med. J.*, **49**, 788–91.

71. Gardner, P.S., Court, S.D.M., Brocklebank, J.T., Downham, M.A.P.S. and Weightman, D. (1973) Virus cross-infection in paediatric wards. *Br. Med. J.*, **2**, 571–5.

72. Gething, M.J., Bye, J., Skehel, J. and Wakefield, M. (1980) Cloning and DNA sequence of double-stranded copies of haemagglutinin genes from H2 and H3 strains elucidates antigenic shift and drift in human influenza virus. *Nature*, **287**, 301–6.

73. Gill, P.W. and Murphy, A.M. (1985) Naturally acquired immunity to influenza type A. Lessons from two coexisting subtypes. *Med. J. Aust.*, **142**, 94–8.

74. Glezen, W.P. (1977) Pathogenesis of bronchiolitis – epidemiologic considerations. *Pediatr. Res.*, **11**, 239–43.

75. Glezen, W.P. (1980) Consideration of the risk of influenza in children and indications for prophylaxis. *Rev. Infect. Dis.*, **2**, 408–20.

76. Glezen, W.P. and Denny, F.W. (1973) Epidemiology of acute lower respiratory disease in children. *N. Engl. Med. J.*, **288**, 498–505.

77. Glezen, W.P., Frank, A.L., Taber, L.H. *et al.* (1984) Parainfluenza virus type 3: seasonality and risk of infection and reinfection in young children. *J. Infect. Dis.*, **150**, 851–7.

78. Glezen, W.P., Loda, F.A. and Denny, F.W. (1982) Parainfluenza viruses, in *Viral Infections of Humans. Epidemiology and Control*, (ed. A.S. Evans), Plenum Press, New York, pp. 441–54.

79. Glezen, W.P., Paredes, A., Allison, J.E. *et al.* (1981) Risk of respiratory syncytial virus infection for infants from low-income families in relationship to age, sex, ethnic group and maternal antibody level. *J. Pediatr.*, **98**, 708–15.

80. Glezen, W.P., Taber, L.H., Frank, A.L. and Kasel, J.A. (1986) Risk of primary infection and reinfection with respiratory syncytial virus. *Am. J. Dis. Child.*, **140**, 543–6.

81. Gold, E. and Nankervis, G.A. (1982) Cytomegalovirus, in *Viral Infections of Humans: Epidemiology and Control*, 2nd edn, (ed. A.S. Evans), Plenum Press, New York, pp. 167–86.

82. Greenberg, M., Jacobziner, H., Pakter, J. and Weisl, B.A.G. (1958) Maternal mortality in the epidemic of Asian influenza, New York City, 1957. *Am. J. Obstet. Gynecol.*, **76**, 897–902.

83. Groothuis, J.R., Gutierrez, K.M. and Lauer, B.A. (1988) Respiratory syncytial virus infection in children with bronchopulmonary dysplasia. *Pediatrics*, **82**, 199–203.

84. Gross, P.A., Green, R.H. and Curnen, M.G.M. (1973) Persistent infection with parainfluenza type 3 virus in man. *Am. Rev. Respir. Dis.*, **108**, 894–8.

85. Gump, D.W., Phillips, C.A., Forsyth, B.R. *et al.* (1976) Role of infection in chronic bronchitis. *Am. Rev. Respir. Dis.*, **113**, 465–74.

86. Gwaltney, J.M. (1982) Rhinoviruses, *in Viral Infections of Humans: Epidemiology and Control*, 2nd edn, (ed. A.S. Evans), Plenum Press, New York, pp. 491–517.

87. Gwaltney, J.M., Hendley, J.O., Simon, G. and Jordan, W.S. Jr (1966) Rhinovirus infections in an industrial population. I. The occurrence of illness. *N. Engl. J. Med.*, **275**, 1261–8.

88. Gwaltney, J.M. Jr and Jordan, W.S. Jr (1966) Rhinoviruses and respiratory illnesses in university students. *Am. Rev. Respir. Dis.*, **93**, 363–71.

89. Gwaltney, J.M. Jr, Moskalski, P.B. and Hendley, J.O. (1978) Hand-to-hand transmission of rhinovirus colds. *Ann. Intern. Med.*, **88**, 463–7.

90. Hall, C.B. (1981) Nosocomial viral respiratory infections: Perennial weeds on pediatric wards. *Am. J. Med.*, **70**, 670–6.

91. Hall, C.E., Cooney, M.K. and Fox, J.P. (1973) The Seattle Virus Watch. IV. Comparative epidemiologic observations of infections with influenza A and B viruses, 1965–1969, in families with young children. *Am. J. Epidemiol.*, **98**, 365–80.

92. Hall, C.B. and Douglas, R.G. (1981) Modes of transmission of respiratory syncytial virus. *J. Pediatr.*, **99**, 100–3.

93. Hall, C.B. and Douglas, R.G. (1981) Nosocomial respiratory syncytial virus infections: should gowns and masks be used? *Am. J. Dis. Child.*, **135**, 512–15.

94. Hall, C.B., Douglas, R.G., Geiman, J.M. *et al.* (1975) Nosocomial respiratory syncytial virus infections. *N. Engl. J. Med.*, **293**, 1343–6.

95. Hall, C.B., Geiman, J.M. and Douglas, R.G. (1980) Possible transmission by fomites of respiratory syncytial virus. *J. Infect. Dis.*, **141**, 98–102.

96. Hall, C.B., Kopelman, A.E., Douglas, R.G. *et al.* (1979) Neonatal respiratory syncytial virus infection. *N. Engl. J. Med.*, **300**, 393–6.

97. Hall, C.B., Powell, K.R., MacDonald, N.E. *et al.* (1986) Respiratory syncytial viral infection in children with compromised immune function. *N. Engl. J. Med.*, **315**, 77–81.

98. Hall, C.E., Brandt, C.D., Frothingham, T.E. *et al.* (1971) The Virus Watch program: A continuing surveillance of viral infections in metropolitan New York families. IX. A comparison of infections with several respiratory pathogens in New York and New Orleans families. *Am. J. Epidemiol.*, **94**, 367–85.

99. Hall, C.E., Cooney, M.K. and Fox, J.P. (1970) The Seattle Virus Watch Program. I. Infection and illness experience of Virus Watch families during a community-wide epidemic of echovirus type 30 aseptic meningitis. *Am. J. Public Health*, **60**, 1456–65.

100. Hamparian, V.V., Colonno, R.J., Cooney, M.K. *et al.* (1987) A collaborative report: rhinoviruses–extension of the numbering system from 89 to 100. *Virology*, **159**, 191–2.

101. Hamre, D. (1968) *Rhinoviruses*, in Monographs in Virology series, (ed. J.L. Melnick), Karger, Basle, Switzerland.

102. Hamre, D. and Beem, M. (1972) Virologic studies of acute respiratory disease in young adults. V. Coronavirus 229E infections during six years of surveillance. *Am. J. Epidemiol.*, **96**, 94–106.

103. Hamre, D., Cooney, A.P. Jr and Procknow, J.J. (1966) Virologic studies in acute respiratory disease. IV. Virus isolation during four years of surveillance. *Am. J. Epidemiol.*, **83**, 238–49.

104. Hamre, D. and Procknow, J.J. (1963) Virologic studies on common colds among young adult medical students. *Am. Rev. Respir. Dis.*, **88**, 277–89.

105. Hasony, H.J. and MacNaughton, M.R. (1982) Prevalence of human coronavirus antibody in the population of Southern Iraq. *J. Med. Virol.*, **9**, 209–16.

106. Henderson, F.W., Collier, A.M., Clyde, W.A. *et al.* (1979) Respiratory syncytial virus infections: reinfections and immunity: a prospective, longitudinal study in young children. *N. Engl. J. Med.*, **300**, 530–4.

107. Henderson, F.W., Clyde, W.A., Collier, A.M. *et al.* (1979) The etiologic and epidemiologic spectrum of bronchiolitis in pediatric practice. *J. Pediatr.*, **95**, 183–90.

108. Hendley, J.O., Gwaltney, J.M. Jr and Jordan, W.S. Jr (1969) Rhinovirus infections in an industrial population. IV. Infection within families of employees during two fall peaks of respiratory illness. *Am. J. Epidemiol.*, **89**, 184–96.

109. Hendley, J.O., Wenzel, R.P. and Gwaltney, J.M. Jr (1973) Transmission of rhinovirus colds by self-inoculation. *N. Engl. J. Med.*, **288**, 1361–4.

110. Hendry, R.M., Talis, A.L., Godfrey, E. *et al.* (1986) Concurrent circulation of antigenically distinct strains of respiratory syncytial virus during community outbreaks. *J. Infect. Dis.*, **153**, 291–7.

111. Hilleman, M.R. (1957) Epidemiology of adenovirus respiratory infection in military recruit populations. *Ann. N.Y. Acad. Sci.*, **67**, 262–72.

112. Hilleman, W.R., Werner, J.H., Dascomb, H.E. *et al.* (1955) Epidemiology of RI (RI-67) group respiratory virus infections in recruit populations. *Am. J. Hyg.*, **62**, 29–43.

113. Hillis, W.D., Cooper, M.R., Bang, F.B., Dey, A.K. and Shah, K.V. (1971) Respiratory syncytial virus infection in West Bengal. *Indian J. Med. Res.*, **59**, 1354–64.

114. Hinshaw, V.S. and Webster, R.G. (1982) The natural history of influenza A viruses, in *Basic and Applied Influenza Research*, (ed. A.S. Beare), CRC Press, Boca Raton, Florida, pp. 79–104.

115. Holmes, M.J., Reed, S.E., Stott, E.J. and Tyrrell, D.A.J. (1976) Studies of experimental rhinovirus type 2 infections in polar isolations and in England. *J. Hyg. (Lond.)*, **76**, 379–93 .

116. Horn, M.E.C., Brain, E., Gregg, I. *et al.* (1975) Respiratory viral infection in children: a survey in general practice, Roehampton 1967. *J. Hyg. (Camb.)*, **74**, 157–68.

117. Horn, M.E.C., Brain, E.A., Gregg, I. *et al.* (1979) Respiratory viral infection and wheezy bronchitis in childhood. *Thorax*, **34**, 23–8.

118. Horn, M.E.C., Reed, S.E. and Taylor, P. (1979) Role of viruses and bacteria in acute wheezy bronchitis in childhood: a study of sputum. *Arch. Dis. Child.*, **54**, 587–92.

119. Hudgel, D.W., Langston, L., Selner, J.C. and McIntosh, K. (1979) Viral and bacterial infections in adults with chronic asthma. *Am. Rev. Respir. Dis.*, **120**, 393–7.

120. Hutto, S.C., Ricks, R., Garvie, M. and Pass, R.F. (1985) Epidemiology of cytomegalovirus infections in young children: day care vs home care. *Pediatr. Infect. Dis.*, **4**, 149–52.

121. Ireland, D.C., Kent, J. and Nicholson, K.G. (1993) Improved detection of rhinoviruses in nasal and throat swabs by seminested RT-PCR. *J. Med. Virol.*, **40**, 96–101.

122. Jennings, L.C., Barns, G. and Dawson, K.P. (1987) The association of viruses with acute asthma. *N. Z. Med. J.*, **100**, 488–90.

123. Johnson, P.R., Feldman, S., Thompson, J.M., Mahoney, J.D. and Wright, P.E. (1986) Immunity to influenza A virus infection in young children. A comparison of natural infection, live cold-adapted vaccine, and inactivated vaccine. *Am. J. Infect. Dis.*, **154**, 121–7.

124. Johnston, S.L., Bardin, P.G. and Pattemore, P.K. (1993) Viruses as precipitants of asthma symptoms III. Rhinoviruses: molecular biology and prospects for future intervention. *Clin. Exp. Allergy*, **23**, 237–46.

125. Johnston, S.L., Pattemore, P.K., Campbell, M.J. *et al.* (1995) The association of upper respiratory infections with hospital admissions for asthma in adults and children: a time trend analysis. *Am. J. Respir. Crit. Care Med.* (in press).

126. Johnston, S.L., Pattemore, P.K., Lampe, F. *et al.* (1994) Community study of role of virus infections in exacerbations of asthma in 9–11 year old children. *Br. Med. J.*, **310**, 1225–8.

127. Johnston, S.L., Sanderson, G., Pattemore, P.K. *et al.* (1993) Use of polymerase chain reaction for diagnosis of picornavirus infection in sub-

jects with and without respiratory symptoms. *J. Clin. Microbiol.*, **31**, 111–17.

128. Jordan, W.S. Jr (1961) The mechanism of spread of Asian influenza. *Am. Rev. Respir. Dis.*, **83**, 29–40.

129. Jordan, W.S. Jr, Badger, G.F., Curtiss, C. *et al.* (1956) A study of illness in a group of Cleveland families. X. The occurrence of adenovirus infections. *Am. J. Hyg.*, **64**, 336–48.

130. Kapikian, A.Z., Bell, J.A., Mastrota, F.M. *et al.* (1961) An outbreak of febrile illness and pneumonia associated with respiratory syncytial virus infection. *Am. J. Hyg.*, **74**, 234–8.

131. Kapikian, A.Z., Bell, J.A., Mastrota, F.M. *et al.* (1963) An outbreak of parainfluenza 2 (croup associated) virus infection. *JAMA.*, **183**, 324–30.

132. Kapikian, A.Z., Chanock, R.M., Reichelderfer, T.E. *et al.* (1961) Inoculation of human volunteers with parainfluenza virus type 3. *JAMA.*, **18**, 537–41.

133. Kapikian, A.Z., Conant, R.M., Hamparian, V.V. *et al.* (1971) Collaborative report: rhinoviruses—extension of numbering system. *Virology*, **43**, 524–6.

134. Kapikian, A.Z., James, H.D. Jr, Kelly, S.J. *et al.* (1969) Isolation from man of 'avian infectious bronchitis-like' viruses (coronaviruses) similar to 229E virus, with some epidemiological observations. *J. Infect. Dis.*, **119**, 282–90.

135. Kaye, H.S., Marsh, H.B. and Dowdle, W.R. (1971) Seroepidemiologic survey of coronavirus (strain OC43) related infections in a children's population. *Am. J. Epidemiol.*, **94**, 43–9.

136. Kellner, G., Popow-Kraupp, T., Kundi, M., Binder, C. and Kunz, C. (1989) Clinical manifestations of respiratory tract infections due to respiratory syncytial virus and rhinoviruses in hospitalized children. *Acta Paediatr. Scand.*, **78**, 390–4.

137. Ketler, A., Hall, C.E., Fox, J.P. *et al.* (1969) The Virus Watch program: a continuing surveillance of viral infections in metropolitan New York families. VIII. Rhinovirus infections: observations of virus excretion, intrafamilial spread and clinical response. *Am. J. Epidemiol.*, **90**, 244–54.

138. Kida, H., Kawaoka, Y., Naeve, C.W. and Webster, R.G. (1987) Antigenic and genetic conservation of H3 influenza in wild ducks. *Virology*, **159**, 109–19.

139. Kim, H.W., Arrobio, J.O., Brandt, C.D. *et al.* (1973) Epidemiology of respiratory syncytial virus in Washington D.C. I. Importance of the virus in different respiratory tract disease syndromes and temporal distribution of infection. *Am. J. Epidemiol.*, **98**, 216–25.

140. Kim, H.W., Brandt, C.D., Arrobio, J.O., Murphy, B., Chanock, R.M. and Parrott, R.H. (1979) Influenza A and B infection in infants and young children during the years 1957–76. *Am. J. Epidemiol.*, **109**, 464–79.

141. Klemola, E., von Essen, R., Wager, O. *et al.* (1969) Cytomegalovirus mononucleosis in previously healthy individuals. *Ann. Intern. Med.*, **71**, 11–19.

142. Kogon, A., Spigland, I., Frothingham, T.E. *et al.* (1969) The Virus Watch program: a continuing surveillance of viral infections in metropolitan New York families. VII. Observations on viral excretion, seroimmunity, intrafamilial spread and illness association in coxsackie and echovirus infections. *Am. J. Epidemiol.*, **89**, 51–61.

143. Krilov, L., Pierik, L., Keller, E. *et al.* (1986) The association of rhinoviruses with lower respiratory tract disease in hospitalized patients. *J. Med. Virol.*, **19**, 345–52.

144. Kumar, M.L., Nankervis, G.A. and Gold, E. (1973) Inapparent congenital cytomegalovirus infection: a follow-up study. *N. Engl. J. Med.*, **288**, 1370–2.

145. Lang, W.R., Howden, C.W., Laws, J. and Burton, J.F. (1961) Bronchopneumonia with serious sequelae in children with evidence of adenovirus type 21 infection. *Br. Med. J.*, **1**, 73–9.

146. Larson, H.E., Reed, S.E. and Tyrrell, D.A.J. (1980) Isolation of rhinoviruses and coronaviruses from 38 colds in adults. *J. Med. Virol.*, **5**, 221–9.

147. Las Heras, J. and Swanson, V.L. (1983) Sudden death of an infant with rhinovirus infection complicating bronchial asthma. *Paediatr. Pathol.*, **1**, 319–23.

148. Laver, W.G., Air, G.M., Webster, R.G. and Markoff, L.J. (1982) Amino acid sequence changes in antigenic variants of type A influenza virus N2 neuraminidase. *Virology*, **122**, 450–60.

149. LeClair, M.M., Freeman, J., Sullivan, B.F., Crowley, C.M. and Goldmann, D.A. (1987) Prevention of nosocomial respiratory syncytial virus infections through compliance with glove

and gown isolation precautions. *N. Engl. J. Med.*, **317**, 329–34.

150. Lemanske, R.F., Dick, E.C., Swenson, C.A., Virtis, R.F. and Busse, W.W. (1989) Rhinovirus upper respiratory tract infection increases airway hyperreactivity and late asthmatic reactions. *J. Clin. Invest.*, **83**, 1–10.

151. Liu, K.-J. and Kendal, A.P. (1987) Impact of influenza epidemics on mortality in the United States from October 1972 to May 1985. *Am. J. Public Health*, **77**, 712–16.

152. MacDonald, N.E., Hall, C.B., Suffin, S.C. *et al.* (1982) Respiratory syncytial virus infection in infants with congenital heart disease. *N. Engl. J. Med.*, **307**, 397–400.

153. McIntosh, K., Ellis, E.F., Hoffman, L.S. *et al.* (1973) The association of viral and bacterial respiratory infections with exacerbations of wheezing in young asthmatic children. *J. Pediatr.*, **82**, 578–90.

154. McIntosh, K., Kapikian, A.Z., Turner, H.C. *et al.* (1970) Seroepidemiologic studies of coronavirus infections in adults and children. *Am. J. Epidemiol.*, **91**, 585–92.

155. McMillan, J.A., Weiner, L.B., Higgins, A.M. and MacKnight, K. (1993) Rhinovirus infection associated with serious illness among pediatric patients. *Pediatr. Infect. Dis. J.*, **12**, 321–5.

156. Madge, P., Paton, J.Y., McColl, J.H. and Mackie, P.L. (1992) Prospective controlled study of four infection-control procedures to prevent nosocomial infection with respiratory syncytial virus. *Lancet*, **340**, 1079–83.

157. Melnick, J.L. (1965) Echoviruses, in *Viral and Rickettsial Infections of Man*, 4th edn, (eds F.L. Hosfall Jr and W. Tamm), Lippincott, Philadelphia, pp. 513–45.

158. Melnick, J.L. (1980) Taxonomy of viruses. *Prog. Med. Virol.*, **26**, 214–32.

159. Mertsola, J., Ziegler, T., Ruuskanen, O. *et al.* (1991) Recurrent wheezy bronchitis and viral respiratory infections. *Arch. Dis. Child.*, **66**, 124–9.

160. Milner, M.E., de la Monte, S.M. and Hutchins, G.M. (1985) Fatal respiratory syncytial virus infection in severe combined immunodeficiency syndrome. *Am. J. Dis. Child.*, **135**, 1111–14.

161. Minor, T.E., Dick, E.C., DeMeo, A.N. *et al.* (1974) Viruses as precipitants of asthmatic attacks in children. *JAMA.*, **227**, 292–8.

162. Mitchell, I., Inglis, H. and Simpson, H. (1976) Viral infections in wheezy bronchitis and asthma in children. *Arch. Dis. Child.*, **51**, 707–11.

163. Mitchell, I., Inglis, J.M. and Simpson, H. (1978) Viral infection as a precipitant of wheeze in children: combined home and hospital study. *Arch. Dis. Child.*, **53**, 106–11.

164. Mogabdab, W.J. (1968) Acute respiratory illnesses in university (1962–1966), military and industrial (1962–1963) populations. *Am. Rev. Respir. Dis.*, **98**, 359–79.

165. Mogabdab, W.J. (1975) Antigenic relationships of common rhinovirus types from disabling upper respiratory illnesses. *Dev. Biol. Stand.*, **28**, 400–11.

166. Monto, A.S. (1968) A community study of respiratory infection in the tropics. I. Introduction and transmission of infection without families. *Am. J. Epidemiol.*, **88**, 69–79.

167. Monto, A.S. (1974) Coronaviruses. *Yale J. Biol. Med.*, **47**, 234–51.

168. Monto, A.S., Bryan, E.R. and Ohmit, S. (1987) Rhinovirus infections in Tecumseh, Michigan: frequency of illness and number of serotypes. *J. Infect. Dis.*, **156**, 43–9.

169. Monto, A.S. and Johnson, K.M. (1968) A community study of respiratory infection in the tropics. II. The spread of six rhinovirus isolates within the community. *Am. J. Epidemiol.*, **88**, 55–68.

170. Monto, A.S. and Lim, S.K. (1974) The Tecumseh study of respiratory illness. VI. Frequency of and relationship between outbreaks of coronavirus infection. *J. Infect. Dis.*, **129**, 271–6.

171. Monto, A.S. and Ullman, B.M. (1974) Acute respiratory illness in an American community: The Tecumseh study. *JAMA.*, **227**, 164–9.

172. Moore, M. (1982) From the Centers for Disease Control: Enteroviral disease in the United States, 1970–79. *J. Infect. Dis.*, **146**, 103–8.

173. Mufson, M.A., Belshe, R.B., Örvell, C. and Norrby, E. (1987) Subgroup characteristics of respiratory syncytial virus strains recovered from children with two consecutive infections. *J. Clin. Microbiol.*, **25**, 1535–9.

174. Mufson, M.A., Levine, H.D., Walsh, R.E., Mocega-Gomzales, H.E. and Krause, H.E. (1973) Epidemiology of respiratory syncytial virus infections among infants and children in Chicago. *Am. J. Epidemiol.*, **98**, 88–95.

175. Mufson, M.A., Webb, P.A., Kennedy, H., Gill, V. and Chanock, R.M. (1966) Etiology of upper

respiratory tract illness among civilian adults. *JAMA.*, **195**, 1–7.

176. Myint, S.H., Johnston, S.L., Sanderson, G. and Simpson, H. (1994) Evaluation of nested polymerase chain methods for the detection of human coronaviruses 229E and OC43. *Mol. Cell. Probes* , **8**, 357–64.

177. Nakajima, K., Desselberger, U. and Palese, P. (1978) Recent human influenza A (H1N1) viruses are closely related genetically to strains isolated in 1950. *Nature*, **274**, 334–9.

178. Nasciemento, J.P., Siqueira, M.M., Sutmolleri, F. *et al.* (1991) Longitudinal study of acute respiratory disease in Rio de Janeiro: occurrence of respiratory viruses during four consecutive years. *Rev. Inst. Med. Trop. Sao Paulo*, **33**, 287–96.

179. Nicholson, K.G., Kent, J. and Ireland, D.C. (1993) Respiratory viruses and exacerbations of asthma in adults. *Br. Med. J.*, **307**, 982–6.

180. Noble, G.R. (1982) Epidemiological and clinical aspects of influenza, in *Basic and Applied Influenza Research*, (ed. A.S. Beare), CRC Press, Boca Raton, Florida, pp. 11–50.

181. Nolan, T.F., Goodman, R.A., Hinman, A.R. *et al.* (1980) Morbidity and mortality associated with influenza B in the United States, 1979–1980. A report from the Centers for Disease Control. *J. Infect. Dis.*, **142**, 360–2.

182. Orstavik, I., Carlsen, K.-H. and Halvorsen, K. (1980) Respiratory syncytial virus infection in Oslo 1972–78. I. Virological and epidemiological studies. *Acta Pediatr. Scand.*, **69**, 717–22.

183. Oxford, J.S., Abbo, H., Corcoran, T. *et al.* (1983) Antigenic and biochemical analysis of influenza B viruses from a single epidemic. *J. Gen. Virol.*, **64**, 2367–77.

184. Paniker, C.K. (1968) Serologic relationships between the neuraminidases in influenza viruses. *J. Gen. Virol.*, **2**, 385–94.

185. Parrott, R.H., Kim, H.W., Arrobio, J.O. *et al.* (1973) Epidemiology of respiratory syncytial virus infection in Washington, D.C. II. Infection and disease with respect to age, immunologic status, race and sex. *Am. J. Epidemiol.*, **98**, 289–300.

186. Parrott, R.H., Vargosko, A., Kim, H.W. *et al.* (1962) Myxoviruses. III. Parainfluenza. *Am. J. Public Health*, **52**, 907–17.

187. Parrott, R.H., Vargosko, A., Luckey, A. *et al.* (1959) Clinical features of infection with hemadsorption viruses. *N. Engl. J. Med.*, **260**, 731–8.

188. Pass, R.F., August, A.M., Dworsky, M.E. and Reynolds, D.W. (1982) Cytomegalovirus infection in a day care center. *N. Engl. J. Med.*, **307**, 477–9.

189. Pass, R.F. and Kinney, J.S. (1985) Child care workers and children with congenital cytomegalovirus infection. *Pediatrics*, **75**, 971–3.

190. Pass, R.F., Little, E.A., Stagno, S. *et al.* (1987) Young children as a probable source of maternal and congenital cytomegalovirus infection. *N. Engl. J. Med.*, **316**, 1366–70.

191. Pattemore, P.K., Johnston, S.L. and Bardin, P.G. (1992) Viruses as precipitants of asthma. I. Epidemiology. *Clin. Exp. Allergy*, **22**, 325–36

192. Pelon, W., Mogabdab, W.J., Phillips, I.A. and Pierce, W.E. (1957) A cytopathic agent isolated from naval recruits with mild respiratory illness. *Proc. Soc. Exp. Biol. Med.*, **94**, 262–7.

193. Perrotta, D.M., Decker, M. and Glezen, W.P. (1985) Acute respiratory disease hospitalizations as a measure of impact of epidemic influenza. *Am. J. Epidemiol.*, **122**, 468–76.

194. Potter, C.W. and Oxford, J.S. (1979) Determinants of immunity to influenza infection in man. *Br. Med. J.*, **35**, 69–75.

195. Price, W.H. (1956) The isolation of a new virus associated with respiratory clinical disease in volunteers. *Proc. Natl Acad. Sci. USA*, **42**, 892–6.

196. Ravenholt, R.T. and Foege, W.H. (1982) Before our time. 1918 influenza, encephalitis lethargica, Parkinsonism. *Lancet*, ii, 860–4.

197. Reed, S.E. (1984) The behaviour of recent isolates of human respiratory coronaviruses *in vitro* and in volunteers: evidence of heterogeneity among 229E–related strains. *J. Med. Virol.*, **13**, 179–92.

198. Reilly, C.M., Hoch, S.M. and Stokes, J. (1962) Clinical and laboratory findings in cases of respiratory illness caused by coryzaviruses. *Ann. Intern. Med.*, **57**, 515–25.

199. Reynolds, D.W., Stagno, S., Hosty, T.S. *et al.* (1973) Maternal cytomegalovirus excretion and perinatal infection. *N. Engl. J. Med.*, **289**, 1–5.

200. Roldaan, A.C. and Masural, N. (1982) Viral respiratory infections in asthmatic children staying in a mountain resort. *Eur. J. Respir. Dis.*, **63**, 140–50.

201. Schleible, J.H., Fox, V.L. and Lennette, E.H. (1967) A probable new human picornavirus associated with respiratory disease. *Am. J. Epidemiol.*, **85**, 297–310.

202. Schmidt, H.J. and Fink, R.J. (1991) Rhinovirus as a lower respiratory tract pathogen in infants. *Pediatr. Infect. Dis. J.*, **10**, 700–2.

203. Shortridge, K.F., Webster, R.G., Butterfield, W.K. and Campbell, C.H. (1977) Persistence of Hong Kong influenza virus variants in pigs. *Science*, **196**, 1454–5.

204. Simila, S., Ylikorkala, O. and Wasz-Hockert, O. (1971) Type 7 adenovirus pneumonia. *J. Pediatr.*, **79**, 605–11.

205. Sims, D.G., Downham, M.A.P.S., McQuillin, J. *et al.* (1976) Respiratory syncytial virus infection in north-east England. *Br. Med. J.*, **2**, 1095–8.

206. Six, H.R., Webster, R.G., Kendal, A.P. *et al.* (1983) Antigenic analysis of H1N1 viruses isolated in the Houston metropolitan area during four successive seasons. *Infect. Immunol.*, **42**, 453–8.

207. Smith, C.B., Golden, C.A., Kanner, R.E. and Renzetti, A.D. Jr (1980) Association of viral and *Mycoplasma pneumoniae* infections with acute respiratory illness in patients with chronic obstructive pulmonary diseases. *Am. Rev. Respir. Dis.*, **121**, 225–32.

208. Smith, C.B., Purcell, R.H., Bellanti, J.A. and Chanock, R.M. (1966) Protective effect of antibody to parainfluenza type 1 virus. *N. Engl. J. Med.*, **275**, 1145–52.

209. Snydman, D.R., Greer, C., Meissner, H.C. and McIntosh, K. (1988) Prevention of nosocomial transmission of respiratory syncytial virus in a newborn nursery. *Infect. Control Hosp. Epidemiol.*, **9**, 105–8.

210. Spence, L. and Barratt, N. (1968) Respiratory syncytial virus associated with acute respiratory infections in Trinidadian patients. *Am. J. Epidemiol.*, **88**, 257–66.

211. Spigland, I., Fox, J.P., Elveback, L.R. *et al.* (1966) The Virus Watch program: a continuing surveillance of viral infections in metropolitan New York families. II. Laboratory methods and a preliminary report on infections revealed by virus isolation. *Am. J. Epidemiol.*, **83**, 413–35.

212. Stagno, S., Pass, R.F., Dwosky, M.E. *et al.* (1982) Congenital cytomegalovirus infection: the relative importance of primary and recurrent maternal infection. *N. Engl. J. Med.*, **306**, 945–9.

213. Stevens, D.J., Douglas, A.R., Skehel, J.J. and Wiley, D.C. (1987) Antigenic and amino acid sequence analysis of the variants of H1N1 influenza virus in 1986. *Bull. WHO*, **65**, 177–80.

214. Storr, J. and Lenney, W. (1989) School holidays and admissions with asthma. *Arch. Dis. Child.*, **64**, 103–7.

215. Sung, R.Y.T., Murray, H.G.S., Chan, R.C.K., Davies, D.P. and Franch, G.L. (1987) Seasonal patterns of respiratory syncytial virus infection in Hong Kong: a preliminary report. *J. Infect. Dis.*, **156**, 527–8.

216. Thwing, C.J., Arruda, E., Vieira Filho, J.P.B., Filho, A.C. and Gwaltney, J.M. Jr (1993) Rhinovirus antibodies in an isolated Amazon indian tribe. *Am. J. Trop. Med. Hyg.*, **48**, 771–5.

217. Torphy, D.E., Ray, C.G., Thompson, R.S. and Fox, J.P. (1970) An epidemic of aseptic meningitis due to echovirus type 30: epidemiological features and clinical laboratory findings. *Am. J. Public Health*, **60**, 1447–55.

218. Tyrrell, D.A. (1977) Aspects of infection in isolated communities. Health and disease in tribal societies. Ciba Foundation Symposium, 49 (new series), pp. 137–53.

219. Tyrrell, D.A.J. and Bynoe, M.L. (1969) Studies on parainfluenza type 2 and 4 viruses obtained from patients with common colds. *Br. Med. J.*, **1**, 471.

220. Tyrrell, D.A.J., Bynoe, M.L., Birkum, K., Petersen, S. and Perreira, M.S. (1959) Inoculation of human volunteers with parainfluenza viruses 1 and 3 (HA2 and HA1) *Br. Med. J.*, **2**, 909.

221. Tyrrell, D.A.J., Bynoe, M.L., Buckland, F.E. and Hayflick, L. (1962) The cultivation in human embryo cells of a virus (D.C.) causing colds in man. *Lancet*, **ii**, 320–2.

222. Tyrrell, D.A.J. and Parsons, R. (1965) Some virus isolations from common colds. III. Cytopathic effects in tissue cultures. *Lancet*, **i**, 239–42.

223. Uncapher, C.R., DeWitt, C.M. and Colonno, R.J. (1991) The major and minor group receptor families contain all but one human rhinovirus serotype. *Virology*, **180**, 814–17.

224. Vihma, L. (1969) Surveillance of acute respiratory disease in children. *Acta Pediatr. Scand.*, **92**(suppl.), 8–52.

225. Walters, J.H. (1978) Influenza 1918: The contemporary perspective. *Bull. N.Y. Acad. Med.*, **54**, 855–64.

226. Wang, M.-L., Skehel, J.J. and Wiley, D.C. (1986) Comparative analyses of the specificities of anti-influenza haemagglutinin antibodies in human sera. *J. Virol.*, **57**, 124–8.

227. Warshauer, D.M., Dick, E.C., Mandel, A.D. *et al.* (1989) Rhinoviruses in an isolated antarctic station. Transmission of viruses and susceptibility of the population. *Am. J. Epidemiol.*, **129**, 319–40.

228. Webster, R.G. and Berton, M.T. (1981) Analysis of antigenic drift in the haemagglutinin molecule of influenza B virus with monoclonal antibodies. *J. Gen. Virol.*, **54**, 243–51.

229. Webster, R.G., Campbell, C.H. and Granoff, A. (1971) The 'in vivo' production of 'new' influenza A viruses. *Virology*, **44**, 317–28.

230. Wenzel, R.P., Hendley, J.O., Davies, J.A. and Gwaltney, J.M. Jr (1974) Coronavirus infections in military recruits. *Am. Rev. Respir. Dis.*, **109**, 621–4.

231. Winter, G.F. and Inglis, J.M. (1987) Respiratory viruses in children admitted to hospital in Edinburgh 1972–1985. *J. Infect.*, **15**, 103–7.

232. Wright, P.F., Thompson, J. and Karzon, D.T. (1980) Differing virulence of H1N1 and H3N2 influenza strains. *Am. J. Epidemiol.*, **112**, 814–19.

THE CLINICAL SPECTRUM OF DISEASE IN ADULTS

Elliott Larson

2.1 INTRODUCTION

Since the various viruses are discussed in detail in other chapters, this is an overview of the clinical findings associated with respiratory viruses, a re-look at the role bacteria play in respiratory infections, viral respiratory infections in patients with airways damaged by tobacco smoke or other toxic chemicals, previous chronic infection, antibody deficiency, or inherited disease, and pneumonia due to viruses usually restricted to the upper airways in patients with severe immunodeficiency.

Viral and Other Infections of the Human Respiratory Tract.
Edited by S. Myint and D. Taylor-Robinson. Published in 1996 by Chapman & Hall. ISBN 0 412 60070 6

2.2 COMMON COLDS

What is condescendingly described as the common cold has proven to be a complex melange of symptoms, agents and therapeutic challenges. Indeed, this medical syndrome has become a symbol for the limits of scientific competence: '...but you can't cure the common cold'. Nevertheless, the five decades since World War II have witnessed a very large expansion of what is known about colds. Previous to that time a great deal of effort had resulted in a few small steps forward. There was increasing assurance that filterable agents rather than bacteria were primarily involved [12]. Colds may indeed be common, the average adult experiencing two to four episodes a year, but the investigations over this period have revealed literally a world of viral agents. Causes for colds encompass both DNA and RNA viruses, membrane-bound as well as unenveloped. Beyond this there is a small group of tantalizing illnesses for which no agent can be detected, some of them having a glimmer of evidence that they are transmissible, and some whose explanation in terms of infectivity continues to elude the most vigorous attempts at definition.

2.2.1 SYMPTOMATOLOGY

The hallmark of a cold is nasal discharge and obstruction. This can vary considerably in degree, however. Some colds qualify as 'dry' because there is very little discharge whereas others are truly 'streaming'. Sneezing is a harbinger of increased inflammation in nasal passages and may be associated with infection as much as allergy. Some viral upper respiratory infections begin with a sore throat. When the patient is examined all that can be seen is the cobble stoning produced by hyperplastic lymphatic tissue on the posterior pharyngeal wall. Subsequently nasal obstruction and streaming begin. Hoarseness is not uncommonly associated with a cold and, of course, this can be either the initial symptom or the symptom which dominates the entire illness.

Sir William Osler's description of a cold still deserves mention: '... the patient feels indisposed, perhaps chilly, has slight headache, and sneezes frequent .... There is usually a slight fever .... At first the mucus membrane of the nose is swollen, 'stuffed up', and the patient has to breathe through the mouth. A thin, clear irritating secretion flows, and makes the edges of the nostrils sore .... Usually, within thirty-six hours the nasal secretion becomes turbid and more profuse, the swelling of mucosa subsides ... and gradually, within four or five days the symptoms disappear'.

Useful distinctions are observed in the symptoms produced by the different agents causing colds. Clinical studies of rhinovirus infections show that after a period of incubation that varies from 2–6 days, a sore throat occurs, followed by cough 2 days later [1]. The most constant measure of the severity of a rhinovirus infection is the amount of nasal discharge. In volunteers at the Common Cold Unit a wet cold might cause the use of up to 26 paper handkerchiefs on the worst day. However, the amount of discharge is highly variable and some colds are stuffy and relatively dry. The amount of nasal discharge varies throughout the day in a diurnal fashion: discharge is greatest during the morning and tapers through the day with a slight increase during the late evening [15]. When cough frequency was assessed by tape recording, a diurnal variation in frequency was also observed [9]. Here, coughs were more frequent between noon and 6:00 p.m. and less frequent between midnight and 6:00 a.m. There was marked variability in the frequency of coughing between patients; many patients coughed as often as 300 to 400 times in a 6-hour period and occasional patients coughed as many as 800 to 1300 times. The median range was between 12 and 377 for a 6-hour period. Patients were not aware of a variation in cough frequency throughout the day.

Constitutional symptoms can be associated with all agents causing colds. Fever, chills, malaise and myalgia, along with a deep cough are the hallmark of influenza, but can occur in varying degrees with the other agents. Fever is not common in rhinovirus infection, but a few patients complain of chills and myalgias. Rhinoviruses are isolated from some patients presenting with clinical syndromes more severe than a cold: children with bronchitis or bronchopneumonia, adults with fever and laryngitis plus nasal discharge. Contrariwise, nasal stuffiness may be a small part of the illness or acknowledged only on direct questioning in patients with influenza. Virus strains associated with a particular set of symptoms when isolated from one individual do not necessarily produce the same set of symptoms when experimentally inoculated into a volunteer [18].

Where predominating symptoms are pain around the eyes and discomfort over the bridge of the nose and in the face, some patients complain that instead of a cold, they have sinusitis. A study using magnetic resonance imaging (MRI) detected abnormalities in the ethmoid or antral sinuses associated with acute experimental rhinovirus infection [19]. All subjects were young adults between the ages of 18 and 40 years. Sinus involvement was associated with increased levels of nasal secretion. Three subjects had mucosal thickening and one an air–fluid level. One had right ethmoid thickening, one bilateral ethmoid thickening, one right antral thickening and one right antral fluid. Three additional subjects had right antral thickening on a pre-challenge MRI. Three of the four subjects had a follow-up MRI done 37 days after challenge. This showed that the sinuses had returned to normal. No antimicrobial treatment was given. There is, therefore, an overlap between the signs and symptoms of viral colds and those of what is usually presumed to be acute bacterial sinusitis.

2.2.2 PHYSIOLOGY

Physiological measurements in volunteers experimentally infected with coronavirus 229E showed increases in nasal airway resistance and mean nasal mucosal temperature whether or not there was clinical evidence of infection [3]. Mean ear temperature (recorded near the drum by a probe) also increased in both the clinically and subclinically infected groups. Blood flow in the nasal mucosa, measured by the ^{133}Xe washout method, increased in the clinically infected group only. Nasal airway resistance and blood flow correlated with the severity of the symptoms. Increase in secretion followed the changes in the other parameters.

2.2.3 THE ROLE OF BACTERIA

The evolution of the thick, turbid nasal discharge from one that is clear and thin is not necessarily associated with the presence of pathogenic bacteria [24]. More than half of 55 volunteers developed purulent discharge during the course of a naturally acquired cold. Despite repeated sampling, however, only 16% became colonized with *Staphylococcus aureus*, *Haemophilus influenzae*, *Streptococcus pneumoniae* or β-haemolytic streptococci. Light and scanning microscopy of nasal biopsies from 29 student volunteers with naturally acquired colds showed sloughing of epithelial cells, but preservation of the epithelial lining and structurally normal cell borders. The nasal epithelium of infected and normal individuals could not be distinguished by light microscopy in biopsies taken at 2 and 14 days after the onset of symptoms. The epithelium and lamina propria showed an increase in the number of neutrophils by the second day of illness. The acute stage was also characterized by an increase in the extravasation of erythrocytes [23]. In none of these studies, however, was the infecting virus identified. Colds are, therefore, associated with non-destructive infection of the nasal epithelium with subsequent polymorphonuclear leucocyte infiltration and exudation even in the absence of complicating bacterial colonization.

2.2.4 AGENTS ASSOCIATED WITH COLDS

Using sensitive tissue and organ culture methods, agents can be cultivated from two-thirds to three-quarters of adults with colds [10,20]. Rhinoviruses are the single most common type of agent, accounting for 40–50% of isolates. Organ cultures enhance the recovery of rhinoviruses. Beyond the identification of agents that are clearly rhinoviruses, however, lie additional agents which can be established as chloroform stable by means of further testing in human volunteers. It is not completely clear whether these replicate in any *in vitro* system, but it is possible that they represent a further group of rhinoviruses causing a proportion of those colds from which no agent is routinely recoverable.

Coronaviruses cause approximately one-fifth of colds, although their identification in survey studies is very laborious. However, when organ and tissue culture-negative specimens were subsequently passaged in human volunteers, only one additional agent was found to be chloroform sensitive and capable of belonging to this group [10]. It is likely that coronaviruses cause colds less frequently than rhinoviruses, but are still deserving of the term 'common'. Infections with influenza and parainfluenza viruses, respiratory syncytial virus and adenoviruses are also associated with cold syndromes. However, the frequency with which they produce common colds relative to rhinoviruses and coronaviruses is highly variable and depends upon the year in question. Influenza A and B viruses may be responsible for up to 10% of respiratory illness in epidemic years. Parainfluenza viruses, adenoviruses and respiratory syncytial virus share about 10% of the burden of illness in adults.

2.3 VIRAL PNEUMONIA

The viruses which have concerned us in causing illness of the upper respiratory tract may on occasion also involve the lower respiratory tract. The viruses most likely to produce these effects in adults are influenza, parainfluenza, respiratory syncytial (RS) virus and adenoviruses. The occurrence of severe pneumonia in immunosuppressed adults is a special case to be considered in more detail below. However, viral pneumonia occurs even in previously healthy adults. Marrie, Durant and Yates [11] prospectively surveyed 719 cases of both community-acquired and nursing home originating pneumonia over a 5-year period. Viruses were the cause in 72 cases (10%). No aetiology was apparent in 47%. Influenza A and B viruses were responsible for 57 of these, parainfluenza 3 for 18 cases and parainfluenza 1 and 2 for seven cases.

Because adult experience with RS virus is almost entirely due to re-infection, it is often presumed that these infections are clinically mild. Routine testing for RS virus in 2400 patients admitted to hospital with pneumonia in Sweden revealed 73 whose clinical diagnosis had been considered to be RS virus pneumonitis [21]. Diagnosis was based upon detection of RS virus in nasopharyngeal aspirates or a four-fold rise is specific antibody titre or the presence of anti-RS virus IgM serum antibody was present. Altogether, 36 patients had documented RS virus infection together with clinical signs of pneumonia and infiltrates on chest X-ray. Of these, 16 had distressed respiration, a preponderance being exacerbation of underlying obstructive lung disease.

Ruben and Nguyen [14] reviewed the subject of RS virus pneumonia occurring in adults and noted that 44% had respiratory distress. The clinical features are not specific and knowledge of an RS virus outbreak in the community might suggest the diagnosis.

Adenoviruses are some of the most frequent causes of lower respiratory tract disease in adults. These infections can be clinically severe, and even fatal in immunosuppressed bone marrow transplant (BMT) patients. There is usually an upper respiratory prodrome for 4 to 7 days preceding presentation with pneumonia.

2.4 RESPIRATORY VIRAL INFECTIONS IN CHRONIC LUNG DISORDERS

Stenhouse showed in a controlled, prospective study that rhinovirus infection was more likely to be associated with acute lower respiratory tract symptoms even while it was less common in patients with bronchitis than in a comparison group [17]. Virus infections are associated with approximately one-third of exacerbations of chronic obstructive pulmonary disease [4]. Rhinoviruses were found in 2.7% of stored, frozen, sputum specimens from episodes of exacerbation, in contrast to 0.55% of remission interval sputum specimens. Single-agent antibody titre rises to viruses (influenza A and B, parainfluenza virus types 1, 2 and 3, RS virus, adenoviruses, coronaviruses) or *M. pneumoniae* in 24.7% of exacerbations and in 13.8% of remission specimens. Coronavirus and influenza virus A were more often associated with exacerbations than remissions. In this study virus infection was not consistently associated with worsening of respiratory function. One-third of all isolations occurred during a period when there was no compromise of respiratory status. Titre rises to *M. pneumoniae* occurred in association with titre rises to one or more respiratory viruses. A low recovery rate for rhinoviruses may have been due to the extended period of frozen storage.

Smith *et al.* [16] found rhinoviruses, influenza viruses, parainfluenza viruses and coronaviruses significantly associated with acute respiratory illness in patients with chronic obstructive pulmonary disease (COPD). There were 272 detections in 1030 follow-up intervals over an 8-year period. Fifty-three of the 272 infections studied were due to influenza viruses A and B, 17 of which were of the 'influenzal-type' clinical syndrome; 56 of the 272 infections were due to rhinoviruses, only one of which was an 'influenzal-type' syndrome. Patients with an influenzal-type clinical syndrome were most likely to have viruses detected (28.9%). Illnesses with cough and sputum were next most likely (20%). Illnesses confined to the upper respiratory tract were least likely (11.4%). Between 40% and 80% of the acute respiratory illness in COPD patients had no discernible aetiology, a proportion similar to the rest of the population. In this study there was no evidence that patients with COPD were more susceptible to virus infections. However, patients with COPD were more likely to develop increased cough and lower respiratory tract symptoms during rhinovirus infections than were normal subjects or those with mild COPD.

Considerable evidence has accumulated to show that cold viruses are associated with exacerbations of asthma in adults as well as children. The prevalence with which viruses can be identified in such attacks is lower than in children, but there is a correlation between the frequency of virus detection and the severity of the illness. Increases in the frequency of asthma during winter months correlate with the prevalence of common cold viruses during those periods [2]. The exposure of patients with chronic chest disease to other individuals with colds can result in the development of respiratory symptoms and exacerbation of the underlying illness [25]. In patients with asthma, 34% of the exposures resulted in symptoms. Lower respiratory tract involvement persisted for up to 33 days in some patients. It was possible to identify a viral pathogen in one-third of the exposure episodes; rhinoviruses, coronaviruses and respiratory syncytial virus were implicated. Paranasal sinus infection commonly coexists with asthma in adults. In a study where radiographically abnormal sinuses were aspirated and cultured, one-third had viruses detected where the lavage fluid obtained was abnormal [13].

Rhinovirus was administered experimentally to a group of 21 adult patients with asthma. In four of these patients histamine responsiveness was increased and forced expiratory volume (FEV$_1$) was significantly reduced subsequent to infection [6]. None of a group of volunteers without asthma showed such changes, even though they developed symptomatic colds after inoculation. It is possible that viruses other than rhinoviruses are more apt to be associated with airway hyperresponsiveness and likely that adults are less susceptible than children. Nevertheless, virus infection did trigger wheezing in some adult patients with asthma under well-controlled conditions.

2.5 RESPIRATORY VIRAL INFECTIONS IN IMMUNOCOMPROMISED ADULTS

Influenza, parainfluenza, respiratory syncytial and adenoviruses have all been shown to cause serious lower respiratory illness in immunocompromised adults. The main subjects for systematic study have been patients about to undergo or who have undergone BMT. For example, severe RS virus disease was reported in 11 immunocompromised adults [5]. Six had had BMT, four organ transplantation, and one T-cell lymphoma. Patients presented with fever, cough, rhinorrhoea, nasal congestion and otalgia. Chest radiography showed diffuse interstitial pneumonia; there was also radiographic evidence of sinusitis. Half of the BMT patients died while 36% of all the patients died.

In a larger series of 74 BMT patients from March 1987 to April 1988, there were eight cases of acute lung injury due to RS virus. In all cases there were initial upper respiratory tract symptoms (cough) followed by evidence of lower respiratory tract involvement. Six of eight patients had pulmonary infiltrates on chest X-ray. Six of the eight cases became ill before engraftment (<10 days after BMT) of whom four had severe diffuse pneumonia and died despite treatment with ribavirin. Six of the eight patients had abnormal sinus radiographs and went on to have drainage procedures. Autopsy findings in all patients who died showed organizing diffuse alveolar damage, with lung disease being the predominant contributor to death. This was also true clinically, three of the patients dying of respiratory failure, while one died of refractory hypotension after 3 days on the ventilator. Failure of lung repair occurred even though there was evidence that the treatment had eliminated the virus from the lower respiratory tract in three of the four patients who died. It was suggested that the chemotherapy and radiation treatment given before BMT rendered the patients incapable of generating a normal reparative response [8].

The role of engraftment in susceptibility to severe RS virus infection was evaluated in 31 of 199 BMT patients who developed RS virus infection over a period of 3 months, beginning January, 1990 [7]. Eighteen patients developed pneumonia, of whom 14 died. Pneumonia was more common in patients who became ill before engraftment had occurred. Early therapy with ribavirin was thought to be helpful in patients with pneumonia. Seventeen of 18 pneumonia patients were positive on broncheoalveolar lavage (BAL) for RS virus. The other patient had an endotracheal tube aspirate that was positive. Eleven of 14 pre-engraftment patients developed pneumonia and seven of 17 post-engraftment patients did so. One patient with *Pneumocystis carinii* pneumonia who had been ill for some weeks had a superinfection with RS virus. Two other patients had both RS virus and CMV in BAL specimens. Fourteen of 18 pneumonia patients had diffuse, bilateral infiltrates. Autopsy histology showed diffuse alveolar damage. RS virus was able to be detected by immunofluorescence staining in bronchiolar and alveolar epithelium and in cells sloughed into alveolar air spaces. Patients dying of pneumonia shed RS virus from 11 to 22 days. Nine of 11 pre-engraftment pneumonia patients died as compared with five of seven post-engraftment pneumonia patients, not a meaningful difference. Ribavirin treatment was not given to five patients who died of pneumonia, and on autopsy three of these had high rates of RS virus positivity, (>35% of cells per high-power field (HPF) positive for RS virus). Two patients receiving ribavirin for RS virus pneumonia who died had <5% of cells positive for RS virus. Pre-engraftment patients are more highly immunosuppressed. Survival appears to depend on confining RS virus infection to the upper respiratory tract.

An outbreak of parainfluenza type 3 in a renal transplant unit was associated with an increase in the frequency of acute allograft rejection during the period of infection [22]. Some 27 patients, of whom 12 were adults, were found to have developed parainfluenza infection between 1974 and 1990. One-third of the patients with lower respiratory tract involvement developed respiratory failure and all died. Ribavirin was given by aerosol but evidently it was not helpful. During the 1991–1992 influenza epidemic season, 25% of adult BMT patients at the MD Anderson Cancer Center had acute respiratory symptoms with influenza A infection confirmed by culture. Two-thirds of these patients had lower respiratory tract disease. Some 10% of adults with leukaemia with acute respiratory symptoms had influenza A viruses; of these infections, 75% were complicated by pneumonia. During the epidemic, 20% of all pneumonia and nosocomial respiratory infections were associated with influenza A viruses.

2.6 REFERENCES

1. Andrewes, C.A. (1965) *The Common Cold.* W.W. Norton & Co., New York.
2. Ayres, J.G., Noah, N.D. and Fleming, D.M. (1993) Incidence of episodes of acute asthma and acute bronchitis in general practice 1976–87. *Br. J. Gen. Pract.*, **43**, 361–4.
3. Bende, M., Barrow, I., Heptonstall, J. *et al.* (1989) Changes inhuman nasal mucosa during experimental coronavirus common colds. *Acta Otolaryngol. (Stockh.)*, **107**, 262–9.
4. Buscho, R.O., Saxtan, D., Shultz, P.S., Finch, E. and Mufson, M.A. (1978) Infections with viruses and *Mycoplasma pneumoniae* during exacerbations of chronic bronchitis. *J. Infect. Dis.*, 137, 377–83.
5. Englund, J.A., Sullivan, C.J., Jordan, M.C., Dehner, L.P., Vercellotti, G.M. and Balfour, H.H. Jr (1988) Respiratory syncytial virus infection in immunocompromised adults. *Ann. Intern. Med.*, 109, 203–8.
6. Halperin, S.A., Eggleston, P.A. and Beasley, P. (1985) Exacerbations of asthma in adults during experimental rhinovirus infection. *Am. Rev. Respir. Dis.*, 132, 976–80.
7. Harrington, R.D., Hooton, T.M., Hackman, R.C. *et al.* (1992) An outbreak of respiratory syncytial virus in a bone marrow transplant center. *J. Infect. Dis.*, 165, 987–93.
8. Hertz, M.I., Englund, J.A., Snover, D., Bitterman, P.B. and McGlave, P.B. (1989) Respiratory syncytial virus-induced acute lung injury in adult patients with bone marrow transplant: a clinical approach and review of the literature. *Medicine*, **68**, 269–81.
9. Kuhn, J.J., Hendley, J.O., Adams, K.F., Clark, J.W. and Gwaltney, J.M. Jr (1982) Antitussive effect of guaifenesin in young adults with natural colds. *Chest*, 82, 713–18.
10. Larson, H.E., Reed, S.E. and Tyrrell, D.A.J. (1980) Isolation of rhinoviruses and coronaviruses from 38 colds in adults. *J. Med. Virol.*, 5, 221–9.
11. Marrie, T.J., Durant, H. and Yates, L. (1989) Community-acquired pneumonia requiring hospitalization: 5 year prospective study. *Rev. Infect. Dis.*, 11, 586–99.
12. Mills, K.C., Shibley, G.S. and Dochez, A.R. (1928) Studies in the common cold. II. A study of certain Gram-negative filter-passing anaerobes of the upper respiratory tract. *J. Exp. Med.*, 47, 193–206.
13. Rossi, O.V.J., Pirila, T., Laitinen, J. and Huhti, E. (1994) Sinus aspirates and radiographic abnormalities in severe attacks of asthma. *Int. Arch. Allergy Immunol.*, **103**, 209–13.
14. Ruben, F.L. and Nguyen, M.L.T. (1991) Viral pneumonitis. *Clin. Chest Med.*, **12**, 223–35.
15. Smith, A., Tyrrell, D., Coyle, K., Higgins, P. and Willman, J. (1988) Diurnal variation in the symptoms of colds and influenza. *Chronobiology International*, 5, 411–16.
16. Smith, C.B., Golden, C.A., Kanner, R.E. and Renzetti, A.D. Jr (1980) Association of viral and *Mycoplasma pneumoniae* infections with acute respiratory illness in patients with chronic obstructive pulmonary disease. *Am. Rev. Respir. Dis.*, 121, 225–32.
17. Stenhouse, A.C. (1967) Rhinovirus infection in acute exacerbation of chronic bronchitis: a controlled prospective study. *Br. Med. J.*, 3, 461–3.
18. Stott, E.J. and Tyrrell, D.A.J. (1971) Recent advances in the study of virus diseases of the respiratory tract. *Abstracts World Med.*, **45**, 301–6.
19. Turner, B.W., Cail, W.S., Hendley, J.O. *et al.* (1992) Physiological abnormalities in the paranasal sinuses during experimental rhinovirus colds. *J. Allergy Clin. Immunol.*, **90**, 474–8.
20. Tyrrell, D.A.J. and Bynoe, M.L. (1966) Cultivation of viruses from a high proportion of patients with colds. *Lancet*, **i**, 76–7.
21. Vikerfors, T., Grandien, M. and Olcen, P. (1987) Respiratory syncytial virus infections in adults. *Am. Rev. Respir. Dis.*, **136**, 561–4.
22. Whimbey, E. and Bodey, G.P. (1992) Viral pneumonia in the immunocompromised adult with neoplastic disease: the role of common community respiratory viruses. *Semin. Respir. Infect.*, **7**, 122–31.
23. Winther, B., Brofeldt, S., Christensen, B. and Mygind, N. (1984) Light and scanning electron microscopy of nasal biopsy material from patients with naturally acquired common colds. *Acta Otolaryngol.*, **97**, 309–18.
24. Winther, B., Brofeldt, S., Gronborg, H., Mygind, N., Pedersen, M. and Vejlsgaard, R. (1984) Study of bacteria in the nasal cavity and nasopharynx during naturally acquired common colds. *Acta Otolaryngol.*, 98, 315–20.
25. Wiselka, M.J., Kent, J., Cookson, J.B. and Nicholson, K.G. (1993) Impact of respiratory virus infection in patients with chronic chest disease. *Epidemiol. Infect.*, **111**, 337–46.

CLINICAL SPECTRUM OF DISEASE IN CHILDREN

3

David Isaacs

3.1 DEFINITIONS

There have always been problems with the definitions of acute childhood respiratory infections. In 1965, when embarking on a Medical Research Council survey of childhood respiratory infections involving a number of hospitals, Donald Court noted that there was no generally accepted clinical classification of acute respiratory infections in children [11]. Paediatricians thought that they were all diagnosing the same disease, when making a clinical diagnosis, but there was no evidence that this was true. Indeed quite the opposite appeared to be the case when the study was carried out. There was great variation in the diagnostic labels given to children with similar clinical presentations in different institutions that were geographically quite close. In four hospitals, for example, the proportion of all respiratory illnesses diagnosed as either bronchitis or bronchiolitis was almost constant, varying from 59% in Glasgow to 70% in Manchester. However, in Glasgow 52% of all infections were called bronchitis and only 7% called bronchiolitis, whereas in Birmingham the reverse was true: 21% were called bronchitis and 45% bronchiolitis [11]. It seemed highly likely that different diagnostic labels were being applied to infected children in different centres, and that this reflected differing interpretation of signs and local diagnostic preferences.

To an extent, Court himself rectified this situation, by developing clear-cut definitions of acute childhood respiratory infections. These were scientifically based, in that the given frequency of signs and symptoms in each diagnostic category was based on the observed frequency during the survey (see Table 3.1). This was a seminal piece of work, and one that greatly facilitated later clinical studies. Like Scadding before him, Court appreciated the importance of clear definitions of diagnostic terms.

Viral and Other Infections of the Human Respiratory Tract.
Edited by S. Myint and D. Taylor-Robinson. Published in 1996 by Chapman & Hall. ISBN 0 412 60070 6

Table 3.1 Definitions of acute respiratory infections. (From [11])*

Upper respiratory infection syndrome
   Affected children are very frequently febrile with some or all of the following symptoms; red pharynx, red tonsils, with or without exudate, red ear drums, cough and nasal discharge. Abnormal signs in the lungs are infrequent and chest radiographs consistently normal.

Individual categories
   Colds The principal symptom is excessive, mucoid or purulent, nasal discharge. This is very frequently associated with a red pharynx and with cough. From 6 months to 5 years, the illness is very frequently febrile.
   Pharyngitis The principal symptom is redness of the pharynx. Children of all ages are very frequently febrile.
   Tonsillitis The principal symptom is tonsillar redness, frequently with exudate, and very frequently with surrounding redness of the pharynx. Children of all ages are very frequently febrile.
   Otitis media The principal symptom is marked redness of the ear drum, infrequently associated with perforation and discharge. A red pharynx is very frequent in children at all ages and nasal discharge frequent in children under 5. Fever is very frequent in children under 5 but less so between 5 and 14.

Middle respiratory infections
   Croup (acute laryngitis or laryngotracheitis) Illness mainly affecting children between 6 months and 4 years. Stridor is the constant symptom, very frequently associated with hoarseness. Cough, breathlessness and chest recession are also very frequent. Upper respiratory features are generally present; red pharynx very frequently and nasal discharge frequently.

Lower respiratory infections
   Acute bronchitis Illness affecting children mainly in the first 5 years of life. Cough is a constant symptom, and wheezing and breathlessness very frequent. Rhonchi are very frequently, and rales frequently, present. Upper respiratory symptoms, mainly red pharynx and nasal discharge, are frequent. The illness is very frequently febrile but high fever is infrequent. Except in a small minority chest radiographs are normal.
   Acute bronchiolitis Illness mainly affecting infants, especially in the first 6 months of life. Rapid respiration, dyspnoea, wheezing, chest recession, cough, rhonchi and rales are very frequent. Visible distension of the chest and increased pulmonary translucency on the chest radiograph are frequent and of high diagnostic significance. Upper respiratory features, especially nasal discharge and a red pharynx are frequent. Fever is very frequent, but high fever uncommon.
   Pneumonia Illness in which the essential feature is lobular segmental or lobar radiographic shadows, interpreted as pulmonary consolidation. Cough, rapid respiration, dyspnoea and rales are very frequent. Wheezing and chest recession are frequent in children under 5, especially in the first year. Cyanosis is frequent in children under 5. Upper respiratory signs are frequent, especially nasal discharge and a red pharynx. In children aged from 5 to 14, chest pain, abdominal pain and headache are frequent, and they may also complain of body aches and shivering. Fever is very frequent at all ages.

* This table is based primarily on the frequency of each symptom. *Very frequent* indicates that the symptom was present in at least 50% of children in one or more of the four age groups in any particular category: *frequent* symptoms were present in 25–50%; *infrequent* symptoms were present in 10–25%.

3.2 DIAGNOSIS

The development and general application of rapid techniques for viral detection, and in particular immunofluorescence for respiratory syncytial (RS) virus, has been of major importance in resolving diagnostic differences between paediatricians. No longer did the paediatrician making a presumptive diagnosis of bronchiolitis have to wait 2 weeks for virological confirmation that the child, long since discharged home, was RS virus-positive. Instead his diagnosis could be lent substantial weight within a few hours of admission. Rapid diagnosis of RS virus was important, not so much because it helped paediatricians to avoid the unnecessary use of antibiotics, but because it helped them to refine their diagnostic accuracy.

Rapid viral diagnosis is available in only a few centres worldwide for other respiratory viruses. In Dr Pekka Halonen's laboratory in Turku, Finland, each specimen of nasopharyngeal secretion is tested by ELISA against a range of respiratory viruses, including RS virus, adenovirus, herpes simplex virus, parainfluenza viruses 1, 2 and 3, and influenza viruses A and B. Suspected adenovirus tonsillitis or adenoviral pharyngoconjunctival fever can be confirmed rapidly. Viral croup can be shown to be due to a parainfluenza virus in most cases. However, in these conditions there is not as much diagnostic confusion as in bronchiolitis, and the respiratory illnesses may be caused by a wider spectrum of respiratory viruses. Nevertheless, the more ready availability of rapid viral diagnosis would almost certainly help refine our clinical acumen and accuracy and would be beneficial for management.

3.3 CLINICAL SPECTRUM OF DISEASE

Most respiratory viruses are capable of causing a range of different clinical illnesses in children. Almost all can cause asymptomatic infection and some, like adenoviruses, may be carried in the nasopharynx and excreted for several months, without causing clinical illness. Many viruses can cause illnesses ranging from a mild cold or upper respiratory tract illness, through otitis media, laryngotracheobronchitis, to severe lower respiratory tract infection with bronchiolitis or pneumonia. In the cases of some viruses, like RS virus, re-infections of the same child are progressively less severe. Others, like rhinoviruses, may cause a mild cold in one child but precipitate a severe asthma attack in an asthmatic child of the same age.

The spectrum of respiratory disease in children can be considered primarily in one of two ways. The first is *virus-orientated*, and considers the spectrum of respiratory illness that can be caused by a single virus or group of viruses. This may be expressed in tabular form, as in Table 3.2, which gives the range of respiratory infections caused by adenoviruses.

The second approach is *disease-orientated*. In this approach, a clinical diagnosis is assigned to each child with a respiratory illness, and the viruses isolated in association with this illness are plotted. An example of this is given in Figure 3.1, which comes from a longitudinal study of children with recurrent respiratory infections ('index' group) and their siblings who did not have recurrent infections ('control' group) [22]. This study was undertaken at Northwick Park Hospital, and it was Dr David Tyrrell, one of my supervisors, who suggested the 'spectrum of illness' presentation of these

Table 3.2 Respiratory infections caused by adenoviruses

Illness	Comments	Commonest types
Tonsillitis or pharyngitis	Common; epidemics	1, 2, 3, 5, 7
Pharyngoconjunctival fever	Common; epidemics	2, 3, 4, 5, 7, 14
Common cold	Rare; sporadic	1, 2, 3, 5, 7
Pneumonia	Common; epidemics	3, 7, 21
Acute respiratory disease (ARD)	Epidemics in military recruits	4, 7
Laryngotracheitis (croup)	Rare; sporadic	1, 2, 3, 5, 6, 7
Bronchiolitis	Rare; sporadic	3, 6, 21
Bronchiectasis	Rare complication	3, 7, 21
Bronchiolitis obliterans	Very rare; late complication	7, 21

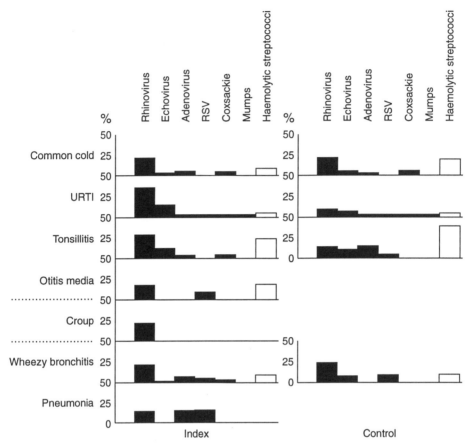

Figure 3.1   Clinical spectrum of disease in children: clinical diagnoses in children with recurrent infections (index) and their siblings (control) in association with different organisms.

data. The editors of the British Medical Journal disagreed and published the data tabulated numerically [27], but the graphic presentation in Figure 3.1 is, I believe, far more visually accessible to the reader.

3.4 THE COMMON COLD

The common cold, an illness characterized by mucoid or purulent nasal discharge, is the most common infectious disease of childhood, so well-deserves the adjective 'common'. However, more than 200 years ago, Benjamin Franklin recognized that colds were caught by exposure to other people and not to cold weather.

More than 50% of pre-school-age children with colds have fever, and pharyngitis and cough are also common features. Pre-school children have three to eight colds a year on average in developed countries [4,7,13]. These are generally caught within families, although the advent of child care facilities has revealed the latter as a potent source of respiratory infections.

The Common Cold Research Unit at Salisbury, England was at the forefront of early post-war studies on the spread of natural colds, even before cold viruses had been identified. Children from a local school with natural colds were invited to the Unit [32]. Volunteers got colds if they played table games with the children but not if they played cards using playing cards soiled with nasal secretions from the children. Further evidence for droplet spread was obtained by an experiment in which volunteers and children with colds in a hut were separated by a blanket across the hut. A fan from behind the children blew air over the blanket, and the volunteers got colds. Subsequent experiments have shown that rhinovirus-induced colds may be spread directly by contaminated secretions being rubbed into the nasal epithelium or conjunctiva, as well as by droplet spread [8]. Colds can be surprisingly difficult to spread in adults, even those in close contact, whereas there generally seems little problem spreading them among children [12]. This may reflect prior exposure to that strain or to similar strains of virus, and spread is also dependent on viral dose.

A number of different viruses can cause common colds in children, of which some are frequent causes and others are much rarer (see Table 3.3).

3.5 UPPER RESPIRATORY TRACT INFECTIONS (URTIs)

It is common practice in paediatrics to refer to children having URTIs or sometimes URIs (upper respiratory infections). Children with URTIs have something more than a cold but the infection is not sufficiently localized to be classified as pharyngitis, tonsillitis or otitis

Table 3.3 Viruses causing the common cold

Rhinoviruses
Coronaviruses
Parainfluenza viruses
Respiratory syncytial virus
Adenoviruses
Coxsackieviruses
Echoviruses
Influenza viruses
Reoviruses

media. The features of URTIs are fever, coryza, pink eardrums with intact light reflex, some pharyngitis, and cervical lymphadenopathy. URTI is an imprecise term, the use of which has often been discouraged, but it is a term used and understood by many general practitioners and paediatricians. That the term persists is an indication that it occupies a diagnostic niche not adequately covered by other diagnoses. Cervical lymphadenopathy is probably the most useful distinguishing clinical feature between URTIs and common colds. Viruses are the main cause of URTIs, and the responsible viruses are the same as those causing common colds [27] (Table 3.3).

Viral URTIs have been shown to be an important prodrome in 50% of episodes of invasive *Haemophilus influenzae* type b disease in pre-school children [38]. However, it has been consistently shown that prescribing oral antibiotics for children with URTIs does not prevent bacterial superinfection or other complications.

3.5.1 SINUSITIS

Infection of the paranasal sinuses is more difficult to diagnose clinically in children than adults, because facial pain and headache are rare. Ellen Wald, who has performed several important studies on sinusitis in children, describes 10 and 30 days of upper respiratory symptoms as important demarcation times. Most viral upper respiratory infections last 5–7 days, and the persistence of symptoms for 10 days is suggestive of the possibility of acute sinusitis. If symptoms persist for 30 days, subacute or chronic sinusitis should be considered [40].

Clinical features

The main features of sinusitis are persistent nasal discharge and cough [39]. The nasal discharge can be thin or thick, and can be clear or mucopurulent. The cough can be either dry or moist, and occurs in the day as well as at night.

Clearly these non-specific symptoms may also occur in children with persistent respiratory infections without sinusitis, and also in atopic children with allergic rhinitis and asthma, although children with allergic rhinitis are likely to have an itchy nose. Other features which may suggest sinusitis are bad breath, and painless periorbital swelling in the morning. Fever is generally low-grade or absent, although occasionally high (>39°C) with purulent nasal discharge. Children rarely complain of frontal headache, although older children may be tender over the frontal area.

Mild changes in sinus radiographs are common even in normal children, but complete opacification, dense mucosal thickening or air–fluid levels are strongly predictive of sinus infection. Computed tomography (CT) scans of the sinuses give better definition than plain radiographs in chronic sinusitis, but are unnecessary in acute sinusitis.

Isolates

Respiratory viruses are isolated from aspirated sinuses in about 10% of children with acute sinusitis, those most commonly detected being adenoviruses, parainfluenza viruses, influenza viruses and rhinoviruses. Bacteria, notably *Streptococcus pneumoniae*, untypable *H. influenzae*, and *Branhamella catarrhalis*, are isolated in about 75% of cases with clinical and radiological evidence of acute sinusitis. It is likely that respiratory virus infection predisposes to bacterial sinusitis, and that respiratory viruses are not cultured more often because viral specimens are collected late in the course of infection and the collection technique is often suboptimal.

3.5.2 THE CATARRHAL CHILD

John Fry, a general practitioner, wrote about *The Catarrhal Child*, the child with a perpetual mucopurulent nasal discharge, hanging like candlesticks from each nostril [16]. The incidence of chronic mucopurulent nasal discharge is higher in indigenous children from poor, crowded family environments, such as many native American, African and Australian Aboriginal children. This suggests that recurrent or chronic infection might be an important aetiological factor in chronic catarrh. Indeed, in a study of preschool children with recurrent respiratory infections and healthy children, catarrhal children were found to have rhinitis most, but not all, of the time. Their nasal discharge cleared for a short time but then recurred, and different respiratory viruses, predominantly rhinoviruses, could be isolated at the start of each new episode [27]. What is more, the children were being infected more often but with the same spectrum of respiratory viruses as their 'healthy' siblings. Catarrh in these children, it seemed, was due to recurrent respiratory viral infections.

Sinusitis is another possible cause of chronic catarrh, and it is difficult to estimate what proportion of John Fry's catarrhal children [16] actually had chronic sinusitis.

A third possibility is that the child with chronic nasal discharge has perennial allergic rhinitis [30]. This is suggested clinically by a history of nasal stuffiness, itchiness and nasal blockage. The child continually rubs his or her nose, the 'allergic salute', a lateral crease may develop across the nose, the 'allergic crease', and dark rings under the eyes, 'allergic shiners'. Nasal smears show eosinophils and/or mast cells in allergic rhinitis [30].

3.5.3 PHARYNGITIS AND TONSILLITIS

Pharyngitis is inflammation of the mucous membranes of the pharynx, whereas tonsillitis involves predominantly tonsillar tissue. However, the two frequently coexist, *viz.* tonsillopharyngitis, and there may also be inflammation of the nose (nasopharyngitis), palate and uvula. Acute nasopharyngitis, in which there is high fever, sore throat and mucopurulent nasal discharge, is nearly always viral, with adenoviruses by far the most important cause. Other viral causes of nasopharyngitis are influenza viruses, parainfluenza viruses, rhinoviruses and RS virus. Adenoviruses are particularly likely to cause follicular pharyngitis or pharyngotonsillitis with exudate, an appearance identical to group A streptococcal tonsillitis [36]. The main distinguishing feature is age: most adenoviral pharyngotonsillitis occurs in children under 3 years of age, whereas group A streptococcal tonsillitis is rare under 5 years of age. Adenovirus infection often results in high fever and adenoviruses are an important cause of febrile convulsions [34].

Other important viral causes of exudative tonsillopharyngitis are Epstein–Barr virus, the so-called anginose form of glandular fever, primary herpes simplex virus infection, and parainfluenza viruses [33].

Pharyngoconjunctival fever, caused by adenoviruses 3 and 7, is a febrile pharyngitis with an intense conjunctivitis, which is usually unilateral or more severe in one eye if it is bilateral [36].

3.5.4 ACUTE OTITIS MEDIA

Acute otitis media (AOM) is a common infection, with about 60% of children experiencing at least one episode before a year of age, 20% in their second year of life, and 10–12% each year until 6 years of age. Thereafter the incidence falls to less than 1% by the age of 12 years.

Clinical features

AOM is characterized clinically by fever, irritability and otalgia, and otoscopically by redness and bulging of the tympanic membrane and middle ear effusion. More than 90% of patients have concurrent upper respiratory signs such as rhinitis or cough, which precede the onset of acute otitis media by 3 to 7 days [1].

Isolates

Since bacteria are isolated from about 75% of episodes of AOM, and with refined bacteriological techniques from over 90%, this condition is generally treated with antibiotics.

However, recent studies have emphasized the importance of viruses in childhood AOM, with viral antigens being detected or viruses isolated from middle ear fluid or nasopharyngeal aspirates in 30–60% of episodes of AOM [1,9,18,35]. This may be an underestimate of the prevalence of respiratory virus infection in AOM, because in many studies very few rhinoviruses are isolated. However, when Ohio HeLa cells are used, rhinoviruses are one of the most common viruses to be isolated from middle ear fluid [2,3]. Viruses appear to initiate swelling of the middle ear mucosa, and Eustachian tube obstruction, and predispose to superinfection with nasopharyngeal bacteria.

Episodes of acute otitis media that do not respond rapidly to antibiotics are more likely to have both bacteria and a virus present in middle ear fluid. It has been suggested that combined viral and bacterial infection may be important in the pathogenesis of chronic otitis media with effusion, sometimes called 'glue-ear', a common and important cause of conductive deafness in childhood [9].

3.6 MIDDLE RESPIRATORY TRACT INFECTIONS

3.6.1 CROUP (ACUTE LARYNGOTRACHEITIS)

Croup is a term used to describe respiratory illnesses characterized by inspiratory stridor, a barking cough like a seal, and a hoarse voice due to tracheal or laryngeal obstruction. It was originally an old English verb, 'to croup' meaning to croak. There are various causes of laryngotracheal obstruction, including foreign body and intubation trauma. Diphtheritic croup has been recognized since at least the fifteenth century. However, by convention the term croup is often now used as synonymous with viral laryngotracheitis.

Croup is not a particularly common respiratory infection, affecting less than 5% of children, predominantly those between 1 and 5 years of age. It is more common in boys.

Aetiology

Parainfluenza virus type 1 is the most common cause of viral laryngotracheitis, tending to cause outbreaks in autumn or winter. Sometimes these epidemics occur every second year, but this pattern may change to annual outbreaks. Not uncommonly parainfluenza virus croup is also associated with wheezing, and the illness is best described as acute laryngotracheobronchitis.

Parainfluenzal virus type 2 virus can cause croup, but less frequently and less severely than type 1, and most type 2 infections are mild upper respiratory infections. Parainfluenza virus type 3 causes bronchiolitis and pneumonia in infants, but can cause croup in older children, which it does throughout the year without clear epidemics. The parainfluenza viruses, like RS virus, cause frequent successively milder re-infections, and older children and adults with colds can infect younger children.

A number of other viruses can cause croup. Influenza A virus can cause a severe laryngotracheitis, and children with such croup are more likely to require endotracheal intubation than those with croup due to parainfluenza viruses. Influenza B virus is a rarer and milder cause of croup. Other viral causes of croup include those viruses which can cause upper respiratory infections, such as rhinoviruses, adenoviruses, enteroviruses and RS virus. Coronavirus antigens have been detected in nasopharyngeal aspirates from children with croup [28]. Measles virus may also cause a significant tracheitis as part of acute measles.

3.6.2 SUPRAGLOTTIS

Over 95% of cases of epiglottis are caused by *Haemophilus influenzae* type b. However, there have been occasional cases of supraglottic inflammation where children have presented identically to epiglottis with acute laryngeal obstruction, but the only positive isolates have been viruses, parainfluenza or influenza virus being the usual cause [16a].

3.7 LOWER RESPIRATORY TRACT INFECTIONS

3.7.1 BRONCHITIS

Acute bronchitis is a febrile illness with cough, and usually with wheeze, which follows an upper respiratory infection. It has a number of synonyms and there is little agreement either within or between countries about a clear definition of acute bronchitis. For a number of years in the UK the condition was called wheezy bronchitis, although there is overlap with asthmatic or asthmatoid bronchitis. In America the name wheeze-associated respiratory illnesses (WARIs) was at one time proposed, but it never caught on. Unfortunately, as discussed previously, the term bronchiolitis is sometimes, and erroneously, used as synonymous with bronchitis.

Clinical features

Upper respiratory signs include rhinitis and often pharyngitis. The child is often tachypnoeic and dyspnoeic. Wheezing is usually audible on auscultation, and crackles are also audible in up to half the children. The chest radiograph must be normal or show only minor inflammatory changes, otherwise the condition is better classified as pneumonitis or pneumonia.

Acute bronchitis is extremely common in pre-school children. More than 20% of all children will have at least one episode by the age of 5 years [10,31]. Of these, approximately half will go on to have asthma, with recurrent wheezing. The younger the child with acute bronchitis the less likely they are to develop asthma. Boesen found that 3–7% of infants under 1 year of age who were admitted to hospital with a first attack of acute bronchitis developed asthma, compared with 18% of children 1–3 years old, and 42% of children over 3 [6]. Children are more likely to develop asthma if they are atopic or have a family history of atopic disease. However, not all asthmatics are atopic, and Frick among others has suggested that virus infections may cause bronchial mucosal damage allowing entry of inhaled allergens, and sensitization to these allergens [15]. His theory is that virus infections may themselves cause asthma, rather than that just trigger wheezing in atopic children. Certainly, Frick may be correct regarding non-atopic children, and it may also be that a significant respiratory virus infection is the main trigger for an atopic child to become asthmatic.

Acute bronchitis in infants under 1 year of age, which rarely leads to asthma, is probably caused predominantly by oedema and debris from exfoliated cells, resulting in obstruction of small bronchi and bronchioles. In asthmatic children with acute bronchitis the narrowing is also caused by constriction of bronchial wall smooth muscle, and is reversible by bronchodilator therapy.

Viruses

Most viruses that cause upper respiratory symptoms can also cause acute bronchitis. RS virus and parainfluenza virus type 3 are the major causes in many studies. However, where optimal techniques are used to isolate rhinoviruses, particularly the use of Ohio HeLa cells, these viruses are the main ones found, being responsible for more than half of all episodes in which a virus is detected [20–22,27]. Moreover, rhinoviruses can be isolated in high titre from sputum of older children with acute rhinovirus bronchitis [21].

Older children with acute bronchitis may well have asthma, and recurrent bronchitis is highly suggestive of asthma. Acute bronchitis does not respond to antibiotics, but may respond to anti-asthma treatment if there is a significant component of bronchospasm. Clearly, it is important always to consider asthma in children with acute bronchitis.

During a study on a highly selected group of pre-school children with recurrent respiratory infections, a small subgroup of four children was identified with frequent episodes of febrile cough and wheeze. These children did not produce interferon-α in response to virus infection *in vitro* [26]. The children were found

not to lack major genes for interferon-α, and indeed subsequently began to produce interferon-α normally. Other workers have subsequently confirmed these findings, identifying occasional children with impaired or delayed interferon-α production.

3.7.2 BRONCHIOLITIS

Bronchiolitis is a pathological description, otherwise inflammation of the bronchioles, that has come to be used as a clinical diagnosis, in the process engendering much confusion. One problem is that there may be inflammation of the bronchioles as well as the small bronchi during acute bronchitis and in asthmatics with wheezing secondary to respiratory viral infection. Nonetheless, the tendency to use the terms bronchiolitis and bronchitis interchangeably is an unfortunate one, and has contributed to varying reports of the rate of subsequent asthma in children with acute bronchiolitis.

Clinical features

The term bronchiolitis should be reserved for infants under 12 months of age who develop an acute respiratory infection characterized by cough, fever, tachypnoea, dyspnoea, hyperinflation and crackles. There may be sternal, subcostal and intercostal recession. Wheeze may occur but is by no means invariable. Cyanosis indicates severe disease. It can be seen, by comparing this clinical description with that for acute bronchitis (see Table 3.1), that there is overlap between the definitions and the potential for confusion. This can be minimized by restricting the diagnosis of bronchiolitis to infants under 1 year of age, and making a clinical diagnosis of bronchiolitis mainly during the epidemic season for respiratory syncytial virus (RS virus), which is the cause of over 70% of cases.

RS virus

RS virus causes an annual winter epidemic in temperate climates, and although the size and severity of the epidemic may vary from year to year, there is always an epidemic. In the tropics, in contrast, epidemics are much more variable in size and timing, tending to occur in the rainy season, if at all. There are two major subtypes of RS virus, A and B, and one or other subtype may predominate or both may circulate simultaneously.

Maternal antibody is poorly protective against RS virus, and babies may be infected shortly after birth. However, there is a 'honeymoon period' in the first 2 months of life when, although babies can be infected, this occurs less frequently than from 2–6 months of age. It is not clear whether this relative protection is due to a protective effect of passively acquired maternal antibody before this wanes, or because very young babies are cosseted and hence less likely to be exposed to infection.

Babies who are born pre-term and are infected with RS virus, even if they do not require artificial ventilation in the neonatal period and are some weeks old, may present initially with apnoeic episodes, before the illness evolves into florid bronchiolitis.

RS virus is ubiquitous. By the end of their second winter over 95% of children have been infected. Approximately 40% of infants infected with RS virus develop lower respiratory involvement and 1% are hospitalized. Reinfections with RS virus are common throughout life, and indeed 70% of adults can be infected by intranasal inoculation of RS virus [29]. Infections are successively milder, but neither specific serum nor secretory antibody to RS virus protects against re-infection. Thus, adults can be an important source of RS virus infection for babies at home and in hospital, although within families most RS virus is introduced by school-aged children.

Infection

RS virus is spread mainly by direct contact with nasal secretions, usually via hands though also by fomite spread on inanimate objects where the virus can survive for several hours. Hall and Douglas conducted a simple, elegant experiment on the mode of transmission of RS virus [17]. Adult volunteers were divided into three groups. The 'cuddlers' changed nappies, cuddled and fed RS virus-positive babies for a short period: five of seven of the adults acquired RS virus infection. The 'touchers' touched surfaces such as cot-sides and chest of drawers from an empty room recently vacated by a baby with RS virus, then rubbed their nose and eyes: four of 10 became infected. The 'sitters' sat 1.8 metres away from a baby with RS virus bronchiolitis for several hours, but touched nothing: none of 14 sitters was infected. Thus for RS virus, droplet spread appears to be an uncommon mode of transmission.

RS virus is by far the most common cause of bronchiolitis, being responsible for at least 70% of cases. However, outside the epidemic period for RS virus, an identical clinical picture may be caused by parainfluenza viruses, particularly type 3, by rhinoviruses, adenoviruses, influenza viruses, enteroviruses, and rarely by sporadic RS virus (Table 3.4). Children hospitalized with RS virus-induced bronchiolitis have an increased risk of developing asthma or at the very least have more reactive airways. In most studies approximately half of the children have subsequently developed recurrent wheezing.

Recovery from RS virus bronchiolitis depends mainly on T-cell function, and children with impaired T-cell function, either congenital or acquired due to HIV infection or cytotoxic therapy for leukaemia, are at risk of developing persistent, severe, symptomatic infection with wheeze and pulmonary infiltrates. In experiments carried out at Salisbury shortly before the Common Cold Research Unit closed, 70% of adult volunteers infected with RS virus were shown to develop a specific cytotoxic T-cell response directed against RS virus fusion protein or nucleoprotein [29]. The Salisbury studies and previous studies on babies with RS virus bronchiolitis [25] suggested that individuals who mounted a detectable specific cytotoxic T-cell response to RS virus tended to have milder infections.

Table 3.4   Relative frequency of virus isolation in acute respiratory infections of childhood

	Adenovirus	Enteroviruses	Coronaviruses	Influenza virus	Parainfluenza viruses	Rhinoviruses	RS virus	Epstein–Barr virus	Herpes simplex virus	Cytomegalovirus	Reoviruses	Measles virus
Common cold	+	++	+++	+	+	++++	+				+	
Sinusitis	+			+	+	+						
Pharyngitis	++++	+++	++	++	++	+	+	++	++	+	+	++
Otitis media	++	+	+	+	+	+++	++		+			++
Croup	++	+	+	+++	++++	+	++		+		+	+
Bronchiolitis		+		+	++	+	++++					
Bronchitis	+	+	++	++	+++	++++	++++					++
Pneumonia	++	+	+	+	++	+++	++++		+			++

Children and adults infected with RS virus produce very little nasopharyngeal interferon-α, in contrast to the situation with rhinovirus infections [24–26]. As RS virus is very sensitive to interferon *in vitro*, it was suggested that interferon might be used therapeutically in RS virus infections. However, studies at Salisbury showed that interferon given intranasally before infection could protect adult volunteers against RS virus infection, but was ineffective in treating volunteers with RS virus-induced colds [19].

3.7.3 PNEUMONIA

Clinical features

Pneumonia is a condition in which radiological features of pulmonary consolidation are present in conjunction with a clinical picture characterized by cough and fever. Tachypnoea and dyspnoea are common, and crackles are usually audible. Viral pneumonia is virtually impossible to distinguish clinically, radiographically or by simple laboratory tests from bacterial pneumonia [23]. Some authors have stated that wheeze is more likely in viral pneumonia, but formal studies have not confirmed this assertion [23].

Histopathological types

There are four major histopathological expressions of pneumonia: acute bronchiolitis, necrotizing bronchiolitis, interstitial pneumonia and alveolar pneumonia.

Acute bronchiolitis.

In this condition there is destruction of the bronchiolar ciliated respiratory epithelium. A degree of pulmonary shadowing, which could be called consolidation, is extremely common in acute bronchiolitis, and RS virus is a common cause of pneumonia as well as bronchiolitis in infancy. There is almost certainly great variation in the clinical and radiological interpretation of which babies have RS virus bronchiolitis, which have RS virus pneumonia, and which have both bronchiolitis and pneumonia. Indeed, in a number of studies in which a comparison has been made of the radiological interpretation of 'normal' and 'pneumonia' radiographs by experienced radiologists, there have been worrying discrepancies in interpretation of what constitutes consolidation [5,37], as well as an inability to differentiate viral from bacterial pneumonia [5,23,37].

Necrotizing bronchiolitis.

In this condition there is deep destruction of epithelial cells and submucosa lining the respiratory tract. It is most often seen in severe, often fatal pneumonia, usually caused by adenovirus types 3, 7 or 21.

Interstitial pneumonia.

This involves diffuse inflammation of the peribronchial alveolar septae with mononuclear cell infiltration. Radiographically there may be a diffuse, hazy pattern, sometimes described as interstitial pneumonitis, although this radiographic pattern of 'interstitial pneumonitis' does not necessarily coincide with a histopathological interstitial pneumonia. Fatal viral pneumonia usually has the histopathological features of bronchiolitis and interstitial pneumonia. Interstitial pneumonia or pneumonitis is more typical of non-bacterial causes of pneumonia, such as viruses, *Pneumocystis carinii*, *Mycoplasma pneumoniae*, and *Chlamydia pneumoniae*, than bacterial causes.

Alveolar pneumonia.

This is characterized by an intense alveolitis, with polymorphonuclear and mononuclear cell infiltrate, degeneration of alveolar cells and the formation of hyaline membranes. This pathological appearance is seen in fatal bacterial pneumonia.

Clinical

In children with viral pneumonia, there is generally a history of upper respiratory tract symptoms, such as coryza or sore throat, preceding the onset of respiratory distress by 1–2 days. Cough and fever are usual, although children with chlamydial pneumonitis are characteristically afebrile. Tachypnoea and dyspnoea may be associated with subcostal and intercostal recession, nasal flaring and sometimes grunting. Cyanosis is relatively uncommon. Young babies may develop apnoeic episodes.

Aetiology

Most viral pneumonias result from an initial upper respiratory infection, as suggested by the symptoms, followed by spread to the lower respiratory tract. The viruses involved include RS virus, parainfluenza viruses, influenza viruses, adenoviruses, measles and rhinoviruses. Some viruses, which are not usually classified as respiratory viruses, can cause pneumonia by haematogenous spread: these include the herpesviruses, varicella zoster virus (VZV), cytomegalovirus (CMV) and Epstein–Barr virus, all of which are most likely to cause pneumonia in immunocompromised children with impaired T-cell function. Children with impaired T-cell function also get more severe or atypical pneumonia with RS virus, parainfluenza viruses, influenza viruses and measles virus. Viruses acquired in the neonatal period may cause an intense, severe pneumonitis. Herpes simplex virus acquired from the maternal vaginal tract at delivery can cause a severe pneumonia at 3–7 days of age, which is almost always fatal unless treated early with acyclovir. Pregnant women who develop chickenpox 5 days before to 2 days after delivery, may transmit VZV transplacentally to the fetus. The baby is born well but, unless varicella–zoster immune globulin (VZIG) is given to the baby soon after birth, 30% will develop a fatal pneumonitis at about 7 days of age.

Incidence

Viral pneumonia is predominantly a disease of pre-school children. The annual incidence of pneumonia under 5 years of age was 42 per 1000 in one study in the USA, compared with five per 1000 from 15–29 years of age [14]. Most pneumonias in children in developed countries are viral [23], perhaps as many as 90%, whereas in developing countries bacterial pneumonia is much more common. Boys are affected 1.5 to 2 times more often than girls. Maternal smoking and air pollution in the urban environment are risk factors for pneumonia in infancy, and breast-feeding is protective. Pneumonia is most common in winter months in temperate climates. RS virus is the commonest cause of pneumonia, and winter peaks of pneumonia often coincide with the RS virus season. There may also be peaks when there are epidemics of other respiratory viruses such as parainfluenza virus 1 or influenza A virus. *M. pneumoniae* affects mainly school-age children in epidemics which occur every 3 to 4 years.

3.8 SUMMARY

Upper respiratory syndromes in childhood are relatively well defined. In contrast, there is considerable potential for diagnostic confusion between acute bronchitis, bronchiolitis and pneumonia. Respiratory viruses are generally capable of causing both upper and lower respiratory tract infections in children (see Table 3.4). Childhood respiratory virus infections are of major significance in their frequency of occurrence, morbidity and mortality, and their effect on family function.

ACKNOWLEDGEMENTS

I am privileged to have worked with David Tyrrell. I would also like to acknowledge my debt to Bernard Valman and David Webster at Northwick Park Hospital for initiating in me a love of the biology and immunology of respiratory viruses. At the John Radcliffe Hospital, Oxford my teachers in the immunology of respiratory viruses were Richard Moxon, Andrew McMichael, Charles Bangham, Frances Gotch and Alain Townsend, to all of whom I owe a great debt.

3.9 REFERENCES

1. Arola, M., Ruuskanen, O., Ziegler, T. *et al.* (1990) Clinical role of respiratory virus infection in acute otitis media. *Pediatrics*, **86**, 848–55.
2. Arola, M., Ziegler, T., Ruuskanen, O. *et al.* (1988) Rhinovirus in acute otitis media. *J. Pediatr.*, **113**, 693–5.
3. Arola, M., Ziegler, T. and Ruuskanen, O. (1990) Prolonged respiratory virus infection as a cause of prolonged symptoms in acute otitis media. *J. Pediatr.*, **116**, 697–701.
4. Badger, G.F., Dingle, J.H., Feller, A.E. *et al.* (1953) A study of illness in a group of Cleveland families. II. Incidence of the common respiratory diseases. *Am. J. Hyg.*, **58**, 31–40.
5. Bettenay, F.A., de Campo, J.F. and McCrossin, D.B. (1988) Differentiating bacterial from viral pneumonia in children. *Pediatr. Radiol.*, **18**, 453–4.
6. Boesen, I. (1953) Asthmatic bronchitis in children; prognosis for 162 cases, observed 6–11 years. *Acta Paediatr. (Uppsala)*, **42**, 87–96.
7. Brimblecombe, F.S.W., Cruickshank, R., Masters, P.L. *et al.* (1958) Family studies of respiratory infections. *Br. Med. J.*, **1**, 119–28.
8. Bynoe, M.L., Hobson, D., Horner, J., Kipps, A., Schild, G.C. and Tyrrell, D.A.J. (1961) Inoculation of human volunteers with a strain of virus isolated from a common cold. *Lancet*, **i**, 1194–6.
9. Chonmaitree, T., Owen, M.J., Patel, J.A. *et al.* (1992) Effect of viral respiratory tract infection on outcome of acute otitis media. *J. Pediatr.*, **120**, 856–62.
10. Colley, J.R.T. and Reid, D.D. (1970) Urban and social origins of childhood bronchitis in England and Wales. *Br. Med. J.*, **2**, 213–17.
11. Court, S.D.M. (1973) The definition of acute respiratory illnesses in children. *Postgrad. Med. J.*, **49**, 771–6.
12. Editorial (1988) Splints don't stop colds – surprising. *Lancet*, **i**, 277–8.
13. Fox, J.P., Hall, C.E., Cooney, M.K. *et al.* (1972). The Seattle virus watch. II. Objectives, study population, and its observation, data processing and summary of illnesses. *Am. J. Epidemiol.*, **96**, 270–85.
14. Foy, H.M., Cooney, M.K., McMahon, R. *et al.* (1973) Viral and mycoplasmal pneumonia in a prepaid medical care group during an eight year period. *Am. J. Epidemiol.*, **97**, 93–102.

15. Frick, O.L., German, D.F. and Mills, J. (1979) Development of allergy in children. I. Association with virus infections. *J. Allergy Clin. Immunol.*, **63**, 228–41.

16. Fry, J. (1971) *The Cattarrhal Child.* Butterworth, London.

16a. Grattan-Smith, T., Forer, M., Kilham, H. and Gillis, J. (1987) Viral supraglottis. *Pediatrics*, **110**, 434–5.

17. Hall, C.B. and Douglas, R.G. Jr (1981) Modes of transmission of respiratory syncytial virus. *J. Pediatr.*, **99**, 100–3.

18. Henderson, F.W., Collier, A.M., Sanyal, M.A. *et al.* (1982) A longitudinal study of respiratory viruses and bacteria in the etiology of acute otitis media with effusion. *N. Engl. J. Med.*, **306**, 1377–83.

19. Higgins, P.G., Barrow, G.I., Tyrrell, D.A.J., Isaacs, D. and Gauci, C.L. (1990) The efficacy of intranasal interferon a-2α in respiratory syncytial virus infection in volunteers. *Antiviral Research*, **14**, 3–10.

20. Horn, M.E.C., Brain, E., Gregg, I., Yelland, S.J. and Inglis, J.M. (1975) Respiratory viral infection in childhood. A survey in general practice. Roehampton 1967–1972. *J. Hyg. (Camb)*, **74**, 157–68.

21. Horn, M.E.C., Reed, S.E. and Taylor. P. (1979) Role of viruses and bacteria in acute wheezy bronchitis in childhood: a study of sputum. *Arch. Dis. Child.*, **54**, 587–92.

22. Isaacs, D. (1984) The epidemiology and immunology of the syndrome of recurrent respiratory infections in pre-school children. MD Thesis. Cambridge University, England.

23. Isaacs, D. (1989) Problems in determining the etiology of community-acquired pneumonia. *Pediatr. Infect. Dis. J.*, **8**, 143–8.

24. Isaacs, D. (1989) Production of interferon in respiratory syncytial virus bronchiolitis. *Arch. Dis. Child.*, **64**, 92–5.

25. Isaacs, D., Bangham, C.R.M. and McMichael, A.J. (1987) Cell-mediated cytotoxic T-cell response to respiratory syncytial virus. *Lancet*, **ii**, 769–71.

26. Isaacs, D., Clarke, J.R., Tyrrell, D.A.J., Webster, A.D.B. and Valman, H.B. (1981) Deficiency of production of leucocyte interferon (interferon-a) in children with recurrent respiratory tract infections. *Lancet*, **ii**, 950–2.

27. Isaacs, D., Clarke, J.R., Tyrrell, D.A.J. and Valman, H.B. (1982) Selective infection of lower respiratory tract by respiratory viruses in children with recurrent respiratory tract infections. *Br. Med. J.*, **284**, 1746–8.

28. Isaacs, D., Flowers, D., Clarke, J.R., Valman, H.B. and Macnaughton, M.R. (1983) Epidemiology of coronavirus respiratory infections. *Arch. Dis. Child.*, **58**, 500–3.

29. Isaacs, D., MacDonald, N.E., Bangham, C.R.M., McMichael, A.J., Higgins, P.E. and Tyrrell, D.A.J. (1991) The specific cytotoxic T-cell response of adult volunteers to infection with respiratory syncytial virus. *Immunol. Infect. Dis.*, **1**, 5–12.

30. Kemp, A. and Bryan, L. (1984) Perennial rhinitis. A common childhood complaint. *Med. J. Aust.*, **141**, 640–3.

31. Leeder, S.R., Corkhill, R.T., Irwig, L.M., Holland, W.W. and Colley, J.R.T. (1976) Influence of family factors on asthma and wheezing during the first five years of life. *Br. J. Prevent. Soc. Med.*, **30**, 213–18.

32. Lovelock, J.E., Porterfield, J.S., Roden, A.T., Somerville, T. and Andrewes, C.H. (1952) Further studies on the natural transmission of the common cold. *Lancet*, **ii**, 657–60.

33. Putto, A. (1987) Febrile exudative tonsillitis: viral or streptococcal? *Pediatrics*, **80**, 6–12.

34. Rantala, H., Uhari, M. and Tuokko, H. (1990) Viral infections and recurrences of febrile convulsions. *J. Pediatr.*, **116**, 195–9.

35. Ruuskanen, O., Arola, M., Putto-Laurila, A. *et al.* (1989) Acute otitis media and respiratory virus infections. *Pediatr. Infect. Dis. J.*, **8**, 94–9.

36. Ruuskanen, O., Meurman, O. and Sarkkinen, H. (1985) Adenoviral disease in children: a study of 105 hospital cases. *Pediatrics*, **76**, 79–83.

37. Stickler, G.B., Hoffman, A.D. and Taylor, W.F. (1984) Problems in the clinical and roentgenographic diagnosis of pneumonia in young children. *Clin. Pediatr.*, **23**, 398–9.

38. Takala, A.K., Meurman, O., Kleemola, M. *et al.* (1993) Preceding respiratory infection predisposing for primary and secondary invasive Haemophilus influenzae type b disease. *Pediatr. Infect. Dis. J.*, **12**, 189–95.

39. Wald, E.R. (1992) Sinusitis in children. *N. Engl. J. Med.*, **326**, 319–23.

40. Wald, E.R., Guema, N. and Byers, C. (1991) Upper respiratory infections in young children: duration of and frequency of complications. *Pediatrics*, **87**, 129–33.

Steven Myint

Viral and Other Infections of the Human Respiratory Tract.
Edited by S. Myint and D. Taylor-Robinson. Published in
1996 by Chapman & Hall. ISBN 0 412 60070 6

4.1 INTRODUCTION

Developments in the diagnosis of respiratory virus infections have often led to advances in other areas of diagnostic virology. Despite this technology most hospital laboratories are less confident in the detection of respiratory viruses as they are, for example, blood-borne viruses. This is often not only due to greater difficulties with getting adequate clinical samples but also because of the greater breadth of resources required confidently to detect the range of respiratory viruses. Most of these methods will be discussed in other chapters. This chapter will give an overview of the technologies and their use in respiratory virology.

4.2 SPECIMEN COLLECTION

It can be argued that this is the most critical step in making a diagnosis of illness due to respiratory viral infection.

4.2.1 GENERAL PRINCIPLES

As with the whole of diagnostic virology, it is better to be able to find a virus at the site of disease than elsewhere if one is trying to associate cause and effect. The respiratory tract can be sampled by nasal swab, throat swab, nasal washings, nasopharyngeal aspirate, nasal aspirate, sputum collection or biopsy. The first four of these have proved to be the most practical.

4.2.2 NASAL SWABS

A nasal swab should be taken from the turbinates for maximum virus recovery and normally requires a little discomfort to the patient to get an adequate sample. The patient should be seated with the head tilted backwards and the swab moistened with viral transport medium (VTM) before introduction along the floor of the nasal passage. It should then be rotated against the turbinates to coat it in nasal secretion. Throat swabs should also be taken with similar meticulousness from the posterior pharyngeal wall or wherever there is obvious inflammation; a 'gag' reflex may be elicited so the patient should be warned. Plain sterile cotton wool swabs should be used and the tip broken off into a container of VTM. Alginate and charcoal swabs can inactivate some viruses and chlamydiae [10,17,48] so should not be used.

4.2.3 NASAL WASHINGS

Nasal washings are a more reliable means of recovering virus than a swab, but are also less convenient for both operator and patient. Either phosphate-buffered saline (PBS) or Hank's buffered saline solution with 0.5% gelatine (BSS) should be used. In either case, the solution should be cold enough to preserve viral infectivity but not ice-cold as this is uncomfortable for the patient. The patient is seated with a towel placed around the neck and requested to inspire submaximally, holding the head backwards, while 1–2 ml of PBS/BSS is inoculated, using a sterile dropper, into each nostril. The patient is then requested to lean forward over a sterile wide-mouthed specimen container and to breathe gently out through the nose. This is repeated until 10 ml of PBS/BSS has been used with a final 'snort' to expel the final drops. With a practised technique it should be possible to minimize the loss of PBS/BSS to 1–2 ml.

4.2.4 NASOPHARYNGEAL ASPIRATES

Nasopharyngeal aspirates are collected in sterile plastic disposable mucus extractors using dis-

posable suction catheters attached to a vacuum pump. These are commercially available and the manufacturers' instructions should be followed [31]. After aspiration, a volume of VTM greater than the dead space of the catheter should be aspirated; about 4–5 ml should suffice. Like nasal washings, nasopharyngeal aspirates tend to lead to greater virus recovery rates than either nose or throat swabs. An alternative method is nasal aspiration using the same equipment. This is suitable for patients with obvious rhinorrhoea and is less uncomfortable. For common cold viruses it is, at least, as good as nasopharyngeal aspirates (unpublished data from the author's laboratory). Thick secretions may have to be diluted a further 1:2 or 1:3 in VTM or PBS and may also require low-speed centrifugation (2000 r.p.m. for 5 min) to remove excess cellular debris.

4.2.5 SPUTUM

It is not always necessary to collect sputum from patients with viral pneumonia as the clinical picture involves usually both the upper and lower respiratory tracts and a positive diagnosis based on a nasopharyngeal aspirate may be sufficient. In any case, the collection of bronchial secretions for virus analysis is best done by bronchoalveolar lavage, which reduces the risk of contamination by microbial flora from the upper airways. Lavage fluid should be collected into VTM. Thick secretions should be clarified in the same manner as for nasopharyngeal aspirates.

4.2.6 BIOPSY MATERIAL

Biopsy material should be collected under sterile conditions directly into VTM. Material should be ground with sterile washed sand as a 10% w/v solution in VTM and then clarified by low-speed centrifugation (e.g. 3000 r.p.m. in a bench-top centrifuge for 5 min) to remove debris. As biopsy material is usually taken post mortem there is likely to be heavy bacterial contamination and further antibiotics are often added, for example, penicillin at a final concentration of 400 units/ml and streptomycin at 400 μg/ml.

4.2.7 TIMING

Whatever specimen is taken, the timing of the sample is important. Specimens should be taken in the symptomatic period, preferably within the first 2–3 days, as virus titres fall significantly during this period. They should be collected into sterile containers containing sufficient fluid to prevent evaporation. This VTM should be of neutral pH, contain stabilizing proteins to maintain infectivity, and anti-biotics to kill other microorganisms. Many different recipes exist for VTM [64]. Respiratory viruses are particularly labile and are best trans-ported at 4°C to the laboratory and processed within 24 hours. If this is impractical, then slow freezing of specimens to –70°C or lower in the presence of 10% glycerol or 10% dimethylsulphoxide (DMSO) should be used to maintain infectivity. Multiple freeze–thawing should, however, be avoided at all costs, and if multiple testing is envisaged aliquots of the specimen should be made before storage. Self-defrosting freezers should not be used.

4.3 DIAGNOSTIC APPROACHES

The diagnosis of a specific viral infection can be made by one of five principal means: detection of virus by electron microscopy; detection of the toxic effects of a virus; detection of viral antigens; detection of viral genome; and detection of antibody to the virus.

4.4 DETECTION OF VIRUS BY ELECTRON MICROSCOPY

This is accomplished with or without immunological enhancement. Electron microscopy (EM) is, however, a fairly unreliable method for the diagnosis of respiratory viruses in clinical material because of low virus titres, except at the peak of infection, and is more often used to confirm

the cause of a particular cytopathic effect in cell culture. It has, however, been used for primary detection of coronaviruses, influenza and paramyxoviruses: a drop of nasopharyngeal secretion is inoculated onto a specimen grid but this may need dilution with an equal volume of sterile distilled water if viscous. Grids are ideally Formvar/carbon coated and staining is with phosphotungsten or uranyl acetate; the latter tends to result in a greater apparent size of the virus. For detecting virus in cell culture supernatant fluid, prior concentration by ultracentrifugation (90 000*g* for 30 min) improves the sensitivity of the method. The reader is referred elsewhere for methodological detail [4,19].

The other main diagnostic use of EM in respiratory virology is the detection of viruses (principally herpesviruses and measlesvirus) in lung biopsies. The biopsy is ground or freeze–thawed several times, after which a suspension is prepared in sterile water for inoculation onto grids.

The sensitivity of EM can be enhanced several hundred-fold by immunoelectron microscopy (IEM). There are various methods employed (direct, solid-phase, immunogold, immunosorbent, agar diffusion) but they all require the use of specific antibody. As EM is not a diagnostic method of choice for the detection of respiratory viruses, few laboratories use any of these methods.

4.5 DETECTION OF TOXIC EFFECTS OF A VIRUS

Common cold viruses were first identified by the use of organ culture and this is a method still used by some researchers. Tracheal organ culture is described in Chapter 10.

4.5.1 CELL CULTURE

The 'gold standard' for detecting most respiratory viruses is still accepted to be cell monolayer culture. This is despite other methods, particularly that of gene amplification, being shown to be more sensitive. Most hospital laboratories use cell culture and it has been a reliable screening method for the presence of viruses in respiratory secretions and tissues for several decades.

Many cell lines are available for the isolation of respiratory viruses (Table 4.1) but clearly, it is not practical for any diagnostic laboratory to carry the complete range. In general a lung fibroblast cell line and a primate kidney

Table 4.1 Cell lines for the propagation and identification of important viral respiratory pathogens

Virus	Cell lines/type	Diagnostic features	Comments
Adenovirus	Human diploid fibroblast (HELF, WI-38, MRC-5)	Large, round cell clusters with refractile intranuclear inclusion bodies in 5–10 days	Can be confirmed by EM
	Primary human embryonic kidney (HEK) Malignant/continuous (Hep-2, HeLa, KB)		Cannot be maintained for longer than 10–14 days; blind passage may be required for some serotypes.
	Monkey kidney (PMK, SMK, TMK, 293)		Useful for serotypes 40/41
Coronaviruses	Human diploid fibroblasts (MRC-C, WI-38, C16)	Some rounding up of cells after 7–14 days	Primary isolation unreliable
Enteroviruses	Monkey kidney (PMK, BGM)	Rounded small cells with lysis after 2–5 days	Some Coxsackie A strains will not grow
	Human diploid fibroblast (MRC-5)		Less sensitive
	Malignant/continuous (Hep-2, HeLa)		Less sensitive
Influenza viruses	Monkey kidney (PMK, GMK, LLC-MK2, BSC-1)	Vacuolation and lysis after 3–5 days with influenza B but not A	
Measles	Monkey kidney (PMK)	Multinucleate giant cells after 7–10 days	
	Human kidney (HEK)		
Parainfluenza viruses	Monkey kidney (PMK, LLC-MK2)	Dark granular cells in syncytia after 7–10 days	Poor CPE
Respiratory syncytial virus	Monkey kidney (PMK)	Syncytia after 5–14 days	
	Malignant/continuous (Hep-2, HeLa) Human diploid fibroblasts (MRC-5, HEL)		Less distinct syncytia
Rhinovirus	Monkey kidney (PMK)	Rounded, refractile cells after 5–10 days	
	Human diploid fibroblasts		CPE may appear earlier than with PMK cells

CPE, cytopathic effects, EM, electron microscopy

cell line are essential for routine use with the occasional use of others as required.

Roller culture tubes are used most often because they tend to enhance any cytopathology, although the enhancement may be due more to better growth of the cell line than increasing infectivity of the virus [22]. Specimens should be inoculated into duplicate tubes with cell monolayers of 70–90% confluence. A 0.2-ml specimen is used for each tube and appropriate uninoculated controls must also be set up. Roller culture tubes should be incubated at 33°C for specimens from the upper respiratory tract and at both 33°C and 37°C for those from the lower respiratory tract. All tubes should be examined at least twice weekly, with changes of medium weekly or more frequently. Cultures may need to be kept for 3 weeks before being deemed 'negative'. Identification of respiratory syncytial (RS) virus and adenoviruses (and herpes simplex virus which may also be found) can be made by the recognition of a characteristic cytopathic effect (CPE). Haemadsorption may be used to identify influenza virus and paramyxoviruses: a CPE is seen with the former, but is usually absent with the latter. Rhinoviruses, enteroviruses and coronaviruses often require confirmatory tests such as neutralization, EM and acid lability. Differential growth at 33°C and 37°C is a helpful indicator of the presence of rhinoviruses and coronaviruses. The addition of trypsin to the medium for isolation of influenza virus or calcium [67] and glutamine [49] for RS virus may be required and can be used as a guide to identification. Details of specific cell culture methods can be found in specific chapters and in standard laboratory manuals [65, 71].

4.5.2 SUPPORTIVE FORMATS FOR CELL CULTURE

Neutralization tests are often required to confirm the presence of a suspected virus in cell culture. If virus is mixed with known dilutions of antibody an estimate of the amount of virus can be ascertained from the minimal titre of antibody necessary to neutralize the CPE of the virus. A haemagglutination–inhibition test is useful for determining the presence of influenza virus; this uses the same principle as the neutralization test in that neutralized virus does not agglutinate a suspension of erythrocytes. The capacity of the virus to agglutinate added erythrocytes is evaluated after 2–3 hours.

Cell culture methods are often too slow to be of practical benefit to the management of patients but the use of centrifugation followed by immunofluorescence to detect viral antigens enhances the speed of diagnosis. These so-called shell-vial cultures (because the cells are on cover slips contained within 1-dram shell vials) have been applied to influenza viruses grown in Madin–Derby canine kidney (MDCK) cells [8,26,30,44,54,59,66,72,82,83], RS virus [36,68,82], adenoviruses [18,25,47,50,81, 84], measles and parainfluenza viruses [55]. The sensitivity of this approach, however, varies enormously from 10–80%, possibly because of a lack of standardization. Moreover, the use of immunofluorescence combined with conventional culture, and without centrifugation, has been shown in some studies to be just as sensitive .

4.5.3 EGG CULTURE

This is more sensitive than cell culture for isolating influenza A and C viruses but is too specialized for routine diagnostic laboratories to be used other than rarely. It is, however, useful for growing large amounts of virus required for vaccines and immunological assays. Typically, 13 to 15-day-old eggs are used for influenza A virus, and younger ones for influenza C virus [35]. Aliquots (0.1 ml) of specimen are inoculated into the amnion and incubated at 37°C for 3 days. The eggs are then chilled to 4°C for 4–6 hours and the allantoic and amniotic fluids harvested separately. Identification of influenza virus is confirmed by haemagglutination with either human (influenza A at 4°C and 20°C), chicken (influenza C at 4°C) or both (influenza B at 4°C and 20°C) erythrocytes.

4.6 DETECTION OF VIRAL ANTIGENS

Many of the test formats which will be described for detecting anti-viral antibody have been applied to detect respiratory virus antigens and these are discussed in the chapters on specific viruses. Only two methods, immunofluorescence and enzyme-linked immunoassay, are, however, commonly used in routine diagnostic laboratories.

4.6.1 IMMUNOFLUORESCENCE

Immunofluorescence to detect viral antigens was the original rapid diagnostic method [28,29]. Many laboratories still only use cell culture and immunofluorescence in their diagnostic armamentarium for respiratory viruses.

Direct immunofluorescence.

In this assay, cell deposit (either specimen or cell-culture) is fixed on a microscope slide with acetone. Virus-specific serum which has been labelled with a fluorescent dye is added and any unbound antibody is washed away; bound antibody is then detected by fluorescence microscopy. Commercial polyclonal sera are available for most respiratory viruses but these are being replaced by monoclonal antibodies which have more consistent specificity and therefore are less likely to require absorption before use. Labelling may be with fluoroscein isothiocyanate (FITC) which emits maximally at 517 nm and rhodamine B which emits maximally at 595 nm. The appearance of the former is, therefore, green and the latter, red.

Indirect immunofluorescence.

This has three instead of two basic steps. Specific unlabelled antiviral antibody is first bound to any antigen and then a second labelled anti-antibody, from another species, is used to detect the presence of the first. This increases the number of bound antibodies

and, therefore, increases sensitivity. The increase in the number of steps still allows a result within 2–3 hours of taking the sample. It also has the added advantage that only one type of labelled antibody need be used on one specimen compared with a number of virus-specific labelled antibodies in the direct assay. Although simple in principle, immunofluorescence has some pitfalls: the specimen needs to contain an adequate number of cells; the antibodies used need to be specific; and background fluorescence occurs commonly. Improvement in the quality of specimens can be made by using procedures such as 20% Percoll centrifugation [78] but a trained and experienced microscopist is still vital for the success of the immunofluorescence technique.

4.6.2 ENZYME-LINKED IMMUNOSORBANTASSAY (ELISA)

The principle of this method is illustrated in Figure 4.1. The wells of a microtitre plate are coated with virus-specific antibody. Any virus or viral antigen present in the specimen which is added is then bound and unbound material is washed away. A second antiviral antibody of another species which is labelled with an enzyme is added and binds to any antigen now present. The addition of a substrate for the enzyme then produces a colour reaction. It is possible by this method to detect as little as 0.1 ng of viral protein/ml of secretion. This is usually well above the sensitivity required, as up to 100 ng/ml of nasopharyngeal secretion are produced at the peak of most respiratory virus infections.

As with immunofluorescence, polyclonal sera have been used but are being replaced by monoclonal antibodies. Many enzyme–substrate pairs are used but horseradish peroxidase with 0-phenyldiamine and alkaline phosphatase–BCIP/NBT remain the most popular combinations.

1. Specific antibody bound
   to solid phase

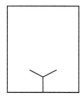

2. Sample added

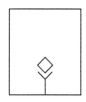

3. Unbound sample washed off

4. Add second specific antibody
   conjugated with enzyme

5. Add substrate for enzyme

6. Measure colorimetric reaction

Figure 4.1   Principles of enzyme-linked immunoassay for antigen/virus detection.

4.6.3 OTHER METHODS

Immunoperoxidase staining.

This is the enzymatic equivalent of immuno-fluorescence. Instead of a fluorochrome dye being conjugated to antibody, the latter is labelled with peroxidase enzyme which in the presence of a substrate produces a colour reaction. Both direct and indirect formats have been described and they have the inherent advantage that only light microscopy is needed. The technique has not, however, gained a foothold in respiratory virology, although in one study it was shown to be more sensitive than indirect immunofluorescence for the detection of RS virus in nasopharyngeal secretions [34].

Radioimmunoassay (RIA).

This utilizes the same format as the ELISA but the antibody used for detection is radioactively labelled. The method has marginally greater sensitivity than the ELISA but has the problems inherent in manipulating radioactive substances.

Time-resolved fluoroimmunoassay.

The format is also that of the ELISA. In this method the detector antibody has a fluorescent label, Europium-chelate, which is activated by the addition of an 'enhancer' solution [69]. Fluorescence is measured immediately by a single photon fluorimeter. It has been shown to be more sensitive than the ELISA for influenza virus [80].

Flow cytometry.

This is a rapid method but appeared to lack sensitivity when applied to rhinovirus detection [37].

4.6.4 PREPARATION OF ANTISERA

Polyclonal antibodies raised in animals are most often used and are adequate for most purposes. All the respiratory viruses are good antigens when given parenterally to animals. Antibodies to subunits, such as the common hexon protein of adenovirus, are however produced by intramuscular injection with Freund's adjuvant. Intranasal inoculation of ferrets [73] and guinea pigs [79] has been used to raise antisera against influenza viruses and paramyxoviruses, respectively.

The use of monoclonal antibodies for respiratory virus diagnosis is identical to that for other viruses and is reviewed elsewhere [85].

4.7 DETECTION OF VIRAL GENOME

Viral genomes can be detected directly or indirectly. Direct detection comprises the isolation of nucleic acid and then visualization in agarose or polyacrylamide gels by staining with a suitable dye. It is only practical if there is sufficient nucleic acid in the clinical specimen and a characteristic pattern of genome nucleic acid, or its fragments, can be identified. In diarrhoeal stools, virus is produced in such vast amounts that it is possible to detect the segmented RNA of rotavirus or restriction enzyme products of adenoviruses. For respiratory virus infections, however, the amount of virus – and hence nucleic acid – tends to be too small and indirect detection of nucleic acid with probes is necessary. Methods for specific viruses are given in other chapters, but the principles involved are outlined here.

4.7.1 PRINCIPLES OF NUCLEIC ACID HYBRIDIZATION

All replicative organisms, whether prokaryotes or eukaryotes, contain deoxyribonucleic acid (DNA) or ribonucleic acid (RNA) or both. The basic building bricks of these nucleic acids are the four nucleosides: adenosine (A), cytidine (C), guanosine (G) and either thymidine (T) in DNA or uridine (U) in RNA. RNA is, with few exceptions, made up of a single chain of phosphorylated nucleosides (nucleotides). DNA usually exists as a duplex with one chain being 'complementary' to the other. This complementarity arises from the specific binding that exists between A and T (or U) and C and G. Gene probes are nucleotide (nt) sequences, which have been labelled to enable their detection, that are complementary to a sought nucleotide sequence. Hybrids can be formed

that consist of two strands of DNA, of RNA, or one of each. The temperature at which only half of the strands are in the form of a strand is known as the melting temperature (T_m). The T_m of a hybrid is dependent on the base composition of the hybrid, the ionic strength of the hybridization solution and the concentration of agents such as formamide. This is represented in the equation:

$$T_m = \frac{81.5 + 16.6\log M + 0.41(\%G + C) - 500}{n - 0.61(\% \text{ formamide})}$$

where M is the ionic strength in moles/l and n is the length of the shorter chain of a duplex. A simplified formula which more accurately reflects hybrids where n is shorter than 50 bases is:

$$T_m = 4(G + C) + 2(A + T)$$

The T_m of an RNA:RNA hybrid is 9–12°C higher than the equivalent DNA:DNA duplex [11]; RNA:DNA hybrids may have a higher or lower T_m [12,15,16].

There are five principal components of the hybridization reaction: the probe, the conditions of hybridization, the hybridization format, the labelling or detection system, and the target.

Probes and probe selection.

Probes are usually designed from known nucleic acid sequences. They may be many hundreds of nucleotides in length or short sequences of only tens of nucleotides. The former tend to be products of cloned nucleic acid, the latter synthesized chemically. In either case it is a prerequisite that the target sequence is known and unique to the agent sought.

Long or 'full-length' probes have greater specificity because of their length and complexity. They can also be labelled more extensively, thus increasing the level of sensitivity. Synthetic oligonucleotides are, however, easier and less costly to generate. They can also be synthesized in an already modified form, for example with a detector molecule(s), and have a much faster rate of hybridization. Most useful oligonucleotide probes are 18–30 bases in length, have 40–60% G+C, are without intra-complementary regions and have less than 70% homology with non-target regions. Computer software is now freely available to aid the design of oligonucleotide probes.

Hybridization conditions.

These determine both the kinetics and stringency of hybridization:

1. Temperature. Standard hybridization for full-length probes takes place most readily at T_m –25°C. In practice this will be between 60 and 75°C. Temperatures closer to T_m can be used if greater stringency is required. There is less flexibility when employing oligonucleotide probes, and typically hybridization would be allowed to take place at T_m –5°C.
2. Salt concentration. This affects predominantly the rate of hybridization [63]. Typically a sodium concentration between 0.4 and 1.0 M is used.
3. Formamide concentration. The use of formamide reduces the T_m of the duplex [46] and is particularly useful for RNA:RNA hybridizations. Dimethyl-sulphoxide (DMSO) can be used in place of formamide, particularly in the polymerase chain reaction as Taq polymerase is inhibited by formamide.
4. Hybridization accelerators. A number of inert substances have been employed to speed up hybridization reactions with large probes. The most commonly used is dextran sulphate in final concentrations of 5–10%. Similar final concentrations are used for polyethylene glycol [3]. These substances are thought to act by decreasing the volume that nucleic acids can occupy thus speeding up the interaction process. Polyacrylic acid (2–4% final concentration)

[53], phenol [39] and 4 M guanidine thiocyanate [76] are also used under appropriate circumstances. The last of these is, particularly, useful in RNA:RNA hybridization reactions. These hybridization accelerators are not effective with probes less than 150 base pairs (bp) long but then are generally not needed with these short probes.

Hybridization format.

Essentially, there are two formats in use. Single-phase hybridization implies a liquid environment, and two-phase (or mixed-phase) hybridization has either the probe or target bound to a solid matrix. Liquid phase hybridization is used in the technologies employing gene or probe amplification. Most non-amplification-based probe methods use a two-phase system.

There are several advantages in fixing either the probe or target to a solid support [51]. The time for hybridization can be reduced, multiple targets or probes can be incorporated into a single hybridization and the solid phase can be recovered to enable detection of hybrids. Also, bound strands of nucleic acid will not self-anneal. The main problem is that of non-specific binding to the solid phase producing so-called 'background'. This can be reduced by pre-hybridizing the solid phase with components such as non-target nucleic acid (e.g. salmon sperm). Additionally, the hybridization mix may include blocking agents such as Denhardt's solution (0.2% w/v bovine serum albumen, Ficoll, polyvinylpyrrolidine) or detergents such as sodium dodecyl sulphate (SDS, usually 0.1–0.5%). A washing step with a detergent is also usually incorporated post-hybridization to remove non-specifically bound components. The stringency of this wash can be adjusted by varying both the temperature of the wash and the amount of detergent.

Two types of solid phase are most often used. These are nitrocellulose and nylon membranes [51]. Single-stranded DNA is bound by its sugar–phosphate backbone non-covalently to nitrocellulose by heat applied under a vacuum. Nitrocellulose membranes have been used reliably for many years and provide an inexpensive way of immobilizing nucleic acid. They cannot, however, bind RNA or short DNA strands (less than 300 bases) effectively. There are several other limitations: the formation of complete duplexes results in loss of the DNA from the filter; prolonged hybridization reactions at high temperature tend to lead to loss of bound nucleic acid; there is often batch-to-batch variation in the binding properties for nucleic acid, and the filters are physically fragile. Alternatively, nucleic acids can be linked covalently to nylon with the use of ultraviolet light. Nylon has all the advantages of nitrocellulose apart from its low cost. Moreover, the binding capacity of nylon is said to be three to five times that of nitrocellulose. The main problem with the use of nylon membranes is non-specific binding of probe, which is particularly troublesome when some non-radioactive labels, such as biotin, are used. A number of commercially prepared nylon-based membranes are available which differ slightly in their precise characteristics.

Other solid supports which are being used increasingly are magnetic beads [77] and polystyrene microtitre plates [32,38,58]. The use of magnetic beads allows easy separation of hybridized products. The microtitre plate methodologies, when coupled with fluorimetric or colorimetric detection methods, enable more easily the processing of the large numbers of specimens that is necessary in a routine diagnostic laboratory.

Test formats.

The now classical method is that developed by Southern [70]. The target DNA is separated by gel electrophoresis and transferred to a membrane by capillary action. The membrane is then hybridized to probes in solution. Northern blotting uses the same methodology

for detecting target RNA species [2]. Reverse hybridization has the probes on the filter and the targets in solution.

Target nucleic acid can be applied directly to a filter either as a 'dot' or a 'slot' for subsequent probing. The slot–blot method has the added advantage of being semi-quantitative if a densitometer is used. Several manufacturers produce manifolds that allow nucleic acids in solution to be applied under vacuum to a membrane, thus ensuring uniformity of size of dots or slots.

In situ hybridization allows the localization of nucleic acids within cells and tissues. Intact cells are fixed on a solid phase, either a membrane or microscope slide, probed, and examined under a microscope. Non-isotopic methods allow direct visualization of target sites.

Sandwich hybridization uses two probes, one that is unlabelled to capture target sequences and the other, labelled, to detect the captured target. This can be easily adapted to the microtitre plate format [38,58].

Labels and labelling methods.

Probes can be labelled with isotopic (radioactive) or non-isotopic (non-radioactive) markers. For research purposes isotopes such as ^{32}P, ^{35}S, or ^{125}I have been used to label probes and allow sensitive detection by autoradiography. For safety reasons the routine use of radioactivity to detect infectious agents has largely been superseded by a number of non-radioactive methods. The first of these utilized biotin–avidin [1,14,43]. Biotin is a small water-soluble vitamin of molecular weight 244 Da which has a very high affinity for the much larger glycoprotein, avidin. Biotinylated nucleotides are incorporated into the probe and then avidin coupled to an enzyme, such as alkaline phosphatase, is used as a secondary detector molecule. After the addition of an appropriate substrate for the enzyme, a colorimetric or chemiluminescence reaction can be detected. The sensitivity is near to that of using ^{32}P, and less than 1 pg of target DNA can be detected. The main technical difficulty tends to be a high level of background, particularly for tissue in situ hybridization. Alternatively, the second detector molecule can be an antibody labelled with such an enzyme. Alternatives to biotin–streptavidin, such as digoxigenin, have since been introduced [21]. It is also possible to label the nucleotide probe directly with an enzyme [33] or use labelled antibodies directed against nucleic acid duplexes [13]. These are all modifications on a theme and have roughly equivalent sensitivity. Chemiluminescent labels have the added advantage that the results can be conveniently recorded on an X-ray film. Details of methods can be found in the review by Kricka [40].

Labelling methods.

Oligonucleotide probes are easily labelled at the time of synthesis by the incorporation of modified nucleotides. Long probes can be labelled by a variety of techniques.

Nick translation is a well-tested and relatively inexpensive method of generating DNA probes of high specific activity [60]. The method relies on DNase I to introduce random single-stranded breaks (nicks) in both strands of a DNA duplex and then for DNA polymerase I to remove one or more bases at the exposed 5' phosphoryl end and to replace them with radiolabelled substitutes at the corresponding 3' terminus. The nick would have been shifted ('translated') in a 3' direction. In this way probes can be produced with an activity of 10^6–10^9 c.p.m./μg DNA.

Greater specific activities can be achieved using the random priming method [27]. Oligonucleotides are used to act as primers for the synthesis of new radiolabelled copies of probe sequences, using the Klenow fragment of DNA polymerase I. Less than 200 ng of input DNA is required and the probe can be used without the removal of unincorporated nucleotides (unlike nick-translated probes). The method also has the advantage, compared

with nick-translation, of being applicable to single-stranded DNA templates and DNA templates of less than 500 bp.

It is also possible to synthesize probes from templates. Specific single-stranded DNA probes can be made by cloning and inserting the sequence of interest into the multiple cloning site (MCS) of a vector such as M13mp19. Then with a known M13 primer and Klenow DNA polymerase it is possible to generate strand-specific probes. A modification of this approach is the use of the polymerase chain reaction (PCR, see below). Similarly single-stranded RNA probes can be generated from Riboprobe vectors such as pGEM^R [52]. These incorporate promoters either side of a MCS for SP6, T7 or T3 RNA polymerases so that both complementary and anti-complementary single-stranded RNA probes can be synthesized. The advantages of RNA probes are that RNA:RNA and RNA:DNA hybrids tend to be more stable than DNA:DNA ones, they can be used without denaturation, they do not self-anneal and unhybridized probe can be removed by RNase digestion. The major problem is that an RNase-free environment is required.

The target.

Sample preparation is as important a part of the procedure as the other steps. For most biological samples some degree of processing is required to enable the extraction and concentration of target nucleic acid. The possible exception to this is in situ hybridization where the nucleic acid target may be already concentrated within cells.

Cellular material may have to be lysed to release nucleic acid. A combination of physical methods (shearing, ultrasound) and detergent is usually sufficient for this purpose. The principal material that may then interfere with hybridization is protein. This may be partially digested at first with proteases and then separated from the nucleic acids. The most common procedure used depends on the differential solubility of nucleic acids and proteins in organic solvents such as phenol and chloroform. Phenol extraction of proteins results in nucleic acids separating into the aqueous component, while the phenol phase contains the lipids and some proteins. The interphase will also contain some proteinaceous material. The DNA can then be recovered by precipitating in salt and ethanol (or isopropanol).

RNA preparation is technically more demanding than that for DNA. This is because of the ubiquitous nature of RNases – many cell types contain high levels of RNase activity – which makes it a particular problem with secretions from the respiratory tract. Thus an RNase-free environment must be attained. This involves treating glassware and solutions with diethylpyrocarbonate and extensive autoclaving of materials. It is often necessary to incorporate inhibitors of RNase activity into the extraction mix. Guanidinium isothiocyanate rapidly reduces RNase activity and standard procedures utilize this. These procedures are effective but labour-intensive. Alternatively more specific inhibitors such as vanadyl–ribonucleoside complex (VRC) and human placental or recombinant RNasin can be used. These latter methods are technically less demanding but have other inherent disadvantages. VRC also inhibits a number of other enzymes and will therefore interfere with subsequent enzymatic manipulations. Specific methods can be found in gene cloning and PCR manuals [9,57,61,63].

4.7.2 GENE AND PROBE AMPLIFICATION

Gene amplification technologies, principally the PCR, have found widespread application in the few years since their introduction [23,24,62]. Two oligonucleotide primers are annealed to regions flanking the target region in duplex DNA and then extended by DNA polymerase (Figure 4.2). Using a thermostable DNA polymerase it is possible to cycle this many times

without the addition of new enzyme at each cycle and obtain a million-fold amplification of known target sequences. Typically 30 to 50 cycles are used. This entire process is usually automated. The products of a PCR reaction can then be visualized by gel electrophoresis. The length of the product is that of the two primers plus the intervening region. It is then usual to confirm these products as being the required target region by probing with an oligonucleotide probe. 'Nested PCR' involves two rounds of amplification with the second round utilizing primers complementary to sequences within the primer product of the first. This may increase the amount of amplification achievable and has increased specificity.

Most respiratory viruses have RNA genomes. Amplification of RNA target sequences by PCR requires the prior step of making cDNA copies. This reverse transcription can be done with a separate enzyme such as avian myeloblastosis virus (AMV) reverse transcriptase, or with one of the newly available thermostable enzymes such as rTth polymerase which has both reverse transcriptase and DNA polymerase properties. There are a number of potential problems in the application of the methods. Currently, thermostable polymerases have a higher misincorporation than, for example, the thermolabile T4 DNA polymerase. This is because of the absence of 3' to 5' exonuclease ('proof-reading') activity in the former; it is calculated that the misincorporation rate is of the order of 10^{-5} nucleotides per cycle. Another thermostable enzyme, VENT DNA polymerase, from *Thermococcus litoralis*, has proof-reading activity and is therefore likely to have a lower misincorporation rate, but has drawbacks for sequence specific priming reactions.

The specificity of priming has also been a problem with priming of non-target sequences resulting in primer artefacts. Modifications such as the addition of the thermostable polymerase at high temperature only, the so-called 'hot start', and the addition of DNA binding

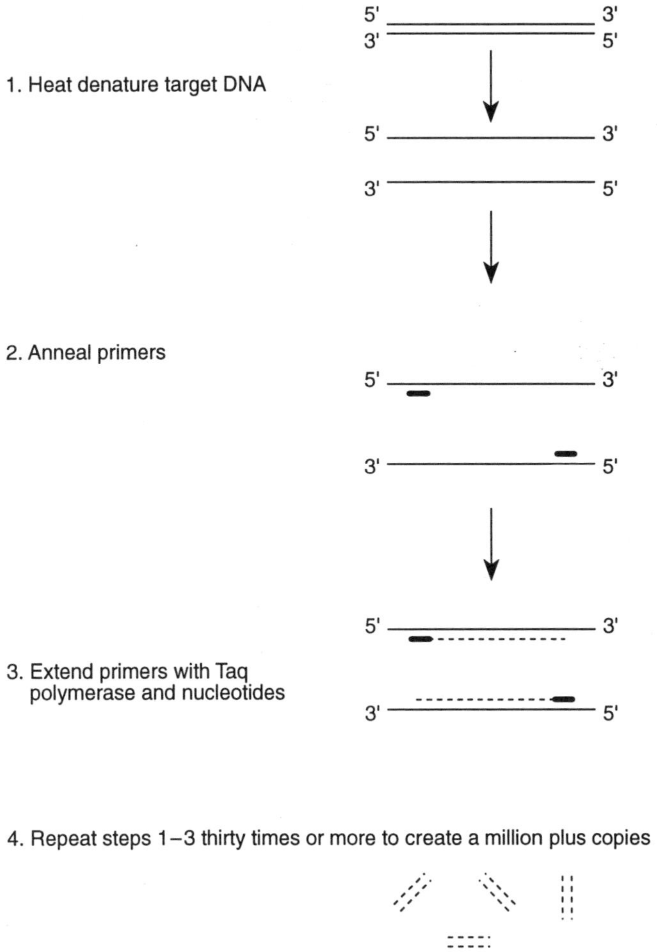

1. Heat denature target DNA

5' ════════════ 3'
3' ──────────── 5'

↓

5' ──────────── 3'

3' ──────────── 5'

↓

2. Anneal primers

5' ──────────── 3'
 ▬

3' ──────────── 5'
             ▬

↓

5' ──────────── 3'
 ▬▬------------

3. Extend primers with Taq
   polymerase and nucleotides

3' ──────────── 5'
   ------------▬▬

4. Repeat steps 1–3 thirty times or more to create a million plus copies

Figure 4.2 Principles of the polymerase chain reaction (PCR).

proteins may be the solution to this. The hot start also seems to reduce the incidence of 'primer–dimers', a product which consists essentially of primer and complementary sequences only.

The major problem with such an exquisitely sensitive technique is that of contamination. This can arise from previous PCR reactions (product carryover) or stray DNA, particularly plasmids. The most important measure to prevent this is attention to strict handling procedures. These include separation of preparation areas from those for analysis of reaction prod-

ucts and clinical material or plasmid DNA, the use of dedicated and positive-displacement pipettes and pre-aliquoting of reagents. In addition to this one can use short-wave UV irradiation of the PCR reaction mix before addition of the template.

Another commercially available (Perkin Elmer Cetus) strategy uses substituted nucleotides. Deoxyuridine triphosphate (dUTP) is used in place of deoxythymidine triphosphate (dTTP). In subsequent PCR mixes, the enzyme uracil–N-glycosylase (UNG) enables the excision of uracil from any contaminating DNA containing uracil. The resultant abasic polynucleotides cannot function as templates and are susceptible to alkaline hydrolysis at the high temperatures used in a PCR reaction. The enzyme does not excise uracil from RNA so that this can still be used as template.

The PCR can also be used to produce vast quantities of probes for amplification. This method is akin to the M13 template methodology already described. An excess of one primer and labelled nucleotides are used in a standard reaction to produce an excess of one strand ('asymmetric PCR') which is labelled.

There have been a number of modifications of the basic PCR protocol including quantitation, adaptation to microtitre plates, and modification of primers so that colorimetric detection is possible [38,56]. Commercial kits for PCR diagnosis are now available based on the last two modifications. Perhaps the biggest obstacles that still need to be overcome are the cost and labour intensiveness of these methods; some of these modifications may help in this regard.

Another gene amplification technique is transcription-based [41]. TAS, or transcription-based amplification systems, are applicable to RNA targets. A version of this, self-sustained sequence replication or 3SR, is illustrated in Figure 4.3. The sensitivity of the method is of the same order as that of the PCR but currently the system is difficult to automate.

An alternative to amplifying genes is amplifying the probes themselves. There are two systems that show promise. The first of these is based on the ability of an RNA-dependent RNA polymerase, Qβ replicase, to synthesize 10^6–10^9 copies of a template sequence [45]. The principles are illustrated in Figure 4.4. The probe sequences are inserted into the MCS of Qβ phage-derived plasmid containing a T7 promoter. T7 RNA polymerase is then added to produce RNA transcripts containing probe sequence. These are hybridized to their target, and after elution, amplified by the addition of Qβ replicase. This is several times more rapid

than the thermal cycling of a PCR reaction. The other major advantage is that because the amount of amplified RNA depends on the

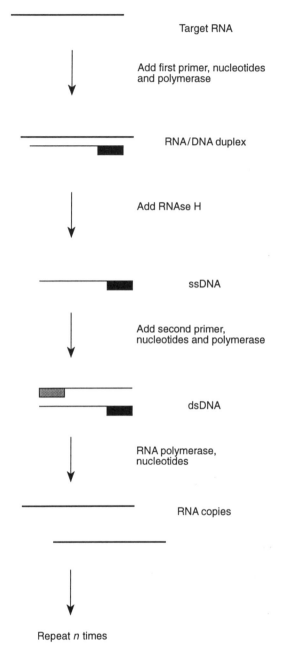

Figure 4.3 Transcription-based amplification: self-sustained sequence replication (3SR).

amount hybridized, the method is quantifiable. The major difficulty is non-specific hybridization. Once this is overcome, the Qβ replicase system may offer an alternative to PCR.

Perhaps the most exciting development since that of automated PCR is the Ligase Chain Reaction (LCR). This technique has been devel-oped consequent to the cloning of the thermostable enzyme, Taq ligase [5, 6]. The method (illustrated in Figure 4.5) relies on their being complete complementarity at the junction of two contiguous oligonucleotides hybridized to a target. A single base mismatch means that ligation will not occur. The use of four oligonu-

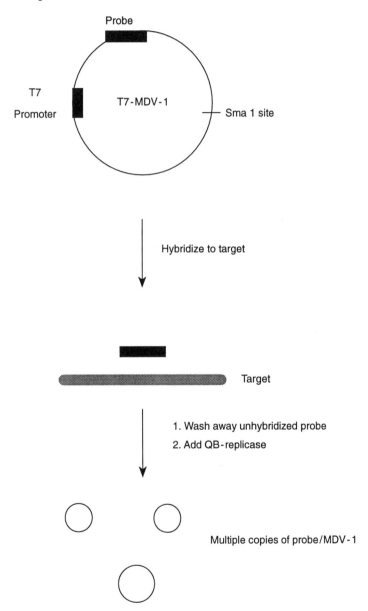

Figure 4.4 Principles of Qβ-replicase system.

cleotides, two complementary to each target strand, and repeated cycles of denaturation, hybridization and ligation results in multiple copies of ligated oligonucleotides. It has been shown that sub-picomolar quantities of lambda phage DNA can be detected. Developments have paralleled those of the PCR, but have been even more rapid. Labelling of one oligonucleotide with biotin, and the other with a reporter molecule allows non-isotopic detection without electrophoresis [42]. 'Nested LCR'

utilizes, as in its PCR equivalent, two rounds of amplification. This may improve the signal-to-background ratio. Perhaps the most significant development for the detection of infectious disease is a combination of PCR and LCR. Primary amplification of RNA or DNA is with PCR/RT–PCR with subsequent LCR. The exquisite sensitivity of PCR is combined with the specificity of LCR. Automation of this procedure has much potential for the routine detection of infectious agents.

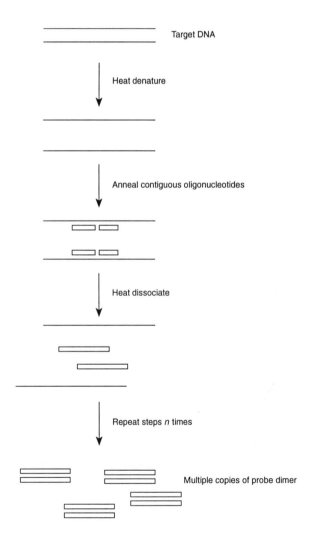

Figure 4.5 Principles of ligase chain reaction (LCR).

4.8 DETECTION OF ANTIVIRAL ANTIBODY

Detection of IgG, IgM, IgA and IgE antibody to respiratory viruses has been attempted. The detection of virus-specific IgE and IgA in nasal secretions has been used as a research tool, but its diagnostic significance remains unclear. Serodiagnosis has, usually, been based on the ELISA format or complement fixation although any of the test formats used to detect viral antigen can be adapted. The principle of the ELISA method is the same as that described already for antigen detection with the exception that an antigen is first applied to the plate to bind any antibody present in serum. The use of antibody to the class of immunoglobulin sought is a further refinement. Most commercial assays are based on this format.

Complement fixation is still a standard means of screening for respiratory viral antibodies despite well-recognized drawbacks [73,74]. Serum is mixed with antigen in the presence of complement. If specific antibody (IgG and/or IgM) is present in the mixture then the complement will be bound ('fixed'). Thus, if sheep erythrocytes which have been coated with anti-sheep antibody are then added to the initial mixture, there will not be any complement left to cause lysis of the sheep red cells. The test requires careful titration of all components which, in turn, have to be prepared freshly. Controls, including one for complement as well as standard negative and positive controls, must also be set up to allow interpretation. The complexity and the requirement for rigid attention to standardization of methodology have led to its abandonment in many laboratories. It has also been suggested that a cost–benefit analysis for respiratory tract diagnosis would favour the use of other tests [7] but it is still a useful screening test when other, simpler formats, do not exist. With the exception of influenza virus infection outside an epidemic season (high single titres

of influenza virus antibody at non-epidemic times are diagnostic as titres fall rapidly) and *Mycoplasma pneumoniae*, a four-fold rise (or fall) in paired sera taken 10–14 days apart is required to ascertain recent infection.

Radioimmunoassay, haemagglutination–inhibition and other tests are used for some viruses and are described in specific chapters.

4.9 WHICH TEST?

The tests most commonly used in the author's laboratory for the detection of specific respiratory viruses are summarized in Table 4.2. Immunofluorescence and ELISA tests for antigen detection of respiratory viruses tend to lack sensitivity (though not specificity) when compared with cell culture and such tests are best used in combination to maximize the positivity rates.

There is now an ever-increasing choice of commercial assays. A study examining commercial kits to detect RS virus and influenza A virus infections in nasal wash–throat swab specimens showed 71% sensitivity with the best ELISA kit (for influenza A virus) and 87% sensitivity for the best immunofluorescence kit (for RS virus) [20]. Even better sensitivity has been shown with such assays when nasopharyngeal secretions have been used [75].

PCR and RT–PCR in our hands are more sensitive than cell culture for the detection of most respiratory viruses and are equal to a 'gold standard' of cell culture plus serology. Many of these methods are described in specific chapters.

It is clear that rapid methods using direct antigen or genome detection will become more common for diagnosing virus infections. Gene amplification offers significant advantages in sensitivity but may be detecting non-viable virus and will thus have to be supported by antigen detection or, preferably, cultural methods.

Table 4.2   Recommended tests for diagnosis of specific respiratory viruses

Virus	Diagnostic test	Notes
Adenovirus	Cell culture	Nose and throat swabs provide adequate specimens for cell culture.
	ELISA (antigen)	Commercial ELISA assays are available
	PCR	PCR provides more rapid, but not more sensitive diagnosis than cell culture
Coronavirus	PCR	Nested PCR is more sensitive than cell culture.
	Cell culture	Primary isolation in C16 or WI-38 cells may be unreliable
Enterovirus	PCR	Cell culture limited in its range
Influenza virus	Cell culture	Commercial IF assays available.
	Immunofluorescence	PCR offers little over a combination
	PCR	of culture and immunofluorescence
Measles	Cell culture	
Parainfluenza virus	Cell culture	Nested with pan-paramyxovirus
	PCR	degenerate outer primers and parainfluenza virus inner primers
Respiratory Hantann (HPS virus, Four corners virus)	PCR	Cell culture not yet available
Respiratory syncytial virus	Cell culture	
	Immunofluorescence	Nested PCR with outer pan-
	PCR	paramyxovirus degenerate outer primers and inner RSV-specific primers
Rhinovirus	Cell culture	Ohio Hela cells most reliable for
	PCR	culture. Nested PCR with pan-picornavirus outer primers and rhinovirus inner primers

IF, immunofluorescence; PCR, polymerase chain reaction

4.10 REFERENCES

1. Al-Hakim, H. and Hull, R. (1986) Studies towards the development of chemically synthesized non-radioactive biotinylated nucleic acid hybridization probes. *Nucleic Acids Res.*, **14**, 9965–72.
2. Alwine, J.C., Kemp, D.J. and Stark, G.R. (1977) Method for the detection of specific RNA in agarose gels by transfer to diazobenzoxylmethyl paper and hybridization with DNA probes. *Proc. Natl Acad. Sci. USA*, **74**, 5350–4.
3. Amasino, R.M. (1986) Acceleration of nucleic acid hybridisation rate by polyethylene glycol. *Anal. Biochem.*, **152**, 304–7.
4. Balows, A., Hausler, W.J. Jr, Lennette, E.H. (eds) (1988) *Laboratory Diagnosis of Infectious*

Diseases: Principles and Practice. Vol. II, Viral, Rickettsial and Chlamydial Diseases. Springer-Verlag, New York.

5. Barany, F. (1991) Genetic disease detection and DNA amplification using cloned thermostable ligase. *Proc. Natl Acad. Sci. USA*, **88**, 189–93.

6. Barany, F. (1991) The ligase chain reaction (LCR) in a PCR world. *PCR Methods Applications*, **1**, 5–17.

7. Bassett, D.C.J., Tam, J.S., McBride, G.A., Leung, K.T. and Cheng, A.F.B. (1994) Viral, mycoplasmal and chlamydial lower respiratory tract infections in Hong Kong: cost and diagnostic value of serology. *Br. J. Biomedical. Sci.*, **51**, 9–13.

8. Bartholoma, N.Y. and Forbes, B.A. (1989) Successful use of shell vial centrifugation and 16–18 hour immunofluorescent staining for the detection of influenza A and B in clinical specimens. *Am. J. Clin. Pathol.*, **92**, 487–90.

9. Berger, S.L. and Kimmel, A.R. (1987) *Guide to Molecular Cloning Techniques. Methods in Enzymology*, Vol. 152. Academic Press Inc., San Diego.

10. Bettoli, E.J., Brewer, P.M., Oxtoby, M.J., Zaidi, A.A. and Guinan, M.E. (1982) The role of temperature and swab material in the recovery of herpes simplex virus from lesions. *J. Infect. Dis.*, **145**, 399.

11. Billeter, M.A., Weissman, C. and Warner, R.C. (1966) Properties of double-stranded RNA from *Escherichia coli* infected with bacteriophage MS2. *J. Mol. Biol.*, **17**, 145–73.

12. Bishop, J.A. (1972) Molecular hybridisation of ribonucleic acid in a large excess of deoxyribonucleic acid. *Biochem. J.*, **126**, 171–85.

13. Boguslawski, S.J., Smith, D.E., Michalak, M.A. *et al.* (1986) Characterisation of a monoclonal antibody to DNA:RNA and its application to the immunodetection of hybrids. *J. Immunol. Methods*, **89**, 123–30.

14. Brigati, D.J., Myerson, D., Leary, J.J. *et al.* (1983) Detection of viral genomes in cultured cells and paraffin-embedded tissue sections using biotin-labelled hybridisation probes. *Virology*, **126**, 32–50.

15. Casey, J. and Davidson, M. (1977) Rate of formation and thermal stabilities of RNA:DNA and DNA:DNA duplexes at high concentrations of formamide. *Nucleic Acids Res.*, **4**, 1539–52.

16. Chamberlin, M. and Berg, P. (1964) Mechanism of action of RNA polymerase, formation of

DNA:RNA hybrids with single-stranded templates. *J. Mol. Biol.*, **8**, 297–313.

17. Crane, L.R. , Gutterman, P.A., Chapel, T. and Lerner, A.M. (1980) Incubation of swab material with herpes simplex virus. *J. Infect. Dis.*, **141**, 531.

18. Darougar, S., Walpita, P., Thaker, U., Viswalingam, N. and Wishart, M.S. (1984) Rapid culture test for adenovirus isolation. *Br. J. Ophthalmol.*, **68**, 405–8

19. Doane, F.W. and Anderson, N. (1987) *Electron Microscopy in Diagnostic Virology: A Practical Guide and Atlas*. Cambridge University Press, Cambridge, England.

20. Dominguez, E.A., Taber, L.H. and Couch, R.B. (1993) Comparison of rapid diagnostic techniques for respiratory syncytial and influenza A virus respiratory infections in young children. *J. Clin. Microbiol.*, **31**, 2286–90.

21. Dooley, S., Radtke, J., Blin, N. and Unteregger, G. (1988) Rapid detection of DNA-binding factors using protein blotting and digoxigenin-dUTP marked probes. *Nucleic Acids Res.*, **16**, 1183–9.

22. Earle, W.R., Schilling, E.L. and Bryant, J.C. (1954) Influence of tube rotation velocity on proliferation of strain L cells in surface substrate roller tube culture. *J. Natl Cancer Inst.*, **14**, 853–64.

23. Ehrlich, H. (1989) *PCR Technology, Principles and Applications for DNA Amplification*. Stockton Press, New York.

24. Ehrlich, H.A., Gelfand, D. and Sninsky, J. (1991) Recent advances in the polymerase chain reaction. *Science*, **252**, 1643–51.

25. Espy, M.J., Hierholzer, J. and Smith, T. (1987) The effect of centrifugation on the rapid detection of adenovirus in shell vials. *Am. J. Clin. Pathol.*, **88**, 358–60.

26. Espy, M.J., Smith, T.F., Harmon, W.M. and Kendal, A.P. (1986) Rapid detection of influenza virus by shell vial assay with monoclonal antibodies. *J. Clin. Microbiol.*, **24**, 677–9.

27. Feinberg, A.P. and Vogelstein, B. (1983) A technique for radiolabelling DNA restriction endonuclease fragments to high specific activity. *Anal. Biochem.*, **132**, 6–13.

28. Gardner, P.S and McQuillin, J. (1968) Application of the immunofluorescent antibody technique in the rapid diagnosis of respiratory syncytial virus infection. *Br. Med. J.*, **3**, 340–3.

29. Gardner, P.S. and McQuillin, J. (1980) *Rapid Virus Diagnosis: Applications of Immunofluorescence.* Butterworths, London.

30. Guenthner, S.H. and Linnemann, R. (1988). Indirect immunofluorescence assay for rapid diagnosis of influenza virus. *Lab. Med.,* **19,** 581–3.

31. Huntoon, C.J., House, R.F. and Smith, T.F. (1981) Recovery of viruses from three transport media incorporated into culturettes. *Arch. Pathol. Lab. Med.,* **105,** 436–7.

32. Inouye, S. and Hondo, R. (1991) Microplate hybridisation of amplified viral DNA segment.. *J Clin. Microbiol.,* **28,** 1469–72.

33. Jablonski, E., Moomaw, E.W., Tullis, R.H. and Ruth, J.L. (1986) Preparation of oligodeoxynucleotide-alkaline phosphatase conjugates and their use as hybridisation probes. *Nucleic Acids Res.,* **14,** 6115–28

34. Jalowayski, A.A., England, B.L., Temm, C.J. *et al.* (1987) Peroxidase-antiperoxidase assay for rapid detection of respiratory syncytial virus in nasal epithelial specimens from infants and children. *J. Clin. Microbiol.,* **25,** 722–5.

35. Jennings, R. and Freeman, M.J. (1972) Studies on type C influenza virus in the chick embryo. *J. Hyg. (Camb),* **70,** 1–4.

36. Johnson, S.L.G and Siegel, C.S. (1990) Evaluation of direct immunofluorescence, enzyme immunoassay, centrifugation culture, and conventional culture for the detection of respiratory syncytial virus. *J. Clin. Microbiol.,* **28,** 2394–7.

37. Johnstone, S., Wilson, J., Mason, S., Forsyth, M. and Holgate, S.T. (1990) Flow cytometry of cultured cells – a model for rapid diagnosis of respiratory viral and mycoplasmal infections in clinical samples. Abstracts of VIIIth International Congress of Virology, Berlin.

38. Keller, G.H, Huang, D.P. and Manak, M.M. (1991) Detection of human immunodeficiency type 1 DNA by polymerase chain amplification and capture hybridisation in microtitre wells. *J. Clin. Microbiol.,* **29,** 638–41.

39. Kohne, D.E., Levinson, S.A. and Byers, M.J. (1977) Room temperature method for increasing the rate of DNA reassociation by many thousandfold, the phenol re-emulsion technique. *Biochemistry,* **16,** 5329–41.

40. Kricka, L.J. (ed.) (1992) *Nonisotopic DNA Probe Techniques.* Academic Press, San Diego.

41. Kwoh, D.Y., Davis, G.R., Whitfield, K., Chapelle, H., DiMichele, L. and Gingeras, T.R. (1988) Transcription-based amplification system and detection of amplified HIV-1 sequences in infected cells using a bead-based sandwich hybridisation format. Practical Aspects of Molecular Probes. 3rd San Diego Conference.

42. Landegren, U., Kaiser, R., Sanders, J. and Hood, L. (1988) A ligase-mediated gene detection technique. *Science,* **241,** 1077–80

43. Leary, J.J., Brigati, D.J. and Ward, D.C. (1983) Rapid and sensitive colorimetric method for visualising biotin-labelled probes hybridised to DNA or RNA immobilised on nitrocellulose bio-blots. *Proc. Natl Acad. Sci. USA,* **80,** 4045–9.

44. Ling, A.E. and Dresingham, S. (1988) Comparison of tube cultures of Madin Darby canine kidney cells with shell vial cultures after low speed centrifugation for influenza virus isolation. *Pathology,* **20,** 346–8.

45. Lizardi, P.M., Guerra, C.E., Lomeli, H., Tussie-Luna, I. and Kramer, F.R. (1988) Exponential amplification of recombinant RNA hybridisation products. *Biotechnology,* **6,** 1197–202.

46. McConaughy, B.L., Laird, C.L. and McCarthy, B.J. (1969) Nucleic acid reassociation in formamide. *Biochemistry,* **8,** 3289–95.

47. Mahafzah, A.M. and Landry, M.L. (1989) Evaluation of immunofluorescent reagents, centrifugation, and conventional cultures for the diagnosis of adenovirus infection. *Diagn. Microbiol. Infect. Dis.,* **12,** 407–11.

48. Mardh, P.-A., Westrom, L., Colleen, S. and Wolner-Hanssen, P. (1981) Sampling, specimen handling, and isolation techniques in the diagnosis of chlamydial and other genital infections. *Sex. Transm. Dis.,* **8,** 280–5.

49. Marquez, A. and Hsuing, G.D. (1967) Influence of glutamine on multiplication and cytopathic effect of respiratory syncytial virus. *Proc. Soc. Exp. Biol. Med.,* **124,** 95–9.

50. Matthey, S., Nicholson, D., Rush, S. *et al.* (1992) Rapid detection of respiratory viruses by shell vial culture and direct staining by using pooled and individual monoclonal antibodies. *J. Clin. Microbiol.,* **30,** 540–4.

51. Meinkoth, J. and Wahl, G.M. (1984) Hybridisation of nucleic acids immobilised on solid supports. *Anal. Biochem.,* **138,** 267–84.

52. Melton, D.A., Krieg, P.A., Rebagliati, M.R., Maniatis, T., Zinn, K. and Gren, M.R. (1984)

Efficient *in-vitro* synthesis of biologically active RNA and RNA hybridization probes from plasmids containing a bacteriophage SP6 promoter. *Nucleic Acids Res.*, **12**, 7035–56.

53. Miller, C.A., Patterson, W.L., Johnson, P.K. et al. (1988) Detection of bacteria by hybridisation of rRNA with DNA latex and immunodetection of hybrids. *J. Clin. Microbiol.*, **26**, 1271–6.

54. Mills, R.D., Cain, K.J. and Woods, G.L. (1989) Detection of influenza virus by centrifugal inoculation of MDCK cells and staining with monoclonal antibodies. *J. Clin. Microbiol.*, **27**, 2505–8.

55. Minnich, L.L., Goodenough, F. and Ray, C.G. (1991) Use of immunofluorescence to identify measles virus infections. *J. Clin. Microbiol.*, **29**, 1148–50.

56. Olson, J.D., Panfili, P.R., Zuk, R.F. and Sheldon, E.L. (1991) Quantitation of DNA hybridisation in a silicon sensor-based system: application to PCR. *Mol. Cell. Probes*, **5**, 351–8.

57. Persing, D.H., Smith, T.F., Tenover, F.C. and White, T.J. (eds) (1993) *Diagnostic Molecular Microbiology: Principles and Applications*. American Society for Microbiology, Herndon, USA.

58. Polsky-Cynkin, R., Parsons, G.H., Allerdt, L. *et al.* (1985) Use of DNA immobilized on plastic and agarose supports to detect DNA by sandwich hybridisation. *Clin. Chem.*, **31**, 1438–43.

59. Rabalais, G.P., Stout, G.G., Ladd, K.L and Cost, K.M. (1992) Rapid diagnosis of respiratory virus infections by using shell vial assay and monoclonal antibody pool. *J. Clin. Microbiol.*, **28**, 1505–8.

60. Rigby, P.W.S., Dieckmann, M., Rhodes, C. and Berg, P. (1977) Labelling deoxyribonucleic acid to high specific activity *in-vitro* by nick translation with DNA polymerase I. *J. Mol. Biol.*, **113**, 237–51.

61. Rolfs, A., Schuller, I., Finckh, U. and Weber-Rolfs, I. (1992) *PCR: Clinical Diagnostics and Research*. Springer-Verlag, Berlin.

62. Saiki, R.K., Gelfand, D.H., Stoffel, S. *et al.* (1988) Primer directed enzymatic amplification of DNA with a thermostable polymerase. *Science*, **239**, 487–94.

63. Sambrook, J., Fritsch, E.F. and Maniatis, T. (1989) *Molecular Cloning: A Laboratory Manual*. Cold Spring Harbour Laboratory Press, Cold Spring Harbor, New York.

64. Schildkraut, C. and Leifson, S. (1965) Dependence of the melting temperature of DNA on salt concentration. *Biopolymers*, **3**, 195–208.

65. Schmidt, N.J. and Emmons, R.W. (eds) (1989) *Diagnostic Procedures for Viral, Rickettsial and Chlamydial Infections*, 6th edn, American Public Health Association, Washington DC.

66. Seno, M., Takao, S., Fukuda, S. and Kanamoto, Y. (1990) Enhanced isolation of influenza virus in conventional plate cell cultures by using low-speed centrifugation from clinical specimens. *Am. J. Clin. Pathol.*, **95**, 765–8.

67. Shahrabadi, M.S. and Lee, P.W.K. (1988) Calcium requirements for syncytium formation in HEp-2 cells by respiratory syncytial virus. *J. Clin. Microbiol.*, **26**, 139–41.

68. Smith, M.C., Creutz, C. and Huang, Y.T. (1991) Detection of respiratory syncytial virus in nasopharyngeal secretions by shell vial technique. *J. Clin. Microbiol.*, **29**, 463–5.

69. Soini, E. and Kojola, H. (1983) Time-resolved fluorimeter for lanthanide chelates – a new generation of nonisotopic immunoassays. *Clin. Chem.*, **29**, 65–8.

70. Southern, E.M. (1975) Detection of specific sequences among DNA fragments separated by gel electrophoresis. *J. Mol. Biol.*, **98**, 503–17.

71. Specter, S. and Lancz, G. (eds) (1992). *Clinical Virology Manual*, 2nd edn, Elsevier Science Publishing Co. Inc, New York

72. Stokes, C.E., Bernstein, J.M., Kyger, S.A. and Hayden, F.G. (1988) Rapid detection of influenza A and B by 24-hour fluorescent focus assays. *J. Clin. Microbiol.*, **26**, 1263–6.

73. Stott, E.J. and Tyrrell, D.A.J. (1986) Applications of immunological methods in virology, in *Handbook of Experimental Immunology*, vol. 4, (eds D.M. Weir, L.A. Herzenberg and C. Blackwell), Blackwell, Oxford, England.

74. Swack, N.S., Gahan, T.F. and Hausler, W.J. (1992) The present status of the complement fixation test in viral serodiagnosis. *Infect. Ag. Dis.*, **1**, 219–24.

75. Takimoto, S., Grandien, M., Ishida, M.A. *et al.* (1991) Comparison of enzyme-linked immunosorbent assay, indirect immunofluorescence assay, and virus isolation for detection of respiratory viruses in nasopharyngeal secretions. *J. Clin. Microbiol.*, **29**, 470–4.

76. Thompson, J. and Gillespie, D. (1987) Molecular hybridisation with RNA probes in

concentrated solutions of guanidine thiocyanate. *Anal. Biochem.*, **163**, 281–91.

77. Uhlen, M. (1989) Magnetic separation of DNA. *Nature*, **340**, 733–4.

78. Ukkonen, P. and Julkunen, I. (1987) Preparation of nasopharyngeal secretions for immunofluorescence by one-step centrifugation through Percoll. *J. Virol. Methods*, **15**, 291–301.

79. Van der Veen, J. and Sonderkamp, H.J.A. (1965) Secondary antibody response of guinea pigs to parainfluenza virus and mumps viruses. *Arch. Gesamte Virusforsch.*, **15**, 721–34.

80. Walls, H.H., Johanssen, K.H., Harmon, M.W., Halonen, P.E. and Kendal, A.P. (1986) Time-resolved fluoroimmunoassay with monoclonal antibodies for rapid diagnosis of influenza infections. *J. Clin. Microbiol.*, **24**, 907–12.

81. Walpita, P. and Darougar, S. (1989) Double-label immunofluorescence method for simultaneous detection of adenovirus and herpes simplex virus from the eye. *J. Clin. Microbiol.*, **27**, 1623–5.

82. Waris, M., Ziegler, T., Kivivirta, M. and Ruuskanen, O. (1990) Rapid detection of respiratory syncytial virus and influenza virus in cell cultures by immunoperoxidase staining with monoclonal antibodies. *J. Clin. Microbiol.*, **28**, 1159–62.

83. Woods, G.L. and Johnson, A.M. (1989) Rapid 24-well plate centrifugation assay for detection of influenza A in clinical specimens. *J. Virol. Methods*, **24**, 35–42.

84. Woods, G.L., Yamamoto, M. and Young, A. (1988) Detection of adenovirus by rapid 24-well plate centrifugation and conventional cell culture with dexamethasone. *J. Virol. Methods*, **20**, 109–14.

85. Yolken, R.H. (1983) Use of monoclonal antibodies for viral diagnosis. *Curr. Top. Microbiol. Immunol.*, **104**, 177–95.

Anders Akerlund and Mats Bende

5.1 INTRODUCTION

The common cold is the cause of the great majority of respiratory tract infections. It seems well established that this is a viral infection in the upper airways. In the early 1950s symptoms and signs of a common cold were induced in isolated healthy individuals for the first time by nasal inoculation of virus cultured *in vitro* [11]. There is a widespread traditional belief that the symptoms of the common cold are related to cooling of the body. Although cooling, e.g. of the feet, induces physiological alterations in the nasal mucosa [69,82,130], local (or general) cooling has failed to result in disease in volunteers, or even to make them more prone to catch a common cold infection in controlled experiments [10,66]. Common colds are brief, self-limited infections that do not affect the life expectancy of otherwise healthy and well-fed subjects. However, they can be a threat to patients who are immunocompromised, are malnourished or have severe cardiac or pulmonary disease. Malnourishment is a contributing factor in developing countries where the children affected suffer more persistent airway infections and more complications, mainly bronchopneumonia [106], which may lead to considerable mortality [51]. There are vast social, as well as economic consequences, as illustrated by an annual loss of 100 million working days in the USA where each episode of a common cold costs an estimated $15–40 for medications and diagnostic medical consultations [117].

The signs and symptoms of the common cold are the principal tools for identifying clinical infections. The viral aetiology is often verified by virological examinations of the nasal discharge and serological analyses. In practice,

Viral and Other Infections of the Human Respiratory Tract.
Edited by S. Myint and D. Taylor-Robinson. Published in 1996 by Chapman & Hall. ISBN 0 412 60070 6

symptom scoring systems have been widely used to monitor the course of the infection [21,106]. Knowledge of the pathophysiology has been further extended by access to increasingly sophisticated objective physiological, biochemical and morphological methods. The nose is the outermost part of the respiratory tract and the port of entry for infective microbes; hence the nose is the part of the airways most frequently affected by viral infections [58]. The reactions in the airway mucosa are to some extent similar throughout the respiratory tract. Hence, nasal mucosal events also reflect the reactions of other parts of the airway mucosa to some extent [142]. The nose is easily accessible and therefore has been studied extensively. This chapter focuses on the nasal physiology and nasal pathophysiology in viral infections.

5.2 APPLIED PHYSIOLOGY OF THE RESPIRATORY TRACT

The anatomy and physiology relevant to the pathophysiology here presented is based on an extensive volume of reviews edited by Proctor and Andersen [147]. The air-conditioning function of the nose is essential to the whole of the respiratory system. The respiratory air is modified during its passage through the nose, according to the needs of the lower airways. Humans are exposed to ambient air with extreme temperatures that can range from –40°C to +40°C and humidity extremes that range from dry to saturated moist air. An adult human breathes 400 litres of air every hour at rest and considerably more during physical activity. The upper airways have the capacity to clean, warm and humidify the inspiratory air to 32°C in an ambient temperature of –20°C. The complex split-like lumen in the nose and the narrowing at the nasal valve cause turbulent airflow, which offers the ideal conditions for a high air-conditioning capacity [108] (Figure 5.1).

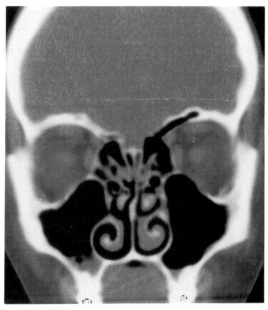

Figure 5.1 Computed tomographic image (coronal projection) showing a cross-section of the nasal cavity of a normal individual. Note the asymmetry in mucosal swelling which is due to the nasal cycle.

5.2.1 NASAL STRUCTURE

The rigid bony nasal cavity is covered by erectile tissue, the nasal mucosa, which is lined by a ciliated epithelium. This pseudostratified ciliated epithelium also lines the paranasal sinuses, the Eustachian tube, the middle ear, the nasopharynx and the larynx (except the vocal cords). In the anterior part of the nose and towards the oropharynx, the epithelium gradually changes to squamous epithelium. The epithelium is covered by a secretion layer, which consists of a low-viscosity periciliary fluid layer under a high-viscosity mucous blanket. The turbulent airflow facilitates the deposition of particles inhaled from the ambient air on the mucous blanket. The latter is the first mechanical barrier which an infective microbe has to overcome. The activity of the epithelial cell cilia is co-ordinated with ciliary beats 10–20 times per second and the cooperation of many cells results in a continuous systematic propulsion of the mucous blanket. The mucociliary transport system thus clears the airway from dust, pollen and microbes. In the mucus, secretory antibodies – mainly IgA – are present as the first line of immunological defence. The mucus thus has characteristics important for defence against viruses and other microbes deposited in the nose. Additional defence capacity is available in the tissue where inflammatory cells play a major role. The healthy nasal mucosa contains a small number of cells which, if activated, may participate in and escalate an inflammatory reaction. The lymphocyte, the most common cell type, is mainly located in the subepithelial layer. T-cells are the most numerous, T-helper cells outnumbering T-suppressor cells and are, primarily, periglandular. The B-cells are located mainly in the lamina propria, occasionally in clusters [188].

5.2.2 NASAL FUNCTION

The air-conditioning function of the nose is regulated mainly by adjustments in the vasculature of the nasal mucosa. The abundant veins form networks of large vessels, sinusoids, which can pool considerable volumes of blood and, because of their swelling and shrinkage capacity, can regulate the patency of the nose (Figure 5.2). Under the effect of neurogenic mechanisms, the blood volume in the mucosa can change rapidly, and thereby within minutes adjust the nasal airway resistance and airflow pattern [53,147]. The turbulence of the airflow is essential for bringing the respiratory air into contact with the mucosa.

Nasal blood flow

Arteriovenous shunts allow the mucosa to be heated by blood at core temperature when

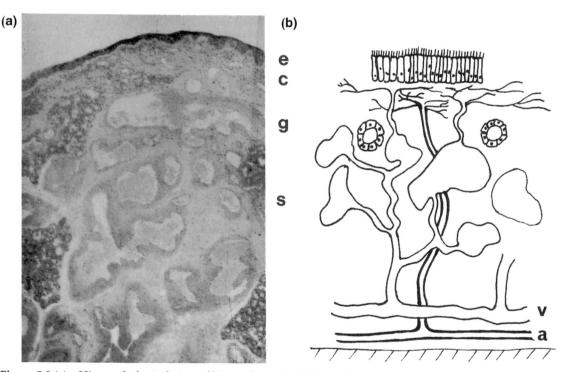

(a)

(b)

e
c

g

s

v
a

Figure 5.2 (a) Histopathological view of biopsy from a healthy nasal mucosa and (b) schematic drawing of nasal mucosal vasculature. a, arteries; c, capillaries; e, epithelium; g, submucosal glands; s, sinusoids; v, veins.

required. An extensive system of fenestrated capillaries delivers water to the inspired air. The blood flow in the nasal mucosa, which is regulated by the tone of the arterioles, is about 35 ml/min/100 g tissue [28]. Such a blood flow, which is 10 times that of the skin, exceeds the nutritive demands of the mucosa. The high blood flow in the nasal mucosa can rapidly activate the humoral systems for defence against viruses and bacteria. To understand the function of the nose it is important to distinguish between blood flow and blood content. Sometimes these parameters change in the same direction: e.g. in the common cold where there is an increase in both the blood flow and the blood content. However, in physiological conditions, blood flow and blood content can change in opposite directions [2,138]. The blood content and the amount of oedema in the nasal mucosa are responsible for the mucosal thickness and thereby regulate nasal patency. This nasal patency is proportional to the nasal airway resistance. Consequently, the blood content of the mucosa can be assessed objectively by rhinomanometry. The temperature of the nasal mucosa is a crucial factor for the regulation of humidity and heat from both inspired and expired air [147]. This is regulated by the blood flow in the mucosa and it is also influenced by the temperature and volume of the airflow through the nose. During normal conditions, the temperature of the nasal mucosa is several degrees below the body core temperature, about 30°C, and about 2°C higher in the expiratory breathing phase [2,69,182].

Nasal patency

Under physiological conditions there exists a circadian variation in nasal secretions [131]. There is, also, a periodic variation in nasal patency, a physiological phenomenon referred to as 'the nasal cycle'. This is regulated by the sympathetic nervous system whereby the blood content in the mucosa

changes asymmetrically, causing cyclical and reciprocal changes in the resistance to airflow every 2–4 hours. This phenomenon is present in about 80% of healthy subjects [81,147]. However, the subjective feeling of nasal patency can be illusory and does not always correspond to the objectively measured nasal airway resistance. It is a common experience, for instance, that the inhalation of cold air causes a sensation of increased patency, while the exposure to heat causes a feeling of stuffiness. Cold sensation in the nose is accompanied by increased airflow sensation [73] although inhalation of cold air actually causes nasal mucosal congestion [25,69].

An increase in sympathetic activity decongests the nasal mucosa by constricting the capacitance vessels [126]. Powerful decongestion – 'maximal decongestion' – is achieved after physical exercise at pulse rates over 120 beats/min. This is physiological and supplies a maximally widened airway for the best achievable ventilation. Exercise-induced decongestion can be used to distinguish between nasal obstruction due to structural defects and to obstruction caused by mucosal swelling. Mucosal swelling depends on the body position and increases in the supine position, a phenomenon which is accentuated in infectious rhinitis [54,133,158]. Some of the various reflexes that involve the nasal mucosa have a pathophysiological significance. For example, cooling of the extremities induces a reflectory rapid decrease in the nasal mucosal blood flow, a reflex mediated by sympathetic nerve fibres [134]. Irritants deposited in the nose, e.g. dust in the ambient air, initiate the protective and airway-clearing sneezing reflex.

Sensory function

The senses of smell and hearing and the ability to speak are perceptive and communicative functions associated with the upper airway. Odours are perceived via the olfactory epithelium, which is located in the superior part of the nasal passage. A turbulent airflow through the nose, an adequate nasal patency in the upper nasal cavities and the ability to sniff, are necessary prerequisites for an optimal function of the sense of smell. The sensation of smell can be assessed by the detection of odour intensity – for example, by threshold tests or odour quality – by identification tests.

Speech.

This is executed through the airways where the larynx is the source of sound. During the phonation a pulsating expiratory airstream flows through the oscillating vocal cords, generating the basic sound for speech. The sound is modified by the resonating rooms in the upper airways and by modifications of the tongue and lips to characterize the sophisticated details of speech and the subject's distinctive voice.

The ear.

This converts sound waves to nerve signals which are processed in the central nervous system. The middle ear transfers the sound waves as vibrations to the fluids of the inner ear. Effusion in the middle ear, as in secretory otitis media, leads to a conductive hearing loss. The middle ear and the Eustachian tubes should not be forgotten as parts of the upper respiratory tract. The Eustachian tube maintains a middle ear pressure equivalent to the ambient atmospheric pressure but it is closed to the rapid changes in air pressure in the nasopharynx that are induced by normal breathing. If the closing is suboptimal, a tuba aperta will lead to an inability to close off the sounds of breathing and a disturbed feed-back hearing of the subject's own voice. The Eustachian tube is the common route of entry for the infective microbes that cause acute middle-ear infections.

The paranasal sinuses are anatomical structures of the respiratory tract, which have an obscure function (Figure 5.1). The mucociliary system in the sinus mucosa is similar to that in the nose, but the vascular arrangement in the mucosa is different. There are few, if any, sinusoidal vessels in the sinus mucosa, but the blood flow is as high as that in the nose. The patency of the sinus ostium is essential for adequate sinus ventilation [147]. During nasal breathing, there is a continuous exchange of air in the paranasal sinuses and half of the maxillary sinus volume of air is exchanged in about 30 minutes [137]. The size of the maxillary sinus ostium varies with body posture as in the case of posture effects on the nasal mucosa [15,107,125].

5.3 PATHOPHYSIOLOGY OF COMMON COLDS

The patient history and the clinical investigation form the basis for diagnosing common colds in clinical practice. The signs found in the clinical investigation include changes in the airway mucosa, swelling, reddening and pathological discharge, sometimes antral sinus tenderness, watery eyes, cervical lymphadenitis or secretory otitis media. The symptoms of the common cold are mainly nasal – blockage and discharge (Figure 5.3) [178]. However, symptoms from all parts of the respiratory tract, such as the paranasal sinuses, ears, throat, larynx, trachea and bronchi, are seen. More general symptoms, like headache, malaise, chills and sometimes fever, are also present. The clinical signs and symptoms are not specific for inflammations caused by infections and the patient history is important for distinguishing toxic or allergic inflammation and non-specific hyperreactivity.

5.3.1 EPITHELIAL CHANGES

The viruses mainly infect the ciliated epithelium in the nose. They damage the cells and various viral pathogens are associated with different degrees of destruction of the respiratory epithelium [102]. Histopathological damage to the

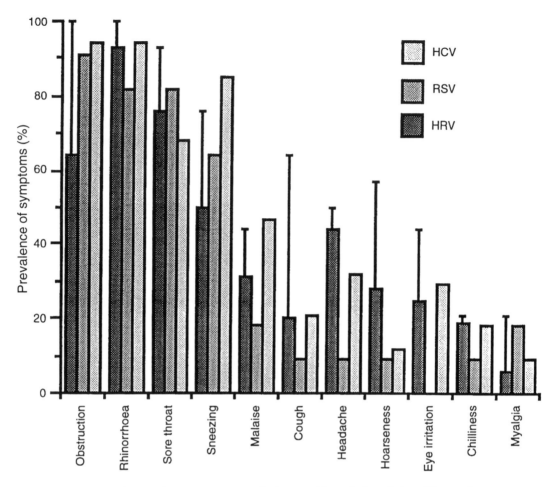

Figure 5.3 The prevalence of symptoms reported after nasal inoculation of virus. The most common are the nasal symptoms and a sore throat; the severity of symptoms is similarly distributed. The error bars indicate the range of the prevalence of symptoms with rhinoviruses types 2, 9 and 14. HRV, rhinovirus; HCV, coronavirus; RSV, respiratory syncytial virus. (From [178].)

epithelium results from experimental rhinovirus infections [64,103]. Investigations in subjects with natural colds revealed no pathological changes on light microscopy but, with the scanning electron microscope patchy damage to the epithelium was seen [185]. In interpreting these findings it should be held in mind that viral infection has a preference for the nasopharyngeal mucosa but can have a patchy distribution [187] and single biopsies may not be representative for the entire mucosa. Virus-infected epithe-

lial cells are shed during colds [176] but, even if a viral infection causes damage to the epithelium, this may not necessarily be great enough to disturb the epithelial function as a mechanical barrier. This aspect has been investigated in coronavirus colds, when the ability of the mucosa to absorb a luminally deposited macromolecular tracer was not significantly altered [89]. An impaired barrier function certainly cannot be ruled out with infection due to other viruses, bearing in mind the differences between the effects of various viral pathogens on the morphology of the mucosal epithelium [102].

In colds, the function of the nasal mucociliary clearing system is impaired [86,159,184]. Impairment of this system is significant already when the symptoms begin, and is most pronounced when the symptoms reach peak scores. The impairment lasts for 2–3 weeks, but can persist for several weeks [139]. During common colds, the volume of nasal discharge increases and its composition changes [155]. The composition and viscosity of the mucus and the relation between the volumes of mucus and periciliary fluid are essential to an optimal function of the mucociliary clearing system. Changes during viral infection that affect these factors impair the clearance rate. Besides changes in the mucus there is a loss of ciliated cells in upper respiratory tract infections [176,184]. Also noted during common colds is a reduction in the number of ciliated cells, an asynchronous and reduced ciliary beat frequency, and an increase in the proportion of living ciliated cells with immotile cilia [139].

Lymphocytes, as well as poly- and mononuclear leucocytes increase in numbers in the nasal secretions during upper airway infections and the numbers are directly related to the severity of the symptoms. White cells also increase in the epithelium and in the lamina propria of the mucosa [114,185,186]. In virus infections the peripheral blood leucocyte population is altered, there being an increase in neutrophil granulocytes and a decrease in lymphocyte counts [44,47,65,101] due to a decrease in T-cells [113]. The presence of leucocytes at these sites in the mucosa, and in secretions, indicates that they actively contribute to the local inflammatory process and antiviral defence [95].

5.3.2 VASCULAR REACTIONS

Nasal blockage is the main symptom reported subjectively by patients [178] (Figure 5.3) and an increase in the nasal airway resistance has been confirmed objectively [5,27] (Figure 5.4).

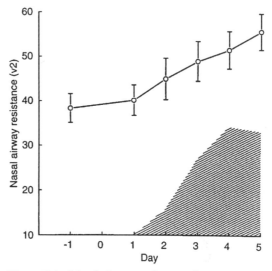

Figure 5.4 Nasal airway resistance changes (v2-values) after inoculation of coronavirus in healthy subjects in relation to symptom scores (shaded). (From [27].)

The mechanisms causing nasal blockage in the common cold are multifactorial. There is congestion of blood in the erectile nasal mucosa; the relevance of this pooling of blood in sinusoidal capacitance vessels is illustrated by the prompt decongestant effect produced by vasoconstricting drugs [6]. Interstitial oedema usually explains the cardinal sign of inflammation – *tumor*. However, no data have been presented that convincingly show oedema in the nasal mucosa during the common cold. Obstruction by a large volume of highly viscous discharge is another obvious cause. The mucosal swelling, although uncomfortable, may be beneficial in the common cold. A reduced airflow, due to nasal obstruction, contributes to an increase in the temperature of the nasal mucosal surface, since it eliminates the cooling effect of the inspiratory nasal airflow (on the nasal mucosa). An increase in the local blood flow is a general phenomenon observed in inflammatory conditions responsible for the cardinal signs, *calor* and *rubor*. The nasal mucosal blood flow increases during a common cold [27] (Figure 5.5) and, with the reduction of airflow, will contribute to an increase in nasal mucosal temperature. The increase in the mucosal surface temperature is 3–5°C, while the increase in body temperature is measured only in tenths of a degree centigrade [2,27] (Figure 5.6). The increase in the local temperature is, theoretically, a defence mechanism *per se* since the replication rate of many respiratory viruses is reduced above an optimal 33°C [136]. A further potentially beneficial effect is that other defence mechanisms (such as leukotaxis, phagocytosis and the production of interferon) are facilitated at higher temperatures [153]. The facilitation of cell migration is especially relevant because it may counteract the impairment of chemotaxis induced by viral infection [111].

5.3.3 RHINORRHOEA

Rhinorrhoea is another common phenomenon accompanying the common cold [178] (Figure 5.3). The volume of discharge collected in the nose increases due to incapacity of the mucociliary clearance and the discharge accu-

mulates in the nose and requires clearance in alternative ways (Figure 5.7). The composition of the discharge may differ between individuals and it changes during an infection [104]. Furthermore, the composition and characteristics of the discharge may vary as a result of factors such as biorhythms, climate and the presence of non-specific irritants. Potential sources of the nasal discharge include the water condensed from humid expired air, secretions from serous and seromucous glands, epithelial cells, goblet cells and fluids from the mucosal microcirculation. The origin of the increased nasal discharge in viral infections is still a matter of speculation. There is no reason to believe that the condensation from humidity will increase significantly in the common cold. Active transport of chloride ions causes a flow of water from epithelial cells in response to mediators. The density of goblet cells in the nasal mucosa increases transiently during viral infections [173]. Glandular secretions contribute to the rhinorrhoea stimulated by mediators and neural reflexes [37,104]; these secretions are mainly derived from sero-

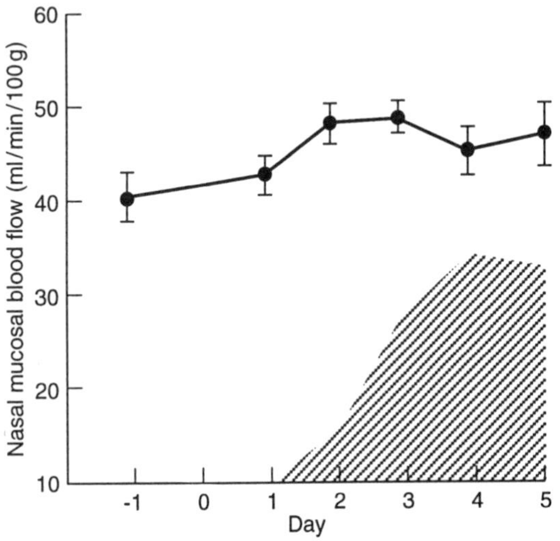

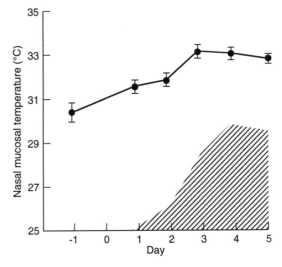

Figure 5.5 Nasal mucosal blood flow changes (ml/min/100 g tissue) after inoculation of coronavirus on day 0 in healthy subjects in relation to symptom scores (shaded). (From [27].)

Figure 5.6 Nasal mucosal temperature changes (°C) after inoculation of coronavirus on day 0 in healthy subjects in relation to symptom scores (shaded). (From [27].)

mucous glands in the mucosa, with a minor contribution from intraepithelial and anterior serous glands. The significance of the glandular contribution, however, can be questioned in the common cold, since anticholinergic drugs which block glandular secretion, far from abolish the increased secretion [38,62,68,84]. It seems contradictory that anticholinergics appear to be effective during the first days of infection [38], while the glandular secretory contribution to the rhinorrhoea tends to predominate in a later phase of the infection [104].

The presence of plasma proteins in the discharge during a common cold indicates that the nasal mucosal microcirculation contributes to the discharge in inflammatory conditions [7,40,132,155] and that an inflammatory exudate, rather than a transudate, is the origin of the fluid. Plasma exudation across the endothelial and epithelial barriers is thought to be a specific feature of inflammation [141]. It is a finely regulated and controlled process with a selective outward bulk flux of plasma solutes through morphologically intact barriers. The

first barrier is the vascular endothelium. Inflammatory mediators cause endothelial cell contraction in the postcapillary venules, allowing bulk plasma to extravasate through the gaps in the endothelium driven by the hydrostatic pressure [57,121,172]. The second barrier is the airway mucosal epithelium. It has been suggested that the extravasated plasma molecules, via osmotic forces, increase the subepithelial hydrostatic pressure. The pressure forces the epithelium to widen its intracellular spaces and this opens a transepithelial paracellular route for the flux of macromolecules into the airway lumen. This flux is unidirectional and outwardly directed because of the pressure gradients [140–142]. The exudates in the common cold occur concominantly with the major symptoms and may be a crucial defence mechanism against infection by virtue of providing humoral antibodies and other potent plasma proteins access to the mucosal surface [5]. This mechanism may also be of pathogenic significance, enabling these substances to contribute to symptoms as a result of the induction of inflammatory cascades. Besides the mechanical flushing effect, the non-specific effects of the plasma exudate include inactivation of microbes by enzymatic degradation and binding to plasma proteins. The amount of total protein in nasal secretions has been found to be positively related to resistance to rhinovirus infection [156]. Specific immunity develops and the amounts of secretory IgA and humoral IgG are of importance in mucosal infections [43]. Together with products from the glandular secretions, several molecules with antimicrobial activity are present in human discharges, including glycoproteins showing antiviral activity [123], the antibacterial glandular protein lactoferrin [157] and lysozyme [39,151,154].

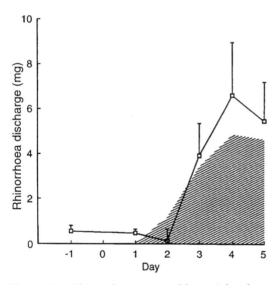

Figure 5.7 Rhinorrhoea assessed by weight of used paper handkerchiefs after inoculation of coronavirus on day 0 in healthy subjects in relation to symptom scores (shaded). (From [27].)

5.3.4 THE INFLAMMATORY PROCESS

The inflammatory process in the nasal mucosa that results from a viral infection is complex.

The multiple mechanisms and mediators involved interact in a cascade of events resulting in overt symptomatic disease. Parts of this process are known, but the knowledge is still to some extent incomplete. Specific immunity will not be discussed as it is presented in chapters on individual viruses.

The innate non-specific mechanisms include exudation and binding by plasma proteins. Four distinct enzyme systems participate to modulate the inflammation. These are the clotting and fibrinolytic systems, and the kinin and the complement systems [122]. The prerequisite for their action in the first line of defence is their presence on the mucosal surface, which is achieved through the inflammatory exudation of unfiltered bulk plasma. *Fibrinogen* is found on the mucosal surface in the common cold [5]. The fibrinopeptides participating in the inflammatory process induce leukotaxis [163]. Fibrin and fibrin–fibronectin gels can repair epithelial damage in the airway. Furthermore, fibrin can facilitate the migration of immunopotent cells by supplying a matrix for inflammatory cell traffic and for re-epithelialization and revascularization of the airway mucosa [59]. The precursor of the kinin system, kininogen, is converted to *bradykinin* through the enzymatic action of kallikrein. The biological effects of bradykinin include smooth muscle contraction, vasodilatation and an increase in vascular permeability [77,146], and they contribute to nociceptive pain in inflammatory conditions [170]. Characteristic symptoms of the common cold, such as nasal obstruction, rhinorrhoea and sore throat (Figure 5.3), can be induced by nasal bradykinin provocation [150]. Bradykinin is found on the nasal mucosal surface during rhinovirus colds and the concentrations of kinins found in nasal lavages are related to the severity of symptoms [132,149]. *Complement* factors play an important role in the local inflammatory response. The presence of complement fractions C3b and C5a in lavage fluids has likewise been shown to be related to the symptoms encountered with a nasal influenza virus infection [34].

Cytokines.

These are small peptide regulatory molecules that participate in the modulation of inflammation. Interferons (IFNs) are a group of quickly mobilized endogenous proteins and glycoproteins and they are one of the cytokines which participate in modulation of the viral airway inflammation. Interferons also have an immediate inhibitory effect on virus replication. Two types have been identified: type 1 (IFN-α from lymphocytes and macrophages; IFN-β from fibroblasts and epithelial cells) and type 2 (IFN-γ from activated T-cells) [18]. Both types are present in nasal lavages from subjects with viral upper airway infections [48,96,109,116,125]. Another multipotent cytokine is interleukin-1 (IL-1β), which is also found in nasal lavages during common colds [116,148]. IL-1β and IFN-γ are of special significance, since they can up-regulate the cellular surface receptor intercellular adhesion molecule-1 (ICAM-1), which is the receptor most rhinoviruses use to adsorb to epithelial cells in order to infect [90,169]. Knowledge of the role of cytokines in viral airway infection remains fragmentary.

Nervous mechanisms

Nervous mechanisms also participate in the pathophysiology of airway inflammation. The airway mucosa is densely innervated by sensory and efferent nerves involved in nociception, protective reflexes and vascular and glandular control. In brief, cholinergic stimulation results in vasodilatation and glandular secretion, and adrenergic stimulation in vasoconstriction and cessation of glandular secretion. Neuropeptides are transmitter substances involved in the control of smooth muscle, blood flow, vascular permeability and glandular secretion in the airways. Tachykinins participate in the neurogenic inflammation, e.g. the release of the sensory neuronal mediator *substance P* results in smooth muscle contraction, mucus secretion and increased microvascular permeability [118]. *Vasoactive intestinal polypeptide* (VIP) in parasympathetic nerves can induce vasodilatation, smooth muscle relaxation and possibly hypersecretion in the airways [17]. Consequently, nervous mechanisms can initiate symptoms indistinguishable from, and possibly be partly responsible for, symptoms of the common cold. Little is known about the role of these transmitters in viral respiratory airway inflammation.

It seems clear that the mediators of a viral infective inflammation differ from those of allergic inflammation, which is orchestrated mainly by mast cells, eosinophilic and basophilic granulocytes and the main mediator, histamine. In viral rhinitis basophils and mast cells do not appear to participate in the inflammatory process [74,132,185].

Inflammatory mediators

Inflammatory mediators are relevant because of their contribution to the development of disease. *Histamine* concentrations do not increase in rhinoviral rhinitis [132]. However, other viruses (RS virus and parainfluenza virus) seem to be able to induce histamine release in the nasal mucosa [180,181]. The failure of antihistamines alone to counteract the symptoms of viral rhinitis indicates that histamine is of no major significance in this inflammation.

Several arachidonic acid metabolites contribute to the pathogenesis of inflammation. Arachidonic acid can be transformed to bioactive molecules which have implications on inflammatory conditions via two major enzymatic pathways. The cyclooxygenase pathway leads to the formation of prostaglandins and thromboxanes, the lipoxygenase pathway to the formation of leukotrienes, hydroxy-eicosatetraenoic acids (HETEs) and lipoxin [115]. Such arachidonic acid derivatives seem to be of no major importance in viral airway inflammation and attempts to identify leukotrienes in nasal lavages from subjects with viral rhinitis have failed [45]. There is no convincing evidence of an active role of prostaglandins in viral rhinitis and furthermore, prostaglandin inhibitors have no obvious effect on the local inflammatory symptoms [87].

5.3.5 EFFECTS ON OTHER FUNCTIONS IN THE UPPER AIRWAYS

Olfaction

Upper respiratory tract infections are often associated with an altered ability to smell and there is an impairment of the ability to detect odours during a common cold. This impairment correlates with the degree of nasal blockage and, to a lesser extent, the volume of nasal discharge, which suggests that the alteration in the characteristics of the mucus and in the pattern of airflow result in an impaired access of the odour to the olfactory sensory epithelium [4]. Some upper airway infections – e.g. influenza – can induce long-lasting, and even permanent anosmia [60,63]. This anosmia is less likely to be due to a persistent nasal congestion, more to a damage of the olfactory epithelium. Moreover, a change in the perception of odours occurs in patients with rhinitis, for example previously pleasant odours being experienced as unpleasant. This phenomenon may be explained by the effects of a viral infection on the nervous system. Related to the impairment of smell is a loss of the appreciation of the taste of food which emphasizes the connection between taste and smell. In terms of food tasting, the olfactory function is of major importance. Only the basic distinguishable taste qualities – sweet, salty, bitter and sour – are identified by the taste buds in the mouth. A normal sense of smell is important for a good quality of life and impairment may be responsible for mood changes, as shown by the increased incidence of mental depression in people who become anosmic [129].

Voice

When the vocal cords are involved in upper airway infections, local inflammatory oedema interferes with the vibratory characteristics of the vocal cords and results in hoarseness. The voice is also changed by alterations in resonance. The resonance of the upper airway is altered in common colds because of mucosal swelling causing nasal obstruction, leading to a hyponasal voice.

Middle ear

Common colds are associated with transient middle ear pressure abnormalities [75,124] and otitis media with effusion [41,174]. The mucosa of the middle ear is subject to infections by respiratory viruses and they can be isolated from this locus [14,50,92]. Pressure abnormalities and otitis media with effusion produce a temporary conductive hearing loss and supposedly predispose to bacterial purulent otitis media [110,161,174]. Acute purulent otitis media is the major complication of the common cold in children.

Paranasal sinuses

Sinus pain is experienced subjectively by one in three subjects with a common cold. Transient sinus mucosal abnormalities are at least as frequent [56,94,175]. Mucosal swelling of the maxillary and ethmoidal sinuses are the most common findings [94]. It must be held in mind, however, that sinus abnormalities are regularly noted on magnetic resonance imaging, and mucosal thickening and sinus fluid are frequent findings in subjects without any symptoms from the upper respiratory tract [56]. In sinusitis, there is viral infection of the sinus mucosa and respiratory viruses can be found in sinus aspirates [78,99]. Viral infection in acute rhinitis frequently affects the patency of the sinus ostium [71], an obstruction of which is considered to predispose to the development of sinusitis [70,189]. Viral infec-

tion of the upper respiratory tract is the principal underlying reason for the development of acute bacterial sinusitis [147].

5.3.6 HYPERREACTIVITY AND VIRAL EFFECTS ON THE LOWER AIRWAYS

Recurrent colds are related to lower respiratory tract symptoms such as chronic cough, persistent wheezing or asthma in children [19]. Lower airway symptoms do not necessarily indicate an infection of the bronchial mucosa, even if virus can be found at this site [97], but could be induced by an upper airways infection through nervous reflexes or inflammatory mediators. Viral upper airway infections induce hyperreactivity in all parts of the respiratory mucosa. Nasal hyperreactivity is present in viral upper airway infections since the exudative response is enhanced by histamine challenge in coronavirus common colds [89]. In upper airway infections there is also a bronchial hyperreactivity to cold air and exercise [13]. In normal subjects, bronchial reactivity is unchanged during and after rhinovirus infections [171] but it increases temporarily in association with *symptomatic* viral airway infections [12,76]. Subjects with allergic rhinitis show bronchial hyperreactivity during common colds when this is provoked pharmacologically [42]. When, on the other hand, asymptomatic subjects with an existing bronchial hyperresponsiveness were investigated for a possible viral aetiology, a previous viral infection did not seem to be a likely cause [16]. Viral airway infections can provoke bronchial obstruction, with wheezing in children, as well as obstructive episodes in persons with asthma [33,105,127,128]. The relation between infections and asthma remains controversial [52] and any link between infection and asthma remains speculative. In addition to bronchial hyperreactivity, peripheral pulmonary functions are also affected in viral upper respiratory tract infections [49,145].

5.3.7 CIRCADIAN VARIATIONS

In viral upper airway infections the most severe symptoms are found in the mornings [165]. This is in accordance with the finding that the most intense inflammatory activity, seen as plasma exudation, occurs during the morning [88]. Other periodic variations include an enhancement of the nasal cycle in the common cold, a greater asymmetry in nasal airway resistance accompanying the increased symptoms [27,183]

5.4 PATHOPHYSIOLOGICAL ASPECTS OF SYMPTOMATIC TREATMENT

Up till now, no drugs have been used regularly in clinical practice to treat the causes of the common cold and only symptomatic treatment is available, many approaches having been tried for generations. High doses of vitamin C have been thought to have a prophylactic effect on the duration of symptoms [100]. However, several investigators have failed to show any positive effect of vitamin C as a means of treating or preventing colds [46,179]. The inhalation of aromatic vapours, such as menthol induces a sensation of increased nasal patency and relief of nasal blockage, even if it has no effect on objective nasal airway resistance in healthy subjects [73] or in subjects with colds [72]. It seems plausible that the temperature of the nasal mucosa is significant for the pathogenesis of colds. Studies of the propagation and basic biology of rhinoviruses reveal that they grow optimally at 33°C and that their growth is impaired at higher temperatures [136]. Consequently, by increasing the nasal mucosal temperature the viral replication can be inhibited and in accordance, inhalation of hot steam seems to decrease the severity and duration of symptoms [120,135,177].

5.4.1 ANTI-INFLAMMATORY DRUGS

Anti-inflammatory drugs, such as *glucocorticoids* and *nedocromil sodium* nasal spray, commonly used in treating allergic rhinitis, appear to improve significantly the symptoms of rhinovirus colds [20,79]. The mechanisms of action of these drugs need further study but effects on cytokines are likely to contribute. Non-steroidal anti-inflammatory drugs (NSAIDs) are frequently used in upper airway infections and can reduce fever, myalgia and malaise [167]. These drugs affect virus shedding and immune system functions in common colds. In one study, aspirin was reported to increase virus-shedding in rhinovirus infections [168] but this could not be confirmed in a study where, on the other hand, aspirin was associated with the suppression of the serum neutralizing antibody response [87]. Although, this suppressed response theoretically may be important, normally aspirin should not have a significant negative clinical effect [87].

5.4.2 ANTI-MEDIATOR DRUGS

Anticholinergic drugs.

These have minor but statistically significant effects on rhinorrhoea [38,62,84]. These will inhibit the glandular secretion from the cholinergically innervated submucous glands but are not likely to affect goblet cells, fenestrated capillaries or plasma exudation. The blocking of glandular secretion could be less favourable for the course of the infection, since the nasal secretions contain components which contribute to the mucosal defence. Such aspects, however, have not been elucidated satisfactorily.

Histamine is not a mediator of viral upper airway inflammation and not surprisingly, *antihistamines* have little or no effect on symptoms in experimental rhinovirus infections [67,83] and in naturally acquired colds [32,85].

5.4.3 DECONGESTANTS

Oral sympathomimetics.

These have an effect on subjective as well as objectively measured nasal congestion [112], but the effect after a single dose is not sufficient for symptomatic relief in the common cold [91]. The oral sympathomimetics are often used in combinations with other drugs. However, additive effects by combinations are rarely reported even if the individual effects of the drug combinations together can reduce the total symptoms [93].

Topical nasal decongestants (nosedrops).

These fulfil most of the criteria for an ideal symptomatic treatment, namely that it should be effective, easily accessible, well tolerated, inexpensive and should not counteract the defence mechanism of the host. Decongestive nosedrops fall into two categories – the sympathomimetic amines (e.g. phenylephrine hydrochloride) and the imidazoline derivatives (e.g. oxymetazoline hydrochloride and xylometazoline hydrochloride). Congestion of the nasal mucosa in acute rhinitis is due to vasodilatation of the sinusoids and to interstitial oedema. Nosedrops act by constricting the sinusoids in the nasal mucosa, the onset of action occurs within 10 minutes and the duration of the decongestive effect is about 4–8 hours [6,31,55,98]. These positive effects have led to the widespread use of these drugs, especially since they are accessible as over-the-counter drugs. The proper indication for the use of nosedrops is nasal blockage in the common cold. If the use of nosedrops continues for weeks or even months, a rebound congestion develops [3,35,80,119,152].

Imidazoline derivatives.

These are selective α_2-adrenoceptor agonists, while sympathomimetic amines act on α_1-adrenoceptors [8]. This different mode of action may be important because in humans the sinusoids in the nasal mucosa are regulated by α_1-, as well as α_2-adrenoceptors, while the nasal mucosal blood flow is not significantly affected by α_1-adrenoceptor agonists [9,24]. Along with their desirable decongestive effect, treatment with nasal decongestants of the imidazoline type leads to a decreased blood flow which has a duration of about 6

hours [31]. In clinical doses the reduction in the blood flow after oxymetazoline nosedrops is about 50% in healthy subjects and about 60% of the increased blood flow in patients with the common cold [22]. In common colds the mucociliary function, as previously discussed, is inhibited. Nosedrops can possibly promote [160, 164], as well as inhibit the mucociliary activity and the difference may be due to different effects by different doses [144] or to added preservatives [61]. An eventual promoting effect does not diminish the impairment of activity seen in common colds [143]. Even protracted use of nosedrops in healthy individuals does not seem to affect mucociliary clearance [143].

Calor and rubor.

These are cardinal signs of infection, caused by an increase in the local blood flow. Drugs interfering with blood flow interfere with the local defence reaction and may counteract an adequate inflammatory response. The basic studies in humans regarding the effects of topical nasal decongestants, other than decongestion, have so far been insufficient. In the rabbit, topical oxymetazoline reduces the nasal mucosal blood flow dose-dependently, with similar effects as in humans [1,30]. Furthermore, the application of oxymetazoline nosedrops in the nasal cavity induces a decrease in the mucosal blood flow in the maxillary sinus [1]. Experiments in this species, however, indicate the existence of less desirable effects on the inflammatory mechanisms. The reduction in blood flow may induce ischaemia in the mucosa, as illustrated by the presence of a microcirculatory phenomenon called vasomotion [30] – i.e. periodic oscillations in blood flow, a mechanism that is thought to optimize tissue perfusion during low-flow conditions. Vasomotion is mobilized when the reduction in local blood flow causes ischaemia in the tissue [36,162]. Nosedrops also inhibit the leucocyte migration during the first 2 hours of a provoked inflammatory reaction but whether this is due to ischaemia or other causes [23] associated with a reduced vascular permeability [29] is not clear. Oxymetazoline nosedrops seem to prolong the infection in nasal and sinus mucosa. In an experimental bilateral bacterial sinusitis nosedrops induce a more pronounced inflammatory reaction 2 days after treatment with oxymetazoline [26] (Figure 5.8).

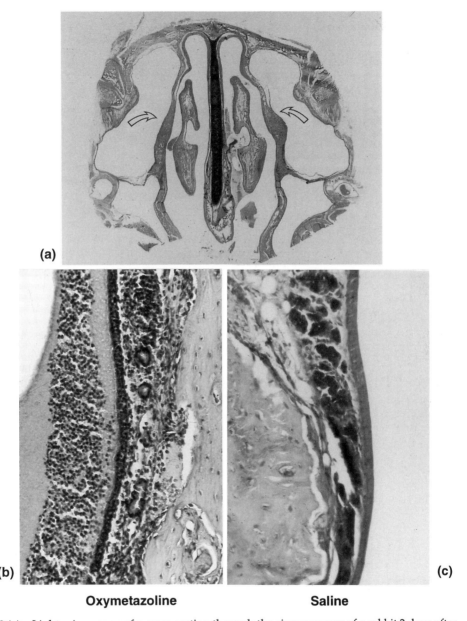

(a)

(b)   **(c)**

Oxymetazoline   **Saline**

Figure 5.8 (a)   Light microscopy of a cross-section through the sinus mucosa of a rabbit 2 days after induction of bilateral sinusitis. **(b)** One side was treated with topically applied decongestant nosedrops (oxymetazoline) and the other with placebo (saline). Note the increased inflammatory cell invasion on the actively treated side, indicating a more pronounced inflammation due to treatment.

In conclusion, the extensive use of topically administered decongestive nosedrops in upper respiratory tract infections contrasts with the lack of studies concerning the side effects of these drugs [166].

5.5 REFERENCES

1. Akerlund, A., Arfors, K.-E., Bende, M. *et al.* (1993) Effect of oxymetazoline on nasal and sinus mucosal blood flow in the rabbit as measured with laser-Doppler flowmetry. *Ann. Otol. Rhinol. Laryngol.*, **102**, 123–6.

2. Akerlund, A. and Bende, M. (1989) Nasal mucosal temperature and the effect of acute infective rhinitis. *Clin. Otolaryngol.*, **14**, 529–34.

3. Akerlund, A. and Bende, M. (1991) Sustained usage of xylometazoline nose-drops aggravates vasomotor rhinitis. *Am. J. Rhinol.*, **5**, 157–60.

4. Akerlund, A., Bende, M. and Murphy, C. (1994) Olfactory threshold shift during common cold is due to rhinorrhea and nasal blockage. *Acta Otolaryngol. (Stockh).*, **115**, 88–92.

5. Akerlund, A., Greiff, L., Andersson, M. *et al.* (1993) Mucosal exudation of fibrinogen in coronavirus-induced common colds. *Acta Otolaryngol. (Stockh)*, **113**, 642–8.

6. Akerlund, A., Klint, T., Olen, L. *et al.* (1989) Nasal decongestant effect of oxymetazoline in the common cold: a dose–response study in 106 patients. *J. Laryngol. Otol.*, **103**, 743–6.

7. Anderson, T., Riff, L. and Jackson, G. (1962) Immunoelectrophoresis of nasal secretions collected during a common cold: observations which suggest a mechanism of seroimmunity in viral respiratory infections. *J. Immunol.*, **89**, 691–7.

8. Andersson, K.-E. (1980) Adrenoceptors – classification, activation and blockade by drugs. *Postgrad. Med. J.*, **56**, 7–16.

9. Andersson, K.-E. and Bende, M. (1984) Adrenoceptors in the control of human nasal mucosal blood flow. *Ann. Otol. Rhinol. Laryngol.*, **93**, 179–82.

10. Andrewes, C. (1959) Adventures among viruses. Ill The puzzle of the common cold. *N. Engl. J. Med.*, **242**, 235–40.

11. Andrewes, C., Chaproniere, D., Gompels, A. *et al.* (1953). Propagation of common-cold virus in tissue cultures. *Lancet*, **ii**, 546–7.

12. Annesi, I., Oryszcyn, M., Neukirch, F. *et al.* (1992) Relationship of upper airways disorders to FEV1 and bronchial hyperresponsiveness in an epidemiological study. *Eur. Respir. J.*, **5**, 1104–10.

13. Aquilina, A., Hall, W., Douglas, G. *et al.* (1980) Airway reactivity in subjects with viral upper respiratory tract infections: the effects of exercise and cold air. *Am. Rev. Respir. Dis.*, **122**, 3–10.

14. Arola, M., Ziegler, T., Lehtonen, O. *et al.* (1990) Rhinovirus in otitis media with effusion. *Ann. Otol. Rhinol. Laryngol.*, **99**, 451–3.

15. Aust, R. and Drettner, B. (1975) The patency of the maxillary ostium in relation to body posture. *Acta Otolaryngol. (Stockh)*, **80**, 443–6.

16. Backer, V., Ulrich, C., Bach-Mortensen, N. *et al.* (1993) Relationship between viral antibodies and bronchial hyperresponsiveness in 495 unselected children and adolescents. *Allergy*, **48**, 240–7.

17. Baraniuk, J., Lundgren, J., Okayama, M. *et al.* (1990) Vasoactive intestinal peptide in human nasal mucosa. *J. Clin. Invest.*, **86**, 825–31.

18. Baron, S., Stanton, G., Fleischmann, W.J. *et al.* (1987) Introduction: general considerations of the interferon, in *The Interferon System. A Current Review to 1987*, (eds S. Baron, F. Dianzani, G. Stanton *et al.*), University of Texas Press, Austin, Texas, pp. 1–17.

19. Barr, M., Weiss, S., Segal, M. *et al.* (1992) The relationship of nasal disorders to lower respiratory tract symptoms and illness in a random sample of children. *Pediatr. Pulmonol.*, **14**, 91–4.

20. Barrow, G., Higgins, P., Al-Nakib, W. *et al.* (1990) The effect of intranasal nedocromil sodium on viral upper respiratory tract infections in human volunteers. *Clin. Exp. Allergy*, **20**, 45–51.

21. Beare, A. and Reed, S. (1977) The study of antiviral compounds in volunteers, in *Chemoprophylaxis and Virus Infections of the Respiratory Tract*, (ed. J. Oxford), CRC Press, Cleveland, pp. 27–55.

22. Bende, M. (1983) The effect of topical decongestant on blood flow in normal and infected nasal mucosa. *Acta Otolaryngol(Stockh)*, **96**, 523–7.

23. Bende, M., Akerlund, A., Intaglietta, M. *et al.* (1992) The effect of nose drops on an acute sinusitis; an experimental study in the rabbit. *Am. J. Rhinol.*, **6**, 55–8.

24. Bende, M., Andersson, K.-E. and Johansson, C. (1985) Vascular effects of phenylpropanolamine on human nasal mucosa. *Rhinology*, **23**, 43–8.

25. Bende, M., Angesjo, P., Hultsten, K. *et al.* (1990) Effect of cold dry air on nasal airway resistance. *Am. J. Rhinol.*, **4**, 65–7.

26. Bende, M., Arfors, K.-E., Stierna, P. *et al.* (1995) Effect of oxymetazoline nose drops on acute sinusitis in the rabbit. *Ann. Otol. Rhinol. Laryngol.* (In press).

27. Bende, M., Barrow, I., Heptonstall, J. *et al.* (1989) Changes in human nasal mucosa during experimental coronavirus common colds. *Acta Otolaryngol. (Stockh)*, **107**, 262–9.

28. Bende, M., Flisberg, K., Larsson, I. *et al.* (1983) A method for determination of blood flow with 133Xe in human nasal mucosa. *Acta Otolaryngol. (Stockh)*, **96**, 277–85.

29. Bende, M., Hansell, P., Intaglietta, M. *et al.* (1992) Effect of oxymetazoline nose drops on vascular permeability of the nasal mucosa in the rabbit after provocation with leukotriene B4. *J. Otorhinolaryngol.*, **54**, 270–4.

30. Bende, M., Intaglietta, M. and Arfors, K.-E. (1993) Nose drops induce vasomotion in the microcirculation of the sinus mucosa of the rabbit. *J. Otorhinolaryngol.*, **55**, 110–13.

31. Bende, M. and Loth, S. (1986) Vascular effect of oxymetazoline on human nasal mucosa. *J. Laryngol. Otol.*, **100**, 285–8.

32. Berkowitz, R. and Tinkelman, D. (1991) Evaluation of terfenadine for treatment of the common cold. *Ann. Allergy*, **67**, 593–7.

33. Bjornsdottir, U. and Busse, W. (1992) Respiratory infections and asthma. *Clin. Allergy*, **76**, 895–915.

34. Bjornson, A., Mellencamp, M. and Schiff, G. (1991) Complement is activated in the upper respiratory tract during influenza virus infection. *Am. Rev. Respir. Dis.*, **143**, 1062–6.

35. Blue, J. (1968) Rhinitis medicamentosa. *Ann. Allergy*, **26**, 425–9.

36. Borgstrom, P., Schmidt, J., Bruttig, S. *et al.* (1992) Slow-wave flow motion in rabbit skeletal muscle after acute fixed-volume hemorrhage. *Circ. Shock*, **36**, 57–61.

37. Borson, D. and Nadel, J. (1990) Regulation of airway secretions: role of peptides and proteases, in *Rhinitis and Asthma: Similarities and Differences*, (eds N. Mygind, U. Pipkorn and R. Dahl), Munksgaard, Copenhagen, pp. 76–99.

38. Borum, P., Olsen, L., Winther, B. *et al.* (1981) Ipratropium nasal spray: a new treatment for rhinorrhea in the common cold. *Am. Rev. Respir. Dis.*, **123**, 418–20.

39. Bowes, D., Clark, A. and Corrin, B. (1981) Ultrastructural localization of lactoferrin and glycoprotein in human bronchial glands. *Thorax*, **36**, 108–15.

40. Butler, W., Waldmann, T., Rossen, R. *et al.* (1970) Changes in IgA and IgG concentrations in nasal secretions prior to the appearance of antibody during viral respiratory infection in man. *J. Immunol.*, **105**, 584–91.

41. Bylander, A. (1984) Upper respiratory tract infection and Eustachian tube function in children. *Acta Otolaryngol. (Stockh)*, **97**, 343–9.

42. Calhoun, W., Swenson, C., Dick, E. *et al.* (1991) Experimental rhinovirus 16 infection potentiates histamine release after antigen bronchoprovocation in allergic subjects. *Am. Rev. Respir. Dis.*, **144**, 1267–73.

43. Callow, K. (1985) Effect of specific humoral immunity and some non-specific factors on resistance of volunteers to respiratory coronavirus infection. *J. Hyg. (Camb)*, **95**, 173–89.

44. Callow, K., Parry, H., Sergeant, M. *et al.* (1990) The time course of the immune response to experimental coronavirus infection of man. *Epidemiol. Infect.*, **105**, 435–46 .

45. Callow, K., Tyrrell, D., Shaw, R. *et al.* (1988) Influence of atopy on the clinical manifestations of coronavirus infection in adult volunteers. *Clin. Allergy*, **18**, 119–29 .

46. Carson, M., Cox, H., Cobett, M. *et al.* (1975) Vitamin C and the common cold. *J. Soc. Occup. Med.*, **25**, 99–102.

47. Cate, T., Couch, R. and Johnson, K. (1964) Studies with rhinoviruses in volunteers: production of illness, effect of naturally acquired antibody, and demonstration of a protective effect not associated with serum antibody. *J. Clin. Invest.*, **43**, 56–67.

48. Cate, T., Douglas, R. Jr and Couch, R. (1969) Interferon and resistance to upper respiratory virus illness. *Proc. Soc. Exp. Biol. Med.*, **131**, 631–6.

49. Cate, T., Roberts, T., Russ, M. *et al.* (1973) Effect of common cold on pulmonary function. *Am. Rev. Respir. Dis.*, **108**, 858–65.

50. Chonmaitree, T., Howie, V. and Truant, A. (1986) Presence of respiratory viruses in middle ear fluids and nasal wash specimens from children with acute otitis media. *Pediatrics*, **77**, 698–702.

51. Chretien, J., Holland, W., Macklem, P. *et al.* (1984) Acute respiratory infections in children. A global public-health problem. *N. Engl. J. Med.*, **310**, 982–4.

52. Clarke, C. (1979) Relationship of bacterial and viral infections to exacerbations of asthma. *Thorax*, **34**, 344–7.

53. Cole, P. (1993) Nasal and oral airflow resistors. Site, function, and assessment. *Arch. Otolaryngol. Head Neck Surg.*, **118**, 790–3.

54. Cole, P. and Haight, J. (1984) Posture and nasal patency. *Am. Rev. Respir. Dis.*, **129**, 351–4 .

55. Connell, J. (1969) Effectiveness of topical nasal decongestants. *Ann. Allergy*, **27**, 541–6 .

56. Cooke, L. and Hadley, D. (1991) MRI of the paranasal sinuses: incidental abnormalities and their relationship to symptoms. *J. Laryngol. Otol.*, **105**, 278–81.

57. Cotran, R. and Majno, G. (1964) The delayed and prolonged vascular leakage in inflammation. I. Topography of the leaking vessels after thermal injury. *Am. J. Pathol.*, **45**, 261–81.

58. Couch, R., Cate, T., Douglas, R. *et al.* (1966) Effect of inoculation on experimental respiratory viral disease in volunteers and evidence for airborne transmission. *Bacteriol. Rev.*, **30**, 517–27.

59. Crouch, E. (1990) Pathobiology of pulmonary fibrosis. *Am. J. Physiol.*, **259**, L159–84 .

60. Davidson, T., Jalowayski, A., Murphy, C. *et al.* (1987) Evaluation and treatment of smell dysfunction. *West. J. Med.*, **146**, 434–8.

61. Deitmer, T. and Scheffler, R. (1993) The effect of different preparations of nasal decongestants on ciliary beat frequency in vitro. *Rhinology*, **31**, 151–3.

62. Dockhorn, R., Grossman, J., Posner, M. *et al.* (1992) A double-blind, placebo controlled study of the safety and efficacy of ipratropium bromide nasal spray versus placebo in patients with the common cold. *J. Allergy Clin. Immunol.*, **90**, 1076–82.

63. Doty, R. (1979) A review of olfactory dysfunctions in man. *Am. J. Otolaryngol.*, **1**, 57–79 .

64. Douglas, R., Alford, B. and Couch, R. (1968) Atraumatic nasal biopsy for studies of respiratory virus infection in volunteers. *Antimicrob. Agents Chemother.*, **8**, 340–3.

65. Douglas, R., Alford, R., Cate, T. *et al.* (1966) The leukocyte response during viral respiratory illness in man. *Ann. Intern. Med.*, **64**, 521–30.

66. Dowling, H., Jackson, G., Spiesman, I. *et al.* (1958) Transmission of the common cold to volunteers under controlled conditions. III. The effect of chilling of the subjects upon susceptibility. *Am. J. Hyg.*, **68**, 59–65.

67. Doyle, W., McBride, T., Skoner, D. *et al.* (1988) A double-blind, placebo controlled clinical trial of the effect of chlorpheniramine on the response of the nasal airway, middle ear and Eustachian tube to provocative rhinovirus challenge. *Pediatr. Infect. Dis.*, **7**, 229–38.

68. Doyle, W., Riker, D., McBride, T. *et al.* (1993) Therapeutic effects of an anticholinergic–sympathomimetic combination in induced rhinovirus colds. *Ann. Otol. Rhinol. Laryngol.*, **102**, 521–7.

69. Drettner, B. (1961) Vascular reactions of the human nasal mucosa on exposure to cold. *Acta Otolaryngol. (Stockh)*, Suppl., 166.

70. Drettner, B. and Aust, R. (1977) The patency of the maxillary ostium in relation to body posture. *Acta Otolaryngol. (Stockh)*, **83**, 16–19.

71. Drettner, B. and Lindholm, C.-E. (1967) The borderline between acute rhinitis and sinusitis. *Acta Otolaryngol. (Stockh)*, **64**, 508–13.

72. Eccles, R., Jawad, M. and Morris, S. (1990) The effect of oral administration of menthol on nasal resistance to airflow in subjects suffering from nasal congestion associated with the common cold. *J. Pharm. Pharmacol.*, **42**, 652–4.

73. Eccles, R. and Jones, A. (1983) The effect of menthol on nasal resistance to air flow. *J. Laryngol. Otol.*, **97**, 705–9.

74. Eggleston, P., Hendley, J. and Gwaltney, J. Jr (1984) Mediators of immediate hypersensivity in nasal secretions during natural colds and rhinovirus infection. *Acta Otolaryngol. (Stockh)*, Suppl. 413, 25–35.

75. Elkhatieb, A., Hipskind, G., Woerner, D. *et al.* (1993) Middle ear abnormalities during natural rhinovirus colds in adults. *J. Infect. Dis.*, **168**, 618–21.

76. Empey, D., Laitinen, L., Jacobs, L. *et al.* (1976) Mechanisms of bronchial hyperreactivity in normal subjects after upper respiratory tract infection. *Am. Rev. Respir. Dis.*, **113**, 131–9.

77. Erjefalt, I. and Persson, C. (1991) Allergen, bradykinin, and capsaicin increase outward but not inward macromolecular permeability of guinea-pig tracheobronchial mucosa. *Clin. Exp. Allergy*, **21**, 217–24.

78. Evans, F., Sydnor, J., Moore, W. *et al.* (1975) Sinusitis of maxillary antrum. *N. Engl. J. Med.*, **293**, 735–9.

79. Farr, B., Gwaltney, J. Jr, Hendley, J. *et al.* (1990) A randomized controlled trial of glucocorticoid prophylaxis against experimental rhinovirus infection. *J. Infect. Dis.*, **162**, 1173–7.

80. Feinberg, A. and Feinberg, S. (1971) The 'nose drop nose' due to oxymetazoline (Afrin) and other topical vasoconstrictants. *Illinois Med. J.*, **140**, 50–2.

81. Fischer, E., Scadding, G. and Lund, V. (1993) The role of acoustic rhinometry in studying the nasal cycle. *Rhinology*, **31**, 57–61.

82. Flisberg, K. and Ingelstedt, S. (1962) Vascular responses to feet cooling in normal and allergic nose. *Acta Otolaryngol. (Stockh)*, **55**, 457–66.

83. Gaffey, M., Gwaltney, J. Jr, Sastre, A. *et al.* (1987) Intranasally and orally administered antihistamine treatment of experimental rhinovirus colds. *Am. Rev. Respir. Dis.*, **136**, 556–60.

84. Gaffey, M., Hayden, F., Boyd, J. *et al.* (1988) Ipratropium bromide treatment of experimental rhinovirus infection. *Antimicrob. Agents Chemother.*, **32**, 1644–7.

85. Gaffey, M., Kaiser, D. and Hayden, F. (1988) Ineffectiveness of oral terfenadine in natural colds: evidence against histamine as a mediator of common cold symptoms. *Pediatr. Infect. Dis.*, 7, 223–8.

86. Ginzel, A. and Illum, P. (1981) Messung des Schleimtransportes in der Nase an Gesunden und im Anschluss an eine Erkaltung mittels des Saccharin-sky-blue-tests. *HNO-Praxis*, **6**, 31–7.

87. Graham, N., Burrell, C., Douglas, R. *et al.* (1990) Adverse effects of aspirin, acetaminophen, and ibuprofen on immune function, viral shedding, and clinical status in rhinovirus-infected volunteers. *J. Infect. Dis.*, **162**, 1277–82.

88. Greiff, L., Akerlund, A., Andersson, M. *et al.* (1994) Day–night differences in mucosal exudation of bulk plasma in coronavirus-induced common cold. (submitted).

89. Greiff, L., Andersson, M., Akerlund, A. *et al.* (1994) Microvascular exudative hyperresponsiveness in human coronavirus-induced common cold. *Thorax*, **49**, 121–7.

90. Greve, J., Davis, G., Meyer, A. *et al.* (1989) The major human rhinovirus receptor is ICAM-1. *Cell*, **56**, 839–47.

91. Groenborg, H., Winther, B., Brofeldt, S. *et al.* (1983) Effects of oral norephedrine on common cold symptoms. *Rhinology*, **21**, 3–12.

92. Gwaltney, J. (1971) Virology of the middle ear. *Ann. Otol.*, **80**, 365–70.

93. Gwaltney, J. (1992) Combined antiviral and antimediator treatment of rhinovirus colds. *J. Infect. Dis.*, **166**, 776–82.

94. Gwaltney, J., Phillips, C., Miller, R. *et al.* (1994) Computed tomographic study of the common cold. *N. Engl. J. Med.*, **330**, 25–30.

95. Hackemann, M., Denman, A. and Tyrrell, D. (1974) Inactivation of influenza virus by human lymphocytes. *Clin. Exp. Immunol.*, **16**, 583–91.

96. Hall, C., Douglas, R. Jr, Simons, R. *et al.* (1978) Interferon production in children with respiratory syncytial, influenza, and parainfluenza virus infections. *J. Pediatr.*, **93**, 28–32.

97. Halperin, S., Eggleston, P., Beasley, P. *et al.* (1985) Exacerbations of asthma in adults during experimental rhinovirus infection. *Am. Rev. Respir. Dis.*, **132**, 976–80.

98. Hamilton, L. (1981) Effect of xylometazoline nasal spray on nasal conductance in subjects with coryza. *J. Otolaryngol.*, **10**, 109–16.

99. Hamory, B., Sande, M., Sydnor, A. *et al.* (1979) Etiology and antimicrobial therapy of acute maxillary sinusitis. *J. Infect. Dis.*, **139**, 197–202.

100. Hemila, H. (1992) Vitamin C and the common cold. *Br. J. Nutr.*, **67**, 3–16.

101. Henderson, F., Dubovi, E., Harder, S. *et al.* (1988) Experimental rhinovirus infection in human volunteers exposed to ozone. *Am. Rev. Respir. Dis.*, **137**, 1124–8.

102. Hoorn, B. and Tyrrell, D. (1966) Effects of some viruses on ciliated cells. *Am. Rev. Respir. Dis.*, **93**, 156–61.

103. Hoorn, B. and Tyrrell, D. (1969) Organ cultures in virology. *Prog. Med. Virol.*, **11**, 408–50 .

104. Igarashi, Y., Skoner, D., Doyle, W. *et al.* (1993) Analysis of nasal secretions during experimental rhinovirus upper respiratory infections. *J. Allergy Clin. Immunol.*, **92**, 722–31.

105. Isaacs, D., Clarke, J., Tyrrell, D. *et al.* (1982) Selective infection of lower respiratory tract by respiratory viruses in children with recurrent respiratory tract infections. *Br. Med. J.*, **284**, 1746–8.

106. Jackson, G., Dowling, H., Spiesman, I. *et al.* (1958) Transmission of the common cold to volunteers under controlled conditions. *Arch. Intern. Med.*, **101**, 267–78 .

107. James, J. (1972) Longitudinal study of the morbidity of diarrheal and respiratory infections in malnourished children. *Am. J. Clin. Nutr.*, **25**, 690–4.

108. Jannert, M., Andreasson, L., Ivarsson, A. *et al.* (1984) Patency of the maxillary sinus ostium in healthy individuals. *Acta Otolaryngol. (Stockh)*, **97**, 137–49 .

109. Jao, R., Wheelock, E. and Jackson, G. (1970) Production of interferon in volunteers infected with Asian influenza. *J. Infect. Dis.*, **121**, 419–26.

110. Knight, L. and Eccles, R. (1993) The relation between nasal airway resistance and middle ear pressure in subjects with acute upper respiratory tract infection. *Acta Otolaryngol. (Stockh)*, **113**, 196–200.

111. Larson, H., Parry, R. and Tyrrell, D. (1980) Impaired polymorphonuclear leucocyte chemotaxis after influenza virus infection. *Br. J. Dis. Chest*, **74**, 56–62.

112. Lea, P. (1984) A double-blind controlled evaluation of the nasal decongestant effect of Day Nurse^R in the common cold. *J. Int. Med. Res.*, **12**, 124–7.

113. Levandowski, R., Ou, D. and Jackson, G. (1986) Acute-phase decrease of T lymphocyte subsets in rhinovirus infection. *J. Infect. Dis.*, **153**, 743–8.

114. Levandowski, R., Weaver, C. and Jackson, G. (1988) Nasal-secretion leukocyte populations determined by flow cytometry during acute rhinovirus infection. *J. Med. Virol.*, **25**, 423–32.

115. Lewis, R., Austen, K. and Soberman, R. (1990) Leukotrienes and other products of the 5-lipoxygenase pathway. Biochemistry and relation to pathobiology in human disease. *N. Engl. J. Med.*, **323**, 645–55.

116. Linden, M., Greiff L, Andersson, M. *et al.* (1995) Nasal cytokines in common cold and allergic rhinitis. *Clin. Exp. Allergy*, **25**, 166–72.

117. Lowenstein, S. and Parrino, T. (1987) Management of the common cold. *Adv. Intern. Med.*, **32**, 207–34.

118. Lundblad, L. (1990) Neuropeptides and autonomic nervous control of the respiratory mucosa, in *Rhinitis and Asthma: Similarities and Differences*, (eds N. Mygind, U. Pipkorn and R. Dahl), Munksgaard, Copenhagen, pp. 65–75.

119. Mabry, R. (1982) Rhinitis medicamentosa: the forgotten factor in nasal obstruction. *South. Med. J.*, **75**, 817–19.

120. Macknin, M., Mathew, S., vander Brug, S. *et al.* (1990) Effect of inhaling heated vapor on symptoms of the common cold. *JAMA*, **264**, 989–91.

121. Majno, G., Shea, S. and Leventhal, M. (1969) Endothelial contraction induced by histamine-type mediators. *J. Cell Biol.*, **42**, 647–72.

122. Male, D. and Roitt, I. (1989) Adaptive and innate immunity, in *Immunology*, 2nd edn, (eds I. Roitt, J. Brostoff and D. Male), Gower Medical Publishing, London, pp. 1.1–1.10.

123. Matthews, T., Nair, C., Lawrence, M. *et al.* (1976) Antiviral activity in milk of possible clinical importance. *Lancet*, **ii**, 1387–9.

124. McBride, T., Doyle, W., Hayden, F. *et al.* (1989) Alterations of the Eustachian tube, middle ear, and nose in rhinovirus infection. *Arch. Otolaryngol. Head Neck Surg.*, **115**, 1054–9.

125. McIntosh, K. (1978) Interferon in nasal secretions from infants with viral respiratory tract infections. *J. Pediatr.*, **93**, 33–6.

126. Melen, I., Andreasson, L., Ivarsson, A. *et al.* (1986) Effects of phenylpropanolamine on ostial and nasal airway resistance in healthy individuals. *Acta Otolaryngol. (Stockh)*, **102**, 99–105.

127. Minor, T., Dick, E., Baker, J. *et al.* (1976) Rhinovirus and influenza type A infections as precipitants of asthma. *Am. Rev. Respir. Dis.*, **113**, 149–53.

128. Minor, T., Dick, E., DeMeo, A. *et al.* (1974) Viruses as precipitants of asthmatic attacks in children. *JAMA*, **227**, 292–8.

129. Moore-Gillon, V. (1989) Olfactometry and the sense of smell, in *Rhinitis. Mechanisms and Management*, (ed. I. Mackay), Royal Society of Medicine Services Ltd, London, pp. 69–79.

130. Mudd, S., Goldman, A. and Grant, S. (1921) Reactions of the nasal cavity and postnasal space to chilling of the body surface. *J. Exp. Med.*, **34**, 11–45.

131. Mygind, N. and Thomsen, J. (1976) Diurnal variation of nasal protein concentration. *Acta Otolaryngol. (Stockh)*, **82**, 219–21.

132. Nacleiro, R., Proud, D., Lichtenstein, L. *et al.* (1988) Kinins are generated during experimental rhinovirus colds. *J. Infect. Dis.*, **157**, 133–42.

133. O'Flynn, P. (1993) Posture and nasal geometry. *Acta Otolaryngol. (Stockh)*, **113**, 530–2 .

134. Olsson, P. and Bende, M. (1987) Sympathetic neurogenic control of blood flow in human nasal mucosa. *Acta Otolaryngol. (Stockh)*, **102**, 482–7.

135. Ophir, D. and Elad, Y. (1987) Effects of steam inhalation on nasal patency and nasal symptoms in patients with the common cold. *Am. J. Otolaryngol.*, **3**, 149–53.

136. Parsons, R. and Tyrrell, D. (1961) A plaque method for assaying some viruses isolated from common colds. *Nature*, **189**, 640–2.

137. Paulsson, B., Bende, M., Larsson, I. *et al.* (1992) Ventilation of paranasal sinuses studied with dynamic emission computer tomography. *Laryngoscope*, **102**, 451–7.

138. Paulsson, B., Bende, M. and Ohlin, P. (1985) Nasal mucosal blood flow at rest and during exercise. *Acta Otolaryngol. (Stockh)*, **99**, 140–3.

139. Pedersen, M., Sakakura, Y., Winther, B. *et al.* (1983) Nasal mucociliary transport, number of ciliated cells, and beating pattern in naturally acquired common colds. *Eur. J. Respir. Dis.*, **64**, 355–64.

140. Persson, C. (1991) Plasma exudation in the airways: mechanisms and function. *Eur. Respir. J.*, **4**, 1268–74.

141. Persson, C., Erjefalt, I., Alkner, U. *et al.* (1991) Plasma exudation as a first line respiratory mucosal defence. *Clin. Exp. Allergy*, **21**, 17–24.

142. Persson, C., Svensson, C., Greiff, L. *et al.* (1992) The use of the nose to study the inflammatory response of the respiratory tract. *Thorax*, **47**, 993–1000.

143. Petruson, B. and Hansson, H.-A. (1982) Function and structure of the nasal mucosa after 6 weeks' use of nose-drops. *Acta Otolaryngol. (Stockh)*, **94**, 563–9.

144. Phillips, P., McCaffrey, T. and Kern, E. (1990) The in vivo and in vitro effect of phenylephrine (Neo Synephrine) on nasal ciliary beat frequency and mucociliary transport. *Otolaryngol. Head Neck Surg.*, **103**, 558–65.

145. Picken, J., Niewoehner, D. and Chester, E. (1972) Prolonged effects of viral infections of the upper respiratory tract upon small airways. *Am. J. Med.*, **52**, 738–46.

146. Polosa, R. (1993) Role of the kinin–kallikrein pathway in allergic disease. *Allergy*, **48**, 217–25.

147. Proctor, D. and Andersen, I. (1982) *The Nose – Upper Airway Physiology and the Atmospheric Environment*, Elsevier Biomedical Press, Amsterdam, New York, Oxford.

148. Proud, D., Gwaltney, J., Hendley, J. *et al.* (1994) Increased levels of interleukin-1 are detected in nasal secretions of volunteers during experimental rhinovirus colds. *J. Infect. Dis.*, **169**, 1007–13.

149. Proud, D., Nacleiro, R., Gwaltney, J. *et al.* (1990) Kinins are generated in nasal secretions during natural rhinovirus colds. *J. Infect. Dis.*, **161**, 120–3.

150. Proud, D., Reynolds, C., Lacapra, S. *et al.* (1988) Nasal provocation with bradykinin induces symptoms of rhinitis and a sore throat. *Am. Rev. Respir. Dis.*, **137**, 613–16.

151. Raphael, G., Jeney, E., Baraniuk, J. *et al.* (1989) Pathophysiology of rhinitis. Lactoferrin and lysozyme in nasal secretions. *J. Clin. Invest.*, **84**, 1528–35.

152. Rijntjes, E. (1985) Nose-drops abuse: a functional and morphological study. (Thesis), University of Leiden.

153. Roberts, N. (1979) Temperature and host defense. *Microbiol. Rev.*, **43**, 241–59.

154. Rossen, R., Alford, R., Butler, W. *et al.* (1966) The separation and characterization of proteins intrinsic to nasal secretion. *J. Immunol.*, **97**, 369–78.

155. Rossen, R., Butler, W., Cate, T. *et al.* (1965) Protein composition of nasal secretion during respiratory virus infection. *Proc. Soc. Exp. Biol. Med. N.Y.*, **119**, 1169–76.

156. Rossen, R., Butler, W., Waldman, R. *et al.* (1970) The proteins in nasal secretion. II. A longitudinal study of IgA and neutralizing antibody levels in nasal washings from men infected with influenza virus. *JAMA*, **211**, 1157–61.

157. Rossen, R., Schade, A., Butler, W. *et al.* (1966) The proteins in nasal secretion: a longitudinal study of the gamma-A-globulin, gamma-G-globulin, albumin, siderophilin and total protein concentrations in nasal washings from adult male volunteers. *J. Clin. Invest.*, **45**, 768–76.

158. Rundcrantz, H. (1969) Postural variations of nasal patency. *Acta Otolaryngol. (Stockh)*, **68**, 434–43.

159. Sakakura, Y., Sasaki, Y., Togo, Y. *et al.* (1973) Mucociliary function during experimentally induced rhinovirus infection in man. *Ann. Otol. Rhinol. Laryngol.*, **82**, 203–11.

160. Saketkhoo, K., Yergin, B., Januszkiewicz, A. *et al.* (1978) The effect of nasal decongestants on nasal mucus velocity. *Am. Rev. Respir. Dis.*, **118**, 251–4.

161. Schilder, A., Zielhuis, G., Straatman, H. *et al.* (1992) An epidemiological approach to the etiology of middle ear diseases in The Netherlands. *Eur. Arch. Otorhinolaryngol.*, **249**, 370–3.

162. Schmidt, J., Borgstrom, P. and Intaglietta, M. (1993) The vascular origin of slow-wave flow motion in skeletal muscle during local hypotension. *Int. J. Microcirc. Clin. Exp.*, **12**, 287–97.

163. Senior, R., Skogen, W., Griffin, G. *et al.* (1986) Effects of fibrinogen derivatives upon the inflammatory response. Studies with human fibrinopeptide. *Br. J. Clin. Invest.*, **77**, 1014–19.

164. Simon, H., Drettner, B. and Jung, B. (1977) Messung des schleimhauttransportes in menschlichen Nase mit ^{51}Cr markierten harzkugeln. *Acta Otolaryngol. (Stockh)*, **83**, 378–90.

165. Smith, A., Tyrrell, D., Coyle, K. *et al.* (1988) Diurnal variation in the symptoms of colds and influenza. *Chronobiol. Int.*, **5**, 411–16.

166. Smith, M. and Feldman, W. (1993) Over-the-counter cold medications; a critical review of clinical trials between 1950 and 1991. *JAMA*, **269**, 2258–63.

167. Sperber, S., Sorrentino, J., Riker, D. *et al.* (1989) Evaluation of an alpha agonist alone and in combination with a nonsteroidal antiinflammatory agent in the treatment of experimental rhinovirus colds. *Bull. N. Y. Acad. Med.*, **65**, 145–60.

168. Stanley, E., Jackson, G., Panusarn, C. *et al.* (1975) Increasing virus shedding with aspirin treatment of rhinovirus infection. *JAMA*, **231**, 1248–51.

169. Staunton, D., Merluzzi, V., Rothlein, R. *et al.* (1989) A cell adhesion molecule, ICAM-1, is the major surface receptor for rhinoviruses. *Cell*, **56**, 849–53.

170. Steranka, L., Manning, D. and DeHaas, C. (1988) Bradykinin as a pain mediator: Receptors are localized to sensory neurons, and antagonists have analgesic actions. *Proc. Natl Acad. Sci. USA*, **85**, 3245–9.

171. Summers, Q., Higgins, P., Barrow, I. *et al.* (1992) Bronchial reactivity to histamine and bradykinin is unchanged after rhinovirus infection in normal subjects. *Eur. Respir. J.*, **5**, 313–17.

172. Svensjo, E., Arfors, K.-E., Arturson, G. *et al.* (1978) The hamster cheek pouch preparation as a model for studies of macromolecular permeability of the microvasculature. *Ups. J. Med. Sci.*, **83**, 71–9.

173. Tos, M. (1982) Goblet cells and glands in the nose and paranasal sinuses, in *The Nose. Upper Airway Physiology and the Atmospheric Environment*, (eds D. Proctor and I. Andersen), Elsevier Biomedical Press, Amsterdam, New York, Oxford, pp. 99–144.

174. Tos, M., Poulsen, G. and Borch, J. (1979) Etiologic factors in secretory otitis. *Arch. Otolaryngol. Head Neck Surg.*, **105**, 582–8.

175. Turner, B., Cail, W., Hendley, J. *et al.* (1992) Physiologic abnormalities in the paranasal sinuses during experimental rhinovirus colds. *J. Allergy Clin. Immunol.*, **90**, 474–8 .

176. Turner, R., Hendley, J. and Gwaltney, J. Jr (1982) Shedding of infected ciliated epithelial cells in rhinovirus colds. *J. Infect. Dis.*, **145**, 849–53.

177. Tyrrell, D., Barrow, I. and Arthur, J. (1989) Local hyperthermia benefits natural and experimental common colds. *Br. Med. J.*, **298**, 1280–3.

178. Tyrrell, D., Cohen, S. and Schlarb, J. (1993) Signs and symptoms in common colds. *Epidemiol. Infect.*, **111**, 143–56.

179. Walker, G., Bynoe, M. and Tyrrell, D. (1967) Trial of ascorbic acid in prevention of colds. *Br. Med. J.*, **1**, 603–6.

180. Welliver, R., Wong, D., Middleton, E. *et al.* (1982) Role of parainfluenza virus-specific IgE in pathogenesis of croup and wheezing subsequent to infection. *J. Pediatr.*, **101**, 889–96.

181. Welliver, R., Wong, D., Sun, M. *et al.* (1981) The development of respiratory syncytial virus-specific IgE and the release of histamine in nasopharyngeal secretions after infection. *N. Engl. J. Med.*, **305**, 841–6.

182. Willatt, D. (1993) Continuous infrared thermometry of the nasal mucosa. *Rhinology*, **31**, 63–7.

183. Williams, R. and Eccles, R. (1992) Nasal airflow asymmetry and the effects of a topical nasal decongestant. *Rhinology*, **30**, 277–82.

184. Wilson, R. (1987) Upper respiratory tract viral infection and mucociliary clearance. *Eur. J. Respir. Dis.*, **70**, 272–9.

185. Winther, B., Brofeldt, S., Christensen, B. *et al.* (1984) Light and scanning electron microscopy of nasal biopsy material from patients with naturally acquired common colds. *Acta Otolaryngol. (Stockh)*, **97**, 309–18.

186. Winther, B., Farr, B., Turner, R. *et al.* (1984) Histopathologic examination and enumeration of polymorphonuclear leukocytes in the nasal mucosa during experimental rhinovirus colds. *Acta Otolaryngol. (Stockh)*, Suppl. 413, 19–24.

187. Winther, B., Gwaltney, J., Mygind, N. *et al.* (1986) Sites of rhinovirus recovery after point inoculation of the upper airway. *JAMA*, **256**, 1763–7.

188. Winther, B., Innes, D., Mills, S. *et al.* (1987) Lymphocyte subsets in normal airway mucosa of the human nose. *Arch. Otolaryngol. Head Neck Surg.*, **113**, 59–62.

189. Zeiger, R. (1992) Prospect for ancillary treatment of sinusitis in the 1990s. *J. Allergy Clin. Immunol.*, **90**, 478–95.

TAXONOMY OF RHINOVIRUSES – A HISTORY

6

Albert Kapikian

.1 Introduction
6.2 The early days
6.3 The rhinovirus numbering system
6.4 References

6.1 INTRODUCTION

It is well recognized that the most important known aetiological agents of the 'common cold' in adults are the rhinoviruses which are associated with 15–40% of such illnesses [7,11,23]. These viruses are also implicated as important causes of acute respiratory illnesses in paediatric populations [3,9]. The course of scientific inquiry leading to the discovery of the major known causes of the common cold was indeed tortuous, with numerous erroneous suppositions regarding the aetiology of this condition [1,10,43]. For example, it was thought: that: (i) chilling and not infection caused the illness; (ii) nasal discharge (coryza) was caused by a secretion from the brain, which passed through perforations in the skull on its way to the nose; (iii) nasal discharge consisted of serum that could not be released through the skin because of the constriction of the pores during cold weather; and (iv) bacteria were the major cause of this malady

Viral and Other Infections of the Human Respiratory Tract. Edited by S. Myint and D. Taylor-Robinson. Published in 1996 by Chapman & Hall. ISBN 0 412 60070 6

because they could be recovered readily from the nose and throat during illness.

6.2 THE EARLY DAYS

Most of the folklore regarding the common cold was laid to rest at the beginning of this century when intranasal instillation of bacteria-free filtrates derived from nasal secretions from 'common cold' patients was shown to induce a similar illness in volunteers [1,7,10,11,43]. Volunteer studies with bacteria-free filtrates derived from nasal washings from individuals with common colds gained further momentum in the middle of this century and established the properties of the 'common cold virus' or viruses [1,7,11,43]. Nonetheless, in spite of the availability of known infectious filtrates, these fastidious non-bacterial agents defied cultivation reproducibly in any tissue culture system. Beginning in the 1950s and early 1960s, several distinct viruses were isolated in tissue culture from common cold patients, but because of the fastidious growth requirements of these agents, it continued to be difficult to isolate common cold viruses from clinical specimens [1,2,4,7,11,43]. However, with the introduction of semi-continuous human diploid fibroblast cell culture systems, notably the WI26 cell strain of human fetal lung, the ability to cultivate common cold viruses changed rapidly [20]. These cells were extremely sensitive and efficient for the growth of these viruses and proved to be the long-sought after substrate required for the rapid and efficient detection of the major known aetiological agents of common colds [16,17,27]. What had previously been an arduous task was now relatively simple. With the conditions for optimizing the propagation of common cold viruses established as well as a new sensitive cell culture system that was readily available, the number of strains described in the literature increased so rapidly that laboratories could not characterize a newly detected strain with regard to serotype by neutralization assay because antisera for many of these described strains were not generally available. The numbering of serotypes was disorganized, confusing, and not uniform as arbitrary numbers were assigned to isolates by various groups of investigators.

In addition, various names were being assigned to these common cold viruses by their discoverers such as coryzaviruses, enterovirus-like viruses, Echo 28-rhinovirus-coryzavirus (ERC) viruses, respiroviruses, muriviruses (mild upper respiratory illness viruses), rhinoviruses and Salisbury strains [30,44]. It is of interest that the first one to be isolated and characterized serotypically was classified as a new Echo virus with serotype 28 specificity [30,32]. However, in 1963, Tyrrell and Chanock at the request of the Virus Subcommittee of the International Committee on the Nomenclature of Bacteria and Viruses composed a paper to clarify the decision of this subcommittee that these common cold viruses be included in a subgroup of picornaviruses that was called *rhinoviruses* [29,44]. The rhinoviruses were classified as a subgroup of the picornaviruses because of certain properties including: (i) small size (15–30 nm); (ii) RNA core; (iii) ether resistance; and (iv) complete or almost complete inactivation at pH 3.0. It was this last property that distinguished the rhinoviruses from the enterovirus subgroup of picornaviruses [29,44].

In a collaborative effort, leadership was provided by the two World Health Organization (WHO) International Reference Centres for Respiratory Diseases other than Influenza, one in Bethesda, Maryland (R.M. Chanock, Director), at the Laboratory of Infectious Diseases (LID), National Institute of Allergy and Infectious Diseases (NIAID), National Institutes of Health (NIH), and the other in Salisbury, Wiltshire (D.A.J. Tyrrell, Director) at the Common Cold Research Unit (CCRU), Harvard Hospital. Inquiries were made to laboratories engaged in common cold virus studies regarding their interest in participating in a collaborative programme aimed at organizing the numerous candidate rhinovirus serotypes, which numbered almost 90, into a cohesive, scientifically-based system. It was envisaged that the actual number of serotypes could be elucidated by neutralization assay with the goal of adopting an internationally recognized numbering system. This was essential because the number of serotypes described was increasing rapidly and the ability to test each isolate against other described strains by neutralization in tissue culture was becoming less feasible. This chaotic situation made it difficult to interpret epidemiological data and also hindered epidemiological investigations. In preparation for the cross-testing of rhinovirus candidate strains, large lots of hyperimmune rhinovirus antisera were prepared in cows and goats. Antisera to 26 strains were prepared in cows by Abbott Laboratories, Inc. under contract to the Vaccine Development Branch (VDB) of NIAID, 26 strains were prepared in goats under the auspices of the LID and seven strains were prepared in goats at the CCRU.

To enable the execution of the collaborative study, the VDB of the NIAID awarded a contract to the Children's Hospital Research Foundation, Columbus, Ohio (V.V. Hamparian). The laboratory was to act as a Reference Laboratory with the responsibility of carrying out reciprocal neutralization tests with the candidate rhinoviruses and sera submitted to the programme. Thus, each candidate virus was to be tested by the submitting laboratory against every other virus submitted to the programme in reciprocal neutralization tests with confirmation by the Reference Laboratory [25].

A plan was devized to compare existing prototype strains. At a workshop held at the NIH on January 25, 1965, investigators from various laboratories submitted a list of candidate rhinoviruses which they considered to be distinct antigenically on the basis of neutralization tests carried out against all available rhinovirus antisera. It was required that any prototype strain submitted to the programme must: (a) be purified by three terminal dilutions in tissue culture or by three single plaque passages; (b) be acid labile and ether resistant; (c) have an RNA core; (d) be less than 50 nm in diameter; (e) be distinct by neutralization assay in tests against all other candidate antisera available at the time of submission; and (f) be of human origin as evidenced by a four-fold or greater increase in neutralizing antibody between acute and convalescent paired sera from at least one individual from whom the virus had been recovered or by isolation of the virus from two or more individuals [25]. A total of 68 viruses were included in this initial phase of the programme [25].

6.3 THE RHINOVIRUS NUMBERING SYSTEM

When all the data were amassed, a meeting was held in Bethesda, Maryland, on June 22, 1966 (A.Z. Kapikian, Chairman), with all of the collaborating laboratories with the exception of the CCRU in attendance [25]. Data from the latter were presented at the meeting by the chairman. Each laboratory or its representative presented the cross-neutralization data on the viruses each had submitted, and these results were compared with the findings of the Reference Laboratory. The collaborators and Reference Laboratory were in total agreement with regard to major cross-reactions. A candidate rhinovirus strain was considered to be a distinct serotype if it met two criteria: (i) at

least 20 times the limiting concentration of a specific antiserum raised to each of the other viruses submitted to the programme and which neutralized 32–320 TCID$_{50}$ of the homotypic strain (i.e. 20 antibody units) failed to neutralize 32–320 TCID$_{50}$ of the candidate virus; and (ii) at least 20 antibody units of serum to the candidate virus did not neutralize 32–320 TCID$_{50}$ of each of the other viruses submitted to the programme. The numbers were assigned chronologically with precedence given to the virus described earlier in the literature or if not yet published, according to the date of submission to the Reference Laboratory. Among the 68 viruses submitted to the first phase of the programme, six pairs of viruses and three groups of three viruses were identical by the criteria outlined above, thus decreasing the number of candidate rhinoviruses by 12. In addition, two strains were found to be related but not identical, resulting in the only subtype designation (1A and 1B). In this way, as shown in Table 6.1, a numbering system of 1 to 55 serotypes with one subtype was unanimously agreed upon at the Bethesda meeting [25].

Subsequently, two representatives from the Bethesda meeting (A.Z. Kapikian and V.V. Hamparian) had a follow-up meeting in Salisbury, England, at the CCRU on July 13, 1966, at which all of the data and the numbering system were presented to C.H. Andrewes, D.A.J. Tyrrell and P.C. Chapple. There was unanimous agreement among the Salisbury collaborators on the recommendations made at the Bethesda meeting [25]. The only modification was a suggestion by D.A.J. Tyrrell that for clarity the subtype designations of 1 and 1A adopted at the Bethesda meeting be changed to 1A and 1B. This was agreed. The numbering system was also presented at the meeting of the Directors of the WHO Respiratory and Enterovirus Reference Centres in Moscow, USSR (July 19–22, 1966), attended by D.A.J. Tyrrell and V.V. Kapikian where it was unanimously approved, was suggested for publica-

Table 6.1   Rhinovirus numbering system

Rhinovirus *number*	Prototype *strain*	*Other strains*	*Reference*
1A	Echo 28		[5,33,38]
1B	B632	K779*	[5,30,42]
2	HGP		[42]
3	FEB		[42]
4	16/60		[42]
5	Norman		[42]
6	Thompson		[42]
7	68-CV11	BU-109†	[15,16,37]
8	MRH-CV12		[16]
9	211-CV13	DC*	[2,5,16]
10	204-CV14		[16]
11	1-CV15		[16]
12	181-CV16		[16]
13	353	5007-CV23*	[21,22,27]
14	1059		[21,22]
15	1734		[21,22]
16	11757		[21,22]
17	33342	1376–64†	[15,21,22,41]
18	5986-CV17		[27]
19	6072-CV18		[27]
20	15-CV19	4462–63*	[24,27,41]
21	47-CV21		[27]
22	127-CV22	203F*; 1321–62†§	[15,18,19,27,40,41]
23	5124-CV24	100319†	[5,26,27]
24	5146-CV25	147H*	[18,19,27]
25	5426-CV26	K2218†; 55216†	[5,26,27]
26	5660-CV27	127–1*	[5,18,19,27]
27	5870-CV28		[27]
28	6101-CV29	113E†	[5,18,19,27]
29	5582-CV30	179E†	[5,18,19,27]
30	106F		[6,19]
31	140F		[6,19]
32	363		[45]
33	1200		[45]
34	137-3	6692-CV42*	[16,18,24]
35	164A		[5,18]
36	342H		[18]
37	151-1	1770-CV36*	[16,18]
38	CH79	201–3C*	[13,14,18,19]
39	209	00052*	[19,31]
40	1794	184E*	[18,19,31]
41	56110	137F*	[18,19,31]
42	56822	248A*	[19,31]
43	58750	E2No. 133†; WIS258E†; 04374*; 1936-CV43*	[5,9,17,18,24,31]
44	71560		[31]
45	Baylor 1-Tippett	037211†; E2No. 46†	[5,19,25,35]
46	Baylor 2-Crell	477-CV50*; CH202*	[5,13,17,24,35]
47	Baylor 3-Calvert	1979M-CV46*; CH310*	[5,13,17,24,35]
48	1505		[26]
49	8213		[26]
50	A2No. 58		[25]
51	F01-4081	19143†; 605-CV45*; 313G*	[5,17–19,24–26]

52	F01-3772	16413†; 515-CV34*	[5,17,24–26]
53	F01-3928	252B*; 464-CV53†	[15,16,19,25]
54	F01-3774	2253-CV49*‡	[15,17,25]
55	WIS315E	Baylor 4*	[8,25]
56	CH82	6660-CV38†; Baylor 5	[13,17,36]
57	CH47		[13,14]
58	21-CV20		[27]
59	611-CV35	1833-63†	[17,40,41]
60	2268-CV37		[17]
61	6669-CV39		[17]
62	1963M-CV40		[17]
63	6360-CV41		[17]
64	6258-CV44	1647-63†; Baylor 6†	[17,36,41]
65	425-CV47	143-3†; 4411-65†	[17–19,40]
66	1983-CV48		[17]
67	1857-CV51		[17]
68	F02-2317-Wood	SF23†	[12,39]
69	F02-2513-Mitchinson		[39]
70	F02-2547-Treganza		[39]
71	SF365		[12]
72	K2207	4704-62†; 410A†	[19,24,41]
73	107E		[17,19]
74	328A		[24]
75	328F		[19]
76	H00062	SF123	[12,19,24]
77	130-63		[41]
78	2030-65		[40]
79	101-1		[18,19]
80	277G		[24]
81	483F2		[24]
82	03647		[24]
83	Baylor 7	191-1†	[17,19,36]
84	432D		[19]
85	50-525-CV54		[28]
86	121564-Johnson		[24]
87	F02-3607-Corn		[24]
88	CVD01-0165-Dambrauskas		[24]
89	41467-Gallo		[24]
90	K2305		[15]
91	JMI	321B‡	[15,19]
92	SF-1662		[15]
93	SF-1492		[15]
94	SF-1803		[15]
95	SF-998		[15]
96	SF-1426		[15]
97	SF-1372		[15]
98	SF-4006		[15]
99	604		[15]
100	K6579		[15]

* Indicates any virus not included in a phase or not submitted to programme but found to be identical with the prototype strain by a collaborating laboratory, except as noted below.
† Indicates a virus submitted to programme and found to be identical to a prototype strain.
‡ The prototype strain is the prime strain.
§ This strain is the prime strain.
CV, coryzavirus.
Adapted from [5,15,24,25].

tion, and was recommended to be presented to the International Subcommittee on Virus Nomenclature for their information [25].

This collaborative programme was continued into a second phase (A.Z. Kapikian, Chairman) in which 73 rhinovirus strains were included, 23 were 'held over' from Phase I and 50 were potential new candidate strains that had been submitted. The ground rules for determining a serotype were the same as in Phase I. In this way, as shown in Table 6.1, 34 distinct new prototype strains were recognized and the numbering system was extended from 56 to 89 [24].

A third phase of the programme could not be completed because of lack of financial support. However, evaluations of 25 submitted or held-over strains were completed (V.V. Hamparian, Chairman), of which 11 were identified as distinct new prototype strains and, as shown in Table 6.1, the numbering system was extended from 90 to 100 [15].

This collaborative programme was a superb model of cooperation in the national and international scientific communities as it undertook a difficult task with the emergence of an agreeable, unified solution. Without this spirit of collaboration, the scientific community would have had to wrestle with a chaotic system of identifying rhinovirus serotypes. Not stated in the description above are specific examples of individual laboratories extending this spirit of collaboration by performing tasks on viruses they had not submitted to the programme in order to include these already described viruses in the programme [25]. For example: (a) coryzaviruses 11–18 and 28 were 'purified' by the LID, Bethesda, Maryland, and coryzaviruses 19, 21, 22, 24–27, 29 and 30 were purified by the CCRU, Salisbury, England; (b) antisera to coryzaviruses 11–18 and 28 were prepared by LID and antiserum for coryzavirus 24 was prepared by the CCRU, and antisera for coryzaviruses 19, 21, 22, 25–27, 29 and 30 were prepared by the California State Department of Public Health Laboratory; and

(c) most of the reciprocal neutralization assays for the submitted coryzaviruses were performed by the University of Chicago.

6.4 REFERENCES

1. Andrewes, C. (1965) *The common cold.* W.W. Norton & Co., New York, pp. 13–17.
2. Andrewes, C.H., Chaproniere, D.M., Gompels, A.E.H., Pereira, H.G. and Roden, A.T. (1953) Propagation of common-cold virus in tissue cultures. *Lancet*, ii, 546–7.
3. Cherry, J.D. (1992) The common cold, in *Textbook of Pediatric Infectious Diseases*, 3rd edn, (eds R.D. Feigin and J.D. Cherry), W.B. Saunders Co., Philadelphia, pp. 137–42.
4. Conant, R.M. and Hampanan, VV. (1968) Rhinoviruses: basis for a numbering system. I. Hela cells for propagation and serologic procedures. *J. Immunol.*, **100**, 107–13.
5. Conant, R.M. and Hamparian, V.V. (1968) Rhinoviruses: basis for a numbering system. II. Serologic characterization of prototype strains. *J. Immunol.*, **100**, 114–19.
6. Connelly, A.P. Jr and Hamre, D. (1964) Virologic studies of acute respiratory disease in young adults. II. Characteristics and serologic studies of three new rhinoviruses. *J. Lab. Clin. Med.*, **63**, 30–43.
7. Couch, R.B. (1990) Rhinoviruses, in *Virology*, 2nd edn, (eds B.N. Fields, D.M. Knipe *et al.*), Raven Press., New York, pp. 607–29.
8. Dick, E.C., Blumer, C.R. and Evans, A.S. (1967) Epidemiology of infections with rhinovirus types 43 and 55 in a group of University of Wisconsin student families. *Am. J. Epidemiol.*, **86**, 386–400.
9. Dick, E.C. and Inhom, S.L. (1992) Rhinoviruses, in *Textbook of Pediatric Infectious Diseases*, 3rd edn, (eds R.D. Feigin and J.D. Cherry), W.B. Saunders Co., Philadelphia, pp. 1507–32.
10. Douglas, R.G. Jr, Lindgren, K.M. and Couch, R.B. (1968) Exposure to cold environment and rhinovirus common cold. Failure to demonstrate effect. *N. Engl. J. Med.*, **279**, 742–7.
11. Gwaltney, J.M. Jr (1989) Rhinoviruses, in *Viral Infections of Humans: Epidemiology and Control*, 3rd edn, (ed. A.S. Evans), Plenum Medical Books Co., New York, pp. 593–615.
12. Gwaltney, J.M. Jr, Hendley, J.Q., Simon, G. and Jordan, W.S. Jr (1968) Rhinovirus infections in

an industrial population. III. Number and prevalence of serotypes. *Am. J. Epidemiol.*, **87**, 158–66.

13. Gwaltney, J.M. Jr and Jordan, W.S. Jr (1964) Rhinoviruses and respiratory disease. *Bacteriol. Rev.*, **28**, 409–22.

14. Gwaltney, J.M. Jr and Jordan, W.S. Jr (1966) Rhinoviruses and respiratory illnesses in university students. *Am. Rev. Resp. Dis.*, **93**, 362–71.

15. Hamparian, V.V. (Chairman), Colonno, R.J., Cooney, M.K. *et al.* (1987) Rhinoviruses – extension of the numbering system from 89 to 100. *Virology*, **159**, 191–2.

16. Hamparian, V.V., Ketler, A. and Hilleman, M.R. (1961) Recovery of new viruses (coryzavirus) from cases of common cold in human adults. *Proc. Soc. Exp. Biol. Med.*, **108**, 444–53.

17. Hamparian, V.V., Leagus, M.B. and Hilleman, M.R. (1964) Additional rhinovirus serotypes. *Proc. Soc. Exp. Biol. Med.*, **116**, 976–84.

18. Hamre, D., Connelly, A.P. Jr and Procknow, J.J. (1964) Virologic studies of acute respiratory disease in young adults. III. Some biologic and serologic characteristics of seventeen rhinovirus serotypes isolated October, 1960, to June, 1961. *J. Lab. Clin. Med.*, **64**, 450–60.

19. Hamre, D., Connelly, A.P. Jr and Procknow, J.J. (1966) Virologic studies of acute respiratory disease in young adults. IV. Virus isolations during four years of surveillance. *Am. J. Epidemiol.*, **83**, 238–49.

20. Hayflick, L. and Moorehead, P.S. (1961) The serial cultivation of human diploid cell strains. *Exp. Cell. Res.*, **25**, 585–621.

21. Johnson, K.M., Bloom, H.H., Chanock, R.M., Mufson, M.A. and Knight, Y. (1962) VI. The newer enteroviruses. *Am. J. Public Health*, **52**, 933–40.

22. Johnson, K.M. and Rosen, L. (1963) Characteristics of five newly recognized enteroviruses recovered from the human oropharynx. *Am. J. Hyg.*, **77**, 15–25.

23. Kapikian, A.Z. (1992) The common cold, in *Cecil Textbook of Medicine*, 19th edn, (eds J.M. Wyngaarden, L.H. Smith and J.C. Bennett), W.B. Saunders Co., Philadelphia, pp. 1806–10.

24. Kapikian, A.Z. (Chairman), Conant, R.M., Hamparian, V.V. *et al.* (1971) Collaborative report: Rhinoviruses – extension of the numbering system. *Virology*, **43**, 524–6.

25. Kapikian, A.Z. (Chairman), Conant, R.M., Hamparian, V.V. *et al.* (1967) Rhinoviruses, a numbering system. *Nature*, **213**, 761–3.

26. Kapikian, A.Z., Mufson, M.A., James, H.D. Jr, Kalica, A.R., Bloom, H.H. and Chanock, R.M. (1966) Characterisation of two newly recognized rhinovirus serotypes of human origin. *Proc. Soc. Exp. Biol. Med.*, **122**, 1155–62.

27. Ketler, A., Hamparian, V.V. and Hilleman, M.R. (1962) Characterization and classification of ECHO 28 rhinovirus-coryzavirus agents. *Proc. Soc. Exp. Biol. Med.*, **110**, 821–31.

28. Mascoli, C.C., Leagus, M.B., Hilleman, M.R., Weibel, R.E. and Stokes, J. Jr (1967) Rhinovirus infection in nursery and kindergarten children. New rhinovirus serotype 54. *Proc. Soc. Exp. Biol. Med.*, **124**, 845–50.

29. Melnick, J.L., Cockburn, W.C., Dalldorf, G. *et al.* (1963) Picornavirus group. *Virology*, **19**, 114–16.

30. Mogabgab, W.J. (1962) Additional respirovirus type related to GL2060 (ECHO28) virus, from military personnel, 1959. *Am. J. Hyg.*, **76**, 160–72.

31. Mufson, M.A., Kawana, R., James, H.D., Gauld, L.W., Bloom, H.H. and Chanock, R.M. (1965) A description of six new rhinoviruses of human origin. *Am. J. Epidemiol.*, **81**, 32–43.

32. Pelon, W. (1961) Classification of the '2060' virus as ECHO 28 and further study of its properties. *Am. J. Hyg.*, **73**, 36–54.

33. Pelon, W., Mogabgab, W.J., Phillips, I.A. and Pierce, W.E. (1956) Cytopathic agent isolated from recruits with mild respiratory illnesses. *Bacteriol. Proc.* (Abstract), 67.

34. Pelon, W., Mogabgab, W.J., Phillips, I.A. and Pierce, W.E. (1957) A cytopathic agent isolated from naval recruits with mild respiratory illness. *Proc. Soc. Exp. Biol. Med.*, **94**, 262–7.

35. Phillips, C.A., Melnick, J.L. and Grim, C.A. (1965) Characterization of three new rhinovirus serotypes. *Proc. Soc. Exp. Biol. Med.*, **119**, 798–801.

36. Phillips, C.A., Melnick, J.L. and Grim, C.A. (1986) Rhinovirus infection in a student population. Isolation of five new serotypes. *Am. J. Epidemiol.*, **87**, 447–56.

37. Phillips, C.A., Melnick, J.L. and Sullivan, L. (1970) Characterization of four new rhinovirus serotypes. *Proc. Soc. Exp. Biol. Med.*, **134**, 933–5.

38. Price, W.H. (1956) The isolation of a new virus associated with respiratory clinical disease in humans. *Proc. Natl Acad. Sci. USA*, **42**, 892–6.

39. Schieble, J.H., Lennette, E.H. and Fox, V.L. (1968) Rhinoviruses: the isolation and characterization of three new serologic types. *Proc. Soc. Exp. Biol. Med.*, **127**, 324–8.

40. Stott, E.J., Eadie, M.B. and Grist, N.R. (1969) Rhinovirus infections of children in hospital; isolation of three possibly new rhinovirus serotypes. *Am. J. Epidemiol.*, **90**, 45–52.

41. Stott, E.J., Grist, N.R. and Eadie, M.B. (1968). Rhinovirus infections in chronic bronchitis: isolation of eight possible new rhinovirus serotypes. *J. Med. Microbiol.*, **1**, 109–18.

42. Taylor-Robinson, D. and Tyrrell, D.A.J. (1962) Serotypes of viruses (rhinoviruses) isolated from common colds. *Lancet*, **i**, 452–4.

43. Tyrrell, D.A.J. (1965) *Common Colds and Related Diseases*. WIlliams and Wilkins Co., pp. 1–7.

44. Tyrrell, D.A.J. and Chanock, R.M. (1963) A description of rhinoviruses. *Science*, **141**, 152–3.

45. Webb, P.A., Johnson, K.M. and Mufson, M.A. (1964) A description of two newly recognized rhinoviruses of human origin. *Proc. Soc. Exp. Biol. Med.*, **116**, 845–52.

Yuri Pervikov, Harry Campbell and Malcolm Molyneaux

7.1 INTRODUCTION

Viruses are important in the aetiology of both diarrhoeal disease and respiratory disease, both major causes of morbidity and mortality in the tropics. The viruses responsible for most of the respiratory mortality are measles, influenza, parainfluenza and respiratory syncytial (RS) virus. Children suffer the greatest toll from these infections, but persons of all ages, and especially an enlarging population of elderly people, are also at considerable risk.

Viral and Other Infections of the Human Respiratory Tract.
Edited by S. Myint and D. Taylor-Robinson. Published in 1996 by Chapman & Hall. ISBN 0 412 60070 6

Both the incidence and the case-fatality of pneumonia are greater among children in the tropics than among children in industrialized countries. In the first section of this chapter, we outline the clinical problem of childhood respiratory infections in the tropics. There are many potential avenues for reducing both the incidence and the case-fatality of these diseases. Better birth-weights, better nutrition, breast-feeding, vitamin A supplements, reduced indoor air pollution and malaria control may all help. Improved diagnosis and case management of pneumonia – by people and methods available close to where the patient lives – undoubtedly improve survival. There is a great need for improvement in preventive vaccination against both viral and bacterial respiratory diseases. We need new vaccines – for example against RS virus and parainfluenza viruses – and, of equal importance, we need to improve the delivery of existing vaccines – especially measles vaccine – to those who need them.

We still do not know how commonly viral infections precede or augment bacterial infections, in particular lobar pneumonia, which carry a high mortality in the very young and the very old. We need more information on the extent to which viruses contribute -either as main pathogens or as predisposing factors – to severe and fatal respiratory infections in the tropics. In the second part of this chapter we discuss the facilities available for identifying viruses causing respiratory disease in various parts of the globe.

As the HIV epidemic continues its catastrophic advance in tropical countries, respiratory complications multiply. Mycobacterial and other bacterial agents are the major opportunists, but viruses may pave the way for these in immunocompromised individuals. The recognition and prevention of virus infections may therefore be of increasing importance in the immediate future.

The third part of this chapter outlines the global surveillance programmes employed by the World Health Organization to monitor respiratory infections.

7.2 CLINICAL ASPECTS IN CHILDREN

Acute respiratory infections (ARI) are the most common cause of death in young children in developing countries [12]. According to 1990 official WHO mortality estimates, 4.3 million deaths in young children under the age of 5 years were associated with ARI; two-thirds of these deaths were in infants. Almost all of these deaths are due to acute lower respiratory infections and in particular pneumonia [22,30,40,41].

7.2.1 ACUTE UPPER RESPIRATORY INFECTIONS

Epidemiology

Acute upper respiratory infections (AURI) result in few deaths in children but cause considerable morbidity. The mean incidence of AURI in young children is 3–5 episodes per year in rural areas and 5–7 episodes per year in urban areas.

The high incidence of AURI is reflected in the fact that attendances associated with ARI are one of the three most common causes of out-patient attendances in children, accounting typically for about one-third of all such consultations.

As in developed countries most AURI are viral in origin and the specific viral agents involved are remarkably similar throughout the world. However, there are two forms of AURI which are associated predominantly with bacterial infections and which cause considerable disability. Otitis media is the leading preventable cause of deafness in developing countries and is a significant contributor to developmental and learning problems in children. In addition, acute rheumatic fever may follow streptococcal pharyngitis.

Common cold

Most episodes of AURI present as the common cold and are caused by respiratory viruses. The viral aetiological agents involved are very similar to those found in developed countries and these viral infections are in general mild, self-limited illnesses. These infections show a typical epidemic pattern and there is often a seasonal pattern, with peaks of incidence in cold or rainy seasons [30].

The management of the common cold in young children in developing countries is essentially the same as in developed countries. Controlled trials performed in developing countries, including communities in which there is a significant prevalence of malnutrition, have shown no benefit from antibiotic therapy: this does not shorten the illness episode, prevent complications such as pneumonia or otitis media, nor provide any symptomatic relief [11]. Appropriate supportive measures include giving extra oral fluids, continuing breast-feeding, encouraging the child to eat nutritious food, and keeping it warm (but not overheated). Over-prescription of antibiotics in the treatment of the common cold is very widespread worldwide. This practice is expensive and encourages drug resistance in respiratory bacteria such as *Streptococcus pneumoniae* and *Haemophilus influenzae* which are carried in the nasopharynx of the great majority of young children in developing countries.

Paracetamol can be given when the child's axillary temperature exceeds 38°C. A blocked nose should be gently cleaned out using saline drops if necessary to loosen mucus, and a cough soothed by use of a simple linctus. Adequate oral hydration should loosen respiratory secretions and act as an expectorant.

'Cough and cold' remedies which contain atropine, codeine, alcohol, phenergan, mucolytics or high-dose antihistamines should not be given to young children. Infants in the first few months of life should be exclusively breast-fed and should not receive any form of cough or cold remedy.

Medicated nosedrops or sprays containing sympathomimetic agents such as ephedrine cause rebound congestion ('rhinitis medicamentosa') which may cause difficulty of breathing in infants. Those containing antihistamines are a mainstay of treatment of allergic rhinitis but should not be used for treatment of the common cold as they cause sedation and may make feeding difficult. Mentholated balms or balms containing camphor should not be ingested or applied inside the nose. Camphor is highly toxic if ingested, with a lethal dose of 50–500 /kg.

Topical antiseptics and anaesthetics may cause local sensitization and, in lozenge form, can result in choking. They should not be used. Proprietary drug combinations containing mixtures of antihistamines, mucolytics and expectorants are often found in irrational or counter-productive combinations (for example an expectorant together with a cough suppressant) and should not be prescribed [15,49].

An important general consideration is the very poor evidence for the effectiveness of these agents in the treatment of the common cold in young children in relation to the considerable cost expended by families and health services in their purchase. Local safe home remedies should be encouraged and resources directed to the relatively few children with a lower respiratory infection who require antibiotic treatment or oxygen therapy.

Croup (acute laryngotracheobronchitis)

In developed countries stridor due to ARI is usually caused by parainfluenza viruses, influenza virus or RS virus. In developing countries this is less often the case with the most important causes being measles, diphtheria (in some countries) and bacterial agents such as *Staphylococcus aureus* (bacterial tracheitis) and *H. influenzae* (epiglottitis). When severe stridor occurs in a calm child this is a sign that the child requires hospital admission. Children with severe croup should, where possible, be transferred to a unit with the facilities and expertise to insert an artificial airway should this be required and these children should be closely monitored. Some paediatricians recommend the use of nebulized 1:1000 adrenaline mixed with 1 ml saline every 2 hours under close supervision. Chloramphenicol treatment is indicated and the child should be closely supervised in case their condition deteriorates and a tracheostomy is required. Cooled steam, cough suppressants and mucolytics are not indicated.

7.2.2 ACUTE LOWER RESPIRATORY INFECTIONS (ALRI)

Pneumonia and bronchiolitis

Viral infections have been identified in 30–40% of cases when both cell culture and rapid diagnostic technologies have been applied to nasopharyngeal aspirates in young children with ALRI. The distribution of viral respiratory agents is approximately: RS virus 15–40%; parainfluenza viruses 7–10%; influenza A and B viruses 5%; and adenovirus 2–4%. In almost all studies RS virus is the predominant viral agent. A clear seasonal pattern is usually found, although there is no consistent pattern with respect to climatic factors in tropical areas. In temperate climates there is usually increased activity in the cold season [1,3,9,10,16,18,26,39,42].

Mixed viral and bacterial ALRI have been found more commonly than in developed countries. In Pakistan, 26% of children infected with RS virus also had *S. pneumoniae* or *H. influenzae* bacteraemia, and 54% of the cases of *H. influenzae* and 47% of those with *S. pneumoniae* bacteraemia were associated with a respiratory viral infection [13].

The most important clinical syndromes of ALRI are pneumonia and bronchiolitis. These can be recognized by signs of difficulty with breathing in young children, with the principal signs being fast breathing and lower chest wall indrawing. The current WHO 'standard case management' strategy is directed at identifying these children from among all those who present with cough or difficult breathing. Full details of this approach are given in publications of the WHO Programme for the Control of Respiratory Infections (see later). The basic principles of standard case management can be summarized as follows: young children with certain danger signs such as inability to drink, and those with chest indrawing are classified as 'very severe disease' and 'severe pneumonia' respectively and should be referred for antibiotic treatment in hospital; children with fast breathing but no chest indrawing are classified as 'pneumonia' and should be treated with antibiotic therapy at home; and those with neither chest indrawing nor fast breathing are classified as 'no pneumonia/cough or cold' and should be given supportive care at home. Full details of this classification system are published by WHO [47].

This classification recognizes that a chest X-ray is expensive, unavailable to most young children presenting with cough or difficult breathing, and in any case, rarely influences treatment in uncomplicated pneumonia. In addition, this classification is based on the fact that auscultatory signs are not very reliable in children. The validity of the simple clinical signs of fast breathing and chest indrawing, in contrast, have been shown in studies in a number of developing countries to have a sensitivity and specificity of approximately 60–90% for the diagnosis of pneumonia [4,6,24,37]. Finally, the standard case management approach recognizes that an aetiological diagnosis of pneumonia is very difficult to establish in young children and so treatment decisions must be based on clinical criteria.

In developing countries, and especially those with high infant mortality rates, approximately 50% of pneumonia cases in young children attending hospital may be associated with bacterial infection and in particular with *S. pneumoniae* and *H. influenzae* infection [47]. However, as noted above, many of these may represent mixed respiratory viral/bacterial infections with the most likely explanation being that a primary viral infection predisposed to the secondary bacterial infection [8]. Empirical treatment with an antibiotic is therefore justified in all children presenting with signs of difficulty in breathing, as defined above. Recommended first-line antibiotics for the home treatment of pneumonia are co-trimoxazole, amoxycillin and procaine penicillin. Paracetamol is indicated if the child's axillary temperature exceeds 38°C. Appropriate supportive care includes adequate fluid intake, a simple linctus or safe home remedy to soothe the cough, clearing a blocked nose if it interferes with feeding and dietary advice with encouragement to continue breast-feeding and to provide extra feeding once the child's appetite returns. It is particularly important that the child's mother or guardian be informed of what signs to look for which would indicate that the child is not improving and should be brought back to the health worker.

The recommended antibiotic for the hospital treatment of severe pneumonia in children aged 2 months up to 5 years is benzylpenicillin [34]. If the pneumonia is very severe, chloramphenicol is indicated [38]. In the first 2 months of life benzylpenicillin should be used together with gentamicin [45]. These antibiotics should be given intramuscularly or by intravenous bolus via an indwelling catheter. Intravenous fluids should not be given due to the risk of fluid overload associated with inappropriate secretion of antidiuretic hormone (ISADH) which can be associated with pneumonia.

Oxygen is indicated if the young child has any of the following clinical signs: central cyanosis, inability to drink, severe chest indrawing, restlessness (if it improves with oxygen therapy), grunting in young infants or very fast breathing [23,27,35,46]. Robust oxygen concentrators are now available for use (through the UNICEF UNIPAC supply system) in developing countries if a minimum maintenance and repair capability is present and electricity is available. Low flow oxygen at 0.5–1.0 litres/min can be efficiently delivered by means of a catheter in the nasal cavity or nasopharynx [33,36] or by the use of nasal prongs. These result in acceptable inspired oxygen concentrations ranging from approximately 30–35% with nasal prongs to 45–60% with a nasopharyngeal catheter. Full details are given in a review of oxygen therapy prepared by WHO [50].

If there is associated wheezing an appropriate bronchodilator such as oral or nebulized salbutamol or subcutaneous adrenaline (epinephrine) should be used. Adequate oral fluids (breast milk, clean water and other low-salt fluids) should be given to replace the increased fluid loss due to fever and fast breathing. Recommendations for the fluid management of children presenting with shock or with associated diarrhoea are given in detail in WHO documents [46]. The child should be nursed, lightly clothed, in a warm room (25°C). This is particularly important for neonates. Attention should be given to maintaining an adequate airway by clearing secretions and continued breast-feeding and small frequent meals encouraged. Humidification of the air has not been shown to be beneficial.

It should be stressed, once again that this treatment is appropriate for all young children who have signs of difficult breathing which define pneumonia in the WHO classification. It is recognized that this will include children with bronchiolitis and those with wheeze due to asthma and perhaps triggered by a viral ARI. However, since bronchiolitis may often be complicated by secondary bacterial pneumonia, and since wheeze has been found to be associated with bacterial pneumonia as well as viral ARI in a developing country, empirical antibiotic treatment is justified.

7.2.3 MEASLES

A recent review of community-based studies of ARI mortality reported that a mean of 18% of ARI deaths was associated with measles [12]. The most important complications are pneumonia, stridor, otitis media, severe conjunctivitis, severe stomatitis and enteritis. Important causes of death from measles are pneumonia, stridor and diarrhoea. In addition, the conjunctivitis can lead to corneal ulceration and is an important cause of blindness. Measles often leads to malnutrition and the increased mortality rate following an attack of measles can persist for 2 years.

The child should be admitted to hospital if the rash is haemorrhagic, stridor is present, there is difficulty in eating or drinking, or if pneumonia, dehydration or severe malnutrition is found. An antibiotic is only recommended if pneumonia or otitis media are present. Pneumonia associated with measles should be treated with amoxycillin or benzylpenicillin or, if staphylococcal pneumonia is suspected, cloxacillin plus gentamicin. In areas with known vitamin A deficiency, vitamin A should be given once orally in a dose of 1000 units for infants and 2000 units for children 12 months up to 5 years of age [46]. Careful skin and eye care should be given together with an emphasis on maintaining nutrition. If the child's eye is discharging pus, tetracycline eye ointment should be used. The child's mouth should be cleaned with a clean water and salt solution with 1% gentian violet applied to mouth sores. Herpes stomatitis is common but acyclovir is too expensive to be considered for wide use in developing countries. Severe stomatitis may interfere with breast-feeding and a cup and spoon may be required to deliver expressed breast milk. The management of fever and stridor and supportive care are as described above. It is important to follow up these children after the acute episode with particular attention to their nutritional status. Dehydration associated with diarrhoea should be treated appropriately with oral or intravenous fluids according to severity and following WHO recommendations.

7.2.4 CONCLUSION

The key to reducing mortality from ARI in young children in developing countries is to improve coverage of measles, pertussis and diphtheria vaccines and to ensure better access to and timely use of correct case management of pneumonia. This requires the strengthening of health services to enable them to provide early treatment with antibiotics based on clinical signs that are easily detectable. Standard case management strategies issued by WHO are given in Tables 7.1 and 7.2. Intervention studies which have evaluated the ARI case management strategy have been carried out in India, Nepal, Phillipines, Tanzania, Pakistan and Bangladesh and have reported reductions in ARI mortality of between 25% and 67% [2,7,17,21,28,47]. This impact on mortality was also found in high-risk groups such as low birth-weight infants, in areas with a high prevalence of malnutrition, and in settings in which case management relied almost entirely on community health workers and home treatment because referral to hospital was very problematic.

7.3 LABORATORY ASPECTS

Rapid and accurate laboratory diagnosis of infectious diseases is essential both for the

immediate care of the patients and for the introduction of necessary public health control measures. Rapidity and simplicity of methods make it easier to carry out surveys of infectious diseases, and the information obtained may lead to the implementation of individual or public health action. Early diagnosis of viral respiratory infections will help to prevent hospital cross-infection and its spread to contacts, especially in cases presenting with atypical symptoms. The WHO Scientific Group defined rapid viral diagnosis as a method that would provide acceptable results in a short time and allow successful intervention in the treatment of patients and their contacts, or in the control of disease in communities [43].

Some of the classical techniques currently available for diagnostic purposes (see Chapter 4), though specific and sensitive, are tedious and difficult to carry out. In view of this they cannot be used on a daily basis in laboratories in developing countries, where budget and technical facilities are limited. The availability of simple and inexpensive tests is a crucial factor for involving laboratories in the tropics in surveillance of respiratory viruses. In recent years several simple and rapid methods for diagnosis of viral infections have been devised which might be assessed as appropriate tools for laboratories all over the world. They include the conventional techniques for virus isolation, antigen detection and virus serology.

7.3.1 CELL CULTURE

Isolation of viruses in cell culture is commonly considered as a standard and essential procedure in full-service diagnostic laboratories. This is consequent on the premise that culture is only a dependable system where many viruses, including rhinoviruses, can be detected [20]. Viruses isolated in cell culture should be submitted for further more detailed analysis, including subgroup antigenic or genetic differentiation. This method is also the reference for evaluation of the specificity and sensitivity of all newer techniques. The greatest disadvantage of the isolation technique is that it is applicable only in laboratories with cell culture facilities. Consequently, it is unacceptable for many local laboratories. Moreover, with the traditional version of virus

Table 7.1 Pneumonia management at the small hospital. For the child age 2 months up to 5 years with cough or difficult breathing (who does not have stridor, severe undernutrition, or signs suggesting meningitis)*

Clinical signs	Classify as:[†]	Summary of treatment instructions
● Central cyanosis or ● Not able to drink	VERY SEVERE PNEUMONIA	ADMIT Give oxygen. Give an antibiotic: chloramphenicol. Treat fever, if present. Treat wheezing, if present. Give supportive care. Reassess twice daily.
● Chest indrawing and ● No central cyanosis and ● Able to drink	SEVERE PNEUMONIA If child is wheezing, assess further before classifying	ADMIT[‡] Give an antibiotic: benzylpenicillin. Treat fever, if present. Treat wheezing, if present. Give supportive care. Reassess daily.
● No chest indrawing and ● Fast breathing[§]	PNEUMONIA	ADVISE MOTHER TO GIVE HOME CARE. Give an antibiotic (at home): cotrimoxazole, amoxycillin, ampicillin or procaine penicillin. Treat fever, if present. Treat wheezing, if present. Advise the mother to return in 2 days for reassessment, or earlier if the child is getting worse.
● No chest indrawing and ● No fast breathing	NO PNEUMONIA: COUGH OR COLD	If coughing more than 30 days, assess for causes of chronic cough. ADVISE MOTHER TO GIVE HOME CARE. Assess and treat ear problem or sore throat, if present. Assess and treat other problems. Treat fever, if present. Treat wheezing, if present.

* If the child has stridor, follow the treatment guidelines outlined in section 3.4.
If the child has severe undernutrition, admit for nutritional rehabilitation and medical therapy (see Annex 3). Treat pneumonia with chloramphenicol (see section 3.1).
If the child has signs suggesting meningitis, admit and treat with chloramphenicol (see pages 7–8).
† These classifications include some children with bronchiolitis and asthma – see section 3.3.
‡ If oxygen supply is ample, also give oxygen to a child with:
● restlessness (if oxygen improves the condition),
● severe chest indrawing, or
● breathing rate of 70 breaths per minute or more
§ Fast breathing is: 50 breaths per minute or more in a child aged 2 months up to 12 months;
                    40 breaths per minute or more in a child aged 12 months up to 5 years.

Table 7.2 Management of the young infant with cough or difficult breathing at the small hospital. For the young infant of age less than 2 months

Clinical signs	Classify as:	Summary of treatment instructions
• Stopped feeding well, • Convulsions, • Abnormally sleepy difficult to wake, • Stridor in calm child, • Wheezing, • Fever (38°C or more) or low body temperature (below 35.5°C), • Fast breathing* • Severe chest indrawing, • Central cyanosis, Grunting, • Apnoeic episodes, or • Distended and tense abdomen	SEVERE PNEUMONIA OR VERY SEVERE DISEASE	ADMIT Give oxygen† If: • central cyanosis, • not able to drink. Give antibiotics: benzylpenicillin and gentamicin. Careful fluid management. Maintain a good thermal environment. Specific management of wheezing or stridor.
• No fast breathing, and • No signs of pneumonia or very severe disease.	NO PNEUMONIA: COUGH OR COLD	ADVISE MOTHER TO GIVE THE FOLLOWING HOME CARE: Keep young infant warm. Breast-feed frequently. Clear nose if it interferes with feeding. Return quickly if: Breathing becomes difficult. Breathing becomes fast. Feeding becomes a problem. The young infant becomes sicker

* Fast breathing is 60 breaths per minute or more in the young infant (age less than 2 months); repeat the count.
† If oxygen supply is ample; also give oxygen to a young infant with:
• restlessness (if oxygen improves the condition),
• severe chest indrawing, or
• grunting.

isolation, a virus may be identified only after several days of incubation of specimens in cells. For instance, detection of adenoviruses, parainfluenza viruses and CMV usually takes 1–4 weeks. This means it is not of great value in the rapid detection of viral respiratory infections. However, recent advances in culture methodology provide a basis for the development of simpler and rapid tests. The main improvements are the brief centrifugation of cell and specimens to enhance sensitivity, a short incubation for viral multiplication and antigen development, followed by staining with monoclonal antibodies. For instance, preliminary enhancement of inoculation of shell-vials by centrifugation and the use of mixtures of monoclonal antibodies for staining allows the identification of RS virus, influenza viruses A and B, parainfluenza virus types 1, 2 and 3, adenovirus and CMV in cell cultures after 1 day of incubation [32].

7.3.2 IMMUNOFLUORESCENCE

Of all the methods currently available for rapid laboratory confirmation of viral respiratory infections, the immunofluorescence (IF) test is one of the best and most used. It can be used in local laboratories although, of course, a fluorescence microscope is required. Construction of fluorescence objectives which can convert almost any compound microscope into a fluorescence microscope provides a good approach for the introduction of IF technology to the diagnostic work of local laboratories. Virus antigens in pharyngeal aspirates can be detected in 30 minutes to 2 hours. The reagents are often made commercially and are quality controlled.

Training of laboratory staff is needed. Experience shows that a course of 2–3 weeks may be sufficient for beginners to get familiar with the technique. A further period may make available positive and negative control specimens to provide enough local experience to enable a service to be set up. The test itself does not present many technical difficulties. However, the selection and preparation of specimens for the test are of great importance in obtaining reliable results. Nasopharyngeal secretions offer the best material, and the proper collection of these specimens should yield intact cells containing the viral antigen. Immunofluorescence is preferable to alternative methods because it enables feedback on specimen quality. However, the skill required to recognize genuine viral fluorescence requires experience as well as basic training. The acquisition of this skill is easier and faster if the fluorescence results are validated using other techniques, such as cell culture. In some laboratories IF has been performed with hyperimmune antiviral sera and this technical approach caused excessive amounts of non-specific fluorescence. In view of this, a very careful selection of appropriate hyperimmune serum should be made before testing. Considerable increase in test specificity was obtained when monoclonal antibodies were introduced instead of hyper-immune antisera. As a general rule, a slight loss in sensitivity can be expected with monoclonal pools compared with good polyclonal antisera, but this disad-

vantage is offset by the ease of reading preparations stained with the monoclonal antibodies. Commercially produced kits are now available containing monoclonal antibodies for identification of viruses. They provide sensitivity and specificity at a level of 85–95% compared with those provided by virus isolation in culture [25].

7.3.3 ENZYME IMMUNOASSAY

Similarly to the IF test, enzyme immunoassays (EIAs) are techniques used routinely for the detection of minute amounts of antigens or antibodies. The technology has the following advantages: large numbers of samples may be examined at the same time; reagents are relatively stable; specimens can be easily transported because no intact cells or intact viral particles are required for performance of the test and soluble viral proteins are quite stable. However, the classical procedure of EIA is relatively intensive and time consuming and simplified techniques should be recommended to laboratories in developing countries. Currently, the following simplifications are already available and have been explored in some laboratories: a naked-eye appreciation of results may be made by the use of coloured standards; the different incubation times for immune-complex formation and enzyme activity can be shortened without dramatically decreasing sensitivity by slightly increasing the concentration of the enzyme–conjugate; and the use of nitrocellulose paper in a spot test format.

A study carried out in China has shown that the use of monoclonal antibodies against RS virus antigens in an EIA considerably increases the specificity and sensitivity (82.9% and 94% respectively) when compared with the same method without using monoclonal antibodies (76.9% and 66.7%) [52]. The tests were being evaluated for laboratory confirmation of viral pneumonia in infants and children.

7.3.4 SEROLOGY

Serological methods are an important laboratory tool for diagnosis of respiratory virus infection, because antigen detection and isolation of viruses in culture are only possible during the acute phase of illness and because no single technique can provide 100% sensitivity. Several techniques have been developed for the demonstration of antibodies against respiratory viruses in blood of patients. They can be used for the diagnosis of viral infections if paired serum samples are available. They can also be used to assess the immune status of individuals and for epidemiological purposes. The most commonly used formats are virus neutralization (VN), complement fixation (CF) and enzyme immunoassay. The latter has been developed as full commercial kits but in addition, for the other assays, components can also be obtained from commercial sources. In general, VN and CF tests are relatively time consuming and labour intensive, the former requiring tissue culture facilities and the latter a source of complement. In spite of these disadvantages, they are operational in many local laboratories, and have proven to be useful in epidemiological surveys and vaccine efficacy assessment studies.

Immunoassays for the detection of virus specific antibodies which are associated with ongoing or recent infections are based on the detection of IgM. Such techniques include the IgM-capture ELISA and the IgM immunofluorescence tests. Specificity and sensitivity of these tests is around 90–100% in comparison with the neutralization test [51]. Reliable results depend on the quality of the antigens used, and in this respect the inclusion of synthetic antigens in tests looks very promising. The rapid development of DNA cloning and sequencing technology which, in turn, facilitate protein sequencing, have made the synthesis of antigens, either chemical or by bioengineering, more feasible and practical. In the light of the high thermostability of these antigens it seems likely that they will be especially useful in hot-climate countries.

7.4 THE WHO PROJECT ON GLOBAL SURVEILLANCE OF RESPIRATORY VIRUSES

The main objectives of this project are:

1. To transfer rapid and simple techniques for the diagnosis of respiratory viruses to laboratories in developing countries.
2. To develop a network of laboratories for the rapid diagnosis and epidemiological surveillance of respiratory virus infection.
3. To obtain data on the circulation of respiratory viruses in developing countries and to evaluate the importance of viruses in the aetiology of acute respiratory infections.

For a long while the WHO has been distributing hyperimmune sera for the detection of respiratory viruses in material from patients. In the early 1980s a limited network of laboratories was developed with facilities for the detection of viruses in infected cell culture by an IF test. The first attempts to expand the network were not very successful because the number of laboratories in developing countries with cell culture facilities was scanty. In order to meet requests from countries for a simple and rapid technique to diagnose respiratory virus infections, WHO initiated in 1986 a project to develop a monoclonal antibody diagnostic kit to detect antigens directly in clinical specimens. To promote this aim, WHO began collecting suitable candidate monoclonal antibodies from different collaborating institutions. During 1987–1988 antibodies to RS virus, influenza virus A, influenza virus B, parainfluenza viruses (types 1, 2 and 3) and adenoviruses were chosen as candidate reagents from a larger group that were initially examined. These were then assessed by eight laboratories, mainly in Europe and the USA, all of which were experienced in the use of the IF technique. The results obtained have shown that the pools of antibodies exhibited about

80–100% sensitivity in detecting antigens in clinical specimens, with a specificity of about 95–100% compared with reference methods such as cell culture or IF (with polyclonal antisera or monoclonal antibodies). Several antimouse conjugates were evaluated before one was chosen for inclusion in the kit.

Following the initial evaluation, laboratories in different parts of the world that were already routinely diagnosing respiratory virus infections, were invited to test the kit clinically. These laboratories had been routinely using polyclonal antisera in IF and EIA tests or isolation in cell culture. A workshop for staff from the participating laboratories in South America was organized in Brazil in 1991 to share experiences they had had with the kit. The three WHO Collaborating Centres for Virus Diagnosis provided the expertise, and quality control of the results obtained. These centres are located at The Royal Victoria Infirmary, Newcastle-upon-Tyne, UK, the Centers for Disease Control and Prevention (CDC), Atlanta, USA and the Swedish Institute for Infectious Disease Control, Stockholm, Sweden. Twenty-five laboratories in the six

WHO Regions (Africa, the Americas, South-East Asia, Europe, Eastern Mediterranean and Western Pacific) took part in the study. Sixteen of the laboratories (Table 7.3) have submitted to WHO information obtained with the diagnostic kit in 1990–1991.

Some laboratories (for example, laboratories 15 and 16 in Table 7.3) were unable to carry out more than preliminary testing on small numbers of samples; nevertheless, they found the reagents to be satisfactory. Comparative data from the remaining laboratories on the performance of the monoclonal antibodies are presented in Table 7.4 [48]. In part 1 of Table 7.4 the results are shown comparing monoclonal antibodies or polyclonal antisera in IF or other assays. There are four numbers in each entry. The first and fourth numbers represent the extent of concordance between the two methods, while the second and third indicate the extent of discordance in favour of one reagent or the other; the discrepant results require explanation. One possible reason is that the polyclonal reagents used by participants were prepared commercially either in animals or in the yolk sac of fertile eggs; some reagents pre-

Table 7.3 Sixteen laboratories in the six WHO regions (Africa, The Americas, South-East Asia, Europe, Eastern Mediterranean and Western Pacific)

1. Department of Pathology, Singapore General Hospital, Singapore
2. Department of Medical Microbiology, University of Malaya, Kuala Lumpur, Malaysia
3. Department of Virology, National Institute of Health, Department of Medical Sciences, Bangkok, Thailand
4. Commonwealth Serum Laboratories, Victoria, Australia
5. National Institute of Virology, Indian Council of Medical Research, Pune, India
6. Sendai National Hospital, Clinical Research Division, Virus Research Centre, Sendai, Japan.
7. Department of Microbiology, University of Buenos Aires, Buenos Aires, Argentina
8. Instituto Carlos Malbran, Buenos Aires, Argentina
9. Instituto Adolfo Lutz, Sao Paulo, Brazil
10. Instituto Evandro Chagas, Belem, Brazil
11. Institute of Public Health, Seccion Virologia Clinica, Santiago, Chile
12. Central Public Health Laboratory, Montevideo, Uruguay
13. Centro Nacional de Microbiologia, Virologie e Inmunologia Sanitarias, Madrid, Spain
14. Central Virology Laboratory, Chaim Sheba Medical Center, Tel-Hashomer, Israel
15. Central Pasteur de Cameroun, Yaoundé, Cameroon
16. Hopital Charles Nicolle, Tunis, Tunisia

pared in this way show non-specific reactions and have low sensitivity. In part II of Table 7.4 the results obtained with the WHO monoclonal antibodies are compared with those of isolation in cell culture. A few discordant results were recorded and there are several possible reasons for this: the cultures may vary in their sensitivity to particular viruses; IF results may still be positive with specimens collected during the late phase of infection when locally produced antibody may inhibit virus infectivity; the infectivity in the specimens may have been lost in transit. During the study the monoclonal antibodies were used on a large number of clinical specimens (Table 7.5). The positivity rates are similar to those obtained in other published studies, although the rates for parainfluenza viruses and adenovirus were low.

Additional information about the detection of viruses in nasopharyngeal specimens was obtained from laboratories in 1992–1993, during which period the second batch of kits had been used. A total of 3683 virus-positive specimens were identified by the participants in the project from tropical climate countries during 1990–1993 and 2340 (63.6%) of them contained antigens of RS virus (Table 7.6). These data show that RS virus is the main cause of respiratory viral infection in developing countries,

Table 7.4 Comparison of immunofluorescence results with those obtained using the WHO monoclonal antibody kit, on other specimens

Laboratory No.	RSV	Influenza A	Virus tested Influenza B	Parainfluenza	Adenovirus
Part I*					
2	12/0/0/88†	4/0/3/58	1/0/0/64	0/0/8/57	0/0/2/63
3	18/0/1/119	0/0/0/138	0/0/1/137	2/0/4/132	2/0/0/136
7	21/3/9/56	0/0/3/86	4/0/11/74	2/0/3/84	4/1/0/84
9	6/0/2/39	1/1/7/39	0/2/0/41	0/0/0/8	0/0/1/14
11	69/1/0/228	24/1/4/102	34/0/0/344	14/1/10/320	45/0/0/251
12	96/5/1/138	12/1/0/231	ND	6/0/6/240	7/1/5/250
13	50/2?/2?/82‡	ND	ND	7/0/0/129	2/0/0/134
	44/2?/6/84§	ND	ND	ND	ND
14	8/2/2/26§	ND	ND	ND	2/1/2/12
Part II**					
1	0/0/0/1††	24/0/0/0	14/0/0/0	ND	ND
2	9/0/21/135	7/0/0/58	1/0/0/64	11/0/0/119	1/0/2/121
7	25/5/54	1/3/2/83	5/1/10/73	0/0/5/84	3/1/1/84
8	14/1/8/23	ND	ND	4/0/3/7	8/2/2/12
12	49/4/16/110	7/2/4/368	1/9/1/370	0/0/0/68	3/1/0/64
13	12/0/10/44	0/0/0/66	0/0/0/66	4/0/1/61	2/0/0/64
14	8/3/2/125	0/0/0/110	3/2/0/105	4/2/0/124	3/8/1/126

* Part I shows a comparison between the results obtained with polyclonal antibody immunofluorescence, commercial monoclonal antibody immunofluorescence or enzyme-linked immunosorbent assay (ELISA).
† The four figures represent the number of samples that were P+M+, P+M–, P–M+, P–M–, respectively. P, polyclonal antibody; M, monoclonal antibody; ND, not done.
‡ Other monoclonal antibodies
§ ELISA
** Part II shows a comparison between the results obtained with isolation in cell culture.
‡‡ The four figures represent the number of samples that were C + M +, C + M–, C – M+, C–M–, respectively C, cell culture; M, monoclonal antibody.

Table 7.5 Proportions of positive results obtained with the WHO monoclonal antibody kit on clinical specimens, by laboratory and virus

Laboratory no.	RSV	Influenza A	Virus tested Influenza B	Parainfluenza	Adenovirus
1	649/2086*	46/2167	11/2167	30/2127	24/2118
2	30/165	7/165	1/165	11/165	3/165
3	19/138	0/138	1/138	6/138	2/138
7	30/89	3/89	15/89	5/89	4/89
8	22/46	–	–	7/14	10/24
9	8/47	8/48	0/43	0/8	1/15
11	69/298	28/131	34/378	24/345	45/296
12	65/179	11/381	2/381	0/68	3/68
13	50/136	0/66	0/66	7/136	2/136
14	10/138	0/110	3/110	4/130	4/138
A†	6/230	34/130	0/130	10/130	4/130
B†	58/696	7/696	10/696	15/696	4/696
Total	1016/4148	144/4121	77/4363	119/4046	106/4013
% Positivity	24.5	3.5	1.8	2.9	2.6

* Shown are no. positive/no. of specimens examined.
† Results were obtained in 1991–1992 with the second batch of the WHO kit. A, Dr J. Hang, Folkehelse, Oslo, Norway; B, Dr B. Marten, Institute of Medical Research, Goroka, Papua New Guinea.

Table 7.6 Contribution of RSV and other respiratory viruses among positive specimens

	RSV	Influenza A	Influenza B	Para-influenza	Adeno-virus	Total
No. positive	2340	537	237	374	195	3683
%	63.6	14.5	6.4	10.2	5.3	100

which is concordant with earlier results. However, this is not universal and the aetiological picture of viruses in the pathology of respiratory tract infection can vary from one country to another. For instance, 306 strains of respiratory viruses have been identified in specimens from children tested in the Virus Research Institute, Thailand during 1991–1993. Of this number, 141 (46.1%) and 108 (35.2%) contained antigens of influenza A and influenza B viruses respectively. RS virus was detected only in 27 (8.8%) specimens. The detection rate for parainfluenza viruses and adenoviruses was 4.9% for each. In view of the fact that different drugs have been used for

the treatment of patients infected with influenza or RS viruses, laboratory diagnosis is strongly recommended.

Clinical and laboratory studies carried out in the Singapore General Hospital revealed that two viruses regularly associated with annual epidemics in Singapore are RS virus and prevailing strains of influenza virus. Singapore usually has two influenza seasons: a major one which occurs in April–June, and a smaller one from around November to January the following year. Sporadic cases of influenza are registered throughout the year. An upsurge of RS virus cases occurs from May to September. During these 5 months in 1992,

41.8% of studied specimens contained RS virus antigens. Parainfluenza virus cases usually increase around April, May, June and July. During these 4 months in 1993 the antigens of parainfluenza viruses were detected in 8.9% of specimens studied, whereas the detection rate of parainfluenza viruses during the four months before and after the epidemic period was 3.4% and 1.6%, respectively.

Adenoviruses cause intermittent outbreaks with no definite seasonal pattern. The viruses identified in specimens from patients admitted to the Institute of Medical Research, Goroka, Papua New Guinea from April 1991 to May 1993 were: RS virus 88 (45.1%); parainfluenza virus 43 (22.1%); influenza A virus 22 (11.2%); influenza B virus 26 (13.3%); adenovirus 16 (8.2%). An increase in the detection rate for influenza A virus, influenza B virus and adenoviruses was shown between November (only adenoviruses) and December 1992.

A WHO group of experts analysed the data obtained during 1990–1993 and assessed the current status of the project as promising. A new kit of monoclonal antibodies was prepared and quality controlled in WHO Collaborating Centres. In order to provide data to underpin vaccine development, a specific monoclonal antibody to each of the parainfluenza viruses was included in the kit.

Studies in some countries have emphasized the important role of measles virus in causing disease of the respiratory tract. For example, in Papua New Guinea measles virus was identified in 22% of children admitted to Goroka Hospital during 1983–1985 [29]. In a group of children in the Philippines with acute bronchiolitis and pneumonia, measles infection was identified in 21.4% of cases [31]. In order to provide rapid laboratory methods for the confirmation of measles, a study was begun at Chicago University, USA, with the objective of evaluating monoclonal antibodies for identification of measles antigens in nasopharyngeal specimens. If the results of the study are successful, a diagnostic kit will be developed and proposed to countries where measles infection

is still frequent. To promote quality control, an interchange of specimens (both positive and negative) and information between laboratories has been encouraged. Further training programmes are in progress, mainly for laboratories without extensive experience in immunofluorescence. Intra-regional working groups will also be convened for the discussion of results. The aim is that existing laboratories will be used as reference laboratories to transfer the appropriate diagnostic technology to countries. In order to encourage countries to produce diagnostic kits locally a reagent bank for RS virus and parainfluenza virus type 3 (PIV3) was established which contains monoclonal antibodies for different proteins of RS virus and PIV3, and anti-RS virus hyperimmune antisera with a known titre of antibody in different tests and to prototype strains of viruses.

The laboratory aspects of respiratory virus surveillance are more uniformly covered worldwide than statistical aspects. However, support for increased coverage is needed in some developing countries where researchers have difficulty in obtaining material for respiratory virus diagnosis and standardization of diagnostic methods.

7.5 THE WHO PROGRAMME ON LABORATORY SURVEILLANCE OF INFLUENZA

Influenza continues to be a major problem worldwide and surveillance is an important part of the control of this infection. The main objectives of the surveillance are [14]:

1. Collection of influenza virus isolates and analysis of antigenic characteristics of influenza viruses, including outbreaks of sporadic cases of influenza, so that a decision can be made on which antigenic variant should be used for the production of influenza virus vaccines during the next epidemiological period.

2. Collection and analysis of epidemiological information on influenza morbidity and mortality in order to estimate the impact of the disease.

3. Early detection of influenza epidemics, enabling immunization of persons not previously vaccinated, and notification of health authorities to prepare for the possible impact on clinical workloads and hospital admissions.

4. Identification of high-risk areas and population groups where additional immunization input may be needed.

Information on morbidity and mortality caused by influenza viruses is available in industrially developed countries. However, these infections have not been studied extensively in developing countries, many of which are located in tropical and subtropical areas. In developing countries the methods used to predict baseline mortality may be inaccurate, resulting in an underestimation of mortality associated with influenza virus infection. Moreover, the excess mortality attributable to influenza virus may not be apparent because the baseline mortality rates are so high. Some of the risk factors that may influence the severity and outcome of influenza virus infection are particularly important in developing countries. The following risk factors have been identified as most pertinent to developing countries:

- compromised immune status
- malnutrition
- infection with other microorganisms
- inhalation of indoor pollutants
- and limited access to medical care [19].

The role of these risk factors can vary from one country to another and an accurate epidemiological study should be carried out to determine which risk factors are most prevalent in influencing the outcome of influenza virus infections. Access to medical care is especially important because it should result in early diagnosis, prompt treatment with rimantadine or amantadine in influenza cases, or antibiotic therapy for influenza-associated bacterial infections. In many tropical countries where data on morbidity and mortality are available, the procedure adopted for their collection and analysis varies according to the type of information and source. Epidemiological surveillance would be much improved if all the data from a given country or geographical area were standardized and could be collected by a central sorting office for analysis and distribution.

In contrast to infections caused by most respiratory viruses, those caused by influenza virus can be prevented by immunization with live or inactivated vaccines which have been used for many years. It is clear that the priorities of individual countries in developing a programme of immunization against influenza depend on the impact of influenza in relation to the incidence of other disease problems. National policies on the use of influenza virus vaccines vary widely from country to country. However, in many of them the vaccination of high-risk persons each year before the influenza season is the most important measure in reducing the impact of influenza. Vaccination can be highly effective when it is aimed at individuals who might suffer severe consequences from influenza virus infection [5]. Two main objectives for influenza virus immunization strategy were identified by a WHO group of experts [44]: (i) to protect individuals who are at particular risk of disease, e.g. elderly persons in nursing homes; and (ii) to protect other defined subsets of the populations such as school children (as in Japan) and factory workers (as in the former USSR). In closed or semi-closed communities, maximum benefit from immunization is likely to be achieved when more than about three-quarters of the population are immunized, so that the benefit of herd immunity can be exploited.

A strong bar to protection of human beings with influenza virus vaccine is related to the ability of influenza viruses to change their antigenic character at irregular intervals. If the antigenic changes of haemagglutinin and/or neuraminidase are major, previously formed antibodies do not recognize the new strains and, therefore, do not provide protection against re-infection. Similarly, vaccination may be effective only against re-infection caused by strains which are antigenically similar to the strains used for production of the vaccine.

In order to help countries in the control of influenza virus infection the WHO has developed a programme of laboratory surveillance of influenza which is maintained through the efforts of 110 National Influenza Centres in 77 countries in collaboration with WHO in Geneva, and three WHO Collaborating Centres for Reference and Research on Influenza: the Centers for Disease Control and Prevention (CDC), Atlanta, USA; the National Institute for Medical Research, London, UK, and the Commonwealth Serum Laboratories, Parkville, Australia. The network of national laboratories covers nearly all parts of the world: 46 laboratories are located in 25 developed countries and 64 laboratories are in 52 other countries [19]. The WHO Collaborating Centres obtain and characterize strains from outbreaks in different parts of the world and distribute them to production laboratories. They advise on the strains for inclusion in influenza virus vaccines and act as resources for the development and use of training materials and also provide training sites. The WHO Collaborating Centre for Reference and Research on Influenza in Atlanta produces and provides diagnostic kits for National Influenza Centres, the functions of which are to isolate influenza viruses from patients and to dispatch the isolates from each outbreak to one of the international centres. Furthermore, they perform serological tests for the diagnosis of influenza and provide virological and epidemiological information to WHO. Some National Influenza Centres participate in the above-mentioned WHO project on surveillance of main respiratory viruses and use the kit for rapid diagnosis of influenza viruses, parainfluenza viruses, RS virus and adenoviruses. This allows the differential diagnosis between influenza virus infection and other viral respiratory infections. Directors of National Influenza Centres are encouraged to collaborate with local community hospitals to maintain records of the daily attendance of persons requiring treatment for respiratory diseases, and to collect virus specimens from a predetermined number or proportion of such persons for further laboratory confirmation of viral disease.

The information obtained from all sources is collected in Geneva and published in the WHO Weekly Epidemiological Record, which is distributed to health authorities, influenza institutions and other interested parties. Each year WHO convenes a meeting of Directors of the WHO Collaborating Centres to discuss the information available and produce recommendations on the composition of influenza virus vaccines for the forthcoming epidemiological season. For example, analysis of data obtained during the 1992–1993 season enabled the conclusion to be made that the vaccines for protection against influenza during the next season should contain the following components: A/Beijing/327/92(H3N2); A/Singapore/6/86(H1N1) and B/Panama/45/90.

7.6 REFERENCES

1. Avila, M.M., Carballal, G., Rovaletti, H., Ebekian, B., Cusminsky, M. and Weissenbacher, M. (1989) Viral etiology in acute lower respiratory infections in children from a closed community. *Am. Rev. Respir. Dis.*, **140**, 634–7.
2. Bang, A.T., Bang, R.A., Tale, O., Sontakke, P., Solanki, J., Wargantiwar, R. and Kelzarkar, P. (1990) Reduction in pneumonia mortality and total childhood mortality by means of community-based intervention trial in Gadchiroli, India. *Lancet*, **i**, 202–6.
3. Berman, S., Duenas, A., Bedoya, A., Constain, V., Leon, S., Borrero, I. and Murphy, J. (1983) Acute lower respiratory tract illnesses in Cali, Colombia: a two-year ambulatory study. *Paediatrics*, **71**, 210–18.
4. Campbell, H., Byass, P. and Greenwood, B.M. (1988) Simple clinical signs for diagnosis of acute lower respiratory infections. *Lancet*, **ii**, 742–3.

5. Centers for Disease Control (1989) Prevention and control of influenza. *MMWR*, **38**, 317–25.
6. Cherian, T., Jacob, J.T., Simoes, E., Steinhoff, M.C. and John, M. (1988) Evaluation of simple clinical signs for the diagnosis of acute lower respiratory tract infection. *Lancet*, **ii**, 125–8.
7. Datta, N., Kumar, V., Kumar, L. and Singhi, S. (1987) Application of case management in the control of acute respiratory infections in low-birth-weight infants: a feasibility study. *Bull. WHO*, **65**(1), 77–82.
8. Degré, M. (1986) Interaction between viral and bacterial infections in the respiratory tract. *Scand. J. Infect. Dis.*, **49**, 140–5.
9. Forgie, I.M., O'Neill, K.P., Lloyd-Evans, N. *et al.* (1991) Etiology of acute lower respiratory tract infections in Gambian children I: Acute lower respiratory tract infections in infants presenting at the hospital. *Pediatr. Infect. Dis. J.*, **10**, 33–41.
10. Forgie, I.M., O'Neill, K.P., Lloyd-Evans, N. *et al.* (1991) Etiology of acute lower respiratory tract infections in Gambian Children II: Acute lower respiratory tract infections in children ages one to nine years presenting at the hospital. *Pediatr. Infect. Dis. J.*, **10**, 42–7.
11. Gadomski, A.M. (1993) Potential interventions for preventing pneumonia among young children: lack of effect of antibiotic treatment for upper respiratory. *Paediatr. Infect. Dis. J.*, **12**, 115–20.
12. Garenne, M., Ronsmans, C. and Campbell, C. (1992) The magnitude of mortality from acute respiratory infections in children under 5 years in developing countries. *World Health Stat. Q.*, **45**, 180–9.
13. Ghafoor, A., Nomani, N.K., Ishaq, Z. *et al.* (1990) Diagnosis of acute lower respiratory tract infections in children in Rawalpindi and Islamabad, Pakistan. *Rev. Infect. Dis.*, **12** (suppl. 8), S907–14.
14. Ghendon, Y. (1992) World Health Organization Programme on the Control of Influenza, *in Control of Virus Diseases*, 2nd edn, (ed. E. Kurstak), Revised and Expanded. Marcel Dekker, Inc., New York.
15. Gove, S. (1990) Remedies for young children with coughs and colds. *ARI News*, **18**, 203.
16. Hazlett, D.T.G., Bell, T.M., Tukei, P.M. *et al.* (1988) Viral etiology and epidemiology of acute respiratory infections in children in Nairobi, Kenya. *Am. J. Trop. Med. Hyg.*, **39**, 632–40.
17. Khan, A.J., Khan, J.A., Akbar, M. and Addis, D.G. (1990) Acute respiratory infections in children: a case management intervention in Abbottabad District, Pakistan. *Bull. WHO*, **68**(5), 577–85.
18. Kloene, W., Bang, F.B., Chakraverty, S.H. *et al.* (1970) A two-year respiratory virus survey in four villages in West Bengal, India. *Am. J. Epidemiol.*, **92**, 307–20.
19. Leigh, M.W., Carson, J.L. and Denny, F.W. Jr (1991) Pathogenesis of respiratory infections due to influenza virus: Implications for developing countries. *Rev. Infect. Dis.*, **164** (suppl. 6), S501–8.
20. McIntosh, K., Halonen, P. and Ruuskanen, O. (1993) Report of a workshop on respiratory viral infections: Epidemiology, diagnosis, treatment and prevention. *Clin. Infect. Dis.*, **16**, 151–64.
21. Mtango, F.D.E. and Neuvians, D. (1986) Acute respiratory infections in children under 5 years. Control project in Bagamoyo District, Tanzania. *Trans. R. Soc. Trop. Med. Hyg.*, **80**, 851–8.
22. Mtango, F.D.E., Neuvians, D. and Korte, R. (1989) Magnitude, presentation, management and outcome of acute respiratory infections in children under the age of five in hospitals and rural health centres in Tanzania. *Trop. Med. Parasitol.*, **40**, 97–102.
23. Mulholland, E.K., Olinsky, A. and Shann, F.A. (1990) Clinical findings and severity of acute bronchiolitis. *Lancet*, **335**, 1259–61.
24. Mulholland, E.K., Simoes, E.A.F, Costales, M.O.D. *et al.* (1992) Standardized diagnosis of pneumonia in developing countries. *Pediatr. Infect. Dis. J.*, **11**, 77–81.
25. Olsen, M.A., Shuck, K.M., Sambol, A.R., Bohnert, V.A. and Henery, M.L. (1993) Performance of the Kallestad Pathfinder Enzyme Immunoassay in the diagnosis of respiratory syncytial virus infections. *Diagn. Microbiol. Infect. Dis.*, **16**(4), 325–9.
26. Ong, S.B., Lam, K.L. and Lam, S.K. (1982) Viral agents of acute respiratory infections in young children in Kuala Lumpur. *Bull. WHO*, **60**, 137–40.
27. Onyango, F.E., Steinhoff, M.C., Wafula, E.M, Wariua, S., Musia, J. and Kitonyi, J. (1993) Hypoxaemia in young Kenyan children with acute lower respiratory infection. *Br. Med. J.*, **306**, 612–15.
28. Pandey, M.R., Sharma, P.R., Gubhaju, B.B. *et al.* (1989) Impact of a pilot acute respiratory infection (ARI) control programme in a rural community of the hill region of Nepal. *Ann. Trop. Paediatr.*, **9**, 212–20.

29. Phillips, P.A., Lehmann, D., Spooner, V. *et al.* (1990) Viruses associated with acute lower respiratory tract infections in children from the Eastern highlands of Papua New Guinea (1983–1985). *Southeast Asian J. Trop. Med. Public Health*, **21**(3), 373–82.

30. Pio, A., Leowski, J. and Ten Dam, H.G. (1985) The magnitude of the problem of acute respiratory infections, in *Acute Respiratory Infections in Childhood*, (eds R.M. Douglas and E. Kirby-Eaton), Proceedings of an International Workshop, Sydney, August 1984. University of Adelaide, Australia, pp. 3–16.

31. Ruutu, P., Halonen, P., Meurman, O. *et al.* (1990) Viral lower respiratory tract infections in Filipino children. *J. Infect. Dis.*, **161**, 175–9.

32. Schirm, J., Luijt, D.S., Pastoor, G.W., Mandema, J.M. and Schroder, F.P. (1992) Rapid detection of respiratory viruses using mixtures of monoclonal antibodies on shell vial cultures. *J. Med. Virol.*, **38**, 146–51.

33. Shann, F. (1989) Nasopharyngeal oxygen in children. *Lancet*, **i**, 1077–8.

34. Shann, F., Barker, J. and Poore, P. (1985) Chloramphenicol alone versus chloramphenicol plus penicillin for severe pneumonia in children. *Lancet*, **ii**, 684–6.

35. Shann, F., Barker, J. and Poore, P. (1989) Clinical signs that predict death in children with severe pneumonia. *Pediatr. Infect. Dis.*, **8**, 852–5.

36. Shann, F., Gatchalian, S. and Hutchinson, R. (1988) Nasopharyngeal oxygen in children. *Lancet*, **ii**, 1238–40.

37. Shann, F., Hart, K. and Thomas, D. (1984) Acute respiratory infections in children: possible criteria for selection of patients for antibiotic therapy and hospital admission. *Bull. WHO*, **62**, 749–53.

38. Shann, F., Linnemann, V. and Gratten, M. (1987) Serum concentrations of penicillin after intramuscular administration of procaine, benzyl and benethamine penicillin in children with pneumonia. *J. Pediatr.*, **110**, 299–302.

39. Sobeslavsky, O., Sebikari, S.R.K., Harland, P.S.E.G., Skritie, N., Fayinka, O.A. and Soneji, A.D. (1977) The viral etiology of acute respiratory infections in children in Uganda. *Bull. WHO*, **55**, 625–31.

40. Sunakorn, P., Chunchit, L., Nihawat, S., Wangweerawong, M. and Jacobs, R.F. (1990) Epidemiology of acute respiratory infections in young children from Thailand. *Paediatr. Infect. Dis. J.*, **9**, 873–7.

41. Von Schirnding, Y.E., Yach, D. and Klein, M. (1991) Acute respiratory infections as an important cause of childhood deaths in South Africa. *S. Afr. Med. J.*, **80**, 79–82.

42. Wafula, E.M., Tukei, P.M., Bell, T.M. *et al.* (1985) Aetiology of acute respiratory infections in children aged below 5 years in Kenyatta National Hospital. *East Afr. Med. J.*, **62**, 757–67.

43. WHO (1981) *Rapid laboratory techniques for the diagnosis of viral infections*. WHO Technical Report Series, 661.

44. WHO (1987) Progress in the Development of influenza vaccines: Memorandum from a WHO Meeting. *Bull. WHO*, **65**(3), 289–93.

45. WHO (1990) *Antibiotics in the treatment of acute respiratory infections in young children*. WHO/ARI/90.10, Geneva, Switzerland.

46. WHO (1990) *Acute respiratory infections in children: case management in small hospitals in developing countries. A manual for doctors and other senior health workers*. Document WHO/ARI/90.5. Geneva, Switzerland.

47. WHO (1991) *Technical bases for the WHO recommendations on the management of pneumonia in children at first-level health facilities*. WHO/ARI/91.20. Geneva, Switzerland.

48. WHO (1992) Use of monoclonal antibodies for rapid diagnosis of respiratory viruses: memorandum from a WHO meeting. *Bull. WHO*, **70**(6), 699–703.

49. WHO (1994) *Cough and cold remedies in the treatment of acute respiratory infections in young children*. WHO unpublished document WHO/ARI, Geneva.

50. WHO (1994) *Oxygen therapy for acute respiratory infections in young children in developing countries*. WHO unpublished document WHO/ARI 1994, Geneva, Switzerland.

51. Zhaori Getu and Fu Li-tao (1990). Respiratory syncytial virus infection. A brief review of epidemiology and current research in China. *Chin. Med. J.*, **103**(6), 508–12.

52. Zheng, W.J. *et al.* (1989) Detection of respiratory syncytial virus (RS virus) in nasopharyngeal secretions by indirect immunoperoxidase assay with RS virus monoclonal antibody. *J. Exp. Clin. Virol. (China)*, **3**, 353–72.

RHINOVIRUSES

8

Widad Al-Nakib

8.1 INTRODUCTION

Rhinoviruses are the major cause of the common cold. They are members of the family Picornaviridae. There are some 89 to 100 different serotypes circulating in different geographical areas [28]. Rhinoviruses are about 28–34 nm in diameter with a buoyant density of 1.38–1.42 ml in $CsCl_2$ gradient. Like other picornaviruses, rhinoviruses are resistant to ether and chloroform but differ in being acid labile at pH below 5. The infectivity of rhinoviruses is very much affected by exposure to different temperatures, though this may vary depending on the serotype. Thus, at

Viral and Other Infections of the Human Respiratory Tract.
Edited by S. Myint and D. Taylor-Robinson. Published in 1996 by Chapman & Hall. ISBN 0 412 60070 6

–70°C, generally the virions retain infectivity for long periods of time while at 20–37°C the virions can lose infectivity fairly rapidly; at 56°C most rhinoviruses are totally inactivated. They grow optimally at temperatures ranging from 33°C to 35°C [40]. Infectious particles (termed D particles) are immunologically distinct from non-infectious particles (termed C particles). Empty capsid can have either C or D type antigenicity.

The infectious aetiology of the common cold was first recognized as early as 1914 when cell-free filtrates, obtained from the nasal secretion of a person with acute symptoms of coryza and nasal obstruction, were inoculated intranasally into volunteers of whom about one-third developed identical respiratory symptoms [35]. However, it was not until 1953 that the aetiology of the common cold was first established by successfully isolating the virus after inoculating cell cultures with nasal secretion from a patient with common cold symptoms [5]. The recognition that rhinoviruses can replicate in tissue cultures such as human fibroblast cells facilitated the isolation of a large number of these agents over the following decade-and-a-half and their typing with specific antisera into different serotypes using neutralization assays [27,28,46,57].

8.2 STRUCTURE AND BIOLOGY

Human rhinoviruses (HRV), like other picornaviruses, have a total molecular weight of approximately 8 x 10⁶ Da. The nucleocapsid is icosahedral in structure with a 5:3:2 arrangement of capsomers. Recent studies [48] suggest that the icosahedral capsid of a rhinovirus is composed of 60 identical capsomers of 100 Da, each arranged as 12 groups of five. These are mainly composed of four structural polypeptides (VP1, VP2, VP3, VP4 with molecular mass in HRV-14 of 32 000, 29 000, 26 000 and 7000 Da, respectively). Recent molecular studies [48] on the three-dimensional structure of rhinoviruses suggested that the main immunogenic neutralizing sites reside in four distinct areas on the surface of the viral capsid. These have been designated as NIm-IA, NIm-IB, NIm-II and NIm-III with NIm-A and NIm-IB located in VP1 while NIm-II and NIm-III located on VP2 and VP3, respectively [48] (Figure 8.1). The results of the same studies suggested the presence of a 2.5-nm depression orientated around the five-fold axis in the surface of the capsid. The results also indicated that this depression is potentially the site of virus–receptor interaction (Figure 8.2). Colonno and colleagues [12,13] showed that rhinoviruses can be classified into two receptor groups depending on their receptor binding specificities. Thus 89% of the viruses

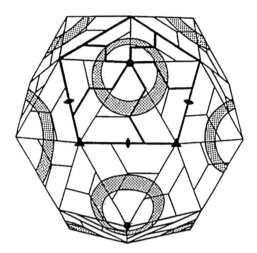

Figure 8.2 The three-dimensional structure of a rhinovirus showing the position of the 2.5-nm deep 'canyons' circulating around each of the 12 pentamer vertices (stippled areas).

studied attached to one receptor on HeLa cells and their attachment could be inhibited by incubating them with a mouse monoclonal antibody against this receptor, while 11% of rhinoviruses (HRV 1A, 1B, 2, 29, 30, 31, 44, 47, 49 and 62) attached to another receptor.

The capsid surrounds a linear segment of a single-stranded RNA of positive polarity with an approximate molecular weight of 2 x 10⁶ Da [38,40]. The viral RNA accounts for some 30% of the total particle mass. The nucleotide sequence of the genome of a number of rhinoviruses has recently been reported (e.g. HRV-1B, -2, -14, -89) [31,49,50]. Generally, the genome of the virus codes for some 7500 nucleotides, is polyadenylated at its 3' terminus and has a small protein (VPg) which is covalently attached to the 5' terminus [50]. The general organization of a rhinovirus genome is shown in Figure 8.3. It is interesting to note that the first 624 nucleotides from the 5' end do not appear to code for any function and are highly conserved among the different rhinoviruses. This non-coding region is followed by an open reading frame of some 6400–6550 nucleotides which comprise about

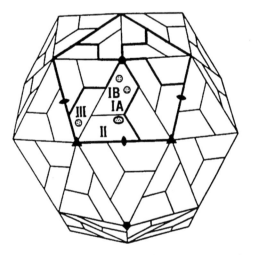

Figure 8.1 The position of the four major neutralizing sites in relation to viral capsid proteins (circled).

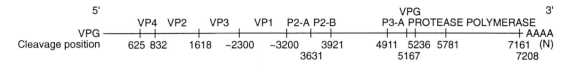

Figure 8.3 The general organization of the human rhinovirus 14 (HRV-14) genome. The P2-A, B and C and P3-A region code for proteases involved in the cleavage of various viral proteins.

90% of the genome and terminates in another non-coding region at the 3' end of approximately 40–44 nucleotides in length [18,49,50].

8.3 VIRUS REPLICATION

Rhinoviruses generally replicate only in cells of primate origin although recent studies suggest that a strain of rhinovirus type 2 has been adapted by serial passage to replicate in a cell line of murine origin. A number of cell cultures have been found to be capable of propagating rhinoviruses – these include human fetal kidney, human diploid lung fibroblasts, human fetal tonsil, monkey kidney, and continuous cell lines such as HeLa, KB, HEp-2. Human diploid lung fibroblasts are most commonly used, although a strain of HeLa cells known as the Ohio strain was routinely employed and found extremely useful for isolation and neutralization of many rhinovirus strains. In the 1960s, human organ cultures such as fetal tonsil and nasal polyp were used extensively in the primary isolation of wild rhinoviruses that were not adaptable for growth in cell culture [53]. A critical stage in the infectious cycle of rhinoviruses is the attachment of the virus to a specific receptor. Although the nature of these receptors is not entirely clear, studies suggest that HeLa cells have at least two different proteins, each of approximately 450 kDa molecular weight [41,54]. Furthermore, Greve *et al.* [24], Staunton *et al.* [51] and Tomassini *et al.* [54] have identified one of these receptors as an intercellular attachment molecule 1(ICAM-l) which is present on the surface of leucocytes and other cells and plays an important role in interleucocyte signalling functions in the immune response.

Following successful attachment of the virus to the cell, it enters the cell and then begins to go through a process of uncoating followed by release of viral RNA. As yet, the mechanism by which the virus uncoats itself is still unclear but appears to be pH-dependent and involves conformational changes in the internalized viral capsid which subsequently result in the release of viral RNA from the endosome. The viral RNA, once uncoated, is released into the cytoplasm of the cell and is first translated to produce an RNA-dependent RNA polymerase(s) which is required to replicate the genomic RNA (plus strands) to produce complementary minus strands. In turn the minus strands replicate to produce progeny plus strands which ultimately become incorporated into progeny virus particles [40]. The entire genome is translated into one polyprotein which is cleaved during translation initially into three proteins (VP0, VP1 and VP3). VP0 then undergoes further cleavages to form VP2 and 4 (Figure 8.4). The virus matures in the cytoplasm of infected cells and is released when a large number of particles begins to accumulate in the cytoplasm, thus resulting in the rupture of the cell membrane. It may take up to 8 hours before new infectious particles appear. Rhinovirus replication in cells will invariably result in the destruction of the cell which is typical for other picornaviruses.

8.4 CLINICAL FEATURES

Rhinovirus infection of humans often results in common cold symptoms which are exemplified by rhinorrhoea, nasal obstruction, sneezing, sore throat and cough. The sore throat often precedes other symptoms. Fever

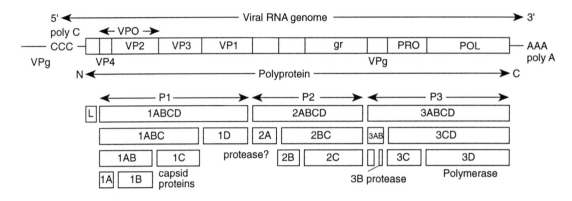

Figure 8.4   Picornaviral processing map. Mature picornaviral peptides are derived by progressive post-translational cleavage of the polyprotein. Peptide 2A is responsible for the first cleavage between P1 and P2 regions. Most of the other cleavages in all polyproteins are carried out by peptide 3C, which is a viral protease. For example, 3ABCD is cleaved to 3AB + 3C + 3D and P1 to VP0 + VP3 + VP1. Finally VP0 is cleaved to VP4 + VP2 (VP0, VP4, VP2, VP3 and VP1 are also called, 1 AB, 1A, 1B, 1C and 1D respectively). The P3 peptides, especially 3D (polymerase), is closely associated with genome replication. Protein 3B is VPg, which is attached to the 5' end of the genome during initiation of RNA synthesis.

or systemic reactions are usually absent. The majority of patients presenting with common cold symptoms recover within 1 week of onset although in a small proportion of individuals these may persist in the form of nasal discharge or cough for periods ranging from 2–4 weeks. In some patients, otitis media and sinusitis may complicate rhinovirus colds although usually these occur in conjunction with bacterial infections. Furthermore, in patients with a history of obstructive airways disease, rhinovirus infection may trigger acute episodes of asthmatic attacks which may be prolonged [23,32]. It is possible that these episodes may be the result of virus spread to the lower respiratory tract, causing an inflammatory reaction [36] and there is some evidence that this is so. Thus, Krilov *et al.* [34], and more recently McMillan *et al.* [39], showed that rhinoviruses can be isolated from a relatively high proportion of paediatric patients with lower respiratory tract symptoms including bronchiolitis. Although the frequency of such infections was less than that caused by respiratory syncytial (RS) virus, the severity of

illness and clinical presentation in these young infants were similar [39]. In addition, Monto and colleagues [42,43], have shown in a prospective study in the USA that 15% to 65% of normal 'healthy' individuals presented with a cough that could be related to a rhinovirus infection and that over an 11-year period rhinoviruses actually produced the greatest number of medical consultations when compared with those related to other respiratory virus infections. Nevertheless, when compared with other respiratory viruses, rhinoviruses do not generally cause serious lower respiratory tract symptoms such as pneumonia, croup or bronchiolitis.

8.5 PATHOGENESIS

Rhinoviruses are transmitted usually by either inhalation of aerosolized droplets or by self-inoculation of virus from contaminated inanimate objects or secretions from infected individuals through contaminated fingers (see below). The virus multiplies initially in the epithelial cells lining the nasopharynx and is detectable usually in nasopharyngeal secretions some 24 hours after inoculation as has been demonstrated in various volunteer studies. Virus titres reach a maximum by the third day and then decline to almost undetectable levels by the fifth to the sixth day after inoculation. However, it has been shown that a proportion of infected individuals with rhinoviruses may continue to excrete virus for periods as long as 3–4 weeks after initial challenge [7]. The virus may replicate initially in the posterior side of the nasopharynx and then spread anteriorly [58]. Infected individuals normally begin to show signs and symptoms of the common cold 1 day after virus becomes detectable in nasal secretions, although the severity of symptoms generally coincides with peak virus excretion in the nose. Furthermore, during the acute phase of illness, marked oedema with infiltration of neutrophils, lymphocytes, plasma cells and eosinophils is often observed. This is usually accompanied by narrowing of the nasal cavities and excessive secretory activities. The involvement of cellular functions in the activation of the inflammatory response is also suggested by the increase of neutrophils and reduction in circulating lymphocytes in the blood [36]. It has been suggested by Turner *et al.* [55] and Hendley [30] that immunomediators are released following rhinovirus infection and these in turn trigger a series of events that ultimately result in cold symptoms. Indeed, nasal obstruction and production of excessive amounts of nasal secretion, that are often seen in the common cold, strongly suggest a role for vasoactive mediators in the development of cold symptoms. Histamines do not appear to play any role in the development of colds although kinins may be important. Thus Naclerio *et al.* [44] have shown that these are normally elevated in nasal secretions of volunteers with rhinovirus infections and that local administration of bradykinins in the nose will result in common cold symptoms that are indistinguishable from those caused by rhinoviruses [47]. Local interferon production, on the other hand, is clearly

very important in the recovery from a rhinovirus infection.

8.6 EPIDEMIOLOGY AND MODE OF TRANSMISSION

8.6.1 INCIDENCE

Rhinovirus infections have worldwide distribution and affect all populations. Most individuals become infected quite early in life and the frequency of infection generally decreases with increasing age, presumably because of, at least partial, immunity from previous infections. Due to the multiplicity of serotypes it is estimated that on average an individual may experience three to five episodes of colds each year. Although infection may take place at any time during a year, there appear to be two major peaks, one in the autumn and another during spring time. It is thought that the autumn peak coincides with the opening of schools. Thus, it has been postulated that children exchange infection at school and later infect their siblings and parents at home. Indeed, schools, nurseries, families and confined groups such as the military provide the major foci of infection within communities. Although there are over 100 different serotypes of rhinoviruses, it has been found that only a few serotypes circulate in a particular community. Indeed, the results of studies in the USA suggested that such small numbers of serotypes may remain within a particular community for many years [21,22,25].

8.6.2 TRANSMISSION PROCESSES

Droplet infection is generally thought to be the major route of transmission of the virus though many well-controlled studies in volunteers failed to show this. However, a recent experiment by Dick *et al.* [12] demonstrated that susceptible volunteers when restrained to preclude hand-to-face contact, became infected as a result of aerosol transmission. Most experiments in volunteers both in the USA and at the

Common Cold Unit in Salisbury, UK, however, demonstrated that virus is transmitted most commonly by direct inoculation of the conjunctiva and/or nasal epithelium with fingers and hands that have been in contact with contaminated objects and fomites [26]. Crowding and close contact, particularly if prolonged, between infected individuals and susceptible persons clearly increases the chance of infection. The amount of virus required to establish an infection, however, is small provided that the virus successfully reaches the target tissues.

8.7 IMMUNITY

Following infection by a rhinovirus, antibodies of both the IgG and IgA class appear in the serum 1 week after the onset of symptoms and reach a peak 3 weeks later. Similarly, virus-specific IgA appears in nasal secretions some 10 days after virus inoculation and reaches a peak within 2 weeks after onset of symptoms [9]. These studies at the MRC Common Cold Unit, Salisbury, also showed that both virus-specific serum IgG and IgA are important in the protection against re-infection and development of common cold symptoms when volunteers were re-challenged with the same virus serotype 1 year later. Thus, individuals who had high levels of both immunoglobulins in the serum resisted challenge and were neither infected nor developed symptoms of colds on re-challenge, while those with intermediate levels became infected, but remained asymptomatic. In contrast, volunteers with low or undetectable levels of virus-specific serum IgA and IgG became infected and developed symptoms of colds when re-challenged [9]. The presence of virus-specific secretory IgA in nasal secretion was also important in protecting volunteers from infection. However, it must be emphasized that this immunity is serotype-specific and does not protect against infection by a heterologous serotype [9]. Circulatory and local cellular immune responses also appear to be impor-

tant in the recovery from and the protection against infection although further studies are required to demonstrate this clearly [37].

8.8 LABORATORY DIAGNOSIS

8.8.1 CELL CULTURE SYSTEMS

The diagnosis of a rhinovirus infection is confirmed by isolating the virus in an appropriate cell culture system such as human lung fibroblasts or Hela cells. In these cultures, the virus may show a cytopathic effect (CPE) as early as 48 hours though cultures are not normally declared negative until 8 days after inoculation. For wild-type rhinoviruses that may be ill-adapted for growth in such cell cultures, it may be necessary to re-passage three times before a CPE can be observed. Like that of enteroviruses, rhinovirus CPE is characterized by rounding of cells which may begin as small foci and then spread to affect the whole cell layer. Alternatively, a rhinovirus infection may be confirmed by showing significant antibody rises (over four-fold) between a sample taken from a patient at the acute phase of the disease and another one 10 to 14 days later using neutralization (NT) or enzyme-linked immunosorbant assay (ELISA) methods [7,8]. However, both NT and ELISA are highly serotype-specific.

8.8.2 OTHER METHODS

Over the past decade a number of new assays have been developed to diagnose rhinovirus infections. Thus, it has been shown that rhinoviruses in nasal samples may be detected by dot–blot DNA hybridization procedures or following virus RNA amplification using polymerase chain reaction (PCR) assays [2,10,45]. Virus antigens in nasal secretions may also be detected reliably using ELISAs [3,14]. Recent field studies showed that the PCR assay may indeed be the test of choice for the detection of wild rhinoviruses since it proved to be far more sensitive than virus isolation [33]. A culture-amplified immunofluorescence test which detects wild rhinoviruses in clinical samples within 48 hours has also been reported recently [1]. This new test is capable of detecting a 'common' rhinovirus epitope which appears to be expressed by all rhinoviruses some 48 hours after infection of Ohio HeLa cells. This test is simple to conduct, appears to be sensitive and reliable and may thus offer an interesting alternative to detection by virus isolation. Unfortunately, none of these new assays is being used routinely or widely for the diagnosis of rhinovirus infections.

8.9 PREVENTION AND TREATMENT

Due to the multiplicity of serotypes (over 100), it has been difficult to develop an effective vaccine against rhinovirus infections. Interferon-α and β have been shown to be effective in preventing rhinovirus colds only when they are administered intranasally before virus challenge. However, these substances can be given only for short periods (5–7 days) since prolonged administration results in local 'toxic' inflammatory changes in the nasal mucosa [4].

Although a number of synthetic compounds such as 4',6-dichloroflavan, Ro O9-0410 (4'ethoxy-2'-hydroxy-4',6'-dimethoxychalcone), R61837 (3-methoxy6-[4-(3-methylphenyl)-1-piperazinyl]-pyridazine) and R77975 (Pirodavir) have undergone clinical evaluation in human volunteers, only the latter two compounds have been shown to suppress colds when given intranasally before or just after virus challenge [4,29]. None of the interferons or the synthetic compounds has proved effective when used to treat a rhinovirus infection. The problem with the synthetic compounds, at least, appears to be that insufficient drug is retained at minimal inhibitory concentrations in the nasal epithelium to sustain effective antiviral activity [6] and hence it is important to find sustained release formulations. It is noteworthy that all these synthetic compounds have similar molecular weights, are hydrophobic and are insoluble in aqueous solutions. They work by binding to the viral capsid and thus render the virus non-infectious, either by stabilizing the capsid and hence preventing disassembly and penetration, or by interfering with virus adsorption to target cells [15]. X-ray crystallography studies have demonstrated that compounds such as WIN 51711, WIN 54954 and R61837 interact with the hydrophobic pocket, sometimes referred to as the 'WIN pocket' just underneath the 'canyons' present on the surface of the virus [11] (Figure 8.5).

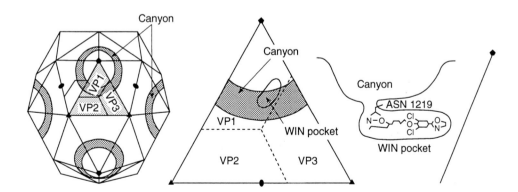

Figure 8.5 The rhinoviral capsid and WIN-pocket binding site are shown schematically in progressively greater detail from left to right. Left: the capsid consists of 60 identical triangular units each containing parts of two protomers. Centre: within one of the 60 units, the location of the drug binding site is showing within VP1, but close to an adjoining VP3. Right: WIN 54954 is shown in the pocket which is underneath the floor of the canyon, the site of receptor attachment. (Figure courtesy of Chapman, Kim and Rossmann [11].)

Local immune modulation has been conceived as a possible method to control rhinovirus infection and this at the present time is under evaluation including treatment directed against mediators [17,20]. Local hyperthermia with fully saturated air at 43°C seems to have both an immediate and delayed effect in the treatment of rhinovirus colds but the mechanism remains obscure [56]. Finally, there remain conflicting opinions about the efficacy of zinc gluconate lozenges for the treatment of rhinovirus colds and their use requires further evaluation [19].

In summary, despite extensive research over the past 10 to 15 years, no satisfactory remedy or treatment has yet been found to relieve the symptoms of colds caused by rhinoviruses.

8.10 REFERENCES

1. Al-Mulla, W., El-Mekki, A. and Al-Nakib, W. (1994) Rapid culture amplified immunofluorescent test for the detection of human rhinoviruses in clinical samples: Evidence of a common epitope in culture. *J. Med. Virol.*, **42**, 182–7.
2. Al-Nakib, W., Stanway, G., Forsyth, M., Hughes, P.J., Almond, J.W. and Tyrrell, D.A.J. (1986) Detection of human rhinoviruses and their molecular relationship using cDNA probes. *J. Med. Virol.*, **20**, 289–96.
3. Al-Nakib, W., Dearden, C.J. and Tyrrell, D.A.J. (1989) Evaluation of a new enzyme-linked immunosorbent assay (ELISA) in the diagnosis of rhinovirus infection. *J. Med. Virol.*, **29**, 268–72.
4. Al-Nakib, W. and Tyrrell, D.A.J. (1992) Drugs against rhinoviruses. *J. Antimicrobial. Chemother.*, **30**, 115–17.
5. Andrewes, C.H., Chaproniere, D.M., Gompels, A.E.H. *et al.* (1953) Propagation of common cold virus in tissue cultures. *Lancet*, **ii**, 546–7.
6. Andries, K., Janssens, M. and Parys, W. (1993) Pirodavir, a broad-spectrum antirhinoviral drug. *Int. Antiviral News*, **1**(4), 52.
7. Barclay, W.S. and Al-Nakib, W. (1987) An ELISA for the detection of rhinovirus specific antibody in serum and nasal secretion. *J. Virol. Methods*, **15**, 53–64.
8. Barclay, W.S., Callow, K., Sergeant, M. and Al-Nakib, W. (1988) Evaluation of an enzyme-linked immunosorbent assay which measures rhinovirus-specific antibodies in human sera and nasal secretions. *J. Med. Virol.*, **25**, 475–82.
9. Barclay, W.S., Al-Nakib, W., Higgins, P.G. and Tyrrell, D.A.J. (1989) The time course of the humoral immune response to rhinovirus infection. *Epidemiol. Infect.*, **103**, 659–69.
10. Bruce, C.B., Al-Nakib, W., Tyrrell, D.A.J. and Almond, J.W. (1988) Synthetic oligonucleotide as diagnostic probes for rhinoviruses. *Lancet*, **ii**, 53.
11. Chapman, M.S., Kim, K.H. and Rossmann, M. (1993) Structural comparisons of several antiviral agents complexed with human rhinoviruses of different serotypes. *Int. Antiviral News*, **1**(4), 53.
12. Colonno, R.J., Callahan, P.L. and Long, W.J. (1986) Isolation of a monoclonal antibody that blocks attachment of the major group of human rhinoviruses. *J. Virol.*, **57**, 7–12.
13. Colonno, R.J., Condra, J.H., Mizutani, S., Callahan, P.L., Davis, M.E. and Murcho, M.A. (1988) Evidence of the direct involvement of the rhinovirus canyon in receptor binding. *Proc. Natl Acad. Sci. USA*, **85**, 5449–53.
14. Dearden, C.J. and Al-Nakib, W. (1987) Direct detection of rhinoviruses by an enzyme-linked immunosorbent assay. *J. Med. Virol.*, **23**, 179–89.
15. Diana, G.D., Treasurywala, A.M., Bailey, T.R., Oglesby, R.C., Pevear, D.C. and Dutko, F.J. (1990) A model for compounds active against human rhinovirus-14. *J. Med. Chem.*, **33**, 1306–11.
16. Dick, E.C., Jennings, L.C., Mink, K.A., Wartgow, C.D. and Inhom, S.L. (1987) Aerosol transmission of rhinovirus colds. *J. Infect. Dis.*, **156**, 442–8.
17. Doyle, W.J., Riker, D.K., McBride, T.P. *et al.* (1993) Therapeutic effects of an anticholinergic-sympathomimetic combination in induced rhinovirus colds. *Ann. Otol. Rhinol. Laryngol.*, **102**(7), 521–7.
18. Duechler, M., Skem, T., Sommergruber, W. *et al.* (1987) Evolutionary relationships within the human rhinovirus genus; comparison of serotypes 89, 2 and 14. *Proc. Natl Acad. Sci. USA*, **84**, 1–5.
19. Eby, G. (1994) *The Zinc Lozenge Story. Handbook for Curing the Common Cold.* George Eby Research, Austin, Texas.

20. Farr, B.M., Gwaltney, J.M., Hendley, J.O. *et al.* (1990) A randomized controlled trial of glucocorticoid prophylaxis against experimental rhinovirus infection. *J. Infect. Dis.*, **162**, 1173–7.

21. Fox, J.P., Cooney, M.K. and Hall, C.E. (1975) Seattle virus watch: V. Epidemiologic observations of rhinovirus infections, 1965–1969, in families with young children. *Am. J. Epidemiol.*, **101**, 122–42.

22. Fox, J.P., Cooney, M.K., Hall, C.E. and Foy, H.M. (1985) Rhinoviruses in Seattle families, 1975–1979. *Am. J. Epidemiol.*, **122**, 830–46.

23. Gregg, I. (1983) Provocation of airflow limitation by viral infections: implication for treatment. *Eur. J. Respir. Dis.*, **64** (suppl. 128), 369–79.

24. Greve, J.M., Davis, G., Meyer, A.M. *et al.* (1989) The major human rhinovirus receptor is ICAM-1. *Cell*, **56**, 839–47.

25. Gwaltney, J.M. Jr, Hendley, J.O., Simon, G. *et al.* (1968) Rhinovirus infections in an industrial population: III. Number and prevalence of serotypes. *Am. J. Epidemiol.*, **87**, 158–66.

26. Gwaltney, J.M. (1982) Rhinoviruses, *in Viral Infections of Humans: Epidemiology and Control*, (ed. A.S. Evans), Plenum, New York, pp. 491–517.

27. Kapikian, A.Z., Hamory, B.H., Hamparian, V.V. *et al.* (1967) Rhinoviruses: a numbering system. *Nature*, **213**, 761–2.

28. Hamparian, V.V., Colonno, R.J., Cooney, M.K. *et al.* (1987) A collaborative report: Rhinoviruses – extension of the numbering system from 89–100. *Virology*, **159**, 191–2.

29. Hayden, F.G., Andries, K. and Janssen, P.A.J. (1992) Safety and efficacy of intranasal Pirodavir (R 77975) in experimental rhinovirus infection. *Antimicrob. Agents Chemother.*, **36**, 727–32.

30. Hendley, J.O. (1983) Rhinovirus colds: immunology and pathogenesis. *Eur. J. Respir. Dis.*, **64** (suppl.), 340–5.

31. Hughes, P.J., North, C., Jellis, C.H., Minor, P.D. and Stanway, G. (1988) The nucleotide sequence of human rhinovirus 1B: Molecular relationship within the rhinovirus genus. *J. Gen. Virol.*, **69**, 49–58.

32. Johnston, S.L., Bardin, P.G. and Pattemore, P.K. (1993) Viruses as precipitants of asthma symptoms. Rhinoviruses: molecular biology and prospects for future intervention. *Clin. Exp. Allergy*, **23**(4), 237–46.

33. Johnston, S.L., Sanderson, G., Pattemore, P.K. *et al.* (1993) Use of polymerase chain reaction for diagnosis of picornavirus infection in subjects with and without respiratory symptoms. *J. Clin. Microbiol.*, **31**(1), 111–17.

34. Krilov, L., Pierik, L., Keller, E. *et al.* (1986) The association of rhinoviruses with lower respiratory tract disease in hospitalized patients. *J. Med. Virol.*, **19**, 345–52.

35. Kruse, W. (1994) Die Erreger von Husten and Schnupfen. *Munch. Med. Wochenschr.*, **61**, 1547.

36. Levandowski, R.A., Ou, D.W. and Jackson, G.G. (1986) Acute phase decrease of T lymphocyte subsets in rhinovirus infection. *J. Infect. Dis.*, **153**, 743–8.

37. Levandowski, R.A. (1991) Rhinoviruses, in *Textbook of Human Virology*, (ed. R.B. Belshe), Mosby-Year Book, St. Louis, MO, pp. 411–26.

38. McGregor, S. and Mayor, H.D. (1968) Biophysical studies on rhinovirus and poliovirus: I. Morphology of viral ribonucleoprotein. *J. Virol.*, **2**, 149–54.

39. McMillan, J.A., Weiner, L.B., Higgins, A.M. and Macknight, K. (1993) The Rhinovirus infection associated with serious illness among paediatric patients. *Paediatr. Infect. Dis.*, **12**(4), 321–5.

40. McNaughton, M.R. (1982) The structure and replication of rhinoviruses. *Curr. Topics Microbiol. Immunol.*, **97**, 1–26.

41. Mischak, H., Neubauer, C., Kuechler, E. *et al.* (1988) Characteristics of the minor group receptor of human rhinovirus. *Virology*, **163**, 19–25.

42. Monto, A.S., Bryan, E.R. and Ohmit, S. (1987) Rhinovirus infections in Tecumseh, Michigan: frequency of illness and number of serotypes. *J. Infect. Dis.*, **156**, 43–9.

43.. Monto, A.S. and Sullivan, K.M.A.D. (1993) Acute respiratory illness in the community. Frequency of illness and the agents involved. *Epidemiol. Infect.*, **110**(1), 145–60.

44. Naclerio, R.M., Proud, D., Lichtenstein, L.M. *et al.* (1988) Kinins are generated during experimental rhinovirus colds. *J. Infect. Dis.*, **157**, 133–42.

45. Olive, D.M., Al-Mufti, S., Al-Mulla, W. *et al.* (1990) Detection and differentiation of picornaviruses in clinical samples following genomic amplification. *J. Gen. Virol.*, **71**, 2141–7.

46. Philipps, C.A., Melnick, J.L. and Grim, C.A. (1968) Rhinovirus infections in a student popu-

lation: Isolation of five new serotypes. *Am. J. Epidemiol.*, **87**, 447–56.

47. Proud, D., Reynolds, C.J., Lacapra, S., Kagey-Sobotka, A., Lichtenstein, L.M. and Naclerio, R.M. (1988). Nasal provocation with bradykinin induces symptoms of rhinitis and a sore throat. *Am. Rev. Respir. Dis.*, **137**, 613–16.

48. Rossmann, M.G., Arnold, E., Erickson, J.W. *et al.* (1985) Structure of a human common cold virus and functional relationship to other picornaviruses. *Nature*, **132**, 145–53.

49. Skern, T., Sommergruber, W., Blaas, D. *et al.* (1985) Human rhinovirus 2: complete nucleotide sequence and proteolytic processing signals in the capsid protein region. *Nucleic Acids Res.*, **13**, 2111–26.

50. Stanway, G., Hughes, P.J., Mountford, R.C., Minor, P.D. and Almond, J.W. (1984) The complete nucleotide sequence of a common cold virus: Human Rhinovirus 14. *Nucleic Acids Res.*, **12**, 7859–77.

51. Staunton, D.E., Merluzzi, V.J., Rothlein, R. *et al.* (1989) A cell adhesion molecule, ICAM-1, is the major surface receptor for rhinoviruses. *Cell*, **56**, 849–53.

52. Stott, E.J. and Killington, R.A. (1972) Rhinoviruses. *Annu. Rev. Microbiol.*, **26**, 503–24.

53. Tomassini, J.E. and Collono, R.J. (1986) Isolation of a receptor protein involved in attachment of human rhinoviruses. *J. Virol.*, **58**, 290–5.

54. Tomassini, J.E., Graham, D., DeWitt, C.M. *et al.* (1989) cDNA cloning reveals that the major group rhinovirus receptor on HeLa cells is intercellular adhesion molecule 1. *Proc. Natl Acad. Sci. USA*, **86**, 4907–11.

55. Turner, R.B., Winther, B., Hendley, J.O., Mygind, N. and Gwaltney, J.M. Jr (1984) Sites of virus recovery and antigen detection in epithelial cells during experimental rhinovirus infection. *Acta Otolaryngol.*, **100** (suppl. 413), 9–14.

56. Tyrrell, D.A.J. and Chanock, R.M. (1963) Rhinoviruses: a description. *Science*, **141**, 152–3.

57. Tyrrell, D.A.J., Barrow, I. and Arthur, J. (1989) Local hyperthermia benefits natural and experimental common colds. *Br. Med. J.*, **198**, 1280–3.

58. Winther, B., Gwaltney, J.M. Jr, Mygind, N., Turner, R.B. and Hendley, J.O. (1986) Sites of rhinovirus recovery after point inoculation of the upper airway. *JAMA*, **156**, 1763–7.

CORONAVIRUSES

Stuart Siddell and Steven Myint

Viral and Other Infections of the Human Respiratory Tract.
Edited by S. Myint and D. Taylor-Robinson. Published in
1996 by Chapman & Hall. ISBN 0 412 60070 6

9.1 INTRODUCTION AND HISTORY

The first report of a human coronavirus (HCV) was in 1965 when Tyrrell and Bynoe isolated a virus from nasal washings that had been collected 5 years earlier from a male child [168] in a boarding school [88]. This child had typical symptoms and signs of a common cold and the washing was found to be able to induce common colds in volunteers challenged intranasally. The virus, termed B814 after the number of the nasal washing, was cultivated in human embryo tracheal organ tissue but not in cell lines used at that time for growing other known aetiological agents of the common cold. The organ culture method was tried because it had already been shown that other respiratory viruses could be propagated in this manner and detected by the cessation of ciliary activity [174]. Hamre and Procknow were simultaneously working on five 'new' agents from the respiratory tract of six medical students with colds, collected in 1962 [55]. One of these, strain 229E, was adapted to grow in WI-38 cells. Almeida and Tyrrell showed that these new viruses were morphologically identical to the viruses of avian bronchitis and mouse hepatitis [4]. McIntosh and colleagues working at the National Institutes of Health in Bethesda, USA then found six morphologically related viruses that could not be adapted to cell monolayer culture but would grow in organ cultures [100]. Two of these isolates, OC (for organ culture) 38 and 43 were then adapted to grow in suckling mouse brain. The term 'coronavirus' that described the characteristic morphology of these agents was accepted in 1968 [169].

9.2 CLASSIFICATION OF CORONAVIRUSES

Even before the term 'coronavirus' was coined, members of the family 'Coronaviridae' were well-recognized veterinary pathogens. Table 9.1 shows those members of the family that have been recognized by the International Committee for the Taxonomy of Viruses (ICTV) and their relatedness to the human viruses; this relatedness has been determined on the basis of antigenic cross-reactivity [12,51,52,58,59,68,83,86,95,109,135,160,163,178] and more recently on gene sequence data (see Structure and molecular biology). Recently it has been accepted that toroviruses should also be included in this family but, as yet, there are no known human pathogens.

There are a number of human respiratory coronaviruses described in the literature but few have been well characterized (Table 9.2). On the basis of serological cross-reactivity, however, it is possible to classify most of them [17,102,110,142]. The two main serogroups are 229E-related and OC43-related and it is the prototype viruses that will be discussed in the rest of this chapter.

Coronavirus-like particles have also been seen in the stools of humans and have become known as human enteric coronaviruses

(HECVs). They are only now being characterized and at present must remain as putative members. Moreover, their role in causing disease is far from proven. In any case, as these are not respiratory pathogens, the reader is referred elsewhere for a review of these agents [28].

Other putative coronaviruses have also been described but not confirmed as coronaviruses. The Tettnang virus was isolated from the cerebrospinal fluid of a 1-year-old female with rhinitis, pharyngitis and encephalitis [116]. It was isolated originally from a suckling mouse brain culture and was shown subsequently to be a murine virus [10].

9.3 STRUCTURE AND MOLECULAR BIOLOGY

9.3.1 ORGANIZATION AND EXPRESSION OF THE HCV 229E GENOME

The hallmark of coronavirus gene expression is the production of a 3' co-terminal set of subgenomic mRNAs in the infected cell. With the exception of the smallest mRNA, all mRNAs are structurally polycistronic but only the information encoded in the 5' unique region of the RNA (i.e. the region absent from the next smallest RNA) is translated into protein. In the majority of cases, a single polypeptide is translated from each mRNA. The mechanism by which coronavirus subgenomic mRNAs are generated is complex and the details are

Table 9.1   Coronaviruses

Natural host	Virus	Acronym	Relatedness to human virus
Chicken	Avian infectious bronchitis virus	IBV	None
Cattle	Bovine coronavirus	BCV	OC43
Dog	Canine coronavirus	CCV	229E
Man	Human coronavirus 229E	HCV 229E	–
Man	Human coronavirus OC43	HCVOC43	–
Cat	Feline infectious peritonitis virus	FIPV	229E
Mouse	Murine hepatitis virus	MHV	OC43
Pig	Porcine transmissible gastroenteritis virus	TGEV	229E
Pig	Porcine haemagglutinating encephalomyelitis virus	HEV	OC43
Turkey	Turkey coronavirus	TCV	OC43

Table 9.2　Classification of human coronaviruses

Serogroup	Prototype virus	Members
A	229E	229E
		LP
		PR
		TO
		KI
		PA
		AD
		Linder
		VH
		Others
B	OC43	OC38
		OC44
		GI
		HO
		RO
Unclassified		B814
		692
		OC16
		OC37
		OC48

beyond the scope of this article. However, clearly, in common with other coronaviruses, the synthesis of HCV 229E subgenomic mRNA involves a process of discontinuous transcription and each mRNA has a leader RNA that is derived from the 5′ end of the genomic RNA [141]. It is also most likely that, as has been shown for mouse hepatitis virus (MHV) [152], HCV 229E-infected cells will contain a 5′ co-terminal set of subgenomic negative strand viral RNAs, each of which is involved in a transcriptionally active subgenomic replicative intermediate (RI) structure [151].

9.3.2 THE GENOMIC RNA OF HCV 229E

The genomic RNA of HCV 229E is a single molecule of positive-strand RNA composed of about 27 300 nucleotides and a 3′ poly-A tract of not less than 50 residues [62]. It is assumed that the 5′ end of the HCV 229E genome is linked to a 'cap' structure but this has not been demonstrated directly. At the 5′ and 3′ ends of

the genome, there are non-translated regions of approximately 300 and 400 nucleotides, respectively. It can be assumed that these regions contain cis-acting elements that are needed for the replication of genomic and, possibly, subgenomic mRNAs. Within the genomic RNA, a common sequence motif, U/A CU C/A AAC (the so-called 'intergenic' region), is located immediately upstream of most HCV 229E genes. These cis-acting elements have an important role in the synthesis of coronavirus mRNAs [77, 15].

The HCV 229E genomic RNA contains eight major open reading frames (ORFs; Figure 9.1). The total number of HCV 229E gene products is, however, unknown. First, two of the ORFs, ORF 1a and ORF 1b, are polycistronic and, almost certainly, encode a number of functionally distinct proteins (see below). Second, there are many redundant ORFs located within the coding regions of larger ORFs and, at least theoretically, these could be expressed by mechanisms such as RNA editing or ribosomal

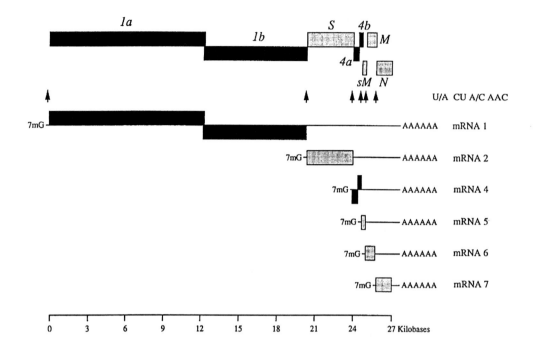

Figure 9.1 The organization and expression of the HCV 229E genome. The HCV genome is represented with eight major open reading frames (ORFs). The ORFs encoding the structural proteins (S, 5/sM, M and N) are shaded. Those encoding the non-structural proteins (polymerase/1a–1b, 4a and 4b) are solid. The position of the 'intergenic' motif (see text) and the relationship of the subgenomic mRNAs to the genomic RNA is illustrated. The transitionally active region of each mRNA is also indicated.

frameshifting. Nevertheless, it is possible to assign the HCV 229E ORFs to specific gene products. In some cases, these correspond to virus structural proteins that have been identified. In other cases, they correspond to putative proteins, the function of which can be deduced by analogy with other virus proteins. In a few cases, they correspond to putative proteins of unknown function. These assignments are listed in Table 9.3.

9.3.3 THE RNA POLYMERASE GENE (ORFS 1A AND 1B)

The HCV 229E RNA-dependent RNA polymerase gene is composed of two overlapping ORFs of 4086 and 2687 codons, respectively. These ORFs, ORF 1a and ORF 1b, encompass approximately 20 kilobases at the 5' end of the

genome. A computer-assisted analysis of the RNA polymerase genes of HCV 229E, MHV and infectious bronchitis virus (IBV) reveals a number of sequence motifs that have been associated with RNA replicative functions and protease activities in a variety of positive strand RNA viruses (Figure 9.2). These include an RNA-dependent RNA polymerase module and an NTP-dependent helicase motif located in ORF 1b and two or three putative protease domains located in ORF 1a. The protease domains belong either to the chymotrypsin/3C-like or papain-like protease groups. In addition, a motif reminiscent of a growth factor-like domain has been located in ORF 1a. The large size of the polymerase gene suggests that it may also encode many, yet unidentified, functions.

Table 9.3 The gene products of HCV 229E

Gene	Bases	Codons	Protein	Molecular mass (Da)
5' NTR	293	–	–	–
ORF 1a	12 258	4 086	[Polymerase]	454 200
ORF 1ab	20 277	6 759	[Polymerase]	754 200
S	3522	1 174	Surface	128 600
ORF 4a	402	134	Unknown	15 300
ORF 4b	267	89	Unknown	10 200
ORF 5	234	78	[Small membrane]	9 100
M	678	226	Membrane	26 000
N	1170	390	Nucleocapsid	43 500
3' NTR	422 +A$_n$	–	–	–

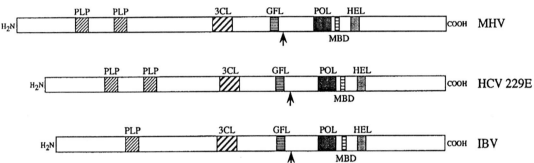

Figure 9.2 Putative functional domains in the HCV 229E polymerase. The position of putative functional domains in the RNA polymerases of HCV 229E, MHV and IBV are shown. The figure is drawn to scale, although the boundaries of the motifs cannot be defined precisely. PLP, papain-like protease; 3CL, 3C-like protease; GFL, growth-like factor/receptor; POL, polymerase module; MBD, metal binding domain; HEL, helicase (NTP binding) domain. The arrows indicate the position of the ORF 1a/1b junctions.

9.3.4 RNA POLYMERASE EXPRESSION

Ribosomal frameshifting

Expression of the proteins encoded in the polymerase gene of HCV 229E is mediated by translation of the genomic RNA, also known as mRNA 1. Translation of ORF 1a is thought to be initiated by a conventional cap dependent event at the 5' end of the mRNA. However, translation of ORF 1b is believed to be accomplished by a (–1) ribosomal frameshifting event in the region of the mRNA that encompasses the ORF 1a/1b overlap [61]. Thus, without any further processing, the polymerase gene could encode an ORF 1a pro-tein of 454 200 Da molecular mass and a carboxy-terminal extended ORF 1ab protein of 754 200 Da molecular mass.

The HCV 229E frameshifting event is mediated by a specific RNA structure composed of a sequence motif, the so-called slippery sequence UUUAAAC, located immediately upstream of a tertiary structure, the RNA pseudoknot [61]. For MHV and IBV, the pseudoknot is a bipartite structure, but for HCV it has been shown that an elaborated, probably a tripartite, quasi-helical pseudoknot, is required for a high frequency of frameshifting (Figure 9.3). The frameshifting event could

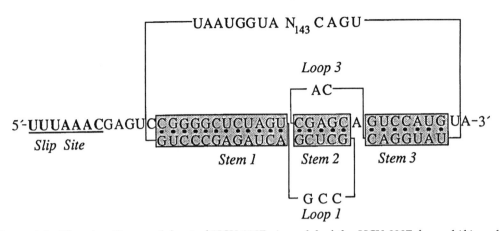

Figure 9.3 The tripartite pseudoknot of HCV 229E. A model of the HCV 229E frameshifting element, including the slippery sequence and a tripartite pseudoknot structure. A number, but not all, of the base-pairing interactions shown have been demonstrated experimentally [61].

be one of the important regulatory mechanisms of polymerase gene expression.

Proteolytic polyprotein processing

Complementation analysis of MHV temperature-sensitive (ts) mutants with an RNA minus phenotype has shown that there are at least five distinct viral functions related to coronavirus RNA synthesis [152]. Analysis of these mutants by genetic recombination allows the different functions to be located and ordered within the RNA polymerase locus [11]. The complementation frequencies of these mutants indicate intergenic, rather than intragenic recombination, so they provide strong evidence for the activity of viral encoded proteases that process the primary translation products of the polymerase gene into smaller, functional proteins. The protease motifs identified in the polymerase locus of HCV 229E (and other coronaviruses) are the most likely candidates for these activities.

Direct evidence of autoproteolytic activity encoded in a coronavirus polymerase gene has, so far, only been obtained for the first papain-like protease encoded in ORF 1a of MHV [9]. This activity has the classical features of a so-called 'leader protease' that cleaves an amino-terminal polypeptide from the primary translation product. The 3C-like protease of coronaviruses is most probably responsible for the processing of the replicative functions located in ORF 1b. There have been some preliminary attempts to analyse the proteolytic processing of the polymerase gene products of MHV [39] and similar studies on the polymerase gene of HCV 229E should follow soon.

9.3.5 THE SURFACE PROTEIN GENE (ORF S)

The HCV 229E surface glycoprotein gene encodes a polypeptide of 1173 amino acids with a molecular mass of 128 600 Da. The polypeptide has 30 potential *N*-glycosylation sites. The difference in the predicted size of the surface (S) protein and its apparent molecular weight in SDS–PAGE suggests that the majority of these sites are used. The coronavirus S protein forms the large peplomers on the surface of the virion and a number of structural features typical of coronavirus S proteins can be recognized in the HCV 229E S protein gene product [141]. These include an amino-terminal signal sequence and a carboxy-terminal membrane anchor and a cytoplasmic domain that displays a characteristic cluster of cysteine residues. Furthermore, the carboxyl half of the HCV 229E S protein contains two regions of hydrophobic residues that are arranged as heptad repeats. These regions are predicted to form a rigid, elongated, intramolecular coiled-coil region that would correspond to the stalk of the peplomer structure. In contrast, the amino-terminal half of the protein is thought to have a globular conformation.

In contrast to the S proteins of MHV and IBV, the HCV 229E S protein does not contain a central basic region with the motif RRXRR or RRAHR (where X is F, S, H or A). This motif has been identified as the sites at which the MHV and IBV S proteins are proteolytically cleaved. Apparently, the HCV S protein is not post-translationally cleaved.

The HCV 229E S protein has at least two major functions. First, it binds to cellular receptor(s) to initiate the infection process. A major class of receptor for HCV 229E has been identified as human aminopeptidase N [181]. Second, the S protein mediates membrane fusion. During the initial stages of the infection this function is responsible for the fusion of viral and cellular membranes at the cell surface. The structural features of the HCV 229E S protein fit remarkably well with the current view of fusogenic viral glycoproteins [94].

9.3.6 THE ORF 4A AND ORF 4B GENES

The proteins encoded by the HCV 229E ORF 4a and ORF 4b have predicted molecular masses of 15 300 and 10 200 Da, respectively. The central region of the ORF 4a gene product displays several long hydrophobic domains that may indicate a membrane-bound protein. To date these proteins have not been identified in the infected cell or in virions and they are considered provisionally as non-structural proteins of unknown function. There have been suggestions that one or both of these proteins have accessory functions and are not essential for replication in tissue culture [40,78]

9.3.7 THE SMALL MEMBRANE PROTEIN GENE (ORF 5)

On the basis of its structural similarity to the small membrane proteins of IBV and transmissible gastroenteritis virus (TGEV), the HCV 229E ORF 5 gene product is most probably a structural protein of the virus. This has not, however, been demonstrated directly. The predicted molecular mass of the HCV sM protein is 9100 Da. Besides its similarity to other coronavirus sM proteins, there is a striking structural resemblance with the M2 protein of influenza virus [98]. The influenza virus protein has an ion channel activity selective for monovalent ions and is believed to have an important role in regulating the flow of H+ ions in and out of viral and cellular compartments.

9.3.8 THE MEMBRANE PROTEIN GENE (ORF M)

The membrane glycoprotein gene of HCV 229E encodes a polypeptide of 225 amino acids with a molecular mass of 26 000 Da. The HCV 229E M protein has several features that are characteristic of coronavirus membrane proteins. First, there are three potential N-linked glycosylation sites, one of which is near the amino terminus. It has been shown that the HCV 229E M protein is N-glycosylated [87]. Second, the polypeptide displays three internal hydrophobic domains within the amino-terminal half and a relatively hydrophilic carboxy terminus. Third, the polypeptide is slightly basic with a net charge of +4 at neutral pH. These data suggest that the membrane topology of the HCV 229E M protein is very similar to that proposed by Rottier *et al.* [146] for the MHV M protein. The amino-terminal ectodomain is, however, shorter than that of MHV (16 residues compared with 25 residues).

The function of the HCV 229E M protein, in common with all coronavirus M proteins, is to bind the nucleocapsid structure to the virus envelope during virus assembly. As the coronavirus M protein is not transported to the plasmalemma but, instead, accumulates in the Golgi apparatus, this interaction may also dictate the intracellular site of coronavirus maturation.

9.3.9 THE NUCLEOCAPSID PROTEIN GENE (ORF N)

The nucleocapsid protein gene of HCV 229E lies at the 3' end of the genome. It encodes a polypeptide of 43 500 Da, which agrees with the apparent molecular weight of the HCV 229E N protein in SDS–PAGE [127]. In common with other coronavirus nucleocapsid proteins, the HCV 229E N protein is a serine-rich, basic protein (net charge +16 at neutral pH). The protein is most probably phosphorylated, although this has not been demonstrated directly. The distribution of basic and acidic residues is compatible with a three-domain structure, as proposed for the MHV N protein by Parker and Masters [132].

The function of the HCV 229E N protein has not been studied in any detail. However, by analogy with other coronavirus N proteins, the major function of the N protein will be to encapsidate the genomic RNA. Unfortunately, just now, there is no information on an N protein:RNA interaction that specifically leads to encapsidation.

9.3.10 THE MORPHOLOGY OF HCV 229E VIRIONS

Coronaviruses are described as roughly spherical, enveloped particles, approximately 100 nm in diameter with a characteristic 'fringe' of 20-nm long surface projections that are round or petal-shaped. The nucleocapsid structure is extended and helical. These features, and the virion proteins that comprise these structures, are represented diagrammatically for 229E virus in Figure 9.4. In this diagram, the ORF 5 gene product, the sM protein, is depicted as a structural component of the virus, although this has not yet been proved experimentally.

9.3.11 MOLECULAR BIOLOGY OF HCV OC43

OC43 virus is even more fastidious in its tissue culture requirements than 229E virus; consequently, analysis of the molecular biology of this virus has been correspondingly slow. It is clear, however, that it belongs to a haemagglutinating group which is distinguished by the presence of an additional 'haemagglutinin–esterase' (HE) gene, located between the polymerase and surface protein genes. The HE protein is a component of the virion and has receptor-binding activity, which is specific for the ligand, *N*-acetyl-9-*O*-acetylneuraminic acid as well as an acetylesterase receptor-destroying activity [158].

At the moment, however, the role of the HE protein in the replication cycle of these viruses is not exactly clear. First, it has been shown that the surface glycoprotein of bovine coronavirus

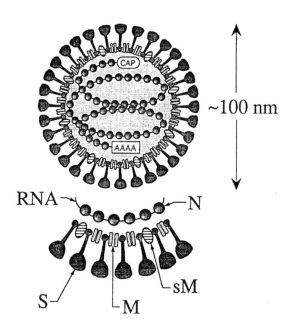

Figure 9.4 The structural components of HCV 229E, illustrated schematically. The model does not try to take account of the stoichiometry of the components nor their precise spatial relationships. RNA, genome; S, surface glycoprotein; M, membrane glycoprotein; sM, small membrane protein; N, nucleocapsid protein.

(BCV) [159] also has a strong binding affinity with *N*-acetyl-9-O-acetylneuraminic acid, i.e. it is also a haemagglutinin. Second, there is good evidence that the specific cellular receptors for MHV are members of the murine carcinoembryonic antigen (CEA) gene family [41]. And third, it has recently been shown that there is a rapid selection of HE-defective mutants during infection of mice with MHV [183]. Taken together, these data suggest that the HE protein may have an accessory function that is not obligatory. Nevertheless, this accessory function could, in certain circumstances, have an important influence on the virulence or tropism of HCV OC43 infections [184].

With the exception of the HE protein gene, the genome of HCV OC43 closely parallels that of HCV 229E regarding the RNA polymerase, surface, small membrane, membrane and nucleocapsid protein genes [155]. It differs quite significantly, however, regarding the putative non-structural genes. Thus, by analogy to MHV and BCV, there is probably an additional ORF upstream of the HE protein gene. Also, between the S and sM protein genes there is one ORF [124] that does not appear to be related to either the ORF 4a or 4b gene products of HCV 229E. In spite of these differences, it appears that the expression of the HCV OC43 genome follows the general pattern of other coronaviruses.

9.4 PHYSICOCHEMICAL PROPERTIES

Coronaviruses are susceptible to inactivation (of infectivity) by heat, acid and ultraviolet irradiation. 229E virus is more sensitive than OC43 virus to these procedures [22]. In the presence of 2% fetal calf serum, 229E virus has a 99% loss of infectivity after 10 hours at 33°C, whereas OC43 virus requires 40 hours to lose the same amount of infectivity; the same effect is produced in half of the time at 37°C. At 33°C, in the presence of 2% fetal calf serum and at pH 3.65 there is 99% loss of 229E virus infectivity after 1 hour, whereas this takes 2 hours

with OC43 virus. Similarly, 229E virus is inactivated 50% more quickly by ultraviolet irradiation than OC43. Viral infectivity of 229E, however, is stable for up to 2 weeks if kept at 4°C in the presence of serum and there is only minimal loss of infectivity after 25 cycles of freeze–thawing [93], though the RNA yield from clinical material shows more rapid decay unless the virus is stored with an RNAase inhibitor (unpublished data). Chloroform (5% for 10 min) and heat (56°C for 1 hour at pH 7.0) completely abolish infectivity of 229E virus [63]. Sodium desoxycholate, sodium dodecyl sulphate, β-proprionolactone and Triton X-100 also abolish infectivity.

229E virus is rapidly inactivated by chemical disinfectants: within 1 minute after exposure to 3% hydrogen peroxide and 0.2% C4 (a quaternary ammonium compound) [120].

Virus growth is sensitive to the presence of actinomycin D, the addition of 0.1 µg/ml to L132 cells causing a 50% reduction in infectious titre of 229E virus [87].

9.5 EPIDEMIOLOGY

Most epidemiological surveys of these viruses as agents of respiratory tract illness (see Disease manifestations, p.150) have been based on serology, using either complement-fixation or haemagglutination-inhibition tests for the two prototype viruses 229E and OC43. As it is clear that serologically unrelated coronaviruses exist, the prevalence of these viruses is likely to be underestimated. Moreover, it is also certain that not all human respiratory coronaviruses have yet been adapted to tissue or organ culture. This is supported by a study based in London, UK, in which coronaviruses were isolated, by tissue or organ culture, from 18.4% of 38 patients with common colds but a further 13% of the isolates, not identifiable as other viruses, were able to induce colds in inoculated volunteers [96].

The most extensive epidemiological survey of OC43 virus infection has been the study of the community of Tecumseh, Michigan, USA [122]. A 4-year study involved looking for serological evidence, by complement fixation or haemagglutination-inhibition, of infection with OC43 virus in 910 persons in 269 families. A mean of 17.1% of individuals showed evidence of infection with OC43 virus in any one year. There was, however, a cycling of the frequency of occurrence: OC43 virus infections occurred in most years but there were peaks of infection every 3 years. Over 80% of infections occurred despite pre-existing antibody, with infections gradually diminishing with age. These peaks took place in the winter–spring months. A study of 229E virus infections in the same community showed a mean annual rate of 7.7% infected [29]. This figure was half that recorded in a 6-year study of Chicago medical students, in which there was also marked year-to-year variation with peaks of 35% incidence [57]. Nearly 97% of 229E virus infections also occurred in the winter and spring months. The Tecumseh study showed that either 229E or OC43 virus was dominant in any one year.

This seasonal pattern was also seen in Seattle during 1975–1979, in a study in which the ELISA format was used to seek antibody rises in sequential sera from 44 adults and children in 10 families [157]. This study also showed that children had almost three times as many infections as adults.

In London, UK, the peak incidence of coronavirus infections in one study was not in winter but in summer [112]. A total of 298 sequential serum samples, collected in the years 1976–1981, were analysed for antibody rises to OC43 and 229E viruses by ELISA. Using this method, antibody could be detected for a mean of 3.5 months. Individuals appeared to get a coronavirus infection at a mean interval of 7.8 months. Another study by the same group in the same locality showed, however, that in children there was still a winter peak. In this latter study, 30% of 108 acute respiratory tract infections, in children under

the age of 6 years, were found to be caused by coronaviruses as detected by ELISA. Interestingly, 30% of index cases also had lower respiratory tract symptoms of wheezy bronchitis, but none of their siblings showed evidence of such symptoms [75].

Studies of coronaviruses as causes of clinical illness have shown that they are second only to rhinoviruses as the causes of the common cold. In the USA and England, 229E and OC43 viruses are responsible for 1–30% of all clinical cases [75,84,104,131,180], with approximately an equal number of subclinical infections. Studies from around the world – Italy [30,48], Finland [143], India [118], Germany [60,148], Czech Republic [72], Japan [119,121], Romania [150] – show that coronavirus infections are just as common.

9.6 DISEASE MANIFESTATIONS

Most coronavirus infections are asymptomatic and this has been seen in seroepidemiological surveys and in human volunteer studies [15,18]. In the latter, only 10–30% of inoculated volunteers have suffered clinical effects.

9.6.1 UPPER RESPIRATORY TRACT INFECTION

The most common clinical presentation is an upper respiratory tract infection (URTI) and, in particular, the common cold. The viruses have been isolated from patients with the common cold and when inoculated intranasally into volunteers they have produced common colds [5,15,18]. There is a mean incubation period of 3 days (range 2–5 days) followed by an illness that lasts a mean of 6–7 days (range 2–18 days). The classical clinical illness is well known and consists of general malaise, headache, nasal discharge, sneezing and a mild sore throat [5]. Approximately one-tenth of individuals will also have a fever, and one-fifth will have a cough. In Table 9.4, the clinical features of rhinovirus type 2-induced common colds, those induced by coronavirus 229E, and the illness

Table 9.4 Clinical features of rhinovirus, coronavirus and influenza A respiratory tract infection

Clinical Feature	Coronavirus 229E (%)	Rhinovirus 2 (%)	Influenza A (%)
Fever	9–23	7–16	98
Rhinorrhoea and/or obstruction	94–100	64–100	20–30
Headache	32–85	28–50	85
General malaise	46–47	28–43	80
Sneezing	85	50	30
Sore throat	54–68	87–93	50–60
Cough	21–31	64–68	90
Hoarseness	12	57	10
Myalgia	9	21	60–75
Watery/sore eyes	29	43	60–70
Chills	18	21	90

caused by the influenza A virus are contrasted (compiled from [5,177]). It is not possible on an individual basis to distinguish rhinovirus colds from coronavirus colds, and although earlier studies suggested that there may be a different incidence of some clinical symptoms following infection with different coronaviruses, this has not been substantiated. In any case, it would not help in identifying the cause in any single patient.

9.6.2 LOWER RESPIRATORY TRACT INFECTION

Less well documented is lower respiratory tract infection associated with coronaviruses. In a seroepidemiological study, using a complement-fixation test, coronaviruses were less likely to be found in hospitalized children with lower respiratory tract infection than in controls with non-respiratory tract disease: 3.5% of 565 children in the former group had evidence of infection with OC38 or OC43 virus compared with 8.2% of 245 children in the latter group [106]. In a later study, however, of 417 hospitalized children under 18 months of age with lower respiratory tract disease, there was serological evidence of either 229E or OC43 virus infection in 8.2% [105]. The incidence of coronavirus infection was higher than that of other respiratory viruses, except parainfluenza virus type 3 and respiratory syncytial (RS)

virus. It is unlikely that there is direct infection of the lower respiratory tract and that any association is through secondary phenomena. Though these secondary phenomena are yet to be defined, a link between coronavirus infection of the upper respiratory tract and wheezing attacks has been shown in several studies over the last 20 years [75,76,105,112]. The basis of this is unclear although a correlation between severity of cold symptoms and IgE levels in nasal secretions was shown in one study [25]; there did not, however, appear to be a correlation with systemic atopy, as assessed by pin-prick to common allergens, or with mediator (leukotriene C4, leukotriene B4 and histamine) release in nasal secretions. In clinical studies, asthmatic children appear to be at particular risk of virus-induced wheezing. In this situation up to 30% of acute wheezing episodes may be due to coronavirus infection. Increased airways resistance has been shown to occur in the upper respiratory tract in non-atopic individuals with colds [2,14] and asthmatic individuals are likely to be at greater risk of this being generalized in the respiratory tract.

9.6.3 NON-RESPIRATORY TRACT INFECTION

Non-respiratory tract infection has also been described with coronaviruses and includes multiple sclerosis, pancreatitis, thyroiditis, pericarditis, nephropathy and infectious mononucleosis when seroconversion has been sought [7,8,145]. The association with multiple sclerosis has been of particular interest since coronavirus-like particles were seen in the post-mortem brain of a patient who died with the disease [164]. Subsequently, OC43-related coronaviruses, SD and SK, were isolated from the brain material of two multiple sclerosis patients [23]. However, it is likely that these isolates were murine coronaviruses present in the mice used for cultivation [44,45,179], although a lack of neutralization, in a plaque assay, of SD or SK with OC43 antiserum was shown in one study [46]. Seroepidemiological studies in which attempts have been made to ascertain an association with multiple sclerosis have been conflicting in that either an association has not been found or when seen, it has been to either OC43 or 229E virus [70,97,114, 148]. Gene detection has failed to detect OC43 virus either by standard probes [161] or polymerase chain reaction (PCR) but has suggested a neurotropism for 229E virus [78]. There is also evidence of coronavirus transcription in the brains of patients with multiple sclerosis [162]. In addition, a five amino-acid motif on mRNA 4 of 229E virus is shared with myelin basic protein [78]. It is clear that further research is needed in this area.

The role of coronaviruses in causing diseases outside the respiratory tract has been doubted mainly because there has been no evidence that the virus spreads from the nasal mucosa. In support of this has been the finding that human serum contains one or more natural non-antibody inhibitors to coronavirus OC43 [36–38]. It is possible that 229E virus evades such mechanisms, however, as it is capable of replicating in macrophages which may not only impair the immune response but may also be a route of spread of the virus [134]. Recently it has also been possible, using the PCR, to show that there is a short viraemic phase with 229E virus in at least some experimentally inoculated volunteers (unpublished data). Allied to this, the identification of the

229E virus receptor as a metalloprotease [181] which is found on cells of many tissue types would also suggest greater possibilities for disease causation than just the common cold.

Although it is likely to be due to secondary phenomena, coronaviruses, in common with influenza viruses and rhinoviruses, have profound and occasionally prolonged effects on the psychology and psychomotor performance of individuals [20,33,166]. It is these effects which may provide an even more important rationale for the development of therapeutic and preventive measures to coronavirus infection as the economic consequences are great.

9.7 PATHOGENESIS AND IMMUNE RESPONSE

Not only the virus but also the host plays an important role in the pathogenesis of common colds. Everyone knows someone who claims that they never suffer from a cold. One factor appears to be the psychological state of the individual [20,33,166], though environmental [90] and genetic factors also play a role; susceptibility to 229E virus infection has been shown to reside on the q11-qter region of human chromosome 15 in a study using murine–human hybrid cell lines [147].

Little is known about the detail of pathogenic mechanisms in human coronavirus infection, principally because man is the only model of infection. Even the predominant mode of transmission is uncertain. Infection can be induced experimentally by direct inoculation of virus into the nose but this is unlikely to be the natural route. By analogy with rhinoviruses it is likely to be either by aerosols or fomites [43] with the former a more likely explanation for the frequent simultaneous transmission that is seen to those exposed. In support of this is the finding that 229E virus survives well in an atmosphere of high humidity and low temperature [74].

Once in the nose the virus is thought to enter by a specific receptor, aminopeptidase N

(CD13) in the case of 229E virus and a sialic acid-containing receptor in the case of OC43 virus (see Structure and molecular biology, p.142). It is suggested by ultrastructural studies that it is ciliated cells, and not goblet cells, that are infected [1]. *In vitro*, 229E virus appears to be randomly distributed on the surface of cells at 4°C but redistributes using an energy-dependent mechanism on heating to 33°C [133]. Replication then takes place over the next 8–24 hours; in MRC-c and WI-38 cells, 229E virus can be found in rough endoplasmic reticulum 12 hours after infection, and in cytoplasmic vesicles and extracellulary 24–36 hours after infection [13,108]. Replication is optimal at 32–33°C, the temperature in the superficial layers of the nasal mucosa. Virus budding then takes place from Golgi apparatus with little loss of cells. Cilia appear to be withdrawn into the cell and it may be this which is associated with rhinorrhoea [1] although increased plasma exudation of proteins such as fibrinogen, which may be histamine-responsive must also play a role [3,54]. Serum antibody levels rise after about a week but it is not clear whether it is this response or cell-mediated mechanisms which clear infection [26]. Certainly there is some, inverse, correlation of the severity and likelihood of disease with pre-existing serum antibody but the mere presence of such antibody is not protective [24]. Serum antibody levels peak about 2 weeks after infection and decline to low or undetectable levels at 12–18 months. The serum antibody response is directed mainly against the surface protein but there is also an, albeit lesser, antibody response directed against the nucleoprotein and membrane proteins [111,154].

Interferon, specific nasal secretory IgA and total nasal secretory protein appear to play a role in protection from infection [172]. The last component may have natural antiviral properties [117]. 229E virus has been shown to be a good inducer of interferon in natural leucocyte cultures derived from healthy children [138]. Other non-specific factors, such as a rise in local temperature that occurs during nasal blockage, may be important not in resistance but in aborting an infection as they may induce the production of heat-shock proteins, activate lymphocytes and inhibit viral replication.

The cell-mediated immune response has not been investigated in natural infections but sera from volunteers has been shown to possess antibody that can elicit antibody-dependent cellular toxicity *in vitro* [69].

Whatever the relative importance of different arms of the immune response in clearing coronavirus infection, re-infections are common and can occur within a year to the same serotype or within 2 months to a heterologous serotype [29,96,157].

9.8 DIAGNOSIS

Because of the trivial and temporary nature of common colds, the detection of coronavirus infections has not been attempted in routine diagnostic laboratories. This situation is unlikely to change unless antiviral therapy becomes available. Although most techniques available to the diagnostic virologist have been used to detect human respiratory coronaviruses, even in research laboratories the range of tests employed by any one centre tends to be limited. A synopsis of the range of techniques used is presented below, but the reader is referred elsewhere for details of methods [128].

Organ culture

The method, or modifications of it, developed by Tyrrell and Bynoe [168] to isolate B819 virus is still used by some laboratories but is hampered by the difficulty of obtaining human embryonic tracheal tissue. It remains, however, the best method available for primary isolation of the broadest range of respiratory coronaviruses.

Tracheal tissue is taken from 14 to 24-week-old embryos and planted in sterile plastic Petri dishes containing 199 medium. The tissue is immersed with the cilia uppermost. Virus can be inoculated onto the cilia and the tissues are then incubated at 33°C for up to 10 days. Viral replication is indicated by cessation of ciliary activity and confirmed by interference with another virus (echovirus, parainfluenza or Sendai). Electron microscopy was used originally to confirm that isolates as coronavirus but other tests such as neutralization are now employed.

Trachea obtained from 5 to 9-month-old fetuses has also been shown to support the growth of coronaviruses, and different media recipes can also be used [55].

Mouse brain culture

OC38 and OC43 viruses (but not B814, OC16, OC37, OC44 or OC48) have been adapted to grow in suckling mouse brains [55,101,103, 170]. The mice can be inoculated intracerebrally or via the peritoneum, with encephalitis occurring some days later. In the initial description by McIntosh and colleagues, CD-1 Swiss mice were used, with encephalitis occurring 11–15 days after inoculation with virus that had been passaged several times in organ culture. After the fourth passage, the time to illness was reduced to 40–60 hours. Evidence of infection was not found in other organs (liver, heart or lungs). Virus can be prepared from brain suspensions by clarification through low-speed centrifugation and then adsorption to and elution from group O erythrocytes or by more elaborate ultracentrifugation-based procedures [82]. Virus may be visualized by electron microscopy. Although this is not used for primary isolation of virus, this method is still a commonly used means of preparing OC43 antigen for serological assays.

Cell culture

Cell cultures have proved unreliable for primary isolation of all human respiratory coronaviruses but certain strains have been adapted to growth in them. The 229E virus and related strains grow well in a continuous heteroploid

cell line termed C16 because they were the sixteenth clone of MRC-c cells that was selected [137]. The original description of these cells from the MRC Common Cold Unit in Salisbury, indicated that morphologically they were a mixture of fibroblastic and epithelioid cells. The former constituted three-quarters of the cell population. These cells were contaminated with organisms detected by Hoechst stain 33258, presumably *Mycoplasma* spp. Interestingly, attempts to remove this contamination with agents, such as ciprofloxacin, have resulted in a poorer yield of virus from the cell line (unpublished data).

C16 would, arguably, be the cell line of choice for the isolation of 229E virus but frequent passage of these cells results in an increasing proportion of the epithelioid content and a consequent reduction in the ability of the cell line to sustain detectable replication of virus. Many laboratories have continued to utilize a cell line that was originally used in the work of Hamre and Procknow [55]: a human diploid cell strain from the Wistar Institute, so-called WI-38 cells. Although 229E virus was readily adapted to this cell line, primary isolation was in human kidney cells and the authors noted that WI-38 cells may not be ideal for primary isolation.

Apart from C16 and WI-38 cells, many other cell lines and strains have been used to grow individual virus strains [16,19,21,31,32,56,79, 153,154,171]. These are summarized in Table 9.5. These viruses do not grow with easily recognizable cytopathic effects in cell types commonly used for the isolation of other respiratory viruses, such as HEp-2, rhesus monkey kidney or MRC-5, which makes routine identification difficult. It is, moreover, clear that the ideal cell line is not yet available and other methods of diagnosis have been applied.

Haemadsorption

OC43 virus, once adapted to BSC-1 cells, can be detected in cell culture by an ability to adsorb rat and mouse erythrocytes.

Table 9.5 Cell lines and strains used for the cultivation of human coronaviruses

Cell type	Virus	Primary isolation or adaptation
Human embryonic kidney (HEK)	229E	Primary
C16 (see text)	229E	Adaptation ≥ primary
Human embryonic lung fibroblast, WI-38	229E	Adaptation ≥ primary
Human embryonic lung fibroblast, MRC-c	229E and OC43	Adaptation > primary (OC43, adaptation only)
Human embryonic lung epithelium, L132	229E and OC43	Adaptation > primary
Human embryonic intestinal fibroblast, MA177	229E	Primary isolation of some strains
Human type II pneumocytes	229E	Primary > adaptation
Human fetal tonsil fibroblast (FT)	229E and OC43	Adaptation
Human embryonic rhabdomyosarcoma (RD)	229E and OC43	Adaptation
Primary monkey kidney (PMK)	OC43	Adaptation
Rhesus monkey kidney epithelium (LLC-MK2)	OC43	Adaptation
Continuous green monkey kidney epithelioid (BSC-1)	OC43	Adaptation

Adsorption of human, monkey and guinea pig cells occurs poorly, if at all [80].

Electron microscopy and immune electron microscopy

Electron microscopy of nasal washings is impractical as the virus load is usually below the level of sensitivity of standard methods. It has been used, however, to detect virus in tissue sections, such as in mouse brain culture of OC43 virus. It is also the means by which HECVs have been detected [27]. Negative staining with tungsten has usually been the method of choice, but molybdenum and uranium salts have also been used. The particles appear larger if uranium salts are used in place of tungsten [35].

An attempt to enhance electron microscopy by utilizing antibody concentration of cultured virus has been used successfully to detect 692 virus in washings from an adult with upper respiratory tract infection [81]. Nasal washings were passaged through both cell culture and tracheal organ culture and the resultant supernatant fluid was then incubated with convalescent serum from the same patient. After centrifugation the pellet was examined on a Formvar–carbon-coated grid.

Aggregates of virus were clearly discernible. Supernatant fluid that had been incubated with phosphate-buffered saline, instead of convalescent serum, was also examined but virus particles were not seen.

Immunofluorescence

An immunofluorescence method has been developed and applied to the detection of 229E and OC43 viruses in nasopharyngeal secretions and washings [107]. Sera were raised against mouse brain-derived OC43 virus and cell culture-grown 229E virus in rabbits and used in an indirect fluorescence assay. This test was able to detect homologous coronavirus antigens in nasal washings from infected volunteers, though cross-reactivity was noted with the 229E virus antiserum in washings from volunteers who had been inoculated with OC43 and OC44 viruses. No nasopharyngeal aspirates from 106 children who were hospitalized with respiratory tract infections had detectable coronavirus antigen by this method. It is difficult to ascertain whether this was due to a lack of sensitivity of this system as paired sera collected from 66 children during the study period did not show evidence of coronavirus infection.

Enzyme-linked immunoassay (ELISA)

An ELISA method based on purified 229E virus and HECV CV-Paris (which has cross-reactivity with OC43 virus) has been used to diagnose infections in children [110,111]. The ELISA method is a modification of that described for antibody detection using rabbit antisera (see below). In a study of 30 children aged 6 months to 6 years, 159 samples were collected: 111 nose swabs, 11 throat swabs and 55 nasopharyngeal aspirates. Some 34.2% of the nose swabs were positive for either 229E or OC43 virus, but only 18.2% of the throat swabs and nasopharyngeal aspirates. No comparison with serology was attempted but the positivity rate would suggest that this ELISA was a sensitive test.

Time-resolved fluoroimmunoassay

Using monoclonal antibodies raised in Balb/c mice, a time-resolved immunoassay has been developed to detect OC43 and 229E antigens. When compared with monoclonal- and polyclonal-based ELISA assays, it was a more sensitive procedure than either: 100% of known positive nasopharyngeal aspirates were identified with the assay whereas only 69% of 229E infections and 90% of OC43 infections were diagnosed by the monoclonal-based ELISA. The sensitivity of the assay was determined as 0.308 ng virus for 229E and 0.098 ng virus for OC43 [65].

Nucleic acid hybridization

A gene detection method for human coronaviruses was first applied to 229E virus using Northern hybridization [126,127]. A cDNA that encoded the entire nucleocapsid gene for 229E was ligated into a Riboprobe (RPromega) vector, pGEM-1, from which ^{32}P-labelled full-length transcripts could be generated. These

transcripts could be made as sense or antisense depending on whether an SP6 or T7 promoter was used. This method has been applied to the detection of 229E virus in nasal washings from inoculated volunteers and has been shown to be at least as sensitive as culture, with the advantage of a diagnostic result being available within 48 hours. An interesting feature was that the probe was able to detect virus for longer than cell culture in sequential samples from the volunteers. This probe method will not detect OC43 virus and attempts to remove the radioactive labelling by incorporating biotin or digoxigenin into transcripts have led to significant loss of sensitivity (unpublished data).

Reverse transcription–polymerase chain reaction (RT–PCR)

With the advent of gene and probe amplification strategies it was to be expected that these methods would be seen as advantageous for these difficult to cultivate viruses. Gene amplification methods based on 'nested' priming have been shown to be a sensitive and specific means of detecting both 229E and OC43 viruses [129]. Serotype-specific 'nested' primers were designed from the known sequences of the nucleocapsid genes of 229E and OC43 viruses. The inner primers were, in particular, chosen to produce a small fragment of about 100 base pairs for maximum sensitivity. Ribonucleic acid (RNA) is extracted using an acid–phenol–guanidium isothiocyanate procedure followed by reverse transcription using murine Moloney leukaemia virus RT. Two 20-cycle amplification steps are then used with the outer and inner sets of primers, respectively. The sensitivity of the assay appears to be much greater than that of cell culture or probe and each primer pair appears to be either 229E or OC43 virus-specific. The use of this method has greatly enhanced the diagnostic yield of coronaviruses in clinical material from asthmatic children [76]: the assay is at least as sensitive as a combination of culture and serology for diagnosing infection and is more specific. RT–PCR is likely to become the method of choice for direct virus detection.

Serological methods

Because of the lack of reliable detection methods before the development of those using gene detection, most epidemiological studies have been based on serological assays to determine evidence of coronavirus infection. The most widely used and sensitive format is the ELISA. The assay was first described for virus strain 229E [92] but has since been adapted for detection of antibodies to OC43 virus [91,113,156]. The 229E virus assay uses antigen that is grown in cell monolayers and then clarified. The OC43 virus test uses mouse brain-derived antigen. Rabbit antisera have been used for both 229E and OC43 tests. The specificities of the assays are similar to those of counter-immunoelectrophoresis, neutralization and complement-fixation, but sensitivity is over 1000-fold greater. In volunteer studies, the 229E virus assay has shown a close correlation between clinical illness and virus shedding. It has been the principal method for determining the occurrence and frequency of coronavirus infections in serological surveys but recent data suggest that some false-positive and false-negative reactions occur [129]. The 229E virus assay also detects antibody rises to some 229E-like viruses (PR, KI and TO). There also appears to be cross-reactivity of 229E virus with MHV-3 in the specified format, due to bilateral cross-reactions between the S protein of MHV-3 and the N protein of 229E virus [91]. The use of recombinant S protein as antigen may overcome some of these problems [139].

OC43 virus will also agglutinate chicken, mouse and rat erythrocytes at 4°C, 20°C and 37°C [82]. Human and monkey cells can be agglutinated at 4°C but not at the other two temperatures. This phenomenon is abolished by trypsin or Tween-80/ether treatment of virus. A haemagglutination-inhibition test based on this phenomenon was specific and did not detect other haemagglutinating viruses such as influenza virus.

Other serological test formats have also been used: indirect haemagglutination [85] and immune-adherence haemagglutination [47] for 229E virus antibody; rapid microneutralization [49] and plaque-reduction [50] for OC43 virus antibody; and immunofluorescence [123], complement-fixation [71] and single radial haemolysis [64,144] for both 229E and OC43 antibody. These test formats have been superseded by the ELISA.

9.9 MANAGEMENT AND THERAPY

The management of common colds due to coronaviruses is the same as that for other causes, essentially aimed at symptomatic relief [99]. Such general remedies are summarized in Table 9.6. The use of local hyperthermia is, particularly, interesting as it has undergone more rigorous scientific evaluation than other measures [130,173,175,182]. An apparatus was developed to deliver hot, humid air into the nares at a temperature of 43°C and a rate of 4 l/min. It was hoped that this would have an antiviral effect in that this temperature should be non-permissive for the replication of either rhinoviruses or coronaviruses; moreover, there might be stimulation of lymphocyte activity or heat-shock proteins. Initial controlled studies were promising, with 55% of volunteers treated daily for 20 minutes being asymptomatic, compared with only 10% of those who had air at 30°C [175]. The existence of an antiviral effect was, however, not shown. The use of intranasal nedocromil, a mediator blocker, on coronavirus colds has, similarly, been tried with some clinical benefit but without any effect on virus shedding [176].

Specific therapy has, however, been tried with intranasal recombinant human interferon-α. In a double-blind placebo-controlled study

Table 9.6   Popular remedies for the common cold

Remedy	Possible effect(s)	Comments
Aspirin*	Antipyretic	Increases viral shedding in experimental rhinovirus colds.
	Analgesic	
		Diminishes efficiency of mucociliary mechanisms
Paracetamol*	Antipyretic	
	Analgesic	
Caffeine*	Analgesic	Alters mood and sleep
	Anti-sedative	Raises blood pressure
		Biochemical side effects
		Side effects outweigh any theoretical benefit
Anticholinergic compounds* (atropine, scopolamine)	Reduce nasal secretions	Contraindicated in glaucoma
		Multiple side effects (urinary retention, constipation, confusion etc.)
		No evidence of efficacy
Alcohol	Sedative	
Antihistamines* (chlorpheniramine, brompheniramine)	Reduce secretions	Probable 'drying' effect due to anticholinergic activity rather than as antihistamines
	Sedative	
Sympathomimetics* (pseudoephredine, ephredine, phenylephrine, phenylpropamine)	Reduce secretions	Have been shown to reduce nasal airways resistance and provide symptomatic relief if applied locally. Rebound hyperaemia, however, occurs
Antitussives* (dextromethorphan)	Reduce cough	Centrally acting
Expectorants* (guaifenesin, ammonium chloride, ipecac, terpin hydrate)	Reduce cough	No evidence of efficacy
Menthol and other inhalations	Centrally acting	Provides symptomatic relief from congestion
		Does not change airways resistance
Garlic	Reduces risk of contact	No large case-controlled studies
	Has antimicrobial properties	
Vitamin C	Inhibits prostaglandin synthesis[140]	Marginal benefit shown in studies so far[53]
	Antiviral (bacteriophages)[125]	Aspirin potentiates uptake of ascorbic acid in common colds
	Antihistamine	
	Immunomodulation[165]	
	Reverses decrease in leukocyte ascorbic acid levels[73]	
Steam inhalation	Loosens tenacious secretions	Symptomatic relief only
Zinc lozenges	Inhibits post-translational cleavage of picornaviral polyprotein[42]	Reduces duration of symptoms if given as one lozenge every 2 hours while awake
Hot soups	Increase nasal mucus velocity	Chicken soup appears to be, particularly, efficacious
Cessation of smoking	Unknown	Reduces duration of symptoms
Local hyperthermia	Antiviral effect	See text

* As part of a proprietary formula

involving 83 volunteers, the incidence and severity of colds was reduced, as was the duration of virus shedding [66]. In those receiving interferon the mean total dose was 3.53×10^7 Units, given as a course three times daily over 4 days. This dose regimen was well tolerated. This was not, however, a truly therapeutic trial in that the virus challenge was not made until 4 hours after the fourth dose of interferon. Similar results were obtained in another placebo-controlled study of 55 volunteers with intranasal recombinant α-2b interferon [167]. In this latter study, 2×10^6 i.u./day were given for 15 days; some 19 of 26 (73%) of the placebo arm had a symptomatic cold, compared with 12 of 29 (41%) of those that received interferon. The severity of colds was also diminished and the mean duration of colds was reduced from over 4 days to 0.5 day. There was a statistically insignificant increase in bloody nasal mucus in the group receiving interferon but more unpleasant side effects were not noted.

Another immunomodulator, a thioguanosine derivative, appeared to enhance the resistance of mice to coronavirus infection but did not appear to have the same effect in man at the doses used [67].

Although no trials have been attempted in humans, there are a number of substances which appear to have *in vitro* activity against coronaviruses. The protease inhibitor, leupeptin, inhibits the growth of 229E virus in MRC-c cells if given within 2 hours of infection. The IC_{50} in a plaque reduction test was shown to be 0.4 μg/ml, whereas the growth of the MRC-c cells was unaffected by 50 μg/ml [6]. Another protease inhibitor, cystatin C, has been shown to inhibit both 229E and OC43 virus [34]. Antipain had a lesser activity but other inhibitors, such as pepstatin A and cathaipsin, had no effect. Protease inhibitors have, until now, found greater use as a means of studying the infectious cycle of 229E virus than as possible antivirals.

It has been a recent vogue to investigate the possibility of natural products as antimicrobial substances, and one such compound has been discovered that has anticoronaviral activity. An extract from the *Mycale* sponge, mycalamide A (Figure 9.5), protected four of eight mice from lethal A59 virus infection, whereas all eight control mice died [136].

It is hoped that some of these substances, or derivatives of them, will be developed in the near future as antivirals. With the advent of rapid diagnostics, their use in patients with chronic lung disease is now practical.

9.10 CONCLUSIONS

The use of animal coronaviruses to study viral pathogenesis has been successful for many years. The human coronaviruses are, however, under-investigated but this may change with the recognition that they may be associated

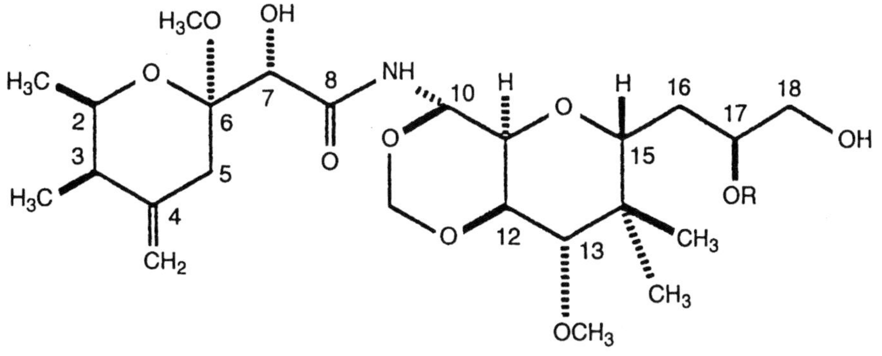

Structure of mycalamide *A* : *R = H*

Figure 9.5 The structure of mycalamide.

with serious disease. The next decade should see more advances in our understanding of this group of viruses.

9.11 REFERENCES

1. Afzelius, B.A. (1994) Ultrastructure of human nasal epithelium during an episode of coronavirus infection. *Virchows Arch.*, **424**, 295–300.
2. Akerlund, A, (1993) Nasal pathophysiology in the common cold. PhD Thesis. University of Lund, Sweden.
3. Akerlund, A., Greiff, L., Andersson, M., Bende, M., Alkner, U. and Persson, C.G. (1993) Mucosal exudation of fibrinogen in coronavirus-induced common colds. *Acta Otolaryngol.*, **113**, 642–8.
4. Almeida, J.D. and Tyrrell, D.A.J. (1967) The morphology of three previously uncharacterised human respiratory viruses that grow in organ culture. *J. Gen. Virol.*, **1**, 175–8.
5. Andrewes, C.H. (1962) The Harben lectures: The Common Cold. *J. R. Inst. Public Health Hyg.*, supplement.
6. Appleyard, G. and Tisdale, M. (1985) Inhibition of the growth of human coronavirus 229E by leupeptin. *J. Gen. Virol.*, **66**, 363–6.
7. Apostolov, K. and Spasic, P. (1975) Evidence of a viral aetiology in endemic (Balkan) nephropathy. *Lancet*, **ii**, 1271–3.
8. Arnold, W., Klein, M., Wang, J.B., Schmidt, W.A.K. and Trampisch, H.J. (1981) Coronavirus-associated antibodies in nasopharyngeal carcinoma and infectious mononucleosis. *Arch. Otorhinolaryngol.*, **232**, 165–7.
9. Baker, S.C., Shieh, C.K., Soe, L.H., Chang, M.F., Vannier, D.M. and Lai, M.M.C. (1989) Identification of a domain required for autoproteolytic cleavage of murine coronavirus gene A polyprotein. *J. Virol.*, **63**, 3693–9.
10. Bardos, V., Schwanzer, V. and Pesko, J. (1980) Identification of Tettnang virus ('possible arbovirus') as mouse hepatitis virus. *Intervirology*, **13**, 275–83.
11. Baric, R.S., Fu, K., Schaad, M.C. and Stohlman, S.A. (1990) Establishing a genetic recombination map for murine coronavirus strain A59 complementation groups. *Virology*, **177**, 646–56.
12. Barlough, J.E., Johnson-Lussenberg, C.M., Stoddart, C.A., Jacobson, R.H. and Scott, F.W. (1984) Experimental inoculation of cats with human coronavirus 229E and subsequent challenge with feline infectious peritonitis virus. *Can. J. Comp. Med.*, **49**, 303–7.
13. Becker, W.B., McIntosh, K., Dees, J.H. and Chanock, R.M. (1967) Morphogenesis of avian infectious bronchitis virus and a related human virus (strain 229E). *J. Clin. Microbiol.*, **1**, 1019–27.
14. Bende, M., Barrow, G.I., Heptonstall, J. *et al.* (1989) Changes in human nasal mucosa during experimental coronavirus common colds. *Acta Otolaryngol.*, **107**, 262–9.
15. Bradburne, A.F., Bynoe, M.L. and Tyrrell, D.A.J. (1967) Effects of a 'new' human respiratory virus in volunteers. *Br. Med. J.*, **3**, 767–9.
16. Bradburne, A.F. (1969) Sensitivity of L132 cells to some 'new' respiratory viruses. *Nature*, **221**, 85–6.
17. Bradburne, A.F. (1970) Antigenic relationships amongst coronaviruses. *Archiv. fur die ges. Virusforschung*, **31**, 352–64.
18. Bradburne, A.F. and Somerset, B.A. (1972) Coronavirus antibody titres in sera of healthy adults and experimentally infected volunteers. *J. Hyg. (Camb)*, **70**, 235–44.
19. Bradburne, A.F. (1972) An investigation of the replication of coronaviruses in suspension cultures of L132 cells. *Arch. Gesamte Virusforsch.*, **37**, 297–307.
20. Broadbent, D.E., Broadbent, M.H.P., Philpotts, R.J. and Wallace. J. (1984) Some further studies on the prediction of experimental colds in volunteers by psychological factors. *J. Psychosom. Res.*, **28**, 511–23.
21. Bruckova, M., McKintosh, K., Kapikian, A.Z. and Chanock, R.M. (1970) The adaptation of two human coronavirus strains (OC38 and OC43) to growth in cell monolayers. *Proc. Soc. Exp. Biol. Med.*, **135**, 431–5.
22. Bucknall, R.A., King, L.M., Kapikian, A.Z. and Chanock, R. (1971) Studies with human coronaviruses II. Some properties of strains 229E and OC43. *Proc. Soc. Exp. Biol. Med.*, **139**, 722–7.
23. Burks, J.S., DeVald, B.L., Jankovsky, L.D. and Gerdes, J.C. (1980) Two coronaviruses isolated from central nervous system tissue of two multiple sclerosis patients. *Science*, **209**, 933–4.
24. Callow, K.A. (1985) Effect of specific humoral immunity and some non-specific factors on resistance of volunteers to respiratory coronavirus infection. *J. Hyg.*, **95**, 173–89.
25. Callow, K.A., Tyrrell, D.A.J., Shaw, R.J., Fitzharris, P., Wardlaw, A.J. and Kay, A.B. (1988) Influence of atopy on the clinical mani-

festations of coronavirus infection in adult volunteers. *Clin. Allergy*, **18**, 119–29.

26. Callow, K.A., Parry, H.F., Sergeant, M. and Tyrrell, D.A.J. (1990) The time course of the immune response to experimental coronavirus infection of man. *Epidemiol. Infect.*, **105**, 435–46.

27. Caul, E.O. and Egglestone, S.I. (1977) Further studies on human enteric coronaviruses. *Arch. Virol.*, **54**, 107–17.

28. Caul, E.O. and Egglestone, S.I. (1982) Coronaviruses in humans, in *Virus Infections of the Gastrointestinal Tract*, (eds D.A.J. Tyrrell and A.Z. Kapikian), Marcel Dekker, New York.

29. Cavallaro, J.J. and Monto, A.S. (1970) Community-wide outbreak of infection with a 229E-like coronavirus in Tecumseh, Michigan. *J. Infect. Dis.*, **122**, 272–9.

30. Cereda, P.M., Pagani, L. and Romero, E. (1986) Prevalence of antibody to human coronaviruses 229E, OC43 and neonatal calf diarrhoea coronavirus (NCDCV) in patients of northern Italy. *Eur. J. Epidemiol.*, **2**, 112–17.

31. Chaloner-Larsson, G. and Johnson-Lussenberg, C.M. (1981) Establishment and maintenance of a persistent infection of L132 cells by human coronavirus strain 229E. *Arch. Virol.*, **69**, 117–30.

32. Chaloner-Larsson, G. and Johnson-Lussenberg, C.M. (1981) Characteristics of a long-term in vitro persistent infection with human coronavirus 229E. *Adv. Exp. Med. Biol.*, **142**, 309–22.

33. Cohen, S., Tyrrell, D.A.J. and Smith, A.P. (1991) Psychological stress and susceptibility to the common cold. *N. Engl. J. Med.*, **325**, 606–12.

34. Collins, A.R. and Grubb, A. (1991) Inhibitory effects of recombinant human cystatin C on human coronaviruses. *Antimicrob. Agents Chemother.*, **35**, 2444–6.

35. Davies, H.A. and MacNaughton, M.R. (1979) Comparison of the morphology of three coronaviruses. *Arch. Virol.*, **59**, 25–33.

36. Debiaggi, M., Perduca, M., Romero, E. and Cereda, P.M. (1985) Phosphatidyl-serine inhibition of OC43 and NCDCV coronavirus infectivity. *Microbiologica*, **8**, 313–17.

37. Debiaggi, M., Luini, M., Cereda, P.M., Perduca, M. and Romero, E. (1986) Serum inhibitor of coronaviruses OC43 and NCDCV: a study in vivo. *Microbiologica*, **9**, 33–7.

38. DeBiaggi, M., Perduca, M., Cereda, P.M. and Romero, E. (1986) Coronavirus inhibitor in human sera: age distribution and prevalence. *Microbiologica*, **9**, 109–12.

39. Denison, M.R., Zoltik, P.W., Hughes, S.A. *et al.* (1992) Intracellular processing of the N-terminal ORF 1a proteins of the coronavirus MHV-A59 requires multiple proteolytic events. *Virology*, **189**, 274–84.

40. Duarte, M., Tobler, K., Bridgen, A., Rasschaert, D., Ackermann, M. and Laude, H. (1994) Sequence analysis of the porcine epidemic diarrhoea virus genome between the nucleocapsid and spike protein genes reveals a polymorphic ORF. *Virology*, **198**, 466–76.

41. Dveksler, G.S., Pensiero, M.N., Cardellichio, C.B. *et al.* (1991) Cloning of the mouse hepatitis virus (MHV) receptor: expression in human and hamster cell lines confers susceptibility to MHV. *J. Virol.*, **65**, 6881–91.

42. Eby, G.A., Davis, D.R. and Halcomb, W.W. (1984) Reduction in duration of common colds by zinc gluconate lozenges in a double-blind study. *Antimicrob. Agents Chemother.*, **25**, 20–4.

43. Editorial (1988) Splints don't stop colds – surprising! *Lancet*, **i**, 277–8.

44. Fleming, J.O., El Zaatari, F.A.K., Gilmore, W. *et al.* (1988) Antigenic assessment of coronaviruses isolated from patients with multiple sclerosis. *Arch. Neurol.*, **45**, 629–33.

45. Gerdes, J.C., Klein, I., DeVald, B.L. and Burks, J.S. (1981) Coronavirus isolates SK and SD from multiple sclerosis patients are serologically related to murine coronaviruses A59 and JHM and human coronavirus OC43, but not to human coronavirus 229E. *J. Virol.*, **38**, 231–8.

46. Gerdes, J.C., Jankovsky, L.D., DeVald, B.L., Klein, I. and Burks, J.S. (1981) Antigenic relationships of coronaviruses detectable by plaque neutralization, competitive enzyme linked immunosorbent assay, and immunoprecipitation. *Adv. Exp. Med. Biol.*, **142**, 29–41.

47. Gerna, G., Achilli, G., Cattaneo, E. and Cereda, P. (1978) Determination of coronavirus 229E antibody by an immune-adherence hemagglutination method. *J. Med. Virol.* **2**, 215–23.

48. Gerna, G., Cattaneo, .E, Cereda, P. and Revello, M.G. (1978) Seroepidemiologic study of human coronavirus OC43 infections in Italy. *Bull. 1st Sieroter. Milanese*, **57**, 535–41.

49. Gerna, G., Cereda, P.M., Revello, M.G., Torsellini Gerna, M. and Costa, J. (1979) A rapid microneutralization test for antibody determination and serodiagnosis of human coronavirus OC43 infections. *Microbiologica*, **2**, 331–44.

50. Gerna, G., Cattaneo, E., Cereda, P.M., Revelo, M.G. and Achilli, G. (1980) Human coronavirus serum inhibitor and neutralizing antibody by a new plaque-reduction assay. *Proc. Soc. Exp. Biol. Med.*, **163**, 360–6.

51. Gerna, G., Cereda, P.M., Revello, M.G., Cattaneo, E., Battaglia, M. and Gerna, M.T. (1981) Antigenic and biological relationships between human coronavirus OC43 and neonatal calf diarrhoea virus coronavirus. *J. Gen. Virol.*, **54**, 91–102.

52. Gerna, G., Battaglia, M., Cereda, P.M. and Passarani, N. (1982) Reactivity of human coronavirus OC43 and neonatal calf diarrhoea coronavirus membrane-associated antigens. *J. Gen. Virol.*, **60**, 385–90.

53. Gotzsche, A.-L. (1989) Pernasal vitamin C and the common cold. *Lancet*, **ii**, 1039.

54. Greiff, L., Andersson, M., Akerlund, A. *et al.* (1994) Microvascular exudative hyperresponsiveness in human coronavirus-induced common cold. *Thorax*, **49**, 121–7.

55. Hamre, D. and Procknow, J.J. (1966) A new virus isolated from the human respiratory tract. *Proc. Soc. Exp. Biol.*, **121**, 190–3.

56. Hamre, D., Kindig, D.A. and Mann, J. (1967) Growth and intracellular development of a new respiratory virus. *J. Virol.*, **1**, 810–16.

57. Hamre, D. and Beem, M. (1972) Virologic studies of acute respiratory disease in young adults. V. Coronavirus 229E infections during six years of surveillance. *Am. J. Epidemiol.*, **96**, 94–106.

58. Hartley, J.W., Rowe, W.P., Bloom, H.H. and Turner, H.C. (1964) Antibodies to mouse hepatitis viruses in human sera. *Proc. Soc. Exp. Biol. Med.*, **115**, 414–18.

59. Hasony, H.J. and MacNaughton, M.R. (1982) Serological relationships of the subcomponents of human coronavirus strain 229E and mouse hepatitis virus strain 3. *J. Gen. Virol.*, **58**, 449–52.

60. Henigst, W. (1974) Occurrence of antibody against coronavirus OC43 in the healthy population and in patients with a disease of the respiratory tract. *Zbl. Bakteriol. I. Abst. Orig. A*, **299**, 150–8.

61. Herold, J. and Siddell, S. (1993) An 'elaborated' pseudoknot is required for high frequency frameshifting during translation of HCV 229E polymerase mRNA. *Nucleic Acids Res.*, **21**, 5838–42.

62. Herold, J., Raabe, T., Schelle-Prinz, B. and Siddell, S.G. (1993) Nucleotide sequence of the human coronavirus 229E RNA polymerase locus. *Virology*, **195**, 680–91.

63. Hierholzer, J.C. (1976) Purification and biophysical properties of human coronavirus 229E. *Virology*, **75**, 155–65.

64. Hierholzer, J.C. and Tannock, G.A. (1977) Quantitation of antibody to non-haemagglutinating viruses by single radial haemolysis: serological test for human coronaviruses. *J. Clin. Microbiol.*, **5**, 613–20.

65. Hierholzer, J.C., Halonen, P.E., Bingham, P.G., Coombs, R.A. and Stone, Y.O. (1994) Antigen detection in human respiratory Coronavirus infections by monoclonal time-resolved fluoroimmunoassay. *Clin. Diagn. Virol.*, **2**, 165–79.

66. Higgins, P.G., Philpotts, R.J., Scott, G.M., Wallace, J., Bernhardt, L.L. and Tyrrell, D.A.J. (1983) Intranasal interferon as protection against experimental respiratory coronavirus infection in volunteers. *Antimicrob. Agents Chemother.*, **34**, 713–15.

67. Higgins, P.G., Barrow, G.I., Tyrrell, D.A.J., Snell, D.J.C., Jones, K. and Jolley, W.B. (1991) A study of the immunomodulatory compound 7-thio-8-oxoguanosine in coronavirus 229E infections in human volunteers. *Antiviral Chemother.*, **2**, 61–3.

68. Hogue, B.G., King, B, and Brian, D.A. (1984) Antigenic relationships among proteins of bovine coronavirus, human respiratory coronavirus OC43, and mouse hepatitis coronavirus A59. *J. Virol.*, **51**, 384–8.

69. Holmes, M.J., Callow, K.A., Childs, R.A. and Tyrrell, D.A.J. (1986) Antibody dependent cellular cytotoxicity against coronavirus 229E-infected cells. *Br. J. Exp. Pathol.*, **67**, 581–6.

70. Hovanec, D.L. and Flanagan, T.D. (1983) Detection of antibodies to human coronaviruses 229E and OC43 in the sera of multiple sclerosis patients and normal subjects. *Infect. Immun.*, **41**, 426–9.

71. Hovi, T. (1978) Nonspecific inhibitors of coronavirus OC43 haemagglutination in human sera. *Med. Microbiol. Immunol.*, **166**, 1773–6.

72. Hruskova, J., Heinz, F., Svandova, E. and Pennigerova, S. (1990) Antibodies to human coronaviruses 229E and OC43 in the population of C.R. *Acta Virologica*, **34**, 346–52.

73. Hume, R. and Weyers, E. (1973) Changes in leucocyte ascorbic acid during the common cold. *Scot. Med. J.*, **18**, 3–8.

74. Ijaz, M.K., Brunner, A.H., Sattar, S.A., Nair, R.C. and Johnson-Lussenberg, C.M. (1985)

Survival characteristics of airborne human coronavirus 229E. *J. Gen. Virol.*, **66**, 2743–8.

75. Isaacs, D., Flowers, D., Clarke, J.R., Valman, H.B. and MacNaughton, M.R. (1983) Epidemiology of coronavirus respiratory infections. *Archiv. Dis. Child.*, **58**, 500–3.

76. Johnston, S.L., Pattemore, P.K., Sanderson, G. *et al.* (In press) The importance of virus infections in asthma-like exacerbations in 9–11 year-old children in a UK community. *Br. Med. J.*

77. Joo, M. and Makino, S. (1992) Mutagenic analysis of the coronavirus intergenic consensus sequence. *J. Virol.*, **66**, 6330–7.

78. Jouvenne, P., Mounir, S., Stewart, J.N., Richardson, C.D. and Talbot, P.J. (1992) Sequence analysis of human coronavirus 229E mRNAs 4 and 5: evidence for polymorphism and homology with myelin basic protein. *Virus Res.*, **22**, 125–41.

79. Kapikian, A.Z., James, H.D., Kelly, S.J. *et al.* (1969) Isolation from man of 'avian infectious bronchitis virus-like' viruses (coronaviruses) similar to 229E virus with some epidemiological observations. *J. Infect. Dis.*, **119**, 282–90.

80. Kapikian, A.Z., James, H.D., Kelly, S.J., King, L.M., Vaughn, A.L. and Chanock, R.M. (1971) Haemadsorption by coronavirus strain OC43. *Proc. Soc. Exp. Med. Biol.*, **139**, 179–86.

81. Kapikian, A.Z., James, H.D., Kelly, S.J. and Vaughn, A.L. (1973) Detection of coronavirus strain 692 by immune electron microscopy. *Infect. Immun.*, **7**, 111–16.

82. Kaye, H.S. and Dowdle, W.R. (1969) Some characteristics of haemagglutination of certain strains of 'IBV-like' virus. *J. Infect. Dis.*, **120**, 576–81.

83. Kaye, H.S., Hierholzer, J.C. and Dowdle, W.R. (1970) Purification and further characterization of an 'IBV-like' virus (coronavirus). *Proc. Soc. Exp. Biol., Med.*, **135**, 457–63.

84. Kaye, H.S., Marsh, H.B. and Dowdle, W.R. (1971) Seroepidemiologic survey of coronavirus (strain OC43) related infections in a children's population. *Am. J. Epidemiol.*, **94**, 43–9.

85. Kaye, H.S., Ong, S.B. and Dowdle, W.R. (1972) Detection of coronavirus 229E antibody by indirect haemagglutination. *Appl. Microbiol.*, **24**, 703–7.

86. Kaye, H.S., Yarbrough, W.B., Reed, C.J. and Harrison, A.K. (1977) Antigenic relationship between human coronavirus strain OC43 and haemagglutinating encephalomyelitis virus strain 67N of swine: antibody responses in human and animal sera. *J. Infect. Dis.*, **135**, 201–9.

87. Kemp, M.C., Hierholzer, J.C., Harrison, A. and Burks, J.S. (1984) Characterization of viral proteins synthesized in 229E infected cells and effect(s) of inhibition of glycosylation and glycoprotein transport. *Adv. Exp. Med. Biol.*, **173**, 65–77.

88. Kendall, E.J.C., Bynoe, M.L. and Tyrrell, D.A.J. (1962) Virus isolations from common colds occurring in a residential school. *Br. Med. J.*, ii, 82–6.

89. Kennedy, D.A. and Johnson-Lussenberg, C.M. (1978) Inhibition of coronavirus 229E replication by actinomycin D. *J. Virol.*, **29**, 401–4.

90. Kingston, D., Lidwell, O.M. and Williams, R.E.O. (1962) The epidemiology of the common cold III. The effect of ventilation, air disinfection and room size. *J. Hyg. (Camb)*, **60**, 341–51.

91. Kraaijeveld, C.A., Madge, M.H. and MacNaughton, M.R. (1980) Enzyme-linked immunosorbent assay for coronaviruses HCV229E and MHV3. *J. Gen. Virol.*, **49**, 83–9.

92. Kraaijeveld, C.A., Reed, S.E. and MacNaughton, M.R. (1980) Enzyme-linked immunosorbent assay for detection of antibody in volunteers experimentally infected with human coronavirus strain 229E. *J. Clin. Microbiol.*, **12**, 493–7.

93. Lamarre, A. and Talbot, P.J. (1989) Effect of pH and temperature on the infectivity of human coronavirus 229E. *Can. J. Microbiol.*, **35**, 972–4.

94. Lamb, R.A. (1993) Paramyxovirus fusion: a hypothesis for change. *Virology*, **197**, 1–11.

95. Lapps, W. and Brian, D.A. (1985) Oligonucleotide fingerprints of antigenically related bovine coronavirus and human coronavirus OC43. *Arch. Virol.*, **86**, 101–8.

96. Larson, H.E., Reed, S.E. and Tyrrell, D.A.J. (1980) Isolation of rhinoviruses and coronaviruses from 38 colds in adults. *J. Med. Virol.*, **5**, 221–9.

97. Leinikki, P.O., Holmes, K.V., Shekarchi, I., Livanainen, M., Madden, D. and Sever, J.L. (1981) Coronavirus antibodies in patients with multiple sclerosis. *Adv. Exp. Med. Biol.*, **142**, 323–6.

98. Liu, D.X. and Inglis, S.C. (1991) Association of the infectious bronchitis virus 3c protein with the virion envelope. *Virology*, **185**, 911–17.

99. Lowenstein, S.R. and Parrino, T.A. (1987) Management of the common cold. *Adv. Intern. Med.*, **32**, 207–23.

100. McIntosh, K., Dees, J.H., Becker, W.B., Kapikian, A.Z. and Chanock, R.M. (1967) Recovery in tracheal organ cultures of novel

viruses from patients with respiratory disease. *Proc. Natl Acad. Sci. USA*, **57**, 933–40.

101. McIntosh, K., Becker, W.B. and Chanock, R.M. (1967) Growth in suckling mice brain of IBV-like viruses from patients with upper respiratory tract disease. *Proc. Natl Acad. Sci. USA*, **58**, 2268–73.

102. McIntosh, K,. Kapikian, A.Z., Hardison, K.A., Hartley, J.W. and Chanock, R.M. (1969) Antigenic relationships among the coronaviruses of man and between human and animal coronaviruses. *J. Immunol.*, **102**, 1109–18.

103. McIntosh, K., Bruckova, M., Kapikian, A.Z., Chanock, R. and Turner, H. (1970) Studies on new virus isolates recovered in tracheal organ culture. *Ann. N. Y. Acad. Sci.*, **174**, 983–9.

104. McIntosh, K., Kapikian, A.Z., Turner, H.C., Hartley, J.W., Parrott, R.H. and Chanock, R.M. (1970) Seroepidemiologic studies of coronavirus infection in adults and children. *Am. J. Epidemiol.*, **91**, 585–92.

105. McIntosh, K., Ellis, E.F., Hoffmann, L.S., Lybass, T.G., Eller, J.J. and Fulginiti, V.A. (1973) The association of viral and bacterial respiratory infections with exacerbations of wheezing in young asthmatic children. *J. Paediatr.*, **82**, 579–90.

106. McIntosh, K., Chao, R.K., Krause, H.E., Wasil, R., Mocega, H.E. and Mufson, M.A. (1974) Coronavirus infections in lower respiratory tract disease of infants. *J. Infect. Dis.*, **130**, 502–7.

107. McIntosh, K., McQuillin, J., Reed, S.E. and Gardner, P.S. (1978) Diagnosis of human coronavirus infection by immunofluorescence: method and application to respiratory disease in hospitalized children. *J. Med. Virol.*, **2**, 341–6.

108. MacNaughton, M.R., Thomas, B., Davies, H.A. and Patterson, S. (1980) Infectivity of human coronavirus 229E. *J. Clin. Microbiol.*, **12**, 462–8.

109. MacNaughton, M.R. (1981) Structural and antigenic relationships between human, murine and avian coronaviruses. *Adv. Exp. Med. Biol.*, **142**, 19–28.

110. MacNaughton, M.R., Madge, M.H. and Reed, S.E. (1981) Two antigenic groups of human coronaviruses detected by using enzyme-linked immunoassay. *Infect. Immun.*, **33**, 734–7.

111. MacNaughton, M.R., Hasony, H.J., Madge, M.H. and Reed, S.E. (1981) Antibody to virus components in volunteers experimentally infected with human coronavirus 229E group viruses. *Infect. Immun.*, **31**, 845–9.

112. MacNaughton, M.R. (1982) Occurrence and frequency of coronavirus infections in humans as determined by enzyme-linked immunosorbent assay. *Infect. Immun.*, **38**, 419–23.

113. MacNaughton, M.R., Flowers, D. and Isaacs, D. (1983) Diagnosis of human coronavirus infections in children using enzyme-linked immunosorbent assay. *J. Med. Virol.*, **11**, 319–26.

114. Madden, D.L., Wallen, W.C., Houff, S.A. *et al.* (1981) Coronavirus antibodies in sera from patients with multiple sclerosis and matched controls. *Arch. Neurol.*, **38**, 209–10.

115. Makino, S. and Joo, M. (1993) Effect of intergenic concensus flanking sequences on coronavirus transcription. *J. Virol.*, **67**, 3304–11.

116. Malkova, D. Holubova, J., Kolman, J.M., Lobkovic, F., Pohlreichova, L. and Zikmundova, L. (1980) Isolation of Tettnang coronavirus from man. *Acta Virol.*, **24**, 363–6.

117. Matthews, T.H.J., Nair, C.D.G., Lawrence, M.K. and Tyrrell, D.A.J. (1976) Antiviral activity in milk of possible clinical significance. *Lancet*, **ii**, 1387–9.

118. Mathur, A., Arora, K.L., Rajvanshi, S. and Chaturvedi, U.C. (1982) Coronavirus in respiratory infection. *Indian J. Med. Res.*, **75**, 323–8.

119. Matsumoto, I. and Kawana, R. (1992) Virological surveillance of acute respiratory tract illnesses of children in Morioka, Japan III. Human respiratory coronavirus. *J. Jap. Assoc. Infect. Dis.*, **66**, 319–26.

120. Mentel, R. and Schmidt, J. (1974) Versuche zur chemischen Inaktivierung von Rhinoviren und Coronaviren. *Z. Gesamte Hyg.*, **26**, 530–3.

121. Miyazaki, K., Tsunoda, A., Kumasaka, M. and Ishida, N. (1971) Presence of neutralising antibody against the 229E strain of coronavirus in the sera of residents of Sendai. *Jpn J. Microbiol.*, **15**, 276–7.

122. Monto, A.S. and Lim, S.K. (1974) The Tecumseh study of respiratory illness. VI. Frequency of and relationship between outbreaks of coronavirus infection. *J. Infect. Dis.*, **129**, 271–6.

123. Monto, A.S. and Rhodes, L.M. (1977) Detection of coronavirus infection of man by immunofluorescence. *Proc. Soc. Exp. Biol. Med.*, **155**, 143–8.

124. Mounir, S. and Talbot, P.J. (1993) Human coronavirus OC43 RNA 4 lacks two open reading frames located downstream of the S gene of bovine coronavirus. *Virology*, **192**, 355–60.

125. Murata, A. (1972) Inactivation of single-stranded DNA and RNA phages by ascorbic

acid and thiol-reducing agents. *Agricultural Biol. Chem.*, **36**, 2597–600.

126. Myint, S., Siddell, S. and Tyrrell, D. (1989) Detection of human coronavirus 229E in nasal washings using RNA:RNA hybridization. *J. Med. Virol.*, **29**, 70–3.

127. Myint, S., Harmsen, D., Raabe, T. and Siddell, S.G. (1990) Characterisation of a nucleic acid probe for the diagnosis of human coronavirus 229E infections. *J. Med. Virol.*, **31**, 165–72.

128. Myint, S. and Tyrrell, D.A.J. (1995) Coronaviruses, in *Diagnostic Procedures for Viral, Rickettsial and Chlamydial Infections*, (eds E.H. Lennette, E.T. Lennette and D.A. Lennette), American Public Health Association, California. (In press)

129. Myint, S., Johnstone, S., Sanderson, G. and Simpson, H. (1994) An evaluation of 'nested' RT–PCR for the detection of human coronaviruses 229E and OC43 in clinical specimens. *Mol. Cell. Probes*, **8**, 357–64.

130. Ophir, D. and Elad, Y. (1987) Effect of steam inhalation on nasal patency and nasal symptoms in patients with the common cold. *Am. J. Otolaryngol.*, **3**, 149–53.

131. Owen Hendley, J.O., Fishburne, H.B. and Gwaltney, J.M. (1972) Coronavirus infections in working adults. *Am. Rev. Resp. Dis.*, **105**, 805–11.

132. Parker, M.M. and Masters, P.S. (1990) Sequence comparison of the N genes of five strains of the coronavirus mouse hepatitis virus suggests a three domain structure for the nucleocapsid protein. *Virology*, **179**, 463–8.

133. Patterson, S. and MacNaughton, M.R. (1981) The distribution of human coronavirus 229E on the surface of human diploid cells. *J. Gen. Virol.*, **53**, 267–73.

134. Patterson, S. and MacNaughton, M.R. (1982) Replication of human respiratory coronavirus strain 229E in human macrophages. *J. Gen. Virol.*, **60**, 307–14.

135. Pensaert, M.B., Debouck, P. and Reynolds, D.J. (1981) An immunoelectron microscopic and immunofluorescent study on the antigenic relationship between the coronavirus-like agent, CV777, and several coronaviruses. *Arch. Virol.*, **68**, 45–52.

136. Perry, N.B., Blunt, J.W., Munro, M.H.G. and Thompson, A.M. (1990) Antiviral and antitumour agents from a New Zealand sponge, *Mycale* sp. 2. Structures and solution conforma-

tions of mycalaimides A and B. *J. Org. Chem.*, **55**, 223–7.

137. Philpotts, R. (1983) Clones of MRC-c cells may be superior to the parent line for the culture of 229E-like strains of human respiratory coronavirus. *J. Virol. Methods*, **6**, 267–9.

138. Pitkaranta, A. and Hovi, T. (1993) Induction of interferon in human leukocyte cultures by natural pathogenic respiratory viruses. *J Interferon Res.*, **13**, 423–6.

139. Pohl-Koppe, A., Raabe, T., Siddell, S.G. and ter Meulen, V. (1994) Detection of human coronavirus 229E specific antibodies using recombinant fusion proteins. *J. Virol. Methods*. (In press).

140. Pugh, D.M., Sharma, S.C. and Wilson, C.W.M. (1975) Inhibitory effect of L-ascorbic acid on the yield of prostaglandin F from the guinea pig uterine homogenates. *Br. J. Pharmacol.*, **53**, 469–73.

141. Raabe, T., Schelle Prinz, B. and Siddell, S.G. (1990) Nucleotide sequence of the gene encoding the spike glycoprotein of human coronavirus HCV 229E. *J. Gen. Virol.*, **71**, 1065–73.

142. Reed, S.E. (1984) The behaviour of recent isolates of human respiratory coronavirus in vitro and in volunteers: evidence of heterogeneity among 229E-related strains. *J. Med. Virol.*, **13**, 179–92.

143. Riski, H. and Estola, T. (1974) Occurrence of antibodies to human coronavirus OC43 in Finland. *Scand. J. Infect. Dis.*, **6**, 325–7.

144. Riski, H., Hovi, T., Vaananen, P. and Penttinen, K. (1977) Antibodies to human coronavirus OC43 measured by radial haemolysis in gel. *Scand. J. Infect. Dis.*, **9**, 75–7.

145. Riski, H. and Hovi, T. (1980) Coronavirus infections of man associated with diseases other than the common cold. *J. Med. Virol.*, **6**, 259–65.

146. Rottier, P.J., Welling, G.W., Welling, W.S., Niesters, H.G., Lenstra, J.A. and van der Zeijst, B. (1986) Predicted membrane topology of the coronavirus protein E1. *Biochemistry*, **25**, 1335–9.

147. Sakaguchi, A.Y. and Shows, T.B. (1982) Coronavirus 229E susceptibility in man-mouse hybrids is located on human chromosome 15. *Somatic Cell Mol. Genet.*, **8**, 83–94.

148. Salmi, A., Ziola, B., Hovi, T. and Reunanen, M. (1982) Antibodies to coronaviruses OC43 and 229E in multiple sclerosis patients. *Neurology*, **32**, 292–5.

149. Sarateneau, D., Bronitki, A., Popescu, A., Teodosiu, O. and Isala, G. (1974) Incidence of

coronavirus OC43 antibodies among the population of Rumania. *Rev. Roum. Virol.*, **25**, 255–8.

150. Sarateanu, D.E. and Ehrengut, W. (1979) A two year serological surveillance of coronavirus infections in Hamburg. *Infection*, **8**, 70–2.

151. Sawicki, S.G. and Sawicki, D.L. (1990) Coronavirus transcription: subgenomic mouse hepatitis virus replicative intermediates function in RNA synthesis. *J. Virol.*, **64**, 1050–6.

152. Schaad, M.C., Stohlman, S.A., Egbert, J., Lum, K., Fu, K., Wei, T. Jr and Baric, R.S. (1990) Genetics of mouse hepatitis virus transcription. Identification of cistrons which may function in positive and negative strand RNA synthesis. *Virology*, **177**, 634–45.

153. Schmidt, O.W., Cooney, M.K. and Kenny, G.E. (1979) Plaque assay and improved yield of human coronaviruses 229E and OC43 in a human rhabdomyosarcoma cell line. *J. Clin. Microbiol.*, **9**, 722–8.

154. Schmidt, O.W. and Kenny, G.E. (1981) Immunogenicity and antigenicity of human coronaviruses 229E and OC43. *Infect. Immun.*, **32**, 1000–6.

155. Schmidt, O.W. and Kenny, G.E. (1982) Polypeptides and functions of antigens from human coronaviruses 229E and OC43. *Infect. Immun.*, **35**, 515–22.

156. Schmidt, O.W. (1984) Antigenic characterisation of human coronaviruses 229E and OC43 by enzyme-linked immunosorbent assay. *J. Clin. Microbiol.*, **20**, 175–80.

157. Schmidt, O.W., Allan, I.D., Cooney, M.K., Foy, H.M. and Fox, J.P. (1986) Rises in titers of antibody to human coronaviruses OC43 and 229E in Seattle families during 1975–1979. *Am. J. Epidemiol.*, **123**, 862–8.

158. Schultze, B., Gross, H.J., Brossmer, R. and Herrler, G. (1991) The S protein of bovine coronavirus is a haemagglutinin recognizing 9-O-acetylated sialic acid as a receptor determinant. *J. Virol.*, **65**, 6232–7.

159. Schultze, B., Wahn, K., Klenk, H.D. and Herrler, G. (1991) Isolated HE-protein from hemagglutinating encephalomyelitis virus and bovine coronavirus has receptor-destroying and receptor-binding activity. *Virology* **180**, 221–8.

160. Small, J.D., Aurelian, L., Squire, R.A. *et al.* (1979) Rabbit cardiomyopathy: associated with a virus antigenically related to human coronavirus strain 229E. *Am. J. Pathol.*, **95**, 709–29.

161. Sorensen, O., Collins, A., Flintoff, W., Ebers, G. and Dales, S. (1986) Probing for the human

coronavirus OC43 in multiple sclerosis. *Neurology*, **35**, 1604–6.

162. Stewart, J.N., Mounir, S. and Talbot, P.J. (1992) Human coronavirus gene expression in the brains of multiple sclerosis patients. *Virology*, **191**, 502–5.

163. Storz, J. and Rott, R. (1981) Reactivity of antibodies in human serum with antigens of an enteropathogenic bovine coronavirus. *Med. Microbiol. Immunol.*, **169**, 169–78.

164. Tanaka, R., Iwasaki, Y. and Koprowski, H. (1976) Intracisternal virus-like particles in brain of a multiple sclerosis patient. *J. Neurol. Sci.*, **28**, 121–6.

165. Thomas, W.R. and Holt, P.G. (1978) Vitamin C and immunity: an assessment of the evidence. *Clin. Exp. Immunol.*, **32**, 370–5.

166. Totman, R., Kiff, J., Reed, S.E. and Craig, J.W. (1980) Predicting experimental colds in volunteers from different measures of recent life stress. *J. Psychosom. Res.*, **24**, 155–63.

167. Turner, R.B., Felton, A., Kosak, K., Kelsey, D.K. and Meschevitz, C.K. (1986) Prevention of experimental coronavirus colds with intranasal α-2b interferon. *J. Infect. Dis.*, **154**, 443–7.

168. Tyrrell, D.A.J. and Bynoe, M.L. (1965) Cultivation of a novel type of common cold virus in organ culture. *Br. Med. J.*, **1**, 1467–70.

169. Tyrrell, D.A.J., Almeida, J.D., Berry, D.M. *et al.* (1968) Coronaviruses. *Nature*, **220**, 650.

170. Tyrrell, D.A.J., Bynoe, M.L. and Hoorn, B. (1968) Cultivation of 'difficult' viruses from patients with common colds. *Br. Med. J.*, **1**, 606–10.

171. Tyrrell, D.A.J., Mika-Johnson, M., Phillips, G., Douglas, W.H.J. and Chapple, P.J. (1979) Infection of cultured human type II pneumocytes with certain respiratory viruses. *Infect. Immun.*, **26**, 621–9.

172. Tyrrell, D.A.J. (1983) Rhinoviruses and coronaviruses-virological aspects of their role in causing colds in man. *Eur. J. Respir. Dis.*, **64**, 332–5.

173. Tyrrell, D.A.J. (1988) Hot news on the common cold. *Annu. Rev. Microbiol.*, **42**, 35–4.

174. Tyrrell, D.A.J. (1988) Common Colds, in *Portraits of Viruses. A History of Virology*, (eds F. Fenner and A. Gibbs), Karger, Basel, pp. 254–66.

175. Tyrrell, D.A.J., Barrow, I. and Arthur, J. (1989) Local hyperthermia benefits natural and experimental common colds. *Br. Med. J.*, **298**, 1280–3.

176. Tyrrell, D.A.J. (1992) A view from the Common Cold Unit. *Antiviral Res.*, **18**, 105–25.

177. Tyrrell, D.A.J., Cohen, S. and Schlarb, J.E. (1993) Signs and symptoms in common colds. *Epidemiol. Infect.*, **111**, 143–56.

178. Weiss, S.R. and Leibowitz, J.L. (1981) Comparison of the RNAs of murine and human coronaviruses. *Adv. Exp. Med. Biol.*, **142**, 245–59.

179. Weiss, S.R. (1983) Coronaviruses SD and SK share extensive nucleotide homology with murine coronavirus MHV-A59, more than that shared between human and murine coronaviruses. *Virology*, **126**, 669–77.

180. Wenzel, R.P., Hendley, J.O., Davies, J.A. and Gwaltney, J.M. (1974) Coronavirus infections in military recruits. *Am. Rev. Resp. Dis.*, **109**, 621–4.

181. Yeager, C.L., Ashmun, R.A., Williams, R.K. *et al.* (1992) Human aminopeptidase N is a receptor for human coronavirus 229E. *Nature*, **357**, 420–2.

182. Yerushalmi, A. and Lwoff, A. (1980) Traitement du coryza infectieux et des rhinites persistantes allergiques par la thermotherapie. *C. R. Acad. Sci.*, **291**, 957–9.

183. Yokomori, K., Stohlman, S.A. and Lai, M.M.C. (1993) The detection and characterization of multiple hemagglutinin-esterase (HE)-defective viruses in the mouse brain during subacute demyelination induced by mouse hepatitis virus. *Virology*, **192**, 170–8.

184. Zhang, X.M., Kousoulas, K.G. and Storz, J. (1992) The hemagglutinin/esterase gene of human coronavirus strain OC43: phylogenetic relationships to bovine and murine coronaviruses and influenza C virus. *Virology*, **186**, 318–23.

ADENOVIRUSES 10

Dick Madeley, Malik Peiris and Joyce McQuillin

10.1 INTRODUCTION

Adenoviruses frequently infect humans, particularly children in whom prolonged infection is common, but only rarely cause severe disease. As a result they have been neglected recently, although they were one of the first viruses to be grown in tissue or cell culture. However, not all infections are benign and fatal childhood pneumonias have been described. They occasionally complicate transplants and they are overdue for a re-appraisal. Some 47 serotypes are now known with five new ones (types 43–47) having been added to the list in the late 1980s, all of them found in the faeces of patients with AIDS. Their significance is unknown, as is type 42, which was a casual finding. This list may not be complete and the problems of growing the fastidious types 40 and 41 in cell culture show that other presently unrecognized serotypes may exist. Documenting their existence may be very difficult but there are many respiratory infections for which no cause is found. Some may be caused by presently unknown adenoviruses but finally identifying them will require new approaches to diagnosis.

10.2 HISTORICAL SURVEY

Adenoviruses were first discovered in adenoid tissue surgically removed and cultured in plasma clots [35]. They were later recovered from tonsils [21]. The epithelial cells growing out from the tissue began to deteriorate after some 8–28 days in culture and this was found to be due to an agent which could then be passed in HeLa cells producing a characteristic cytopathic effect. The agent appeared to be in a latent form as it could not be isolated directly in HeLa cells from pieces of ground lymphoid

Viral and Other Infections of the Human Respiratory Tract.
Edited by S. Myint and D. Taylor-Robinson. Published in 1996 by Chapman & Hall. ISBN 0 412 60070 6

tissue nor isolated from throat swabs taken from the site of removal [36].

It was soon clear that there were several such agents which had a soluble complement-fixing antigen in common [21], but cross-neutralization tests showed at least six serotypes. The prototype of types 1, 2 and 5 came from adenoids and type 6 from a tonsil.

Evidence that these agents were viruses and associated with respiratory disease followed quickly. A prototype of serotype 4 was isolated from the throat washings of a soldier with primary atypical pneumonia during an outbreak of respiratory illness among military recruits [19]. These outbreaks in new recruits to the armed forces were different from other respiratory illnesses in severity and included headache, aches and pains and sore throat. This had been called 'febrile catarrh' pre-World War II [43] and later termed acute respiratory disease (ARD).

Another serotype was then isolated from the faeces of an infant with febrile pharyngitis and roseola infantum [28] and became the prototype of serotype 3. This type was later isolated from an outbreak in a hospital among children and adults with febrile pharyngitis and conjunctivitis [30] and from 80 people during an outbreak of pharyngoconjunctival fever in a children's summer camp and two communities within a 7-mile radius [4]. As a result of their recovery from the adenoids, the nasopharynx and conjunctiva these agents were termed adenoidal-pharyngeal-conjunctival (APC) agents.

During 1953 and 1954, studies on throat washings from army personnel in the USA who had acute respiratory illnesses yielded 51 isolates of APC agents [5]. Seven were type 3, 22 were type 4, particularly associated with lower respiratory tract infections (LRTI) and another 22 were serologically distinct from types 1–6. These were all the same type and became type 7. In this investigation the type 7 infections were mainly confined to the upper respiratory tract.

Studies by Tyrrell and colleagues in 1954 [56] and 1955 [3] confirmed the presence of types 1, 2 and 5 in surgically removed adenoids and demonstrated diagnostic rises in complement-fixing (CF) antibodies in the sera of army personnel with respiratory infections and in three elderly patients with acute exacerbations of chronic bronchitis. These rises in antibody coincided with recovery from the acute episodes in all cases. In a further paper [48] they described an outbreak of respiratory illness with fever, nasal obstruction and sore throat, sometimes with conjunctivitis, which took place in a boys' residential school. Type 7 was recovered from those infected and the report suggested that the gastrointestinal symptoms experienced by some of the boys could have been due to the virus.

In 1956 the APC agents were re-named adenoviruses [14], by which time eleven serotypes had been identified from human sources. Type 8 was isolated from an eye swab from a patient with epidemic keratoconjunctivitis. It was subsequently found in a number of outbreaks associated with ocular first aid in heavy industry and the conjunctivitis was called 'shipyard eye' [23]. Types 9 and 11 were found in stool specimens collected as part of a survey for polioviruses while type 10 came from a further case of conjunctivitis.

The association between types 1, 2 and 5 and respiratory illness was confirmed over the following years by others, with virus isolated either from the throat or from an anal swab. Isolation from the latter, however, made correlation with respiratory disease more difficult. Type 6 had so far only been recovered from an eye swab.

Between 1953 and 1965 the number of serotypes gradually grew to 31 but, other than types 1 to 7, only two (types 14 and 21) were associated with respiratory illness. Type 14 came from nine military recruits with ARD in the Netherlands [50] while type 21, originally isolated from an eye swab from a child with trachoma in Saudi Arabia was later recovered from 118 recruits with 'febrile catarrh', again in the Netherlands [49]. Subsequently types 32–39 were isolated from faeces, eye swabs and the respiratory tract [18,20].

By 1994, the total number of serotypes had reached 47. Most of the later serotypes have been isolated from stool specimens, particularly types 40 and 41 which are now recognized as causing diarrhoea in children and occasionally adults. The role of these higher serotypes in respiratory disease does not at present appear to be significant but should not be dismissed out of hand. All serotypes can infect the gut and there is no fundamental reason to suppose that the respiratory tract, and possibly the eye, are not susceptible to infection with all serotypes. We should keep an open mind, but the great majority of adenovirus respiratory infections are due to types 1–7.

10.3 STRUCTURE

Adenoviruses are the most perfectly icosahedral of viruses. They are 75–80 nm in diameter, non-enveloped and icosahedral in both symmetry and form, with a highly characteristic appearance by negative contrast in the electron microscope (EM) (Figure 10.1). Unlike many other viruses, it is easy both to draw an adenovirus and to build a convincing model of the naked capsid (Figure 10.2), reflecting its straightforward construction, at least to external appearances.

The capsid is assembled from 252 capsomers, 240 with six adjacent neighbours (hexons) and 12 at the apices which have only five neighbours (pentons). The hexons are shown in white in Figure 10.2 and the pentons in black. As shown in the model, the complete penton structure is complex, consisting of a base, fibre and terminal knob. The knobbed fibres are rarely seen *in situ* in the EM. The reason for this is unknown. They may fall off very readily, being lost even in very gentle preparation or, it has been suggested, they may be retractable. If this latter hypothesis were to be true, such movement would require energy

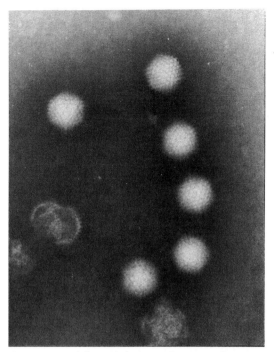

Figure 10.1   Adenovirus (type 5) as seen by negative contrast, showing surface capsomers. (Stain 3% potassium phosphotungstate, pH 7.0; original magnification, × 2000 000.)

systems, something yet to be described in viruses. Occasional adenovirus preparations are found in which detached fibres are present in thousands, more than the number of whole particles (even allowing for 12 from each), but these may be the result of substantial intracellular over-production.

Although adenovirus models are usually made from 252 spheres, the individual capsomers are hollow tubular prisms. The central hollow is not seen clearly when potassium phosphotungstate is used as a negative stain but uranyl salts (particularly uranyl formate) have a smaller effective radius. They can therefore penetrate the hollow (Figure 10.3).

The capsomers, though appearing to be single individual subunits, are complex protein structures with specific antigenic identities (shown in Table 10.1). Their significance is discussed further below.

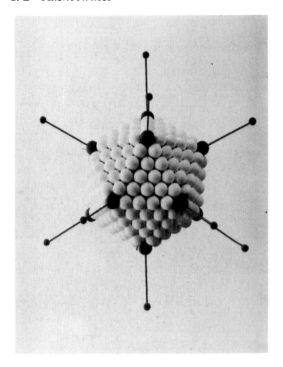

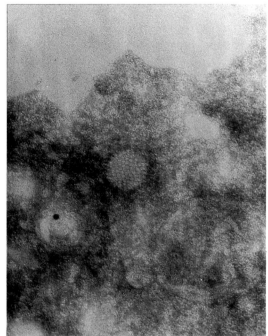

Figure 10.2   Model of adenovirus to show icosahedral symmetry and form. The structure comprises 240 hexons (white) and 12 pentons (black), each with an apical knobbed fibre. The length of the fibres differs between serotypes.

Figure 10.3   Adenovirus seen by negative contrast using uranyl formate. The smaller effective radius of the stain allows it to enter the axial hollow of the hexons. (Original magnification, × 200 000.)

Table 10.1   Antigenic structural proteins of adenovirus

Number	Physical structure	No. of copies of protein/structural unit	Serological specificity
I	Mixture of small proteins	? separate entity	
II	Hexon	3	Genus (specific for all human adenoviruses)
III	Penton base	5	Genus
IV	Fibre	2	Serotype*
V	Core protein	–	
VI	Subsurface connections	–	
VII	Core protein	–	
VIII	Hexon-associated protein	–	
IX	Subsurface connections	–	
X	?Core	–	

* Also has some subgroup specificity with some strains.

Beneath the outer icosahedral surface lie further proteins and the single piece of double-stranded DNA. Subsurface structure is not seen directly in the EM though very occasional 'empty' particles are seen. Preparations of purified particles will degrade rapidly at 4°C. Broken particles show the capsid disintegrating into crown-like arcs releasing amorphous material from the interior. No identifiable nucleic acid has been seen except by using specialist techniques to visualize it.

The capsid may also break down progressively, probably mimicking the normal uncoating process. First the pentons are lost, preceded by their fibres, then the five hexons surrounding each penton (the peripentons). This leaves the main icosahedral faces which separate into 20 groups of nine hexons (Figure 10.4). These in turn break down to three groups of three and finally to individual capsomers.

Antigenically adenoviruses have both group and type specificities (Table 10.1). They are also subgrouped, originally by the species of red cell they would agglutinate, but now on the pattern of fragments produced from the DNA by digestion with restriction endonucleases. Typical subgroup patterns place the serotypes in the same subgroup as found previously (Table 10.2). Some extra digestion sites are found in the DNA from specific isolates, allowing further subdivision. This is important only for epidemiological purposes, showing that isolates from a single ward, for example, are the same and are likely to have come from a common source.

10.4 CLASSIFICATION

Adenoviruses are divided into two genera: *Mastadenoviruses*, infecting mammalian species including man, and *Aviadenoviruses* infecting birds [27]. The 47 human serotypes are divided into six subgenera or groups (both terms are widely used) identified by letters A–F (Table 10.2). Within each subgenus there

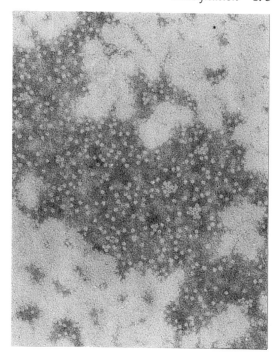

Figure 10.4 Adenovirus which has decayed at 4°C. Two 'groups of nine' hexons are visible and occasional groups of three. Elsewhere, there are a number of single capsomers. Note that all appear hollow when separate from the complete virion, even when using potassium phosphotungstate (pH 7.0) for negative contrast. Note also the absence of any identifiable nucleic acid. (Original magnification, × 200 000.)

is considerable DNA homology (between 50% and 100%) but serotypes are assigned to each on the basis of DNA digestion patterns using the *Sma*1 restriction endonuclease. The same assignments are made when based on the species of erythrocytes agglutinated by each virus. Other groupings have used the varying ability of different adenoviruses to cause tumours in new-born hamsters and on the size patterns of the polypeptides produced in cells infected by each serotype. Fortunately for everyone concerned the groups revealed by each of these methods were essentially the same and clearly reflect basic biological intergroup differences.

Table 10.2   Human adenoviruses – properties and classification. (From [52].)

| Subgroup | Serotypes | DNA | | | Haemagglutination* | Oncogenicity in hamsters |
		Homology within subgroup	Homology with other subgroup	Sma 1 fragments		
A	12, 18, 31	48–69	8–20	4–5	iv	High
B1	3, 7, 16, 21	89–94	9–20	8–10	i	Weak
B2	11, 14, 34, 35					Nil
C	1, 2, 5, 6	99–100	10–16	10–12	iii	Nil
D	8–10, 13, 15, 17, 19, 20, 22–30, 32, 33, 36–39, 42–47	94–99	4–17	14–18	ii	
E	4		4–23	16–19	iii	Nil
F	40, 41	62–69	15–22	9–12	iv	Nil

*Patterns of agglutination: i, complete agglutination of monkey erythrocytes; ii, complete agglutination of rat erythrocytes; iii, partial agglutination of rat erythrocytes (fewer receptors); iv, agglutination of rat erythrocytes visible only after addition of heterotypic antisera.

10.5 REPLICATION

The virus attaches to susceptible cells by the apical fibres, which bind to specific receptors. The virus is taken into endosomes inside the cell, losing the pentons and their fibres in the process. The remaining capsid is carried to the nuclear membrane where the peripentons are removed and the DNA is released into the nucleus. DNA replication, production of viral messenger RNA (mRNA), accumulation of structural proteins and assembly into progeny virions takes place in the nucleus with the mRNA inducing protein production in the cytoplasm. Early proteins take over the cellular functions and induce the cellular DNA replicase which then copies the viral DNA. New viral protein and, later, new virus particles form intranuclear crystals (inclusions) which may represent up to several thousand new progeny.

The DNA is 33–45 kb pairs long, is double-stranded, linear and codes for early (control) and late (structural) proteins. There are over 20 of the former and at least ten of the latter (conventionally notated I–X). Hexons are formed from three copies of II, penton bases from five copies of III, and the fibres from two copies of IV, while IIIa, VI, VIII and IX connect pentons and hexons beneath the outer surface to make a rigid capsid which withstands physical deformation well.

Assembly of new virus leads to rounding up of the host cells, producing a characteristic 'bunches-of-grapes' appearance (Figure 10.5). This is followed by lysis and virus release. However, with highly oncogenic strains (such as those from group A, types 12, 18 and 31) which induce tumours in new-born hamsters, productive virus replication may be inhibited. Instead, tumour (T) antigens are expressed on the surface of infected cells as the viral genome becomes integrated into that of the cell. This releases the cells from normal growth restraints and a malignant tumour is the result. In culture such cells are transformed and immortalized.

A variety of cell types will support adenovirus growth; in particular, continuous cell lines such as HEp2 (epidermoid carcinoma), HeLa (cervical carcinoma) and Graham 293 cells (human embryo kidney cells immortalized by incorporating part of the early region of adenovirus type 5 DNA). The latter cells are

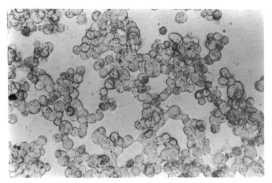

Figure 10.5 Characteristic 'bunch of grapes' cyto-pathic effect of adenoviruses growing in Hep-2 cells. As the cells die they round up and retract into refractile clumps of spheroids.

particularly valuable for growing the fastidi-ous types 40 and 41 which grow poorly if at all in other cell types. Group A viruses and serotype 8 (of group D) are often slow-grow-ing and may require one or more blind pas-sages before the characteristic cytopathic effect appears. Difficulties in recovering type 8 from clinical specimens probably means that it is substantially under-diagnosed.

High-titre virus can be toxic for cells in cul-ture without growth. The cells round up in a matter of hours and fall off the glass or plastic. However, little or no growth occurs and the input virus has to be diluted to allow the cells to survive for long enough for replication to occur.

10.6 CLINICAL FEATURES

Adenoviruses commonly infect the respiratory and gastrointestinal tracts. Infection is often asymptomatic, but once it occurs, the virus may persist for many weeks or even months. This characteristic of adenoviruses makes the clinical significance of a virus 'isolation' from a given patient, especially in children difficult to establish with certainty.

10.6.1 PREVALENCE

The prevalence of adenovirus infections in a community setting has been elegantly investi-

gated in the 'Virus Watch' programmes such as those in New York and Seattle [15]. A similar approach was used to study faecal excretion of adenovirus and diarrhoeal disease in Glasgow [40]. It was noted by Fox and his colleagues [15] that confining diagnostic tests to respiratory specimens meant that a significant number of infections were missed. A considerable number of isolates were only recovered from faeces and this finding focuses attention on the problems of proving disease causation by adenoviruses. Data from the Seattle and New York studies [15] indicate that the infection rate in infants is 179 per 100 person-years for all adenovirus serotypes (57 per 100 person-years for types 1 and 2), with the rate of infection declining in the older child up to the age of 9 years, but increasing again thereafter. The 'epidemic' serotype 7 was an exception to this pattern and infection rates increased progressively with age. Some 36% of infections resulted in persis-tence of virus shedding for longer than a month, 14% lasted longer than 1 year, and occasional patients excreted the same virus serotype for over 2 years. However, it was also noted that although virus excretion could be prolonged, it appeared to be finite (the maxi-mum documented time was 906 days). Thus, life-long persistence of the virus does not occur, and the virus is probably not transmit-ted vertically. There were no clear differences between the common serotypes (type 1, 2, 3 and 5) in their ability to establish this persistent excretion.

10.6.2 SYMPTOMS

Data from the Virus Watch studies also sug-gested that 51–56% of new infections are asymptomatic without clear differences in relation to age or virus type. Isolation of virus from the respiratory tract correlated better with clinical disease in that 65% of such patients were symptomatic in contrast to 31% of patients whose virus excretion was con-fined to the faeces. There was indirect evidence to suggest that less virus was present in the respiratory tract (compared with faeces in which the virus concentration may have been boosted by multiplication in the gut cells). Thus, it was possible that virus isolation from the respiratory tract was successful only from patients with a higher virus load. In those individuals who became ill respiratory symptoms predominated, the majority being confined to the upper respiratory tract. Of 184 episodes of illness documented, only 10 involved the lower respiratory tract [15].

Collaborative studies on the aetiological role of viruses (including adenoviruses) in respiratory disease in the UK were organized by Tyrrell and others [31,46,47]. In a general practice and hospital-based collaborative study carried out on patients under 17 years old in the UK between 1961–1964 [46], adenoviruses were isolated from 6.5% (85 of 1303) of those with respiratory diseases as compared with 1.6% of 'contact controls'. Typically, adenovirus infections caused sore throat and cough

associated with fever, with or without coryzal symptoms. Conjunctivitis was observed in a proportion of cases, and was more common with the 'epidemic' virus serotypes (e.g. type 3). However, as with other respiratory viruses, any combination of signs and symptoms could be associated with an adenovirus infection, and identifying the aetiological agent on clinical features was not possible. Pyrexia was more common in adenovirus, influenza virus and enterovirus infections, and was similar to that seen in streptococcal infection. A subsequent study [31] confirmed these observations and provided comparative data on the clinical manifestations of adenoviral illness in general practice and hospitalized patients (Table 10.3) (Figures 10.6 and 10.7). It was also noted during the latter study that adenoviruses may cause otitis media in a proportion of patients.

10.6.3 VIRUS ISOLATION

In a very large hospital-based study of 18 096 children in Washington, Brandt and colleagues

Table 10.3   Isolation of adenoviruses from patients with respiratory syndromes

	Patient population					
Age group No. of patients Reference	GP and hospital All ages* 1888 [46]		GP All ages 3966 [31]		Hospital All ages 2418 [31]	
	Patients yielding virus isolate (%)					
Clinical syndrome	Adeno- virus	Any virus	Adeno- virus	Any virus	Adeno- virus	Any virus
Coryza	4.4	19	3	20	3	28
Otitis media	0	16.7	0	7	7	23
Pharyngitis	8.2	22.9	3	17	8	41
Croup and laryngo- tracheo bronchitis	3.2	24	1	23	0	31
Bronchitis	4.5	22	3	19	0	26
Bronchiolitis	2.2	41	}4	}32	}0	}30
Pneumonia	2	28				
Influenza	2	28.5	1	34	0	57

* Predominantly patients under 17 years old.

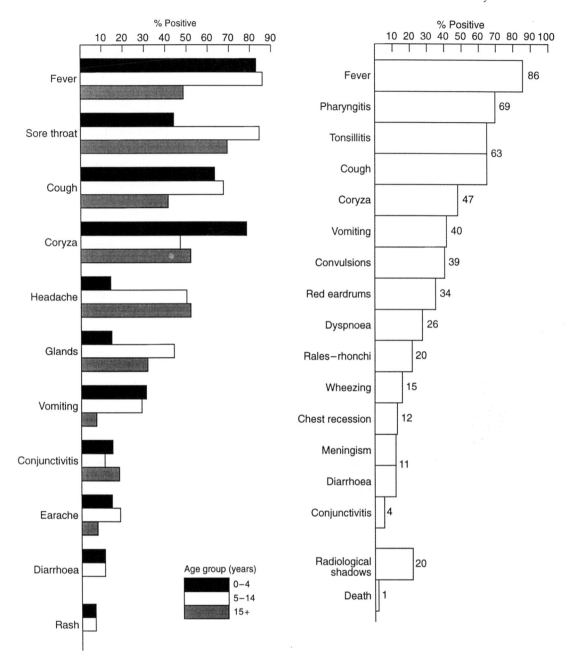

Figure 10.6 Clinical features of adenovirus infections (all serotypes) in different age groups of patients presenting to the general practitioner. (Reproduced from [31], with permission.)

Figure 10.7 Clinical features of adenovirus infections (all serotypes) in hospitalized patients. (Reproduced from [31], with permission.)

[7] isolated adenoviruses from 1792 (9.9%). When compared with the isolation rate in a control population, they estimated that approximately 7% of respiratory diseases in hospital are aetiologically linked to adenoviruses. In over 54% of patients with respiratory disease from whom an adenovirus was isolated from the throat, the virus appeared to be causally related to concurrent illness. However, isolation of virus types 1, 3, 6 or 7 appeared to be more clinically significant, and virus isolation from the throat was more relevant than virus isolation from the faeces. They also observed that when throat swabs were cultured, adenovirus cytopathic effect appeared earlier (mean of 9 days versus 12 days) in patients with disease as compared with controls, suggesting a larger virus load in the former. This observation on the speed of isolation from nasopharyngeal secretions was confirmed in Newcastle-upon-Tyne, UK as discussed later. A concurrent rise in antibody titre by the complement-fixation (CF) test was more often observed when virus isolation was associated with disease. However, the CF test almost never showed a rise in titre in children under 4 months of age [15].

These seminal observations on adenovirus disease have broadly been confirmed and amplified in subsequent studies elsewhere [13,33,34,38]. In a series of studies using antigen detection techniques for diagnosis, Ruuskanen and colleagues [38] have shown that adenoviruses are the commonest cause of exudative tonsillitis in children (particularly children under 3 years of age), and that this is indistinguishable from streptococcal sore throat either clinically or on the basis of WBC/ESR estimations in the laboratory. Adenovirus involvement in otitis media (either with or without concurrent bacterial infection) was confirmed by recovery of virus from the middle ear [10,37].

10.6.3 PNEUMONIA

Pneumonia is a rare, but more serious complication of adenovirus infection. Epidemic aden-

oviral pneumonia in military barracks and other closed communities is well recognized, and is associated with serotypes 4 and 7 (less commonly types 3, 14, 21, intermediate strain 14–11), as reviewed by Liu [26]. Outbreaks of adenovirus respiratory disease occur in other closed-community situations such as college dormitories, but are uncommon.

Pneumonia can occasionally complicate community-acquired infection by 'endemic' serotypes of virus. In hospital-based studies, approximately 7–9.5% of childhood pneumonias were attributed to adenovirus [7]. Conversely, of children hospitalized with adenoviral disease, a similar proportion had LRTIs [38], normally associated with types 1, 3 and 7, while adenovirus type 7b appeared particularly virulent in a number of outbreaks [51]. Rarely, community outbreaks of adenovirus pneumonia in children have been documented. An outbreak in China involved over 3000 children with a fatality rate of 16% [45]. Long-term lung damage (bronchiectasis, pulmonary fibrosis, atelectasis) can follow adenovirus pneumonia [26].

10.6.4 OTHER CONDITIONS

Disseminated adenoviruses disease (sometimes without any obvious respiratory localization) has been reported. The clinical presentation may be easily confused with septic shock (e.g. meningococcaemia or the toxic shock syndrome) [6,29,39]. Disseminated adenovirus infection has also been documented in the new-born (<10 days of age) [1] and must be considered (together with herpes simplex, enteroviruses and bacterial pathogens) in babies developing pneumonitis, hepatitis and neurological symptoms along with a disseminated intravascular coagulation (DIC)-like picture. Premature rupture of the membranes in the mother may be a risk factor, and suggests that an ascending infection may reach the baby *in utero*.

In most parts of the world, adenovirus types 1 and 2 are those most commonly isolated from patients with respiratory disease, followed by types 3, 5, 4 and 7 (Table 10.4).

The overall impression of adenoviruses is that they do not all behave in the same way. Some (types 1, 2, 5 and 6) are endemic, infecting children in whom they behave as opportunist pathogens. They frequently infect but do not always cause disease. They persist in the lymphoid tissue of the nasopharynx in which they behave as if they are latent, though whether their nucleic acid is truly integrated into the cellular genome and, if so, in which cells is unknown. Given the appropriate circumstances they can cause trouble in the form of respiratory disease (usually confined to the upper tract in the form of fever, nasal obstruction, sore throat and general malaise) with a

possible extension to conjunctivitis. Often, however, they remain dormant with isolation only from faeces following leakage into the gut leading to further replication (and therefore amplification of the amount of virus).

Other adenoviruses (typically types 3, 4 and 7) are more epidemic in nature. These are more likely to infect the respiratory epithelium with the development of an acute disease. Diagnosis made directly by taking specimens from the respiratory tract may be easier as more virus is produced (see later). Types 14 and 21 may behave similarly but are rarely seen in civilian communities.

Adenovirus type 8 is slow-growing and difficult to demonstrate. Most of its recorded pathogenesis has been in eye infections, par-

Table 10.4 Adenovirus serotypes isolated from clinical specimens

Reference	[46]	[31]	[31]	[2]	[7]	Unpublished data*	Unpublished data*
Place	UK	UK	UK	UK	USA	UK (Newcastle upon Tyne)	(Newcastle upon Tyne)
Patient type	GP and Hospital	GP	Hospital	Hospital	Hospital	Hospital	Hospital
Years of study	1961–1964	1964–1966	1964–1966	1970	1957–1967	1990–1992	1990–1992
Patient ages	All ages†	All ages	All ages	All ages	Children	All ages	All ages
Specimen type	Throat/ nose	Throat	Throat	All	Throat and faeces	All	Respiratory
No. of isolates	86	93	88	817	1792	422	142
Serotype			Adenovirus isolated belonging to serotype (%)				
1	19	} 66	27	24	26	16	29
2	15		30	23	35	23	26
3	47	20	–	8	10	10	19
4	2	–	–	7	0.2	2.4	5.6
5	13	–	25	16	11	13.5	14
6	1	–	–	1	1	2.1	2.8
7	1	14	} 18	3	3.3	2.1	3.5
>7	2	–		17	11	2.6	2.8
40	Not done	Not done	Not done	Not done	Not done	11.6	0
41	Not done	Not done	Not done	Not done	Not done	11.1	0
Untyped	0	0	0	0	2.8		
Total	100	100	100	100	100	100	100

* Authors' unpublished data;

ticularly in common-source outbreaks. Several of these were in clinics which re-used eye baths and other equipment on successive patients. With an awareness of the mechanism and the use of single-use drug preparations, adenovirus type 8 has largely disappeared. Nonetheless, it must still be at large – out of sight does not mean it has gone for good.

The higher serotypes have been isolated mostly from faeces or the eye. Their role, if any, in respiratory disease is unknown at present. Indeed their presence in the upper respiratory tract is undocumented, but they may be waiting for the right circumstances to make a contribution to disease.

Adenoviruses can cause life-threatening infection in the immunocompromised host [55], but this is an infrequent occurrence. They only rarely seem to take advantage of the reduced immunity (in contrast to other latent/persistent viruses, e.g. herpes viruses), but when infection does occur it may be of increased severity. For example, in a study of 1051 bone marrow transplant patients [41], adenovirus infections were detected in 51 (4.9%) of whom 10 (1%) had invasive disease leading to a fatal outcome in five (0.5%). In this study, pneumonia was the most frequent disease associated with adenovirus infection. Hepatitis (with or without respiratory involvement) is seen [55] particularly in liver transplants [24], but also in AIDS [25] and congenital immunodeficiencies [53]. Other manifestations of adenovirus infections in the immunocompromised include disseminated disease leading to a 'septic' patient with DIC and adenoviral colitis in patients with AIDS [22].

We have seen two children with persistent adenovirus type 2 (nasopharynx) and type 41 (stool) infections, respectively, given a bone marrow transplant without subsequent virus dissemination and disease. Hence, adenovirus infection pre-transplant is not necessarily a serious problem, though not one to be willingly courted.

The adenovirus serotypes associated with respiratory disease in the immunocompromised are no different to those found in immunocompetent individuals. However, adenovirus type 35 or other group B viruses (types 3, 7, 11) have been isolated from the urine of some patients with AIDS. The clinical significance of these isolates is unknown. On the other hand, similar virus types isolated from urine after bone marrow transplantation have been associated with a haemorrhagic cystitis [55].

The activities of adenoviruses are complex and the possibilities range from completely asymptomatic infection to a fatal pneumonia. Interpreting a positive laboratory test requires all the circumstances to be considered, as indicated below.

10.7 DIAGNOSIS

The general approach to diagnosing viral respiratory infections has been covered in an earlier chapter and it is not intended to re-cap this in detail. However, the method, specimen, and amount of virus recovered are essential ingredients in assessing the significance of the diagnosis of an adenovirus infection.

10.7.1 SPECIMEN COLLECTION

Clinical specimens for the diagnosis of adenoviral respiratory disease include nasopharyngeal secretions (NPS), cough/nasal swabs (a nasal swab combined in transport medium with a throat swab taken thoroughly enough to induce the cough reflex) and sputum. Eye swabs may be useful if conjunctivitis is present. While adenoviruses may be isolated from faeces, their significance in respiratory disease is less certain, as has been discussed above. NPS is the specimen of choice because it allows rapid diagnosis by techniques such as immunofluorescence. Sputum may be useful for investigation of LRTI, but more invasive specimens such as bronchoalveolar lavage or lung biopsy provide a more accurate indication of LRT involvement.

10.7.2 DIAGNOSTIC TECHNIQUES

Fluorescent antibody (FA)

Examination of exfoliated respiratory cells in NPSs with a good quality, high-titre, group-specific antiserum provides the quickest method of diagnosing a respiratory infection. Moreover, it also indicates whether the respiratory epithelium itself is involved. Typical ciliated cells showing positive fluorescence confirm that respiratory cells are infected. However, only about 45% of adenoviruses isolated in cell culture from an NPS can be detected by FA [17]. The significance of this is discussed below.

Cell culture

Isolation in cell cultures confirms the presence of complete infective virus. Adenovirus will multiply in susceptible cell lines such as HEp 2 and HeLa cells. Primary human embryonic kidney (HEK) cells are particularly sensitive for primary isolation but are less widely available. Growth will also occur in other cells such as human diploid cells, but less readily.

Virus isolates can be identified as adenovirus by FA or CF testing, and typed by neutralization or haemagglutination-inhibition tests. Higher concentrations of virus will result in more rapid growth and in a wider spectrum of cells. The more virus there is present the greater the clinical significance.

Enzyme immunoassays

Assuming the same quality of antisera as used in the FA technique, these will detect viral antigen but do not indicate its location and give only a general indication of quantity. Hence, positive results are more difficult to interpret.

Serology

This provides evidence for an antigenic stimulus. There is some evidence [15] that reliable seroconversion only follows a significantly widespread infection, but this leaves a grey zone of uncertainty over more minor episodes. Complement fixation (using a group specific antigen) is still the most widely used technique, but is not reliable in young children. Neutralization tests are type-specific but impractical for routine use.

Electron microscopy (EM)

There is no evidence that this has any place in diagnosing respiratory infection with adenoviruses. It has a valuable place in the diagnosis of adenovirus diarrhoea typically caused by types 40 and 41. Types 1–7 rarely reach EM-detectable levels in faeces.

Histology

The histological changes produced by adenoviruses in lung tissue were first related to these viruses by Chany and colleagues in 1958 [9] during an investigation of childhood deaths from adenovirus type 7 pneumonia. Later studies by EM showed that cells with intranuclear inclusions and the 'smudge cells' characteristic of adenovirus infections contained crystalline arrays of adenovirus [32]. Methods (including immunohistology) which allow the localization of virus in tissue confirm the invasion of lungs or other organs. On the other hand, isolation of adenovirus from the nasopharynx or even the lung does not necessarily mean that tissue invasion has taken place.

10.8 INTERPRETING ADENOVIRUS ACTIVITY IN THE RESPIRATORY TRACT

In the field of respiratory virus illness, the isolation of an adenovirus from the respiratory tract presents a challenge in interpretation. The presence of endemic adenoviruses has been demonstrated in a large number of asymptomatic individuals, particularly children. From the results of appropriate diagnostic tests, the extent of the infection can be assessed, where it is located and what serotype is involved. But with the adenoviruses, particularly the endemic types, whether the adenovirus isolated from the patient is causing the symptoms under investigation should be questioned.

It should also be remembered that dual infections with other respiratory viruses can occur, adding to the problems of interpretation. Respiratory syncytial (RS) virus, influenza A or B viruses and parainfluenza viruses are all capable of co- or super-infecting the respiratory mucosa. Whether synergy occurs and makes the overall diseases worse is not clear – all are capable of causing severe disease on their own. Studies in Newcastle-upon-Tyne, UK have shown that in dual infection with adenovirus and another respiratory virus, where the adenovirus overgrows the second virus in culture, only the use of FA diagnosis will bring attention to the presence of another virus.

Considering adenoviruses alone, a strongly positive FA test with numerous infected ciliated cells, complemented by early virus isolation (<8 days) in several cell lines will suggest very strongly that the adenovirus is the cause of a clinically significant infection. Seroconversion is also likely, except in young children. On the other hand, slow isolation (>12 days) from a FA-negative NPS or throat swab in one cell line is probably a casual finding with minimal significance in relation to disease.

Between these extremes are various combinations. Although it is not possible to be totally dogmatic about interpretation, some useful guidance to the clinical virologist and clinician can be derived from the following data, collected in Newcastle over a 20-year period during which childhood respiratory illness has been studied by using FA and cell culture. The guidelines are based on preliminary results first reported in 1972 [16], and confirmed by further results reported in 1980 [17], together

with other unpublished observations. The demonstration of invasion of the respiratory epithelium by adenovirus (FA-positive NPS) is convincing evidence of active infection of the respiratory tract and can be reported as such. Neither the endemic serotypes (1, 2, 5 and 6) or epidemic serotypes (3, 4 and 7) are exclusively pathogenic or non-pathogenic, but some patterns can be seen.

In Figure 10.8 are summarized the FA and cell culture results from NPS specimens taken from 661 patients examined between 1968 and 1981. It may be seen that the proportion of epidemic serotypes (3, 4 and 7) shown positive by FA was consistently greater than that seen with the endemic (1, 2, 5 and 6) serotypes. The latter make a significant contribution to the overall FA positive results because these serotypes occur in much larger numbers. However, the FA results demonstrated that the epidemic types showed a greater tendency to involve the respiratory epithelium.

When the FA results are compared with the speed of virus isolation from 432 NPS from these patients (Figure 10.9) there is a strong correlation. In the FA-positive NPS the virus was isolated rapidly (within 8 days from the majority and by 12 days in almost all of them). In the few FA-negative specimens where the virus was isolated rapidly, the NPS specimens were found to be of poor quality and had insufficient cells for reliable FA diagnosis. In contrast, virus isolation took longer (9–28 days) in the majority of FA-negative NPS specimens. These observations suggest that high virus titres in the NPS correlated with viral invasion of the nasopharynx.

In the case of most of the common respiratory viruses, a significant proportion of clinically important infections would be missed if the NPS was not cultured. Isolation of virus from NPS can exceed that from cough/nasal swabs by as much as 15%, most of these being FA-positive and clinically significant. With adenoviruses this is less clear-cut.

The source of isolation of the adenovirus in relation to the results of FA tests on the NPS of

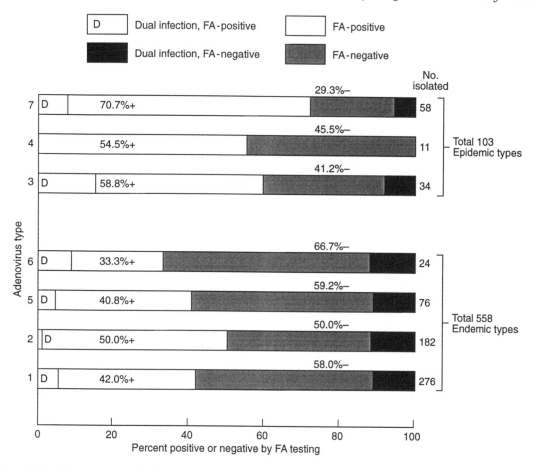

Figure 10.8 Fluorescent antibody (FA) test results on 661 nasopharyngeal secretions infected with epidemic (3, 4, 7) and endemic (1, 2, 5, 6) adenovirus types, Newcastle upon Tyne, December 1968 to March 1981.

192 patients is recorded in Figure 10.10. As with other respiratory viruses, more isolations were made from NPS specimens than from cough/nasal swabs. Of the 192 isolates, 68 were obtained only from NPS. However, the majority of the isolates obtained solely from the NPS were FA-negative, and their clinical significance was doubtful. Conversely, the virus was isolated from cough/nasal swabs from the majority of patients with FA-positive NPS. Thus, unlike other respiratory viruses, the culture of cough/nasal swabs alone will not compromise the detection of clinically significant adenovirus infections.

A second striking difference between adenoviruses and other respiratory viruses concerns the rates of FA positivity in patients with infection. With most of the common respiratory viruses (e.g. RS virus) 90–99% of infections (whether detected by culture from cough/nose swab or NPS) can be diagnosed by FA. The picture with adenoviruses is quite different. Only 45% of patients from whom virus was isolated from similar specimens taken from the upper respiratory tract were FA-positive (Figure 10.10). The FA-negative specimens were predominantly those endemic serotypes which were isolated solely from the NPS; 60 of

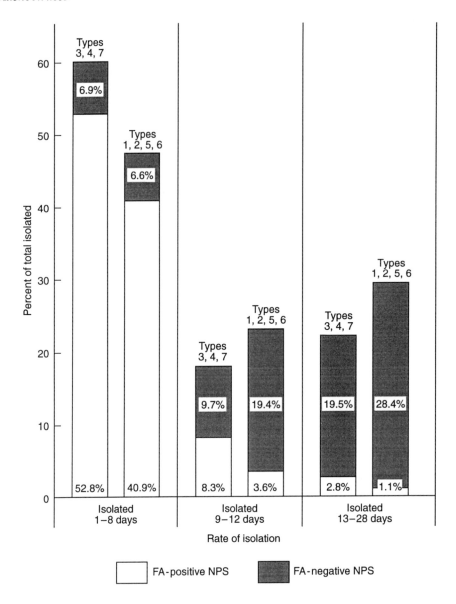

Figure 10.9 The relationship of the rate of isolation to the detection of adenovirus-infected cells, measured by immunofluorescence, in nasopharyngeal secretions, Newcastle upon Tyne. Comparison of epidemic (3, 4, 7) and endemic (1, 2, 5, 6) types.

96 patients with adenovirus infections which were FA-negative yielded the virus solely from the NPS. It is reasonable to suppose that the virus is not replicating in the nasopharyngeal epithelium in these FA-negative patients.

Other studies on the length of excretion of adenoviruses from the respiratory tract showed rapid disappearance from the respiratory epithelium (i.e. patients rapidly became FA-negative), but the virus could be isolated

Percentage distribution

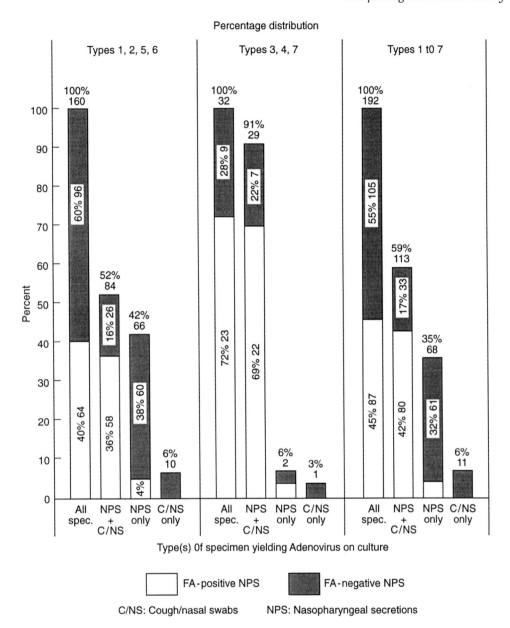

Figure 10.10 Comparison of source(s) of isolation of adenovirus in relation to fluorescent antibody (FA) diagnostic results on nasopharyngeal secretions from 192 patients, Newcastle upon Tyne.

from the upper respiratory tract for long periods. This is also in marked contrast to other respiratory viruses where in the convalescent phase of infection the virus can no longer be isolated, but remains detectable in the nasopharynx by FA for a few days longer. This is presumably because the virus becomes noncultivable as a result of being neutralized by mucosal antibody, although viral antigens are still detectable. Thus, the ability to culture adenovirus for long periods from the NPS without visual evidence of virus infection of nasopharyngeal cells poses an enigma. Where does the virus go during asymptomatic infection of the respiratory tract? Why is it not neutralized by locally produced antibody?

Some light has been thrown on this enigma by studying children with acute lymphoblastic leukaemia (ALL). Such children were found to be susceptible to prolonged infections with respiratory viruses such as RS virus and other viruses, such as measles, sometimes resulting in severe life-threatening illness [12]. Because of the number of children known to have persistent asymptomatic infections with adenoviruses it must be inevitable that children with ALL would frequently come into contact with them. Yet we have never seen a persistent adenovirus infection in a child with ALL, in contrast to the situation with other respiratory viruses.

During the treatment of ALL, the lymphatic tissue is almost completely destroyed, which suggests that in the normal host the virus must be dependent on the presence of this tissue in the nasopharynx for survival during the asymptomatic period of infection. Further, it seems likely that the nasopharyngeal lymphatic tissue is the true reservoir of adenoviruses. The infection of the respiratory tract may be due to a spill-over of virus from the lymphatic tissue rather than vice-versa. The original discovery of adenoviruses in the adenoid does not quite fit with this hypothesis since it was reported that it was difficult to release the virus from the adenoidal tissue without culturing the tissue itself, but perhaps *in vivo* the conditions are different and vary from time to time.

Given the spectrum of presentation there are no clear markers of severity in the lower respiratory tract except as outlined above. However, major upper tract activity should provide a warning to look for evidence of lung disease, although proof would depend on bronchoalveolar lavage and/or lung biopsy. Such episodes are rare but a prudent virologist should be aware of the possibility and ask questions about lower respiratory tract involvement, particularly in infants.

10.9 OTHER ASPECTS

There is no evidence that adenoviruses from other species infect humans at any age. Although domestic animals such as the dog are known to be infected the species barrier appears not to be crossed.

Adenovirus types 12, 18 and 31 are all oncogenic in hamsters and other rodents under very artificial laboratory conditions. Chimaeras with polyoma viruses can also be made but there is again no evidence that they or adenoviruses are oncogenic in humans despite a careful search.

10.10 MANAGEMENT AND PREVENTION

10.10.1 TREATMENT

The vast majority of adenovirus-caused illnesses are brief and self-limited, even if post-illness excretion may be prolonged. These episodes will usually require no treatment other than symptomatic relief with analgesics and maintenance of fluids and calorie intake. Three groups may, however, have more serious illness: the very young, the immunocompromised and the physically exhausted. As indicated above, these may be at serious risk, raising the question of whether antiviral drugs can help.

There is at present, no antiviral therapy of proven efficacy for adenovirus disease. Ganciclovir [44] and ribavirin [42] have activity against adenoviruses *in vitro*. There are anecdotal reports of the use of these antivirals in patients with life-threatening adenoviral disease [8,54], but there is no proof of clinical benefit. Neither of these antivirals is licensed in the UK for treating adenovirus infection and each would have to be used only on a named-patient basis.

10.10.2 PREVENTION

There is no indication for routine immunization against adenoviruses. Most of the associated illness is neither severe nor common enough to justify it. However, the problems of serious disease in their recruit camps led the US Army to develop a vaccine to serotypes 3, 4 and 7. Later, type 21 was added. This was an oral enteric-coated preparation containing virus passed 10 times in WI-38 cells. This produced partly attenuated virus which was released into the gut beyond the acid barrier of the stomach to induce systemic humoral antibody and prevent severe lower respiratory tract infection. The vaccine-induced circulating neutralizing antibody in >95% of the vaccinees and reduced the incidence of disease by two-thirds [11]. For army purposes, this proved useful but evidence that the vaccine virus may spread to contacts has reduced interest in using it more widely.

Severe adenovirus complications in the immunosuppressed are rare. It is certainly not enough to make development of a vaccine for this purpose a viable possibility. The US Army vaccine discussed above is specific only for these serotypes and would not provide the broad spectrum of cover necessary.

10.11 CONCLUSIONS

Adenoviruses have been out of fashion for some time. In the 1950s and 1960s they were the focus of considerable attention following their discovery and as their basic properties and intracellular replication were established. They were again the focus of attention as their oncogenic potential was recognized, especially when combined with their interaction with papovaviruses. There was a hope that further research would throw new light on the mechanisms of cancer. Although widely distributed in humans, there has never been any evidence that they are involved in any aspect of human malignancy, except as opportunist invaders of the respiratory and alimentary tracts of those under treatment.

There was a further period of interest in the 1970s as the contribution of the fastidious types 40 and 41 to diarrhoea, particularly in children, was uncovered and, briefly, as types 43–47 were isolated from patients with AIDS. Latterly, interest has waned and they are perhaps due for another season in fashion. This may come through virologists exploring the value in diagnosis of the polymerase chain reaction (PCR), the sensitivity of which is likely to rediscover evidence for adenovirus persistence. This, seen in the context of the search for viral causes for chronic diseases, such as the chronic fatigue syndrome, multiple sclerosis, Paget's disease of bone, ulcerative colitis, may cause a quickening of interest. Whether this will lead to useful new evidence only time will tell but, even if it does not, it still leaves the evidence of more banal activities to be understood.

Overall, most adenovirus activity involves a small number of serotypes (1–7) which can infect the respiratory tract, particularly in infants, and are confined largely to the lymphoid tissue of the tonsils and adenoids. This can spread more widely to involve the respiratory mucosa and cause respiratory disease, which is unpleasant but not generally life threatening. What precipitates it is unclear – do other viruses or even bacteria lead the way? – but it is often followed (or even preceded?) by prolonged virus shedding which can be at a low level, making isolation variable. However, since this leaks or overflows into the gut, replication in gut cells can amplify the concentration of virus to more readily detectable levels.

On the basis of this interpretation it is clear that not all virus isolations are significant and the challenge is to distinguish the significant from the incidental. Diagnosis by enzyme immunoassay and cell culture will at best be marginally quantitative and cannot reveal the site of infection, however carefully the specimen is taken as secretions and mucus can wash the virus to any part of the nasopharynx.

As has been indicated above, the FA technique is particularly useful in clarifying the situation. It is just as important to know *where* the infection is located as to know whether it is there at all. Other techniques, with the exception of *in situ* DNA detection, are not able to do this. The advent of the PCR offers a new opportunity to monitor long-term excretion of virus. It should detect even latent virus, provided the appropriate specimen can be and is taken. Further tests for the presence of messenger RNA (mRNA) to indicate transcriptional activity for example, may be necessary to understand the significance of any positive results.

What are the unanswered questions about adenoviruses? These include 'What is the difference between latent (?) and/or trivial carriage and a significant infection leading to disease? If there is a clear distinction, what converts one to the other? What leads an otherwise trivial infection with an endemic strain to become more aggressive? Does it require another microorganism or virus – perhaps unrecognized – to promote its activity?

Adenoviruses have been described as 'the weeds in the virological garden'. Perhaps this is unfairly dismissive of a family of viruses whose interaction with their host is more complex than it appears at first glance. They certainly provide a challenge to the virologist to develop appropriate diagnostic techniques and to all concerned to interpret the results.

10.12 REFERENCES

1. Abzug, M.J., Beam, A.C., Gyorkos, E.A. and Levin, M.J. (1990) Viral pneumonia in the first year of life. *Pediatr. Infect. Dis. J.*, **9**, 881–5.
2. Anon. (1971) Epidemiology: adenovirus infections. *Br. Med. J.*, **2**, 719.
3. Balducci, D., Zaiman, E. and Tyrrell, D.A.J. (1956) Laboratory studies of A.P.C. and influenza C viruses. *Br. J. Exp. Pathol.*, **37**, 205–18.
4. Bell, J.A., Rowe, E.P., Engler, J.I. *et al.* (1955) Pharyngoconjunctival fever. Epidemiological studies of a recently recognized disease entity. *JAMA*, **157**, 1083–92.
5. Berge, T.O., England, B., Mouris, C. *et al.* (1955) Etiology of acute respiratory disease among service personnel at Fort Ord, California. *Am. J. Hygiene*, **62**, 283–94.
6. Bojang, K. and Walters, M.D.S. (1992) Toxic shock-like syndrome caused by adenovirus infection. *Arch. Dis. Child.*, **67**, 1112–14.
7. Brandt, C.D., Kim, H.W., Vargosko, A.J. *et al.* (1969) Infections in 18 000 infants and children in a controlled study of respiratory tract disease. I. Adenovirus pathogenicity in relation to serologic type and illness syndrome. *Am. J. Epidemiol.*, **90**, 484–500.
8. Buchdahl, R.M., Taylor, P. and Warner, J.O. (1985) Neubulized ribavirin for adenovirus pneumonia. *Lancet*, **ii**, 1070.
9. Chany, C., Lépine, P., Lelong, M. *et al.* (1958) Severe and fatal pneumonia in infants and young children associated with adenovirus infection. *Am. J. Hygiene*, **67**, 367–78.
10. Chonmaitree, T., Howie, V. and Truant, A.L. (1986) Presence of respiratory viruses in middle ear fluids and nasal-wash specimens from children with acute otitis media. *Pediatrics*, **77**, 698–702.
11. Couch, R.B., Chanock, R.M., Cate, T.R. *et al.* (1963) Immunization with types 4 and 7 adenovirus by selective infection of the intestinal tract. *Am. Rev. Respir. Dis.*, **88**, 394–403.
12. Craft, A.W., Reid, M.M., Gardner, P.S. *et al.* (1979) Virus infections in children with acute lymphoblastic leukaemia. *Arch. Dis. Child.*, **54**, 755–9.
13. Edwards, K.M., Thompson, J., Paolimi, J. and Wright, P.F. (1985) Adenovirus infections in young children. *Pediatrics*, **76**, 420–4.
14. Enders, J.F., Bell, J.A., Dingle, J.H. *et al.* (1956) 'Adenoviruses': group name proposed for new respiratory-tract viruses. *Science*, **124**, 119–20.
15. Fox, J.P., Hall, C.E. and Cooney, M.K. (1977) The Seattle Virus Watch. VIII. Observations of adenovirus infections. *Am. J. Epidemiol.*, **105**, 362–86.
16. Gardner, P.S., McGuckin, R. and McQuillin, J. (1972) Adenovirus demonstrated by immunofluorescence. *Br. Med. J.*, **3**, 175.
17. Gardner, P.S. and McQuillin, J. (1980) *Rapid Virus Diagnosis: application of immunofluorescence*, 2nd edn. Butterworth, London, pp. 197–200.

18. Hierholzer, J.C., Kemp, M.C., Gary, G.W.G. Jr, and Spencer, H.C. (1982) New human adenovirus associated with respiratory illness: candidate adenovirus type 39. *J. Clin. Microbiol.*, **16**, 15–21.

19. Hilleman, M.R. and Werner, J.H. (1954) Recovery of new agents from patients with acute respiratory illness. *Proc. Soc. Exp. Biol. Med.*, **85**, 183–8.

20. Horwitz, M.S. (1990) Adenoviruses, in *Virology*, (eds B.N. Fields and D.M. Knipe), Raven Press, New York, pp. 1723–40.

21. Huebner, R.J., Rowe, W.P., Ward, T.G. et al. (1954) Adenoidal-pharyngeal-conjunctival agents. A newly recognized group of common viruses of the respiratory system. *N. Engl. J. Med.*, **251**, 1077–86.

22. Janoff, E.N., Orenstein, J.M., Manischewitz, J.F. and Smith, P.D. (1991) Adenovirus colitis in the acquired immunodeficiency syndrome. *Gastroenterology*, **100**, 976–9.

23. Jawetz, J. (1959) The story of shipyard eye. *Br. Med. J.*, **i**, 873–8.

24. Koneru, B., Jaffe, R., Esquivel, C.O. et al. (1987) Adenoviral infections in pediatric liver transplant recipients. *JAMA*, **258**, 489–92.

25. Krilov, L.R., Rubin, L.G., Frogel, M. et al. (1990) Disseminated adenovirus infection with hepatic necrosis in patients with human immunodeficiency virus infection and other immunodeficiency states. *Rev. Infect. Dis.*, **12**, 303–7.

26. Liu, C. (1991) Adenoviruses, in *Textbook of Human Virology*, (ed. R.B. Belshe), Mosby Year-Book, Inc., St Louis, pp. 791–803.

27. Mautner, V. (1989) *Andrewes' Viruses of Vertebrates*, (ed. J.S. Porterfield), Baillière Tindall, London, pp. 249–82.

28. Neva, F.A. and Enders, J.F. (1954) Isolation of a cytopathogenic agent from an infant with a disease in certain respects resembling roseola infantum. *J. Immunol.*, **72**, 315–21.

29. Odio, C., McCracken, G.H. and Nelson, J.D. (1984) Disseminated adenovirus infection: a case report and review of the literature. *Pediatr. Infect. Dis. J.*, **3**, 46–9.

30. Parrott, R.H., Rowe, W.P., Huebner, R.J. et al. (1954) Outbreak of febrile pharyngitis and conjunctivitis associated with type 3 adenoidal-pharyngeal-conjunctival virus infection. *N. Engl. J. Med.*, **251**, 1087–90.

31. Pereira, M.S. (1973) Adenovirus infections. *Postgrad. Med. J.*, **49**, 798–801.

32. Pinkerton, H. and Carroll, S. (1971) Fatal adenovirus pneumonia in infants. Correlation of histological and electron microscopic observations. *Am. J. Pathol.*, **65**, 543–6.

33. Putto, A. (1987) Febrile exudative tonsillitis: viral or streptococcal. *Pediatrics*, **80**, 6–12.

34. Putto, A., Ruuskanen, O. and Meurman, O. (1986) Fever in respiratory virus infections. *Am. J. Dis. Child.*, **140**, 1159–63.

35. Rowe, W.P., Huebner, R.J., Gilmore, L.K. et al. (1953) Isolation of a cytopathogenic agent from human adenoids undergoing spontaneous degeneration. *Proc. Soc. Exp. Biol. Med.*, **84**, 570–3.

36. Rowe, W.P., Huebner, R.J., Hartley, J.W. et al. (1955) Studies of adenoidal-pharyngeal-conjunctival (APC) group of viruses. *Am. J. Hygiene*, **61**, 197–218.

37. Ruuskanen, O., Avola, M., Lutto-Lanvilla, A. et al. (1989) Acute otitis media and respiratory virus infections. *Pediatr. Infect. Dis. J.*, **8**, 94–9.

38. Ruuskanen, O., Meurman, O. and Sarkkinen, H. (1985) Adenoviral diseases in children: a study of 105 hospital cases. *Pediatrics*, **76**, 79–83.

39. Sahler, O.J. and Wilfert, C.M. (1974) Fever and petechiae with adenovirus type 7 infection. *Pediatrics*, **53**, 233–5.

40. Scott, T.M., Madeley, C.R., Cosgrove, B.P. and Stanfield, J.P. (1979) Stool viruses in babies in Glasgow. 3. Community studies. *J. Hygiene*, **83**, 469–85.

41. Shields, A.F., Hackman, R.C., Fife, K.H. et al. (1985) Adenovirus infections in patients undergoing bone-marrow transplantation. *N. Engl. J. Med.*, **312**, 529–33.

42. Sidwell, R.W. (1980) Ribavirin: *in vitro* antiviral activity, in Ribavirin: *A Broad Spectrum Antiviral Agent*, (eds R.A. Smith and W. Kirkpatrick), Academic Press, New York, pp. 23–42.

43. Stuart-Harris, C.H., Andrewes, C.H., Smith, W. et al. (1938) *Acute Respiratory Infections*, Medical Research Council Special Report Series.

44. Taylor, D.L., Jeffries, D.J., Taylor-Robinson, D. et al. (1988) The susceptibility of adenovirus infection to the anti-cytomegalovirus drug ganciclovir (DHPG). *FEMS Microbiol. Lett.*, **49**, 337–41.

45. Teng, C.H. (1960) Adenovirus pneumonia epidemic among Peking infants and preschool children in 1958. *Chinese Med. J.*, **80**, 331–9.

46. Tyrrell, D.A.J. (1965) A collabatory study of the aetiology of acute respiratory infections in Britain 1961–64. A report of the Medical Research Council Working party on acute respiratory infections. *Br. Med. J.*, ii, 319–26.

47. Tyrrell, D.A.J. (1973) Concluding remarks. Patterns of respiratory virus infection. *Postgrad. Med. J.*, **49**, 820–1.

48. Tyrrell, D.A.J., Balducci, D. and Zaiman, T.E. (1956) Acute infections of the respiratory tract and the adenoviruses. *Lancet*, ii, 1326–30.

49. van der Veen, J. and Dijkman, J.H. (1962) Association of type 21 adenovirus with acute respiratory illness in military recruits. *Am. J. Hygiene*, **76**, 149–59.

50. van der Veen, J. and Kok, G. (1957) Isolation and typing of adenoviruses recovered from military recruits with acute respiratory disease in The Netherlands. *Am. J. Hygiene*, **65**, 119–29.

51. Wadell, G., Varsany, T.M., Lord, A. and Sutton, R.N.P. (1980) Epidemic outbreaks of adenovirus 7 with special reference to the patho-genicity of adenovirus genome type 7b. *Am. J. Epidemiol.*, **112**, 619–28.

52. Wadell, G. (1994) Adenoviruses, in *Principles and Practice of Clinical Virology*, 3rd edn (eds A.J. Zuckerman, J.E. Banatvala and J.R. Pattison), John Wiley & Sons, Chichester, pp. 287–308.

53. Washington, K., Gossage, D.L. and Gottfried, M.R. (1993) Pathology of the liver in severe combined immunodeficiency and DiGeorge syndrome. *Pediatr. Pathol.*, **13**, 485–504.

54. Wreghitt, T.G., Gray, J.J., Ward, K.N. *et al.* (1989) Disseminated adenoviral infections after liver transplantation and its possible treatment with ganciclovir. *J. Infection*, **19**, 88–9.

55. Zahradnik, J.M., Spencer, M.J. and Porter, D.D. (1980) Adenovirus infection in the immuno-compromised patient. *Am. J. Med.*, **68**, 725–32.

56. Zaiman, E., Balducci, D. and Tyrrell, D.A.J. (1955) A.P.C. viruses and respiratory disease in northern England. *Lancet*, ii, 595–6.

PARAMYXOVIRUSES

11

Craig Pringle

11.1 INTRODUCTION

A paramyxovirus-induced disease, epidemic parotitis or mumps, was one of the first human afflictions to be recognized as a distinct disease entity [39]. It can be traced back as far as

Viral and Other Infections of the Human Respiratory Tract.
Edited by S. Myint and D. Taylor-Robinson. Published in
1996 by Chapman & Hall. ISBN 0 412 60070 6

Hippocrates, who in the fifth century BC described the mild epidemic sickness involving facial swelling with occasional painful enlargement of the testes now known as mumps. Table 11.1 lists some of the milestones in the development of knowledge of paramyxovirus-related disease and of the paramyxoviruses themselves. This chapter is a rather selective look at the state of progress of research in this field of virology up to the end of 1993.

11.2 THE NEGATIVE-STRANDED NON-SEGMENTED GENOME RNA VIRUSES

Three families of viruses, the Filoviridae, the Paramyxoviridae and the Rhabdoviridae, possess non-segmented negative-stranded RNA genomes. These families are now grouped together as the Order Mononegavirales because of their similar patterns of genome organization and morphogenesis [82]. They are differentiated primarily by their biological properties. The filoviruses have been isolated only infrequently and so far always in association with outbreaks of severe haemorrhagic disease in primates. By contrast the rhabdoviruses are abundant and infect a wide range of organisms embracing both the plant and animal kingdoms. Rhabdoviruses are predominantly transmitted by arthropod vectors, and in mammals are associated with mild febrile or silent infections, or in the exceptional case of rabies virus and its close relatives, neurological disease of varying severity. The

Table 11.1   A chronology of the paramyxoviruses

5th century BC	Mumps described by Hippocrates
9th century AD	Measles described by arab physicians
17th century AD	Measles recognized as a distinct exanthem by Sydenham
1846	Panum's study of measles epidemiology in the Faroe Islands
1911	Measles incubation period defined by experiments with volunteers
1918	Prevention of measles by passive immunization
1926	Comprehensive characterization of canine distemper
1927	Fowl pest (Newcastle disease) recognized as disease entity
1928	Use of formalinized and live virus to control canine distemper
1934	Shown that mumps caused by filterable agent Transmission to monkeys
1935	Characterization of mumps virus
1938	Measles virus propagated *in vitro* in chick embryo cells and transmitted to volunteers
1939	Measles transmitted to monkeys experimentally Isolation of pneumonia virus of mice
1944	Old World Newcastle disease and New World Pneumoencephalitis recognized as the same disease
1945	Growth of mumps virus in chick embryos
1948	Attenuation of canine distemper virus by passage in the chorioallantoic membrane of fertile eggs
1950	Introduction of killed mumps virus vaccine
1952	Propagation of NDV and other paramyxoviruses in eggs and cultured cells (D.A.J. Tyrrell and others)
1956	Isolation of chimpanzee coryza agent (respiratory syncytial virus)
1963	Introduction of killed and live attenuated measles virus vaccine.
1965	Growth of paramyxoviruses in human ciliated epithelium organ culture (D.A.J. Tyrrell and M. Hoorn)
1965	Initial reports of unusual illness on exposure of children previously vaccinated with formalin-inactivated measles virus to natural infection. Followed by removal of killed vaccine from the market in 1968.
1967	Introduction of live attenuated mumps virus vaccine Paradoxical disease observed in trials of formalin-inactivated respiratory syncytial virus vaccine Recognition that subacute sclerosing panencephalitis (Dawson's inclusion body encephalitis or von Bogaert's disease) associated with measles virus infection
1971	Introduction of trivalent measles, mumps and rubella (MMR) virus vaccine
1977	Isolation of first avian pneumovirus (turkey rhinotracheitis virus)
1988	Association of paramyxoviruses with outbreaks of disease in marine and freshwater mammals
1993	Trial of attenuated respiratory syncytial virus in volunteers (D.A.J Tyrrell and others)
1995	WHO target for global measles control of 95% reduction in measles deaths and 90% reduction in measles cases compared with pre-immunization levels

paramyxoviruses, on the other hand, have been isolated only from vertebrates, and arthropod vectors play no role in their transmission. They are transmitted by aerosols or contact, and as a consequence are predominantly associated with respiratory infections. The common and distinguishing features of the viruses of the three families included in the

Order Mononegavirales are listed in Table 11.2. The characteristic properties of the viruses that comprise each of the families Filoviridae, Paramyxoviridae and Rhabdoviridae are compared in Table 11.3.

The family Paramyxoviridae is divided into two subfamilies, Paramyxovirinae and Pneumovirinae [42]. The subfamily Paramyxovirinae comprises the three genera *Paramyxovirus*, *Morbillivirus* and *Rubulavirus*, and the subfamily Pneumovirinae contains the single genus *Pneumovirus* [40]. There are, in addition, a few unassigned viruses. In Table 11.4, a listing is given of the currently recognized virus species in the family. It is likely that the genus name *Paramyxovirus* will be replaced in the near future. Use of the term paramyxovirus as the vernacular name referring to any member of the family Paramyxoviridae will become a source of increasing confusion as knowledge of the diversity of the viruses within the family increases.

11.3 GENOME ORGANIZATION

Figures 11.1 and 11.2 illustrate diagrammatically the organization of the genomes of viruses of the Order Mononegavirales. There is a basic pattern of 5–10 genes in a linear array with a unique promoter at the 3' terminus followed by a block of core protein genes, a block of envelope protein genes and finally the viral polymerase gene adjacent to the 5' terminus. There are minor variations from this pattern, but the polymerase (L protein) gene is always located at the 5' extremity and with the exception of the pneumoviruses, the nucleoprotein (N or NP) gene is located at the 3' terminus.

Table 11.2 The Order Mononegavirales

Features common to the three families comprising the order:

Linear undivided single-stranded negative-sense RNA genome
Helical nucleocapsid
Similar gene order (3'-non-coding leader – core protein genes – envelope protein genes – polymerase gene – 5'-non-coding trailer).
Discrete unprocessed messenger RNAs transcribed by sequential interrupted synthesis
Replication by synthesis of a complete positive-sense RNA anti-genome
Virion-associated RNA-dependent RNA polymerase
Maturation by budding through host cell membrane and investment of the nucleocapsid by a host-derived lipid envelope containing transmembrane viral proteins and largely devoid of host proteins

Features distinguishing the three families comprising the order:

Absolute genome size and coding capacity
Virion morphology
    filamentous – the filoviruses
    pleomorphic – the paramyxoviruses
    bullet-shaped and bacilliform – the rhabdoviruses
Pathogenic potential
    haemorrhagic fever – the filoviruses
    mainly respiratory disease – the paramyxoviruses
    mild febrile to fatal neurological disease – the rhabdoviruses
Host range
    primates – the filoviruses
    vertebrates – the paramyxoviruses
    invertebrates, vertebrates and plants – the rhabdoviruses
Several biological properties

Table 11.3 The characteristic properties of the three families of the Order Mononegavirales

---

The family Filoviridae

Morphology: bacilliform or filamentous with branching; sometimes U-shaped, 6-shaped or circular
Uniform diameter of 80 nm and varying in length up to 14 000 nm; infectious particle length estimated to
  be 790 nm for Marburg virus and 970 nm for Ebola virus
Surface spikes of 7–nm length, spaced at 10-nm intervals
Helical nucleocapsid; 50-nm diameter, with an axial space of 20-nm diameter and a helical periodicity of
  about 5 nm
Genome; $M_r = 4.2 \times 10^6$, negative-sense, single-stranded unsegmented RNA, 19.1 kb in length
At least seven proteins; large 267-kDa polymerase, 75-kDa surface glycoprotein, 78-kDa nucleoprotein,
  two nucleocapsid-associated proteins, and at least two other membrane-associated proteins
Biology enigmatic; two antigenically unrelated viruses (Marburg and Ebola) known; blood-borne
  infection of humans from monkeys; tropism for cells of the reticuloendothelial system, fibroblasts and
  interstitial tissue

The family Rhabdoviridae

Morphology; bullet-shaped or bacilliform, 45–100 nm × 100–430 nm
Surface spikes composed of G protein alone, 5–10 nm length × 3 nm diameter
Helical nucleocapsid; 50 nm diameter, unwinding to helical structure of 20 × 700 nm (VSV)
Genomes size; $M_r = 4.2–4.6 \times 10^6$ 11.161 kb for VSV to 11.932 kb for rabies virus
Proteins; 5–6 open reading frames encoding 5–7 polypeptides, 50–62 kDa nucleoprotein (N), 20–30 kDa
  phosphoprotein (NS, P or M1), 70–85 kDa glycoprotein (G), and 150 kDa large (L) polymerase protein
  common to all; variable proteins include one of unknown function encoded in the P gene, and non-
  structural proteins encoded between G and L and M1 and M2
May be vertically transmitted in insects, but otherwise spread horizontally
Five genera infecting vertebrates (*Vesiculovirus, Lyssavirus, Ephemerovirus*), invertebrates (*Vesiculovirus,
  Ephemerovirus*) and plants (*Cytorhabdovirus, Nucleorhabdovirus*).

The family Paramyxoviridae

Morphology; pleomorphic, some quasi-spherical 150 nm or more in diameter, filamentous forms common
Envelope derived from the cell membrane, incorporating one or two viral glycoproteins and one or two
  unglycosylated proteins
Helical nucleocapsid; 13–18 nm diameter, 5.5–7 nm pitch
Genome size uniform; 15.156 kb (Newcastle disease virus) to 15.892 kb (measles virus)
Seven to eight open reading frames (genes) encoding 10–12 polypeptides, of which 4–5 may be derived
  from 2–3 overlapping reading frames in the P gene
Proteins; a nucleoprotein (N or NP), phosphoprotein (P), matrix protein (M), fusion protein (F) and large
  (L) polymerase protein are common to all genera; variable proteins are the non-structural C, D, Y, W,
  NS1, NS2, cysteine-rich (V), small integral membrane (SH) and second inner membrane (M2) proteins
Horizontal transmission; mainly airborne
Found only in vertebrates; no vectors

There are non-coding leader, trailer and intergenic sequences of varying extents which are
characteristic of the different groups, but overall more than 95% of the sequence encodes
polypeptides. The individual genes, with a
few exceptions (see below), do not overlap
and they are transcribed by interrupted synthesis sequentially from the 3'-terminal promoter. Transcripts from the 3'-terminal genes
are the most abundant with progressive attenuation of transcription towards the 5' terminus, such that transcripts of the L protein gene
are the least abundant (Figure 11.3). In the case
of measles virus the rate of attenuation of transcription has been related to pathogenicity
and propensity to become persistent [27]. In

Table 11.4   The currently recognized members of the family Paramyxoviridae

Family Paramyxoviridae
  Subfamily Paramyxovirinae
    Genus *Paramyxovirus*
      Human parainfluenza virus type 1 (type species)
      Human parainfluenza virus type 3
      Bovine parainfluenza virus type 3
      Mouse parainfluenza virus type 1 (also known as Sendai virus)
      Simian parainfluenza virus 10

    Genus *Morbillivirus*
      Measles virus (type species)
      Dolphin morbillivirus
      Canine distemper virus
      Peste-des-petits-ruminants virus
      Phocine distemper virus
      Rinderpest virus

    Genus *Rubulavirus*
      Mumps virus (type species)
      Avian paramyxovirus 1 (also known as Newcastle disease virus)
      Avian paramyxovirus 2 (Yucaipa virus)
      Avian paramyxovirus 3
      Avian paramyxovirus 4
      Avian paramyxovirus 5 (Kunitachi virus)
      Avian paramyxovirus 7
      Avian paramyxovirus 8
      Avian paramyxovirus 9
      Human parainfluenza virus type 2
      Human parainfluenza virus type 4a
      Human parainfluenza virus type 4b
      Porcine rubulavirus (La-Piedad-Michoacan-Mexico virus)
      Simian parainfluenza virus 5
      Simian parainfluenza virus 41

  Subfamily Pneumovirinae

    Genus *Pneumovirus*
      Human respiratory syncytial virus (type species)
      Bovine respiratory syncytial virus
      Pneumonia virus of mice (also known as murine pneumonia virus)
      Turkey rhinotracheitis virus

Unassigned viruses in the family
      Fer-de-Lance virus
      Nariva virus
      Several viruses isolated from penguins

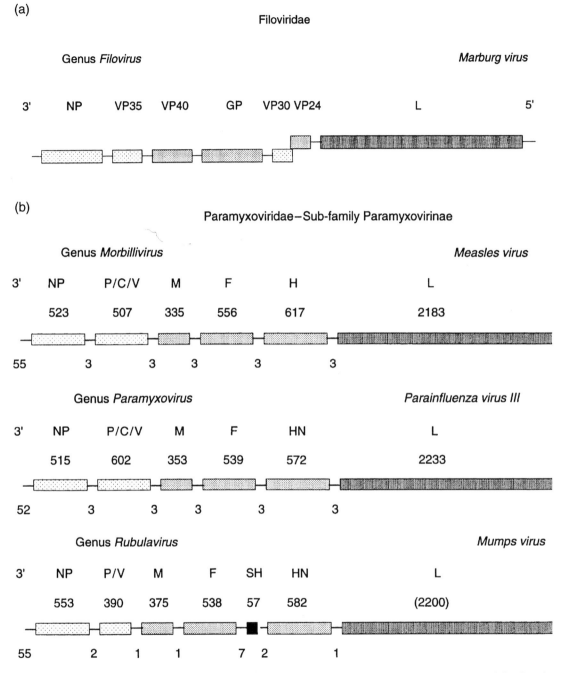

Figure 11.1 (a) Diagrammatic representation of the genome of Marburg virus, type species of the family Filoviridae. The overlap of the VP30 and VP24 genes is indicated by the displacement of the linear sequence. Not to scale. (b) Diagrammatic representation of the genomes of four viruses representing three genera of the family Rhabdoviridae, indicating divergence by expansion of intergenic regions. The figures below the genomes are the number of nucleotides in the intergenic gaps, where known. Approximately to scale.

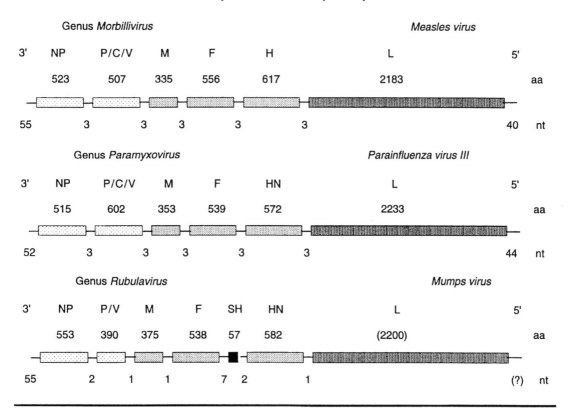

Figure 11.2 (a) Diagrammatic representation of the genomes of the type species of the three genera comprising the subfamily Paramyxovirinae of the family Paramyxoviridae. The figures above the genomes indicate the number of amino acids encoded in the major open reading frames, and the figures below are the number of nucleotides separating the genes. Approximately to scale.

general the genes encode single products, the phosphoprotein (core-associated) protein gene being the principal exception in this respect (see later).

Figure 11.1 emphasizes diagrammatically that despite the morphological resemblance of the filoviruses and the rhabdoviruses with their uniform particle diameter, they are divergent at the genome level. The filovirus genome extends to 19.1 kb and has seven open reading frames (ORFs). A feature of the filovirus genome is overlap of the non-coding regions of some genes. The 3' end of the VP30 mRNA of Marburg virus (shown in Figure 11.1) is overlapped by the 5' coding end of the VP24 mRNA, whereas in Ebola virus, an antigenically distinct filovirus, the overlaps are more extensive involving three other genes.

The five-component genome of the rhabdovirus, vesicular stomatitis virus, probably represents the minimum complement of separately transcribed genes required to constitute an infectious replication competent negative-stranded RNA virus particle.

The existence of six complementation groups in some vesiculoviruses [79], an internal initiation product from the P gene ORF and a small overlapping ORF [97], however, indicates

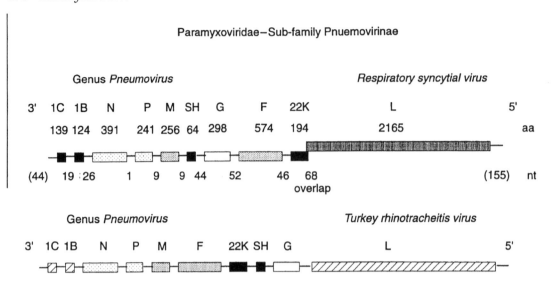

Figure 11.2   (b) Diagrammatic representation of the genomes of a mammalian and an avian virus of the subfamily Pneumovirinae of the family Paramyxoviridae. The figures above the genomes indicate the number of amino acids encoded in the major open reading frames, and the figures below are the number of nucleotides separating the genes. The overlap of the 22K and L protein genes of respiratory syncytial virus is indicated by the displacement of the linear sequence. (The 1C, 1B and L genes of turkey rhinotracheitis virus have still to be sequenced and are conjectural.) Approximately to scale.

that the minimum number of essential gene products is six. Comparison of the genome maps of the four rhabdoviruses in Figure 11.1(b) suggests that the non-segmented genome negative-stranded RNA viruses, which apparently lack either the ability or the opportunity to undergo genetic recombination, may be able to evolve by expansion of their intergenic regions. Bovine ephemeral fever virus, the type species of the genus *Ephemerovirus*, differs from the other rhabdoviruses in possession of two G proteins; these may have evolved by gene duplication as an adaptation of the virus to attachment to different receptors in its vertebrate and invertebrate hosts.

The genome size of the paramyxoviruses is fairly uniform. The genomes of paramyxoviruses sequenced in their entirety so far lie in the range 15.156 kb (Newcastle disease virus) to 15.892 kb (measles virus), whereas the

rhabdoviruses span a wider range from 11.2 kb for vesicular stomatitis virus to 14.8 kb for bovine ephemeral fever virus, with a filovirus reaching 19.1 kb. The encoding of genetic information is sufficiently different in the two subfamilies of the Paramyxoviridae to justify their consideration separately in this chapter. Indeed, in some respects the pneumoviruses show more similarity to the rhabdoviruses than they do to the other paramyxoviruses. The rule of six, enunciated by Calain and Roux [19] from a study of deletions and insertions in an *in vitro* DI (defective interfering) Sendai virus replication system, implies that the nucleocapsid protein molecule contacts six nucleotides and for the efficient replication of the paramyxovirus Sendai virus the genome length must be a multiple of six.

The basic paramyxovirus genome, as exemplified by viruses of the genera *Morbillivirus*

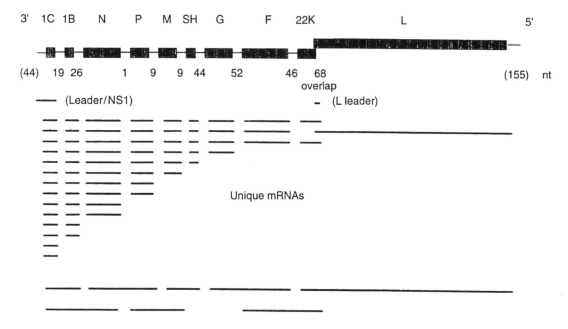

3' 1C 1B   N   P   M SH   G   F   22K   L   5'

(44)   19 26   1   9   9   44   52   46 68   (155)   nt
                                                    overlap

—— (Leader/NS1)   - (L leader)

Unique mRNAs

Polycistronic mRNAs

Figure 11.3   The pattern of transcription of the genome of the pneumovirus, respiratory syncytial virus. The relative abundance of the discrete mRNAs transcribed from the genomic RNA is indicated diagrammatically. The configuration of the polycistronic mRNAs observed is indicated in a non-quantitative manner. The figures below the genome are the number of nucleotides in the intergenic gaps and the non-coding leader and trailer regions. Not to scale. (After [32].)

and *Paramyxovirus*, consists of six transcriptional units (Figure 11.2(a)). The paramyxoviruses and the morbilliviruses are distinguished primarily by the addition of a neuraminidase function to the attachment protein gene product (the haemagglutinin) in the former. The genus *Rubulavirus* exhibits increased complexity; some, but not all rubulaviruses, have an additional transcriptional unit (the SH gene encoding a small hydrophobic protein) located between the F (fusion) and HN (attachment) protein genes. The genus *Pneumovirus* exhibits a further increase in complexity (Figure 11.2(b)). Some of this may be due to dispersion of the multiple coding functions of the P locus (see later) of other paramyxoviruses to separate transcriptional units. The pneumoviruses are unique in the possession of two genes, NS1

(1C) and NS2 (1B), encoding non-structural proteins located between the 3'-terminal region and the nucleocapsid gene, and another two unique genes SH and 22K (or M2) of undetermined function located internally in the genome. Thus the pneumovirus genome comprises 10 separate transcriptional units. The pneumoviruses are also characterized by smaller genes and have intergenic regions of less uniform sequence and extent [32].

11.4 FEATURES OF THE SUBFAMILY PARAMYXOVIRINAE

The attachment proteins (H, HN or G) of paramyxoviruses are type II glycoproteins, i.e. their amino-termini are located internally and their carboxy-termini externally, whereas the other major glycoprotein components of the

Genus (virus)	Unedited/edited transcript	Product	Size (aa)

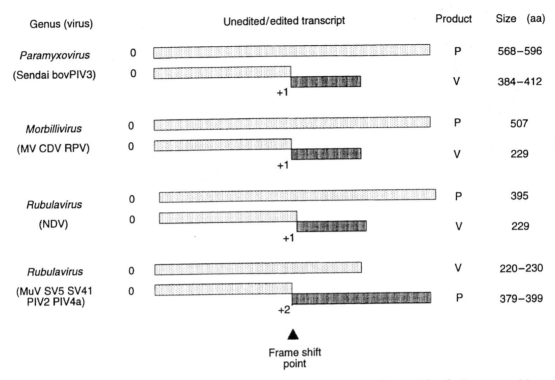

Figure 11.4 Differential editing of transcripts of the P gene of viruses of the subfamily Paramyxovirinae. The transcripts generating the P and V proteins are represented diagrammatically. The transcripts are aligned by their frameshift points, and the unique regions of the edited transcripts are indicated by darker shading. In paramyxoviruses and morbilliviruses the P coding transcript is unedited and non-templated insertion of a single G nucleotide accesses the V protein open reading frame, whereas in the rubulaviruses the V coding transcript is unedited and non-templated insertion of 2G nucleotides accesses the P open reading frame. However, Newcastle disease virus, currently classified as a rubulavirus, does not conform to these generalizations as transcript editing follows the paramyxovirus/morbillivirus pattern. The lengths of the amino acid sequences of the P and V proteins are indicated in the last two columns. NDV, Newcastle disease virus; PIV2, PIV3 and PIV4a, parainfluenzavirus type 2, 3 and 4a, respectively; MV, measles virus; CDV, canine distemper virus; RPV, rinderpest virus; MuV, mumps virus. (Adapted from [100].)

envelope, the fusion (F) proteins, are orientated in the usual manner with their amino-termini external. The 6–7 genes of the viruses belonging to the subfamily Paramyxovirinae encode 10–12 proteins. All the proteins, except those encoded in the P gene, are translated from unique mRNAs. A number of presumptive non-structural proteins, however, are derived from the P gene in addition to the nucleocapsid-associated phosphoprotein from which it takes its name. The C protein, the pre-cise function of which is unknown, is encoded in an ORF which overlaps that of the P protein in viruses classified in the genera *Morbillivirus* and *Paramyxovirus*. Furthermore, several minor truncated C proteins may be generated by internal initiation events. In Sendai virus the C ORF yields four products, C, C′, Y1 and Y2 by internal initiation at four separate codons, one of which in the case of the C′ protein is a non-AUG initiation codon (ACG). In the closely related human parainfluenza virus type 1 the C′ protein is initiated from a GUG start codon. The C, C′, Y1 and Y2 proteins have staggered amino termini and are carboxy co-terminal. A further protein of unknown function, designated X and representing the last 95 amino acids of the P ORF, is derived by internal initiation from the P protein ORF of Sendai virus [53]. Viruses of the genus *Rubulavirus*, however, have no C protein.

The P gene of all three genera encodes an additional non-structural cysteine-rich protein, designated V. The V protein has a zinc-binding motif indicative of a nucleic acid binding function of this protein. It may function as an inhibitor of replication. Both the P and the V protein ORFs are accessed by a process of mRNA editing [55,104,106], whereby one ORF is translated from an unaltered transcript and the other is accessed by the non-templated insertion in the alternate transcript of one or two G residues. A polymerase stuttering model has been proposed [105] to account for this phenomenon. This editing process produces a chimaeric protein which has a shared amino-terminal sequence and a unique carboxy-terminal sequence. In the morbilliviruses and the paramyxoviruses it is the P protein which is translated from an unedited transcript and the V protein from a mRNA with a frameshift created by insertion of a single G residue. With the exception of Newcastle disease virus (avian parainfluenza virus type 1), the inverse is the case in the rubulaviruses, where the P protein is translated from an edited mRNA containing a non-templated insertion of 2G residues, and where the V protein is translated from an unaltered transcript (Figure 11. 4).

The editing process is not always precise and in Sendai virus sometimes two (or more) G residues are inserted, yielding a mRNA expressing a W protein, which represents a truncated form of the P protein with two amino acids from an alternate reading frame at its carboxy-terminus [35]. Further complexity of encoding of information in the P gene occurs in human and bovine parainfluenza

type 3 viruses. Here a fourth ORF designated D, which is in phase with the C ORF, is also accessed by RNA editing. A V and D overlap would be predicted to be accessed by separate mRNA editing sites. Only one site is utilized, however, to generate both proteins; this occurs as a consequence of variable insertion of 1–6 G residues. Bovine parainfluenza virus type 3 is exceptional in that a significant proportion of the P gene is expressed in all three reading frames [75]. Newcastle disease virus, which at present is classified as a member of the genus *Rubulavirus*, lacks a C protein, and is an exception to these generalizations [100]. Human parainfluenza virus type 1 is also unusual in that the V ORF and the editing site are not conserved.

Rima has made a comparison of the amino acid sequences of viruses belonging to the

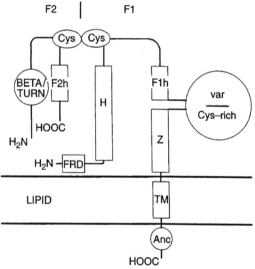

Figure 11.5 A hypothetical model of the arrangement of secondary structure elements in the fusion (F) proteins of paramyxoviruses, rubulaviruses and pneumoviruses. The proposed alpha-helical regions (F2h, H, F1h and Z) are shown as rectangles with beta-sheet and turn-rich regions as circles. FRD represents the possible location of the fusion-related domain and Z a zinc-binding domain. TM is the transmembrane region and Anc the cytoplasmic domain. (Adapted from [28].)

Paramyxovirinae [89]. The pattern of protein variability followed the succession P >H = N > F = M > L.

11.5 FEATURES OF THE SUBFAMILY PNEUMOVIRINAE

The genes encoding the nucleocapsid (N), matrix (M1), phosphorylated (P), fusion (F), attachment (G) and polymerase (L) proteins are common to all members of the family Paramyxoviridae. The pneumoviruses possess four additional genes specifying two non-structural proteins, NS1 (or 1C) and NS2 (or 1B), a small hydrophobic protein SH (or 1A), and a second non-glycosylated membrane-associated protein, M2 (or 22K). The SH gene is presumed to be analogous to the SH gene of rubulaviruses, but the other three appear to be unique to pneumoviruses [9]. NS1 may be a third non-glycosylated membrane-associated protein, whereas NS2 is a rapidly turned-over protein of unknown function (J. Evans, P.A. Cane and C.R. Pringle, unpublished data).

Only three pneumoviruses are known; two, respiratory syncytial (RS) virus and pneumonia virus of mice (PVM), exhibit an identical gene order, whereas the third, turkey rhinotracheitis virus, exhibits an inversion of the SH-G and F-22K gene pairs which restores the attachment protein (G) gene to a location adjacent to the polymerase (L) gene as in all other paramyxoviruses (see Figure 11.2). This inversion suggests that despite the fact that recombination has not been observed in laboratory experiments with any member of the Mononegavirales, recombination events may have occured rarely during the evolution of these viruses in nature [57]. Nonetheless, evolutionary processes in these viruses appear to be mediated mainly by random accumulation of mutations and by expansion of the intergenic regions if the direction of evolution is towards greater complexity, or the converse if the direction of evolution is towards greater economy in the use of genetic information [81].

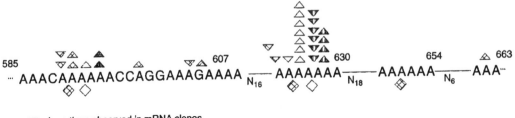

▽ Insertions observed in mRNA clones

▼ Insertions observed in genomic clones

▽ Insertions observed in T3 clones

◈ Sites of insertions observed in clones from escape mutants [7]

△ Deletions observed in mRNA clones

▲ Deletions observed in genomic clones

△ Deletions observed in T3 clones

◇ Sites of deletions observed in clones from escape mutants [7]

Figure 11.6 Schematic diagram of the region of the G gene of RS virus (mRNA sense) indicating oligoA tracts between nucleotides 585 and 663 where frameshift mutations have been observed. The insertion and deletion symbols indicate the number of clones detected with that mutation, and the key gives their origin. The diamonds below the line indicate sites of insertion (filled) and deletion (empty) observed by Melero and colleagues [44] in monoclonal antibody-selected escape mutants. Adapted from [23].

The P protein gene of RS virus encodes a single protein, although an internally initiated product is translated from the P mRNA both *in vitro* and *in vivo* [25]. By contrast two ORFs are present in the P gene of PVM [7]. Both are expressed *in vivo,* and several products of internal initiation derived from the larger ORF have been observed. The P gene of PVM is larger than that of RS virus and the entire length of the second ORF coincides with a region of low homology with the RS virus P gene, suggesting either a separate origin or absence of structural conservation in this region.

The number and nature of the envelope-associated proteins distinguish the pneumoviruses from the remainder of the Paramyxoviridae [66,101]. Only the F proteins have clearly discernible structural features in common, and it has been possible to derive a hypothetical generalized structure [28,29] for the paramyxovirus F protein by comparative nucleotide sequence analysis (Figure 11.5). The attachment (G) protein, on the other hand, is extensively *O*-glycosylated and does not resemble the H and HN proteins of paramyxoviruses. The G gene of RS virus has hypervariable regions where there is a high ratio of amino acid to nucleotide substitutions, suggesting rapid evolution in response to immune pressure [21]. Paradoxically, in experiments with laboratory animals the major neutralization epitopes locate to the F protein. The known G protein epitopes are located towards the carboxy-terminal region of the G protein. Selection of monoclonal antibody escape mutants has resulted in coordinate loss of epitopes as a result of frameshifting [44]. Furthermore, a high frequency of polymerase error during mRNA transcription [44] has been observed in the region of these frameshift sites (Figure 11.6), suggesting that this phenomenon may assist the virus in evading the immune response and allow more efficient attachment. Similar mutations are also generated during replication and during *in vitro* synthesis (Figure 11.6), indicating that the error proneness is template-associated [23]. The syncytial cytopathic effect of these viruses enables them to spread without restraint by the immune response.

The M1 protein seems to play the same role in morphogenesis as the M proteins of other paramyxoviruses [73], but the M2 (or 22K) protein is an enigma. It is not an important inducer of cytotoxic T-cell responses in mice, but on the basis of experiments using recombinant vaccinia viruses expressing RS virus proteins individually, cells expressing the 22K protein have been ranked as the prime target for cytotoxic T-cell-mediated lysis [71]. Assay of cytotoxic T-cells from human volunteers has revealed great variability in the responses of individuals, but overall all the proteins of RS virus are recognized by cytotoxic T-cells [30].

The L proteins of a filovirus (Marburg), four rhabdoviruses (the Indiana and New Jersey serotypes of vesicular stomatitis virus, rabies virus, and sonchus yellow net virus), two paramyxoviruses (Sendai virus and human parainfluenza virus type 3), five rubulaviruses (mumps virus, human parainfluenza virus type 2, SV5 , SV41 and Newcastle disease virus), two morbilliviruses (measles virus and canine distemper virus) and two isolates of one pneumovirus (human RS virus) have been determined and comparison of the inferred amino acid sequences has revealed several highly conserved domains. The most highly conserved central domain, containing an invariant QGDNQ pentapeptide, probably represents the active site for RNA replication (43). Some of these domains share low but consistent relatedness with corresponding domains in the RNA-dependent RNA polymerases of segmented negative-strand RNA viruses, positive-strand RNA viruses, retro-transposons and retroviruses [77]. Poch *et al.* [76] concluded from a five-way comparison of two rhabdovirus and three paramyxovirus L protein sequences that the genus *Lyssavirus* may represent an evolutionary intermediate stage between the vesiculoviruses and the viruses of the family

Paramyxoviridae. The sequences also predict unexpectedly that the L proteins may have short half-lives [4]. Muehlberger *et al.* [68] considered that the L proteins could be partitioned; the N-terminal portion accommodating the enzymatic domains and the C-terminal portion the less-conserved virus-specific regions. The L protein of RS virus seems to have a 70-amino acid extension at the amino terminus not present in the other L proteins, which may be a consequence of the overlap of the L gene with its upstream neighbour, the M2 (22K) protein gene [99]. A phylogenetic tree constructed from the inferred protein sequences of the L proteins reinforced the conclusion that the pneumoviruses represent a distinct lineage within the family Paramyxoviridae.

A comparative analysis of a nuclear inclusion protein of the unclassified Borna virus by McClure *et al.* [60] revealed distant homology with two regions of the L proteins of paramyxoviruses and rhabdoviruses. This led them to hypothesize that Borna virus may in fact be a negative-stranded RNA virus, and that this sequence similarity indicates a previously undetected duplication in these viral polymerases. Phylogenetic analysis suggested that the Borna virus gene shared a common ancestor with the L polymerase genes of paramyxoviruses and rhabdoviruses prior to the duplication event.

11.6 THE INFECTIOUS CYCLE

11.6.1 ATTACHMENT

Molecules containing sialic acid (sialoglycoconjugates) have been considered to be the receptors for paramyxoviruses, rubulaviruses and pneumoviruses, since sialidase (neuraminidase) treatment of susceptible cells prevents infection and serum proteins and gangliosides containing *N*-acetylneuraminic acid have facilitated the attachment of paramyxoviruses to sialidase-treated erythrocyte membranes. Three related gangliosides, GD1a, GT1b and GQ1B, have been identified

as receptors for Sendai virus both in cultured cells and in tissue [59]. The specific sialoglycoconjugates functioning as host cell receptors for other paramyxoviruses have not been defined precisely.

The morbilliviruses, by contrast, do not possess a receptor-destroying sialidase activity. Their cellular receptors do not appear to be sensitive to the action of bacterial and viral sialidases, but are sensitive to proteases. Two molecules have been identified as receptors for measles virus. A study of lymphocyte markers by Dorig *et al.* [36] suggested that the human CD46 molecule (the complement-binding protein MCP) was the measles virus receptor. By way of confirmation, these workers were able to demonstrate that expression of CD46 in hamster cells conferred the ability to bind measles virus, and that polyclonal antiserum against CD46 inhibited binding and infection. CD46 is not present on all cells susceptible to measles virus infection, however, indicating the existence of alternative receptor molecules. Dunster *et al.* [37] using a monoclonal antibody that inhibited measles virus infection of susceptible cells, specifically precipitated a 75-kDa protein from surface-labelled human monocyte U937 cells. The purified protein and a specific antibody against the protein both inhibited measles virus infection of susceptible cells. Amino acid microsequencing of this protein revealed it to be moesin, a heparin-binding protein expressed at high levels on the surface of human and primate cells. It is not clear yet whether CD46 and moesin are alternate receptors for measles virus, or part of the same receptor complex. Measles virus infection results in a down-regulation of both CD46 and moesin, and experiments with recombinant vaccinia viruses have shown that expression of the measles virus H protein is sufficient to trigger down-regulation of CD46.

11.6.2 FUSION AND PENETRATION

Entry into the cytoplasm of the host cell is mediated by fusion with the cell membrane following attachment. Fusion of the viral and cell membranes occurs at neutral pH. Experiments with recombinant SV40 and vaccinia virus vectors have demonstrated that co-expression of both the attachment (HN) and F protein may be necessary for optimal cell-to-cell fusion in the case of some viruses (Sendai virus, NDV, mumps virus, parainfluenza 2 virus and parainfluenza 3 virus), whereas in others (e.g. SV5) co-expression of HN protein may only marginally influence membrane fusion (reviewed in 54). In the case of RS virus there is evidence that the SH protein may also be involved in the fusion process in addition to the F and G proteins (J. Bernstein, personal communication). Activation of biological function by proteolytic cleavage is an important property of the surface glycoproteins of paramyxoviruses and proteolytic cleavage of the fusion protein of viruses of the family Paramyxoviridae is required for biological function. Cleavage activation is not necessary for particle formation and is dependent on the presence of an appropriate protease in the host cell.

The paramyxovirus attachment protein does not normally undergo post-translational cleavage, nor is there cleavage of a signal sequence. In the avirulent Ulster and Queensland strains of NDV, however, a HN precursor protein (HN$_0$) is synthesized which is converted to the active HN form by proteolytic removal of a 90-amino acid C-terminal peptide. The HN gene sequence of NDV has an unusually long non-coding region at the 3' end of the mRNA, and it may be that mutation has altered the length of the reading frame, since in virulent strains HN appears to be synthesized without a precursor [63]. However, as in other paramyxoviruses, cleavage of the F protein plays a more important role in determining the pathogenicity of NDV; for example, the F$_0$ precursor protein of the virulent La Sota and B1 strains of NDV is not cleaved in some cell lines, resulting in decreased infectivity [69]. Proteolytic cleavage of the attachment protein has also been observed in the case of pneumoviruses. Two forms of the attachment protein of bovine and human RS viruses are synthesized from a single mRNA, the smaller form being initiated from a second in-frame methionyl codon. Both forms of the G protein are N-glycosylated and translocated to the surface membrane. The larger G protein is stably incorporated into the plasma/viral membrane, whereas the shorter form is inserted into the membrane and released to the exterior by a process of proteolytic cleavage [58,92]. The functional significance of this shedding of G protein is not known.

The tropism of paramyxoviruses is determined predominantly after attachment, and the role of the F protein has been demonstrated in a series of classic experiments by Schied and Choppin (reviewed in [31]). The F protein is synthesized as an F$_0$ precursor, which requires a cellular protease for its cleavage activation. The two cleaved sub-units (F$_1$ and F$_2$) are held together by disulphide bonds. Cells lacking the appropriate protease are incapable of activating F$_0$ and do not support virus replication. Consequently, proteolytic activation can be a determinant of host range, tissue tropism and pathogenicity. The cleavage of F$_0$ unmasks a highly conserved hydrophobic amino-terminus on F$_1$ (the larger subunit) and a new carboxy-terminal F$_2$ (see Figure 11.5). The hydrophobic terminus is considered to be the region most directly involved in membrane fusion. The sequence of the fusion-associated region is highly conserved within and between genera, reflecting their taxonomic relationships [28]. Synthetic oligopeptides which mimic the amino-terminal region of F$_1$ glycoproteins of several paramyxoviruses and morbilliviruses inhibit virus replication. The apparent sequence specificity of the phenomenon, however, is spurious because a recent study has demonstrated that most enveloped RNA viruses are inhibited, whereas non-enveloped RNA viruses are not [51]. One carbobenzoxylated tripeptide (Z-D-phe-phe-gly) is more potent than the others, and in all cases inhibition is dependent on pre-treatment of virus.

The protease activation (pa) mutants of Sendai virus isolated by Schied and Choppin [95] illustrate the potential role of the F protein in determining tissue tropism and host range. Sendai virus is exceptional in that the F protein in released virus is uncleaved and cleavage of the F protein is necessary for infection to proceed beyond the attachment stage. This cleavage is normally mediated by cellular proteases with trypsin-like specificity. Proteases of other specificities cannot activate the Sendai virus F protein. However, pa mutants were isolated by passaging virus in a non-permissive system in the presence of an added protease. In this way mutants were obtained in which the F protein is cleaved by proteases of different specificities, including chymotrypsin, elastase, plasmin and thermolysin. These mutants may have lost or retained their sensitivity to trypsin. The relevance of such mutations to pathogenicity was shown experimentally by inoculation of chick embryos; e.g. a chymotrypsin sensitive, trypsin-resistant mutant (pa-c1) which failed to multiply in the allantoic sac could be activated by simultaneous injection of chymotrypsin. Other protease mutants responded to the appropriate protease in like manner. Nucleotide sequencing has confirmed that the majority of these mutations are due to single base substitutions resulting in single amino acid replacements. In Sendai virus the sequence of amino acids around the cleavage site is Val-Pro-Gln-Ser-Arg/Phe-Phe-Gly-Ala. Mutant pa-c1, for example, has a single amino acid substitution → at the cleavage site, which renders the F protein resistant to trypsin and sensitive to chymotrypsin (and also elastase). The ease of isolation of protease-activation mutants and the low specificity of the attachment process suggest that variants can arise under natural conditions which are able to infect a normally non-susceptible cell type and alter the disease potential of the virus. It remains to be demon-

strated that the host range and tropism of other members of the Paramyxoviridae are determined by protease activation of infectivity by cleavage of the F protein.

11.6.3 TRANSCRIPTION, REPLICATION AND MATURATION

The paramyxoviruses might be expected to replicate efficiently in the cytoplasm since they closely resemble the rhabdoviruses in genome organization. Indeed the pneumoviruses, RS virus and pneumonia virus of mice, multiplied as efficiently in BS-C-1 cells enucleated by cytochalasin treatment, as in untreated nucleate cells. However, this was not the case with paramyxoviruses and the morbilliviruses. Although synthesis of polypeptides was observed in Sendai virus-infected enucleate BS-C-1 cells, no infectious virus was released. Enucleation of cells as late as 6.5 hours after infection did not release the block [78]. Similar data were reported for paramyxovirus 6/94 [103]. It would appear that a host-specified function is required for normal maturation of paramyxoviruses. Similarly infectious measles virus was not released from enucleated BS-C-1 cells, although polypeptide synthesis reached 20% of normal levels and cell fusion occurred, indicating that proteins were synthesized and transported to the plasma membrane. Identical results were observed with several strains of measles virus, including three strains originally isolated from patients with subacute sclerosing panencephalitis (SSPE), although these strains are now presumed to be contaminating wild-type viruses. In the case of measles virus infection of BS-C-1 cells, the presence of the cell nucleus was only required during the very early stage of the growth cycle, because measles virus yields were increased when enucleation was delayed until 2 hours after infection and were returning to near normal values when enucleation was delayed until 6 hours after infection [41].

The details of the replication of the genome of all members of the family Paramyxoviridae are complex and beyond the scope of this review. The reader is referred to recent reviews by Kolakofsky, Vidal and Curran [53], Blumberg, Chan and Udem [18] and Moyer and Horikami [67]. Briefly, replication starts with a triphosphate at the 3' end of the template and proceeds by uninterrupted synthesis to produce an exact copy of the genome. Transcription, on the other hand, produces a short leader RNA, followed sequentially by a

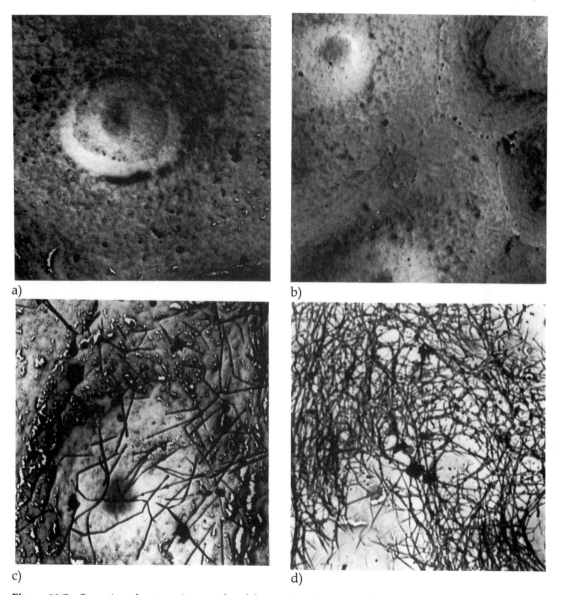

a) b) c) d)

Figure 11.7 Scanning electronmicrographs of the surface of potoroo kidney cells infected with respiratory syncytial virus showing the development of filaments. (a) Uninfected cells; (b) 33 hours; (c) 48 hours; (d) 72 hours after infection. (From [85].).

series of unique mRNAs, and terminates before the precise end of the template. The experimental evidence favours the hypothesis that the multiple transcripts are generated by interrupted synthesis involving polymerase pausing and re-starting at each gene junction, rather than by polymerase detachment and re-attachment. The individual transcripts are capped and polyadenylated during synthesis, whereas the leader RNA remains unmodified at both termini. It is presumed that the junctions contain conserved signals mediating these events. mRNA synthesis in non-segmented genome RNA viruses is primer independent and is thought to take place only on re-initiation of polymerase activity at each junction. The replicative RNAs are encapsidated during synthesis and nucleocapsids assemble independently in the cytoplasm. The nucleocapsids are enveloped at the cell surface at sites containing virus envelope proteins. In all cases the maturation process is relatively inefficient and quasi-spherical pleomorphic particles or long filamentous structures are produced (Figure 11.7).

The principal role of the matrix protein appears to be in promoting assembly, although studies with mutant viruses and addition of M protein to *in vitro* transcription–translation systems indicate that it may play a subsidiary role in the modulation of gene expression. Roux and Waldvogel [93] observed that the viral M protein was labile in cultures of cells persistently infected with Sendai virus and little infectious virus was produced. M protein interacts with itself and other viral components such as nucleocapsids and probably with the cytoplasmic domains of the viral glycoproteins. It also interacts with cellular components such as the plasma membrane and actin. The best evidence for the role of the M protein of measles virus in assembly comes from study of variants obtained from patients with SSPE. Such viruses, which frequently exhibit deficiencies in their M protein, replicate poorly and are unable to spread from

cell to cell efficiently. Two of the three genuine SSPE viruses which have been sequenced have carboxy-terminal deletions of their F proteins. The M gene of a third cloned SSPE strain and the M gene of a virus from a patient with measles inclusion body encephalitis (MIBE) were grossly defective [15]. Consequently, if the M protein interacts with the cytoplasmic tail of the F protein, mutation in either region would affect virion assembly. There is evidence also that there may be direct interaction between the M protein and the attachment protein. However, pseudotype particles form readily on co-infection of cells with different enveloped viruses suggesting that the specificity of the interaction between the M protein and viral membrane glycoproteins is normally not high [109]. The stoichiometry of the components of paramyxovirus virions is difficult to calculate in view of the generally heterogenous morphology of the virions.

However, the M protein of Sendai virus and NDV could also be cross-linked to the nucleocapsid protein in newly released virions, indicating that the M protein may be complexed with the nucleocapsid protein [85]. The M protein of NDV and measles virus appears to associate also with the cell cytoskeleton. Actin has been found as a component of many enveloped viruses, including paramyxoviruses. Electron microscopy of budding measles virus has revealed that many particles contain barbed actin filaments which appear to be growing into the budding virus and may be the motive force in the budding process [98]. Measurement of filament production (see Figure 11.7) at 30°C in Vero cells infected with RS virus, using video-enhanced microscopy, revealed that the process was extremely rapid [3]. Filaments reaching a length of 5–10 μm extended at a rate of 110–250 nm/s. Thus a particle of 150 μm diameter would bud from the cell in less than 2 seconds.

Curiously, Peeples [74] has reported that although polyclonal antisera stain only the cytoplasm of infected cells, staining of infected

Table 11.5 The Paramyxoviridae and their associated diseases

Serotype	Primary host (virus)	Associated disease
Genus *Paramyxovirus*		
Parainfluenza virus type 1	Man	URTI, rarely pneumonia in adults
	Mouse (Sendai virus)	Inapparent, persistent infections in mice
Parainfluenza virus type 3	Man	URTI, laryngitis, bronchiolitis and pneumonia in childen
	Cattle	URTI, shipping fever
SV10	Monkey	URTI
Genus *Rubulavirus*		
Mumps virus	Man	Parotitis, gastroenteritis, oöphoritis, orchitis, pancreatitis, etc.
Parainfluenza virus type 2	Man	URTI; croup mainly in children
Parainfluenza virus type 4a/b	Man	URTI
Porcine rubulavirus	Pig	Nervous system disorders, pneumonia
SV5	Dog (monkey)	URTI
SV41	Monkey	URTI
Avian paramyxoviruses:		
Type 1	Chicken (Newcastle disease)	Inapparent to lethal viscerotropic and neurotropic (meningoencephalitis in mink)
Type 2	Several avian spp. (Yucaipa)	Inapparent
Type 3	Several avian spp.	Inapparent
Type 4	Duck, chicken, geese	Inapparent
Type 5	Budgerigar (Kunitachi)	Inapparent
Type 6	Duck and chicken	Inapparent
Type 7	Dove	–
Type 8	Geese, wild ducks	Inapparent
Type 9	Rock pigeon, domestic duck	–
Genus *Morbillivirus*		
Measles virus	Man	Measles, rarely (1/2000) encephalitis, exceptionally (1/1 000 000) SSPE; giant-cell pneumonia in the immunocompromised
Canine distemper	Dog and mustelides	URTI, skin eruptions, bronchopneumonitis, keratitis of the feet (hard pad), old dog encephalitis, demyelination of neural tissue
Phocine distemper	Seals	Respiratory disease
Dolphin morbillivirus	Dophins	Respiratory disease
Rinderpest virus	Cattle, sheep, goats, pigs, buffalo	Mucosal lesions, diarrhoea, bronchopneumonia
Peste des petits ruminants	Sheep, goats	Mucosal disease, diarrhoea, bronchopneumonia, abortion

Genus *Pneumovirus*

Respiratory syncytial virus	Man	URTI, bronchiolitis and pneumonia in infants, otitis media
	Cattle, sheep, goats	URTI, pneumonia, bronchiolitis and emphysema in calves
Murine pneumonia virus	Mouse, Syrian hamster, cotton rats	Inapparent in mice, rarely disease-producing in guinea-pigs
Turkey rhinotracheitis	Turkey	URTI, ocular and nasal discharge, submaxillary oedema, swollen infraorbital sinuses
	Chicken	Swollen head disease
Ungrouped virus*	Fer de Lance viper	Respiratory disease

* Fer de Lance viper virus is serologically unrelated to other paramyxoviruses; it possesses haemagglutinin and neuraminidase, but the diameter of the nucleocapsid is closer to that of pneumoviruses. Other uncharacterized viruses include Nariva virus from rodents, Mapuera virus from a bat and several viruses from penguins. SSPE, subacute sclerosing panencephalitis; URTI, upper respiratory tract infection.

cells with anti-M monoclonal antibodies reveals that a large proportion of the M protein is present within the nucleus and located in discrete regions (possibly the nucleolus). It appears that the M protein is not very immunogenic and anti-M antibody is not abundant in polyclonal sera. The presence of a cross-reacting host antigen in the nucleus is unlikely as anti-M monoclonals recognizing several different epitopes behaved similarly. M protein in the nucleus can be detected as early as 3.5 hours after infection, which is also the earliest time it can be detected in the cytoplasm. Consequently, the presence of the M protein in the nucleus is not a late event accompanying degradation of cellular structures as a consequence of virus replication. The M protein of NDV contains a four amino acid motif (K-R/K-X-R/K) considered to be a nuclear transport signal. The functional significance of the transport of the NDV M protein to the cell nucleus is not known, although some imaginative but as yet untested hypotheses have been put forward by Peeples [73]. Penetration of the M protein into the nucleus was less evident in Sendai virus-infected cells, and was not observed in measles virus-infected cells, although in the latter case the NP protein was detected in the nucleus.

11.7 PARAMYXOVIRUSES AND DISEASE

The paramyxoviruses are predominantly associated with respiratory infections. The clinical and pathological aspects of paramyxovirus infections of humans and animals are listed in Table 11.5.

11.7.1 DEFECTIVE MEASLES VIRUSES IN THE CENTRAL NERVOUS SYSTEM

SSPE develops in children several years after acute measles virus infection and is associated with cryptic persistence of measles virus in neural tissue. The disease presents first as behavioural and emotional changes, followed by intellectual deterioration and loss of control of motor functions, progressing remorselessly through cerebral degeneration to coma and death. A strong inflammatory response, leucocyte infiltration, and limited demyelination are characteristic features of SSPE, as also are the occurrence of high antibody titres to measles virus antigens in the cerebrospinal fluid (CSF). No infectious virus or budding particles can be detected in the brains of infected individuals, although inclusion bodies containing measles virus ribonucleoprotein are prominent [84]. It is generally accepted that a maturation-defective measles virus is responsible for this disease [15]. The evidence supporting this conclusion is failure to detect antibodies to M protein in CSF (as well as in serum), or M protein in brain tissue, and absence of cytopathology (syncytium formation) typical of lytic infection. However, recent molecular studies indicate that the underlying situation is more complex. Viral nucleic acids derived from the brains of patients with SSPE or MIBE (a SSPE-like condition associated with immunosuppression), show evidence of extensive and systematic mutational change [15]. Attempts to associate measles virus with the aetiology of multiple sclerosis have been inconclusive [34,62].

Persistent infection

Persistent infection may be a normal consequence of acute measles virus infection since this would best account for the life-long immunity that usually follows acute infection or primary vaccination with live virus vaccine. Virus sequestered in some reservoir tissue in rare instances could be transported to the brain by infected lymphocytes. Measles virus RNA can be detected abundantly in the brains of SSPE patients after death, and there is some evidence from analysis of biopsied material that the amount of measles virus RNA increases with progression of the disease. Study of the nucleotide sequences of the RNAs from SSPE brain material has revealed extensive mutational change which may be responsible for the transition from cryptic infection to expansion of the persistent infection in the brain.

Persistently infected cell cultures have been generated by co-cultivation of brain material with susceptible cells. Transcription studied by RNA blot hybridization was found to be 7 to 20-fold lower than in lytic cells and the gradient of attenuation of transcription was steeper. Protein synthesis was more difficult to quantitate; there appeared to be a deficit in the synthesis of the M and H proteins in some cases, but overall there was great variation between individuals [15].

Mutational changes

The mutational changes observed in viruses recovered from SSPE brains were first described comprehensively by Billeter and colleagues (reviewed in [15]). Comparisons were made with a consensus sequence constructed from all known measles virus sequences to circumvent the problem that the sequence of the infecting virus in any SSPE patient was unknown; nucleotides deviating from this consensus sequence were described as 'mutations'. A higher frequency of mutation was observed to occur more or less randomly throughout the genome in RNA cloned from the brains of several SSPE cases post mortem. One MIBE case exhibited a particularly dramatic pattern. The M gene sequence from this MIBE patient exhibited a phenomenon termed biased hypermutation; about 50% of the U residues had undergone transition to C (read in positive-strand polarity). Since independently isolated clones from the same brain did not show significant variation, it was assumed that the hypermutation occurred as a singular event rather than as a result of progressive accumulation of changes. Later Baczko *et al.* [5] cloned M genes from several sites in the brain of a single SSPE patient and their sequence analyses revealed the presence of wild-type measles virus and variants representing five independent hypermutational events of both U → C and A → G specificity (in positive-strand polarity). Analysis of the distribution of these viruses within the brain suggested that the hypermutated viruses had expanded clonally throughout the brain. Curiously, in this brain no M protein could be detected, although M mRNA with an intact ORF was present.

Extensive mutational change has also been observed in the cytoplasmic domain of the F protein of SSPE viruses [96], and more subtle changes in the extracellular domains of the F and H proteins. The U → C or A → G transitions are distributed randomly throughout the M gene which suggests that impairment of assembly is an essential step in the development of persistent infection following penetration of measles virus into the brain. The more subtle changes in the extracellular domains of the F and H proteins may be necessary for spread of virus in the brain and suppression of residual wild-type virus. Hirano [48] has shown that SSPE virus dominantly interferes with the replication of wild-type virus.

Double-stranded RNA unwindase

The phenomenon of biased hypermutation has been attributed to an enzyme activity termed 'double-stranded RNA unwindase', which converts A residues to inosine (I) in double-stranded RNA [12]. It has been hypothesized that occasionally during transcription the transcript may collapse onto the template creating a stretch of double-stranded RNA [13]. The RNA unwindase converts some of the A residues in this region to I. The I residues in the template strand are read usually as G, so that C residues are incorporated into the anti-genome progeny strand. On completion of replication G residues are incorporated in the genome strand where A residues were present originally. Biased hypermutation was later reported in the M gene of several SSPE-derived viruses and Wong *et al.* [108] were able to reproduce the phenomenon *in vitro* by passage of SSPE virus in neuroblastoma cells. RNA unwindase activity has been shown to be high in neurones, oligodendrocytes and neuroblastoma cells, which might explain the association of biased hypermutation with virus persistently infecting neural tissue.

11.7.2 PARAMYXOVIRUSES AND PAGET'S DISEASE OF BONE

Persistent infection of cultured cells by paramyxoviruses can be established with ease, but apart from measles virus persistence in neural tissue in SSPE patients there is little direct evidence for persistent infection *in vivo*. Nonetheless, the ubiquity of antibodies to viruses such as parainfluenza virus type 3 and RS virus in the adult population, coupled with the often seasonal prevalence of overt infection (see Figure 11.8), are difficult to explain without invoking persistence. Parkinson *et al.* [72] have documented an outbreak of upper respiratory tract infection attributed to parainfluenza virus type 3 in a community in Antarctica after a long period of isolation. The high frequency of carriage of unique paramyxoviruses by Antarctic penguins [1], however, suggests that the existence of a cryptic avian reservoir can not be discounted.

Paget's disease

The question of the involvement of paramyxoviruses in Paget's disease of bone, a common disorder of bone metabolism in the middle-aged and elderly due to osteoclast malfunction, illustrates the problems encountered in attempting to investigate the involvement of paramyxoviruses in non-respiratory tract disease. Certain features of Paget's disease of bone suggest a viral aetiology; e.g. the focal nature of the lesions, and the presence in osteoclasts of nuclear and occasionally cytoplasmic inclusions consisting of filaments resembling the nucleoprotein of paramyxoviruses. Evidence both for and against the involvement of RS virus, measles virus, canine distemper virus and an unknown paramyxovirus, together or separately, has been accumulating, but the aetiology of Paget's disease of bone remains an enigma.

The dimensions of the filaments in malfunctioning osteoclasts suggested the presence of a pneumovirus and RS virus antigens were detected by immunofluorescence [50,64]. Serological support for [64] and against [83,87] the involvement of RS virus has been published. Independently, Basle *et al.* [9,11] reported detection of measles virus antigens by immunofluorescence microscopy and measles virus mRNA by *in situ* hybridization [10] using a cDNA copy of the measles virus nucleocapsid protein gene. Others have been unable to confirm these observations, although Mills *et al.* [65] reported the simultaneous presence of RS virus and measles virus antigens in osteoclasts from six patients.

More recently canine distemper virus has been favoured as a possible aetiological agent on the basis of epidemiological data demonstrating a positive correlation between dog ownership and Paget's disease of bone in the north of England [49,70], although the same correlation was not observed elsewhere. Cartwright *et al.* [26] examined bone samples by *in situ* hybridization using probes for the nucleocapsid protein genes of canine distemper virus and measles virus, and the F and HN membrane protein genes of RS virus and SV5, respectively. A large proportion of patients were positive for canine distemper virus and one patient for RS virus; none was positive for measles virus or SV5 virus. The same group reverse-transcribed RNA from pagetic bone tissue and amplified sequences by polymerase chain reaction (PCR) using canine distemper virus and measles virus primers [46]. Eight of fifteen patients were positive for canine distemper virus and one of 10 for measles virus. Ralston *et al.* [88], however, in an independent study failed to find evidence of paramyxovirus sequences by reverse transcription and PCR from RNA obtained from 10 patients, using primers specific for RS virus and degenerate primers recognizing conserved sequences in the genomes of canine distemper virus, measles virus, parainfluenza virus type 3 and Sendai virus. It remains possible, particularly in view of the serological data of Pringle and Eglin [83], that some as yet uncharacterized virus is involved.

Attempts to isolate an infectious agent from pagetic tissue have been consistently negative. It may be that the foregoing observations are either entirely artefactual, or that the structures, antigens and sequences detected in pagetic tissue represent late consequences of osteoclast malfunction rather than their cause. It is possible that viral antigens are scavenged by and become sequestered in osteoclasts, or that cross-reactive antigens are present. Nuclear inclusion bodies very similar to those

observed in osteoclasts have been observed in giant cells in four of eight cases of primary oxalosis [14], suggesting that the filaments are associated with cell fusion and giant cell formation rather than virus infection.

Paramyxovirus-like intranuclear and cytoplasmic filaments have been described in the muscle disease known as inclusion body myositis, and immunoreactivity with mumps virus antiserum has been reported. However, Kallajoki *et al.* [52] were unable to confirm this by using a range of paramyxovirus group antibodies and by *in situ* hybridization using a cRNA probe representing the nucleocapsid gene.

11.8 EPIDEMIOLOGY

Two contrasting patterns are observed which can be explained in terms of the transmissibility of paramyxoviruses as aerosols and the durability of the immune response to infection (reviewed in [17]).

11.8.1 MORBILLIVIRUSES

Immunity

Immunity to measles virus, and probably all members of the genus *Morbillivirus*, and to mumps virus is durable, and individuals once recovered from infection play no further role in the spread of infection. In the pre-vaccination era, measles in the human population exhibited a cyclical pattern of epidemics and periods of quiescence. As Panum first observed in the Faroe Islands in 1846, measles can disappear from small island communities once immunity embraces the majority of the population. The disease cannot re-establish itself until the immune majority has been replaced by a new generation of susceptibles. A minimum population size, estimated at 300 000 from studies of urban populations by Bartlett [8] and of isolated communities by Black [16], is necessary to maintain measles virus as an endemic infection. Since settled communities

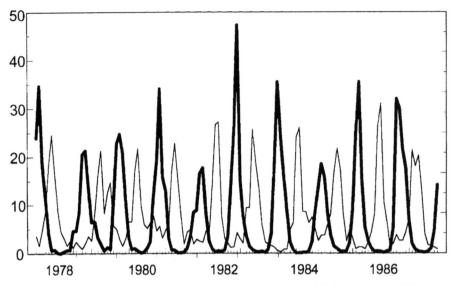

Figure 11.8 Periodicity of epidemics of parainfluenza type 3 virus and RS virus in the UK over a 10-year period from the beginning of 1978 to the end of 1987. The figures are cases reported in 4-weekly periods to the PHLS Communicable Disease Surveillance Centre and expressed as percentages of total annual cases. During this period summer epidemics of parainfluenza virus type 3 infection (thin line) alternated with winter epidemics of respiratory syncytial virus infection (thick line). The number of reports of respiratory syncytial infection exceed those of parainfluenza virus type 3 infection by a factor of ten approximately. (From [38].)

of 300 000 individuals probably did not exist before the establishment of the urban societies of the Middle-Eastern river valleys, it is probable that measles has only existed as an epidemic human disease for less than 5000 years. The morbilliviruses are a tight group of viruses, closely related to each other antigenically and genetically, whereas the other viruses in the Paramyxoviridae are more diverse. It is likely, therefore, that the progenitor of measles virus was not very dissimilar from one of the animal morbilliviruses in current circulation. Mumps exhibits a similar but less obvious epidemic pattern, because only a proportion of those infected exhibit moderate to severe disease. The introduction of effective vaccination has dramatically reduced the incidence of measles in the developed world. However, the high transmissibility of the virus and the uneven uptake of vaccine have conspired to delay eradication of the virus. This may also in part be a consequence of the fact that the propor-

tion of mothers with vaccine-induced immunity rather, than natural immunity, in the population is increasing. This may have retarded the elimination of the virus because maternal immunity provided by a vaccinated mother decays more rapidly than that donated by a naturally infected mother [56]. It is likely that a population immunity rate of about 99% will be necessary to interrupt measles virus transmission and finally eradicate the disease.

Canine distemper virus

Canine distemper virus exhibits many similarities to measles virus except that it has a broader host range. Most of the Order Carnivora are susceptible to canine distemper virus, domestic dogs being the usual reservoir host. Major epizootics of distemper with high mortality were observed in seals in the Baltic, Irish and North seas in 1988 due to both phocine distemper virus, a virus closely

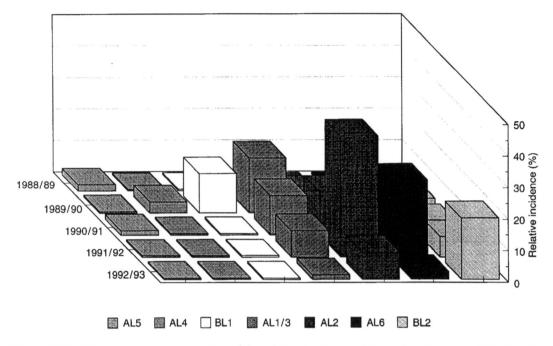

Figure 11.9 Diagrammatic representation of the relative incidence of the various lineages of RS virus isolates compared with total isolates during five successive annual epidemics in the East Birmingham area. AL 1–6 are the six lineages of the A subgroup (1 and 3 have been combined for display purposes), and BL 1–2 are the two lineages of the B subgroup of RS virus currently recognized. Several lineages are present in every epidemic, but there are progressive changes in the frequencies of some lineages during this 5-year period. (From [24].).

related to the canine virus [90,91], and to canine distemper virus itself, illustrating the characteristic devastating effect of morbillivirus infection in populations devoid of immune protection. It has been suggested that the vulnerability of seals to morbillivirus infection is a consequence of the fact that they live at higher population densities than other non-domesticated members of the Carnivora. A further similarity between the biology of canine distemper virus and measles virus is the occurrence in dogs of a rare persistent neurological infection, old dog distemper, bearing resemblance to SSPE in humans. A significant difference is that fully infectious canine distemper virus can be recovered from infected animals. In this case persistent infection may play a role in maintaining the virus in nature.

Rinderpest

The cattle plague, rinderpest, provides a further illustration of the catastrophic effects of morbillivirus infection in susceptible non-immune hosts. Historically, rinderpest has been responsible for serious epizootics of disease in cattle in Europe and in both cattle and game animals in Africa. Indian breeds of cattle, however, are more resistant to infection by rinderpest virus which suggests that the disease may have been prevalent in Asia before it spread west to Europe and Africa. Currently, rinderpest remains a serious problem in Africa. A vaccinia virus vector carrying proteins which induce neutralizing antibody to rinderpest virus may be the first genetically engineered live virus vaccine to be used in control of disease caused by any member of the Paramyxoviridae (T. Yilma, personal communication).

11.8.2 PNEUMOVIRUSES

A different epidemiological behaviour is exhibited by the pneumoviruses and by some of the paramyxoviruses and rubulaviruses, of which parainfluenza virus type 3 is the best example. In contrast to the morbilliviruses, the immune response to these viruses does not interrupt the spread of the virus in the population. Both RS virus and parainfluenza virus type 3 are endemic in the human population, and in temperate climates cause annual epidemics of serious respiratory disease in children (Figure 11.8).

These viruses appear to be ubiquitous and present throughout the world in all populations irrespective of their size, degree of isolation or socioeconomic status. It is not clear whether an inappropriate immune response, age-related host responses or antigenic variability of the viruses is responsible for the recurrent re-infections which are a characteristic feature of the biology of these viruses [94].

RS virus

RS virus, in particular, is not as antigenically homogeneous as once supposed [95]. It has become apparent that the attachment (G) protein of RS virus is highly variable. Since the majority of the genetic changes observed in the hypervariable regions of the G protein gene result in coding changes [21], it is likely that this variability affects transmissibility of the virus. This is notwithstanding the observation that in experimental animals neutralizing antibody induced by the fusion (F) protein appears to be adequate to provide cross-protective immunity. It must be assumed that the genetic hypervariability of the G protein is relevant to the mode of transmission of the virus and its maintenance in the global population. Unfortunately, knowledge of RS virus is restricted to the subset of viruses which are recovered from children sufficiently ill to require hospitalization during the seasonal epidemics (November to March in temperate climates; Figure 11.8) [2], and there is no knowledge of how the virus is maintained in the inter-epidemic period (April to October in temperate climates). It is clear from a 5-year study of a stable population in East Birmingham [24] that several co-circulating lineages (genotypes) are present in each annual episode of severe illness in children, and that the frequency of these lineages changes from year-to-year (Figure 11.9). Currently, six discrete lineages of the A antigenic subgroup of RS virus are recognized and two of subgroup B [20]. There is some evidence that at the local community level (defined by postal codes) virus of a particular lineage may be excluded from an area where it was prevalent in the preceding year [24] (P.A. Cane and C.R. Pringle, unpublished data). Remarkably, viruses recovered from children throughout the world fall into the same lineage classification, although the frequency of isolation of viruses of the different lineages may be different [22].

More significantly there appears to be a gradual global genetic drift in the G protein genes of the RS viruses isolated from children during the four decades since the discovery of the virus, which resembles the pattern of antigenic drift of the haemagglutinin protein of the influenza B viruses [20a].

11.9 CONTROL

Effective vaccines are available for control of the major human diseases caused by paramyxoviruses and rubulaviruses [45]. Currently employed live measles and mumps virus vaccines provide effective protection (>90%), and the incidence of both diseases has declined markedly since the widespread use of these vaccines in the developed regions of the world. An encouraging feature of the widespread use of live measles virus vaccine is that there has been a decline also in the rare conditions, such as SSPE, associated with measles virus persistent infection. The current WHO target is to reduce measles deaths by 95% and measles cases by 90% worldwide by 1995. The current range of empirically derived live measles virus vaccines do not perform well in children less than 9 months of age, which is the period when children in the economically underdeveloped regions of the world are at greatest risk. A new generation of vaccines may be required to achieve the ultimate goal of eradication of these diseases. A detailed account of vaccine development is beyond the scope of this chapter.

A problem peculiar to viruses of the family Paramyxoviridae is the aberrant responses which were observed in children vaccinated with formalin-inactivated preparations when these children were exposed subsequently to natural infection. Paradoxical disease responses were common to independent trials involving formalin-inactivated alum-adsorbed preparations of measles, parainfluenza and RS viruses. Retrospective analyses suggest that unbalanced immune responses were induced as a consequence of subtle changes in the antigenicity of the envelope glycoproteins of these viruses, such that there was a deficit of neutralizing antibody and an excess of non-neutralizing antibody in the serum of vaccinees (reviewed in [45,80]). Since these perceived changes in the antigenicity of the envelope proteins of any of these viruses have not yet been defined in chemical terms, considerable caution is required in the development and evaluation of replacement vaccines.

The difficulties are best exemplified by consideration of RS virus. No vaccine is yet available although this virus is responsible for most of the severe respiratory illness in infancy. The original trials of a killed vaccine soon after recognition of the association of this virus with severe illness in infancy were marred by the occurrence of an exacerbated disease response when vaccinees were exposed to natural infection. Some deaths occurred that might have been avoided in the normal course of events. Empirically derived live virus vaccines developed by Chanock and his colleagues, although not sufficiently immunogenic, did not induce the exacerbated disease response, giving hope that further development of live virus vaccines will provide a means of controlling the disease (reviewed in [80]). A new live temperature-sensitive candidate vaccine displayed adequate immunogenicity in adult volunteers, but residual virulence could not be eliminated [61,86,107]. Although it may not prove suitable for use in the ultimate target population, seronegative children, it may facilitate identification of the genetic determinants of virulence. Towards this end, the complete nucleotide sequence of this candidate vaccine strain, which was derived by controlled sequential mutagenesis, and that of its virulent progenitor are now being determined. Provisional data indicate that mutations in the L and F genes are responsible for the attenuation of this virus (K. Tolley *et al.*, in preparation).

There is now a variety of subunit and single protein vaccines available, as well as several genetically engineered vectors carrying RS virus antigens. An unresolved problem is how to evaluate these vaccines. There is no cheap accessible experimental animal model and the target population for vaccination, infants in their first weeks of life, is too vulnerable to allow direct trials. Perhaps the best hope lies in the field of reverse genetics. In the near future it may be possible to introduce precisely modified nucleic acids into infectious virus particles. The technical problems of achieving this with negative-strand RNA viruses with nonsegmented genomes in general, and RS virus in particular, are close to solution [33,47]. With the information likely to accrue from the study of the genetic determination of attenuation by the comparative sequencing of virulent and modified viruses, it should be possible to design new live vaccines having the beneficial attributes of live immunogens and none of the hazards of genetic instability or insufficient attenuation associated with the empirically derived vaccines currently employed in medical practice.

In the interim, the guanosine analogue, ribavirin, administered as a controlled aerosol over periods up to 72 hours, or passive immunization with high neutralizing titre immunoglobulin, may be treatments of value in moderation of severe illness [94] in individuals at high risk (i.e. with congenital heart and lung conditions). In the future, administration of humanized neutralizing monoclonal antibodies may replace these methods of treatment of severe disease in high risk patients [6,102]. However, the development of safe single-dose vaccines remains the only universally applicable approach to control of respiratory disease caused by the Paramyxoviridae.

11.10 REFERENCES

1. Alexander, D.J., Manvell, R.J., Collins, M.S. *et al.* (1989) Characterisation of paramyxoviruses isolated from penguins in Antarctica and sub-Antarctica during 1976–1979. *Arch. Virol.*, **109**, 135–44.
2. Anderson, L.J., Hendry, R.M., Pierik, L.T., Tsou, C. and McIntosh, K. (1991) Multicenter study of strains of respiratory syncytial virus. *J. Infect. Dis.*, **163**, 687–92.
3. Bachi, T. (1988) Direct observation of the budding and fusion of an enveloped virus by video microscopy. *J. Cell Biol.*, **107**, 1689–95.
4. Bachmair, A., Finley, D. and Varshavsky, A. (1986) *In vivo* half life of a protein is a function of its amino-terminal residue. *Science*, **234**, 179–86.
5. Baczko, K., Lampe, J., Liebert, U.G. *et al.* (1993) Clonal expansion of hypermutated measles virus in SSPE brain. *Virology*, **197**, 188–95.
6. Barbas, C.F. III, Crowe, J.E. Jr, Cababa, D. *et al.* (1992) Human monoclonal Fab fragments derived from a combinatorial library bind to respiratory syncytial virus F glycoprotein and neutralize infectivity. *Proc. Natl Acad. Sci. USA*, **89**, 10164–8.
7. Barr, J., Chambers, P., Harriott, P., Pringle, C.R. and Easton, A.J. (1994) Sequence of the phosphoprotein gene of pneumonia virus of mice: expression of multiple proteins from two overlapping reading frames. *J. Virol.*, **68**, 5330–4.
8. Bartlett, M.S. (1957) Measles periodicity and community size. *J. R. Stat. Soc. A*, **120**, 40–70.
9. Basle, M., Rebel, A., Pouplard, A., Koutoumdjian, S., Filmon, R. and Lepatezour, A. (1979) Mise en evidence d'antigens vivaux de rougeole dans les osteoclasts de la maladie osseuse de Paget. *C. R. Acad. Sci. (Paris)*, **289**, 225–8.
10. Basle, M.F., Fournier, J.G., Rozenblatt, S., Rebel, A. and Bouteille, M. (1986) Measles virus RNA detected in Paget's bone tissue by *in situ* hybridization. *J. Gen. Virol.*, **67**, 907–13.
11. Basle, M.F., Russell, W.C., Goswami, K.K.A. *et al.* (1985) Paramyxovirus antigens from Paget's bone tissue detected by monoclonal antibodies. *J. Gen. Virol.*, **66**, 2103–10.
12. Bass, B.L. and Weintraub, H. (1988) An unwinding activity that covalently modifies its double-stranded RNA substrate. *Cell*, **55**, 1089–98.
13. Bass, B.L., Weintraub, B.L., Cattaneo, R. and Billeter, M.A. (1989) Biased hypermutation of viral RNA genomes could be due to unwinding/modification of double-stranded RNA. *Cell*, **56**, 331.
14. Bianco, P., Silvestrini, G., Ballanti, P. and Bonucci, E. (1992) Paramyxovirus-like nuclear inclusions identical to those of Paget's disease of bone detected in giant cells of primary oxalosis. *Virchows Arch. Path. Anat. Histopathol.*, **421**, 427–33.
15. Billeter, M.A. and Cattaneo, R. (1991) Molecular biology of defective measles viruses persisting in the human nervous system, in *The Paramyxoviruses*, (ed. D.W. Kingsbury), Plenum Press, New York, pp. 323–46.
16. Black, F.L. (1965) Measles endemicity in insular populations: critical community size and its evolutionary implication. *Theor. Biol.*, **11**, 207–11.
17. Black, F.L. (1991) Epidemiology of Paramyxoviridae, in *The Paramyxoviruses*, (ed. D.W. Kingsbury), Plenum Press. New York, pp. 509–36.
18. Blumberg, B.M., Chan, J. and Udem, S.A. (1991) Function of paramyxovirus 3' and 5' end sequences: In theory and practice, in *The*

Paramyxoviruses, (ed. D.W. Kingsbury), Plenum Press, New York, pp. 235–48.

19. Calain, P. and Roux, L. (1993) The rule of six, a basic feature for efficient replication of Sendai virus defective interfering RNA. *J. Virol.*, **67**, 4822–30.

20. Cane, P.A. and Pringle, C.R. (1992) Molecular epidemiology of respiratory syncytial virus: rapid identification of subgroup A lineages. *J. Virol. Methods*, **40**, 297–306.

20a. Cane, P. A. and Pringle, C. R. (1995) Evolution of subgroup A respiratory syncytial virus: Evidence for progressive accumulation of amino acid changes in the attachment protein. *J. Virol.*, **69**, 2918–25.

21. Cane, P.A., Matthews, D.A and Pringle, C.R. (1991) Identification of variable domains of the attachment protein (G) of subgroup A respiratory syncytial viruses. *J. Gen. Virol.*, **72**, 2091–6.

22. Cane, P.A., Matthews, D.A. and Pringle, C.R. (1992) Analysis of relatedness of subgroup A respiratory syncytial viruses isolated worldwide. *Virus Res.*, **25**, 15–22.

23. Cane, P.A., Matthews, D.A. and Pringle, C.R. (1993) Frequent polymerase errors observed in a restricted area of clones derived from the attachment (G) protein gene of respiratory syncytial virus. *J. Virol.*, **67**, 1090–3.

24. Cane, P.A., Matthews, D.A. and Pringle, C.R. (1994) Analysis of respiratory syncytial virus strain variation in successive epidemics in one city. *J. Clin. Microbiol.*, **32**, 1–4.

25. Caravokyri, C. and Pringle, C.R. (1992) Effect of changes in the nucleotide sequence of the P gene of respiratory syncytial virus on the electrophoretic mobility of the P protein. *Virus Genes*, **6**, 53–62.

26. Cartwright, E.J., Gordon, M.T., Freemont, A.J., Anderson, D.C. and Sharpe, P.T. (1993) Paramyxoviruses and Paget's disease. *J. Med. Virol.*, **40**, 133–41.

27. Cattaneo, R., Rehmann, G., Baczko, K., ter Meulen, V. and Billeter, M.A. (1987) Altered ratios of measles virus transcripts in diseased human brains. *Virology*, **160**, 523–6.

28. Chambers, P., Pringle, C.R. and Easton, A.J. (1990) Heptad repeat sequences are located adjacent to hydrophobic regions in several types of virus fusion glycoproteins. *J. Gen. Virol.*, **71**, 3075–80.

29. Chambers, P., Pringle, C.R. and Easton, A.J. (1992) Sequence analysis of the gene encoding the fusion glycoprotein of pneumonia virus of mice suggest possible conserved secondary structure elements in paramyxovirus fusion glycoproteins. *J. Gen. Virol.*, **73**, 1717–24.

30. Cherrie, A.H., Anderson, K., Wertz, G.W. and Openshaw, P.J.M. (1992) Human cytotoxic T cells stimulated by antigen on dendritic cells recognize the N, SH, F, M, 22K and 1b proteins of respiratory syncytial virus. *J. Virol.*, **66**, 2102–10.

31. Choppin, P.W. and Scheid, A. (1980) The role of viral glycoproteins in absorption, penetration, and pathogenicity of viruses. *Rev. Infect. Dis.*, **2**, 40–61.

32. Collins, P. L. (1991) The molecular biology of human respiratory syncytial virus (HRSV) of the genus *Pneumovirus*, in *The Paramyxoviruses*, (ed. D.W. Kingsbury), Plenum Press, New York, pp. 103–62.

33. Collins, P.L., Mink, M.A. and Stec, D.S. (1991) Rescue of synthetic analogs of respiratory syncytial virus genomic RNA and effects of truncations and mutations on the expression of a foreign reporter gene. *Proc. Natl Acad. Sci. USA*, **88**, 9663–7.

34. Cosby, S.l., McQuaid, S., Taylor, M.J. *et al.* (1989) Examination of eight cases of multiple sclerosis and 56 neurological and non-neurological controls for genomic sequences of measles virus, canine distemper virus, simian virus 5 and rubella virus. *J. Gen. Virol.*, **70**, 2027–36.

35. Curran, J., Boeck, R. and Kolakofsky, D. (1991) The Sendai virus P gene expresses both an essential protein and an inhibitor of RNA synthesis by shuffling modules via mRNA editing. *EMBO J.*, **10**, 3079–85.

36. Dorig, R.E., Marcill, A., Chopra, A. and Richardson, C.D. (1993) The human CD46 molecule is a receptor for measles virus (Edmonston strain). *Cell*, **75**, 295–305.

37. Dunster, L.M., Schneider-Schaulies, J., Loffler, S. *et al.* (1994) Moesin: a cell membrane protein linked with susceptibility to measles virus infection. *Virology*, **198**, 265–74

38. Easton, A.J. and Eglin, R.P. (1989) Epidemiology of parainfluenza virus type 3 in England and Wales over a ten year period. *Epidemiol. Infect.*, **102**, 531–5.

39. Enders, J.F. (1959) Mumps, in *Viral and Rickettsial Infections of Man*, 3rd edn, (eds. T.M. Rivers and F.L. Horsfall), Pitman Medical, London, pp. 780–9.

40. Fauquet, C.M., Murphy, F.A., Bishop, D.H.L. *et al.* (1995) Classification and Nomenclature of Viruses. *Sixth Report of the International Committee on Taxonomy of Viruses.* Springer-Verlag (in press).

41. Follett, E.A.C., Pringle, C.R. and Pennington, T.H. (1976) Events following the infection of enucleate cells with measles virus. *J. Gen. Virol.,* **32,** 163–5.

42. Francki, R.I.B., Fauquet, C.M., Knudsen, D.L. and Brown, F. (1991) Classification and Nomenclature of Viruses. Fifth Report of the International Committee on Taxonomy of Viruses. *Archiv. Virol.,* Suppl. 2, 1–450, Springer-Verlag, Vienna, New York.

43. Galinkski, M.S. and Wechsler, S.L. (1991) The molecular biology of the *Paramyxovirus* genus, in *The Paramyxoviruses,* (ed. D.W. Kingsbury), Plenum Press, pp. 41–82.

44. Garcia-Barreno, B., Portela, A., Delgado, T., Lopez, J.A. and Melero, J.A. (1990) Frame shift mutations as a novel mechanism for the generation of neutralisation resistant mutants of human respiratory syncytial virus. *EMBO J.,* **9,** 4181–7.

45. Gluck, R. and Just, M. (1990) Vaccination against measles, mumps and rubella, in *Vaccines and Immunotherapy,* (ed. S.J. Cryz Jr), Pergamon Press, New York, pp. 282–91.

46. Gordon, M.T., Mee, A.P., Anderson, D.C. and Sharpe, P.T. (1992) Canine distemper virus transcripts sequenced from pagetic bone. *Bone and Mineral,* **19,** 159–74.

47. Grosfeld, H., Kuo, L., Mink, M., Stec, D. and Collins, P.L. (1993) Rescue of synthetic analogs of respiratory syncytial virus (RSV) genomic RNA (vRNA). *Proceedings of the 9th International Congress of Virology,* Glasgow, Scotland, 8–13 August, 1993. Abstract no. W39–5.

48. Hirano, A. (1992) Subacute sclerosing panencephalitis virus dominantly interferes with replication of wild-type measles virus in a mixed infection: Implication for viral persistence. *J. Virol.,* **66,** 1891–8.

49. Holdaway, I.M., Ibbertson, H.K., Wattie, D., Scraag, R. and Graham, P. (1990) Previous pet ownership and Paget's disease. *Bone and Mineral,* **8,** 53–8.

50. Howatson, A.F. and Fournier, V.L. (1982) Microfilaments associated with Paget's disease of bone: Comparison with nucleocapsids of measles virus and respiratory syncytial virus. *Intervirology,* **18,** 150–9.

51. Inocencio, N.M., Gotoh, B., Toyoda, T., Kitada, C. and Nagai, Y. (1990) Evaluation of the antiviral effect of synthetic oligopeptides whose sequences are derived from paramyxovirus F1 amino-termini. *Med. Microbiol. Immunol.,* **179,** 87–94.

52. Kallajoki, M., Hyypia, T., Halonen, P., Orvell, C., Rima, B.K. and Kalimo, H. (1991) Inclusion body myositis and paramyxoviruses. *Human Pathol.,* **22,** 29–32.

53. Kolakofsky, D., Vidal, S. and Curran, J. (1991) Paramyxovirus RNA synthesis and P gene expression, in *The Paramyxoviruses,* (ed. D.W. Kingsbury), Plenum Press, New York, pp. 215–34.

54. Lamb, R.A. (1993) Minireview. Paramyxovirus fusion: a hypothesis for changes. *Virology,* **197,** 1–11.

55. Lamb, R.A. and Paterson, R.G. (1991) The nonstructural proteins of paramyxoviruses in *The Paramyxoviruses,* (ed. D.W. Kingsbury), Plenum Press, New York, pp. 181–214.

56. Lennon, J.L. and Black, F.L. (1986) Maternally derived measles immunity in the era of vaccine-protected mothers. *J. Paediatr.,* **108,** 671–6.

57. Ling, R., Easton, A.J. and Pringle, C.R. (1992) Sequence analysis of the 22K, SH and G genes of turkey rhinotracheitis virus and their intergenic regions reveals a gene order different to that of other pneumoviruses. *J. Gen. Virol.,* **73,** 1709–15.

58. Mallpeddi, S.K. and Samal, S.K. (1993) Sequence variability of the glycoprotein gene of bovine respiratory syncytial virus. *J. Gen. Virol.,* **74,** 2001–4.

59. Markwell, M.A.K. (1991) New frontiers opened by the exploration of host cell receptors, in *The Paramyxoviruses,* (ed. D.W. Kingsbury), Plenum Press, New York, pp. 407–25.

60. McClure, M.A., Thibault, K.J., Hatalski, C.G. and Lipkin, W.I. (1992) Sequence similarity between Borna disease virus p40 and a duplicated domain within the paramyxovirus polymerase proteins. *J. Virol.,* **66,** 6572–7.

61. McKay, E., Higgins, P., Tyrrell, D. and Pringle, C. (1988) Immunogenicity and pathogenicity of temperature-sensitive modified respiratory syncytial virus in adult volunteers. *J. Med. Virol.,* **25,** 411–21.

62. McLean, B.N. and Thompson, E.J. (1989) Antibodies against the paramyxovirus SV5 are not specific for cerebrospinal fluid from multiple sclerosis patients. *J. Neurol. Sci.,* **92,** 261–6.

63. Millar, N.S., Chambers, P. and Emmerson, P.T. (1986) Nucleotide sequence analysis of the haemagglutinin-neuraminidase gene of Newcastle disease virus. *J. Gen. Virol.*, **67**, 1917–27

64. Mills, B.G., Singer, F.R., Weiner, L.P. and Holst, P.A. (1981) Immunohistological demonstration of respiratory syncytial virus antigens in Paget's disease of bone. *Proc. Natl Acad. Sci. USA*, **78**, 1209–13.

65. Mills, B.G., Stabile, E., Holst, P.A. and Graham, C. (1982) Antigens of two different viruses in Paget's disease of bone. *J. Dent. Res.*, **61**, 347–51.

66. Morrison, T. and Portner, A. (1991) Structure function and intracellular processing of the glycoproteins of Paramyxoviridae, in *The Paramyxoviruses*, (ed. D.W. Kingsbury), Plenum Press, New York, pp. 347–82.

67. Moyer, S.A. and Horikami, S.M. (1991) The role of viral and host cell proteins in paramyxovirus transcription and replication, in *The Paramyxoviruses*, (ed. D.W. Kingsbury), Plenum Press, New York, pp. 249–74.

68. Muehlberger, E., Sanchez, A., Randolf, A. *et al.* (1992) The nucleotide sequence of the L gene of Marburg virus, a filovirus: homologies with paramyxoviruses and rhabdoviruses. *Virology*, **187**, 534–47.

69. Nagai, Y., Klenk, H.-D. and Rott, R. (1976) Proteolytic cleavage of the viral glycoproteins and its significance for the virulence of Newcastle disease virus. *Virology*, **72**, 494–508.

70. O'Driscoll, J.B., Buckler, H.M., Jeacock, J. and Anderson, D.C. (1990) Dogs, distemper, and osteitis deformans. *Bone and Mineral*, **11**, 209–16.

71. Openshaw, P.J.M., Anderson, K., Wertz, G.W. and Askonas, B.A. (1990) The 22-kilodalton protein of respiratory syncytial virus is a major target for K^d-restricted cytotoxic T lymphocytes from mice primed by infection. *J. Virol.*, **64**, 1683–9.

72. Parkinson, A.J., Muchmore, H.G., McConnell, T.A., Scott, L.V. and Miles, J.A.R. (1980) Serologic evidence for parainfluenza virus infection during isolation at South Pole Station, Antarctica. *Am. J. Epidemiol.*, **112**, 334–40.

73. Peeples, M.E. (1991) Paramyxovirus M proteins: putting it all together and taking it on the road, in *The Paramyxoviruses*, (ed. D.W. Kingsbury), Plenum Press, New York, pp. 427–56.

74. Peeples, M.E. (1988) Differential detergent treatment allows immunofluorescent localiza-tion of the Newcastle disease virus matrix protein within the nuclei of infected cells. *Virology*, **162**, 255–9.

75. Pelet, T., Curran, J. and Kolakofsky, D. (1991) The P gene of bovine parainfluenza virus 3 expresses all three reading frames from a single mRNA editing site. *EMBO J.*, **10**, 443–8.

76. Poch, O., Blumberg, B.M., Bougueleret, L. and Tordo, N. (1990) Sequence comparison of the five polymerases (L proteins) of unsegmented negative-strand RNA viruses: theoretical assignment of functional domains. *J. Gen. Virol.*, **71**, 1153–62.

77. Poch, O., Sauvaget, I., Delarue, M. and Tordo, N. (1989) Identification of four conserved motifs among the RNA-dependent polymerase encoding elements. *EMBO J.*, **8**, 3867–74.

78. Pringle, C.R. (1977) Enucleation as a technique in the study of virus–host interactions. *Curr. Top. Microbiol. Immunol.*, **76**, 50–82.

79. Pringle, C.R. (1987) Rhabdovirus genetics, in *The Rhabdoviruses*, (ed. R.R. Wagner), Plenum Press, New York, pp. 167–244.

80. Pringle, C.R. (1990) Respiratory syncytial virus vaccine, in *Vaccines and Immunotherapy*, (ed. S.J. Cryz Jr), Pergamon Press, New York, pp. 357–72.

81. Pringle, C.R. (1991) The genetics of Paramyxoviruses, in *The Paramyxoviruses*, (ed. D.W. Kingsbury), Plenum Press, New York, pp. 1–39.

82. Pringle, C.R. (1991) The order *Mononegavirales*. *Archiv. Virol.*, **117**, 137–40.

83. Pringle, C.R. and Eglin, R.P. (1986) Murine pneumonia virus: seroepidemiological evidence of widespread human infection. *J. Gen. Virol.*, **67**, 975–82.

84. Pringle, C.R. and Heath, R.B. (1990) Paramyxoviridae, in *Topley and Wilson's Principles of Bacteriology, Virology and Immunity, Vol. 4 Virology*, (ed. L.H. Collier and M.C. Timbury), Edward Arnold, London, pp. 273–90.

85. Pringle, C.R. and Parry, J.E. (1980) Location and quantitation of antigen on the surface of virus-infected cells by specific bacterial adherence and scanning electron microscopy. *J. Virol. Methods*, **1**, 61–75.

86. Pringle, C.R., Filipiuk, A.H., Robinson, B.S., Watt, P.J., Higgins, P. and Tyrrell, D.A.J. (1993) Immunogenicity and pathogenicity of a triple temperature-sensitive modified respiratory syncytial virus in adult volunteers. *Vaccine*, **11**, 473–8.

87. Pringle, C.R., Wilkie, M.L. and Elliott, R.M. (1985) A survey of respiratory syncytial virus and parainfluenza virus type 3 neutralising and precipitating antibodies in relation to Paget's disease. *J. Med. Virol.*, **17**, 377–86.

88. Ralston, S.H., Digiovine, F.S., Gallacher, S.J. and Duff, G.W. (1991) Failure to detect paramyxovirus sequences in Paget's disease of bone using the polymerase chain reaction. *J. Bone Mineral Res.*, **6**, 1243–8.

89. Rima, B.K. (1989) Comparison of amino acid sequences of the major structural proteins of the paramyxo- and morbilliviruses, in *Genetics and Pathogenicity of Negative Strand Viruses*, (eds D. Kolakofsky and B.J. Mahy), Elsevier, Amsterdam, pp. 254–63.

90. Rima, B.K., Cosby, S.L., Duffy, N. *et al.* (1990) Humoral immune responses in seals infected by phocine distemper virus. *Res. Vet. Sci.*, **49**, 114–16.

91. Rima, B.K., Curran, M.D. and Kennedy, S. (1992) Phocine distemper virus, the agent responsible for the 1988 mass mortality of seals, in *The Science of the Total Environment*, Elsevier Science Publishers B.V., Amsterdam, pp. 45–55.

92. Roberts, S.R., Lichtenstein, D., Ball, L.A. and Wertz, G.W. (1994) The membrane-associated and secreted forms of the respiratory syncytial virus attachment glycoprotein G are synthesized from alternative initiation codons. *J. Virol.*, **68**, 4538–46.

93. Roux, L. and Waldvogel, F.A. (1982) Instability of the viral M protein in BHK-21 cells persistently infected with Sendai virus. *Cell*, **28**, 293–302.

94. Ruuskanen, O. and Ogra, P.L. (1993) Respiratory syncytial virus. *Curr. Probl. Pediatr.*, **23**, 50–79.

95. Scheid, A. and Choppin, P.W. (1976) Protease activation mutants of Sendai virus. Activation of biological properties by specific proteases. *Virology*, **69**, 265–77.

96. Schmid, A., Spielhofer, P., Cattaneo, R., Baczko, K., ter Meulen, V. and Billeter, M.A. (1992) Subacute sclerosing panencephalitis is typically characterized by alterations in the fusion protein cytoplasmic domain of the persisting measles virus. *Virology*, **188**, 910–15.

97. Spiropoulou, C.F. and Nichol, S.T. (1993) A small highly basic protein is encoded in overlapping frame within the P gene of vesicular stomatitis virus. *J. Virol.*, **67**, 3103–10.

98. Stallcup, K.C., Raine, C.S. and Fields, B.N. (1983) Cytochalasin B inhibits the maturation of measles virus. *Virology*, **124**, 59–74.

99. Stec, D.S., Hill, M.G. III and Collins, P.L. (1991) Sequence analysis of the polymerase L gene of human respiratory syncytial virus and predicted phylogeny of nonsegmented negative-strand viruses. *Virology*, **183**, 273–87.

100. Steward, M., Vipond, I.B., Millar, N.S. and Emmerson, P.T. (1993) RNA editing in Newcastle disease virus. *J. Gen. Virol.*, **74**, 2539–47.

101. Sullender, W.M. and Wertz, G.W. (1991) The unusual attachment glycoprotein of respiratory syncytial viruses: structure, maturation, and role in immunity, in *The Paramyxoviruses*, (ed. D.W. Kingsbury), Plenum Press, pp. 383–406.

102. Taylor, G., Furze, J., Tempest, P.R. *et al.* (1991) Humanised monoclonal antibody to respiratory syncytial virus. *Lancet*, **337**, 1411–12.

103. ter Meulen, V., Koprowski, H., Iwasaki, I., Kackell, Y.M. and Muller, D. (1972) Fusion of cultured multiple sclerosis brain cells with indicator cells; presence of nucleocapsids and virions and isolation of parainfluenza type virus. *Lancet*, **ii**, 1–5.

104. Thomas, S.M., Lamb, R.A. and Paterson, R.G. (1988) Two mRNAs that differ by two non-templated nucleotides encode the amino co-terminal proteins P and V of the paramyxovirus SV5. *Cell*, **54**, 891–902.

105. Vidal, S., Curran, J. and Kolakofsky, D. (1990) A stuttering model for paramyxovirus mRNA editing. *EMBO J.*, **9**, 2017–22.

106. Vidal, S., Curran. J. and Kolakofsky, D. (1990) Editing of the Sendai virus P/C mRNA by G insertion occurs during mRNA synthesis via a virus-coded activity. *J. Virol.*, **64**, 239–46.

107. Watt, P.J., Robinson, B.S., Pringle, C.R. and Tyrrell, D.A.J. (1990) Determinants of susceptibility to challenge and the antibody response of adult volunteers given experimental respiratory syncytial virus vaccine. *Vaccine*, **8**, 231–36.

108. Wong, T.C., Ayata, M., Hirano, A., Yoshikawa, Y., Tsuruoka, H. and Yamanouchi, K. (1989) Generalized and localized biased hypermutation affecting the matrix gene of a measles virus strain that causes subacute sclerosing panencephalitis. *J. Virol.*, **63**, 5464–8.

109. Zavada, J. (1982) The pseudotype paradox. *J. Gen. Virol.*, **63**, 3–24.

RESPIRATORY SYNCYTIAL VIRUS

Peter Watt and Paul Lambden

12.1 INTRODUCTION

Respiratory syncytial virus (RS virus) is a highly successful pathogen, being the most important cause of lower respiratory tract infections in infancy and early childhood. The efficiency of transmission and the lack of sustained immunity is reflected in the unique ability of the RS virus to cause annual outbreaks of disease in all communities. Such outbreaks last 2–5 months, occurring in winter in temperate climates and during the rainy season in some tropical countries. The incidence of severe disease is highest in urban areas. Most commonly, infection is introduced into the household by school-age children with an RS virus 'cold'. Dissemination through the family unit occurs by inhalation of large droplets or contact with contaminated secretions. Thus the risk factors for infection with RS virus in early infancy include the number of older siblings and the level of overcrowding in the household. The incidence of life-threatening narrowing of the terminal airways, termed bronchiolitis, and viral pneumonia peaks at a time when maternal IgG to the RS virus persists in the infant's circulation. Levels of IgG in bronchoalveolar lavage fluids are approximately equal to the serum value, suggesting that maternal IgG to the RS virus may be involved in the pathogenesis of bronchiolitis. Infants with bronchopulmonary dysplasia or congenital heart disease with pulmonary hypertension show enhanced susceptibility to life-threatening RS virus infection. A decade ago mortality rates of 30–70% were reported in such children. Despite intensive virological research the morbidity of RS virus infection in these children remains unchanged. Indeed, it is advances in intensive care management and surgical correction procedures which are responsible for the markedly decreased mortality rate of RS virus infection in these compromised children [28].

In older children and healthy adults RS virus infection typically presents as a 'heavy cold'. That upper-respiratory tract infections with RS virus recur throughout life suggests that protective immunity in the nasopharynx is short-lived but useful immunity in the lower

Viral and Other Infections of the Human Respiratory Tract.
Edited by S. Myint and D. Taylor-Robinson. Published in 1996 by Chapman & Hall. ISBN 0 412 60070 6

airways is more lasting. However, this protection is not life-long since RS virus pneumonia is an important problem in the elderly. A recent analysis suggests RS virus is as important as influenza viruses in causing morbidity and excess deaths among elderly people [18].

12.2 INFECTION WITH RS VIRUS

Studies on outbreaks of RS virus infection in closed communities and experimentally challenged adult volunteers indicate an incubation period of 4–6 days. Initially, RS virus replication is confined to the nasopharynx, reaching titres of 10^4–10^6 TCID$_{50}$/ml secretions. In severe disease viral shedding persists for many days. How RS virus spreads within the airways to cause disease at distant sites is unclear. However, because RS virus-infected cells fuse to form syncytia, direct mucosal cell-to-cell spread is an attractive option. Such a mechanism could, via the Eustachian tube, introduce RS virus into the middle ear. In an important study by Ogra's group [31], RS virus genome was demonstrated in 82% of middle ear effusions from those children in whom infectious virus was detected in the nasopharynx. Around 40% of infants infected with RS virus develop symptoms of lower airways disease. Usually, clinical evidence of lung infection appears 1–3 days after the onset of nasopharyngeal infection. Such a time scale suggests that inhalation of virus and/or infected cells with embolization to the terminal airways is the likely route of spread.

The initial docking of RS virus to the mucosal surface of respiratory tract epithelial cells is mediated by a specific viral attachment protein designated the G protein. Fusion of the RS virus with the host cell membrane occurs at the cell surface and is mediated by a dimerized protein termed F. The glycoproteins are seen as club-shaped spikes protruding 12 nm from the surface of the RS virus envelope (Figure 12.1). Video-enhanced microscopy and digital image processing have been used to analyse the assembly, budding and fusion of RS virus [2]. Filamentous viral processes bud from the cytoplasmic membrane of infected cells protruding to a final length of 10 μm. The rapid and synchronized budding from localized regions of the membrane indicates a directed process of recruitment of viral components to areas selected for virus maturation overlying regions where the cytoplasm is undergoing hectic motion. Filaments extend at a rate of 110 to 250 nm/s and shedding of a complete virus takes less than 1 minute. An interesting feature of RS virus replication is that host cell protein synthesis is not down-regulated.

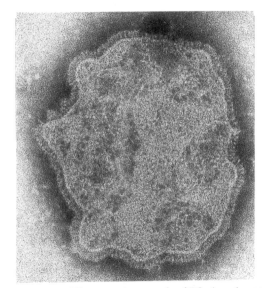

Figure 12.1   Electronmicrograph of RS virus (overall magnification × 140 000). In electromicrographs RS virus appears as pleomorphic spheres of 120–300 nm diameter or as filaments 10 μm long and 60–100 nm diameter. Club-shaped spikes or peplomers (1) 12 nm long and spaced at regular 10-nm intervals protrude from the host cell-derived lipid outer membrane. The nucleocapsid (2) consisting of protein and single-stranded RNA has diameter of 13.5 nm and a helical pitch of 6.5 nm. The fact that the RS virus nucleocapsid diameter is intermediate between the larger paramyxoviruses and the smaller influenza viruses led to the positioning of RS virus in a separate genus designated *Pneumovirus*.

Several features in the natural history of RS virus infections suggest that severe disease has an immunopathological basis. Thus, bronchiolitis occurs while maternal IgG to RS virus persists and enhanced disease was seen in children vaccinated with formalin-inactivated virus when subsequently exposed to natural infection (see Chapter 13). Further, RS virus replication *in vitro* does not induce dramatic cell pathology, and syncytial formation is reputedly rare in the lungs of infants dying of RS virus infection. Taken together these facts suggest that tissue destruction as a direct consequence of virus replication is not the major cause of disease in RS virus infection. However, a definitive analysis of the immunopathology of terminal airways disease in infants dying of RS virus infection is still awaited. In the meantime, conflicting theories (immune-complex disease, IgE-mediated hypersensitivity reaction and T-cell attack on infected cells) can neither be supported nor refuted.

Perhaps the most important determinant of the age-related severity of RS virus bronchiolitis is the incomplete development of the terminal airways in infancy. At age 3 months the terminal two generations of bronchioles are still developing and the numbers of alveoli are only one-quarter of the adult value [17]. The implication is that partial bronchiolar obstruction due to filament production by RS virus-infected cells, the sloughing of dead cells into the lumen plus syncytial formation cause a marked increase in terminal airways resistance given the highly cellular, immature lungs of infants. A recent autopsy study of infants dying of RS virus infection has confirmed the shedding of viral-infected epithelial cells into the bronchial lumen and the presence of giant syncytial cells in the alveoli [30]. Further, the presence of exudate and inflammatory cells at the site of RS virus replication is not proof that antigen-specific immune reactions are involved in the pathogenesis of airways disease. The mRNA for pro-inflammatory cytokines (IL-1β, IL-6, TNF-α), together with

adhesion molecules (ICAM-1, ELAM-1 and VCAM-1) are present in the middle ear exudates of children infected with RS virus but these mediators may have been induced by the direct interaction of virus with epithelial cells and/or mucosal lymphocytes and macrophages [31]. Support for this pathogenic process, comes from studies on bronchoalveolar lavage fluids from RS virus-infected children. Alveolar macrophages were shown to be infected with RS virus and to co-express class II HLA DR, IL-1β and TNF-α proteins [26]. Thus the key features of RS virus-mediated damage to the terminal airways may be explained as viral induced pathology without invoking antigen-specific immunological reactions.

12.3 THE RS VIRUS GENOME

The genome of RS virus is a negative polarity single-stranded RNA molecule with a coding potential of approximately 15 kb and an estimated molecular weight of 5×10^6 Da. The genome is transcribed into 10 discrete polyadenylated mRNAs from a single promoter at the 3′ terminus. During replication of the virus a high-molecular weight, positive polarity, genome-length RNA is produced which serves as the template for the generation of new negative-strand genomes [21]. The genome organization in the Pneumoviruses (Figure 12.2) differs from the typical paramyxoviruses in that the genes for two non-structural proteins immediately follow the promoter and precede the nucleocapsid gene (N). The NS1 and NS2 genes in RS virus subgroups A and B are identical in length and share 78% nucleotide sequence identity. In NS2 the C-terminal amino acids 78–124 are conserved while in NS1 it is the N-terminus which is highly conserved [22]. Human parainfluenza virus and Sendai virus both produce a non-structural protein (C), but this protein is produced from a second overlapping reading frame present in the phosphoprotein gene. The relative order of the fusion protein (F) and attachment glycoprotein (G) genes are reversed in RSV compared with other paramyxoviruses and are placed immediately upstream of a gene encoding a second matrix protein M_2. The presence of a small hydrophobic (SH) protein gene between the M and G genes is also unique to RS virus, but a structural counterpart is produced by SV5 [13].

The complete nucleotide sequence of RS virus has been determined and has confirmed the gene order to be 3'-NS1-NS2-N-P-M-SH-G-F-M2-L-5' (Figure 12.2). Analysis of the sequences in the intergenic regions has revealed conserved sequences which probably function as initiation and termination signals during transcription of the viral genome. The nucleotide sequences of the 3'-leader and 5'-trailer regions have been determined for the A_2 strain genomic RNA [27]. In this strain the 3'-leader consists of a 44-nucleotide extracistronic tract whereas the 5'-trailer sequence is 155 nucleotides and is considerably longer than in other paramyxoviruses. A significant feature is that 16 of the 22 terminal nucleotides in the leader and trailer sequences are complementary (Figure 12.3), presumably serving a role as recognition sequences for the RS virus

RNA polymerase (L). Studies on the involvement of the leader and trailer sequences in transcriptional control have been performed using cDNA constructs joined to a bacterial chloramphenicol acetyltransferase (CAT) reporter gene [15]. Such synthetic constructs expressed CAT and could be rescued by wild-type virus following transfection into virus-infected cells. The 44-nucleotide leader sequence could be replaced by a 50-nucleotide sequence complementary to the 5'-terminus without affecting rescue. However, the addition of 11 heterologous nucleotides to the intact 3' end of the leader sequence abolished rescue, suggesting that the 3' proximal domain is essential for transcription. The first nine genes start with the sequence CCCCGU-UUA- and all genes end with the motif -UCA$^A/_u$UN$^A/_u$$^A/_u$$^A/_u$UUU. The tenth gene (L-polymerase) starts with a modification (CCCUGUUUU) of the consensus sequence. However, a surprising finding was that the L-gene start overlaps with the M2 gene terminus by 68 nucleotides [16]. Since transcription of the L-gene initiates within the M2 gene and the M2 termination signal lies downstream of the L-gene initiation signal, full-length

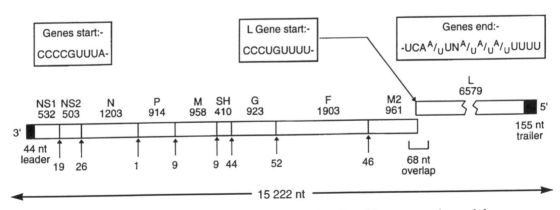

Figure 12.2 RS virus genome structure and organization. The M2 and L genes overlap and the consensus gene start and end sequences are shown in boxes above the genetic map. The size of the mRNAs are indicated below the gene segment name and the number of nucleotides in the intergenic region are arrowed below the genome map. The leader and trailer sequences are indicated by shading at the ends of the map.

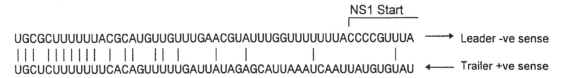

NS1 Start

UGCGCUUUUUUACGCAUGUUGUUUGAACGUAUUUGGUUUUUUUACCCCGUUUA $\longrightarrow$ Leader -ve sense

UGCUCUUUUUUUCACAGUUUUUGAUUAUAGAGCAUUAAAUCAAUUAUGUGUAU $\longleftarrow$ Trailer +ve sense

Figure 12.3 Complementarity in the RS virus leader and trailer sequences. RS virus 3' leader and 5' trailer sequences aligned to show terminal homologies. The leader sequence is in the genomic sense and the trailer sequence is the complementary positive-strand sequence.

L-mRNA can be produced only if the viral polymerase transcribes across the M2 termination signal without pausing or adding a polyadenylate tail. Analysis of cDNA clones of RS virus transcripts revealed that the polymerase occasionally fails to recognize termination signals and produces bicistronic messages. This mechanism accounts for the production of an abundant small 68-base polyadenylated L-leader RNA and an attenuated synthesis of full-length L-mRNA and hence viral polymerase.

Another interesting feature of the RS virus genome is that the intergenic regions are of variable length and are not conserved, as is the case with the paramyxoviruses. The intergenic regions vary in length from one to 54 nucleotides in subgroup A and B strains. The lack of conserved sequences or secondary structure in the intergenic regions probably means that these sequences do not contain termination and re-initiation signals. These RS viral gene end sequences differ from the typical paramyxovirus (AUUC) and rhabdovirus (AUAC) sequences adjacent to the poly-A tail of each gene transcript. However, all newly forming mRNA transcripts in both groups of viruses are thought to polyadenylate by re-iterative copying of the poly-U tracts at the end of the gene [13].

12.4 RS VIRUS GLYCOPROTEINS

12.4.1 STRUCTURE AND FUNCTION

RS virus attachment by the G glycoprotein to receptors on the respiratory epithelial cell sur-

face is the primary event in the fusion process. This initial docking via the viral glycoprotein spikes leaves the viral and host cell membranes some 10 nm apart. How the lipid bilayers of the two membranes are brought into contact despite strong short-range repulsive forces remains unclear. In influenza virus, the haemagglutinin is a trimeric protein with the fusion domain a loop structure at neutral pH. At the pH of the endosome this peptide segment assumes an extended trimeric α-helical coiled-coil formation displacing the fusion region 10 nm from its buried position in the trimer to an exposed fusion competent position [12]. However, the fusion process with RS virus must involve a pH-independent conformational change because it is highly efficient at extracellular pH, not only permitting viral entry into the target cells but also fusion of infected cells with their near-neighbour cells to form giant syncytia.

12.4.2 RS VIRUS FUSION PROTEIN

The structure of the RS virus F glycoprotein is illustrated schematically in Figure 12.4. The F0 precursor protein is synthesized on membrane-bound polysomes and co-translationally inserted into the rough endoplasmic reticulum (ER). Post-translational modifications including *N*-linked glycosylation are essential for the creation of the bioactive conformation. The precursor oligosaccharides (three glucose, nine mannose, two *N*-acetyl-glucosamine) are linked to a lipid dolichol carrier by a pyrophosphate bond. The entire oligosaccharide chain is then transferred to specific asparagine residues (Asn-X-Ser/Thr) on the growing polypeptide. After transportation to the Golgi apparatus the high mannose precursor is trimmed by glycosidases to a core oligosaccharide composed of two N-acetylglucosamine and three mannose residues. In the Golgi cisternae various specific transferases build up the complex oligosaccharide by the addition of N-acetylglucosamine, galactose, fucose and finally sialic acid residues to the core sugars. In the mature protein there are four N-linked oligosaccharides on the F2 subunit and a single chain on the F1 subunit. Also within the Golgi the precursor F protein is modified further by acylation of the single cysteine residue (Cys^{550}) located at the cytoplasmic face of the carboxy terminus [1]. This linkage via a thiol ester bond of palmitate to Cys^{550} plays a major role in anchoring the F protein

to the viral envelope but may also be involved in the fusion process.

During or immediately after translation the hydrophobic 22-residue signal peptide is cleaved from the amino terminus by a specific peptidase. The F protein anchorage domain [Ile^{525} → Cys^{550}] is rich in hydrophobic amino acids and typical of the stop transfer signals which terminate the passage of the nascent polypeptide chain through the ER membrane. Thus membrane-anchored F protein has the carboxy terminal [Lys^{551} → Cys^{574}] free within the host cell cytoplasm. These residues may interact with the M protein during virus assembly.

All functional viral fusion proteins are oligomers, the dimerization of the RS virus F protein occurring during or shortly after translation. Highly conserved in paramyxovirus

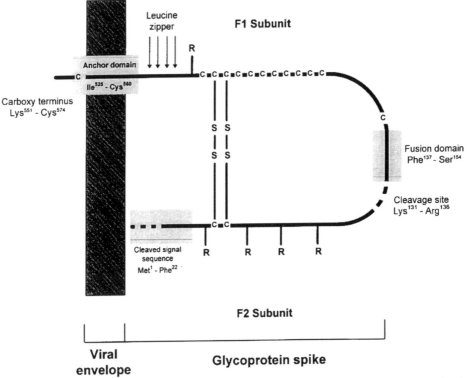

Figure 12.4 Schematic diagram of RS virus fusion glycoprotein. The shaded area indicates hydrophobic regions. R, N-linked oligosaccharides; C, conserved cysteine residues.

fusion proteins, including RS virus, is a leucine zipper motif consisting of five heptad repeats situated 4 to 11 amino acids from the transmembrane anchor. This may serve as the linkage joining F protein monomers [7]. Presumably the highly hydrophobic fusion domains lie buried in the oligomeric glycoprotein spike. Activation of the F0 protein requires proteolytic cleavage releasing the new amino terminal fusion domain. The oligobasic cleavage motif KKRKRR is cut by Furin, a subtilisin-like endoprotease localized on the Golgi membrane [29]. Furin is ubiquitously expressed in a wide range of tissues and cell lines. Subsequent to initial cleavage the remaining basic amino-acids are removed by membrane-bound carboxypeptidases. For some respiratory viruses the enzymatic cleavage of the F0 protein is the determinant of organ tropism. One example is Sendai virus, which has a single arginine cleavage motif. In the case of Sendai virus, the F0 protein is cleaved by an arginine-specific serine protease located exclusively in, and secreted by, the Clara cells of the bronchial epithelium [39]. Such a mechanism cannot explain RS virus localization to the respiratory tract since the six basic amino acids forming the F0 proteolytic site are readily cleaved by Furin. That immunological factors are important in confining RS virus to the airways is suggested by the dissemination of infection in severely immunocompromised infants.

In all fusion proteins activated by proteolytic cleavage the newly exposed fusion domain is located on the same peptide segment (F1 in RS virus) as the viral membrane anchor domain. Presumably, following initial interaction of the fusion domain with the target host cell membrane, the virus remains anchored by the F1 polypeptide to the target cell. The fact that the newly created amino terminus of the fusion domain has the conserved motif Phe-X-Gly (Figure 12.5) in all paramyxoviruses and HIV implies a key role in the induction of fusion [19]. The 19 amino acids

constituting the fusion domain have a marked propensity to α-helix formation in a lipid environment. How the intercalation of a helical peptide causes disorganization of the host cell membrane is not known.

The nucleotide and imputed amino-acid sequences of the F proteins of prototype strains of both subgroup A and B RS virus are highly conserved [23,35]. The proteolytic cleavage site (KKRKRR), the first 15 amino acids of the fusion domain (FLGFLLGVGSAIASG) and the number and position of the cysteine residues are identical across all known RS virus fusion proteins implying essential roles in F protein function.

12.4.3 THE ATTACHMENT GLYCOPROTEIN, G

The attachment protein, G, is the translation product of the seventh cistron of the RS virus genome. Both monoclonal and polyclonal antibodies reactive to G block the attachment of RS virus to the target surface, thus neutralizing virus and protecting against experimental challenge in animal models. Antibodies to G do not inhibit syncytial formation. The inability of an excess of purified G protein to block RS virus infectivity for tissue culture cells implies that the virus receptor is a ubiquitous membrane component [40]. However, RS virus lacks haemagglutinating properties, implying that the host-cell receptor is distinct from Ortho- and Paramyxovirus receptors.

A schematic diagram of G protein is shown in Figure 12.6. An unusual feature is that the protein lacks an N-terminal hydrophobic signal sequence and a C-terminal transmembrane anchor domain. A conserved region (amino acids 38–66) near the N-terminus may serve the dual role of signal sequence and membrane anchor site. The primary translational product of 32 kDa is co-translationally inserted into the ER where N-linked, mannose-rich, oligosaccharides are attached to give an intermediary product (M_r 45 kDa). Within the Golgi this product is extensively

RSV

NDV

HPIV3

Sendai

Mumps

Measles

HIV-1

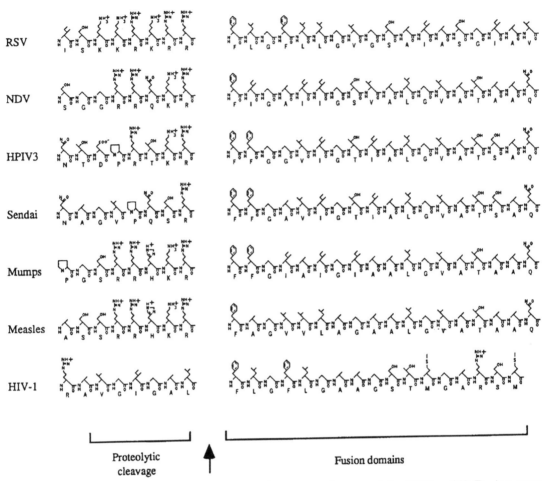

Proteolytic
cleavage

Fusion domains

Figure 12.5 Comparison of the fusion domains of paramyxoviruses and the HIV1 gp160. During proteolytic cleavage of the fusion protein precursors, the cleavage site, rich in basic amino acids (KKRKRR in RS virus), is removed to create a new amino terminus. The now exposed fusion domains begin with the motif *Phe-X Gly* conserved in paramyxoviruses and many retroviruses.

modified by maturation of the N-linked oligosaccharides and the addition of O-linked oligosaccharides [43]. The G protein ectodomain (amino acids 67–298) is remarkably rich in serine and threonine residues (approximately 30%). The proportion of these residues which have O-linked side chains is unclear but the mature G protein has an apparent size of 84–90 kDa on SDS–PAGE gels. One region of the molecule, His[164] to Cys[176], is conserved in all RS virus strains and is remote from N- and most O-linked glycosyla-tion sites. This region is remarkably rich in hydrophobic residues (three Phe, two Val, two Cys) and computer predictions suggest that this peptide forms an internal segment of the protein. One possibility is that this segment is the attachment cleft of the G protein docking onto host-cell receptors. The results of studies on escape mutants and short synthetic peptides reactive with neutralizing monoclonal antibodies suggest that these antibodies bind to a G protein segment (amino acids 190–210) adjacent to the putative receptor [5]. Such

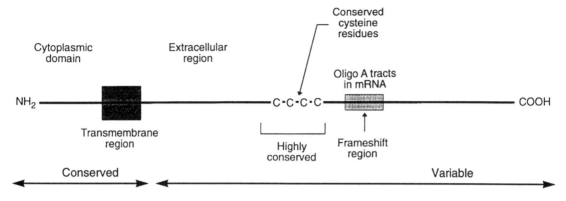

Figure 12.6 Schematic diagram of RS virus attachment glycoprotein.

antibodies could neutralize RS virus by steric interference with the attachment event.

A remarkable feature of RS virus G protein is the extensive diversity between subgroup A and B isolates, with the polypeptide segments either side of the putative receptor site showing only 40% conservation of amino acids [24]. Studies on laboratory animals show that immunization with recombinant vaccinia virus expressing subgroup A G protein confers little or no protection against challenge with a subgroup B isolate [37]. However, both subgroups co-circulate in outbreak situations, suggesting that G protein variation between A and B subgroups is not the critical determinant of the ability of RS virus to cause annual epidemics of infection. Further, extensive genetic variations of the G protein within the subgroups have been identified both in outbreak situations and between successive outbreaks. A study of outbreaks in Birmingham, UK revealed that six genetic lineages of subgroup A virus co-circulated within the city [10]. Comparable changes in the G protein of subgroup B have been recorded in the USA [38]. The molecular basis for these variations was demonstrated in a study of G protein monoclonal antibody escape mutants by Garcia-Barreno and colleagues [6]. This established that antigenic variation was generated by a

novel mechanism involving frameshift mutations. Immediately following the conserved region of the RS Virus G protein gene is a segment (nucleotides 588–654) which contains three runs of six or seven adenosine residues. Insertion or deletion of nucleotides within these gene segments can result in dramatic changes in the amino-acid sequence of the carboxy terminal one-third of the G protein. Downstream of the changed segment, a single nucleotide insertion or deletion may restore the original open reading frame (ORF). Such frameshift mutations lead to many changed amino acids and generate new termination codons resulting in G proteins of variable length. The molecular basis of these frameshifts is unclear but may involve polymerase errors caused by a slippage mechanism [8]. What is certain is that the frameshift mechanism enables RS virus to escape neutralization by anti-G protein antibodies and makes G protein a poor candidate for vaccine development.

12.4.4 SH PROTEIN

Immediately preceding the genes for the G and F proteins is a gene coding for a 64-amino acid polypeptide (7.5 kDa) designated the small hydrophobic (SH) protein. An interesting feature of the SH protein is that it also occurs as a 4.8-kDa species generated by translation initiation at the second AUG in the sequence. The SH protein has a hydrophobic core; residues 14 to 41 contain 21 hydrophobic amino acids and only a single charged residue. There are two acceptor sites for *N*-linked sugars either side of this hydrophobic domain. Antisera raised against a synthetic carboxy-terminus peptide were used to demonstrate that the predominant 7.5-kDa unmodified protein, a 13 to 15-kDa mannose-rich glycosylated form and a 21 to 31-kDa complex carbohydrate form are present at the surface of the infected cell [32]. The role of this unusual integral membrane protein in the pathogenesis of RS virus infection remains unclear. The fact that there is only 75% deduced amino acid sequence identity between the SH protein of subgroups A and B indicates that the SH protein is remarkably tolerant of structural variation [9]. Within the subgroups, 93–99% homology was found; this suggests that SH protein structural variations may not be induced as a response to immune surveillance.

12.5 MATRIX PROTEIN

The matrix protein is non-glycosylated, lies external to the nucleocapsid and elicits a strong immune response in natural RS virus infection [41]. This 28.7-kDa protein is relatively rich in basic amino acids and has two clusters of hydrophobic residues (amino acids 185–206 and 213–238) in the C-terminal third of the protein [34]. These hydrophobic segments of the M protein may interact with lipid structures of the cytoplasmic membrane of the RS virus-infected cell. The matrix protein is solubilized from infected cells using non-ionic detergents and does not co-purify with the nucleocapsids. Although RS virus M protein lacks significant homology with other negative-strand matrix proteins it may well serve a comparable function. During virus maturation matrix proteins associate with the host cell membrane and may be involved in targeting viral glycoproteins and nucleocapsids to the site of viral budding. A second matrix protein,

the M2 or 22K protein is also selectively extracted from RS virus-infected cells by non-ionic detergents. The use of immunofluorescence to study both unfixed and acetone-treated cells showing extensive cytopathic effect has enabled the association of M2 protein with both the viral envelope and capsid to be demonstrated [33]. There is a high degree of conservation of the M2 gene in RS virus of subgroups A and B [14]. A second ORF coding for some 90 amino acids is seen at the carboxy terminus of M2 but it is not known if the translation product is produced in RS virus-infected cells. A further interesting feature is a region of 68 bp overlap between the start of the L gene and the carboxy terminus of the M2 protein gene.

12.6 RS VIRUS CAPSID PROTEINS AND VIRAL REPLICATION

The nucleocapsid core of RS virus consists of genomic RNA and three proteins; nucleoprotein (N, 43.5 kDa), phosphoprotein (P, 27 kDa) and the polymerase (L, 250 kDa). The nucleocapsid protein is an abundant structural protein and forms a complex with the viral RNA. The phosphoprotein is less abundant and is probably involved in the formation of replication complexes with the viral polymerase. The P-protein is extensively phosphorylated, relatively acidic and contains no cysteine or tryptophan residues [25]. Thermosensitive mutants of RS virus belonging to the complementation group E have been mapped to the phosphoprotein gene [11]. Viral protein synthesis in the tsN19 mutant was completely restricted at 39°C and the mutant P protein failed to react with a specific monoclonal antibody at 33°C. Reversion of the mutation identified as a Gly→Ser at position 172, restored reactivity with the monoclonal antibody. In challenge studies of human volunteers tsN19 virus produced fewer colds than wild-type RS virus but did induce neutralizing antibody responses [42]. The P-proteins of paramyxoviruses have been shown to be essential components of the polymerase complex, so that mutations in the P gene would be expected to have major consequences for viral replication. The L-protein or viral polymerase is a minor component of the viral nucleocapsid and is similar in size to the analogous protein in other paramyxoviruses.

Studies with ribonucleoprotein complexes purified from RSV-infected Hep-2 cells have demonstrated that these structures are inactive in *in vitro* transcription. However, addition of extracts from uninfected Hep-2 cells restored the ability to synthesize mRNAs *in vitro* [3]. The transcription process is tightly coupled to capping and methylation of the 5'-terminal nucleotide. The cap structure synthesized *in vitro* in the presence of S-adenosyl-methionine consists of a 7-methyl guanosine linked to a second guanosine residue via a 5'-5' pyrophosphate bridge. The complete cap structure is $^mG(5')$-P-P-P-Gp and does not include O-methylation of ribose [4].

12.7 SUMMARY AND FUTURE DIRECTIONS

The genus *Pneumovirus* is the most divergent member of the Paramyxoviridae. Genome organization in RS virus is distinctive with the presence of 10 cistrons including additional genes coding for NS1, NS2 and a second matrix protein. The major sequence variation between subgroups A and B implies divergence in an early stage of RS virus evolution. Surprisingly, both subgroups co-circulate in most outbreak situations causing comparable clinical disease. Thus the significant differences between subgroups A and B may be irrelevant to the ability of RS virus to cause annual outbreaks of infection and repeated episodes of disease in all age groups.

Despite detailed understanding of the RS virus genome, the structure and functional organization of RS virus proteins and the mechanisms important in the causation of disease are largely not understood. However, it is now feasible to construct full-length cDNA

clones of the RS virus genome to allow genetic manipulation of the viral structural genes and control sequences. The major obstacle to this approach is that the negative-sense RNA genome is not infectious and that the viral polymerase must be provided in the infecting capsids. However, analogous systems are being established in the study of influenza virus by reconstruction of the ribonucleoprotein complex *in vitro* to allow expression and rescue *in vivo* of the genetically engineered RNA [36]. By site-specific modification of RS virus genes it will be possible to gain new insights into the role of different proteins in RS virus virulence. Further, it offers the opportunity to create 'designer' live vaccines; one obvious approach is to replace the cleavage site on F0 susceptible to the ubiquitous protease Furin with a highly specific site. Such a virus should have significantly reduced ability to replicate within the body.

12.8 REFERENCES

1. Arumugham, R.G., Seid, R.C., Doyle, S. *et al.* (1989) Fatty acid acylation of the fusion protein of human respiratory syncytial virus. *J. Biol. Chem.*, **264**, 10339– 42.
2. Bachi, T. (1988) Direct observation of the budding and fusion of an enveloped virus by video microscopy of viable cells. *J. Cell Biol.*, **107**, 1689–95.
3. Barik, S. (1992) Transcription of human respiratory syncytial virus genome RNA *in vitro*: Requirement of cellular factor(s). *J. Virol.*, **66**, 6813–18.
4. Barik, S. (1993) The structure of the 5' terminal cap of the respiratory syncytial virus mRNA. *J. Gen. Virol.*, **74**, 485–90.
5. Barreno, G., Delgado, T., Akerlind Stopner, B. *et al.* (1992) Location of the epitope recognised by monoclonal antibody 63G on the primary structure of human respiratory syncytial virus G glycoprotein and the ability of synthetic peptides containing this epitope to induce neutralising antibodies. *J. Gen. Virol.*, **73**, 2625– 30.
6. Barreno, G., Portela, A., Delgado, T. *et al.* (1990) Frameshift mutations as a novel mechanism for the generation of neutralization resistant

mutants of human respiratory syncytial virus. *EMBO J.*, **9**, 4181–7.

7. Buckland, R. and Wild, F. (1989) Leucine zipper motif extends. *Nature*, **338**, 547.

8. Cane, P.A. (1993) Frameshifting and antigenic variation in respiratory syncytial virus. *Trends Microbiol.*, **1**, 156–9.

9. Cane, P.A. and Pringle, C.R. (1991) Respiratory syncytial virus heterogeneity during an epidemic: analysis by limited nucleotide sequencing (SH gene) and restriction mapping (N gene). *J. Gen. Virol.*, **72**, 349–57.

10. Cane, P.A., Matthews, D.A. and Pringle, C.R. (1991) Identification of variable domains of the attachment (G) protein of subgroup A respiratory syncytial virus. *J. Gen. Virol.*, **72**, 2091–6.

11. Caravokyri, C., Zajac, A.J. and Pringle, C.R. (1992) Assignment of mutant tsN19 (Complementation group E) of respiratory syncytial virus to the P protein gene. *J. Gen. Virol.*, **73**, 865–73.

12. Carr, C.M. and Kim, P.S. (1993) A spring-loaded mechanism of the conformational change of Influenza hemagglutinin. *Cell*, **73**, 823–32.

13. Collins, P.L., Dickens, L.E., Buckler-White, A.J. *et al.* (1986) Nucleotide sequences for the gene junctions of human respiratory syncytial virus reveal distinctive features of intergenic structure and gene order. *Proc. Natl Acad. Sci. USA*, **83**, 4594–8.

14. Collins, P.L., Hill, M.G. and Johnson, P.R. (1990) The two open reading frames of the 22K mRNA of human respiratory syncytial virus: sequence comparison of antigenic subgroups A and B and expression *in vitro*. *J. Gen. Virol.*, **71**, 3015–20.

15. Collins, P.L., Mink, M.A. and Stec, D.S. (1991) Rescue of synthetic analogs of respiratory syncytial virus genomic RNA and effect of truncations and mutation on the expression of a foreign reporter gene. *Proc. Natl Acad. Sci. USA.*, **88**, 9663–7.

16. Collins, P.L., Olmstead, R.A., Spriggs, M.K. *et al.* (1987) Gene overlap and site-specific attenuation of transcription of the viral polymerase L gene of human respiratory syncytial virus. *Proc. Natl Acad. Sci. USA*, **84**, 5134–8.

17. Dunnill MS (1962) Postnatal growth of the lung. *Thorax*, **17**, 329–33.

18. Fleming, D.M. and Cross, K.W. (1993) Respiratory syncytial virus or influenza? *Lancet*, **342**, 1507–10.

19. Gallaher, W.R. (1987) Detection of a fusion peptide sequence in the transmembrane protein of human immunodeficiency virus. *Cell*, **50**, 327–8.

20. Gruber, C. and Levine, S. (1985) Respiratory syncytial virus polypeptides IV: the oligosaccharides to the glycoproteins. *J. Gen. Virol.*, **66**, 417–32.

21. Huang, Y.T. and Wertz, G.W. (1982) The genome of respiratory syncytial virus is a negative-stranded RNA that codes for at least seven mRNA species. *J. Virol.*, **43**, 150–7.

22. Johnson, P.R. and Collins, P.L. (1989) The IB(NS2), IC(NS1) and N proteins of human respiratory syncytial virus (RSV) of antigenic subgroups A and B: sequence conservation and divergence within RSV genomic RNA. *J. Gen. Virol.*, **70**, 1539–47.

23. Johnson, P.R. and Collins, P.L. (1988) The fusion glycoprotein of respiratory syncytial virus of subgroups A and B: sequence conservation provides a structural basis for antigenic relatedness. *J. Gen. Virol.*, **69**, 2623–8.

24. Johnson, P.R., Spriggs, M.K., Olmsted, R.A. *et al.* (1987) The G glycoprotein of human respiratory syncytial virus of subgroups A and B. Extensive sequence divergence between antigenically related proteins. *Proc. Natl Acad. Sci. USA*, **84**, 5625–9.

25. Lambden, P.R. (1985) Nucleotide sequence of the respiratory syncytial virus phosphoprotein gene. *J. Gen. Virol.*, **66**, 1607–12.

26. Midulla, F., Villani, A., Panuska, J.R. *et al.* (1993) Respiratory Syncytial Virus lung infection in infants: immunoregulatory role of infected alveolar macrophages. *J. Infect. Dis.*, **168**, 1515–19.

27. Mink, M.A., Stec, D.S. and Collins, P.L. (1991) Nucleotide sequences of the 3' leader and 5' trailer regions of human respiratory syncytial virus genomic RNA. *Virology*, **185**, 615–24.

28. Moler, F.W., Khan, A.S., Meliones, J.N. *et al.* (1992) Respiratory Syncytial Virus morbidity and mortality estimates in congenital heart disease patients: a recent experience. *Crit. Care Med.*, **20**, 1406–13.

29. Nagai, Y. (1993) Protease-dependent virus tropism and pathogenicity. *Trends Microbiol.*, **1**, 81–7.

30. Neilson K.A. and Yunis, E.J. (1990) Demonstration of Respiratory Syncytial Virus

in an autopsy series. *Paediatr. Pathol.*, **10**, 491–502.

31. Okamoto, Y., Kudo, K., Ishikawa, K. *et al.* (1993) Presence of Respiratory Syncytial Virus genomic sequences in middle ear fluid and its relationship to expression of cytokines and cell adhesion molecules. *J. Infect. Dis.*, **168**, 1277–81.

32. Olmsted, R.A. and Collins, P.L. (1989) The 1A protein of respiratory syncytial virus is an integral membrane protein present as multiple, structurally distinct species. *J. Virol.*, **63**, 2019–29.

33. Routledge, E.G., Willcocks, M.M., Sampson, A.C.R. *et al.* (1988) Heterogeneity of the respiratory syncytial virus 22K protein revealed by Western blotting with monoclonal antibodies. *J. Gen. Virol.*, **68**, 1209–15.

34. Satake, M. and Venkatesan, S. (1984) Nucleotide sequence of the gene encoding respiratory syncytial virus matrix protein. *J. Virol.*, **50**, 92–9.

35. Scopes, G.E., Watt, P.J. and Lambden, P.R. (1990) Identification of a linear epitope on the fusion glycoprotein of respiratory syncytial virus. *J. Gen. Virol.*, **71**, 53–9.

36. Seong, B.L. (1993) Influencing the influenza virus: genetic analysis and engineering of the negative-sense RNA genome. *Infect. Agents Dis.*, **2**, 17–24.

37. Stott, E.J., Taylor, G., Ball, L.A. *et al.* (1987) Immune and histopathological responses in animals vaccinated with recombinant vaccinia viruses that express individual genes of human respiratory syncytial virus. *J. Virol.*, **61**, 3855–61.

38. Sullender, W.M., Mufson, M.A., Anderson, L.J. *et al.* (1991) Genetic diversity of the attachment protein of subgroup B respiratory syncytial virus. *J. Virol.*, **65**, 5425–34.

39. Tashiro, M., Yokogoshi, Y., Tobita, K. *et al.* (1992) Tryptase Clara, an activating protease for Sendai virus in rat lungs, is involved in pneumophogenicity. *J. Virol.*, **66**, 7211–16.

40. Walsh, E.E., Schlesinger, J.J. and Brandriss, M.W. (1984) Purification and characterisation of GP90, one of the envelope glycoproteins of respiratory syncytial virus. *J. Gen. Virol.*, **65**, 761–7.

41. Ward, K.A., Lambden, P.R., Ogilvie, M.M. *et al.* (1983) Antibodies to respiratory syncytial virus; polypeptides and their significance in human infection. *J. Gen. Virol.*, **64**, 1867–76.

42. Watt, P.J., Robinson, B.S., Pringle, C.R. *et al.* (1990) Determinants of susceptibility to challenge and the antibody response of adult volunteers given experimental respiratory syncytial virus vaccines. *Vaccine*, **8**, 231–6.

43. Wertz, G.W., Krieger, M. and Ball, L.A. (1989) Structure and cell surface maturation of the attachment glycoprotein of the human respiratory syncytial virus in a cell line deficient in O glycosylation. *J. Virol.*, **63**, 4767–76.

IMMUNITY TO RESPIRATORY SYNCYTIAL VIRUS INFECTIONS

Jim Stott and Geraldine Taylor

13.1 INTRODUCTION AND HISTORICAL PERSPECTIVE

Respiratory syncytial (RS) virus was first isolated in 1956 [53] and the following year was found to be associated with infant bronchiolitis [12]. Despite almost 40 years of intensive research, no effective vaccine has been developed. Viral bronchiolitis remains the single most common cause of hospitalization of infants in the western world and a significant cause of mortality in the developing world. The costs of hospitalization in the USA alone are estimated to be 300 million dollars annually [38]. Serious lower respiratory tract disease also occurs in the elderly [29,37]. The first vaccine trials were with a parenterally administered formalin-inactivated RS virus preparation. The vaccine not only failed to protect the youngest infants from infection but appeared to enhance the severity of subsequent infection [43]. The precise roles of humoral and

cell-mediated immunity in protection and in the pathogenesis of RS virus infections are not well understood.

Work at the Common Cold Unit, Salisbury, UK in 1963 demonstrated the presence of RS virus inhibitors in calf serum and predicted the existence of bovine strains of RS virus [74,76]. Bovine RS virus was subsequently isolated by Paccaud and colleagues [62]. Another significant observation made at the Common Cold Unit during this period was the identification of an antigenically different strain of RS virus, the 8/60 virus, indicating that there were at least two different subtypes of RS virus [24]. The genetic basis for this antigenic difference was only revealed 28 years later by Wertz and colleagues [69] who demonstrated that the G-protein of 8/60 virus shared only 60% amino acid-identity with that of the prototype A2 virus. It is now known that neutralizing antibodies directed against the fusion protein are cross-reactive between group A and B strains of RS virus and only antibodies directed to the G-protein will distinguish between the two subtypes. The fact that these viruses could be distinguished by simple neutralization tests with polyclonal sera is a testimony to the meticulous care which was used in the preparation of the reagents and the conduct of the assays.

During this period, considerable advances have been made in our understanding of immunity to RS virus. In this chapter the current position of cellular and humoral immu-

Viral and Other Infections of the Human Respiratory Tract. Edited by S. Myint and D. Taylor-Robinson. Published in 1996 by Chapman & Hall. ISBN 0 412 60070 6

nity to RS virus and the role of such immunity in pathogenesis and protection are reviewed.

13.2 STRUCTURE OF THE VIRUS

The genome of RS virus consists of approximately 15 000 nucleotides in a single negative-sense strand of RNA. The single promoter transcribes 10 mRNAs, each coding for a unique protein in the order 1C, 1B, N, P, M, SH, G, F, 22 kDa and L [16]. Only two proteins, 1B and 1C, are non-structural. The fusion protein F, the attachment protein G, and the SH protein are glycosylated and together with the 22 kDa protein they are expressed on the cell membrane. The F protein is synthesized as a 68 kDa precursor molecule (F0) which requires proteolytic cleavage into disulphide-linked fragments of 48 kDa (F1) and 20 kDa (F2) in order to be biologically active and enable the fusion of virus with host-cell membranes. The G protein carries no haemagglutination or neuraminidase activity, unlike other paramyxoviruses. It has an apparent molecular weight of 90 kDa, but this is largely due to O-linked carbohydrate chains. The nucleocapsid of the virus consists of the nucleocapsid protein N, the phosphoprotein P and the large protein L, together with the viral RNA. The functions of the matrix protein M, the 22 kDa, SH, 1B and 1C proteins are not yet fully understood.

13.3 HUMORAL IMMUNITY

Almost half of all infants and over two-thirds of calves are infected with RS virus in the first year of life. Most severe disease in both species occurs in the first 6 months of life when maternally derived neutralizing antibodies are present but re-infection with RS virus occurs throughout life. Since the highest incidence of severe disease occurs at a time when maternal antibody is present it has been proposed that antibodies may not be protective and may, in fact, contribute to the pathogenesis of RS virus-induced disease. The formalin-inactivated vaccine induced high levels of non-

functional antibodies, i.e. antibodies which were poorly neutralizing and did not inhibit virus-induced cell fusion [54,55]. Such antibodies may have contributed to enhanced disease by causing a Type III (Arthus) hypersensitivity reaction in the lungs. Furthermore, there is evidence from *in vitro* studies that antibody can enhance RS virus infection in human and murine macrophages resulting in the release of leukotrienes and platelet-activating factor, which could induce bronchoconstriction [5,30,47]. However, despite the fact that high titres of neutralizing antibodies do not preclude severe infection [63], there is a correlation between the levels of serum antibodies to RS virus and protection against disease (see [44,67]). For example, an analysis of mean titres of neutralizing antibody in acute phase sera from infants less than 2 months of age showed that the mean titre of neutralizing antibodies in control infants was 1:400, compared with 1:200 for infants with RS virus bronchiolitis and 1:100 for infants with RS virus pneumonia [63]. However, a study of humoral immunity to RS virus in the elderly showed that high levels of circulating neutralizing antibodies (approximately 1:2000) are not sufficient to prevent severe respiratory disease in this age group [25]. There is evidence that the presence of mucosal antibody correlates with resistance to reinfection, reduced levels of viral replication and early cessation of virus shedding [13,36,51,52].

A greater understanding of the role of antibody in RS virus infections has been obtained by analysing the antibody response following RS virus infection, examining the effects of passively transferred polyclonal and monoclonal antibodies (MAbs) on RS virus infection and examining the effects of suppression of the antibody response. The serum antibody response after primary RS virus infection is poor [63] and antibody declines to low or undetectable levels within a year [82]. However, higher antibody levels are produced and persist for longer after a second infection.

Thus, there is a pattern of increasing levels of serum antibody which parallels an accumulative acquisition of resistance to severe disease. In infants the antibody responses to the F and G proteins belong to the IgG1 and IgG3 subclasses, whereas in adults the responses to F and G are IgG1 and IgG2 [77,78]. The IgG1:IgG2 ratio to F was four-fold higher than that to the G protein after each of three successive RS virus infections [79]. Although an IgG2 response is characteristic of one to polysaccharide antigens, the immune response to G is different from that to other polysaccharide antigens in that the ratio of IgG1:IgG2 remained constant with age whereas the IgG2 response to a polysaccharide antigen increases with age. The local antibody response after RS virus infection is composed of IgA, IgG and IgM [49,50] and high levels of neutralizing antibody in nasal secretions prevent extensive infection in the upper respiratory tract [51]. However, immunity to re-infection of the nasal passages is less durable than in the lung.

Passive transfer of convalescent sera or human immune globulin, containing high levels of neutralizing antibody to RS virus, to cotton rats, ferrets and owl monkeys protected the lower, but not the upper respiratory tract against RS virus infection [39]. The protective levels of neutralizing antibodies in passively transferred sera were 1:300 to 1:400, which were similar to levels found to be protective in infants [63]. Passive transfer of immune sera did not enhance the severity of disease in any of these studies. Furthermore, cotton rats given immune sera from cotton rats immunized with formalin-inactivated RS virus, failed to develop the enhanced pulmonary pathology observed in vaccinated animals following RS virus challenge [18]. The results of these studies indicate that the phenomenon of enhanced disease is not mediated by serum antibodies.

Studies with monoclonal antibodies (MAbs) to RS virus have also shown that antibody can protect against RS virus infection in mice, cotton rats and calves [42,70,72,81]. Protective MAbs were specific for the F and G proteins of RS virus and were not only able to prevent RS virus infection when administered 24 hours before virus challenge but were also able to clear an established RS virus infection [70]. A comparison of a neutralizing MAb specific for the F protein with a neutralizing MAb specific for the G protein showed that the antibody to the F protein was more effective in clearing RS virus from the lungs of persistently infected nude mice than the antibody to the G protein [70]. Studies in calves, a natural host for RS virus infection, confirmed the findings from experiments with small laboratory animals, namely that antibody did not enhance pneumonic lesions. Thus, passive transfer of protective bovine MAbs to calves before infection with bovine RS virus suppressed the development of lung lesions, and a non-protective, complement-fixing bovine MAb had no effect on the extent of pneumonic lesions [70].

A comparison of the biological properties of MAbs to the F and G proteins has provided information on the possible mechanisms of action of these protective antibodies. The majority of MAbs to the G protein are non-neutralizing, so MAbs to the G protein are able to protect by a mechanism independent of virus neutralization. Furthermore, several protective MAbs to the G protein did not lyse virus-infected cells in the presence of complement. Thus, the mechanism of action of these antibodies may involve cells with Fc receptors [70]. The MAbs to the F protein that were most effective at either preventing or clearing an established RS virus infection were neutralizing and inhibited the fusion of virus-infected cells [71]. MAbs that were neutralizing but did not inhibit fusion were less protective than those with fusion-inhibiting properties and some highly protective antibodies did not lyse virus-infected cells in the presence of complement. The finding that Fab fragments produced by papain digestion of two fusion-inhibiting MAbs were able to protect mice against RS virus infection and clear an established infection when given by the intranasal route [70], indicates that these antibodies mediate their protective effects by a mechanism that is independent of secondary effector mechanisms. Similar findings were reported by Prince *et al.* [65] who demonstrated that F(ab')₂ fragments prepared from human IgG cleared RS virus from the lungs of complement-depleted cotton rats to a similar extent as did intact IgG. In addition, a neutralizing, fusion-inhibiting human Fab fragment produced from a combinatorial antibody phage display library was also effective in clearing RS virus infection from the lungs of mice [22]. Since RS virus can spread by means of cell fusion it is not clear how intracellular virus can be eliminated following the binding of Fab fragments to the F protein expressed on the surface of the cells. However, studies of rabies virus have demonstrated that antibody-mediated inhibition of viral gene transcription may result from endocytosis of antibody bound to virus-infected cells [23].

The antiviral effects of neutralizing antibody observed in a variety of animals challenged with RS virus have stimulated clinical trials of the possible prophylactic and therapeutic effects of human immune globulin (IVIG) on RS virus infections in children. A preparation of IVIG containing high titres of RS virus-neutralizing antibodies was given intravenously to infants hospitalized for RS virus lower respiratory tract disease. A dose of 2 g/kg body weight resulted in significant reductions in virus shedding and improvement in oxygenation when compared with placebo controls [40]. In another clinical trial, 750 mg/kg of IVIG administered intravenously at monthly intervals, was shown to be highly effective in preventing serious RS virus lower respiratory tract illness in high-risk infants and children [35]. In the light of these findings there is considerable interest in using human MAbs for the prophylaxis and therapy of RS virus infections in children and immunosup-

pressed individuals. One of the major advantages of MAbs is that higher concentrations of specific antibody can be obtained than with polyclonal sera and this should reduce the amount of immunoglobulin required to protect against infection to levels that can be administered intramuscularly. A fusion-inhibiting human MAb that is effective in preventing and clearing RS virus in mice, and which recognizes all clinical isolates of RS virus examined so far, has been produced by CDR grafting [75]. The dose of this human MAb required to prevent or clear RS virus infection in mice is 2–5 mg/kg when given parenterally [75]. This is 10–100 times less than the dose of IVIG required to prevent RS virus infection in cotton rats and 4000 times less than the dose required to clear an established infection. A dose of 1 mg/kg of a fusion-inhibiting human Fab fragment cleared RS virus from the lungs of mice infected 3 days previously, when administered intranasally [22]. However, the therapeutic effect was not sustained, and RS virus re-appeared in the lungs of treated mice 2 days after treatment. This rebound was prevented by daily administration of the MAb for 3 days. Although Fab fragments may offer some advantages for the treatment of RS virus infections, there has been no detectable rebound of RS virus in the lungs of persistently infected nude mice after treatment with intact IgG administered either parenterally or intranasally, even when antibody had declined to undetectable levels [70].

The role of antibody in RS virus infections has also been studied in mice depleted of B-cells by treatment with anti-μ [31]. The results of these studies showed that antibody was not required to terminate RS virus replication after a primary infection, but without antibody only partial immunity against re-infection was induced. Passive transfer of RS virus-specific immune serum to anti-μ-treated mice before re-challenge reconstituted complete protection against RS virus infection. Studies in mice vaccinated with recombinant vaccinia viruses expressing the F or G proteins of RS virus and depleted of CD4+ and CD8+ T-cells at the time of RS virus challenge, demonstrated that antibodies are sufficient to mediate the resistance induced by these recombinants [19]. Further, recombinant vaccinia viruses expressing mutant forms of the F protein, in which the proteolytic processing and cell surface transport of the F0 precursor is arrested, induced CD8+ cytotoxic T-lymphocytes (CTLs) in mice but failed to induce RS virus-neutralizing antibodies and failed to protect against RS virus infection when animals were challenged 3 weeks after vaccination [27]. These studies highlight the need for an effective RS virus vaccine to stimulate antibodies.

Despite the wealth of evidence that RS virus-neutralizing antibodies are protective, there remains the dilemma that severe disease can occur in infants and in the elderly despite substantial levels of neutralizing antibody. It may be that the level of fusion-inhibiting antibody is a more accurate reflection of protection than neutralizing antibody, or that the titre of virus that can be neutralized is more important than the titre of neutralizing antibody. Alternatively, the protective capacity of antibodies may be reduced by the presence of competing non-protective antibodies. Non-protective, non-neutralizing and non-fusion-inhibiting MAbs have been identified which map to the same region of the F protein as highly protective, fusion-inhibiting MAbs [71]. Susceptibility to RS virus infection could be related to antigenic differences in the F and G proteins. However, highly protective MAbs to the F protein appear to be cross-reactive among both human and bovine strains of RS virus [72]. Local defence mechanisms may also be critical, since RS virus in the upper respiratory tract is not readily accessible to protective antibodies in the circulation.

13.4 CELLULAR IMMUNITY

The role of antibody in immunity to RS virus infection has been studied extensively and the overwhelming weight of evidence clearly indicates a beneficial role for antibody in both prophylaxis and therapy. In contrast, relatively little attention has been paid to cell-mediated immunity in RS virus infection and disease. Furthermore, the studies which have attempted to assign a protective or pathogenic role to immune cells have yielded equivocal results. Because of the technical difficulties of studying cellular immunity in children and cattle, the primary natural hosts for RS virus, the most extensive studies have been done in a murine model of RS virus infection. These murine studies will be reviewed first and subsequently their relevance to pathogenesis and protection in humans and cattle will be considered.

13.4.1 MURINE STUDIES

Virus-specific CTLs are induced after RS virus infection of the respiratory tract of mice [6,73] and after intraperitoneal inoculation of mice with live RS virus [8]. Cytotoxic T-cells specific for RS virus have also been induced by intraperitoneal inoculation of recombinant vaccinia virus coding for F, N or 22 kDa proteins and these studies showed that intranasal inoculation of mice with recombinant vaccinia virus was less effective than intraperitoneal inoculation for inducing memory CTLs in the spleen [56,57,64]. Interestingly, RS virus inactivated with ultraviolet (UV) light and given intraperitoneally also primed mice for the development of CTLs [8]. RS virus inactivated by UV light can restimulate spleen cells *in vitro* to become cytotoxic and to form a target for the cytotoxic cells. In this respect, RS virus resembles other fusogenic viruses such as Sendai, which can induce CTLs in the absence of viral protein synthesis because it fuses directly with the plasma membrane [46]. Another example of the induction of CTLs by inactivated RS virus antigen was reported by Bangham and Askonas [9]. RS virus-infected mouse cells which had been fixed with glutaraldehyde were inoculated intraperitoneally on two occasions and found to prime mice for memory CTLs. This property may also explain the effectiveness of an inactivated bovine RS virus vaccine composed of virus-infected bovine mucosal cells fixed with glutaraldehyde and inoculated subcutaneously. This vaccine was found to induce more effective immunity than either of two live virus vaccines [68].

Recombinant vaccinia viruses expressing the proteins of RS virus have been used extensively not only to immunize and prime mice for CTL memory but also to generate target cells for CTL assays. Initial studies identified the N protein as one of the target antigens for CTLs induced by RS virus infection in BALB/c mice [64], whereas the G protein was not recognized [11]. Further studies in 1990, showed that the major target for CTLs from RSV-infected BALB/c mice was the 22 kDa protein and identified that the recognition of this protein was restricted by MHC class 1 molecules Kd (but not Dd). This work also showed that recognition of the F and N proteins was comparatively weak and there was little or no detectable recognition of G, P, SH or M proteins [56,60]. Although the CTLs induced by RS virus infection in mice recognized both A and B subtypes of the virus [9], CTLs induced by recombinant vaccinia virus expressing the F protein showed poor cross-reactivity [64].

Recombinant vaccinia virus expressing the F protein appears to induce higher levels of memory CTLs in the mouse than do recombinant vaccinia viruses expressing the N protein [64] or the 22 kDa protein in some studies [60]. In contrast, in other studies, recombinants expressing the 22 kDa protein induced high levels of CTL [56]. Recombinant vaccinia viruses expressing M or SH proteins did not induce significant levels of memory CTLs [64]. The CTLs described in these studies were CD8+; however, repeated *in vitro* re-stimulation of murine CTLs with RS virus resulted in the outgrowth of CD4+ CTLs which eventually became the sole subset in culture. These CD4+ CTLs were highly cytotoxic for RS virus-infected cells, but did not recognize target cells infected with any recombinant vaccinia viruses expressing RS virus structural proteins.

Further studies with recombinant vaccinia viruses expressing RS virus proteins showed that different individual antigens of RS virus stimulate different and quite distinct cellular responses in BALB/c mice [2]. Careful analysis of spleen cell cultures from mice primed with a variety of recombinant vaccinia viruses expressing specific RS virus genes, and restimulated *in vitro* with RS virus, showed that cell cultures gave rise to CD3+, TcR α/β+ T-cell lines with distinct characteristics. Those from mice primed with the F protein contained a mixture of CD8+ and CD4+ T-cells, whereas those specific to G, N and P were mostly CD4+. In contrast, 22 kDa-specific lines were mostly CD8+. The F and 22 kDa-specific lines displayed class I-restricted cytotoxic activity against RS virus infected targets but the lines from mice primed with G, N or P proteins did not. The F-specific lines contained T-helper cells that released high levels of IL-2 and some IL-3, but little IL-4 or IL-5, indicating a Th1 pattern. The G-specific line in contrast released IL-3, 4 and 5, but little IL-2, indicating a Th2 pattern. The 22 kDa line released IL-3. The staining of T-cell subsets for CD45 RB was consistent with the cytokine secretion profiles of the T-helper cells they contained [3]. These patterns of virus-specific immunity appear to be relevant to the pathogenicity of RS virus infection and vaccination.

The role of immune lymphocytes in pathogenesis and protection during RS virus infection has been studied extensively in mice. In normal BALB/c mice only 5% of cells recovered by broncheoalveolar lavage (BAL) are lymphocytes [58]. The majority of cells are macrophages. Following intranasal infection, the proportion of lymphocytes increased to a maximum of about 20% between days 10 and 16. In contrast, histological changes in the lung were maximal at day 7 and had resolved by day 10. Analysis of the lymphocytes recovered from the lungs by flow cytometry revealed that less than 5% were B-cells (surface immunoglobulin positive) and most carried T-cell markers. Of these T-cells, the CD8+ sub-population out-numbered CD4+ cells by 2:1 or 4:1 in the lungs of mice recovering from RS virus infection [58]. Further analysis of lymphocytes in the BAL of RS virus-infected mice also revealed that both the CD4+ CD8– and CD4– CD8+ subpopulations stained uniformly for CD3 and α/β T-cell receptor (TcR) while none stained for TcR γ/δ [59]. These results conflicted with the widely held view that lymphocytes expressing TcR γ/δ predominate in epithelial immune responses to viral infections. A similar increase in the proportion of CD8+ T-cells in BAL and in lymphocytes extracted from lung tissue of BALB/c mice was reported by Kimpen *et al.* [45].

Having identified the T-cell subsets present in the lungs of RS virus-infected mice, this experimental model has also been used to identify the functional significance of these cells during infection or re-infection. Two approaches were used. First, CD4+ or CD8+ lymphocytes (or both) were depleted by injections of specific monoclonal antibodies [32]. These studies showed that both CD4+ and CD8+ lymphocytes were involved in controlling RS virus replication after primary infection. Depletion of either CD4+ or CD8+ T-cells alone had no significant effect on virus infection but when both T-cell subsets were depleted, virus replication was prolonged. Furthermore, mice depleted of both T-cell subsets did not develop signs of illness, indicating that the host immune response, rather than a virus cytocidal effect, was the basic cause of disease in mice. Both CD4+ and CD8+ cells contributed to disease, although CD8+ lymphocytes were dominant. The CD4+ lymphocytes were required for the development of peribroncheovascular lymphocytic aggregates seen after re-challenge of mice and CD8+ T-cells were responsible for alveolar infiltration. The presence of alveolar lymphocytes corre-

lated with illness. Further evidence for the detrimental effect of CD4+ T-cells was obtained by the demonstration that the enhanced disease induced by formalin-inactivated RS virus vaccine could be abrogated by the depletion of CD4+ cells in mice [20]. These cells appear to mediate their effects via IL-4 and IL-10 since depletion of IL-4 and IL-10 immediately before RS virus challenge of mice immunized with formalin-inactivated RS virus completely abrogated the pulmonary histopathological changes [21].

In the second approach, the effect of RS virus-specific T-cell lines, derived from the spleens of mice infected intranasally with RS virus, on RS virus infection of mice was studied [1]. The lines were expanded *in vitro* and separated into CD4+ and CD8+ T-cell-enriched fractions. The cells were then transferred passively into RS virus-infected mice. The most severe immunopathological changes were seen in mice receiving CD4+ cells. However, transfer of CD4+, CD8+ or both cell fractions caused illness and weight loss in RS virus-infected mice. Both cell lines increased the severity of lung pathology with the appearance of haemorrhage and polymorphonuclear cell efflux. In addition, recipients of CD4+ cells developed marked pulmonary eosinophilia. Both T-cell populations decreased the titre of virus in the lungs, but fewer CD4+ than CD8+ cells were required to achieve this effect. The unseparated T-cell line and CD4– cell fraction secreted IL-3, IL-4 and IL-5. High levels of IL-2 were produced only by the unseparated T-cell line and the CD8+ cell fraction secreted only IL-3. In contrast to the first approach, the cell transfer studies showed that CD4 cells were more antiviral and more immunopathogenic than CD8+ cells in RS virus-infected mice [3]. Nevertheless, these studies clearly show that although immune T-cells clear virus, they also augment disease.

Having identified the T-cell subsets responsible for virus clearance and disease, attention then centred on the antigenic specificity of the sensitized T-cells. Mice were immunized by scarification with recombinant vaccinia viruses expressing individual RS virus proteins and then infected intranasally with RS virus; their pulmonary response monitored by analysis of BAL cells [61]. In mice vaccinated with recombinants expressing the G protein, 14–25% of the lavage cells were eosinophils. These comprised less than 3% of lavage cells from mice vaccinated with recombinants expressing either the F, N or SH proteins. Lung haemorrhage developed in mice sensitized to G or F proteins and a pulmonary neutrophil efflux was evident in mice sensitized to G, F or N protein. Most T-cells were CD3+, CD4+, $\alpha/\beta+$, γ/δ + or CD3+, CD8+, $\alpha/\beta+$, $\gamma/\delta-$. Staining for CD45 RB declined rapidly after infection with RS virus on both CD4+ and CD8+ cells. The rate of loss of CD45 RB on CD4 T-cells was accelerated by prior vaccination with G protein, consistent with the conversion to helper T-cell subsets producing eosinophil-promoting cytokines. Thus, the eosinophilic reaction to RS virus infection specifically reflects sensitization to the G protein [61].

As described above, individual RS virus proteins have been shown to generate T-cell lines with quite specific phenotypes and functional characteristics [3]. The *in vivo* function of the T-cell subsets was studied by transfer of the cell lines into mice infected intranasally with RS virus [4]. BALB/c mice showed a mild illness after RS virus infection and recovered fully, but after intravenous injections of T-cells the mice developed respiratory distress. Mice receiving G-specific T-cells, with a Th2 pattern of cytokines, suffered the most severe, sometimes fatal illness characterized by lung haemorrhage, pulmonary neutrophil recruitment (shock lung) and intense pulmonary eosinophilia. This disease was enhanced by the co-injection of 22 kDa-specific CD8+ CTLs which alone caused only mild shock lung without eosinophilia. Cells (CD8+ and CD4+) specific for the F protein caused minimal enhancement of pathology and had little effect on the disease caused by G-specific cells. Each of these three cell lines reduced the titre of virus in the lungs and combined injections of T-cells specific for G and 22 kDa proteins eliminated infection completely [4]. These differing characteristics of antiviral T-cell lines indicate their possible role in protection and pathogenesis of RS virus disease following infection or vaccination. Although these studies show that passive transfer of activated CTLs can protect against RS virus infection in mice, there is evidence that vaccinia recombinants that can prime memory CTLs in mice only induce transient resistance following vaccination. Thus, resistance to RS virus induced by immunization with recombinants expressing the 22 kDa, N or certain mutant forms of the F protein appears to be mainly mediated by primary CTLs [17,48; G. Taylor, unpublished observations]. Activation of memory CTLs in mice does not appear rapidly enough to mediate virus clearance from the lungs.

13.4.2 STUDIES IN CATTLE AND HUMANS

A key question raised by these extensive studies in mice is how far they are applicable to the natural infection in humans and cattle. Although previously infected BALB/c mice are resistant to re-infection with RS virus, the lungs develop a more extensive lymphocytic infiltration than after primary infection [33,45], whereas re-infection of humans results in an accumulative acquisition of resistance to severe lung disease. Furthermore, RS virus disease in immunosuppressed humans and calves is more extensive, whereas that in mice depleted of CD4+ and CD8+ is less severe, than in immunologically normal animals [26,32; G. Taylor *et al.*, in preparation]. Cytotoxic T-cells specific for RS virus have been detected in the peripheral blood in adults [10]. The effector cells were shown to be CD8+ and HLA class I restricted. They were shown to have activity against the N protein but not the G protein [11]. Although, in mice, passively acquired antibodies to RS virus impair the efficiency of priming for CTL memory [7], there is no evidence that maternal antibody suppresses cytotoxic T-cells in infants with RS virus acute bronchiolitis [41]. More recent studies on human cytotoxic T-cells in adults indicate that the six RS virus proteins most strongly recognized were the N protein, the surface proteins SH and F, the matrix protein M and 22 kDa and the non-structural protein 1B. There was no evidence of significant recognition of the major surface glycoprotein G, the internal phosphoprotein P or the non-structural protein 1C [14].

Cattle studies

Recent studies in cattle indicate that CD8+ lymphocytes are important in the clearance of RS virus infection from the bovine respiratory tract. Depletion of these cells results in prolonged and enhanced virus replication and enhanced macroscopic and microscopic pathological changes. The latter are characterized by protracted epithelial proliferation which appears to be a direct result of persistent virus replication. Although depletion of CD4+ lymphocytes also results in enhanced pneumonic lesions, the histological lesions appear to be the result of desquamated and necrotic epithelial cells and a denuded respiratory mucosa (G. Taylor *et al.*, in preparation; L.H. Thomas *et al.*, in preparation). An increase in the proportion of CD8+ T-cells has been demonstrated in the lungs and trachea of calves infected 10 days previously with bovine RS virus (L.H. Thomas *et al.*, in preparation; E. McInness *et al.*, in preparation). Further, CTLs which lyse RS virus-infected targets have been demonstrated in the peripheral blood and lungs of calves infected with bovine RS virus (R. Gaddum *et al.*, in preparation). The effector cells were CD8+ and BoLA restricted.

Human studies

Eosinophil degranulation products have been detected in nasal secretions from infants with RS virus bronchiolitis [28]. The key eosinophil-promoting cytokine is IL-5; this is produced by Th2 cells, which may be activated by the G protein of RS virus. Eosinophil infiltration was observed post mortem in the lungs of RS virus-infected children previously vaccinated with a formalin-inactivated alum-precipitated RS virus vaccine [43] and recipients of this vaccine had an increase in peripheral blood eosinophils at the time of subsequent RS virus infection [15]. It has been demonstrated that BALB/c mice inoculated with RS virus inactivated with formalin or heat, developed a Th2-like pattern of cytokines in the lungs when challenged intranasally with RS virus 4 weeks later, whereas inoculation with live RS virus induced a Th1-like pattern of cytokines [34]. Furthermore, there is evidence that alum, which was used in the formalin-inactivated vaccine given to children, is a good adjuvant for Th2 responses and attracts eosinophils to the injection site [66,80]. These observations, together with the finding that IL-4 and IL-10 were responsible for the inflammatory response in the lungs of mice given formalin inactivated RS virus, suggest that the inactivated vaccine, which failed to induce protective antibodies, induced a predominantly Th2 response in children. This would have resulted in release of IL-4, IL-5 and IL-10 at the sites of RS virus infection in the bronchioles and alveoli causing inflammation and bronchoconstriction. These findings have important implications for vaccine design and suggest that induction of a RS virus-specific Th2 response should be avoided. A knowledge of the pattern of cytokine expression in humans and cattle after RS virus infection and vaccination will provide important information on whether the selective activation of T-cell subsets with different cytokine profiles can influence the pathogenesis of RS virus disease.

13.5 REFERENCES

1. Alwan, W.H., Record, F.M. and Openshaw, P.J.M. (1992) CD4+ T-cells clear virus but aug-

ment disease in mice infected with respiratory syncytial virus. Comparison with effects of CD8+ T cells. *Clin. Exp. Immunol.*, **88**, 527–36.

2. Alwan, W.H. and Openshaw, P.J.M. (1993) Distinct patterns of T and B cell immunity to respiratory syncytial virus induced by individual viral proteins. *Vaccine*, **11**, 431–7.

3. Alwan, W.H., Record, F.M. and Openshaw, P.J.M. (1993) Phenotypic and functional characterisation of T cell lines specific for individual respiratory syncytial virus proteins. *J. Immunol.*, **150**, 5211–18.

4. Alwan, W.H., Kozlowska, W.J. and Openshaw, P.J.M. (1994) Distinct types of lung disease caused by functional sets of antiviral T cell. *J. Exp. Med.*, **179**, 81–9.

5. Ananaba, G.A. and Anderson, L.J. (1991) Antibody enhancement of respiratory syncytial virus stimulation of leukotriene production of a macrophage cell line. *J. Virol.*, **65**, 5054–60.

6. Anderson, J.J., Norden, J., Saunders, D., Toms, G.L. and Scott, R. (1990) Analysis of the local and systemic immune responses induced in BALB/c mice by experimental respiratory syncytial virus infection. *J. Gen. Virol.*, **71**, 1561–70.

7. Bangham, C.R.M. (1986) Passively acquired antibodies to respiratory syncytial virus impair the secondary cytotoxic T-cell response in the neonatal mouse. *Immunology*, **59**, 37–41.

8. Bangham, C.R.M., Cannon, M.J., Karzon, D.T. and Askonas, B.A. (1985) Cytotoxic T cell response to respiratory syncytial virus in mice. *J. Virol.*, **56**, 55–9.

9. Bangham, C.R.M. and Askonas, B.A. (1986) Murine cytotoxic T cells specific to respiratory syncytial virus recognise different antigenic subtypes of the virus. *J. Gen. Virol.*, **67**, 623–9.

10. Bangham, C.R.M. and McMichael, A.J. (1986) Specific human cytotoxic T cells recognize B-cell lines persistently infected with respiratory syncytial virus. *Proc. Natl Acad. Sci. USA*, **83**, 9183–7.

11. Bangham, C.R.M., Openshaw, P.J.M., Ball, L.A., King, A.M.Q., Wertz, G.W. and Askonas, B.A. (1986) Human and murine cytotoxic T cells specific to respiratory syncytial virus recognize the viral nucleoprotein (N) but not the major glycoprotein (G), expressed by vaccinia virus recombinants. *J. Immunol.*, **137**, 3937–77.

12. Chanock, R.M., Roizman, B. and Myers, R. (1957) Recovery from infants with respiratory illness of a virus related to chimpanzee coryza agent (CCA). I. Isolation, properties and characterization. *Am. J. Hyg.*, **66**, 281–90.

13. Chanock, R.M., Kapikian, A.Z., Mills, J., Kim, H.W. and Parrott, R.H. (1970) Influence of immunological factors in respiratory syncytial virus disease. *Arch. Environ. Health*, **21**, 347–55.

14. Cherrie, A.H., Anderson, K., Wertz, G.W. and Openshaw, P.J.M. (1992) Human cytotoxic T cells stimulated by antigen on dendritic cells recognise the N, SH, F, M, 22K and 1B proteins of respiratory syncytial virus. *J. Virol.*, **66**, 2102–10.

15. Chin, J., Magoffin, R.L., Shearer, L.A., Schieble, J.H. and Lennette, E.H. (1969) Field evaluation of a respiratory syncytial virus vaccine and a trivalent parainfluenza virus vaccine in a pediatric population. *Am. J. Epidemiol.*, **89**, 449–63.

16. Collins, P.L. and Wertz, G. (1986) Human respiratory syncytial virus genome and gene products,) in *Concepts in Viral Pathogenesis*, Vol. 2, (eds A.L. Notkins and M.B.A. Oldstone), Springer, Berlin, pp. 40–6.

17. Connors, M., Collins, P.L., Firestone, C.-Y. and Murphy, B.R. (1991) Respiratory syncytial virus (RSV) F, G, M2, 22Kd, and N proteins each induce resistance to RSV challenge, but resistance induced by M2 and N proteins is relatively short-lived. *J. Virol.*, **65**, 1634–7.

18. Connors, M., Collins, P.L., Firestone, C.-Y., Sotnikov, A.V. and Murphy, B.R. (1992) Cotton rats previously immunized with a chimeric RSV FG glycoprotein develop enhanced pulmonary pathology with RSV, a phenomenon not encountered during immunisation with vaccinia-RSV recombinants or RSV. *Vaccine*, **10**, 475–84.

19. Connors, M., Kulkarni, A.B., Collins, P.L. *et al.* (1992) Resistance to respiratory syncytial virus (RSV) challenge induced by infection with a vaccinia virus recombinant expressing the RSV M2 protein (Vac-M2) is mediated by CD8+ T cells, while that induced by Vac-F of Vac-G recombinants is mediated by antibodies. *J. Virol.*, **66**, 1277–81.

20. Connors, M., Kulkarni, A.B., Firestone, C-Y. *et al.* (1992) Pulmonary histopathology induced by RSV challenge of formalin inactivated RSV-immunized BALB/c mice is abrogated by depletion of CD4+ T cells. *J. Virol.*, **66**, 7444–51.

21. Connors, M., Giese, N.A., Kulkarni, A.B., Firestone, C.-Y., Morse, H.C. and Murphy, B.R. (1994) Enhanced pulmonary histopathology

induced by respiratory syncytial virus (RSV) challenge of formalin-inactivated RSV-immunized BALB/c mice is abrogated by depletion of interleukin 4 (IL-4) and IL-10. *J. Virol.*, **68**, 5321–5.

22. Crowe, J.E., Murphy, B.R., Chanock, R.M., Williamson, R.A., Barbas, C.F. and Burton, D.R. (1994) Recombinant human respiratory syncytial virus (RSV) monoclonal antibody Fab is effective therapeutically when introduced into the lungs of RSV-infected mice. *Proc. Natl Acad. Sci. USA*, **91**, 1386–90.

23. Dietzschold, B., Hardwick, J.M., Trapp, B.D. *et al.* (1992) Delineation of putative mechanisms involved in antibody-mediated clearance of rabies virus from the central nervous system. *Proc. Natl Acad. Sci. USA*, **89**, 7252–6.

24. Doggett, J. and Taylor-Robinson, D. (1965) Serological studies with respiratory syncytial virus. *Arch. ges. Virusforsch.*, **15**, 601–8.

25. Falsey, A.R. and Walsh, E.E. (1992) Humoral immunity to respiratory syncytial virus infection in the elderly. *J. Med. Virol.*, **36**, 39–43.

26. Fishaut, M., Tubergen, D. and McIntosh, K. (1980) Cellular response to respiratory viruses with particular reference to children with disorders of cell-mediated immunity. *J. Pediatr.*, **96**, 179–86.

27. Gaddum, R., Taylor, G. and Melero, J. (1993) A vaccinia recombinant expressing a mutant RSV F protein does not protect against RSV despite generating a CTL response. Abstracts IXth International Congress of Virology, Glasgow, A237.

28. Garofalo, R., Kimpen, J.L.L., Welliver, R.C. and Ogra, P.L. (1992) Eosinophil degranulation in the respiratory tract during naturally acquired respiratory syncytial virus infection. *J. Pediatr.*, **120**, 28–32.

29. Garvie, D.G. and Gray, J. (1980) An outbreak of respiratory syncytial virus infection in the elderly. *Br. Med. J.*, **281**, 1253–4.

30. Giminez, H.B., Howland, W.C. and Spiegelberg, H.L. (1989) In vitro enhancement of respiratory syncytial virus infection of U937 cells by human sera. *J. Gen. Virol.*, **70**, 89–96.

31. Graham, B.S., Bunton, L.A., Rowland, J., Wright, P.F. and Karzon, D.T. (1991) Respiratory syncytial virus infection in anti-μ-treated mice. *J. Virol.*, **65**, 4936–42.

32. Graham, B.S., Bunton, L.A., Rowland, J., Wright, P.F. and Karzon, D.T. (1991) Role of T-

lymphocyte subsets in the pathogenesis of primary infection and rechallenge with respiratory syncytial virus in mice. *J. Clin. Invest.*, **88**, 1026–33.

33. Graham, B.S., Bunton, L.A., Wright, P.F. and Karzon, D.T. (1991) Reinfection of mice with respiratory syncytial virus. *J. Med. Virol.*, **34**, 7–13.

34. Graham, B.S., Henderson, G.S., Tang, Y.W., Lu, X., Neuzil, K.M. and Colley, D.G. (1993) Priming immunization determines T helper cytokine mRNA patterns in lungs of mice challenged with respiratory syncytial virus. *J. Immunol.*, **151**, 2032–40.

35. Groothuis, J.R., Simoes, E.A.F, Levin, M.J. *et al.* (1993) Prophylactic administration of respiratory syncytial virus immune globulin to high-risk infants and young children. *N. Engl. J. Med.*, **329**, 1524–30.

36. Hall, C.B., Walsh, E.E., Long, C.E. and Schnabel, K.C. (1991) Immunity and frequency of reinfection with respiratory syncytial virus. *J. Infect. Dis.*, **163**, 693–8.

37. Hart, R.J. (1984) An outbreak of respiratory syncytial virus in an old people's home. *J. Infect.*, **8**, 259–61.

38. Heilman, C.A. (1990) Respiratory syncytial and parainfluenza viruses. *J. Infect. Dis.*, **161**, 402–6.

39. Hemming, V.G. and Prince, G.A. (1992) Respiratory syncytial virus: babies and antibodies. *Infect. Agents Dis.*, **1**, 24–32.

40. Hemming, V.G., Rodriguez, W., Kim, H.W. *et al.* (1987) Intravenous immunoglobulin treatment of respiratory syncytial virus infections in infants and young children. *Antimicrob. Agents Chemother.*, **31**, 1882–6.

41. Isaacs, D., Bangham, C.R.M. and McMichael, A.J. (1987) Cell-mediated cytotoxic response to respiratory syncytial virus in infants with bronchiolitis. *Lancet*, **ii**, 769–71.

42. Kennedy, H.E., Jones, B.V., Tucker, E.M. *et al.* (1988) Production and characterization of bovine monoclonal antibodies to respiratory syncytial virus. *J. Gen. Virol.*, **69**, 3023–32.

43. Kim, H.W., Canchola, J.G., Brandt, C.D. *et al.* (1969) Respiratory syncytial virus disease in infants despite prior administration of antigenic inactivated vaccine. *Am. J. Epidemiol.*, **89**, 422–33.

44. Kimman, T.G., Zimmer, G.M., Westenbrink, F., Mars, J. and van Leeuwen, E. (1988) Epidemiological study of bovine respiratory

syncytial virus infections in calves: influence of maternal antibodies on the outcome of disease. *Vet. Rec.*, **124**, 104–9.

45. Kimpen, J.L.L., Rich, G.A., Mohar, C.K. and Ogra, P.L. (1992) Mucosal T-cell distribution kinetics during infection with respiratory syncytial virus. *J. Med. Virol.*, **36**, 172–9.

46. Koszinowski, U., Gething, M.J. and Waterfield, M. (1977) T-cell cytotoxicity in the absence of viral protein synthesis in target cells. *Nature*, **267**, 160–3.

47. Krilov, L.R., Anderson, L.J., Marcoux, L., Bonagura, V.R. and Wedgewood, J.F. (1989) Antibody-mediated enhancement of respiratory syncytial virus infection in two monocyte/macrophage cell lines. *J. Infect. Dis.*, **160**, 777–82.

48. Kulkarni, A.B., Connors, M., Firestone, C.-Y., Morse, H.C. and Murphy, B.R. (1993) The cytolytic activity of pulmonary CD8+ lymphocytes, induced by infection with a vaccinia virus recombinant expressing the M2 protein of respiratory syncytial virus (RSV), correlates with resistance to RSV infection in mice. *J. Virol.*, **67**, 1044–9.

49. McIntosh, K., Masters, H.B., Orr, I., Chao, R.K. and Barkin, R.M. (1978) The immunologic response to infection with respiratory syncytial virus in infants. *J. Infect. Dis.*, **138**, 24–32.

50. McIntosh, K., McQuillin, J. and Gardner, P.S. (1979) Cell-free and cell-bound antibody in nasal secretions from infants with respiratory syncytial virus infection. *Infect. Immun.*, **23**, 276–81.

51. Mills, J., Van Kirk, J.E., Wright, P.F. and Chanock, R.M. (1971) Experimental respiratory syncytial virus infection of adults. *J. Immunol.*, **107**, 123–30.

52. Mohanty, S.B., Lillie, M.G. and Ingling, A.L. (1976) Effect of serum and nasal neutralising antibodies on bovine respiratory syncytial virus infection in calves. *J. Infect. Dis.*, **134**, 409–13.

53. Morris, J.A., Blount, R.E. and Savage, R.E. (1956) Recovery of cytopathic agent from chimpanzees with coryza. *Proc. Soc. Exp. Biol. Med.*, **92**, 544–9.

54. Murphy, B.R., Prince, G.A. and Walsh, E.E. (1986) Dissociation between serum neutralising and glycoprotein antibody responses of infants and children who received respiratory syncytial virus vaccine. *J. Clin. Microbiol.*, **24**, 197–202.

55. Murphy, B.R. and Walsh, E.E. (1988) Formalin-inactivated respiratory syncytial virus vaccine induces antibodies to the fusion glycoprotein that are deficient in fusion-inhibiting activity. *J. Clin. Microbiol.*, **26**, 1595–7.

56. Nicholas, J.A., Rubino, K.L., Levely, M.E., Adams, E.G. and Collins, P.L. (1990) Cytolytic T-lymphocyte responses to respiratory syncytial virus: effector cell phenotype and target proteins. *J. Virol.*, **64**, 4232–41.

57. Nicholas, J.A., Rubino, K.L., Levely, M.E., Meyer, A.L. and Collins, P.L. (1991) Cytotoxic T cell activity against the 22-kDa protein of human respiratory syncytial virus (RSV) is associated with a significant reduction in pulmonary RSV replication. *Virology*, **182**, 664–72.

58. Openshaw, P.J.M. (1989) Flow cytometric analysis of pulmonary lymphocytes from mice infected with respiratory syncytial virus. *Clin. Exp. Immunol.*, **75**, 324–8.

59. Openshaw, P.J.M. (1991) Pulmonary epithelial T-cells induced by viral infection express T-cell receptors α/β. *Eur. J. Immunol.*, **21**, 803–6.

60. Openshaw, P.J.M., Anderson, K., Wertz, G.W. and Askonas, B.A. (1990) The 22,000-kilodalton protein of respiratory syncytial virus is a major target for Kd-restricted cytotoxic T lymphocytes from mice primed by infection. *J. Virol.*, **64**, 1683–9.

61. Openshaw, P.J.M., Clarke, S.L. and Record, F.M. (1992) Pulmonary eosinophilic response to respiratory syncytial virus infection in mice sensitized to the major surface glycoprotein. *Int. Immunol.*, **4**, 493–500.

62. Paccaud, M.F. and Jacquier, A. (1970) A respiratory syncytial virus of bovine origin. *Arch. ges. Virusforsch.*, **30**, 327–42.

63. Parrott, E.H., Kim, H.W., Arrobio, J.O. *et al.* (1973) Epidemiology of respiratory syncytial virus infection in Washington, D.C. II. Infection and disease with respect to age, immunologic status, race and sex. *Am. J. Epidemiol.*, **98**, 289–300.

64. Pemberton, R.M., Cannon, M.J., Openshaw, P.J.M., Ball, L.A., Wertz, G.W. and Askonas, B.A. (1987) Cytotoxic T-cell specificity for respiratory syncytial virus proteins; fusion protein is an important target antigen. *J. Gen. Virol.*, **68**, 2177–82.

65. Prince, G.A., Hemming, V.G., Horswood, R.L., Baron, P.A., Murphy, B.R. and Chanock, R.M. (1990) Mechanisms of antibody-mediated viral

clearance in immunotherapy of respiratory syncytial virus infection of cotton rats. *J. Virol.*, **64**, 3091–2.

66. Sakata, K.M., Tashiro, K., Hirashima, M. *et al.* (1985) Selective regulation of chemotactic lymphokine production 1. Selective potentiation of eosinophil chemotactic lymphokine production in alum hydroxy gel- and *Bordetella pertussis* vaccine-treated guinea pigs. *J. Immunol.*, **135**, 3463–7.

67. Stott, E.J. and Taylor, G. (1985) Respiratory syncytial virus. Brief review. *Arch. Virol.*, **84**, 1–52.

68. Stott, E.J., Thomas, L.H., Taylor, G., Collins, A.P., Jebbett, J. and Crouch, S. (1984) A comparison of three vaccines against respiratory syncytial virus in calves. *J. Hyg.*, **93**, 251–61.

69. Sullender, W.M., Anderson, K. and Wertz, G.W. (1990) The respiratory syncytial virus subgroup B attachment glycoprotein: analysis of sequence, expression from a recombinant vector, and evaluation as an immunogen against homologous subgroup virus challenge. *Virology*, **178**, 195–203.

70. Taylor, G. (1994) The role of antibody in controlling and clearing virus infections, in *Strategies in Vaccine Design*, (ed. G.L. Ada), R.G. Landes, Georgetown, USA, pp. 17–34.

71. Taylor, G., Stott, E.J., Furze, J. *et al.* (1992) Protective epitopes on the fusion protein of respiratory syncytial virus recognized by murine and bovine monoclonal antibodies. *J. Gen. Virol.*, **73**, 2217–23.

72. Taylor, G., Stott, E.J., Bew, M. *et al.* (1984) Monoclonal antibodies protect against respiratory syncytial virus infection in mice. *Immunology*, **52**, 137–42.

73. Taylor, G., Stott, E.J. and Hayle, A.J. (1985) Cytotoxic lymphocytes in the lungs of mice infected with respiratory syncytial virus. *J. Virol.*, **66**, 2533–8.

74. Taylor-Robinson, D. and Doggett, J. (1963) An assay method for respiratory syncytial virus. *Br. J. Exp. Pathol.*, **44**, 473–80.

75. Tempest, P.R., Bremner, P., Lambert, M. *et al.* (1991) Reshaping a human monoclonal antibody to inhibit human respiratory syncytial virus infection in vivo. *Bio/Technology*, **9**, 266–71.

76. Tyrrell, D.A.J. (1963) Discovering and defining the aetiology of acute respiratory viral disease. *Am. Rev. Respir. Dis.*, **88**, 77–84.

77. Wagner, D.K., Graham, B.S., Wright, P.F. *et al.* (1986) Serum immunoglobulin G antibody subclass responses to respiratory syncytial virus F and G glycoproteins after primary infection. *J. Clin. Microbiol.*, **24**, 304–6.

78. Wagner, D.K., Nelson, D.L., Walsh, E.E., Reimer, C.B., Henderson, F.W. and Murphy, B.R. (1987) Differential immunoglobulin G subclass antibody titres to respiratory syncytial virus F and G glycoproteins in adults. *J. Clin. Microbiol.*, **25**, 748–50.

79. Wagner, D.K., Muelenaer, P., Henderson, F.W. *et al.* (1989) Serum immunoglobulin G antibody subclass response to respiratory syncytial virus F and G glycoproteins after first, second, and third infections. *J. Clin. Microbiol.*, **27**, 589–92.

80. Walls, R. S. (1977) Eosinophil response to alum adjuvants: involvement of T cells in non-antigen dependent mechanisms. *Proc. Soc. Exp. Biol. Med.*, **156**, 431–5.

81. Walsh, E.E., Schlesinger, J.J. and Brandriss, M.W. (1984) Protection from respiratory syncytial virus infection in cotton rats by passive transfer of monoclonal antibodies. *Infect. Immun.*, **43**, 756–8.

82. Welliver, R.C., Kaul, T.N., Putnam, T.I., Sun, M., Riddlesberger, K. and Ogra, P.L. (1980) The antibody response to primary and secondary infection with respiratory syncytial virus: Kinetics of class specific responses. *J. Pediatr.*, **96**, 808–13.

INFLUENZA VIRUSES AND VACCINES

14

Geoffrey Schild, Jim Robertson and John Wood

14.1 INTRODUCTION

Epidemics of a feverish respiratory illness undoubtedly due to influenza virus have been

Viral and Other Infections of the Human Respiratory Tract.
Edited by S. Myint and D. Taylor-Robinson. Published in 1996 by Chapman & Hall. ISBN 0 412 60070 6

described by physicians in letters and medical journals for many centuries and several of these outbreaks were apparently of pandemic significance. Accounts from as early as the fourteenth century used the word 'influence' to suggest that planetary phenomena brought about such pandemics of coughs, colds and fever; thus the origin of the name 'influenza'. It was in the eighteenth century that the disease was recognised as contagious in nature and not due to intrinsic, bodily malfunctions or extrinsic atmospheric factors [29]. However, it was another two centuries before the infectious agent was finally isolated in the MRC laboratories in Hampstead, London [57]. Approximately 10 years later, a vaccine prepared in eggs became available and in 1947, with the recognition of the continually changing antigenic makeup of the virus, the World Health Organization set up a network of laboratories to undertake a worldwide surveillance of influenza to collect data relevant to choosing the most appropriate strains for inclusion in vaccines. Since the onset of molecular biology about 20 years ago, possibly more has been learned about influenza viruses than any other virus of human disease.

There have been many reviews of influenza virus most of which deal with specialized aspects of the virus or its biology. Since the accumulated knowledge of influenza virus is too vast a subject to cover adequately in this single chapter, the reader is directed towards various specialized reviews and specific articles where appropriate [60].

14.2 CLASSIFICATION OF INFLUENZA VIRUSES

Influenza viruses are classified serologically into types A, B and C dependent on the antigenicity of their core proteins. Influenza A viruses are further subdivided antigenically into subtypes dependent on the haemagglutinin (HA) and neuraminidase (NA) surface antigens. Fourteen HA and nine NA subtypes of influenza A virus have been described in nature, although only three of these have been associated with human disease, the H1N1 (or Spanish Flu), the H2N2 (or Asian Flu) and the H3N2 (or Hong Kong Flu) subtypes. All of the subtypes have been isolated from avian species while the H1 and H3 subtypes have been isolated from swine and the H3 and H7 from equine species. Occasional isolates of influenza virus have been made from other mammals, including seals (H4 and H7), whales (H1 and H13), mink (H10) and camels (H1). Considerable evidence, especially at the genetic level, implicates viruses from avian species and possibly other non-human hosts as the origin of pandemic strains of human influenza virus (see Antigenic variation, p. 258).

There are four types/subtypes of influenza virus currently causing respiratory disease in humans: the A(H1N1) and A(H3N2) subtypes and types B and C. Influenza A virus infections are a source of severe morbidity, the symptoms of which include fever, sore throat, myalgia and malaise. Severe epidemics are associated with high levels of morbidity and mortality in the elderly, and in those with chronic debilitating conditions such as respiratory or heart disease, renal failure, endocrine disorders or immunosuppression. Influenza B virus infections are generally less severe while influenza C virus causes only a mild or asymptomatic infection and is generally not considered a sufficiently serious human pathogen to warrant consideration for diagnostic or vaccine purposes. Further details of the clinical manifestations are available in Chapter 15.

14.3 THE VIRUS

14.3.1 VIRION STRUCTURE

Influenza virus is a member of the Orthomyxoviridae family. Virions of influenza consist of a nucleocapsid core in which the single-stranded and negative-sense segmented RNA genome is intimately associated with the nucleoprotein and three polymerase proteins (Figure 14.1). The cores are surrounded by a shell of matrix protein encapsulated within a lipid bilayer membrane derived from the host cell during maturation by budding. Protruding

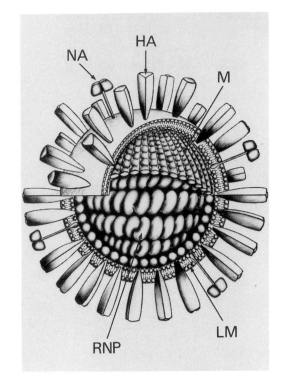

Figure 14.1 Diagram of an influenza virion showing the three-dimensional relationship between the viral lipid membrane (LM), haemagglutinin (HA) and neuraminidase (NA) spikes, the M protein and the ribonucleoprotein (RNP) cores which consist of the genomic RNA segments and nucleoprotein. (Reproduced with permission from Oxford, J.S. and Hockley, D.J. (1987) Orthomyxoviridae, in *Animal Virus Structure*, Elsevier Science Publishers, Amsterdam, New York.)

from the membrane are viral encoded glycoproteins. For influenza A and B viruses the surface glycoproteins are the haemagglutinin (HA) and neuraminidase (NA), while influenza C has a single haemagglutinin–esterase fusion (HEF) glycoprotein. These appear as protrusions or spikes under the electron microscope (Figure 14.2(a)). On the virion surface there are approximately 500 protrusions, with about ten times more HA than NA. In addition, the lipid membrane of influenza A virions contains a few molecules of a small integral membrane protein, M2. A large coiled helical structure can often be visualized within or extruding from slightly disrupted virions (Figure 14.2(b)). Full disruption with mild detergent solubilizes the lipid membrane but leaves the ribonucleoprotein (RNP) cores intact. Electron micrographs of purified RNPs show characteristic structures of strands of subunits of variable length (Figure 14.2(c)). These strands are twisted into a double-helical configuration often opening out

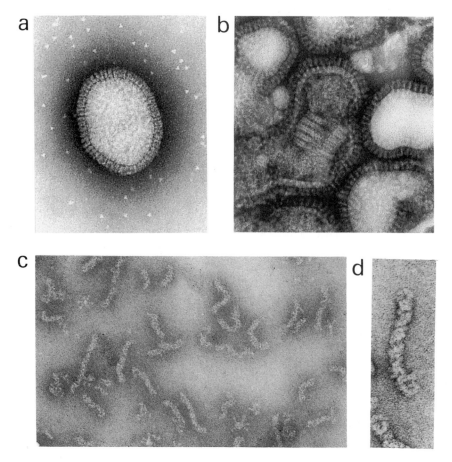

Figure 14.2   Electron micrographs of influenza virus particles. (a) Unfixed virus, negative staining with 4% sodium silicotungstate; original magnification, ×200 000. (b) Minor disruption of lipid bilayer allowing entry of stain shows a large coiled structure within the virion; original magnification, ×200 000. (c) A purified preparation of ribonucleoprotein (RNP) cores consists of short strands of various lengths; original magnification, ×130 500. (d) Close examination of the RNP cores [alpha]-double helical ribbon structure often opening out into a loop at one end. (Reproduced with permission from Oxford, J.S. and Hockley, D.J. (1987) Orthomyxoviridae, in *Animal Virus Structure*, Elsevier Science Publishers, Amsterdam, New York.)

into a loop at one or both ends (Figure 14.2(d)). Each RNP particle contains a single genome segment and the visible subunits correlate with monomers of nucleoprotein. The three virion polymerases are located at one end of these structures.

The genomes of influenza A and B viruses consist of eight segments of single strand RNA of negative polarity, i.e. of opposite sense to mRNA. The entire genome contains approximately 13 500 nucleotides and the eight segments of influenza A and B viruses encode ten polypeptides. For influenza A virus, segments 1–6 encode single polypeptides in large open reading frames (ORFs); these account for the three polymerases, the two surface glycoproteins and the nucleoprotein. The two smallest segments, segments 7 and 8, each encode two polypeptides, one from a large ORF, the other from a mainly non-overlapping smaller reading frame expressed via a spliced mRNA. Segment 7 encodes the matrix protein (M1) and the small membrane protein M2. Segment 8 encodes two non-structural proteins found only within infected cells, NS1 and NS2, which are believed to function in genome replication (for a review, see [30]).

14.3.2 THE REPLICATION CYCLE

During infection of a target cell, the HA attaches virions to receptors on the cell surface and they are then internalized by receptor-mediated endocytosis (see HA function, p. 256). In the mature endosome the HA undergoes a low pH-induced conformational change which induces fusion of the virion and endosome membranes permitting the RNP/transcriptase complex of the virus to enter the cytosol. Recent investigations have centred on the role of the M2 protein on virus entry. M2 is a small (97 amino acids) virion-encoded transmembrane protein. It is a homotetramer composed of two disulphide-linked dimers held together by non-covalent forces, and is found in large quantities within

the plasma and endoplasmic membranes of the infected cell and in small quantities in the membrane of mature virions. Studies with amantadine, an anti-influenza drug, have implicated M2 in having a role during virus entry into the cell. Small concentrations of amantadine block an early stage in the replication cycle and mutants resistant to amantadine have mutations clustered within the transmembrane domain of M2 [20]. This block appears to involve dissociation of the matrix protein from the RNPs and/or transport of the RNPs into the nucleus after internalization and uncoating of the incoming virus. More recently, M2 has been shown to have ion channel activity and it has been hypothesized that, during the period that the incoming virus spends in the acidic environment of the mature endosome, M2 acts to acidify the interior of the virion [22,43]. This treatment appears to be a prerequisite for disassembly of the matrix and RNP cores and in support of this, Zhirnov [75] has demonstrated that the tight interaction between matrix and RNP cores can be disrupted by mildly acidic pH. Much of our current understanding of the function of M2 also derives from studies on the effects of amantadine on late stages of virus replication. Sugrue *et al.* [61] showed that in the presence of amantadine, which inhibits the ion-channel activity of M2, the HA emerges on the cell surface in its low pH-induced conformation. M2, present in large amounts within the membranes of the endoplasmic reticulum and Golgi network, appears to act by de-acidifying the mildly acidic environment of the lumen of the Golgi complex, thus preventing the premature acid-induced conformational changes to the HA during its transport to the cell surface.

After entry into the cytosol and uncoating from matrix, the RNP cores are transported to the nucleus [34] where the negative-sense genome is transcribed and replicated by the three virion RNA-dependent RNA polymerases. During transcription of mRNA, the viral polymerases cannibalize host-cell 5-methylated cap structures for priming viral mRNA synthesis while polyadenylation occurs at a stretch of uridine residues approximately 20 residues from the 5' end of genomic RNA. Thus mRNA is not a complete copy of vRNA. For vRNA replication a full-length complementary RNA is synthesized as an intermediate from which molecules of vRNA are transcribed (for a review, see [28]).

The two surface glycoproteins, HA and NA, are synthesized in the endoplasmic reticulum and transported to the cell surface through the Golgi network where processing takes place (see HA structure, below). Nascent RNP cores associated with matrix protein exit from the nucleus and interact with the cytoplasmic side of the plasma membrane at regions where HA and NA have accumulated. Mature virions are formed when nucleocapsids bud through the plasma membrane at these sites. The NA, by removing any sialic acid which may have been attached to the carbohydrate moieties of the HA, allows release of virions from the cell surface and prevents nascent particles clumping together. In polarized epithelial cells *in vitro*, e.g. the MDCK cell line, the HA, NA and M2 are transported specifically to the apical surface of the cell where budding occurs. Such polarized budding also occurs in epithelial cells *in vivo*, which may contribute to restricting the infection to the respiratory epithelium, although there is no direct evidence for this.

14.4 THE SURFACE GLYCOPROTEINS

Because of their importance in the epidemiology of the disease and as primary vaccine candidates, the HA and NA surface glycoproteins of influenza virus will be considered in greater detail than any other feature of the virus (for a review, see [70]).

14.4.1 HA STRUCTURE

The HA and NA are membrane-bound glycoproteins which are synthesized in the rough endoplasmic reticulum within the infected cell. The HA is a type 1 glycoprotein, i.e. the signal peptide at the N terminus of the nascent polypeptide is cleaved off and the protein remains membrane-associated through a transmembrane anchor region at its carboxy terminus. Considerable processing of the HA occurs during its transport to the cell surface via the Golgi network, including glycosylation, oligomerization, acylation, sulphation and proteolytic cleavage. The HA, initially synthesized as a single polypeptide HA0, is cleaved into two components, HA1 and HA2, by a host-derived, trypsin-like protease at a late stage in its maturation. This event is an absolute requirement for the infectivity (and thus virulence) of virus particles since a free HA2 N-terminal sequence is required for membrane fusion during the early stages of infection. The two components, the HA1 (approximately 330 amino acid residues) and the HA2 (approximately 220 amino acid residues), are held together by a disulphide bridge and other forces. In some cell types *in vitro*, the cleavage step fails to take place and for *in vitro* production of virus, inclusion of trypsin in the growth medium is required. This requirement for cleavage is also a factor affecting tissue tropism *in vivo*.

The HA spike structures observed by electron microscopy are trimers containing three identical molecules of HA (Figure 14.3). The detailed molecular structure of the ectodomain of the HA, both by itself and in association with its receptor sialic acid, has been determined by X-ray crystallography [68,71]. The HA trimer extends 13.5 nm from the surface of the virus and in cross-section has a triangular appearance. Each monomer consists of a globular head distal to the viral membrane made up entirely of the HA1 region and a stem region extending approximately 7.5–8 nm from the viral surface consisting of HA2 and part of HA1. The transmembrane region is located towards the C-terminal end of HA2 and consists of approximately 25 amino acids and located within the virus is the 'cytoplasmic tail' of 10–15 residues. During maturation, palmitic acid is attached to the HA at cysteine residues

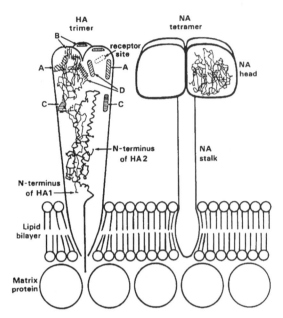

Figure 14.3 Diagram summarising features of the three-dimensional structure of haemagglutinin (HA) and neuraminidase (NA) spikes and their putative interaction with the viral membrane. The [alpha]-carbon backbone of a single HA monomer is shown within the trimeric structure, with the HA1 shown as a thin line and the HA2 as a thick line. The precise folding of the transmembrane region of HA2 and its interaction with Matrix protein are unknown. The N-terminus of HA2, the fusion peptide, is buried within the stem at neutral pH and becomes exposed during the conformational change at acidic pH. The receptor binding site and H3 antigenic sites A–D are indicated on the globular head portion which consists entirely of HA1. The NA is a tetramer with the polypeptide chain of each monomer composed of six identical folding units arranged in a propeller formation when viewed from above. The active sites are situated on the top surface of each monomer and antigenic determinants form a nearly continuous surface across the top of the molecule. (Reproduced with permission from Oxford, J.S. and Hockley, D.J. (1987) Orthomyxoviridae, in *Animal Virus Structure*, Elsevier Science Publishers, Amsterdam, New York.)

within the transmembrane and cytoplasmic domains. It is presumed, although there is no direct evidence, that the matrix protein and possibly the nucleoprotein interact with the cytoplasmic tail, especially during the formation of viral particles by budding from the infected cell plasma membrane.

14.4.2 HA FUNCTION

The function of the HA is in recognition and attachment of virus to the surface of target cells and in fusion of the viral and cellular membranes to allow the inner ribonucleoprotein cores access to the cell in order that the genome may replicate. Fusion takes place within mature endosomes after the virus has penetrated the cell by receptor-mediated endocytosis. The HA is also the major antigenic determinant of influenza virus and induces and binds neutralizing antibodies (see B cell responses, p. 262).

It is well established that the cellular receptor recognized by the virus is sialic acid [17] and the interaction of sialic acid with HA has been defined at the atomic level by X-ray crystallography [68]. The site within the HA which binds sialic acid, the receptor binding site (RBS), is a surface pocket on the globular head containing several amino acid residues conserved through all HA subtypes. Other residues involved to a lesser extent can vary, resulting in subtle variation in the precise specificity of the binding site. Such lesser residues may also be significant antigenically. In comparison with the binding of whole virus to sialic acid bearing cell surfaces, the affinity of individual HA molecules for sialic acid is relatively weak. Thus, it is assumed that there is a cooperative effect of many HA molecules on the binding of whole virus. The precise spatial arrangement of HA over the virus surface is unknown. However, greatly increased binding of a synthetic divalent sialic acid compound to virus particles was obtained when the two sialic acid residues were 5.5 nm apart

[14]. A second, but much weaker, sialic acid binding site has been located at the base of the globular head at the interface between HA1 and HA2 and with interactions to some residues in adjacent monomers within the trimeric structure. Only sialic acid in the configuration α2,3 to galactose binds here and the biological relevance of this finding is unknown [51].

The predominant sialic acid involved in the binding of influenza A and B viruses is 5-*N*-acetyl neuraminic acid which is found ubiquitously in animal species. Sialic acids are found as terminal sugars, generally attached to galactose residues on glycoproteins and glycolipids, and the nature of the linkage between sialic acid and galactose influences the host range of the virus. Human A(H3N2) and post-1977 A(H1N1) viruses attach preferentially to sialic acid with an α2,6 linkage to galactose [47,48]. In contrast, H3 avian and equine strains preferentially attach to sialic acid in α2,3 linkage. For the H3, this linkage preference correlates with HA1 residue 226, which is Leu in the human virus and Gln in avian and equine virus [47]. Human A(H3N2) virus has been shown to bind specifically to the surface of human tracheal epithelium on which the sialic acid is bound α2,6 and not to mucin-producing goblet cells in which sialic acid is bound α2,3 [10].

There is considerable interest in designing an antiviral drug which would inhibit the ability of the virus to attach to cells. In the search for a potent chemical inhibitor of influenza virus which could be used therapeutically, there have been further refinements of the interaction between sialic acid and HA using sialic acid analogues [36,52]. While there are data relating different HA sequences to the nature of the sialic acid species [24] and to the nature of the linkage between the sialic acid and its adjacent sugar moiety [47,48], there is a paucity of information regarding the type of macromolecule which acts as an effective receptor, i.e. whether it is a glycoprotein, a glycolipid, or both, and whether a specific glycoprotein/lipid is involved, although the ganglioside G_{M3} has been identified as a specific receptor for entry and fusion of A(H3N2) virus [62].

Available evidence indicates that after attachment, influenza virus enters the cell via receptor-mediated endocytosis, although our knowledge of the details of this mechanism and its trigger is also minimal [35]. However, the cooperative effect of the attachment of many HA molecules may be a necessary factor for triggering endocytosis.

After internalization by endocytosis, a further stage in virus entry is required for the viral genome to gain access to the cytosol and nucleus of the cell where expression and replication take place. This is achieved by fusion of the viral and endosomal membranes. The drop in pH during maturation of the endosome results in the HA undergoing a conformational change. During this change, which is a requisite for fusion, there is a dissociation of the globular heads of the monomers and exposure of the hydrophobic and highly conserved N-terminal fusion peptide which was previously buried in the stem region at the interface between monomers [58,70].

Recent X-ray crystallographic analysis of a fragment of the HA in the acid-induced conformation has revealed major refolding of the HA2 structure [6]. The fusion peptide is relocated ≥ 10 nm towards the top of the molecule, possibly into the target (endosomal) membrane, by extension of the triple-stranded 8-nm long coil located at the interface of the HA2 subunits. At the same time the bottom section of the coil partially unfolds and relocates, reversing the chain direction. The extent of this molecular movement is quite remarkable. The considerable energy required to drive the process may well be trapped within the HA during its synthesis. Cleavage of the HA during maturation then primes the molecule for release of this energy by a reduction in pH. This dramatic conformational change presumably results in destabilization of the viral and endosome membranes, and hence fusion.

These findings were partially predicted by a model proposed by Carr and Kim [8].

14.4.3 HA ANTIGENIC STRUCTURE

The HA induces and reacts with neutralizing antibodies and as such is described as the main antigenic determinant of influenza viruses. Analysis of the location of amino acid substitutions in naturally occurring variants and in monoclonal antibody escape mutants has identified the immunodominant regions of the HA molecule [9,69,70]. For the H1 and H3 subtypes, four or five overlapping immunodominant regions, or antigenic sites, have been identified (Sa, Sb, Ca and Cb for the H1, and sites A–E for the H3). Changes in glycosylation can mask or uncover antigenically active regions of the molecule and the subtle differences between the H3 and H1 maps are due much in part to differential glycosylation between the two strains. The antigenic sites are located on the surface of the globular head regions and, for the H1 subtype, the regions located on the top of the globular head (Sa and Sb) have been shown to be strain-specific compared with those lower down, which are more cross-reactive. The epitopes for individual antibody molecules are conformational rather than linear, although studies with synthetic peptides have shown that some antipeptide antibodies are capable of reacting with intact virus.

In contrast to influenza A virus HA, the influenza B virus HA appears to have a single immunodominant region composed of several overlapping arrays of epitopes [5]. This region corresponds to the sites on the top of the globular head (sites A and B for the H3, and sites Sa and Sb for the H1). While the atomic structure of the H1 subtype HA is expected to be very similar to that of the H3 subtype, it is difficult to know precisely to what extent the B HA structure will compare with the H3 structure. The antigenic analyses suggest significant differences in the antigenic structure: the HA molecule of influenza B virus is folded more tightly than the sites A and B of influenza A virus.

While the HA is the dominant antigen recognized by the humoral system, it is also a target for T-cell recognition, mainly class II restricted T-helper cells (see T-cell responses, p. 264). Determinants on the HA recognized by murine T-helper cells have been defined in various studies and map to linear epitopes over most of the HA1, although some regions may be more immunodominant than others. It is less clear if non-linear sites or sites on the HA2 are also recognized (see [72]).

14.4.4 NA STRUCTURE AND FUNCTION

The NA is a type 2 glycoprotein, the signal peptide for membrane insertion at the N-terminus acting also as the membrane anchor. It is a tetramer consisting of four identical disulphide-linked monomers (see Figure 14.3). Electron microscopy shows that the NA has a mushroom-like appearance with a box-shaped head measuring 8 nm × 8 nm × 4 nm atop a thin stalk approximately 6–8 nm in length. X-ray crystallography of the heads of both influenza A and B virus neuraminidase has defined their molecular structures [7,65]. The NA has sialidase activity and the sialic acid binding site is a pocket on the top surface of each of the four subunits. This binding site is quite distinct from the sialic acid binding site within the HA molecule by which the virus attaches to cells and which has no enzymatic activity.

The NA is presumed to facilitate the movement of virions through sialic acid-rich mucous layers during virus entry and also the release of mature virions from the cell. By removing sialic acid residues from nascent HA and surrounding cell surface glycoprotein and glycolipid structures, it permits virus release from the cell surface and prevents clumping which occurs if sialic acid is not removed from glycosylated structures on the HA.

14.5 ANTIGENIC VARIATION

Since the isolation of human influenza A virus in 1933, two radical changes in the antigenicity of human influenza viruses have been observed, both of which were associated with pandemic disease; once in 1957 with the outbreak of Asian Flu (when the prevalent subtype changed from H1N1 to H2N2) and again in 1968 with the outbreak of Hong Kong Flu (when the prevalent H2N2 subtype was replaced by the H3N2 subtype) (Figure 14.4). These radical changes in the antigenic characteristics of the causative viruses have been termed 'antigenic shift'. A unique event occurred in 1977 when the A(H1N1) subtype, which had been absent for 20 years, re-emerged but did not displace the prevalent A(H3N2) virus. Historical medical records indicate the occurrence also of periodic pandemic influenza over many centuries and serological analyses implicate the H2, H3 and H1

subtypes in the pandemics of 1889, 1900 and 1918, respectively. Thus, these subtypes appear to be recycled over long periods of time. More subtle changes in the antigenic nature of influenza viruses have occurred during interpandemic periods, often resulting in epidemics, and these changes have been termed 'antigenic drift'.

14.5.1 ANTIGENIC SHIFT

Considerable and detailed molecular and antigenic studies (there are almost 1000 entries for influenza virus in the international nucleotide sequence data banks) have provided a good understanding of the molecular events associated with antigenic shift and drift. Due to the segmented nature of the viral genome,

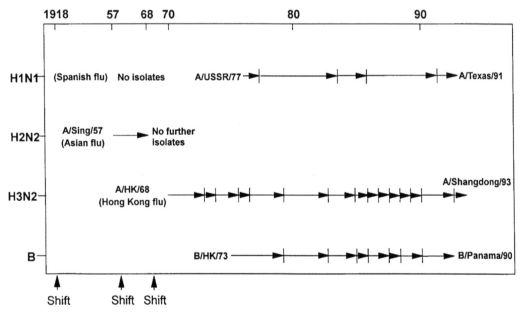

Figure 14.4 Diagram of antigenic shift and drift of human influenza virus. Serological evidence implicates the A(H1N1) subtype in the 1918 'Spanish flu' pandemic. Antigenic shifts occurred in 1957 when the H1N1 subtype was replaced by A(H2N2) A/Singapore/57 'Asian flu' and again in 1968 when the A(H2N2) strain was replaced by the A(H3N2) 'Hong Kong' strain. In 1977, the H1N1 strain reappeared but did not displace the H3N2 strain. The arrowheads indicate significant antigenic drift events when vaccine strains were updated. A/Shangdong/93 and B/Panama/90 are recommended strains for vaccine for 1994–1995. While the current recommended strain for A(H1N1) remains the 1986 strain A/Singapore/6/86, the antigenic variant A/Texas/36/91 is used by several vaccine manufacturers. (Modified from Wood, J.M. (1991) Influenza vaccines, in *Vaccines and Immunotherapy*, Pergamon Press, Oxford.)

genotypic re-assortment readily occurs during a mixed infection. This has readily been observed *in vitro,* and evidence also exists that it has occurred *in vivo* between swine and human strains and between swine and avian strains. Antigenic shift is the result of a re-assortment event in which one or more of the genome segments, including those coding for the surface glycoproteins of the prevalent strain, is replaced by the corresponding segment(s) from the avian reservoir of influenza viruses. In 1957, the H2N2 viruses obtained their HA, NA and PB1 genome segments from an avian virus, with the other five segments being derived from the pre-existing H1N1 viruses. Then in 1968, the H2N2 viruses obtained novel HA and PB1 segments, again from the avian pool, generating the H3N2 strains.

However, sequencing studies have revealed an additional mechanism for the appearance of 'new' human strains and that is direct adaptation of an avian virus to man [13,15]. Phylogenetic analyses of all gene products, with the exception of those replaced by re-assortment during 1957 and 1968, provide convincing evidence that all current human influenza viruses, and also the classical H1N1 swine viruses, were derived directly from an avian virus approximately 80–100 years ago. Indeed, the infamous 'Spanish flu' pandemic of 1918 may have been the result of this adaptation of a non-human strain to man with a concurrent increase in virulence and subsequent devastating effects on human health. However, inaccuracies in extrapolating backwards cannot pinpoint the exact year of origin of the H1N1 virus at the beginning of this century [56,67].

Considerable interest has always been attached to the observation that the two recent pandemic strains arose in south-east Asia and a hypothesis has been proffered that due to the lifestyles and agricultural practices in this part of the world, with close proximity of the population to ducks and pigs, there is a distinct possibility of a mixed infection occurring between these species. Generally, influenza virus is species-specific, and there is little evidence that avian species will infect humans and attempts to do so have generally been unsuccessful [4]. However, limited cross-species infections do occur, and there are several documented cases of swine influenza viruses infecting farm workers and *vice versa.* Also, avian-like viruses have been isolated from infected pigs, and the current H1N1 virus circulating in pigs in Europe since 1979 appears to be derived from an avian virus. Thus, pigs appear to occupy a central role in the evolution of human pandemic strains and the hypothesis is that pigs act as 'mixing vessels' for avian and human strains [54].

The emergence of a new subtype is believed to occur as a result of the human population becoming increasingly immunologically naive to past subtypes and increasingly immune to the current circulating strain which may be approaching the limits of its antigenic evolution. The emergence of a virus in man derived entirely from an avian strain is probably a rare event and maintaining the virulence of circulating human strains by re-assortment events as has occurred in 1957 and 1968 is more likely.

While the NP has been considered to be a major determinant of host range, all of the internal gene products may also contribute and host range should be considered a polygenic trait. However, since all current human viruses are presumed to derive from an avian precursor, nucleotide sequences recognized as host range-specific must in reality be a result of the independent evolution in man of these genes, although some unknown substitution(s) is presumably responsible for the initial establishment of the avian strain in humans and swine. In contrast to the evolving nature of human viruses, the avian genetic pool in aquatic wild birds appears to be close to its evolutionary limits and is fully adapted to its host. Influenza B type viruses are found only in humans and do not undergo re-assortment with non-human strains. It is surmised that the

B lineage evolved also from a common (avian) ancestor a very long time ago and has evolved in humans to the extent that it can no longer interact biologically with non-human viruses. Similarly, the divergence of influenza C which causes little disease must have occurred a very long time ago, possibly thousands of years [56,67].

Although there are two less than 8^8 possible combinations of re-assortment between two influenza viruses, not every combination is necessarily viable and limitations on gene constellations have been observed in the laboratory, dependent on the two viruses under study [31]. In addition, re-assortment is not restricted to viruses from different species and has also occurred in nature between co-circulating human strains [18].

14.5.2 ANTIGENIC DRIFT

Antigenic drift describes the lesser changes in antigenicity which occur within a subtype during interpandemic periods. It results from the substitution of a small number of critical amino acid residues within antigenic sites of the HA (Figure 14.5). Many, but not all, interpandemic strains arising through drift appear to originate also in the Far East with no plausible explanation, although the density of the population in that region may be an important factor. Some epidemic-causing variants have been isolated initially elsewhere, e.g. A/England/72.

Why does influenza virus undergo such rapid variation? In common with all RNA viruses, influenza virus populations are described as quasispecies in that they consist

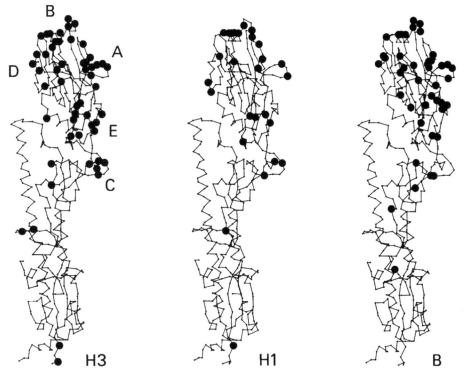

Figure 14.5 Location of naturally occurring amino acid substitutions on the (α–carbon backbone of the (H3) haemaglutinin (HA) molecule where the same substitution has occurred in two or more strains. Substitutions on the H3 HA are from its appearance in 1968 to 1991; antigenic sites A–E are indicated. Substitutions on the H1 subtype are from its reappearance in 1977 until 1991. Substitutions on influenza B cover a 40-year span from 1954 to 1993.

of a complex mixture of microvariants due to the inherent error rate of the viral polymerase [59]. Any variant virus can readily become dominant if an appropriate selection pressure is applied. It is assumed that variant viruses cause epidemics when the antigenicity of the HA has altered sufficiently for the virus to escape neutralization by a sufficient proportion of the population and sequencing studies indicate that in general, epidemic strains have amino acid changes in at least two antigenic sites of the HA molecule.

However, there are likely to be other factors involved including the immune status of the host. It has been observed that new variants are occasionally isolated sporadically and are followed in the next season by an epidemic caused by the same variant [12]. The explanation for this 'herald wave' phenomenon is likely to be that within a population a limited number of individuals will have sufficiently little protection against a new variant and there will be some, but limited, spread of the variant. In the following year, a greater proportion of the population will be at risk of infection by this new variant due to waning of pre-existing protective antibody, occasionally to the extent that an epidemic occurs. Thus, there exists a fine balance between the protective status of the population which is diminishing with time and an evolving virus which is constantly probing our defences. Other viral factors, e.g. mutation in other gene products which affect virulence, may exist about which very little is known at present.

14.6 HOST CELL VARIATION

There is now a considerable body of evidence that genetically distinct influenza viruses are derived in the laboratory from human clinical material depending on the substrate used for isolation and propagation. Sequence analysis of the HA genes of virus grown on Madin Darby canine kidney (MDCK) cells or in eggs identified single amino acid substitutions in A(H1N1), A(H3N2) and B viruses which are associated with the growth of virus in eggs (Figure 14.6). Use of the polymerase chain reaction (PCR) to analyse the HA sequence of virus present in clinical material indicated that the virus derived in tissue culture is representative of the natural virus and confirmed that the viruses isolated in eggs are variants (for a review, see [46]).

The substitutions found in egg-adapted viruses cluster around the receptor binding site of the HA and suggest that variants are selected in the egg due to altered receptor binding properties. The restricted growth of non-egg-adapted virus in eggs appears to be due to a lack of internalization of MDCK-grown virus, although the virus is able to bind specifically to sialic acid residues on the surface of the allantois.

Many of the substitutions occur in antigenically significant regions and in many cases pairs of viruses isolated in cell cultures and in eggs can be distinguished antigenically by monoclonal antibodies and polyclonal sera by a variety of assays. In general, MDCK-derived viruses are more antigenically homogeneous than their egg-grown counterparts. This was true for virus obtained from an epidemic, a single infected individual and from plaque-purified MDCK-grown virus. In addition, tissue culture grown virus has been shown to detect antibody in human sera more frequently and to higher titre than egg-adapted virus [53].

14.7 INFECTION AND CONTROL

14.7.1 B-CELL RESPONSES

Following infection or immunization with influenza virus, B-cells respond by producing antibody to HA, NA and some internal virus proteins. The role of serum antibody to HA in protection against infection has been well documented by observations that resistance to infection correlates with serum anti-HA antibody levels and also by the demonstration of

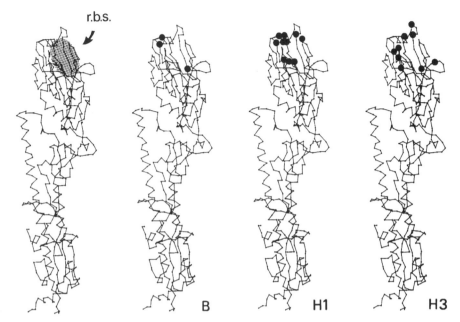

Figure 14.6 Location of HA1 amino acid substitutions on the ([alpha]-carbon backbone of the (H3) haemagglutinin (HA) molecule associated with egg adaptation of human influenza virus. Generally only one substitution is found in an individual egg-adapted virus. r.b.s., receptor binding site. (Reproduced with permission from [46].)

protection of mice by passive transfer of immune sera. The relationship between humoral antibody to HA and protection was first established at the MRC Common Cold Unit at Salisbury, UK by Hobson and colleagues [25]. They came to the conclusion that a haemagglutinating-inhibition (HI) titre of 1:40 was indicative of protection and this value is now accepted universally. In humans, HI antibodies are detectable within 4–7 days of infection, yet after immunization with inactivated vaccine, antibody can be detected as early as 2 days. Peak antibody levels after infection occur from 14–21 days and can persist for months or even years.

The serum antibody response to HA is influenza subtype-specific and includes antibodies directed to cross-reactive as well as strain-specific determinants of the HA molecule. In mice, strain-specific antibody is more protective than cross-reacting antibody [66] and *in vitro* strain-specific antibody is more

effective at virus neutralization [19]. The nature of the humoral response in humans depends upon age and previous experience of influenza virus infection or immunization. In 1977, when the influenza A(H1N1) subtype re-emerged after an absence of 25 years, the antibody response to A/USSR/77 (H1N1) vaccine in young people was much weaker than the corresponding response in adults over the age of 25 years [40]. This was due to anamnestic responses in the older people. Children, naturally infected with influenza virus, respond by producing serum antibodies that are predominantly strain-specific, yet in adults the predominant antibodies to infection or vaccination are cross-reactive or strain-specific to earlier virus strains. This is the phenomenon of 'original antigenic sin' [12] whereby the humoral response is directed to conserved regions of the HA molecule which are shared by early and more modern virus strains. Although this cross-reacting antibody is inefficient at virus neutralization it still appears to play a role in immunity by enhancing uptake of virus by Fc receptor-bearing cells such as antigen-presenting cells.

The most extensive studies of the class specificity of antibody responses to HA, have been performed by Murphy and colleagues in studies with live attenuated vaccines [38]. IgM, IgA and IgG antibodies appeared in serum within 2 weeks of infection. The IgG response peaked at about 6 weeks, whereas the IgM and IgA antibodies declined after 2 weeks. In nasal secretions, IgA was abundant at 2 weeks but trace amounts of IgG and IgM were also present. Influenza virus is neutralized by IgG, IgA, secretory IgA and IgM and each class of antibody probably contributes to protection.

The mechanism of virus neutralization is extremely complex and appears to be affected by the concentration of antibody, and whether the antibody is in the monomeric or polymeric form. Somewhat surprisingly IgG does not inhibit virus attachment but affects a later stage in virus replication, possibly transcription. IgM in high concentrations interferes efficiently with attachment and also internalization of virus, but at low concentrations (less than seven molecules of IgM per virion) IgM does not neutralize virus activity [1]. Polymeric IgA and secretory IgA at high concentrations interfere with virus attachment and possibly internalization, whereas lower concentrations are thought to inhibit the low pH-mediated cell fusion of virions. Monomeric IgA, however, appears to act, as does IgG, by inhibiting virus transcription.

Antibody to NA has a less well-defined role in protection than HA. It can inhibit virus release from infected cells and passive immunization of mice with anti-NA antibody will reduce the size of lung lesions following challenge infection. In humans, virus infection and immunization have been shown to induce serum antibody to NA and in certain studies, protection has been demonstrated in patients with no antibody to HA but detectable levels

of anti-NA antibody. It is likely that serum antibody to NA does play a part in immunity to influenza virus but its importance is probably less than that of serum HI antibody [45].

Antibody to NP is frequently detected following influenza virus infection and after vaccination with whole virus vaccine [40]. In contrast, M protein is not very immunogenic and antibody in humans is rarely detected. It is not certain whether anti-M or NP antibody has a role to play in immunity as passive transfer of antibody to mice has shown no evidence of protection [41] although it is known that NP does in fact vary antigenically.

14.7.2 T-CELL RESPONSES

Infection with influenza virus causes a significant T-cell response involving both helper T (Th) cells and cytotoxic T (Tc) cells [74]. When stimulated with non-infectious antigen, e.g. killed vaccine, there is a Th-cell response but a very poor Tc-cell response. These features are a result of the differing manner in which antigen-presenting cells process and present viral antigen to the two classes of T-cell. Th-cells recognize antigen in association with class II molecules on the surface of antigen-presenting cells primarily of the lymphatic system. Such antigen is derived exogenously by proteolysis of viral components within lysosomes after internalization of viral antigen by endocytosis. In contrast, Tc-cells generally recognize most somatic cells expressing endogenously derived viral antigen. Newly synthesized viral polypeptides are degraded within the cytoplasm of infected cells and are presented on the surface in association with class I molecules of the major histocompatibility complex via a pathway yet to be defined.

Thus Th-cells have a pivotal role in the development of the protective antibody response both post-infection and post-vaccination with inactivated vaccine, and also appear to function in aiding the Tc response. On the other hand, since Tc-cells are stimulated by and function in the destruction of virus-infected cells, they are not induced by inactivated vaccine but have been shown to aid recovery from virus infection [33].

The human Th response is directed mainly at the major structural components of the virus, viz., the HA, NA, NP and Ml and is fairly evenly divided between them. Th clones against the surface glycoproteins tend to be strain-specific while clones which recognize the internal components are generally cross reactive. The regions of the HA recognized by Th clones have been examined in fine detail and appear to be sequential epitopes, although there is evidence that conformational epitopes also exist. Several specific HA peptides which stimulate Th-cells have been defined for human and for murine Th-cells and are located on both surface and internal regions of the HA1. Regions within HA2 also stimulate Th-cells, although these are less well defined.

The Tc response is cross-reactive among influenza A viruses, but not influenza B viruses, and is directed primarily towards the NP [16]. Studies using cells transfected with the NP gene indicated that such cells could be recognized by Tc-cells although there was no detectable expression on the surface of a serological determinant for the NP. Further studies involving synthetic peptides delineated the NP epitopes involved in Tc-cell recognition. Such epitopes, e.g. NP amino acid residues 335–349, have been identified for human Tc-cells but are likely to vary between individuals [64].

The other internal proteins of the virus (Pa, Pb1, Pb2, M, NS1 and NS2], especially the Pb2 polymerase, are to varying but lesser degrees also targets for Tc-cells [16]. The HA, NA and M2, which are synthesized on the endoplasmic reticulum, are seldom targets for the human Tc response and strain-specific Tc-cells recognizing the surface glycoproteins HA and NA are found only at low levels. This may be due to a requirement for degradation of newly synthesized viral polypeptides in the cytosol before presentation on the cell surface with

class I antigens. In animal experiments, murine Tc-cells recognize the various internal viral proteins to a greater degree including the non-structural protein, NS1, and also show a greater recognition of the HA and NA.

14.7.3 INACTIVATED VACCINE

The first attempts at immunization using inactivated influenza vaccines were made in the 1930s using crude preparations of homogenized lungs from infected mice. These preparations were inactivated with formaldehyde and, when injected into animals by subcutaneous or intraperitoneal routes, they were immunogenic and protective. Such vaccines were clearly unsuitable for human use and the breakthrough came in 1937 when it was demonstrated that influenza virus could be grown successfully in the embryonated hen's egg. Eggs are still used to the present day although many aspects of vaccine production and standardization have changed over the years. The early vaccines were only partially purified and gave rise to significant adverse reactions. In the mid 1960s the introduction of zonal centrifugation and column chromatography ensured that most of the contaminating egg proteins were removed. A further improvement came after the discovery that virus particles disrupted by detergent treatment are less reactogenic. In the 1970s and 1980s, a third type of vaccine was developed: the subunit or surface antigen vaccine. This was a split vaccine from which the core proteins had been removed and under the electron microscope the HA and NA structures resembled characteristic rosette patterns (Figure 14.7). Thus, there are three types of vaccine currently available worldwide: whole virus, split and subunit [44].

An additional feature of vaccine manufacture is the use of high-growth re-assortant viruses. Newly isolated virus generally grows poorly in the laboratory and high-growth re-assortants (HGRs) containing the surface antigens of the appropriate recommended vaccine

a   b   c

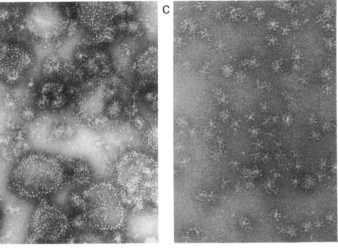

Figure 14.7 Electron micrographs of inactivated influenza vaccines composed of (a) whole-virus particles, (b) split-virus particles and (c) subunit particles. The split-virus vaccine contains all the viral components whereas the subunit vaccine contains only haemaglutinin arranged in rosettes and the neuraminidase arranged in cartwheel formations; original magnification, ×130 000. (Reproduced with permission from Wood, J.M. (1991) Influenza vaccines, in *Vaccines and Immunotherapy*, Pergamon Press, Oxford.)

strain are prepared for industry by co-infecting eggs with the vaccine strain and a laboratory adapted high-growth strain. Due to the segmented nature of the influenza virus genome, genetic re-assortment readily occurs in a mixed infection and a virus with the relevant surface antigens and high-growth characteristics can be selected by the use of appropriate antiserum. In these circumstances it is important to demonstrate that HGRs for vaccine use have the correct antigenic phenotype.

14.7.4 SURVEILLANCE

Control of influenza is currently based primarily on vaccination. Vaccination is not used for controlling the spread of the disease but for protection of certain risk groups, including the elderly and those with chronic illness who may be at risk if they contract the disease. The antigenic characteristics of the virus are continually undergoing change and in order to derive the most benefit from vaccination it is important that the strains included in vaccines are a good antigenic match to those causing

influenza outbreaks. In order to achieve this, an elaborate system of surveillance for both the disease and the virus has been established by the World Health Organization (WHO). Information is gathered on a world scale on the prevalence of the disease, on the antigenic characteristics of prevalent viruses and on the ability of current vaccine to raise antibodies against the prevalent strains (by a haemagglutination-inhibition assay). The WHO has designated three laboratories 'World Influenza Centres' to which viruses isolated in all parts if the world are sent for full characterization. Many other laboratories are designated WHO collaborating centres and assist in the characterization of viruses isolated in their local areas.

Each year, usually in mid-February, the WHO assemble a group of influenza experts including epidemiologists, representatives from surveillance centres, vaccine control authorities and vaccine manufacturers to discuss the vaccine composition for the forthcoming winter (in the Northern hemisphere). A decision to replace an existing strain is based upon several factors including the appearance of a new epidemiologically significant variant, coupled with low-level protection against the new variant by the previous season's vaccine. It is important that these data derive from several international sources. As only three strains currently cause human disease, influenza A(H1N1), A(H3N2) and B, only one strain from each of these subtypes is incorporated into a vaccine. In the past 9 years there have been seven changes of the A(H3N2) strain, one change of the A(H1N1) strain (a reflection possibly of it nearing the end of its evolutionary potential), and three changes of the B strain.

When the selected vaccine strains are compared with viruses causing influenza in the world (Table 14.1), there is a good match in 79% of cases. This is an excellent record in view of the variability of influenza virus. While the major analytical approach remains the antigenic characterization of these viruses, sequence analysis of the HA genes of recent isolates, especially now that the technology can be performed easily and rapidly, has an important role to play in studying antigenic drift and of assessing the impact of laboratory propagation on the selection of variants. Thus, vaccine production is an annual event and

Table 14.1 Comparison of vaccine strains with concurrent epidemic strains

Virus subtype	Year	Vaccine recommendation	Epidemic strain
A(H1N1)	1986–87	Chile/1/83 + Sing/6/86	Sing/6/86
	1987–88	Sing/6/86	Sing/6/86
	1988–89	Sing/6/86	Sing/6/86*
	1989–90	Sing/6/86	Sing/6/86*
	1990–91	Sing/6/86	Sing/6/86
	1991–92	Sing/6/86	Sing/6/86
	1992–93	Sing/6/86	Sing/6/86*
	1993–94	Sing/6/86	Sing/6/86*
	1994–95	Sing/6/86	?
A(H3N2)	1986–87	Mississippi/1/85	Leningrad/360/86*†
	1987–88	Leningrad/360/86	Sichuan/2/87†
	1988–89	Sichuan/2/87	Sich/2/87 + Shang/11/87
	1989–90	Shanghai/11/87	Shanghai/11/87
	1990–91	Guizhou/54/89	Beijing/353/89†
	1991–92	Beijing/353/89	Beijing/353/89
	1992–93	Beijing/353/89	Beijing/32/92*†
	1993–94	Beijing/32/92	Beijing/32/92
	1994–95	Shangdong/9/93	?
B	1986–87	Ann Arbor/1/86	Ann Arbor/1/86
	1987–88	Ann Arbor/1/86	Beijing/1/87†
	1988–89	Beijing/1/87	Beijing/1/87
	1989–90	Yamagata/16/88	Yamagata/16/88
	1990–91	Yamagata/16/88	Yamagata/16/88
	1991–92	Yam/16/88 or Pan/45/90	Yam/16/88 + Pan/45/90
	1992–93	Yam/16/88 or Pan/45/90	Panama/45/90
	1993–94	Panama/45/90	Panama/45/90
	1994–95	Panama/45/90	?

* Little influenza activity
† Poor matches between epidemic and vaccine strains

takes place between the announcement of the WHO recommendation and the autumn of that year, with the intention of placing the vaccine on the market by October in time for vaccination to take place before the influenza season begins. Each year, the vaccine is based on the prevalent virus from the previous season or on a strain which is believed to have the potential to cause disease in the future.

14.7.5 EFFECT OF CELL SUBSTRATE ON VACCINE

Considerable attention has been given in recent years to the most appropriate host cell type in which to cultivate influenza viruses for surveillance and vaccine ever since the demonstration of host cell-mediated selection of influenza virus antigenic variants [53]. Since surveillance studies are performed with, and vaccine prepared from, egg-grown viruses the recognition that during egg-adaptation, viruses with variant HAs are selected has potentially serious implications on the choice of the vaccine strain.

No proper clinical trial has been conducted with cell- versus egg-prepared vaccine; however, animal studies with killed vaccines indicated that mammalian cell-derived virus vaccine induces a greater and more cross-reactive antibody response and gave better protection than the corresponding egg-adapted virus vaccine. When experimental animals were immunized with vaccinia recombinants bearing the HA of MDCK-grown or egg-grown virus, no difference in immunogenicity or protective efficacy was observed. However, no model is ideal for assessing the comparative immunogenicity and protective efficacy that can be achieved in humans, although the current vaccine for human use is based on inactivated virus rather than a live vaccinia recombinant. These studies should not be taken to indicate that the egg-derived influenza virus vaccine currently in use is likely to be irrelevant in providing protection

in humans. Rather, they were designed to determine if HAs which differ by only one or two amino acid residues would invoke a differential immunogenic response and protective efficacy [27,49,73].

The reference strains of human influenza virus chosen by the WHO to represent antigenic phenotypes circulating in a given influenza season are also currently grown in eggs. The exclusive use of egg-adapted viruses in assessing the antigenic nature of epidemiologically significant strains has been shown to overestimate the extent of antigenic diversity among human influenza viruses. Influenza A(H3N2) viruses isolated from around the world over several years exhibited less antigenic drift if isolated in cells than if isolated in eggs [37].

There is now a consensus of opinion that care should be taken in the use of egg-grown viruses for human influenza surveillance and vaccine production. Although all the evidence indicates that the cell-isolated virus is more like, if not identical to, the natural virus, there is not an immediate need to suggest the exclusive use of cell culture for vaccine production, although this is currently being pursued by several vaccine manufacturers. Many egg-adapted influenza A virus isolates are indistinguishable antigenically from their cell-derived counterparts, even though an amino acid within the HA has been substituted and it would be prudent to target such 'cell-like' egg-derived viruses for vaccine manufacture. It is now recognized by WHO surveillance laboratories that the effects of egg-adaptation on the HA must be carefully monitored to avoid diagnosing a new influenza virus strain as epidemiologically significant when it is in fact a laboratory artefact.

14.7.6 ATTENUATION, LIVE VACCINES AND VIRULENCE

Early attempts at attenuating human influenza virus for use as a live vaccine were made by

Russian workers by sequential passaging in eggs. Mass vaccination was conducted with such a vaccine, although the reliability of this method for the preparation of an attenuated strain and the efficacy of the vaccine produced is doubtful.

The MRC Common Cold Unit was the scene of many studies by Beare and colleagues [3] on the attenuation of pathogenic human influenza virus strains for use as live vaccines. These studies made extensive use of the highly passaged and attenuated human strain, A/PR/8/34 (PR8) and re-assortants derived from PR8 which influenza viruses readily produce due to their segmented genome. It was shown that re-assortants of PR8 in which the HA and NA genes were derived from epidemic viruses were indeed attenuated for humans. However, in some instances, virulence was partially restored to PR8 even to the extent that the re-assortant was as virulent as the wild-type parent [2].

Alternative attempts at attenuation by workers at the National Institutes for Health (NIH), USA, involved the generation and use of conditional lethal, temperature-sensitive mutants as the master strain with which to produce vaccine strains by re-assortment. While satisfactory results were obtained in adults with this approach, there were problems with insufficient attenuation for children, possibly due to the lack of anti-NA antibody in the young [39]. This work also demonstrated that mutations in an individual gene coding for an internal protein will attenuate the virus for animals and birds and it was assumed that any gene can serve as a target for attenuation.

Later studies, also by the NIH group, have shown that attenuation of human strains can be achieved by re-assortment with avian viruses. Re-assortment experiments between avian strains and recent pathogenic human strains implicated the internal genes of the avian strain, especially the NP and M, in attenuating virulent human strains. This indicated that internal gene products contribute to viru-

lence. However, the virulence also depended on the precise HA and NA donated from the human strain and this approach was abandoned when some re-assortants made for use as live vaccine were unfortunately, inadequately attenuated for children.

Another attenuated strain being assessed as a suitable donor strain for live vaccine use in man is the cold-adapted strain (ca) of Massab, A/Ann Arbor/6/60, developed by successive *in vitro* passages at gradually lower temperatures [32]. This strain is highly attenuated for humans and sequencing studies have revealed mutations in every gene of the ca master strain [11], although more recently, Herlocher, Maassab and Webster [23] have suggested that attenuation resulted from initial laboratory passage of the human isolate and not from the cold adaptation *per se*.

The only country in which a live attenuated vaccine has been used clinically is in the former USSR, where live vaccines based on an attenuated ts master strain were in use for several years and where ca vaccines were also developed. However, little use, if any, of these live vaccines is now made and most influenza immunization currently utilizes inactivated vaccine.

Many of the above studies on attenuation contributed considerably to our knowledge of the determinants of virulence of influenza virus. Data derived by re-assortment experiments with avian strains also indicated the polygenic nature of virulence. In mixed infections between the highly virulent avian influenza virus strain, fowl plague, and an avirulent strain, progeny re-assortants in which any genes of the fowl plague virus were replaced with the corresponding gene of the avirulent strain had a degree of attenuation dependent on the number of genes replaced [55].

Studies on the pathogenicity of avian strains have also identified the cleavability of the HA as a prime determinant of virulence [50]. Proteolytic cleavage of the HA is an absolute prerequisite for infectivity of the virus; if cleavage fails to take place, newly formed virus is non-infectious. In most avian species the activating protease is restricted to the tissues in which virus replication is found to occur, i.e. the gut where infection is asymptomatic. In the highly pathogenic H5 and H7 strains of fowl plague, a series of basic amino acid residues in the vicinity of the HA are cleaved throughout the organs of the bird and a generalized viraemia occurs. No mammalian HA has ever been found which has the series of basic amino acid residues at the cleavage site found in pathogenic avian strains. A factor restricting the human virus to respiratory epithelium, in addition to polarized budding at the apical surface, may be that the activating protease for cleavage exists only within these cell types. It has also been demonstrated that a concomitant bacterial infection can supply the necessary protease for a more disseminated infection, e.g. co-infection of mice with *Staphylococcus* spp. resulted in an increased and fatal infection in the lungs, and may explain the high fatality rate in humans co-infected with *S. aureus* [63]. In a similar vein, it was possibly co-infection with *Haemophilus influenzae*, isolated regularly during the devastating 1918 pandemic, which was responsible for the increased virulence of the influenza virus during that period.

An additional interesting study on the virulence of influenza virus, comparing viruses which differed by only one or two amino acid residues in the HA, was conducted at the MRC Common Cold Unit [42]. This was designed to study the effect of laboratory passage on pathogenicity. Four groups of volunteers were infected with an influenza B virus; either original infectious clinical material (nasal wash), virus isolated and grown in human embryonic tracheal cultures, egg-adapted virus which had an HA1 substitution affecting a glycosylation site, and a second egg-adapted virus with an HA1 substitution distinct from the previous one (with the latter three viruses all being derived from the clinical sample). Only the volunteers infected with the egg-adapted virus in which the HA had a substitution affecting a glycosylation site failed to become significantly ill. The level of symptoms and the extent of antibody rises and of virus isolation from members of this group were extremely low in contrast to the other three groups. Apparently the loss of the glycosylation site from the HA had considerably attenuated the virulence and infectivity of this virus. This may be why influenza B virus, lacking this site, has not undergone antigenic drift, naturally, in the human population.

The precise genetic basis of the virulence of influenza virus for humans is complex, is not fully understood and remains to be elucidated. It has, however, long been recognized that it is polygenic in nature and a change or mutation in any single gene can result in reduced virulence of a pathogenic strain.

14.7.7 ANTIVIRALS

Two antiviral drugs have been used successfully in the control of influenza – amantadine and more recently, rimantadine. Unfortunately, while good results have been obtained prophylactically, there is minimal advantage in their therapeutic use unless administered very early in infection. The problem of resistant mutants arising must also be considered and the results to date suggest that, although they do occur and are transmissible, their clinical significance is unclear [21].

In the search for an anti-influenza virus compound considerable attention has been given to analysing the structure and function of the viral components in order to identify a suitable target for drug intervention. One such target has been the viral neuraminidase and in the past, analogues of sialic acid have been shown to inhibit virus replication *in vitro*. Recently, starting from these original sialic acid analogues, computer-assisted molecular graphics have been used to design compounds with enhanced inhibitory effects on the NA. Using such approaches an effective and potent neuraminidase inhibitor has been designed at the atomic level which has a greatly decreased K_i (inhibition constant) and which has proved to be effective in inhibiting virus replication *in vitro* and in animal challenge experiments [26]. This compound has a strong potential as an effective drug for human use and is about to enter clinical trials to determine its therapeutic and prophylactic efficacy.

14.8 THE FUTURE

Although there have been significant technical developments since the 1930s to improve the production and quality of influenza virus vaccines, there has been little progress to improve the efficacy of vaccines. Recombinant HA has been produced and is immunogenic; however, the quantities thus synthesized cannot compete economically with the ease of use and yield obtained from embryonated hens' eggs. Novel adjuvants, ranging from tree bark extracts (Saponin) and bacterial cell wall derivatives to recombinant cytokines, are being tried and tested. New methods of antigen delivery are also being investigated such as ISCOMS, liposomes and biodegradable microparticles which may be used for parenteral delivery or by the oral and nasal routes. An entirely novel approach to vaccination, that of genetic immunization, is also under investigation. With genetic immunization, the gene for an antigen is injected and expressed within the host rather than inoculation of the antigen itself. However, infection and immunity involve not only the virus but also the host and ultimately a much greater understanding of the human immune system is required in order to overcome such problems as original antigenic sin and to be able to confer effective cross-protection against a variety of strains.

Influenza virus has been used widely as a model biological system and probably as much is now known about the biology of influenza virus as about any other virus. As the understanding of the biology and molecular biology of the virus increases, so may there be a dramatic improvement also in the prevention and control of influenza through the design of antiviral drugs.

From sequencing studies, it is known that pandemic strains originate from avian species, but exactly how is not understood; however, there are enticing hypotheses involving pigs as intermediaries. Neither is the origin of or what constitutes an epidemic strain fully understood. With this lack of total understanding it is impossible to predict where or when a new strain, especially a pandemic strain, will arise. Already several countries have developed pandemic contingency plans to enable health authorities and vaccine manufacturers to contend with the inevitable frenzy of activity. Whether there will be sufficient time to prepare a vaccine depends upon the place and time of emergence of a pandemic virus.

14.9 REFERENCES

1. Armstrong, S.J., Outlaw, M.C. and Dimmock, N.J. (1990) Morphological studies of the neutralisation of influenza virus by IgM. *J. Gen. Virol.*, **71**, 2313–19.
2. Beare, A.S., Schild, G.C. and Craig, J.W. (1975) Trials in man with live recombinants made from A/PR/8/34 (H0N1) and wild H3N2 influenza viruses. *Lancet*, ii, 729–32.
3. Beare, A.S. (1982) Research into the immunisation of humans against influenza by means of living viruses, in *Basic and Applied Influenza Research*, (ed. A.S. Beare), CRC Press, Boca Raton, Florida, pp. 211–34.
4. Beare, A.S. and Webster, R.G. (1991) Replication of avian influenza viruses in humans. *Arch. Virol.*, **119**, 37–42.
5. Berton, M.T. and Webster, R.G. (1985) The antigenic structure of the influenza B virus hemagglutinin: operational and topological mapping with monoclonal antibodies. *Virology*, **143**, 583–94.

6. Bullough, P.A., Hughson, F.M., Skehel, J.J. *et al.* (1994) Structure of influenza haemagglutinin at the pH of membrane fusion. *Nature*, **371**, 37–43.

7. Burmeister, W.P., Ruigrok, R.W.H. and Cusack, S. (1992) The 2.2Å resolution crystal structure of influenza B neuraminidase and its complex with sialic acid. *EMBO J.*, **11**, 49–56.

8. Carr, C.M. and Kim, P.S. (1993) A spring-loaded mechanism for the conformational change in influenza hemagglutinin. *Cell*, **73**, 823–32.

9. Caton, A.J., Brownlee, G.G., Yewdell, J.W. *et al.* (1982) The antigenic structure of the influenza virus A/PR/8/34 haemagglutinin (H1 subtype) *Cell*, **31**, 417–27.

10. Couceiro, J.N.S.S., Paulson, J.C. and Baum, L.G. (1993) Influenza virus strains selectively recognise sialyloligosaccharides on human respiratory epithelium; the role of the host cell in selection of hemagglutinin receptor specificity. *Virus Res.*, **29**, 155–65.

11. Cox, N.J., Kitame, F., Kendal, A.P. *et al.* (1988) Identification of sequence changes in the cold-adapted, live attenuated influenza vaccine strain, A/Ann Arbor/6/60 (H2N2). *Virology*, **167**, 554–67.

12. Francis, T. Jr (1953) Influenza: the newe acquayantance. *Ann. Intern. Med.*, **39**, 203–21.

13. Gammelin, M., Altmuller, A., Reinhardt, U. *et al.* (1990) Phylogenetic analysis of nucleoproteins suggests that human influenza A viruses emerged from a l9th-century avian ancestor. *Mol. Biol. Evol.*, **7**, 194–200.

14. Glick, G.D., Toogood, P.L., Wiley, D.C. *et al.* (1991) Ligand recognition by influenza virus. *J. Biol. Chem.*, **266**, 23660–9.

15. Gorman, O.T., Donis, R.O., Kawaoka, Y. *et al.* (1990) Evolution of influenza A virus PB2 genes: implications for evolution of the ribonucleoprotein complex and origin of human influenza A virus. *J. Virol.*, **64**, 4893–902.

16. Gotch, F., McMichael, A., Smith, G. *et al.* (1987) Identification of viral molecules recognized by influenza-specific human cytotoxic T lymphocytes. *J. Exp. Med.*, **165**, 408–16.

17. Gottschalk, A. (1959) Chemistry of virus receptors, in *The Viruses*, (eds F.M. Burnet and W.M. Stanley), Academic Press, New York, pp. 51–61.

18. Guo, Y.J., Xu, X.Y. and Cox, N.J. (1992) Human influenza A (H1N2) viruses isolated from China. *J. Gen. Virol.*, **73**, 383–7.

19. Haaheim, L. and Schild, G.C. (1980) Antibodies to the strain specific and cross-reactive determinants of the haemagglutinin of influenza H3N2 viruses. *Acta Pathol. Microbiol. Immunol. Scand. [B]*, **88**, 335–40 .

20. Hay, A.J., Wolstenholme, A.J., Skehel, J.J. *et al.* (1985) The molecular basis of the specific anti-influenza action of amantadine. *EMBO J.*, **4**, 3021–4.

21. Hayden, F.G. and Couch, R.B. (1992) Clinical and epidemiological importance of influenza A viruses resistant to amantadine and rimantadine. *Rev. Med. Virol.*, **2**, 89–96 .

22. Helenius, A. (1992) Unpacking the incoming influenza virus. Cell, 69, 577–8.

23. Herlocher, M.L., Maassab, H.F. and Webster, R.G. (1993) Molecular and biological changes in the cold-adapted 'master strain' A/AA/6/60 (H2N2) influenza virus. *Proc. Natl Acad. Sci. USA*, **90**, 6032–6.

24. Higa, H.H., Rogers, G.N. and Paulson, J.C. (1985) Influenza virus hemagglutinins differentiate between receptor determinants bearing N-acetyl-, N-glycollyl-, and N,O-diacetylneuraminic acids. *Virology*, **144**, 279–82.

25. Hobson, D., Curry, R.L., Beare, A.S. *et al.* (1972) The role of serum haemagglutination inhibition antibody in protection against challenge infection with A2 and B viruses. *J. Hyg. (Camb)*, **70**, 767–77.

26. von Itzstein, M., Wu, W.-Y., Kok, G.B. *et al.* (1993) Rational design of potent sialidase-based inhibitors of influenza virus replication. *Nature*, **363**, 418–23.

27. Katz, J.M. and Webster, R.G. (1989) Efficacy of inactivated influenza A virus (H3N2) vaccines grown in mammalian cells or embryonated eggs. *J. Infect. Dis.*, **160**, 191–8.

28. Krug, R.M., Alonso-Caplen, F.V., Julkunen, I. *et al.* (1989) Expression and replication of the influenza virus genome, in *The Influenza Viruses*, (ed. R.M. Krug), Plenum Press, New York, pp. 89–152.

29. deLacy, M. (1993) The conceptualization of influenza in eighteenth-century Britain. *Bull. History Med.*, **67**, 74–118.

30. Lamb, R.A. (1989) Genes and proteins of the influenza viruses, in *The Influenza Viruses*, (ed. R.M. Krug), Plenum Press, New York, pp. 1–88.

31. Lubeck, M.D., Palese, P. and Schulman, J.L. (1979) Nonrandom association of parental

genes in influenza A virus recombinants. *Virology*, **95**, 269–74.

32. Maassab, H.F. and DeBorde, D.C. (1985) Development and characterisation of cold-adapted viruses for use as live virus vaccines. *Vaccine*, **3**, 355–69.

33. McMichael, A.J., Gotch, F.M., Noble, G.R. *et al.* (1983) Cytotoxic T-cell immunity to influenza. *N. Engl. J. Med.*, **309**, 13–17.

34. Martin, K. and Helenius, A. (1991) Transport of incoming virus nucleocapsids into the nucleus. *J. Virol.*, **65**, 232–44.

35. Matlin, K.S., Reggio, H., Helenius, A. *et al.* (1981) Infectious entry pathway of influenza virus in a canine kidney cell line. *J. Cell Biol.*, **91**, 601–13.

36. Matrosovich, M.N., Gambaryan, A.S., Tuzikov, A.B. *et al.* (1993) Probing of the receptor-binding sites of the H1 and H3 influenza A and influenza B virus hemagglutinins by synthetic and natural sialosides. *Virology*, **196**, 111–21.

37. Meyer, W.J., Wood, J.M., Major, D. *et al.* (1993) Influence of host cell-mediated variation on the international surveillance of influenza A(H3N2) viruses. *Virology*, **196**, 130–7.

38. Murphy, B.R., Nelson, D.L., Wright, P.F. *et al.* (1982) Secretory and systemic immunological response in children infected with live attenuated influenza A virus vaccines. *Infect. Immun.*, **36**, 1102–8.

39. Murphy, B.R., Clements, M.L., Maassab, H.F. *et al.* (1984) The basis of attenuation of virulence of influenza virus for man, in *The Molecular Virology and Epidemiology of Influenza*, (eds Sir C. Stuart-Harris and C.W. Potter), Academic Press, London, pp. 211–25.

40. Nicholson, K.G., Tyrrell, D.A.J., Harrison, P. *et al.* (1979) Clinical studies of monovalent inactivated whole virus and subunit A/USSR/77 (H1N1) vaccine; serological responses and clinical reactions. *J. Biol. Standardisation*, **7**, 123–36.

41. Oxford, J.S. and Schild, G.C. (1976) Immunological and physiochemical studies of influenza matrix (M) polypeptides. *Virology*, **74**, 394–402.

42 Oxford, J.S., Schild, G.C., Corcoran, T. *et al.* (1990) A host-cell-selected variant of influenza B virus with a single nucleotide substitution in HA affecting a potential glycosylation site was attenuated in virulence for volunteers. *Arch. Virol.*, **110**, 37–46.

43. Pinto, L.H., Holsinger, L.J. and Lamb, R.A. (1992) Influenza virus M2 protein has ion channel activity. *Cell*, **69**, 517–28.

44. Potter, C.W. (1982) Inactivated influenza virus vaccine, *in Basic and Applied Influenza Research*, (ed. A. S. Beare), CRC Press, Boca Raton, Florida, pp. 119–58.

45. Potter, C.W. and Oxford, J.S. (1979) Determinants of immunity to influenza infection in man. *Br. Med. Bull.*, **35**, 69–75.

46. Robertson, J.S. (1993) Clinical influenza virus and the embryonated hen's egg. *Rev. Med. Virol.*, **3**, 97–106 .

47. Rogers, G.N., Paulson, J.C., Daniels, R.S. *et al.* (1983) Single amino acid substitutions in influenza haemagglutinin change receptor binding specificity. *Nature*, **304**, 76–8.

48. Rogers, G.N. and D'Souza, B.L. (1989) Receptor binding properties of human and animal H1 influenza virus isolates. *Virology*, **173**, 317–22.

49. Rota, P.A., Shaw, M.W. and Kendal, A.P. (1989) Crossprotection against microvariants of influenza virus type B by vaccinia viruses expressing haemagglutinins from egg- or MDCK cell-derived subpopulations of influenza virus type B/England/222/82. *J. Gen. Virol.*, **70**, 1533–7.

50. Rott, R. (1992) The pathogenic determinant of influenza virus. *Vet. Microbiol.*, **33**, 303–10.

51. Sauter, N.K., Glick, G.D., Crowther, R.L. *et al.* (1992) Crystallographic detection of a second ligand binding site in influenza hemagglutinin. *Proc. Natl Acad. Sci. USA*, **89**, 324–8.

52. Sauter, N.K., Hanson, J.E., Glick, G.D. *et al.* (1992b) Binding of influenza virus hemagglutinin to analogs of its cell-surface receptor, sialic acid: analysis by proton nuclear magnetic resonance spectroscopy and X-ray crystallography. *Biochemistry*, **31**, 9609–21.

53. Schild, G.C., Oxford, J.S., De Jong, J.C. *et al.* (1983) Evidence for host-cell selection of influenza virus antigenic variants. *Nature*, **303**, 706–9.

54. Scholtissek, C. (1987) Molecular aspects of the epidemiology of virus disease. *Experientia*, **43**, 1197–201.

55. Scholtissek, C., Rott, R., Orlich, M. *et al.* (1977) Correlation of pathogenicity and gene constellation of an avian influenza A virus (fowl plague). 1. Exchange of a single gene. *Virology*, **81**, 74–80.

56. Scholtissek, C., Ludwig, S. and Fitch, W.M. (1993) Analysis of influenza A virus nucleoproteins for the assessment of molecular genetic mechanisms leading to new phylogenetic virus lineages. *Arch. Virol.*, **131**, 237–50

57. Smith, W., Andrews, C.H. and Laidlaw, P.P. (1933) A virus obtained from influenza patients. *Lancet*, **ii**, 66.

58. Stegmann, T. and Helenius, A. (1993) Influenza virus fusion: from models toward a mechanism, in *Viral Fusion Mechanisms*, (ed. J. Bentz), CRC Press, Boca Raton, Florida, pp. 89–111.

59. Steinhauer, D.A. and Holland, J.J. (1987) Rapid evolution of RNA viruses. *Annu. Rev. Microbiol.*, **41**, 409–33.

60. Stuart-Harris, C.H., Schild, G.C. and Oxford, J.S. (1985) *Influenza: the viruses and the disease.* Edward Arnold, London.

61. Sugrue, R.J., Bahadur, G., Zambon, M.C. *et al.* (1990) Specific alteration of the influenza haemagglutinin by amantadine. *EMBO J.*, **9**, 3469–76.

62. Suzuki, Y., Matsunaga, M. and Matsumoto, M. (1985) N-acetylneuraminyllactosyl-ceramide, G_{M3}-NeuAc, a new influenza A virus receptor which mediates the adsorption-fusion process of viral infection. *J. Biol. Chem.*, **260**, 1362–5.

63. Tashiro, M., Ciborowski, P., Klenk, H.-D. *et al.* (1987) Role of *Staphylococcus* protease in the development of influenza pneumonia. *Nature*, **325**, 536–7.

64. Townsend, A.R.M., Rothbard, J., Gotch, F.M. *et al.* (1986) The epitopes of influenza nucleoprotein recognised by cytotoxic T lymphocytes can be defined with short synthetic peptides. *Cell*, **44**, 959–68.

65. Varghese, J.N., Laver, W.G. and Colman, P.M. (1983) Structure of the influenza virus glycoprotein antigen neuraminidase at 2.9 resolution. *Nature*, **303**, 35–40.

66. Virelizier, J.L., Allison, A.C. and Schild, G.C. (1979) Immune responses to influenza virus in the mouse. *Br. Med. Bull.*, **35**, 65–8.

67. Webster, R.G., Bean, W.J., Gorman, O.T. *et al.* (1992) Evolution and ecology of influenza A viruses. *Microbiol. Rev.*, **56**, 152–79.

68. Weis, W., Brown, J.H., Cusack, S. *et al.* (1988) Structure of the influenza virus haemagglutinin complexed with its receptor, sialic acid. *Nature*, **333**, 426–31.

69. Wiley, D.C., Wilson, I.A. and Skehel, J.J. (1981) Structural identification of the antibody-binding sites of Hong Kong influenza haemagglutinin and their involvement in antigenic variation. *Nature*, **289**, 373–8.

70. Wiley, D.C. and Skehel, J.J. (1987) The structure and function of the hemagglutinin membrane glycoprotein of influenza virus. *Annu. Rev. Biochem.*, **55**, 365–94 .

71. Wilson, I.A., Skehel, J.J. and Wiley, D.C. (1981) Structure of the haemagglutinin membrane glycoprotein of influenza virus at 3Å resolution. *Nature*, **289**, 366–73.

72. Wilson, I.A. and Cox, N.J. (1990) Structural basis of immune recognition of influenza virus hemagglutinin. *Annu. Rev. Immunol.*, **8**, 737–71.

73. Wood, J.M., Oxford, J.S., Dunleavy, U. *et al.* (1989) Influenza A(H1N1) vaccine efficacy in animal models is influenced by two amino acid substitutions in the hemagglutinin molecule. *Virology*, **171**, 214–21.

74. Yewdell, J.W. and Hackett, C.J. (1989) Specificity and function of T lymphocytes induced by influenza A viruses, in *The Influenza Viruses*, (ed. R.M. Krug), Plenum Press, New York, pp. 361–429.

75. Zhirnov, O.P. (1990) Solubilisation of matrix protein M_1/M from virions occurs at different pH for orthomyxo- and paramyxoviruses. *Virology*, **176**, 274–9.

INFLUENZA VIRUSES: CLINICAL SPECTRUM AND MANAGEMENT

15

Karl Nicholson

Viral and Other Infections of the Human Respiratory Tract.
Edited by S. Myint and D. Taylor-Robinson. Published in 1996 by Chapman & Hall. ISBN 0 412 60070 6

15.1 INTRODUCTION

Influenza is an acute, febrile respiratory illness of global importance caused by influenza A or B virus. Influenza in humans occurs in two epidemiological forms – pandemic influenza which results from the emergence of a new influenza A virus to which the population possesses little or no immunity, so it spreads with a high attack rate in all parts of the world; and interpandemic influenza A or B, occurring as sporadic infections, a localized outbreak, or epidemic, the latter representing an outbreak in a given community which usually occurs abruptly, peaks within 2–3 weeks, lasts 5–6 weeks, and is associated with a significant drift of the surface haemagglutinin and neuraminidase antigens. Epidemics occur virtually every year almost exclusively in the 'winter' months in the northern hemisphere (October to April), and May to September in the southern hemisphere. Explosive epidemics of influenza can exert an enormous toll in terms of morbidity, mortality, and economic and social costs.

Influenza has no pathognomonic features so a precise picture of its impact was impossible until the first isolation of influenza A virus in 1933 [170], and influenza B virus in 1940. Further advances came with the recognition of influenza virus replication in hens' eggs in 1936 [20], discovery of the haemagglutinating properties of influenza virus in 1941, and the subsequent development of serological methods based on haemagglutination-inhibition. In the absence of these tools, a combination of the explosive nature of influenza, its tendency for seasonality, high attack rates, and respiratory and systemic features allow an overview of the disease since ancient times.

15.2 HISTORY

Langmuir and colleagues [111] proposed that the plague of Athens in the years 430–427 BC was caused by influenza associated with toxic shock syndrome and named the association 'the Thucydides syndrome'. The epidemic in 412 BC described by Hippocrates was probably influenza [158]. Clear historical accounts of influenza in Italy, Germany and England can be traced back to the epidemic of 1173 [89]. Hirsch [89] and Thompson [181] provide accounts of epidemics from the twelfth and sixteenth centuries respectively, and Hirsch [89] referred to 'pandemics' occurring over a great part of the globe and noted their occurrence in the years 1510, 1557, 1580, 1593, 1732–1733, 1767, 1781–1782, 1802–1803, 1830–1833, 1836–1837, 1847–1848, 1850–1851, 1855, 1857–1858 and 1874–1875. Interestingly, several of these pandemics, such as the ones in 1781–1782, 1801 (1802–1803), 1830–1833 and 1847–1848, seem to have spread to Europe across Russia from the Far East and Hirsch commented upon the 'regular progress of the disease from east to west'. Modern influenza surveillance has shown that the virus affects much of the globe in most years; 'pandemics' with high attack rates are associated with antigenic 'shift' of the virus rather than antigenic 'drift', and such pandemics occur infrequently with two of the most recent ('Asian' and 'Hong Kong') originating in the Far East. The short intervals between several 'pandemics' in Hirsch's review question whether they were associated with antigenic shift.

The origin of the term 'influenza' is uncertain, although the chronicles of a Florentine family used it in reference to the possible 'influence' of the planets at times of respiratory epidemics [178]. 'Influenza' was used in England during the outbreak of 1743 [50], but earlier references to the malady included 'newe acquayantance', the gentle correction, epidemic catarrh, or catarrhal fever [32, 50].

15.3 PANDEMIC INFLUENZA

15.3.1 INFLUENZA PANDEMICS – 1847 TO 1977

The influenza pandemic of 1847–1848

Review of mortality statistics for 'influenza' in London since 1840 reveals a 39-fold rise in deaths during the fourth quarter of 1847 as compared with the average for the same quarter during the 7-year period 1840–1846 [151]. The increase was 11-fold for the first quarter of 1848 as compared with the average for the same period during the eight preceding years. Such abrupt increases in mortality, with outbreaks occurring in both eastern and western hemispheres, suggest the occurrence of a pandemic associated with antigenic shift. London mortality statistics and the 1847–1848 pandemic enabled William Farr to provide the first estimate of influenza-associated mortality. Farr noted that 'the epidemic was most fatal to adults and the aged' ... 'the mortality in childhood was raised 83%, in manhood 104%, in old age 247%'... 'from the age 4 to 25, however, the mortality was comparatively not much increased'.

The influenza pandemic of 1889–1890

The number of influenza deaths in London gradually decreased from 1739 during the winter of 1847–1848 to less than 10 during each winter from 1884 to 1889 [151]. Some 558 'influenza' deaths were recorded during the first quarter of 1890, and outbreaks of respiratory disease with high attack rates were noted throughout the United Kingdom [151]. The first cases were described in May 1889 in North America and in Bokhara, and the Russian doctors at that time called the disease

'Chinese' influenza. The outbreak reached Western Europe via Central Asia and European Russia and reached London by November 1889. High attack rates throughout Europe were associated with excess deaths as compared with those in previous years.

Seroprevalence studies carried out during the emergence of 'Asian' influenza in 1957–1958 revealed the presence of pre-existing antibody in people aged ≥71 years of age [139]. This suggested that the viruses responsible for both the 1889–1890 and 1957–1958 pandemics contained an H2 haemagglutinin and led to the proposal that influenza virus might recycle. However, during the Asian influenza pandemic (1957–1958) there was no evident reduction in pneumonia and influenza death rates with increasing age, implying that the elderly had little or no protection following prior exposure to related H2 viruses [166].

The 1889–1890 pandemic is the first pandemic for which there are good public health records. Data collected during this period confirmed and extended those obtained during 1847–1848, notably that during periods of influenza activity, certified influenza deaths represent only a small proportion of the excess. In Paris, the excess mortality between 15th December 1889 and the end of January

1890, beyond that during the three previous years, amounted to at least 5500 cases, but only 213 of all deaths were attributed by the certifying medical attendants to influenza [151]. Table 15.1 shows the relative increase in deaths from chronic conditions including diabetes, 'wasting', alcoholism, heart disease, tuberculosis, and chronic bronchitis during the epidemic as compared with the average during the three previous years. These and more recent observations identify 'risk factors' and provide the basis for current vaccine recommendations. A further interesting observation, which has been noted in more recent pandemics, was the increase in influenza mortality during the second and third waves in 1891 and 1892 as compared with the first. This may reflect an increasing awareness of influenza among attendant physicians at the time or an increase in virulence of the virus.

Possible H3 pandemic during the late nineteenth century

The number of 'influenza' deaths in England and Wales fell after 1892 to about 100 per million of the population in 1896 but exceeded 400 per million in 1900 [178]. Some historians suspect the emergence of an H3 strain at the turn of the century based on the finding by

Table 15.1 Increase in deaths during the 1889–90 epidemic in Paris. (Data abstracted from [151])

Cause of death	Ratio, 1889–1990 (Dec 15–Jan 31) to average number in the same period during 3 previous years*
Pneumonia	3.19
Paralysis	2.50
Pleurisy	1.99
'Wasting'	1.94
'Alcoholism'	1.86
Heart disease	1.78
Acute bronchitis	1.67
Tuberculosis	1.58
Chronic bronchitis	1.44

* Period 15th December to 31st January

independent investigators in Asia, Europe and North America that a large number of elderly persons, principally those over 75 years of age in 1968, had high levels of antibody to the A/Hong Kong (H3) strain of influenza before the pandemic in 1968 [35,55,118,122]. Lower mortality rates during the A/Hong Kong pandemic in 1968–1969 in the over 75s as compared with those aged 65–74 years provides additional evidence for infection with an H3 virus at the turn of the century [96].

The influenza pandemic of 1918

The first 18 years of the twentieth century saw influenza death rates of ≤300 per million of the population in England and Wales, but in 1918 influenza mortality increased more than tenfold to 3129 per million [178]. The pandemic in 1918 was due to the appearance of a new virus which seroepidemiological studies have identified as related to swine influenza (H1). The 1918–1919 pandemic is unequalled in recent history, causing an estimated 20 million deaths worldwide and 675 000 in the USA alone [33]. The reason for the exceptionally high toll is unclear, but unlike other outbreaks during the nineteenth and twentieth centuries high mortality was seen in otherwise healthy young adults. During the 1889–1890 pandemic and those since 1918–1919, age-specific mortality curves showed a U-shaped pattern with higher death rates in the very young and the elderly, but in 1918, the mortality curve was W-shaped, with high fatality in young adults and at the extremes of age. This may not be unique in the history of influenza. During the pandemic of 1782 several observers noted '... I think that the middle age felt it most' ... 'I mean from 16 to 45 or so'; 'children and old people escaped entirely or were affected in a slighter manner' [50].

The influenza pandemic of 1957

The research effort generated by the 1918 pandemic culminated in the successful transmission of human influenza to ferrets using

garglings from patients with influenza in 1933 [170]. The subsequent recognition that influenza virus would replicate in hens' eggs, and the development of serologic techniques provided the means to monitor the spread and impact of influenza. The 'Asian' influenza pandemic was apparently first encountered in an outbreak in the Kweichow province of China in February, 1957 [178] and the causative virus – A/Asian/57 (H2N2) – was first identified in cases in Hong Kong in April 1957. The outbreak spread rapidly to South-East Asia and Australasia. It travelled westwards to the Middle East and Europe and eastwards through Japan to the USA. Influenza reached Holland, apparently by sea, between the end of May and the beginning of July [138]. Sporadic cases occurred in Britain during the summer months, and outbreaks developed in many places in the latter half of August, but October was the peak month and the first wave had virtually disappeared by the end of November [52,148,194].

The attack rate caused by the A/Asian virus was unusually high and varied from about 20% to almost 100% in different communities worldwide, with the higher rates (>50%) occurring among school-age children, the institutionalized, and those living in crowded conditions [45,52,69,138,194]. The attack rate in England and Wales averaged 31% for all age groups, but the highest overall attack rate, 49%, occurred in the group aged 5 to 14 years [194]. Similarly, during the first wave in the USA, the overall incidence of infection assessed serologically was 55%, with rates of 78% among 10 to 14-year-olds and 24% among all adults [103]. The incidence of clinical influenza in Kansas and Louisiana was approximately 60% in high-school children, whereas the attack rate was 10–20% in adults above the age of 30 years [23]. Children were paramount in the spread of infection during the 1957 pandemic. In a Kent general practice, families without school-children suffered about half the incidence of those with school-children (18% versus 33%), and attack rates in the elderly were three times greater in households with school-children (31%) when compared with childless households [194].

From 21st August until the end of the year the number of new claims for national insurance sickness benefit in Britain was about 2.5 million more than average for the same periods during the previous 5 years – that is, an increase equivalent to about 12.5% of the insured population [194]. Data collated by the Ministry of Health in London indicated that 7.5–9 million persons suffered some incapacity from influenza and that at least 5.5 million visited a doctor. To illustrate the effect of the epidemic in general practice, Fry [52] in his practice compared the weekly volume of work for common respiratory infections during September and October 1957 with the corresponding weeks over the previous 5 years. At the peak of the epidemic in 1957 there were 3.6 times as many attendances as previously. Sickness statistics for a large factory in southeast London showed that the proportion of employees who were absent also rose approximately four-fold to 23% during the influenza epidemic compared with a level of about 6% in the recent past [194]. The demand for hospital accommodation appears to have been heavy. In Liverpool during September and October 1957 the number of discharges from general medical departments increased by 27% compared with the average for the same 2 months during the years 1953 to 1956 inclusive [128]. During the period 23rd September to 5th November, the London Emergency Bed Service dealt with 2.8 times as many patients suffering from acute respiratory disease as compared with the same period during the three previous years [128].

The 1957 Asian pandemic was generally regarded – on the basis of complications and deaths – as rather mild. A much higher rate of complications occurred during the 1950–1951 epidemic when 20% of those seen in a general practice developed chest complications [52].

During the influenza A virus epidemic of 1953 and the B virus epidemic of 1955 Fry noted chest complications in 10%, whereas in 1957 only 3% of patients had comparable complications [52]. Woodall, Rowson and McDonald [194] similarly observed an incidence of 3% for pneumonia and 2% for otitis media among 187 patients with serologically confirmed influenza in a general practice.

Although elderly individuals had antibodies to previously circulating H2 viruses in 1957, there was no evident reduction in pneumonia and influenza deaths with increasing age [166]. Indeed in Fry's general practice the incidence of bronchitis and pneumonia actually increased from 2–3% in those aged < 50 years, to 5% in those 50–59 years, 9% in those aged 60–69 years, and 15% in those over 70 [52]. Elsewhere in Britain the incidence of pulmonary complications was 5% of 542 patients in a general practice, [42] and 5% of 391 RAF recruits [59]. Asthma exacerbations occurred in 38% of 124 people with atopy and influenza in a general practice in Barrow in Furness [12].

There is remarkably little published work on influenza as met in hospitals in England and Wales during autumn 1957. Most admissions with influenza were for the complication of pneumonia, but some people with uncomplicated influenza were admitted because of difficulty of treatment at home [100]. During the 6-week period commencing 1st October 1957, 262 (48%) of 541 patients admitted to Dundee hospitals had influenzal pneumonia based on clinical and radiological evidence, and an additional 135 (25%) had abnormal chest signs [100]. On admission *Streptococcus pneumoniae* and *Haemophilus influenzae* were isolated from 13% and 8% of cases, respectively, and staphylococci from only 2.5%. However, after 4 and 7 days of hospitalization staphylococci were isolated from 32% and 36% implying nosocomial acquisition. The overall death rate was 6%, but was 11% in those with pneumonia and 1% in the remainder. Among those aged <20 years the case-fatality rate was

0.5%, rising to 2.2% in those aged 20–49 years, and 14% in those over 50 [100]. Of 30 patients dying from pneumonia, eight (27%) died within 24 hours of admission, 11 (37%) within 48 hours, and 18 (60%) within 96 hours. All of these patients were gravely ill on admission and the part played by the nosocomially acquired staphylococci was difficult to assess.

Elsewhere the presence of staphylococci rendered the prognosis grave. Oswald, Shooter and Curwen [148] compared fatality in relation to age and sex in a series of 155 staphylococcal and 145 non-staphylococcal influenzal pneumonias. The mortality was 28% for staphylococcal pneumonia, 12% for non-staphylococcal pneumonia, and the greater fatality rate of staphylococcal pneumonia was observed at all ages. Among survivors, staphylococcal pneumonia was the more severe with a median stay in hospital of 26 days compared with 16 days for the others. As in Dundee [100], there was evidence of nosocomial acquisition of staphylococci and of multiple antibiotic resistance; 93% of staphylococcal isolates were penicillin-resistant, 74% were resistant to tetracycline, 67% were resistant to sulphonamides, and 62% were resistant to streptomycin. However, most were sensitive to chloramphenicol and erythromycin. Overall, 49% of staphylococci isolated within 4 days of admission – which were presumed to have been acquired outside hospital – were penicillin-resistant [148].

Reports from different parts of the world showed that the case fatality varied greatly from place to place. In Malaya it was about 1 in 40 000 and in Pakistan around 1 in 46 000, whereas in the Philippines it was 1 in 220 [163]. In Abadan, Iran it was estimated at 1 in 9000 [163], in Kuwait 1 per 1000 [69] and in Aden 1 in 350 [45]. Estimates of the case-fatality rate in Britain ranged from 13 to 35 deaths per 10 000 [128]. The most comprehensive review of fatal cases in England and Wales was undertaken by the Public Health Laboratory Service, which received records for

477 fatal cases from over 100 hospitals [159]. The disease progressed more rapidly in the young than the old. From the onset of pneumonia deterioration was usually rapid and 86 (18%) of the 477 patients died before admission to hospital. An outstanding feature was the speed with which patients died after admission – two-thirds died within 48 hours of arrival, implying that in most instances the staphylococci were acquired in the community. In this series virtually no deaths occurred in children or young adults in the absence of *S. aureus*, whereas in the very young and elderly, in whom 'milder' infections might well be fatal, staphylococci were found less often. It is clear that *H. influenzae*, despite its prominence in the past did not play an important part in 1957, even though technical difficulties may have led to some underestimate of frequency.

The number of deaths arising from the influenza pandemic is not known with certainty. In Britain, the average number of deaths for the years 1950–1956 in England and Wales for the September and December quarters was 228 760, whereas during this period in 1957 it was 262 191. The difference between these figures (33 431) has been taken as a rough estimate of the toll in deaths [128]. In the USA the excess deaths from influenza were estimated as 35 400 during October to December 1957, and 34 100 during January to March 1958 [95].

The influenza pandemic of 1968

An increase in influenza was noted in Hong Kong during mid-July 1968. The virus isolate reacted poorly with antiserum to 'Asian' strains and was sent to WHO reference centres in London and Atlanta where it was confirmed that the strain differed substantially from previous isolates. Although it was not the usual season for such outbreaks, the virus spread rapidly and involved about 500 000 persons [189]. The peak incidence in Hong Kong occurred in late July, only a fortnight after the first isolation of the virus, and the whole outbreak was over in 6 weeks.

The new virus appeared in Singapore and the Philippines in August and by the end of the month there had been outbreaks in Taipei, Taipan, Kuala Lumpur, Malaysia and Vietnam [164]. By September it had spread to northern Australia, and also to Madras, India, after arrival of a ship from Singapore with 16 cases of influenza on board. A week later influenza was reported in Bombay. The virus was introduced into Japan in August and September but, as in the USA and Britain, there was little general spread until the onset of cooler weather in the autumn.

The Hong Kong virus was introduced into the USA in September and October by people returning from the Far East and sporadic cases were subsequently reported throughout the country. Outbreaks were first noted in California; there was a gradual spread eastwards throughout November and December and by the end of the year all states were involved. A retrospective questionnaire survey of 6994 students from a high school in Kansas City, Missouri, and their families, for histories of influenza-like illness indicated an overall attack rate of 39% [36]. In contrast to the Asian pandemic in 1957 and studies reported by many other authors, the age-specific attack rates of all age groups were similar and adults were as likely as school-children to have been the first family case. This epidemiological observation has implications for the use of influenza vaccine as it has been suggested that vaccination of school-children should prevent dissemination of influenza virus in the community. Overall, some 27% of the population in the USA were ill during the epidemic and schools had 50% absenteeism [168]. A review of the clinical and pathological features in 127 adults – 93 outpatients, 23 who were more seriously ill and were admitted, and 11 patients who acquired influenza nosocomially in the Mayo Clinic – suggested that the manifestations were similar to Asian influenza in

1957 [112]. Pneumonia was a complication in 20 (16%) of cases. Excess mortality from all causes above that expected from seasonal variation was estimated at 56 300 in the USA during the 12-month period 1st October to 30th September, 1968–1969, but at only 9500 for 1969–1970, and there was a deficit of deaths during 1970–1971.

Hong Kong virus reached Europe in autumn 1968, but no widespread outbreaks were reported until the winter and the disease was generally mild. Poland was probably the hardest hit country with an estimated four million cases of influenza-like illnesses [24]. The virus was isolated in London in August from a solitary family infection, and in September there was a small school outbreak, but then the infection apparently halted [177]. In January 1969 an increase in national insurance claims began which plateaued in March and April [162]. Returns from the Royal College of General Practitioners in England indicated that, unlike the situation in Hong Kong, the epidemic was prolonged, lasting from late December to early April; the incidence did not reach high levels in any one week and there was no great pressure on hospital accommodation. Deaths from pneumonia and other causes were little elevated and only in March 1969 did the mortality curve suggest the existence of an influenza epidemic [177].

Hope-Simpson reported his experience of the first wave of Hong Kong influenza as it presented in a Gloucestershire general practice with a list size of 3620 persons [91]. The outbreak lasted 13 weeks, from 15th January to 15th April 1969, and is estimated to have attacked only 4% of the practice population. Within households, the secondary attack rate was 17%. As in the USA, liability to infection was almost independent of age except that persons over 65 years were less frequently attacked. In household outbreaks neither school-children nor children under school age were more commonly the first case than their elders, and the majority of contacts evidently escaped disease. The difference in attack rates between different countries was remarkable; indeed, during the conference on Hong Kong influenza in October 1969, it was suggested that two virus strains were circulating simultaneously – the more virulent one in the USA, and a less virulent one in Europe. It is now recognized that mutations, including substitutions, deletions and insertions, are extremely important in producing variation in the behaviour of influenza viruses. It is therefore conceivable that geographic variation in strains had occurred and that the virus in Britain became more virulent or pneumotropic over time.

Although the first wave of Asian influenza was mild, infection reappeared simultaneously as a major epidemic in Britain, Spain, Portugal, Switzerland, Norway, Italy, Yugoslavia and France in November, 1969. In contrast to the previous year's experience mortality was heavy, the deaths reported during the first 5 weeks of the epidemic amounted to over 8700 and the total for the epidemic exceeded that of all individual outbreaks since 1951 [177]. The age-specific attack rate was concentrated upon those aged 40–60 years and chest complications occurred at a rate of 25% [53].

The H1N1 'pandemic' in 1977

Significant pandemics have arisen with major changes in both the haemagglutinin and neuraminidase, but an exception to this generalization occurred when A/USSR/77 (H1N1) did not cause a pandemic – this may have been because much of the world's population in 1977 were alive before 1957 during the previous H1N1 era. Most people over the age of 23 years possessed antibodies to its surface haemagglutinin antigen and the new virus was confined almost entirely to children and teenagers. Previously, when new influenza A subtypes emerged the preceding antigenic subtype disappeared rapidly throughout the world, but not on this occasion. Currently both

H1N1 and H3N2 subtypes are circulating throughout the world.

15.4 INTERPANDEMIC INFLUENZA

The interpandemic period is represented by relatively minor, but immunologically significant changes in the haemagglutinin or neuraminidase of influenza virus types A and B. Epidemics occur with cycles every few years, but a recent feature has been the co-circulation of one or more strains of H3N2 and H1N1 subtypes of influenza A virus together with influenza B virus. All three types may be detected nationally in any one year but it is usual that one type predominates and annual outbreaks of influenza are not uncommon. Not only in large countries such as the USA and Australia, but also in smaller ones such as France where there is extensive reporting by medical practitioners, it is not uncommon to observe major outbreaks in some regions while others experience little or no activity – at least for a week or two. The spread of influenza through a community typically produces a bell-shaped curve of the incidence of visits to medical practitioners for influenza and influenza-like illness. It usually lasts 6–8 weeks and depending upon the epidemic's size there are concomitant increases in claims for sickness benefit, hospital admissions, school absenteeism and death registrations. Interpandemic influenza is responsible for considerable morbidity and mortality which exceeds that associated with the introduction of the pandemic strain.

Influenza infections are often identified during years when no excess mortality is detected. Silent transmission of virus for a long period in a small community was demonstrated in 1935, more than 2 months after the last ship left a Greenland village [99]. Influenza occasionally persists beyond its normal season, and in the absence of recognizable community outbreaks of disease [71,82,109,133, 180]. Failure to culture influenza virus from over 20 000 specimens collected during the summer from persons with febrile respiratory disease in Houston over 11 years of community surveillance suggests that persistent infection in a community seldom occurs [31]. Influenza outbreaks occur nearly all months of the year somewhere in the world, and the introduction of influenza virus into one country from another is generally accepted as the origin of epidemics.

15.4.1 HERALD WAVES

In certain years, viruses similar to those appearing in the next year have appeared at the end of outbreaks. These so-called 'herald waves' have been observed in some years in community studies in several geographic locations within the USA [47,49]. For example, in Houston during the influenza A/Victoria epidemic of 1976, 34 infections with influenza B/Hong Kong virus were detected and this virus became epidemic the following year [62]. Similarly a herald wave of 48 A/Texas-A/Victoria (H3N2) virus infections occurred during the latter half of the influenza B/Hong Kong epidemic of 1977 [63]. These H3N2 viruses were epidemic during the next season. In Japan, during the 5-year period 1985–1990, the influenza virus type or subtype isolated in the spring was shown by HA gene sequences to be in most cases genetically close to the epidemic strain of the next influenza season [142]. These same investigators compared the HA gene sequences of influenza B virus isolates during herald waves and later epidemic seasons in the same districts in Japan in 1987 and 1989 [143]. Analysis revealed that the herald viruses in one wave (1987) were genetically close to the winter isolates and were considered to be the parental viruses for the following influenza season, while in the other wave (1989) winter isolates were genetically and antigenically different from the herald viruses.

15.4.2 IMPORTANCE OF CHILDREN IN THE SPREAD OF INFLUENZA

Pre-school and school-age children are generally major vectors of infection during periods of antigenic shift and drift – the following observations serve as examples.

Buchan and Reid [18] describing the impact of influenza A virus on the Scottish island of Lewis in 1969–1970, observed that the presence of school-children in a household almost doubled the chance of infection in the household. Intensive surveillance in Seattle, Washington, of families with school-age children during 1975–1979 encompassed type B virus infections in 1975–1976, type A (H3N2) during 1975–1976 and 1977–1978, and type H1N1 in 1978–1979 [48]. Age-related infection rates were highest in children aged 5 to 9 years for H3N2 viruses and in teenagers for H1N1 and type B viruses. During the 1975–1976 season the overall infection rate was 18.4% for influenza type A (H3N2) and 16.6% for type B virus. The rates fell during 1976–1977 to 5.8% for influenza type A (H3N2) and 1.6% for influenza B virus, but during 1977–1978 they increased to 23.7% for influenza type A (H3N2), 4.8% for influenza type A (H1N1), and 3.2% for influenza B virus. Finally during 1978–1979 the rate for influenza type A (H3N2) fell to 0.4%, but for influenza type A (H1N1) it rose to 30.6%, and was 2.4% for influenza B virus. Thus, during three of four periods of virus activity influenza affected almost one-third of the population, the greatest burden occurring in young children.

Houston families with and without infection were also studied during a period of A/Victoria/75 (H3N2) virus activity in 1976 [179]. The overall frequency of infection in family members was 27.7%. Three determinants characterized those families who were infected with the virus: family size, density, and the presence of school-age children or children in day-care facilities. Indeed, as in the study of Buchan and Reid in Lewis [18] the incidence of infection and illness attributed to influenza among persons in families with school-children (38.5%) was more than double that in families without children attending school or day-care facilities (16.9%). However, during the Hong Kong pandemic it was noted the age-specific attack rates of all age groups were similar and adults were as likely as children to have been the first family case [36,91].

Influenza virus certainly has the potential for rapid spread within schools. During the pandemic in 1957, high attack rates were the rule in the closed communities of residential schools in England and Wales, frequently reaching 90% and often affecting the whole school within a fortnight [128]. Blunting of the epidemic curve and delay in the upswing until classes re-start have been observed when influenza first appeared in the community before the Christmas holidays [60]. During epidemics, a consistent finding has been a shift in the age-distribution of persons with virus-positive cultures [61]. During the early phase of epidemics a disproportionate number of cases have been older school-age children in the 10 to 19-year age range, but as epidemics progress an increasing proportion of infections were detected in adults and pre-school children.

A similar shift in age distribution seen with hospital admissions and influenza deaths during the first wave of Asian influenza further supports the concept that influenza virus first spreads mostly in children and then among younger sibs and adults [100,156]. A review of 541 patients admitted to hospitals in Dundee, revealed that during the first week patients under 20 years of age predominated (52 of 88, 59%); this contrasted with subsequent weeks when the number of patients over 50 years of age increased, reaching a maximum of 55% (21 of 51) in the fifth week [100]. In Holland during the early part of the epidemic 146 of 301 deaths (49%) occurred in under-29-year-olds; during the mid-phase 213 of 767 deaths (28%) occurred in this age group; and during the late epidemic period 25 of 137 deaths (18%) were among those aged up to 29 years [156].

15.4.3 INFLUENZA IN RESIDENTIAL HOMES

Influenza A and B viruses are important causes of infection within nursing homes and chronic care facilities. Outbreaks affecting over 60% of residents have been reported and many die or require hospitalization for severe complications [4]. During the winter of 1982–1983 in Genesee County, Michigan, Patriarca *et al.* [153] compared the characteristics of seven nursing homes with outbreaks caused by A/Bangkok/1/79 (H3N2)-like viruses with those of six homes in which there were sporadic infections only. The homes were similar in many respects including the physical characteristics of the facilities, visiting and staffing patterns, infection control practices and demography of the residents. However, homes with outbreaks were almost twice the size of those without and a lower proportion of residents were vaccinated. When assessed by multivariate analysis, the number of susceptible individuals was the most important factor in predicting the occurrence of an outbreak. Arden *et al.* [5] conducted a similar study in 64 nursing homes in Michigan. Preliminary analysis revealed a mean vaccination rate of 71% in 22 outbreak homes and 79% in 39 homes that did not experience an outbreak (non-significant difference). Overall, 25% of homes with vaccination rates of 80% or more experienced an outbreak of pyrexial influenza-like illness, implying that target vaccination rates of 80% or greater can not guarantee herd immunity.

15.5 TRANSMISSION AND PATHOGENESIS

15.5.1 TRANSMISSION

Influenza is spread via virus-laden respiratory secretions from an infected to a susceptible person. It is generally accepted that influenza viruses are transmitted by droplets several microns in diameter that are expelled during coughing and sneezing, rather than fine-droplet nuclei. Influenza virus can evidently survive drying at room temperature for some days and has been demonstrated in dust for as long as 2 weeks [43]. The human infectious dose 50 (HID_{50}) of influenza A virus is 127–320 tissue culture infectious dose 50s ($TCID_{50}$) for virus administered by nosedrops [29,30] but is significantly less, 0.6–3.0 TCID$_{50}$, for virus delivered by aerosol [2]. Pathological evidence suggests initial or early involvement of pulmonary alveolar cells, a site only accessible to droplets up to 5 μm in diameter.

15.5.2 INFECTION AND REPLICATION

The binding of influenza virus to erythrocytes and to host cells is mediated by interaction of the viral haemagglutinin with cell surface oligosaccharides containing sialic acid. Human H3 isolates have been shown to bind almost exclusively to sialyloligosaccharide structures terminated by the SA α-2, 6-Gal sequence and are sensitive to inhibition of binding and infection by a glycoprotein present in non-immune horse serum. Histological studies of fatal cases, nasal exudate cells, and tracheal biopsies indicate that virus replication may occur throughout the entire respiratory tract [87,112,114,116], the principal site of infection occurring in ciliated columnar epithelial cells [86,187]. It has been suggested that infection begins in the tracheobronchial epithelium and then spreads. Indeed, lesions in the tracheobronchial mucosa have been identified in bronchoscopic biopsies from young adults with uncomplicated Asian influenza [187] which correspond with, but are less severe than, those found in the trachea and bronchi of fatal cases [86].

In vitro studies suggest that the cycle of replication takes about 4–6 hours [165]. Thereafter virus is released for several hours before cell death and progeny virions initiate infection in adjacent cells, so that within a short period many cells in the respiratory tract are either infected, releasing virus, or dying. The pattern of virus replication in relation to clinical symptoms and immune responses has been studied by several investigators. Virus can be detected shortly before the onset of illness, usually within 24 hours, rises to a peak of 10^3–10^7 TCID$_{50}$/ml of nasopharyngeal wash, remains elevated for 24–72 hours, and falls to low levels by the fifth day [140]. In young chil-

dren virus shedding at high titres is generally more prolonged and virus can be recovered up to 6 days before and 21 days after the onset of symptoms [51,72,74]; in adults the quantity of virus shedding is related to the severity of illness and temperature elevation [140]. Because of the generalized symptoms present in uncomplicated influenza, viraemic spread from the respiratory tract has been suspected but attempts to demonstrate viraemia have been inconclusive – a few investigators have demonstrated viraemia, even before the onset of symptoms [107,141,174] but other attempts have been unsuccessful [108,131,136].

The incubation period of influenza ranges from 1 to 7 days but is commonly 2 to 3 days. This short period, coupled with the relatively high titres in nasopharyngeal secretions, the fairly lengthy periods of virus shedding (especially in children), and the relatively small amounts necessary to initiate infection, explain the explosive nature of influenza outbreaks.

15.6 CLINICAL FEATURES

15.6.1 ADULTS

Infection with influenza viruses can result in a wide spectrum of clinical responses ranging from asymptomatic infection to fulminant primary viral pneumonia (Table 15.2). The onset of illness is typically abrupt after an incubation period of several days which may depend in part on the severity of exposure and the immune status of the individual. Review of the symptoms and signs of illnesses due to HON1, H1N1, H2N2, and H3N2 viruses reveals no important differences [19,52,90,112, 164,176]. However, there have been differences in the proportion and character of the complications that have been noted.

Influenza B virus can also cause the same spectrum of disease as that seen following influenza A virus infection, and the frequency of severe influenza B virus infections requiring hospital admission may be several-fold less than that for influenza A [123] or may be similar

Table 15.2   Clinical features of influenza illness in serologically confirmed family outbreaks. (Data abstracted from [194]).

	Asian influenza (% of 186 cases with feature)
Onset	
Gradual	54
Sudden	46
Symptoms	
Cough	87
Headache	70
Sneezing	67
Nasal symptoms	65
Sore throat	57
Sweating	48
Shivers	46
Aches and pains	42
Malaise	15
Prostration	14
Drowsy	11
Delirium	10
Nose bleeds	9
Faint and giddy	7
Hoarseness	5
Vomiting	18
Abdominal pain	7
Diarrhoea	3
Complications	
Pneumonia	3
Otitis media	2
Most frequent initial symptom	Headache, sore throat
Most troublesome symptoms	Headache, cough
Days fever (median)	3
Days in bed (median)	3

[154]. In Tecumseh, Michigan, longitudinal studies of influenza in the community showed that type B influenza was associated with illnesses intermediate in severity compared with those associated with influenza A viruses and rhinoviruses [133]. This pattern was unrelated to age. During the period 1976–1981 influenza A H3N2 viruses produced the most severe illnesses, H1N1 viruses the mildest, and type B infections were intermediate [134].

Systemic features are often evident initially and include shivering, feverishness, headache, myalgia affecting the back and limbs, malaise and anorexia. Headache and sore throat are the most frequent initial symptoms; headache, myalgia and cough are usually the most troublesome. The early features are almost invariably accompanied by a non-productive cough, sneezing, nasal discharge or obstruction and less frequently by productive cough and substernal soreness. Photophobia and other ocular symptoms including lacrimation, burning and pain on moving the eyes and nausea occur in up to 20% of cases, hoarseness and abdominal pain are reported by less than 10%, and diarrhoea is a feature in less than 5%.

Fever is the most prominent sign of infection and although it is as high as 41°C in about one-quarter of cases it is more commonly 38–40°C. The pyrexia peaks at the height of systemic features and is typically of 3 days' duration, but may last for 1–5 days. Clinical examination often reveals a toxic appearance early in the course of the illness, the skin is often hot and moist, the face appears flushed, the eyes may have a glistening, injected, sometimes weepy appearance, the mucous membranes of the nose and pharynx are hyperaemic and devoid of an exudate, and breathing may appear laboured with a clear nasal discharge or blocked nose. Small, tender, cervical lymph nodes are palpable in up to one-quarter of cases and crackles and wheezes may be heard in a similar proportion. This constellation of symptoms and signs typically persists for 3–4 days, but cough, lassitude, and malaise may persist for 1–2 weeks after the fever settles. These features relate to a typical case, but during an outbreak cases may be either subclinical or present with other acute respiratory syndromes such as the common cold, pharyngitis, tracheobronchitis, or a systemic illness without respiratory features.

15.6.2 INFANTS AND CHILDREN

Although the manifestations of influenza A virus infection in infants and children are similar to those in adults there are some differences. Maximal temperatures tend to be higher among children, and below the age of 5 years febrile convulsions are prominent, occurring in 35–40% of hospitalized patients [16,60,64,104,157,195,196]. There is also a relative prominence of coryza, otitis media and gastrointestinal manifestations, notably vom-iting, abdominal pain and diarrhoea [60,84,85,157]. Infection in neonates may manifest itself as a pyrexial condition of uncertain aetiology [126]; it may present as croup or bronchiolitis, and lower respiratory tract complications seem more frequent than in adults [16,109,195]. Finally there is a much higher incidence of drowsiness and delirium than in adults.

Infants beyond the age when passively acquired antibody would provide protection represent an especially susceptible group. The risk of hospital admission for infants from low-income families in Harris County, Texas, was estimated to be 4 per 1000 during the A/Victoria influenza outbreak, and the mortality rate for hospitalized infants was 7.5% [64]. Review of three later epidemics (H1N1, H3N2 and influenza B viruses) in Harris County identified hospitalization rates of at least 5 per 1000, regardless of the type of influenza virus [154]. Both influenza A and B viruses represent a significant cause of serious lower respiratory tract infection in children. In one prospective study of 121 young seronegative children during an H3N2 virus outbreak in 1974, 60 children were infected and five developed clinical and X-ray evidence of pneumonia [195]. Kim and colleagues associated influenza virus infection with more than 5% of admissions to a children's hospital for acute respiratory tract illness [109] and showed that 14.3% of 860 patients with croup were infected with influenza virus. The mean period of hospitalization for either influenza A or B was about 8 days, but serious infections with influenza A virus were five times more common than with influenza B virus.

In a review of five studies of manifestations of influenza infections in hospitalized children it was reported that about one-half had complications affecting systems other than the respiratory tract [16,21,106,150,157]. Abdominal pain mimicking acute appendicitis was seen in patients infected with influenza B virus, and mortality rates of 1–4% were reported in four of the five studies. Other reports refer to the occurrence of acute myositis in association with both influenza A and B viruses [9,38,127].

15.7 THE COMPLICATIONS OF INFLUENZA

15.7.1 INFECTION IN THE ELDERLY AND EXCESS DEATHS

Analyses of mortality data have shown consistently that the elderly, especially those with certain chronic medical conditions, are at greatest risk from death during outbreaks of influenza. In temperate climates the incidence of deaths from cerebrovascular, cardiovascular and respiratory diseases increases during winter, and this seasonal variation is influenced by low temperature and respiratory viruses such as influenza and respiratory syncytial (RS) viruses. Additional deaths above the normal winter increase are recorded regularly in association with influenza epidemics and about 90% of these excess deaths are among people aged 65 years and over [6,173,182,183].

Over 10 000 excess deaths were documented in each of seven epidemics in the USA from 1977 to 1988, and more than 40 000 excess deaths occurred during each of two of these epidemics [132]. About 120 000 excess deaths were attributable to influenza in England and Wales during the 10 winters after influenza A/Hong Kong (H3N2) virus first arrived [182]. None was recognized in the UK during seven consecutive winters from 1978–1979 to 1984–1985 [22], but there were almost 30 000 excess deaths in Great Britain during the 56 days between 17th November, 1989 and 11th January, 1990 associated with the influenza epidemic at that time [6]. Generally about half the excess deaths during influenza epidemics are attributed to influenza, bronchitis and pneumonia, and many of the remainder to cerebrovascular and cardiovascular disease – implying that influenza is responsible for many hidden deaths.

In the 3 months subsequent to the 1989–1990 epidemic in Great Britain there was a deficit of almost 11 000 deaths [6], indicating that almost one-third of the deaths would have occurred in the absence of influenza. A small but consistent deficit of observed deaths was also noted by Eickhoff, Sherman and Serfling [44], in relation to mortality in the USA after the 1957–1958 pandemic, principally from June to August 1958. This reinforces a generally held opinion that influenza often kills those individuals who would die of their primary disease alone, in the absence of epidemic influenza, within 6 to 12 months.

Using a computer model, Serfling *et al.* [166] estimated excess pneumonia-influenza deaths during three major influenza epidemics in the USA in late 1957, early 1958, 1960 and 1963. Mortality among infants aged less than 12 months ranged from 3.7–11.6 per 100 000 (mean 7.3/100 000), but fell to 0.8–3.1 per 100 000 (mean 1.8/100 000) among those aged 1 to 4 years, and to 0.1–1.4 per 100 000 (mean 0.5/100 000) among 5 to 14-year-olds. Mortality then increased steadily, to 0.2–2.5 per 100 000 (mean 0.95/100 000) among 15 to 19-year-olds, to 0.9–2.7 per 100 000 (mean 1.6/100 000) among 20 to 44-year-olds, to 5–8.9 per 100 000 (mean 7/100 000) among 45 to 64-year-olds, and to 16.7–33.1 per 100 000 (mean 25/100 000) among 65 to 74-year-olds. Mortality was maximal in the over 75s, at 51.3–101.6 per 100 000 (mean 73.1/100 000).

Comparable rates for deaths certified as due to influenza were seen in England and Wales during the 15 years between 1974 and 1988. The lowest mortality (0.04 per 100 000) was again noted in 5 to 14-year-olds; mortality increased in successive 10-year age bands by three-fold, four-fold, six-fold, 11-fold, 32.5-fold, 100-fold and 765-fold to a mean of 30.6 per 100 000 (range 3.4–190.2) in those aged 75 years and over. Examination of 1989–1990 mortality data for England and Wales [34] showed that certified influenza deaths represent fewer than 10% of the excess mortality associated with influenza, so it is possible that mortality from influenza in the over 75s could be nearer 2% during the most severe epidemics. This high estimate is supported by mortality data for non-vaccinated, elderly, people in nursing homes. In Cardiff, for example, mortality rates attributable to influenza were 4–8% during a 3-year period [98]; rates of 3.7% and 0.2% were seen in Paris during A/Victoria/75 (H3N2) and B/Hong Kong/73 virus outbreaks [167]; a rate of 4.5% was seen in seven nursing homes in Michigan when A/Bangkok/1/79 (H3N2) virus was circulating [152]; and the mortality was 17.7 % during an influenza A/Arizona/80 (H3N2) virus outbreak in a New York home for the elderly [67].

Barker and Mullooly [8] studied pneumonia and influenza deaths among a population of 230 000 in relation to age and presence of chronic medical conditions during two epidemics (1968–1969 and 1972–1973). The death rate was 9 per 100 000 for the over 65s without high-risk conditions and increased to 217 per 100 000 for the presence of one high-risk condition and 306 per 100 000 for two or more. The death rate among the fit elderly was comparable with that of 10 per 100 000 among 45 to 64-year-olds with one high-risk condition, and higher than that in 15 to 44-year-olds with one high-risk condition (no deaths among 6260 subjects). Nguyen-Van-Tam and Nicholson examined certified influenza deaths among a population of 892 000

in Leicestershire during the 1989–1990 epidemic [145]. Influenza was reported as the cause of death for 18 men and 29 women aged 67 to 97 years (mean age 84 years); 41(87%) were aged ≥75 years. There were no certified influenza deaths in younger subjects with chronic medical conditions, and the estimated mortality for the fit elderly was approximately 7 per 100 000 (Table 15.3), i.e. similar to the rate observed by Barker and Mullooly [8]. Among non-residential subjects the rate for influenza-certified deaths was 11.6 and 23.1 per 100 000 for those with lung and heart disease, respectively. The major impact of influenza was seen in the residential care facilities where the rates were 343, 499 and 2703 per 100 000, respectively, for people with one, two and three or more medical conditions. On the basis of these mortality data it appears that pneumonia and influenza deaths in Oregon, and influenza deaths in Leicestershire, are more numerous in the fit elderly than in younger subjects with a single 'high-risk' medical condition.

15.7.2 INFLUENZA DURING PREGNANCY AND THE EFFECTS ON THE FETUS AND OFFSPRING

Women during the second and third trimesters of pregnancy are at increased risk of

Table 15.3 Influenza-associated death rates (certified as influenzal) during the 1989–1990 influenza A epidemic in Leicestershire. (Data abstracted from [145].)

Status	No. of influenza deaths	Estimated no. in the population	Deaths per 100 000
Residential patients			
With 1 medical condition	8	2334	343
With 2 medical conditions	5	1002	499
With ≥3 medical conditions	6	222	2703
With lung disease	5	1590	314
With heart disease	12	1668	719
Non-residential patients			
Without medical condition	3	45399	6.6
With lung disease	6	51534	11.6
With heart disease	15	65031	23.1

hospital admission, severe pulmonary complications of influenza and death. The Public Health Laboratory Service Report on the influenza epidemic in England and Wales during 1957–1958 notes that 12 of 103 fatal cases in females aged 15 to 44 years of age were pregnant, which is about double the expected proportion for this age group [159]. In Holland, during the first wave of Asian influenza, 1230 influenzal deaths were reported among a population of 11 million, and 11 deaths occurred during pregnancy [156]; in the group aged 20–39 years the mortality of pregnant women was twice that of non-pregnant women. Similarly, 10 of 24 females of childbearing age among an American series of 91 patients with Asian influenzal pneumonia were pregnant, all but one in the last trimester or at term, and this was three to four times the expected proportion [155]. During the Asian influenza epidemic in New Orleans, Burch, Walsh and Mogabgab [19] admitted 34 women of whom 13 were pregnant – three in the second trimester and 10 in the third. Here again, the proportion of admissions who were pregnant was several times higher than expected. The evident increase in mortality associated with pregnancy has also been recognized during an interpandemic period. Ashley, Smith and Dunell [6] compared cause of death among a 1 in 15 random sample of deaths that occurred between 25th November 1989 and 8th January 1990 with deaths occurring during the same period in 1985–1986. Analysis revealed a fourfold increase in deaths associated with pregnancy during the 1989–1990 epidemic, and the data suggest that the epidemic in the UK accounted for about 90 excess maternal deaths during pregnancy.

Most of the above reports do not make it clear whether severe influenza during pregnancy was associated with a chronic medical condition. In the series reported by Oswald, Shooter and Curwen [148], seven of 379 patients with influenzal pneumonia were pregnant and two died, one having mitral stenosis in addition. In the series reported by Petersdorf *et al.*, one of 10 pregnant women with influenza went into premature labour, delivered a dead fetus, and died 15 hours later [155]. Neither she nor the other nine pregnant women had any other risk factor. Ramphal, Donnelly and Small [160] describe a fatal A/Texas/77 virus infection in a woman with ventricular septal defect and term pregnancy. Two of 15 women with Asian influenzal pneumonia reported by Louria and colleagues were pregnant [116]; one had rheumatic heart disease with mitral stenosis but both recovered. During the same epidemic Martin *et al.* reported 10 fatal cases among women aged 15 to 44 years, of whom four were pregnant [119]. The number who were pregnant was several times higher than expected, and only one of the four pregnant women had a chronic medical condition, myasthenia gravis. Giles and Shuttleworth [58] reported the death of a 17-year-old pregnant girl among 14 women who died from influenza in Stoke on Trent; post-mortem examination revealed cardiac hypertrophy but she was not known previously to have had cardiac disease. From this small sample it would appear that about 75% of the deaths during pregnancy occur in women without a chronic medical risk factor.

Influenza complicating heart disease in pregnancy appears especially serious. Govan and Macdonald identified nine such patients, of whom four died [65]. They presented with an acute febrile illness and within 12 hours of onset developed fulminating bronchopneumonia. Ketosis was a prominent finding. Five patients went into labour at this stage, of whom three died; two of the babies were stillborn and three died within 48 hours. Post-mortem examinations revealed extensive haemorrhagic bronchopneumonic consolidation.

Few attempts have been made to demonstrate transplacental passage of influenza virus to the fetus. Martin *et al.* [119] and Ramphal, Donnelly and Small [160] failed to identify the presence of influenza antigen in fetuses from four mothers with fatal influenza. Yawn *et al.* [198], however, recovered the virus from the extrapulmonary tissues of a mother who died in the third trimester and from the amnion and myocardium of the fetus, although there were no abnormalities in the tissues from which the virus was isolated. Influenza A/Bangkok (H3N2) virus was isolated from the amniotic fluid of a mother with 'amnionitis' and influenza infection at 36 weeks of gestation. The infant, who was born at 39 weeks, had evidence of infection but remained well [125].

There have been several reports of a slight increase in congenital abnormalities following influenza virus infections during pregnancy [25,66,70] but there is no consistent association between specific defects and influenza, and the virus has not been conclusively implicated. Hardy *et al.* [76] noted major congenital abnormalities in 5.3% of 80 women whose influenza infections occurred during the first trimester compared with 2.1% of 183 women infected during the second trimester and 1.1% infected during the third. Wilson and Stein [191], during the 1957 outbreak, noted no increase in congenital abnormalities among infants conceived during the 3-month period when influenza was epidemic and whose mothers had serologic evidence of infection with influenza virus. Other studies in the USA and the UK have been reported which indicate a possible relationship between maternal influenza and childhood leukaemia [7] and schizophrenia.

Review of influenza deaths during the period 1921–1932 revealed an increase in mortality from premature births in association with influenza [175]. Wynne-Griffith *et al.* [197] showed that early neonatal mortality in the second quarter of 1970 after a major influenza epidemic was slightly but significantly higher (from 9.88 per 1000 live births in 1969 to 10.77 in 1970) than in the corresponding quarter of the previous year. Analysis of infant mortality over the previous 25 years indicated that similar increases occurred in relation to four of the other five major

influenza epidemics during the period, notably in 1951,1953,1959 and 1961; the exception was the Asian flu epidemic of the autumn of 1957. Ashley, Smith and Dunell [6] showed that perinatal deaths increased 1.6-fold during the 1989–1990 epidemic, namely an increase of 255 from approximately 465 to approximately 720, as compared with a similar period in 1985–1986. Hardy *et al.* [76] reported that the incidence of stillbirths was higher in 332 symptomatic pregnant women with serologically confirmed influenza than in 206 asymptomatic women with serologically confirmed influenza or in 73 uninfected women [72].

15.7.3 RESPIRATORY TRACT COMPLICATIONS

Pneumonia

The H3N2 influenza virus epidemic in 1989–1990 provided Connolly, Salmon and Williams with the opportunity of reviewing 342 episodes reported as influenza in the general practice surveillance of infectious diseases in Wales [26]. Pneumonia was the second most common respiratory complication, occurring at a rate of 2.9%. Comparable figures for pneumonia complicating Asian influenza in the primary care setting are 4.8% of 391 [59] 1.8% of 170 [90], 3% of 187 [194], 1.9% of 930 [52], 1.5% of 1990 [69], 3.5% of 750 [54], 4.9% of 161 [12], and 5% of 542 [42]. Overall, the incidence of pneumonia complicating influenza appears to be approximately 3%. In the hospital setting pneumonia was seen in 11.1% of 494 people with severe or complicated Asian influenza [163], 13% of 77 children admitted with Hong Kong influenza [16], 19.7% of 76 subjects with Asian influenza [19], and 48% of 541 cases with Asian influenza [100].

The course and outcome of influenzal pneumonia is influenced by the presence of pre-existing disease, secondary staphylococcal infection and age. The series of patients reported by Jamieson *et al.* [100] exemplifies the importance of pneumonia, age and pre-existing disease. Of 541 hospitalized cases of

Asian influenza, a worse prognosis was noted in those with pneumonia, and of the 262 patients with pneumonia the course was particularly severe among those with pre-existing disease, mostly chronic respiratory or cardiac disease [100]. Case-fatality rates increased from 1% (3 of 279) among admissions with uncomplicated influenza to 11% (30 of 262) among those with pneumonia; from 0.5% (1 of 200) in those aged <20 years, through 2.2% (3 of 135) in those aged 20–49 years, to 14% in those 50 years and over; and from 3% in pneumonia patients aged <50 years to 22% in those aged ≥ 50 years. The overall mortality among those with pre-existing disease was 51%. Giles and Shuttleworth [58] found that 35 of their 46 fatal cases had pre-existing disease.

Children with congenital malformations and chronic diseases are also at increased risk of severe lower respiratory tract disease. Among 77 children hospitalized with Hong Kong influenza, lower respiratory tract infection occurred in 13 (56%) of 23 with pre-existing conditions, compared with 10 (19%) of 54 without [16]. Oswald, Shooter and Curwen [148] compared staphylococcal and non-staphylococcal influenzal pneumonias; in the staphylococcal group, 71 (46%) of the patients were severely ill, compared with 34 (23%) of those with non-staphylococcal pneumonia, many of whom were elderly and had associated chronic respiratory or cardiac disease. The overall mortality rate of 28% for staphylococcal pneumonia was higher than that of non-staphylococcal pneumonia (12%). Moreover, the mortality in the staphylococcal group was similar at all ages, whereas that in the non-staphylococcal group was concentrated in those aged over 55 years.

Influenza complicated by pneumonia generally presents with influenzal symptoms and although there are no clear distinguishing features, cough and chest pain were observed more frequently than in uncomplicated influenza – 54% versus 79%, and 31% versus 51%, respectively, in the series reported by Jamieson *et al.* [100]. The chest pain is of two types – substernal, which presumably reflects the tracheitis commonly seen at post mortem, and pleuritic. During the Asian influenza outbreak the onset of illness in fatal cases was typically abrupt and within 24 hours of onset of symptoms pneumonic or other serious symptoms were observed in one-third of patients [159]. Deterioration was much more rapid in the young than in the elderly with 56% of patients under 5 years of age gravely ill within the day of onset compared with 16% of those > 64 years [159]. In fatal cases deterioration was particularly rapid; in the Public Health Laboratory Service series of 477 deaths, no fewer than 86 occurred before admission to hospital and two-thirds died within 48 hours of admission [159].

Primary viral pneumonia

Louria *et al.* [116] and Martin *et al.* [119] provided the first detailed accounts of primary influenza virus pneumonia that occurred during the Asian influenza pandemic in 1957–1958. The incidence of primary influenzal pneumonia is unclear, but Louria *et al.* saw six cases among 33 with pulmonary complications of influenza that were sufficiently severe to be admitted. All six had underlying cardiac disease. Since the overall incidence of pneumonia during influenza is about 3%, viral pneumonia uncomplicated by secondary bacterial infection evidently occurs in at least 1 in 200 infections. Hers, Masurel and Mulder [88] isolated influenza virus from the lungs of 72% of 148 virologically confirmed fatal cases. Overall, 20% were identified as having influenza virus pneumonia without secondary bacterial infection. These observations indicate that primary viral pneumonia, without secondary bacterial infection, is not uncommon and can be life-threatening, and that the majority of bacterial pneumonias complicating influenza occur in lungs infected with the virus.

After a brief period of typical influenzal symptoms each of the patients in Louria's series [116] became severely dyspnoeic (making speech difficult) and cyanosed. None had chest pain, four produced bloody sputum and two had mucoid sputum. Auscultation revealed fine inspiratory crackles, inspiratory and expiratory wheezes, and no signs of consolidation. The total leucocyte counts were elevated with a polymorphonuclear leucocyte predominance despite the absence of concurrent bacterial infection. Chest radiography mostly revealed a diffuse perihilar infiltrate. Profound hypoxia was resistant to oxygen administered by face-mask and oxygen tent but none of the patients was ventilated. Influenza virus was recovered from the lungs of each of five patients who died. Histological studies revealed tracheitis, bronchitis, bronchiolitis with haemorrhagic areas, and loss of ciliated epithelial cells, especially in the lower lobes. The alveolar spaces contained neutrophils, mononuclear cells, fibrin and fluid, and intra-alveolar haemorrhage was common in the lower lobes.

Combined bacterial and viral pneumonia

Louria *et al.* [116] report nine patients in whom concomitant viral and bacterial pneumonia was considered likely, and 15 in whom the bacterial infection was a 'late' secondary complication. All nine patients with concomitant viral and bacterial pneumonia were severely ill on admission; eight were cyanosed, all had purulent or bloody sputum with signs of consolidation, crackles or wheezes. Six of the nine had underlying chronic disease and one was 6 months pregnant. Chest radiography revealed lobar infiltrates and/or bilateral nodular perihilar infiltrates; sputum examination revealed large numbers of polymorphonuclear leucocytes, even in patients with leukopaenia, and large numbers of bacteria. Influenza virus was isolated from the lungs of three of the four patients who died. Seven of the 15 patients in whom bacterial pneumonia was considered a 'late' event had chronic underlying disease and one was 8 months pregnant. Ten experi-

enced an improvement in influenzal symptoms before pulmonary involvement, and pulmonary symptoms blended with the initial illness, but occurred several days after its onset in the remainder. Chest X-radiography revealed lobar or lobular involvement; and only one of the 15 patients died. Angeloni and Scott [3] reported nine deaths among 41 adult patients who developed pneumonia from 1–14 days after onset of influenza. Post-influenzal pneumonia caused a higher mortality in patients with long-standing chest disease (six deaths in 13 cases) than in those who were previously fit (three deaths in 28 cases).

In the cases with secondary bacterial infections reported by Louria *et al.* [116] *S. aureus* was recovered from four of five fatal cases and from eight of 19 who survived (NS). Although fatal influenzal–staphylococcal pneumonia had been recognized previously, the Asian influenza epidemic highlighted its importance in fatal cases. It was commonly encountered in pathological material – in 62% of 467 deaths in the UK [159], 71% of 62 deaths in London hospitals [148], 59% of 148 fatal cases in Holland [88], 45% of 11 post mortems reported by Martin *et al.* [120], 33% of 46 fatal cases in Stoke on Trent [58], and in eight of nine cases of fulminating influenzal pneumonia [161]. Two large studies of hospitalized patients in Dundee [100] and London [148] illustrate that many of these staphylococcal infections were acquired nosocomially. This fact was again emphasized when the bacterial flora of the lung or sputum was related to the interval between admission to hospital and death [159]. The proportions of patients over 65 years who yielded staphylococci were 13% among those who died on the day or day after admission, 20% after 2 or 3 days, 45% after 5 to 7 days, and 80% after 8 days or more.

Lung abscess

Comparison of chest radiographs and necropsy specimens indicates that radiology does not reflect the true incidence of abscess formation as a complication of influenza [148].

Of 300 influenza-associated pneumonias reported by Oswald, Shooter and Curwen [148] radiological cavitation was observed in 14% of secondary staphylococcal pneumonias, and in 2% of the others. Lung abscess was diagnosed radiologically in eight (29%) of 28 previously fit adults who developed symptoms of pneumonia up to a fortnight after an influenzal illness [3] and frank abscess formation was found at post mortem in four of 41 patients by Giles and Shuttleworth [58].

Croup

Influenza viruses are among the aetiological agents associated with acute laryngotracheobronchitis (croup), and croup associated with influenza A virus appears to be more severe, resulting in more frequent hospitalizations and tracheostomies [97] but less frequent than that associated with parainfluenza virus or RS virus infections. Kim *et al.* [109] showed that whereas influenza viruses (particularly influenza A) ranked second overall after the parainfluenza viruses as a cause of croup, influenza A virus infection was detected in proportionately more croup patients than the most important croup-producing virus, parainfluenza type 1, when the peak months of influenza A virus and parainfluenza virus activity were compared. H3N2 viruses appear to have a greater predilection to produce croup than H2N2 viruses. During the years 1957–1976, influenza A virus infection was detected in approximately 8% of hospitalized children with croup during the H2N2 virus era and 24% of such patients during the H3N2 era [109].

Asthma

In normal individuals with uncomplicated influenza, pulmonary function tests have revealed frequent airway hyperreactivity, peripheral airway dysfunction, and abnormalities in gas exchange that can be prolonged for some weeks after clinical recovery [73,93,102].

Bendkowski [12], a general practitioner, compared the incidence of respiratory complications in three groups of patients, 124 with allergies (of whom 55 had asthma), 59 who were non-allergic but came from allergic families, and 161 who were non-allergic. He noted that asthma complicated 38% of influenzal infections in the allergic group and clinical improvement was slow. Influenza virus infections have consistently precipitated attacks of wheezing in both asthmatic children and adults. For example, in a study by Minor *et al.* [129] which involved 41 children aged 3–17 years and eight adults with a history of 'infectious' asthma, 55% of all respiratory infections precipitated asthma. Five patients had influenza A virus infections and four of these were associated with asthma. In adult asthmatics, almost 50% of upper respiratory tract viral infections are followed by asthma and a similar percentage of asthmatic episodes are associated with a virus, chiefly rhinoviruses and coronaviruses [147]. Severe epidemics of influenza result in a small but significant excess mortality attributed to asthma. Housworth and Langmuir [95] studied asthma excess mortality during seven periods of influenza activity between 1957 and 1966. Asthma deaths increased by 19–46% during influenza A virus outbreaks in 1957, 1958, 1960 and 1963, but were either insignificant or barely significant during the milder epidemics with some influenza B virus activity. During the worst period excess mortality from asthma increased by an estimated 3.8 per million population.

Bronchitis

Acute bronchitis is the most common lower respiratory tract complication of influenza. Bendkowski [12] studied 161 subjects without a history of allergy and noted acute bronchitis during Asian influenza in 30%. Among 350 admissions with Asian influenza in Kuwait, Guthrie, Forsyth and Montgomery [69] diagnosed bronchitis in 22%. Among 76 children admitted to hospital during the Hong Kong epidemic, Brocklebank *et al.* [16] found bronchitis in 12%. During the H3N2 virus outbreak in 1989–1990, Connolly, Salmon and Williams [26] found evidence of acute bronchitis in 19% and noted the risk of bronchitis complicating influenza to be greater in patients with pre-existing diseases, regarded as an indication for vaccination, and also in the elderly.

Studies of chronic bronchitics have shown that one-quarter to two-thirds of exacerbations are associated with viruses, and data collated from several studies have revealed serological evidence of influenza A and B virus infection in approximately 4% of exacerbations. In Stock's review of influenza epidemics, mortality from both acute and chronic bronchitis was noted to have increased during the years 1921 to 1933 [175]. During the worst periods deaths from acute bronchitis increased by 93% and from chronic bronchitis by 52%.

15.7.4 COMPLICATIONS IN DIABETICS

As long ago as 1889-1890 it was noted that there were twice as many 'diabetes' deaths in Paris during the pandemic as compared with the average number during the three previous years [151]. In Holland in 1957 the first wave of the 'Asian' influenza pandemic saw an increase of 61 'diabetes' deaths among a 14.9 million population, representing a rise of 25% compared with the same period in previous years [156]. The actual increase was 4.6-fold lower than excess respiratory deaths and 15.5-fold lower than excess cardio- and cerebrovascular-related deaths. Almost 90% of 'diabetes' deaths occurred in the over 70s, and 99% were over 50 years of age. The same pandemic in the USA saw 'diabetes' deaths increase by an estimated 3.62 per million population (by 9.2%) in late 1957, and by 3.25 per million (7.9%) in early 1958 [95].

The interpandemic years also see increased death rates among diabetics. Thus, 'diabetes' deaths increased by approximately 10% during the five most severe epidemics of 1921 to 1932 [175] and by 5–12% during four of five USA epidemics during 1960–1966 [95]. Endocrine deaths (mostly diabetic) increased by about 1350 (i.e. by 30%) in England, Wales and Scotland during the 1989–1990 epidemic, as compared with 1985–1986, but the actual rise was eight-fold lower than excess circulatory deaths and approximately 17.5-fold lower than excess respiratory deaths [6]. Older-onset diabetics are 1.7 times more likely to die from pneumonia and influenza as compared with the general population, and 1 in 33 die from these conditions overall [137]. A similar but non-significant trend for increased pneumonia and influenza deaths has also been observed among younger, insulin-dependent diabetics [137].

Hospitalization rates for 'diabetes' were marginally increased (by 4%) during the 1989–1990 epidemic in the USA, but mortality among 'diabetes' admissions increased by two-thirds to 6% [123]. During 1976 and 1978, hospital admissions in Holland for influenza were six times more common among diabetics than controls with duodenal ulcer, and admissions for ketoacidosis increased by 50%, as compared with years with low influenza activity, from 78 in 1976 to 152 in 1978 [15]. There were four times as many pneumonia deaths among diabetics in 1976 and 1978 (57 versus 15) compared with years with little influenza activity, and twice the number of ketoacidosis deaths (63 versus 29). Bouter *et al.* [15] estimated that during the 1978 epidemic 1 in 1300 patients with diabetes mellitus were hospitalized for pneumonia, 1 in 260 with insulin dependence were hospitalized for ketoacidosis, and 1 in 1000 with insulin dependence died in hospital. The combination of pneumonia and diabetes appears especially serious; in the series of influenzal pneumonias reported by Oswald, Shooter and Curwen [148] six of nine diabetics died.

15.7.5 MYOCARDITIS AND PERICARDITIS

Myocarditis and pericarditis occasionally complicate fatal influenza, and electrocardiographic abnormalities have been reported in some uncomplicated cases of influenza A virus infection [19,57,105]. Louria *et al.* [116] studied the course of influenza in 15 patients with underlying heart disease and found no electrocardiographic evidence of myocarditis. However, autopsies on eight patients revealed one with a widespread, acute, necrotizing myocarditis, and the striking polymorphonuclear leucocyte infiltration of the myocardium appeared to be of recent origin. Martin *et al.* [119] described one case among 32 who died from influenza. The diagnosis was not suspected antemortem and no signs of myocarditis other than tachycardia were observed. Histological studies revealed marked muscle fibre necrosis with a cellular infiltrate of mononuclear cells, lymphocytes and occasional plasma cells. Focal interstitial myocarditis of low degree and questionable significance was observed in nine of the remaining 31 cases. Attempts at virus isolation from cardiac tissue of patients with myocarditis associated with fatal influenzal pneumonia have generally been unsuccessful. Virus has been recovered from pericardial fluid and both influenza A and B virus infections have been associated with myopericarditis [1]. Post-mortem studies of cardiac tissue of 46 patients dying from Asian influenza revealed myocardial oedema as the only constant finding [58].

15.7.6 NEUROLOGICAL COMPLICATIONS

Neurological complications of influenza have long been described but appear to be rare [41]. Apart from Reye's syndrome, a range of neurological complications have been reported in association with influenza, including irritability, confusion, convulsions, psychoses, neuritis, Guillain–Barré, syndrome, coma, transverse myelitis and encephalomyelitis. Their pathogenesis remains unclear but it is likely that the pyrexia, hypoxia and pH abnormalities that accompany influenza are responsible for a toxic encephalopathy in many cases.

The Asian epidemic appears to have stimulated an interest in the neurological complications of influenza, or conceivably H2N2 virus was particularly neurotropic. Dubowitz [41] observed two children who became comatose and recovered rapidly. Horner [92] described four children and a young woman who developed neurological disorders several weeks after an influenzal illness. Three were described as having encephalopathies; one recovered completely except for a mononeuritis, one recovered but was left with severe sequelae, and one died and showed extensive but patchy demyelinization of the central nervous system. A fourth patient developed acute cerebellar ataxia which improved completely within 12 days, and the final patient developed a disseminated encephalomyelitis with prominent features of a transverse myelitis. Four cases were reported by McConkey and Daws [124]; all four patients developed neurological features 4 or 5 days after onset of the respiratory illness. Three became drowsy or semi-comatose and one developed a vertical gaze palsy. Electro-encephalograms (EEGs) in two of three patients were grossly abnormal; all recovered and the EEGs soon reverted to normal.

Flewett and Hoult [46] in an analysis of neurological complications in 18 cases seen during the 1957–1958 outbreak, categorized patients into those with 'encephalopathy' seen at the height of influenza, post-influenzal encephalitis which developed 3 days to 2 weeks after onset of influenza, and Guillain–Barré, syndrome. Fatal encephalopathy and encephalitis appears to be rare. In the UK during the four seasons 1955–1956 to 1958–1959 there were 19 622 certified influenza deaths with 52 fatal cases of 'encephalitis' – a rate of 2.6 fatal cases of 'encephalitis' per 1000 fatal cases of influenza. Virus has occasionally been recovered from the cerebrospinal fluid (CSF) or brain of patients with 'encephalopathy' [148]. Hoult and Flewett [94] described the histological features in the brain of five fatal cases of encephalitis. No gross abnormalities were found except in one case in which there was acute haemorrhagic leucoencephalitis. Other investigators have also commented on the haemorrhagic leucoencephalitis and noted vessel necrosis, perivascular oedema, inflammatory infiltrates, haemorrhage and demyelination. Cerebral abscess formation and bacterial meningitis are also recognized [148]. According to Osler, 'almost every form of disease of the central nervous system may follow influenza', and others have also commented on a variety of mental disturbances associated with the 1918–1919 pandemic.

Acute psychosis developing 2 to 10 days after the onset of Asian influenza possibly represents another manifestation of encephalitis; electroencephalograms in three patients with psychoses were all diffusely abnormal and improved slowly over a period of weeks [13]. Lloyd-Still [115] noted the onset of acute psychoses, with auditory and visual hallucinations, in 19 patients during or after an attack of presumed influenza. All recovered quickly.

Subtle changes in brain function

Influenza B virus was one of several viruses studied at the MRC Common Cold Unit in Salisbury for its ability to affect simple measures of performance [171]. Volunteers infected with influenza B viruses had significant impairments of reaction times as measured by the ability to press a button as soon as possible after seeing a black spot appear on a television screen. A visual search task involving five possible target letters was also significantly impaired in the group infected with influenza B virus. Various tests were applied but only the attention tasks were impaired. Similar tasks were assessed in students infected naturally with influenza B virus. Comparison of baseline and symptomatic periods revealed a 37.5% increase in the variable fore-period simple reaction time, a 13% reduction in response time to repeated numbers detection task, and a significant reduction in a categorical search task [172].

Reye's syndrome

Reye's syndrome, a multisystem disorder which is characterized by encephalopathy and fatty liver degeneration, typically follows an acute viral infection in young children, most commonly influenza B and varicella, but also gastrointestinal infections. Clustering of cases occurs during the winter months and outbreaks of Reye's syndrome have been associated with epidemics of both influenza A and B and salicylate use. During the 1974 influenza outbreak, Reye's syndrome occurred with an estimated rate of 31 to 58 cases per 100 000 influenza B virus infections in children [27,28]. Otherwise, the evaluated incidence of Reye's syndrome has been approximately 0.5 per 100 000 population aged less than 18 years in the USA. An increasing body of data has revealed a strong association between the use of salicylates and Reye's syndrome and, possibly because of reduced salicylate usage, recent trends in the USA and UK indicate a decreased incidence of cases. In the UK, a recent active surveillance of Reye's syndrome revealed an annual incidence of 1 case per million population aged <16 years [144]. The pathogenesis of the condition remains obscure. Isolation of influenza virus from the liver, muscle biopsy specimens, CSF and muscle from a survivor of Reye's syndrome has led to the proposal that viral dissemination may play a role.

15.7.7 TOXIC SHOCK SYNDROME

Toxic shock syndrome, an illness characterized by fever, hypotension and erythroderma followed by desquamation, is found in association with TSST-1- or enterotoxin-producing *S. aureus* infection. MacDonald *et al.* [117] identified nine cases of severe hypotension or death compatible with toxic shock syndrome as a complication of influenza or influenza-like illness during an epidemic of influenza in Minnesota. Four of the patients had a proven influenza B virus infection, and *S. aureus* was cultured from two individuals; one isolate produced TSST-1 and both produced enterotoxin. The authors concluded that toxic shock syndrome may be a rare complication of influenza, and that it may occur more frequently than Reye's syndrome. The mortality rate from toxic shock syndrome complicating influenza is 60%. It has been proposed that a plague in Athens in 430 BC was due to the complications of a toxin-producing *S. aureus* during an influenza epidemic. Interestingly, Jaimovich *et al.* [101] recently reported four cases of influenza B virus infection associated with a non-bacterial septic shock-like illness. Altogether 14 cases of toxic shock syndrome following influenza or influenza-like illness have been described, but influenza A virus infection has been documented serologically in only two cases [184].

15.7.8 MYOSITIS, MYOGLOBINURIA AND RENAL FAILURE

Myalgia affecting the legs and back is a well-recognized feature of influenza and occurs early during the course of the illness. In contrast, myositis (and myoglobinuria with or without renal failure) usually occurs in the recovery phase from a typical influenzal illness.

Middleton, Alexander and Szymanski [127] reported 26 children who developed bilateral lower-limb myositis, characteristically after a period of rest and at a time when the respiratory symptoms and signs began to subside. Influenza B virus infection was proved in 20 cases and type A infection in one. In a further series of paediatric cases, similar evidence of influenza B virus infection was found in 11 of 17 cases [39]. In Middleton's series, leg pains and muscle tenderness lasted 1–5 days and no true muscle weakness was apparent, though

the children often refused to walk, or did so with a bizarre gait. Muscle enzymes were elevated in two-thirds of the cases, but attempts to isolate influenza virus from blood were unsuccessful. In adults, the myositis tends to be more diffuse. Muscle cell necrosis with and without cellular infiltrates can occur, but the pathogenesis of the condition remains obscure. The condition is usually benign and of short duration, but rhabdomyolysis with myoglobinaemia, myoglobinuria and acute renal failure have been described in severe adult cases occurring in association with influenza A virus infection [130,135,169,199]. During outbreaks influenza is probably the leading cause of acute myoglobinuric renal failure. Histological examination reveals normal glomeruli with focal tubular necrosis and pigmented casts in some tubules.

Other renal disease

Glomerulonephritis and Goodpasture's syndrome have been documented in several patients with influenza infections [188,192]. A lesion in the kidney with tubular changes was reported by Beswick and Finlayson [14]. Judging by the absence of clinically important renal complications during influenza and the dearth of pathological abnormalities in renal tissue, it seems that influenza rarely affects the kidneys.

15.8 ANTIVIRALS

15.8.1 ADAMANTANES

Amantadine and other adamantanes such as rimantadine were first shown to inhibit influenza A virus in the 1960s. Amantadine and rimantadine inhibit H1N1, H2N2 and H3N2 strains of influenza type A virus [68,80] including 'new' epidemic strains, so it is to be expected that future variants, including pandemic strains, will be similarly inhibited. Neither agent inhibits influenza B virus strains and higher concentrations than can be achieved safely in humans are required to inhibit parainfluenza and RS viruses. Drug-resistant strains of influenza A virus, which exhibit complete cross-resistance between amantadine and other adamantanes [10,77] can be readily produced in the laboratory and are being recovered increasingly from humans [11,81,83,110,121]. The molecular basis of resistance involves a single amino acid change in one of four amino acid residues in the M2 membrane protein [10,77,78]. Rapid selection and apparent transmission of drug-resistant influenza A viruses have resulted in treatment failures when index cases and family contacts were given rimantadine [11,81] and the emergence of drug resistance was implicated in occasional treatment failures when amantadine prophylaxis was initiated during a nursing home outbreak of influenza [121]. Illnesses caused by resistant strains are probably no more severe than those caused by wild strains of the virus, and although the emergence of virus drug resistance is of paramount concern, no reduction of drug efficacy was observed in the Soviet Union over a 20-year period when 142 227 patients were treated with rimantadine [110].

Currently, only amantadine is licensed for use in the UK. Amantadine is almost completely absorbed after oral administration and is excreted unchanged in the urine. Because of reduced renal clearance in the elderly and in patients with impaired renal function, care should be taken to ensure that the drug does not accumulate to toxic levels [146]. Minor neurological symptoms including insomnia, light-headedness, difficulty in concentration, nervousness, dizziness and headaches develop in 5–20% of individuals receiving 200 mg daily [17]. Other adverse effects include anorexia, nausea, vomiting, dry mouth, constipation and urinary retention. They arise mostly during the first few days of medication and disappear quickly when amantadine is discontinued. Because embryotoxic and teratogenic effects have been described in rats,

but not rabbits, receiving 15 times the usual human dose [149], treatment with amantadine is not justified in women of childbearing potential, except perhaps in life-threatening influenzal pneumonia. Amantadine is contraindicated in epilepsy and gastric ulceration, and should be used cautiously in patients with cardiovascular or renal disorders or those with cerebral atherosclerosis, i.e. individuals who are at increased risk from influenza.

Before amantadine is prescribed, laboratory and epidemiological evidence of an outbreak of influenza A in the community should exist. Taken prophylactically amantadine is about as effective as influenza virus vaccine in preventing influenza A virus infection. Controlled trials suggest that prolonged administration of amantadine, 200 mg daily, during community outbreaks of influenza is about 50% effective in the prevention of infection but 70–100% effective in the prevention or amelioration of illness [40,185]. This distinction may be a desirable feature of prophylaxis since subclinical infection could confer immunity against reinfection. If amantadine or rimantadine is used for both treatment and post-exposure prophylaxis in homes, however, rapid selection and apparent transmission of drug-resistant influenza A viruses can occur, and the drug may provide little or no protection [11,56,81,121]

Because of minor CNS side effects produced by amantadine, prophylaxis is most appropriate for persons in high-risk groups for whom vaccination is indicated [37,190] Indeed, WHO recommends amantadine or rimantadine prophylaxis for elderly and high-risk people in institutional settings to augment protection afforded by vaccination [190]. If vaccination has been overlooked high-risk individuals should still be vaccinated after an outbreak appears locally, but the development of an immune response usually takes several weeks and this vulnerable period (and beyond) can be covered by chemoprophylaxis. When vaccine is unavailable, or the influenza

A strain causing an epidemic differs markedly from the vaccine strain, amantadine should be given for the entire duration of the outbreak, a period of about 4–8 weeks. Chemoprophylaxis should be considered for all unvaccinated household members and medical and paramedical workers in frequent contact with high-risk persons in the home, hospital or institutional setting. It is also advocated to control established outbreaks in facilities that care for high-risk persons, regardless of vaccination status, but in this setting the rapid emergence of resistance may be a problem.

Treating established influenza A virus infections with amantadine, when started within the first 48 hours of symptoms, reduces the duration of fever and other effects by 1–2 days and accelerates the resolution of the peripheral airways abnormalities that usually accompany influenza [113,146,186,193]. The reduction in symptoms far outweighs the drug's toxic effects [186,193]. Early treatment, i.e. before laboratory confirmation of the diagnosis is generally available, seems essential. Treatment for several days is usually effective and short courses may lessen the selection of resistant strains of virus [75]. During a known influenza outbreak most people with acute onset of nasal symptoms, feverishness, shivering, cough, headache, myalgia or anorexia, without vomiting or diarrhoea, will have influenza [79,186] and can be considered for treatment, particularly those in high-risk groups, in whom complications can be expected.

The recommended prophylactic and therapeutic dose of amantadine is 200 mg daily, reduced to 100 mg in those 10–15 years of age or over 65; the suggested dose in children aged 1–9 years is 2–4 mg/kg. The likelihood that drug resistance will increase with extensive use of amantadine, together with minor CNS adverse effects and the logistic difficulties in organizing timely prophylaxis and treatment, underscore the importance of immunization. On balance the adamantanes are still useful clinically and deserve wider distribution as an adjunct to (not a substitution for) vaccination, but doctors must continue to monitor efficacy and the emergence of resistant strains in formal clinical trials.

15.9 REFERENCES

1.   Adams, C.W. (1959) Postviral myopericarditis associated with the influenza virus; report of eight cases. *Am. J. Cardiol.*, **4**, 56–67.
2.   Alford, R.M., Kasel, J.A., Gerone, P.J. *et al.* (1966) Human influenza resulting from aerosol inhalation. *Proc. Soc. Exp. Biol. Med.*, **122**, 800–4.
3.   Angeloni, J.M. and Scott, G.W. (1958) Lung abscess and pneumonia complicating influenza. *Lancet*, **i**, 1255–6.
4.   Arden, N.H., Patriarca, P.A. and Kendal, A.P. (1986) Experiences in the use and efficacy of inactivated influenza vaccine in nursing homes, in *Options for the Control of Influenza*, (eds A.P. Kendal and P.A. Patriarca), A.R. Liss, New York, pp. 155–68.
5.   Arden, N., Bidol, S., Ohmit, S. and Monto, A. (1993) Effect of nursing home size and influenza vaccination rates on the risk of institutional outbreaks during an influenza (H3N2) epidemic, in *Options for the Control of Influenza II*, (eds C. Hannoun, A.P. Kendal, H.D. Klenk, A. McMichael, K.G. Nicholson and A. Oya), Excerpta Medica, Amsterdam, pp. 161–5.
6.   Ashley, J., Smith, T. and Dunell, K. (1991) Deaths in Great Britain associated with the influenza epidemic of 1989/90. *Population Trends*, **65**, 16–20.
7.   Austin, D.F., Karp, S., Dworsky, R. and Henderson, B.E. (1975) Excess leukemia in cohorts of children born following influenza epidemics. *Am. J. Epidemiol.*, **101**, 77–83.
8.   Barker, W.H. and Mullooly, J.P. (1982) Pneumonia and influenza deaths during epidemics. *Arch. Intern. Med.*, **142**, 85–9.
9.   Barton, L.L. and Chalhub, E.G. (1975) Myositis associated with influenza A infection (letter). *J. Pediatr.*, **87**, 1003–4.
10. Belshe, R.B., Smith, M.H., Hall, C.B., Betts, R. and Hay, A.J. (1988) Genetic basis of resistance to rimantadine emerging during treatment of influenza virus infection. *J. Virol.*, **62**, 1508–12.
11. Belshe, R.B., Burk, B., Newman, F., Cerruti, R.L. and Sim, I.S. (1989) Resistance of influenza A virus to amantadine and rimantadine: results

of one decade of surveillance. *J. Infect. Dis.*, **159**, 430–5.

12. Bendkowski. B. (1958) Asian influenza (1957) in allergic patients. *Br. Med. J.*, **2**, 1314–15.

13. Bental, E. (1958) Acute psychoses due to encephalitis following Asian influenza. *Lancet*, **ii**, 18–20.

14. Beswick, I.P. and Finlayson, R. (1959) A renal lesion in association with influenza. *J. Clin. Pathol.*, **12**, 280–5.

15. Bouter, A.M., Diepersloot, R.J.A., van Romunder, L.K.J. *et al.* (1991) Effect of epidemic influenza on ketoacidosis, pneumonia and death in diabetes mellitus: a hospital register survey of 1976–1979 in the Netherlands. *Diab. Res. Clin. Prac.*, **12**, 61–8.

16. Brocklebank, J.T., Court, S.D.M., McQuillin, J. and Gardner, P.S. (1972) Influenza A infections in children. *Lancet*, **ii**, 497–500.

17. Bryson, Y., Monahan, C., Pollack, M. and Shields, W.D. (1980) A prospective double-blind study of side-effects associated with the administration of amantadine for influenza A prophylaxis. *J. Infect. Dis.*, **141**, 543–7.

18. Buchan, K.A. and Reid, D. (1972) Epidemiological aspects of influenza in Lewis. *Health Bulletin*, **30**, 183–6.

19. Burch, G.E., Walsh, N. and Mogabgab, W.J. (1959) Asian influenza – clinical picture. *Arch Intern. Med.*, **103**, 696–707.

20. Burnet, F.M. (1936) Influenza virus on the developing egg. I. Changes associated with the development of an egg-passage strain of virus. *Br. J. Exp. Pathol.*, **17**, 282–93.

21. Caul, E.O., Waller, D.K. and Clarke, S.K.R. (1976) A comparison of influenza and respiratory syncytial virus infections among infants admitted to hospital with acute respiratory infections. *J. Hyg. (Camb)*, **77**, 383–92.

22. Chakraverty, P., Cunningham, P., Shen, G.Z. and Pereira, M.S. (1986) Influenza in the United Kingdom 1982–85. *J. Hyg. (Camb)*, **97**, 347–58.

23. Chin, T.D.Y., Foley, J.F., Doto, I.L., Gravelle, C.R. and Weston, J. (1960) Morbidity and mortality characteristics of Asian strain influenza. *Publ. Health Rep.*, **75**, 148–58.

24. Cockburn, W.C., Delon, P.J. and Ferreira, W. (1969) Origin and progress of the 1968–69 Hong Kong influenza epidemic. *Bull. WHO*, **41**, 345–8.

25. Coffey, V.P. and Jessup, W.J.E. (1959) Maternal influenza and congenital deformities. *Lancet*, **ii**, 935–8.

26. Connolly, A.M., Salmon, R.L. and Williams, D.H. (1993) What are the complications of influenza and can they be prevented? Experience from the 1989 epidemic of H3N2 influenza A in general practice. *Br. Med. J.*, **306**, 1452–4.

27. Corey, L., Rubin, R.J., Thompson, T.R. *et al.* (1976) A nationwide outbreak of Reye's syndrome: its epidemiologic relationship to influenza B. *Am. J. Med.*, **61**, 204–8.

28. Corey, L., Rubin, R.J., Thompson, T.R. *et al.* (1977) Influenza B- associated Reye's syndrome: incidence in Michigan and potential for prevention. *J. Infect. Dis.*, **135**, 398–407.

29. Couch, R.B., Douglas, R.G., Fedson, D.S. *et al.* (1971) Correlated studies of a recombinant influenza-virus vaccine. III. Protection against experimental influenza in man. *J. Infect. Dis.*, **124**, 473–80.

30. Couch, R.B., Kasel, J.A., Gerin, J.L. *et al.* (1974) Induction of partial immunity to influenza by a neuraminidase-specific vaccine. *J. Infect. Dis.*, **129**, 411–20.

31. Couch, R.B., Kasel, J.A., Glezen, W.P. *et al.* (1986) Influenza: its control in persons and populations. *J. Infect. Dis.*, **153**, 431–40.

32. Creighton, C. (1965) *A History of Epidemics in Britain*, Vols 1 and 2, 2nd edn, Frank Cass, London, and Barnes & Noble, New York.

33. Crosby, A.W. (1989) *America's Forgotten Pandemic*, Cambridge University Press, Cambridge.

34. Curwen, M., Dunnell, K. and Ashley, J. (1990) Hidden influenza deaths. *Br. Med. J.*, **300**, 896.

35. Davenport, F.M., Minuse, E., Hennessy, A.V. and Francis, T. (1969) Interpretations of influenza antibody patterns in man. *Bull. WHO*, **40**, 453–60.

36. Davis, L.E., Caldwell, G.G., Lynch, R.E., Bailey, R.E. and Chin, T.D.Y. (1974) Hong Kong influenza: the epidemiologic features of a high school family study analysed and compared with a similar study during the 1957 Asian influenza epidemic. *Am. J. Epidemiol.*, **92**, 240–7.

37. Department of Health, Welsh Offlce, Scottish Home and Health Department. (1992) *Immunization against infectious diseases, Influenza*, London HMSO, pp. 95–9.

38. DiBona, F.J. and Morens, D.M. (1977) Rhabdomyolysis associated with influenza. *Am. J. Pediatr.*, **91**, 943–5.

39. Dietzman, D.E., Schaller, J.G., Ray, C.G. et al. (1976) Acute myositis associated with influenza B infection. *Pediatrics*, **57**, 255–8.

40. Dolin, R., Reichman, R.C., Madore, H.P., Maynard, R., Linton, P.M. and Webber-Jones, J. (1982) A controlled trial of amantadine and rimantadine in the prophylaxis of influenza A infection. *N. Engl. J. Med.*, **307**, 580–4.

41. Dubowitz, V. (1958) Influenzal encephalitis. *Lancet*, **i**, 140–1.

42. Edmundson, P. and Hodgkin, K. (1957) Protean symptomatology. *Br. Med. J.*, **2**, 1058.

43. Edward, D.G. (1941) Resistance of influenza virus to drying and its demonstration in dust. *Lancet*, **ii**, 664–6.

44. Eickhoff, T.C., Sherman, I.L. and Serfling, R.E. (1961) Observations on excess mortality associated with epidemic influenza. *JAMA*, **176**, 776–82.

45. Fawdry, A.L. (1957) Influenza epidemic in Aden. *Lancet*, **ii**, 335–6.

46. Flewett, T.H. and Hoult, J.G. (1958) Influenza encephalopathy and postinfluenza encephalitis. *Lancet*, **ii**, 11–15.

47. Fox, J.P., Cooney, M.K., Hall, C.E. et al. (1982) Influenza virus infections in Seattle families, 1975–79: I. Study design, methods and the occurence of infections by time and age. *Am. J. Epidemiol.*, **116**, 212–27.

48. Fox, J.P., Cooney, M.K., Hall, C.E. et al. (1982) Influenza virus infections in Seattle families, 1975–79: II. Pattern of infection in invaded households and relation to age and prior antibody to occurrence of infection and related illness. *Am. J. Epidemiol.*, **116**, 228–42.

49. Foy, H.M., Hall, C.E., Cooney, M.K. et al. (1983) Influenza surveillance in the Pacific Northwest, 1976–80. *Int. J. Epidemiol.*, **12**, 353–6.

50. Francis, T. (1953) Influenza: the newe acquayantance. *Ann. Intern. Med.*, **39**, 203–21.

51. Frank, A.L., Taber, L.H., Wells, C.R., Wells, J.M., Glezen, W.P. and Paredes, A. (1981) Patterns of shedding of myxoviruses and paramyxoviruses in children. *J. Infect. Dis.*, **144**, 433–41.

52. Fry, J. (1958) Clinical and epidemiological features in a general practice. *Br. Med. J.*, **i**, 259–61.

53. Fry, J. (1970) A report on the influenza epidemic (A/Hong Kong) 1969–70. *Update*, **2**, 369.

54. Fry, J. and Hume, E.M. (1957) Contrasts with previous epidemics. *Br. Med. J.*, **ii**, 1057.

55. Fukumi, H. (1969) Interpretation of influenza antibody patterns in man. *Bull. WHO* **41**, 469–73.

56. Galbraith, A.W., Oxford, J.S., Schild, G.C. and Watson, G.I. (1969) Study of amantadine hydrochloride used prophylactically during the Hong Kong influenza epidemic in the family environment. *Bull. WHO*, **41**, 677–82.

57. Gibson, T.C., Arnold, J., Craige, E. and Curnen, E.C. (1959) Electrocardiographic studies in Asian influenza. *Am. Heart J.*, **57**, 661–8.

58. Giles, C. and Shuttleworth, E.M. (1957) Postmortem findings in 46 influenza deaths. *Lancet*, **ii**, 1224–5.

59. Gilroy, J. (1957) Asian influenza. *Br. Med. J.* **2**, 997–8.

60. Glezen, W.P. (1980) Consideration of the risk of influenza in children and indications for prophylaxis. *Rev. Infect. Dis.*, **2**, 408–20.

61. Glezen, W.P. (1982) Serious morbidity and mortality associated with influenza epidemics. *Epidemiol. Rev.*, **4**, 25–44.

62. Glezen, W.P. and Couch, R.B. (1978) Interpandemic influenza in the Houston area, 1974–76. *N. Engl. J. Med.*, **298**, 587–92.

63. Glezen, W.P., Couch, R.B., Taber, L.H. et al. (1980) Epidemiologic observations of influenza B virus infections in Houston, Texas, 1976–77. *Am. J. Epidemiol.*, **111**, 13–22.

64. Glezen, W.P., Paredes, A. and Taber, L.H. (1980) Influenza in children: relationship to other respiratory agents. *JAMA*, **243**, 1345–9.

65. Govan, A.D.T. and Macdonald, H.R.F. (1957) Influenza complicating heart disease in pregnancy. *Lancet*, **ii**, 891.

66. Griffiths, P.D., Ronalds, C.J. and Heath, R.B. (1980) A prospective study of influenza infections during pregnancy. *J. Epidemiol. Commun. Health*, **34**, 124–8.

67. Gross, P.A., Quinnan, G.V., Rodstein, M. et al. (1988) Association of influenza immunization with reduction in mortality in an elderly population. *Arch. Intern. Med.*, **148**, 562–5.

68. Grunert, R.R. and Hoffman, C.E. (1977) Sensitivity of influenza A/New Jersey/8/76 (Hsw1N1) virus to amantadine HCl. *J. Infect. Dis.*, **136**, 297–300.

69. Guthrie, J., Forsyth, D.M. and Montgomery, H. (1957) Asiatic influenza in the middle east: an

outbreak in a small community. *Lancet*, **ii**, 590–2.

70. Hakosalo, J. and Saxen, L. (1971) Influenza epidemics and congenital defects. *Lancet*, **ii**, 1346–7.

71. Hall, C.E., Cooney, M.K. and Fox, J.P. (1973) The Seattle virus watch. IV Comparative epidemiological observations of infections with influenza A and B viruses., 1965–1969, in families with young children. *Am. J. Epidemiol.*, **98**, 365–80.

72. Hall, C.B. and Douglas, R.G. (1975) Nosocomial influenza infection as a cause of intercurrent fevers in children. *Pediatrics*, **55**, 673–7.

73. Hall, W.J., Douglas, R.G., Hyde, R.W., Roth, F.K., Cross, A.S. and Speers, D.M. (1976) Pulmonary mechanics after uncomplicated influenza A infection. *Am. Rev. Respir. Dis.*, **113**, 141–7.

74. Hall, C.B., Douglas, R.G., Geiman, J.M. and Meagher, M.P. (1979) Viral shedding patterns of children with influenza B infection. *J. Infect. Dis.*, **140**, 610–13.

75. Hall, C.B., Dolin, R., Gala, C.L. *et al.* (1987) Children with influenza A infection: treatment with rimantadine. *Pediatrics*, **80**, 275–82.

76. Hardy, J.M.B., Azarowicz, E.N., Mannini, A., Medearis, D.N. and Cooke, R.E. (1961) The effect of Asian influenza on the outcome of pregnancy, Baltimore 1957–1958. *Am. J. Publ. Health*, **51**, 1182–8.

77. Hay, A.J., Wolstenholme, A.J., Skehel, J.J. and Smith, M.H. (1985) The molecular basis of the specific anti-influenza action of amantadine. *EMBO J.*, **4**, 3021–4.

78. Hay, A.J., Zambon, M.C., Wolstenholme, A.J., Skehel, J.J. and Smith, M.H. (1986) Molecular basis of resistance of influenza A viruses to amantadine. *J. Antimicrob. Chemother.*, **18** (suppl. B), 19–29.

79. Hayden, F.G., Hall, W.J. and Douglas, R.G. (1980) Therapeutic effects of aerosolized amantadine in naturally acquired infection due to influenza A virus. *J. Infect.*, **141**, 535–42.

80. Hayden, F.G., Cote, K.M. and Douglas, R.G. (1980) Plaque inhibition assay for drug susceptibility testing of influenza viruses. *Antimicrob. Agents Chemother.*, **17**, 865–70.

81. Hayden, F.G., Belshe, R.B., Clover, R.D., Hay, A.J., Oakes, M.G. and Soo, W. (1989) Emergence and apparent transmission of rimantadine-resistant influenza A virus in families. *N. Engl. J. Med.*, **321**, 1696–702.

82. Hayslett, J., McCarroll, J., Brady, E., Deuschle, K., McDermott, W. and Kilbourne, E.D. (1962) Endemic influenza. I. Serologic evidence of continuing and subclinical infection in disparate populations in the post-pandemic period. *Am. Rev. Respir. Dis.*, **85**, 1–8.

83. Heider, H., Adamczyk, B., Presber, H.W., Schroeder, C., Feldblum, R, and Indulen, M.K. (1981) Occurrence of amantadine and rimantadine resistant influenza A virus strains during the 1980 epidemic. *Acta Virol.*, **25**, 395–400.

84. Heikkinen, T., Waris, M. and Ruuskanen, O. (1993) Efficacy of influenza vaccination in reducing otitis media in children, *in Options for the Control of Influenza II*, (eds C. Hannoun, A.P. Kendal, H.D. Klenk, A. McMichael, K.G. Nicholson and A. Oya), Excerpta Medica, Amsterdam, pp. 25–37.

85. Henderson, F.W., Collier, A.M., Sanyal, M.A. *et al.* (1982) A longitudinal study of respiratory viruses and bacteria in the etiology of acute otitis media with effusion. *N. Engl. J. Med.*, **306**, 1377–83.

86. Hers, J.F.P. (1966) Disturbances of the ciliated epithelium due to influenza virus. *Am. Rev. Respir. Dis.*, **93**, 162–71.

87. Hers, J.F., Gosling, W.R.O., Masurel, N. and Mulder, J. (1957) Death from Asiatic influenza in the Netherlands. *Lancet*, **ii**, 1164–5.

88. Hers, J.F.P., Masurel, N. and Mulder, J. (1958) Bacteriology and histopathology of the respiratory tract and lungs of fatal asian influenza. *Lancet*, **ii**, 1141–3.

89. Hirsch, A. (1883) Influenza, in *Handbook of geographical and historical pathology, Vol 1, Acute infective diseases*, New Sydenham Society, London, pp. 7–54.

90. Holland, W.W. (1957) A clinical study of influenza in the Royal Air Force. *Lancet*, **ii**, 840–1.

91. Hope-Simpson, R.E. (1970) First outbreak of Hong Kong influenza in a general practice population in Great Britain. A field and laboratory study. *Br. Med. J.*, **3**, 74–7.

92. Horner, F.A. (1958) Neurologic disorders after Asian influenza. *N. Engl. J. Med.*, **258**, 983–5.

93. Horner, G.J. and Gray, F.D. (1973) Effect of uncomplicated, presumtive influenza on the diffusing capacity of the lung. *Am. Rev. Respir. Dis.*, **108**, 866–9.

94. Hoult, J.G. and Flewett, T.H. (1958) Influenzal encephalopathy and postinfluenzal encephalitis. Histological and other observations. *Br. Med. J.*, **i**, 1847–50.

95. Housworth, J. and Langmuir, A.D. (1974) Excess mortality from epidemic influenza, 1957–1966. *Am. J. Epidemiol.*, **100**, 40–8.

96. Housworth, W.J. and Spoon, M.M. (1971) The age distribution of excess mortality during A2 Hong Kong influenza epidemics compared with earlier A2 outbreaks. *Am. J. Epidemiol.*, **94**, 348–50

97. Howard, J.B., McCracken, G.H. and Luby, J.P. (1972) Influenza A2 virus as a cause of croup requiring tracheotomy. *J. Pediatr.*, **81**, 1148–50.

98. Howells, C.H.L., Vesselinova-Jenkins, C.K., Evans, A.D. and James, J. (1975) Influenza vaccination and mortality from bronchopneumonia in the elderly. *Lancet*, **i**, 381–3.

99. Høygaard, A. (1939) Acute epidemic diseases among Eskimos in Angmagssalik. *Lancet*, **i**, 245–6.

100. Jamieson, W.M., Kerr, M., Green, D.M. *et al.* (1958) Some aspects of the recent epidemic of influenza in Dundee. *Br. Med. J.*, **1**, 908–13.

101. Jaimovich, D.G., Kumar, A., Shabino, C.L. and Formoli, R. (1992) Influenza B virus infection associated with non-bacterial septic shock like illness. *J. Infect.*, **25**, 311–15.

102. Johanson, W.G., Pierce, A.K. and Sanford, J.P. (1969) Pulmonary function in uncomplicated influenza. *Am. Rev. Respir. Dis.*, **100**, 141–6.

103. Jordan, W.S. (1961) The mechanism of spread of Asian influenza. *Am. Rev. Respir. Dis.*, **83**, 29–35

104. Jordan, W.S., Denny, F.W., Badger, G.F. *et al.* (1958) A study of illness in a group of Cleveland families. XVII. The occurrence of Asian influenza. *Am. J. Hyg.*, **68**, 190–212.

105. Karjalainen, J., Nieminen, M.S. and Heikkila, J. (1980) Influenza A1 myocarditis in conscripts. *Acta Med. Scand.*, **207**, 27–30.

106. Kerr, A.A., Downham, M.A.P.S., McQuillin, J. and Gardner, P.S. (1972) Gastric 'flu influenza B causing abdominal symptoms in children. *Lancet*, **i**, 291–5.

107. Khakpour, M., Saidi, A. and Naficy, K. (1969) Proved viraemia in Asian influenza (Hong Kong variant) during incubation period. *Br. Med. J.*, **4**, 208–9.

108. Kilbourne, E.D. (1959) Studies on influenza in the pandemic of 1957–1958. III Isolation of Influenza A (Asian strain) viruses from influenza patients with pulmonary complications. Details of virus isolation and characterization of isolates, with quantitative comparison of isolation methods. *J. Clin. Invest.*, **38**, 266–74.

109. Kim, H.W., Brandt, C.W., Arrobio, J.O., Murphy, B., Chanock, R.M. and Parrott, R.H. (1979) Influenza A and B infections in infants and young children during the years 1957–1976. *Am. J. Epidemiol.*, **109**, 464–79.

110. Kubar, O.I., Brjantseva, E.A., Nikitina, L.E. and Zlydnikov, D.M. (1989) The importance of drug-resistance in the treatment of influenza with rimantadine. *Antiviral Res.*, **11**, 313–16.

111. Langmuir, A.D., Worthen, T.D., Slomon, J., Ray, C.K. and Patersen, E. (1985) The Thucydides syndrome: a new hypothesis for the cause of the plague of Athens. *N. Engl. J. Med.*, **313**, 1027–30.

112. Lindsay, M.I., Herrmann, E.C., Morrow, G.W. and Brown, A.L. (1970) Hong Kong influenza: clinical, microbiologic, and pathologic features in 127 cases. *JAMA*, **214**, 1825–32.

113. Little, J.W., Hall, W.H., Douglas, R.G., Mudholkar, G.S., Speers, D.M. and Patel, K. (1978) Airway hyperreactivity and peripheral airway dysfunction in influenza A infection. *Am. Rev. Respir. Dis.*, **118**, 298–303.

114. Liu, C. (1956) Rapid diagnosis of human influenza infection from nasal smears by means of fluorescein-labelled antibody. *Proc. Soc. Exp. Biol. Med.*, **92**, 883–7.

115. Lloyd-Still, R.M. (1958) Psychosis following Asian influenza. *Lancet*, **ii**, 20–1.

116. Louria, D.B., Blumenfeld, H.L., Ellis, J.T., Kilbourne, E.D. and Rogers, D.E. (1959) Studies on influenza in the pandemic of 1957–1958. II. Pulmonary complications of influenza. *J. Clin. Invest.*, **38**, 213–65.

117. MacDonald, K.L., Oserholm, M.T., Hedberg, C.W. *et al.* (1987) Toxic shock syndrome. A newly recognised complication of influenza and influenza-like illness. *JAMA*, **257**, 1053–8.

118. Marine, W.M. and Workman, W.M. (1969) Hong Kong influenza immunologic recapitulation. *Am. J. Epidemiol.*, **90**, 406–15.

119. Martin, C.M., Kunin, C.M., Gottlieb, L.S., Barnes, M.W., Liu, C. and Finland, M. (1959) Asian influenza A in Boston, 1957–1958. *Arch. Intern. Med.*, **103**, 515–31.

120. Martin, C.M., Kunin, C.M., Gottlieb, L.S. and Finland, M. (1959) Asian influenza A in Boston,

1957–1958: Severe staphylococcal pneumonia complicating influenza. *Arch. Intern. Med.*, **103**, 532–42.

121. Mast, E.E., Davis, J.P., Harmon, M.W., Arden, N.H., Circo, R. and Tyszka, G.E. (1989) Emergence and possible transmission of amantadine-resistant viruses during nursing home outbreaks of influenza A(H3N2), in *Program and Abstracts of the 29th Interscience Conference on Antimicrobial Agents and Chemotherapy*, September 18, 1989, Houston, Washington, DC. American Society for Microbiology, pp. 111 (abstract).

122. Masurel, N. (1969) Relation between Hong Kong virus and former A2 isolates and the A/Equi 2 virus in human sera collected before 1957. *Lancet*, **i**, 907–10.

123. McBean, A.M., Babish, J.D., Warren, J.L. and Melson, E.A. (1993) The effect of influenza epidemics on the hospitalisation of persons 65 years and older, in *Options for the Control of Influenza II*, (eds C. Hannoun, A.P. Kendal, H.D. Klenk, A. McMichael, K.G. Nicholson and A. Oya), Excerpta Medica, Amsterdam, pp. 25–37.

124. McConkey, B. and Daws, R.A. (1958) Neurological disorders associated with Asian influenza. *Lancet*, **ii**, 15–17.

125. McGregor, J.A., Burns, J.C., Levin, M.J. *et al.* (1984) Transplacental passage of influenza A/Bangkok/(H3N2) mimicking amniotic fluid infection syndrome. *Am. J. Obstet. Gynecol.*, **149**, 856–9.

126. Meibalane, R., Sedmak, G.V., Sasisharan, P., Garg, P. and Grausz, J.P. (1977) Outbreak of influenza in a neonatal intensive care unit. *J. Infect. Dis.*, **91**, 974–6.

127. Middleton, P.J., Alexander, R.M. and Szymanski, M.T. (1970) Severe myositis during recovery from influenza. *Lancet*, **ii**, 533–5.

128. Ministry of Health. (1960) *The influenza epidemic in England and Wales, 1957–1958.* HMSO, London.

129. Minor, T.E., Dick, E.C., Baker, J.W., Ouellette, J.J., Cohen, M. and Reed, C.E. (1976) Rhinovirus and influenza type A infections as precipitants of asthma. *Am. Rev. Respir. Dis.*, **113**, 149–53.

130. Minow, R.A., Gorbach, S., Johnson, B.L. et al. (1974) Myoglobinuria associated with influenza A infection. *Ann. Intern. Med.*, **80**, 359–361.

131. Minuse, E., Willis, P.W.. Davenport, F.M. and Francis, T. (1962) An attempt to demonstrate viremia in cases of Asian influenza. *J. Lab. Clin. Med.*, **59**, 1016–19.

132. Center for Disease Control. Editorial. (1992) Prevention and control of influenza. Recommendations of the immunization practices advisory committee. *MMWR*, **41**, No. RR–9, 1–17.

133. Monto, A.S. and Kioumehr, F. (1975) The Tecumseh study of respiratory illness. IX Occurrence of influenza in the community, 1966–1971. *Am. J. Epidemiol.*, **102**, 553–63.

134. Monto, A.S., Koopman, J.S. and Longini, I.M. (1985) The Tecumseh study of illness. XIII influenza infection and disease, 1976–1981. *Am. J. Epidemiol.*, **121**, 811–22.

135. Morgensen, J.L. (1974) Myoglobinuria and renal failure associated with influenza. *Ann. Intern. Med.*, **80**, 362–3.

136. Morris, J.A., Kasel, J.A., Saglam, M., Knight, V. and Loda, F.A. (1966) Immunity to influenza as related to antibody levels. *N. Engl. J. Med.*, **274**, 527–35.

137. Moss, S.E., Klein, R. and Klein, B.E.K. (1991) Cause-specific mortality in a population-based study of diabetes. *Am. J. Publ. Health*, **81**,1158–62.

138. Mulder, J. (1957) Asiatic influenza in the Netherlands. *Lancet*, **ii**, 334.

139. Mulder, J. and Masurel, N. (1958) Pre-epidemic antibody against 1957 strain of Asiatic influenza in serum of older people living in the Netherlands. *Lancet*, **i**, 810–14.

140. Murphy, B.R., Baron, S., Chalhub, E.G., Uhlendorf, C.P. and Chanock, R.M. (1973) Temperature-sensitive mutants of influenza virus. IV. Induction of interferon in the nasopharynx by wild-type and a temperature-sensitive recombinant virus. *J. Infect. Dis.*, **128**, 488–93.

141. Naficy, K (1963) Human influenza infection with proved viraemia. *N. Engl. J. Med.*, **269**, 964–6.

142. Nakajima, S., Nakamura, K., Nishikawa, F. and Nakajima, K. (1991) Genetic relationship between the HA genes of type A influenza viruses isolated in off-seasons and later epidemic seasons. *Epidemiol. Infect.*, **106**, 383–95.

143. Nakajima, S., Nishikawa, F., Nakamura, K. and Nakajima, K. (1992) Comparison of the HA genes of type B influenza viruses in herald

waves and later epidemic seasons. *Epidemiol. Infect.*, **109**, 559–68.

144. Newton, L. and Hall, S.M. (1993) Reye's syndrome in the British Isles: report for 1990/91 and the first decade of surveillance. *Comm. Dis. Rep. Rev.*, **3**, R11–16.

145. Nguyen-Van-Tam, J.S. and Nicholson, K.G. (1992) Influenza deaths in Leicestershire during the 1989–90 epidemic: implications for prevention. *Epidemiol. Infect.*, **108**, 537–45.

146. Nicholson, K.G. (1984) Antiviral agents in clinical practice: properties of antiviral agents. *Lancet*, **ii**, 562–4.

147. Nicholson, K.G., Kent, J. and Ireland, D.C. (1993) Respiratory viruses and exacerbations of asthma in adults. *Br. Med. J.*, **307**, 982–6.

148. Oswald, N.C., Shooter, R.A. and Curwen, M.P. (1958) Pneumonia complicating Asian influenza. *Br. Med. J.*, **ii**, 1305–11.

149. Oxford, J.S. and Galbraith, A. (1980) Antiviral activity of amantadine: a review of laboratory and clinical data. *Pharmacol. Ther.*, **11**, 181–262.

150. Paisley, J.W., Bruhn, F.W., Lauer, B.A. and McIntosh, K. (1978) Type A2 influenza viral infections in children. *Am. J. Dis. Child.*, **132**, 34–6.

151. Parsons, H.F. (1891) *Report on the influenza epidemic of 1889–90*. Local Government Board, HMSO, London.

152. Patriarca, P.A., Weber, J.A., Parker, R.A. *et al.* (1985) Eficacy of influenza vaccine in nursing homes. Reduction in illness and complications during an influenza A (H3N2) epidemic. *JAMA*, **253**, 1136–9.

153. Patriarca, P.A., Weber, J.A., Parker, R.A. *et al.* (1986) Risk factors for outbreaks of influenza in nursing homes. *Am. J. Epidemiol.*, **124**, 114–19.

154. Perrotta, D.M., Decker, M. and Glezen, W.P. (1985) Acute respiratory disease hospitalizations as a measure of impact of epidemic influenza. *Am. J. Epidemiol.*, **122**, 468–76.

155. Petersdorf, R.G., Fusco, J.J., Harter, D.H. and Albrink, W.S. (1959) Pulmonary infections complicating Asian influenza. *Arch. Intern. Med.*, **103**, 262–72.

156. Polak, M.F. (1959) Influenzasterfte in de herfst van 1957. *Ned. Tijdschr. Geneeskd.*, **103**, 1098–109.

157. Price, D.A., Postlethwaite, R.J. and Longson, M. (1976) Influenza virus A2 infections presenting with febrile convulsions and gastrointestinal symptoms in young children. *Clin. Pathol.*, **15**, 361–7.

158. Pryor, H.B. (1964) Influenza: that extraordinary malady. Notes on its history and epidemiology. *Clin. Pediatr.*, **3**, 19–24.

159. Public Health Laboratory Service. (1958) Deaths from Asian influenza, 1957. *Br. Med. J.*, **i**, 915–19.

160. Ramphal, R., Donnelly, W.H. and Small, P.A. (1980) Influenza pneumonia in pregnancy: failure to demonstrate transplacental transmission of influenza virus. *Am. J. Obstet. Gynecol.*, **138**, 347–8.

161. Roberts, G.B.S. (1957) Fulminating influenza. *Lancet*, **ii**, 944–5.

162. Roden, A.T. (1969) National experience with Hong Kong influenza in the United Kingdom, 1968–69. *Bull. WHO*, **41**, 375–80.

163. Rowland, H.A.K. (1958) The influenza epidemic in Abadan. *Br. Med. J.*, **i**, 422–5.

164. Saenz, A.C., Assad, F.A. and Cockburn, W.C. (1969) Outbreak of A2/Hong Kong/68 influenza at an international medical conference. *Lancet*, **i**, 91–3.

165. Scholtissek, C., Rott, R. and Klenk, H.-D. (1975) Two different mechanisms of the multiplication of enveloped viruses of glucosamine. *Virology*, **63**, 191–4.

166. Serfling, R.E., Sherman, I.L. and Housworth, W.J. (1967) Excess pneumonia-influenza mortality by age and sex in three major influenza A2 epidemics, United States, 1957–58, 1960 and 1963. *Am. J. Epidemiol.*, **86**, 433–41.

167. Serie, C., Barme, M., Hannoun, C., Thibon, M., Beck, H. and Aquino, J.P. (1977) Effects of vaccination on an influenza epidemic in a geriatric hospital. *Dev. Biol. Stand.*, **39**, 317–21.

168. Sharrar, R.G. (1969) National influenza experience in the USA, 1968–69. *Bull. WHO*, **41**, 361–6.

169. Simon, N.M., Rovner, R.N. and Berlin, B.S. (1970) Acute myoglobinuria associated with type A2 (Hong Kong) influenza. *JAMA*, **212**, 1704–5.

170. Smith, W., Andrewes, C.H. and Laidlaw, P.P. (1933) A virus obtained from influenza patients. *Lancet*, **ii**, 66–8.

171. Smith, A.P., Tyrrell, D.A.J., Al-Nakib, W., Barrow, G.I., Higgins, P.G., Leekam, S. and Tricket, S. (1989) Effects and after effects of the common cold and influenza on human performance. *Neuropsychobiology*, **21**, 90–3.

172. Smith, A.P.,Thomas, M., Brockman, P., Kent, J. and Nicholson, K.G. (1993) Effect of influenza B infection on human performance. *Br. Med. J.,* **306**, 760–1.

173. Sprenger, M.J.W., Beyer, W.E.P., Kempen, B.M. and Mulder, P.G.H. (1993) Risk factors for influenza mortality? in *Options for the Control of Influenza II,* (eds C. Hannoun, A.P. Kendal, H.D. Klenk, A. McMichael, K.G. Nicholson and A. Oya), Excerpta Medica, Amsterdam, pp. 15–23.

174. Stanley, E.D. and Jackson, G.G. (1966) Viraemia in Asian influenza. *Trans. Assoc. Am. Physicians,* **1**, 376–87.

175. Stocks, P. (1935) The effect of influenza epidemics on the certified causes of death. *Lancet,* ii, 386–95.

176. Stuart-Harris, C.H. (1961) Twenty years of influenza epidemics. *Am. Rev. Respir. Dis.,* **83**, 54–61.

177. Stuart-Harris, C.H. (1970) Pandemic influenza: an unresolved problem in prevention. *J. Infect. Dis.,* **122**,108–15.

178. Stuart-Harris, C.H. and Schild, G.C. (eds) (1976) *Influenza, the Viruses and Disease,* Edward Arnold, London.

179. Taber, L.H., Paredes, A., Glezen, W.P. and Couch, R.B. (1981) Infection with influenza A/Victoria virus in Houston families, 1976. *J Hyg. (Camb),* **86**, 303–13.

180. Thacker, S.B. (1986) The persistence of influenza in human populations. *Epidemiol. Rev.,* **8**, 129–42.

181. Thompson, T. *Annals of Influenza in Great Britain,* Sydenham Society, London 1852.

182. Tillett, H.E., Smith, J.W.G. and Clifford, R.E. (1980) Excess morbidity and mortality associated with influenza in England and Wales. *Lancet,* i, 793–5.

183. Tillett, H.E., Smith, J.W.G. and Gooch, C.D. (1983) Excess deaths attributable to influenza in England and Wales: age at death and certified cause. *Int. J. Epidemiol.,* **12**, 344–52.

184. Tolan, R.W. (1993) Toxic shock syndrome complicating influenza in a child: case report and review. *Clin. Infect. Dis.,* **17**, 43–5.

185. Tominack, R.L. and Hayden, F.G. (1987) Rimantadine hydrochloride and amantadine hydrochloride use in influenza A infections. *Infect. Dis. Clin. North Am.,* **1**, 459–78.

186. Van Voris, L.P., Betts, R.F., Hayden, F.G., Christmas, W.A. and Douglas, R.G. (1981) Successful treatment of naturally occurring influenza A/USSR/77 H1N1. *JAMA,* **245**, 1128–31.

187. Walsh, J.J., Dietlein, L.F., Low, F.N., Burch, G.E. and Mogabgab, W.J. (1961) Bronchotracheal response in human influenza. *Arch. Intern. Med.,* **108**, 376–88.

188. Whitaker, A.N., Bunce, I. and Graeme, E.R. (1974) Disseminated intravascular coagulation and acute renal failure in influenza A2 infection. *Med. J. Aust.,* **2**, 196–201.

189. WHO. (1968) Influenza: The Hong Kong virus. *WHO Chron.,* **22**, 528–9.

190. WHO. (1985) Current status of amantadine and rimantadine as anti-influenza-A agents: memorandum from a WHO meeting. *Bull. WHO,* **63**, 51–6.

191. Wilson, C.B. and Stein, A.M. (1969) Teratogenic effects of Asian influenza. An extended study. *JAMA,* **210**, 336–7.

192. Wilson, C.B. and Smith, R.C. (1972) Goodpasture's syndrome associated with influenza A2 virus infection. *Ann. Intern. Med.,* **76**, 91–4.

193. Wingfield, W.L., Pollack, D. and Grunert, R.R. (1969) Therapeutic efficacy of amantadine HCl and rimantadine HCl in naturally occurring A2 respiratory illness in man. *N. Engl. J. Med.,* **281**, 579–84.

194. Woodall, J., Rowson, K.E.K. and McDonald, J.C. (1958) Age and Asian influenza. *Br. Med. J.,* **4**, 1316–18.

195. Wright, P.F., Ross, K.B., Thompson, J. and Karzon, D.T. (1977) Influenza A infections in young children. Primary natural infection and protective efficacy of live-vaccine induced or naturally-acquired immunity. *N. Engl. J. Med.,* **296**, 829–34.

196. Wright, P.F., Thompson, J., McKee, K.T., Vaughn, W.K., Sell, S.H.W. and Karzon, D.T. (1981) Patterns of illness in the highly febrile young child; epidemiologic, clinical, and laboratory correlates. *Pediatrics,* **67**, 694–700.

197. Wynne-Griffith, G., Adelstein, A.M., Lambert, P.M. and Weatherall, J.A.C. (1972) Influenza and infant mortality. *Br. Med. J.,* **3**, 553–6.

198. Yawn, D.H., Pyeatte, J.C., Joseph, J.M., Eichler, S.L. and Bunuel, R.G. (1971) Transplacental transfer of influenza virus. *JAMA,* **216**, 1022–3.

199. Zamkoff, K. and Rosen, N. (1979) Influenza and myoglobinuria in brothers. *Neurology,* **29**, 340–5.

RESPIRATORY HANTAAN VIRUS 16

Steven Myint

16.1 INTRODUCTION

Hantaviruses were first recognized as a cause of serious disease in 1951 when 3000 United Nations troops in Korea developed an illness characterized by fever, headache, back and abdominal pain, with haemorrhage in many cases. This disease had a case-fatality of 5–10% and was named Korean haemorrhagic fever [18]. The causative agent was, putatively, termed 'Hantaan' after the river that now divides North and South Korea. Examination of

Viral and Other Infections of the Human Respiratory Tract.
Edited by S. Myint and D. Taylor-Robinson. Published in 1996 by Chapman & Hall. ISBN 0 412 60070 6

historical records would, however, suggest that the illness has been recognized in many parts of the world for decades [4], if not centuries [37]. The disease course is characterized by increased capillary permeability and acute shock, resulting in renal failure. The term 'haemorrhagic fever with renal syndrome (HFRS)' has now been adopted to describe the illness.

On 14th May, 1993, the New Mexico Department of Health in the USA was notified of two persons living in the same household who died of respiratory failure within 5 days of each other. By 7th June, 22 other cases had been reported to other near-by state departments and the Centers for Disease Control (CDC), Atlanta, Georgia recognized an outbreak of unexplained respiratory illness in this part of the South-western USA. Those affected were all adults living in rural areas. There was a case-fatality of over 50% but neither post-mortem findings nor microbiological analysis could identify a known agent for 3 weeks from the onset of investigations. This changed, however, when an enzyme immunoassay based on prototypical hantaviral agents was used to show an elevated titre of IgM antibody or rise in IgG titre in six people who died with the illness [5]. Subsequent developments have identified a new hantavirus in a rodent reservoir.

16.2 CLASSIFICATION OF HANTAVIRUSES

Hantaviruses are enveloped, segmented RNA viruses that belong to the Bunyaviridae family.

Before the recognition of the respiratory isolates there were five recognized members of the genus, named geographically. The characteristics of these non-pulmonary viruses are summarized in Table 16.1. On the basis of serology, and S and M gene sequences, it is possible to derive a phylogenetic tree (Figure 16.1).There have also been a number of putative, serologically related and distinct viruses described; these also tend to be named geographically (Table 16.2).

16.3 BIOLOGICAL CHARACTERISTICS OF HANTAVIRUSES

Hantaan virus is the best studied of this group and the biology of this virus is taken as representative of the genus.

16.3.1 STRUCTURE AND REPLICATION

Hantaviruses are small (95 nm in diameter), lipid-enveloped viruses with a helical nucleocapsid. The genome is single-stranded, negative-sense RNA and is present in a large (L), medium (M) and small (S) segment: the relative molecular weights are 2.7, 1.2 and 0.6 x 10^6 Da, respectively. These three strands have a common 3'-terminal sequence, AUCAUCAUCUG, which is unique to hantaviruses [41]. The L segment encodes a virion-associated polymerase (L), the M segment encodes two structural envelope glycoproteins G1 and G2 and the S segment a nucleocapsid protein, N [34]. The N, G1 and G2 proteins are 48 kDa, 64 kDa and 53.7 kDa, respectively. It appears that

Table 16.1 Characteristics of prototype non-pulmonary hantaviruses

Virus	Distribution	Reservoir	Disease	Mortality (%)
Hantaan	Former USSR, South-East Asia, Balkans	Striped field mouse (*Apodemus agrarius*)	HFRS	5–15
Seoul	Worldwide	Norwegian rat (*Rattus norvegicus*)	HFRS	1
Puumula	Scandinavia, Former USSR, Balkans	Bank vole (*Clethrionomys glariolus*)	Nephropathia epidemica	Rare
Prospect Hill	Eastern and North-Western USA	Meadow vole (*Microtus pennsylvanicus*)	None	
Dobrava	Balkans	Yellow-necked field mouse (*Apodemus flavicollis*)	HFRS	5–35

HFRS, haemorrhagic fever with renal syndrome

Table 16.2 Named non-prototype hantaviruses

Virus	Location	Serological relationship	Reservoir
Belgrade virus	Balkans	Dobrava	*Apodemus* spp. [46]
Thottapalayam virus	India	?Unrelated	*Suncus marinus* [3]
Fojnica virus	Balkans	Hantaan	*Apodemus flavicollis* [20]
Leaky virus	USA	Prospect Hill	*Microtus* spp.
Porogia virus	Balkans	Dobrova	*Apodemus* spp.
Thailand	Thailand	?Unrelated ?Seoul	*Bandicota indica* [15]

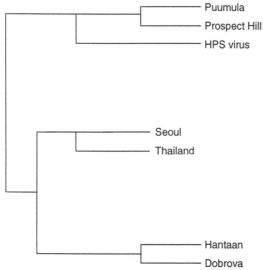

Figure 16.1 Phylogenetic tree of hantaviruses. (Adapted from [33,43].)

there is considerable conservation of both G1 and G2 within serogroups [42], and there also appears to be extensive serological cross-reactivity between serogroups with non-typing sera [17,30,45]. Epitopes responsible for eliciting neutralizing antibodies reside within the two envelope proteins.

A receptor for hantaviruses is yet to be recognized, but replication is known to take place in the cytoplasm. Virus assembly and release of virions takes place on the Golgi apparatus [38].

16.3.2 PHYSICOCHEMICAL PROPERTIES

Being an enveloped virus, infectivity is reduced by 70% ethanol, ether and chloroform [29]. It is also rapidly abolished at pH 5.0 and at 37°C. Virus suspensions can be stored, however, for years at –60°C in the presence of 1% bovine serum albumen [29]. Infectivity has also been reported to persist for 2 days in dried cell culture medium [7].

16.3.3 EPIDEMIOLOGY

Humans contract infection with hantaviruses from close contact with infected rodents.

There are four main rodent reservoirs: *Apodemus*, *Rattus*, *Clethrionomys* and *Microtus* species. The infected rodent is viraemic for a week and the virus is spread to lungs, liver and kidneys [31]. Saliva is also virus-positive for weeks and appears to be important for rodent-to-rodent transmission in that intramuscular passage through bites appears to be the main route [19]. Vertical transmission does not appear to occur. Despite the development of neutralizing antibodies, virus persists for life in infected organs of the rodent [53]. It is thought that humans become infected via the respiratory route. Evidence for this comes from laboratory outbreaks where aerosols are thought to have been the only possible means of transmission [13,26,29,36,47]. Person-to-person transmission has not been recorded in 266 health care workers [2].

Serological evidence for exposure to hantaviruses has been found worldwide [49]. Seroprevalence appears to be 0.25% in Baltimore [10], but is much higher (1–30%) in other parts of the USA [50], in blood donors in Belgium (16%) [51] and in certain occupational groups, such as farm workers in Britain (21.5%) [35]. Endemic and sporadic disease has been described in Scandinavia, the Balkans, North America, the Far East, the former USSR and less commonly in western Europe, the Middle East and west Africa [37,49]. A recent epidemic of HFRS, due to a Puumula serotype, has been reported in France and Belgium; it affected 133 people living near densely forested areas [11].

In South-east Asia, there appears to be seasonality of infection with November and December being the peak months [28].

16.4 HANTAVIRUS PULMONARY SYNDROME

The initial outbreak of unidentified pulmonary disease has identified a new syndrome, which because of the evidence for a hantavirus aetiology, has become known as 'Hantavirus pulmonary syndrome' (HPS).

16.4.1 EPIDEMIOLOGY

The original outbreak that appeared to be centred on the south-western USA was found to have 'spread' to 12 states, in the western half of the USA, within months [6]. Since then cases have been reported also in British Columbia, Canada [21]. Although cases have not been confirmed elsewhere, the report of a retrospective diagnosis on an earlier case in the US in 1983 [52] would suggest that this is not an entirely novel disease and that it is the epidemiology that has changed.

The initial two cases of HPS notified to the New Mexico Office of the Investigator were a 21-year-old female and a 19-year-old male. On the basis of early clinical descriptions, screening criteria were drawn up for identifying patients with possible HPS (Table 16.3). By 28th July, 1994, 83 confirmed cases had been notified to the CDC [8]. Almost all of these

(94%) occurred west of the Mississippi River. The age ranged from 12 years to 69 years (median age 31 years), with 61% in the 20 to 39-year-old age group. There were equal proportions of women and men. About half of the patients were non-Hispanic whites with the others being Hispanic (8%), American Indian (40%) and non-Hispanic blacks (3%). The high percentage of American Indian is likely to represent their lifestyle (see below) rather than racial susceptibility.

The search for a possible reservoir of HPS virus identified the deer mouse, *Peromyscus maniculatus*. Of nearly 1700 small mammals captured in the same geographical locations as the HPS cases, 48% were deer mice. Serological testing for Prospect Hill/Puumula antibodies showed that 30% of these were seropositive [9]. Using nested reverse transcriptase–polymerase chain reaction (RT–PCR) assays, with primers initially based on conserved sequences in the G2 coding region of Prospect Hill, Puumula, Seoul and

---

Table 16.3   CDC screening criteria for hantavirus pulmonary syndrome [6]

Potential case-patients must have one of the following
- a febrile illness (temperature $\geq 101°F$ ($\geq 38.3°C$) occurring in a previously healthy person characterized by unexplained adult respiratory distress syndrome, *or* bilateral intersitial pulmonary infiltrates developing within 1 week of hospitalization with respiratory compromise requiring supplemental oxygen, *or*
- an unexplained respiratory illness resulting in death in conjunction with an autopsy examination demonstrating non-cardiogenic pulmonary oedema without an identifiable specific cause

Potential case-patients are to be *excluded* if they have any of the following
- a predisposing underlying medical condition (e.g. severe underlying pulmonary disease, solid tumours of haematological malignancies, congenital or acquired immunodeficiency disorders), or medical conditions (e.g. rheumatoid arthritis or organ transplant recipients) requiring immunosuppressive drug therapy (e.g. steroids or cytotoxic chemotherapy)
- an acute illness that provides a likely explanation for the respiratory illness (e.g. recent major trauma, burns or surgery; recent seizures or history of aspiration; bacterial sepsis; another respiratory disorder such as respiratory syncytial virus in young children; influenza; or legionella pneumonia)

Confirmed case-patients must have the following
- at least one specimen (i.e. serum and/or tissue) available for laboratory testing for evidence of hantavirus infection *and*
- in a patient with a compatible clinical illness, either serology is positive (presence of hantavirus-specific immunoglobulin M or rising titres of immunoglobulin G), or polymerase chain reaction for hantavirus ribonucleic acid, or immunohistochemistry for hantavirus antigen is positive

---

Hantaan viruses, it was possible to identify identical (or near identical) sequences in all of 10 patients who had died of HPS and in 82% of 54 seropositive deer mice [39]. Cases of HPS have occurred in areas of the USA, such as Louisiana, where the deer mouse is not found, however, and it is possible that there is another vector. Bats have been found to harbour other hantaviruses and are another possible vector [32].

In a case-control study of residents of the south-western USA, risk factors associated with acquiring HPS included cleaning food storage areas, cleaning barns and other outbuildings, ploughing with hand tools, herding animals and trapping rodents [2].

The epidemiological evidence has been overwhelming in linking a hantavirus with HPS although there are some who believe that, because the newly identified virus does not behave like other hantaviruses, it is merely a bystander [12]. In this regard, it is of interest to note that well-documented cases have been described which appear not to have had an identifiable exposure to areas inhabited by deer mouse [2].

16.4.2 CLINICAL MANIFESTATIONS

The onset of illness with HPS is characterized by a prodromal phase lasting 3–6 days. There may be upper respiratory tract symptoms such as cough but these are not marked. Gastrointestinal symptoms (nausea, vomiting and abdominal pain), headache and dizziness appear quite frequent. During this phase, physical examination and laboratory investigation is usually unhelpful in distinguishing the condition from other acute non-specific febrile disorders.

The prodrome heralds the abrupt onset of respiratory distress. There is progressive cough and shortness of breath, with tachypnoea, tachycardia, fever and hypotension. Chest auscultation may reveal widespread crackles. Chest radiography has shown bilateral pulmonary infiltrates in all cases within 2 days of admission to hospital. This is typical of that seen in any case of adult respiratory distress syndrome except that interstitial oedema is more frequent [27]. Haemoconcentration and thrombocytopaenia have been noted in the majority of patients. Leucocytosis, a prolonged partial thromboplastin time, an increased proportion of immature granulocytes, elevated serum lactate dehydrogenase and aspartate aminotransferase, and mild-to-moderate proteinuria have also been common findings. Metabolic acidosis occurs in severe cases [6,14]. The frequency of clinical features in 17 patients on admission to hospital is shown in Table 16.4.

Over 80% of cases have required management of hypoxaemia with mechanical ventilation. The majority have died in association with intractable hypotension and cardiac dysrhythmias. In those that survive, recovery appears to be complete.

16.4.3 DIAGNOSIS

Confirmation of clinical suspicion is made by the detection of hantavirus antibodies in serum, either IgM or a four-fold, or greater, rise in IgG antibodies. Serological testing has relied previously on the use of heterologous antigens, particularly those of Prospect Hill virus. Recombinant homologous N protein antigen has, however, been produced in HeLa cells using a vaccinia-T7 RNA polymerase system [16]. HPS virus RNA was rescued as cDNA from lung post-mortem tissue by using the PCR. The use of homologous antigens appears to improve sensitivity as well as specificity, whether in an enzyme immunoassay or indirect immunofluorescence assay, and should replace heterologous antigens.

RT–PCR assays and immunohistochemistry (IHC) have been used to find hantavirus antigens in post-mortem tissue. In 16 case-patients, heterologous serology, RT–PCR and IHC showed concordance in 15 patients. One patient was antibody-negative [6]. RT–PCR

Table 16.4   Frequency of clinical features on admission to hospital in 17 patients with hantavirus pulmonary syndrome (Adapted from [14].)

Clinical feature	Proportion of cases (%)
Fever	100
Myalgia	100
Respiratory rate ≥20/min	100
Heart rate ≥100/min	94
Temperature ≥38.1°C	75
Headache	71
Cough	71
Nausea/vomiting	71
Chills	65
Malaise	59
Diarrhoea	59
Shortness of breath	53
Systolic blood pressure ≤100 mmHg	50
Dizziness/light-headedness	41
Crackles on lung auscultation	31
Arthralgia	29
Back pain	29
Abdominal pain	24
Abdominal tenderness	24
Chest pain	18
Sweats	18
Cool, clammy skin	18
Conjunctival injection	18
Rhinorrhoea	12
Sore throat	12

has also enabled virus in peripheral blood mononuclear cells to be detected [23]. Virus appeared to disappear over a period of 9 to 123 days in convalescent patients.

16.4.4 MANAGEMENT

Supportive therapy to maintain cardiovascular and pulmonary function in an intensive care setting has proved essential. Allied to this, the use of antiviral therapy is being investigated. The use of ribavirin for HFRS in prospective, placebo- and case-controlled trials in China has shown a significant reduction in disease severity [24]. The use of ribavarin for HPS still needs to be defined. Similarly, the use of corticosteroids in HFRS [40] has shown promise but their role in HPS is undefined.

16.4.5 CONTROL

Control of the rodent population in areas close to centres of human population would help prevent large outbreaks. This can be done by common-sense measures, such as not leaving food available, keeping waste in rodent-proof containers as well as the use of rodenticides. The control of all rodents in rural areas is, of course, an impossibility. Identification of risk activities and education of the public with regard to these is, however, more practical.

Prevention of illness in laboratory workers, and others at high risk, depends on strict containment facilities when handling known, or potentially, infected material. Vaccines to other hantaviruses have been developed [34] and may be useful against all the major serogroups if based on conserved antigens which appear to be present.

16.4.6 PATHOLOGY

Post-mortem examination of HPS patients has revealed serous pleural effusions and oedematous lungs. Microscopic examination of lung tissue has shown mononuclear infiltrates, septal and alveolar oedema, focal hyaline membranes and less frequently, alveolar haemorrhage [6]. Mild splenomegaly is also often found. Macroscopically, lymph nodes, liver, kidneys, heart and brain are normal. Immunohistochemistry, however, has revealed hantavirus antigens in endothelial cells in many organs.

16.5 HANTAVIRUS PULMONARY SYNDROME VIRUSES

The initial isolates from deer mice were found to be related to another hantavirus known to be present in the USA, namely Prospect Hill virus, and Puumula virus. This was based on an antigenic evaluation and by sequence analysis of G2 proteins [22]. There appears to be, however, high sequence diversity (up to 14%) in hantavirus isolates in the USA [43]. The original isolate has gone under the names of Four Corners Virus and Muerto Canyon virus, which represent the geography of the initial isolate. Sequence analysis of an isolate from a case of HPS in Louisiana has shown it to be different from the Muerto Canyon virus [44] and it is likely that the HPS viruses do not represent a single virus but a closely related group. Phylogenetic analysis based on RNA segment sequences would suggest that the HPS viruses are not new, and that there has not been re-assortment to produce a new pathogenic virus [43].

Much of the L, S and M genes that have now been sequenced for different HPS isolates confirm a hantavirus gene structure, with an anomaly that there appears to be an unusually long (728 nucleotide) predicted N mRNA 3' non-coding region. There appears to be no obvious reason for the increased virulence of this virus when compared with other hantaviruses [43].

Epitope mapping of an HPS virus using recombinant N, G1 and G2 proteins in a Western blot assay showed an immunodominant epitope on the N gene protein which elicited antibodies that cross-reacted with Puumula and Prospect Hill viruses. No cross-reactivity was found with G1 and G2 antibodies [25].

16.6 FUTURE PROSPECTS

Modern disease surveillance and molecular biological methods have enabled a 'new' disease entity to be rapidly characterized. Over the next few years, more should be learned of the molecular biology of an important, but understudied, group of pathogens and the mechanisms by which they cause disease. A vaccine appears, with current knowledge, to be more realistic than with most of the other respiratory viruses.

16.7 REFERENCES

1. Antoniadis, A., Grekasm, D., Rossi, C.A. *et al.* (1987) Isolation of a Hantavirus from a severely ill patient with haemorrhagic fever with renal syndrome in Greece. *J. Infect. Dis.*, **156**, 1010–13.
2. Butler, J.C. and Peters, C.J. (1994) Hantaviruses and Hantavirus Pulmonary Syndrome. *Clin. Infect. Dis.*, **19**, 387–95.
3. Carey, D.E., Reuben, R., Panicker, K.N., Shope, R.E. and Myers, R.M. (1971) Thottapalayam virus: a presumptive arbovirus isolated from a shrew in India. *Indian J. Med. Res.*, **59**, 1758–60.
4. Casals, J., Henderson, B.E., Hoogstraal, H. *et al.* (1970) A review of Soviet viral haemorrhagic fevers 1969. *J. Infect. Dis.*, **122**, 437–53
5. Centers for Disease Control and Prevention (1993) Outbreak of acute illness– south-western United States 1993. *MMWR*, **42**, 421–2.
6. Centers for Disease Control and Prevention (1993) Update: Hantavirus Pulmonary Syndrome–United States, 1993. *MMWR*, **42**, 816–20.

7. Centers for Disease Control and Prevention (1994) Laboratory management of agents associated with hantavirus pulmonary syndrome: interim biosafety guidelines. *MMWR*, **43**, 1–7.

8. Centers for Disease Control and Prevention (1994) Hantavirus pulmonary syndrome: north-eastern United States, 1994. *MMWR*, **43**, 548–9, 555–6.

9. Childs, J.E., Ksiazek, T.G., Spiropoulou, C.F. *et al.* (1994) Serologic and genetic identification of *Peromyscus maniculatus* as the primary rodent reservoir for a new hantavirus in the south-western United States. *J. Infect.*, **169**, 1271–80.

10. Childs, J.E. and Rollin, P.E. (1994) Emergence of hantavirus disease in the USA and Europe. *Curr. Opin. Infect. Dis.*, **7**, 220–4.

11. Clement, J.P., McKenna, P., Colson, P. *et al.* (1994) Hantavirus epidemic in Europe, 1993. *Lancet*, **1**, 114.

12. Denetclaw, W.F. Jr and Denetclaw, T.H. (1994) Is 'south-west US mystery disease' caused by hantavirus?. *Lancet*, **1**, 53–4.

13. Desmyter, J., LeDuc, J.W., Johnson, K.M. *et al.* (1983) Laboratory rat associated outbreak of haemorrhagic fever with renal syndrome due to Hantaan-like virus in Belgium. *Lancet*, **2**, 1445–8.

14. Duchin, J.S., Koster, F.T., Peters, C.J. *et al.* (1994) Hantavirus pulmonary syndrome: a clinical description of 17 patients with a newly recognized disease. *N. Engl. J. Med.*, **14**, 949–55.

15. Elwell, M.R., Ward, G.S., Tingpalapong, M. and LeDuc, J.W. (1985) Serological evidence of Hantaan-like virus in rodents and man in Thailand. *Southeast Asian J. Trop. Med. Public Health*, **16**, 3129–32.

16. Feldmann, H., Sanchez, A., Morzunov, S. *et al.* (1993) Utilization of autopsy RNA for the synthesis of the nucleocapsid antigen of a newly recognized virus associated with a hantavirus pulmonary syndrome. *Virus Res.*, **30**, 351–67.

17. Friman, G., French, G.R., Hambraeus, L. *et al.* (1980) Scandinavian epidemic nephropathy and Korean haemorrhagic fever. *Lancet*, **2**, 100–4.

18. Gadjusek, D.C. (1989) Introduction, in *Manual of Haemorrhagic Fever with Renal Syndrome*, (eds H.W. Lee and J. Dalrymple), Korea University, Seoul, Korea.

19. Glass, G.E., Childs, J.E., Korch, C.W. and LeDuc, J.W. (1988) Association of intraspecific wounding with Hantaviral infection in wild rats (*Rattus norvegicus*). *Epidemiol. Infect.*, **101**, 459–72.

20. Gligic, A., Frusic, M., Obradovic, M. *et al.* (1989) Haemorrhagic fever with renal syndrome in Yugoslavia: antigenic characterization of Hantaviruses isolated from *Apodemus flavicollis* and *Clethrionomys glareolus*. *Am. J. Trop. Med. Hyg.*, **41**, 109–15.

21. Health Canada (1994) First reported cases of hantavirus pulmonary syndrome in Canada. *Canada Commun. Disease Report*, **20**, 121–5.

22. Hjelle, B., Jenison, S., Torrez-Martinez, N. *et al.* (1994) A novel hantavirus associated with an outbreak of fatal respiratory disease in the south-western United States: evolutionary relationships to known hantaviruses. *J. Virol.*, **68**, 592–6.

23. Hjelle, B., Spiropoulou, C.F., Torrez-Martinez, N. *et al.* (1994) Detection of Muerto Canyon Virus RNA in peripheral blood mononuclear cells from patients with hantavirus pulmonary syndrome. *J. Infect. Dis.*, **170**, 1013–17.

24. Huggins, J.W., Hsiang, C.M., Cosgriff, T.M. *et al.* (1991) Prospective, double blind, concurrent, placebo-controlled clinical trial of intravenous ribavirin therapy of haemorrhagic fever with renal syndrome. *J. Infect. Dis.*, **164**, 1119–27.

25. Jenison, S., Yamada, T., Morris, C. *et al.* (1994) Characterisation of human antibody responses to Four Corners Hantavirus infections among patients with hantavirus pulmonary syndrome. *J. Virol.*, **68**, 3000–6.

26. Kawamata, J., Yamanouchi, T., Dohmae, *et al.* (1987) Control of laboratory acquired haemorrhagic fever with renal syndrome (HFRS). *Lab. Anim. Sci.*, **37**, 431–6.

27. Ketai, L.H., Williamson, M.R., Telepak, R.J. *et al.* (1984) Hantavirus pulmonary syndrome: radiographic findings in 16 patients. *Radiology*, **191**, 665–8.

28. Lee, H.W. and van der Groen, G. (1989) Haemorrhagic fever with renal syndrome. *Prog. Med. Virol.*, **36**, 62–102.

29. Lee, H.W. and Johnson, K.M. (1982) Laboratory-acquired infections with Hantaan virus, the etiologic agent of Korean haemorrhagic fever. *J. Infect. Dis.*, **146**, 645–51.

30. Lee, H.W., Lee, P.W., Lahdevirta, J. *et al.* (1979) Aetiological relation between Korean haemorrhagic fever and nephropathia epidemica. *Lancet*, **i**, 186–7.

31. Lee, P.W., Yanagihara, R., Gibbs, C.J. and Gadjusek, D.C. (1986) Pathogenesis of experimental Hantaan virus infection in laboratory rats. *Arch. Virol.*, **88**, 57–66.

32. Lee, Y.T. and Park, C.H. (1994) A new natural reservoir of hantavirus: isolation of hantaviruses from lung tissues of bats. *Arch. Virol.*, **134**, 82–95.

33. Liang, M., Li, D., Xiao, S.-Y., Rossi, C.A. and Schmaljohn, C.S. (1994) Antigenic and molecular characterization of hantavirus isolates from China. *Virus Res.*, **31**, 219–33.

34. Lloyd, G. (1990) Aetiology of hantavirus infections. *Antiviral Chemistry Chemother.*, **1**, 227–31.

35. Lloyd, G. (1991) Hantaviruses, in *Current Topics in Clinical Virology*, (ed. P. Morgan-Capner), Public Health Laboratory Service, London.

36. Lloyd, G., Bowen, E.T.W., Jones, N. *et al.* (1984) HFRS outbreak associated with laboratory rats in UK. *Lancet*, **1**, 1175–6.

37. McKee, K.T. Jr, LeDuc, J.W. and Peters, C.J. (1991) Hantaviruses, in *Textbook of Human Virology*, (ed. R.B. Belshe), Mosby Year Book, St. Louis.

38. Matsuoka, Y., Chen, S.Y. and Compans, R.W. (1991) Bunyavirus protein transport and assembly, in (ed. D. Kolakofsky), *Bunyaviridae. Curr. Top. Microbiol. Immunol.*, 169, 161–79.

39. Nichol, S.T., Spiropoulou, C.F., Morzunov, S. *et al.* (1993) Genetic identification of a hantavirus associated with an outbreak of acute respiratory illness. *Science*, **262**, 914–17.

40. Sayer, W.J., Entwhistle, G., Uyeno, B. and Bignall, R.C. (1955) Cortisone therapy of early epidemic haemorrhagic fever: a preliminary report. *Ann. Intern. Med.*, **42**, 839–51.

41. Schmaljohn, C.S. and Dalrymple, L.M. (1983) Analysis of Hantaan virus RNA: evidence for a new genus of *Bunyaviridae*. *Virology*, **131**, 482–91.

42. Schmaljohn, C.S., Arikawa, J., Hasty, S.E. *et al.* (1988) Conservation of antigenic properties and sequences encoding the envelope proteins of prototype Hantaan virus and two virus isolates from Korean haemorrhagic fever patients. *J. Gen. Virol.*, **69**, 1949–55.

43. Spiropoulou, C.F., Morzunov, S., Feldmann, H., Sanchez, A., Peters, C.J. and Nichol, T. (1994) Genome structure and variability of a virus causing hantavirus pulmonary syndrome. *Virology*, **200**, 715–23.

44. Steier, K.J. and Clay, R. (1993) Hantavirus pulmonary syndrome (HPS): report of first case in Louisiana. *J. Am. Osteopath. Assoc.*, **93**, 1252, 1255.

45. Svedmyr, A., Lee, H.W., Berglund, A. *et al.* (1979) Epidemic nephropathy in Scandinavia is related to Korean haemorrhagic fever. *Lancet*, **1**, 100.

46. Taller, A.M., Xiao, S.-Y., Godec, M.S. *et al.* (1993) Belgrade Virus, a cause of haemorrhagic fever with renal syndrome in the Balkans, is closely related to Dobrava virus of field mice. *J. Infect. Dis.*, **168**, 750–3.

47. Umenai, T., Lee, H.W., Lee, P.W. *et al.* (1979) Korean haemorrhagic fever in an animal laboratory. *Lancet*, **1**, 1314–16.

48. World Health Organization (1983) Haemorrhagic fever with renal syndrome: memorandum from a WHO meeting. *Bull. WHO*, **61**, 257–69.

49. World Health Organization (1986) Global survey of antibody to Hantaan-related viruses among perdomestic rodents. *Bull. WHO*, **64**, 139–44.

50. Yanagihara, R. (1993) Hantavirus infections in the United States: epizootiology and epidemiology. *Rev. Infect. Dis.*, **12**, 449–57.

51. van Ypersele de Strihou, C. and Mery, J.P. (1989) Hantavirus-related acute interstitial nephritis in Western Europe: expansion of a worldwide zoonosis. *Q. J. Med.*, **270**, 941–50.

52. Zaki, S.R., Albers, R.C., Greer, P.W. *et al.* (1994) Retrospective diagnosis of a 1983 case of fatal hantavirus pulmonary syndrome. *Lancet*, **1**, 1037–8.

53. Zhang, X.K., Takashima, I. and Hashimoto, N. (1989) Characteristics of passive immunity against hantavirus infection in rats. *Arch. Virol.*, **105**, 235–46.

MYCOPLASMAS AND THEIR ROLE IN HUMAN RESPIRATORY TRACT DISEASE

David Taylor-Robinson

17.1 INTRODUCTION

Almost 60 years ago a mycoplasma was isolated for the first time from humans [29].

Viral and Other Infections of the Human Respiratory Tract.
Edited by S. Myint and D. Taylor-Robinson. Published in
1996 by Chapman & Hall. ISBN 0 412 60070 6

Subsequently, there have been many landmark events in the field of mycoplasmology. An account is given in this chapter of the mycoplasmas that have been found in the human respiratory tract during this period and their role in causing disease, together with comments on diagnosis, treatment and prevention. First, however, as a background, the characteristics and taxonomy of mycoplasmas are discussed briefly.

17.2 CHARACTERISTICS OF MYCOPLASMAS

The first mycoplasma ever to be isolated, just before the turn of the century, came from cattle with pleuropneumonia [83]. Organisms of the same kind isolated subsequently from other animal species were called, therefore, pleuropneumonia-like organisms (PPLO), a term that is now found only in the earlier literature and out-dated textbooks. The name 'mycoplasma' was introduced first in 1929 [84] but did not gain acceptance until it was re-proposed more than 25 years later [31]. Mycoplasmas have been derived from bacteria by gene deletion and are the smallest free-living microorganisms, the smallest viable cells measuring no more than 300 nm in diameter. These are sufficiently small to penetrate agar medium and their growth within the medium forms the central dark area of the 'fried egg'-like colonies which are characteristic of most mycoplasmas. Genetic depletion is such that mycoplasmas lack the ability to produce a rigid cell wall. As a consequence, they are resistant to penicillins and other antibiotics which act on this structure. Instead of a rigid wall, the cytoplasm is limited by a pliable unit membrane which also encloses the DNA, RNA, and other metabolic components necessary for propagation in cell-free media.

Apart from their importance in humans, certain mycoplasmal species are of economic importance because of the pneumonia, arthritis, keratoconjunctivitis and mastitis they cause among livestock and poultry in Africa, Australia and other parts of the world. Furthermore, naturally occurring infections of laboratory animals may affect the results of experimental procedures and a number of mycoplasmal species are a laboratory nuisance

as occult contaminants of eukaryotic cell cultures [3].

17.3 TAXONOMY OF MYCOPLASMAS

Despite the general characteristics that mycoplasmas have in common, they comprise a heterogeneous group of microorganisms which differ from one another in DNA and antigenic composition, nutritional requirements, metabolic reactions and host specificity. Taxonomically, mycoplasmas are members of the class Mollicutes which is divided into three orders (Table 17.1). Almost all of the mycoplasmas isolated from humans belong to the order Mycoplasmatales and to the family Mycoplasmataceae which is one of two families in the order. This family com-

Table 17.1 Taxonomy of the class Mollicutes

Classification		Genome size (kbp)	Mol % G+C of DNA	Other characteristics
Order I	Mycoplasmatales			
Family I	Mycoplasmataceae			
Genus I	*Mycoplasma*	580–1380	23–41	
	About 100 species			
Genus II	*Ureaplasma*	730–1160	27–30	Urea metabolized
	Five species; *U. urealyticum* has at least 14 serotypes			
Order II	Entomoplasmatales			
Family I	Entomoplasmataceae			
Genus I	*Entomoplasma*	790–1140	27–29	
	Five species			
Genus II	*Mesoplasma*	825–1100	27–30	0.04% Tween 80 required
	Four species			
Family II	Spiroplasmataceae			
Genus I	*Spiroplasma*	970–1970	25–31	Helical structure
	Eleven or more species including *S. citri*			
Order III	Acholeplasmatales			
Family I	Acholeplasmataceae			
Genus I	*Acholeplasma*	1230–1690	27–36	Sterol not required
	Twelve species			
Order IV	Anaeroplasmatales			Obigate anaerobes
Family I	Anaeroplasmataceae			
Genus I	*Anaeroplasma*	About 1600	29–33	
	Four species			
Genus II	*Asteroleplasma*	About 1600	40	Sterol not required
	One species			

prises two genera. The first is the genus *Mycoplasma*, which contains the largest number of species (about 100) within the class that metabolize glucose and/or arginine but not urea; these microorganisms are referred to trivially as mycoplasmas, a term often extended to all organisms in the class. The second genus, *Ureaplasma*, contains microorganisms which hydrolyse urea only and are referred to trivially as ureaplasmas. The latter organisms originally were termed T-strains or T-mycoplasmas (T, representing 'tiny') because of the tiny colonies they form on agar medium [95].

17.4 OCCURRENCE OF MYCOPLASMAS IN HUMANS

Twelve *Mycoplasma* species and *Ureaplasma urealyticum* constitute the normal flora or are pathogens of humans [107] (Table 17.2). Two *Acholeplasma* species have also been isolated. Most of these various microorganisms are found predominantly in the oropharynx or infecting the respiratory tract. The suggestion that there may be a thirteenth *Mycoplasma*

species, *M. pirum*, of human origin comes from the fact that this mycoplasma was isolated originally [28] and subsequently [64] from human lymphoblastoid cell lines, and its recent reported isolation from cultures of lymphocytic cells from AIDS patients [77] offers further support for the notion, assuming that the organisms came from the patients and were not contaminants of the cell cultures. More information is required about the distribution of the two species reported most recently, namely *M. spermatophilum* and *M. penetrans*. The former was isolated from human sperm and cervical specimens [46] and the latter from the urine of HIV-positive homosexual men [68] and has been associated serologically with Kaposi's sarcoma [113].

17.5 MYCOPLASMAS OTHER THAN M. *PNEUMONIAE* AND ACUTE RESPIRATORY DISEASE

The dominant role of *M. pneumoniae* in causing acute respiratory disease has not been superseded by any of the other mycoplasmas.

Table 17.2 Mycoplasmas of human origin: primary site of colonization, metabolism and pathogenicity

Genus and species	Primary site of colonization		Metabolism of		Cause of disease
	Oropharynx	Genitourinary tract	Glucose	Arginine	
M. salivarium	+	−	−	+	−
M. orale	+	−	−	+	−
M. buccale	+	−	−	+	−
M. faucium	+	−	−	+	−
M. lipophilum	+	−	−	+	−
M. fermentans	+	−	+	+	+
M. pneumoniae	+	−	+	−	+
M. hominis	−	+	−	+	+
M. genitalium	−	+	+	−	+
M. primatum	−	+	−	+	−
M. spermatophilum	−	+	−	+	−
M. pirum*	?	?	+	+	?
M. penetrans	−	+	+	+	?
U. urealyticum	−	+	−	−	+
A. laidlawii	+	−	+	−	−
A. oculi	?	?	+	−	−

* Possibly of human origin

However, existing knowledge about their involvement is commented on first as a background to a discussion of *M. pneumoniae*.

17.5.1 MYCOPLASMAS BEHAVING APPARENTLY AS COMMENSALS

The species found most commonly in the oropharynx are *M. salivarium* and *M. orale* (Table 17.2); one or both have been isolated from as many as 84% of individuals in one study [62] and both can probably be found as members of the resident flora in all adults; in other words, behaving as commensals. The latter is a judgement based on the fact that these mycoplasmas have not been associated with disease. There are several other mycoplasmas which come into the same category but are isolated less frequently. These *are M. buccale, M. faucium, M. lipophilum, M. primatum, Acholeplasma laidlawii* and, in adults, *Ureaplasma urealyticum*. There may be several reasons to account for infrequent detection. The oropharynx may be the natural habitat for some but they occur infrequently at this site, or their natural habitat is the genital tract and their presence in the oropharynx is intermittent or transient following orogenital contact. In addition, some of the mycoplasmas may appear to occur infrequently, simply because they have fastidious nutritional requirements that limit successful culture. In any event, these mycoplasmas should be regarded as commensals unless proved otherwise.

17.5.2 MYCOPLASMAS EXHIBITING SIGNS OF PATHOGENICITY

M. hominis

During a study of patients with pneumonia, in the 1960s, several mycoplasmal species were recovered from the oropharynx. One of these was *M. hominis* (strain DC63) to which the individual developed complement-fixing and fluorescent-stainable antibodies [80]. Subsequently, rises in antibody titre were detected in the sera of other patients with pneumonia. Furthermore, whereas 3.2% of 346 patients with pneumonia developed complement-fixing antibody to strain DC63, only 0.3% of 939 patients without respiratory disease did so (P <0.001). These findings prompted a volunteer experiment [80] in which 50 men were given a large number of organisms of strain DC63 oropharyngeally via a nebulizer and nasally via a pipette. The organisms were recovered from 42 of the men and 38 developed a four-fold or greater rise in indirect haemagglutinating antibody. An afebrile exudative pharyngitis developed in 21 men and an afebrile non-exudative pharyngitis in four. Half of those with pharyngeal involvement had cervical adenopathy and one-quarter complained of a sore throat. In addition, pharyngitis occurred more often in men who did not have pre-existing antibody than in those who did. All of this amounts to irrefutable evidence for *M. hominis* having caused the pharyngitis. However, attempts subsequently to demonstrate that this mycoplasma is a cause of naturally occurring pharyngitis in children and adults have been unsuccessful [79]. This is probably because contact occurs with smaller numbers of organisms under natural conditions than experimentally, although the volunteer study does illustrate the potential pathogenicity of *M. hominis*.

M. genitalium

M. genitalium was found originally in the male genitourinary tract [109] but subsequently it was isolated from a small proportion of respiratory tract specimens which also contained *M. pneumoniae* [4]. The difficulty experienced in detecting *M. genitalium*, at least by culture, was overcome eventually with the development of the polymerase chain reaction (PCR) [54,87]. Use of this technique has resulted in this mycoplasma being associated strongly with acute non-gonococcal urethritis [47,52]. Indeed, the evidence would tend to indicate that *M. genitalium* occurs more often in the genitourinary tract than the respiratory tract. Nevertheless, its significance at the latter site remains to be determined. Furthermore, because of the close antigenic similarity between *M. genitalium* and *M. pneumoniae*, as mentioned later, caution should be exercised when using the complement-fixation test to diagnose an *M. pneumoniae* respiratory infection.

M. fermentans

M. fermentans was isolated first from the genitourinary tract [91] and, until quite recently, was regarded as occurring infrequently at this site. It is known to contaminate cell cultures and in the late 1960s was suggested, but never proven, to be a cause of rheumatoid arthritis [115]. However, whatever tenuous association *M. fermentans* was thought to have had with human disease needs to be reconsidered in the light of recent events. Using immunohistochemical, DNA hybridization and electron microscopical techniques, Lo and co-workers [67] found *M. fermentans*, at first erroneously called 'M. incognitus' [92], in various tissues of AIDS patients. Also, by means of a PCR technique, it has been detected in the throat of about 20%, the blood of 10% and the urine of 5% of HIV-positive homosexual men, as well as in about the same proportions of specimens from HIV-negative, mainly homosexual, individuals [55]. From the viewpoint of respiratory disease, *M. fermentans* has been recovered from the throats of about 16% of children with community-acquired pneumonia, two-thirds of whom apparently had no other respiratory pathogen [15]; its occurrence in healthy children is not known. In addition, it was detected in a few adults who presented with an acute influenza-like illness, which sometimes deteriorated rapidly with development of an often fatal respiratory distress syndrome [69]. It is clear from these cases that infection by the mycoplasma and associated disease is not linked necessarily with prior immunosuppression of the patient. Most recently, however, it

was detected by use of the PCR in broncheoalveolar lavage specimens from 25% of AIDS patients with pneumonia [1]. Whether *M. fermentans* makes a significant contribution to the latter is still not clear, but there seems no doubt that it is more associated with the respiratory tract than the genitourinary tract and that its significance at the former site needs to be fully evaluated.

The ways of detecting the mycoplasmas mentioned, that is those for which there are some indications of pathogenicity for the respiratory tract, are mainly molecular and have been reviewed recently [104]. The means of treating infection by these mycoplasmas is mentioned later in the section concerned with the treatment of *M. pneumoniae* infection.

17.6 M. PNEUMONIAE AND ACUTE RESPIRATORY DISEASE

17.6.1 HISTORICAL ASPECTS

In the late 1930s, non-bacterial pneumonia was recognized as being distinct from typical lobar pneumonia and the term primary atypical pneumonia (PAP) was coined. Gradually PAP was found to be aetiologically heterogeneous. In one variety of disease, in which cold haemagglutinins often developed, an infectious agent was isolated in embryonated eggs [30]. This microorganism, the 'Eaton agent', produced pneumonia in cotton rats and hamsters, and for a number of years was thought to be a virus. However, the inhibitory effects that chlortetracycline and gold salts had on the agent [75] raised serious doubts about its viral nature and these were justified when it was shown to be a mycoplasma by cultivation on a cell-free agar medium [18]. It was subsequently called *M. pneumoniae* [17] and its ability to cause respiratory disease was established fully by studies based on isolation, serology, volunteer inoculation, and vaccine protection [16].

17.6.2 EPIDEMIOLOGICAL ASPECTS

Transmission

Spread of *M. pneumoniae* from person-to-person occurs slowly, usually where there is continual or repeated close contact, for example in a family, a boarding school or military camp, rather than where there is only casual contact [35].

Geographical and seasonal distribution of infection

M. pneumoniae infections have been reported from every country where appropriate diagnostic tests have been undertaken. Infection is endemic in most areas so that disease is seen during all months of the year with a slight preponderance in the late summer and early autumn, at least in countries where seasonal climatic changes are striking. Apart from such

seasonal variation, epidemic peaks have been observed about every 4–7 years in some countries, and those recorded for England and Wales over the period 1975–1993 are shown in Figure 17.1. A similar cyclical variation was seen in Denmark until 1972. Then, possibly as a consequence of sociological changes, through children being enrolled in day-care at an earlier age, the regular epidemics ceased until 1986 when there was a major epidemic [66].

Relative importance of *M. pneumoniae*

This mycoplasma causes inapparent and mild upper respiratory tract infections (coryza and wheezing) more commonly than severe disease but, nevertheless, it is responsible for only a small proportion of all upper respiratory tract disease, most of it being of viral aetiology.

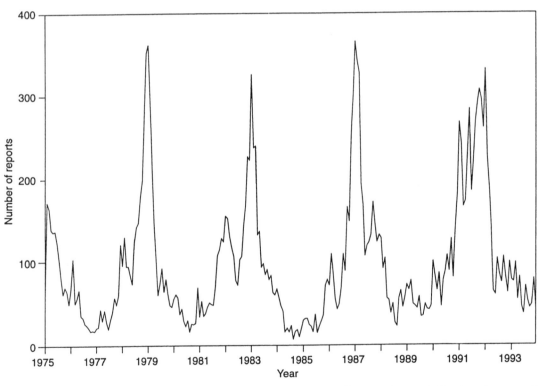

Figure 17.1 *M. pneumoniae* infection in England and Wales during the period 1975–1993. Note the 4-yearly periodicity. Graph based on 4-weekly totals of laboratory reports supplied to and by courtesy of the Communicable Disease Surveillance Centre, Colindale London.

Acute pharyngitis, occurring mostly in adolescents and younger persons, is likely to be due mainly to group A streptococci and rarely (about 5%) to *M. pneumoniae* [41] which plays a relatively greater part in producing lower respiratory tract disease. Even here, *M. pneumoniae* is only one of several microorganisms that are recognized as being important; *Chlamydia pneumoniae* is the respiratory pathogen added most recently to the list of those bacteria causing atypical pneumonia [42]. Nevertheless, the significance of *M. pneumoniae* should not be played down because its contribution as a cause of pneumonia in certain groups of people ranks high and may outweigh that of respiratory viruses and other bacteria. Thus, in the USA, it has been calculated that in a large general population, *M. pneumoniae* has accounted for about 15–20% of all pneumonias [19,35] and in certain populations, for example military recruits, it has been be responsible for as much as 40% of acute pneumonic illness [19]. *Haemophilus influenzae* pneumonia occurring soon after infection by *M. pneumoniae* has been seen [100] but, otherwise, there is very little evidence that infection with *M. pneumoniae* predisposes to infection by this or other microorganisms, or vice versa.

Relationship of infection to age

M. pneumoniae affects children and adults, the consequence of infection depending upon age and immune status [26,35]. Infection is common in children under 5 years of age and it seems that most of the infections are symptomatic, although they tend to be mild and non-pneumonic, usually in the form of coryza and wheezing without fever. Infection rates are greatest in school-aged children and teenagers (5–15 years) and the risk of *M. pneumoniae* pneumonia is maximum (possibly 30% or more), such disease constituting about half of all cases of pneumonia in this age group. In the 25 to 45-year age group, the incidence of pneumonia is five-fold less than at 10 years of age and it diminishes further thereafter; in adults, perhaps only about 5% of *M. pneumoniae* infections manifest as pneumonia. However, when it occurs, such disease in the middle-aged and elderly is often more severe than in the young.

17.6.3 DISEASE MANIFESTATIONS

M. pneumoniae may produce only an inapparent infection. On the other hand, following an incubation period of 2–3 weeks, a spectrum of effects may occur, ranging from mild afebrile upper respiratory tract disease, tracheobronchitis, to severe pneumonia [22,26,35]. Clinical manifestations often are insufficiently distinct to permit an early definitive diagnosis of *M. pneumoniae* pneumonia. Indeed, the latter shares the features of non-bacterial pneumonias in that general symptoms, such as malaise and headache (both common) and occasionally fever, often precede the development of a cough by 1–5 days. Cough is usually non-productive and becomes a prominent feature, its absence making the diagnosis of *M. pneumoniae* pneumonia unlikely. Patients usually do not appear seriously ill and the term 'walking pneumonia' has been used because far less than 10% of cases of pneumonia, in one study [36] no more than 2% (the very young and elderly), are severe enough to warrant admission to hospital. Radiographic examination frequently reveals evidence of pneumonia before physical signs, such as rales, become apparent. Usually, only one of the lower lobes is involved and the radiograph most often shows segmental patchy opacities. About 20% of patients suffer bilateral pneumonia, but pleurisy and large pleural effusions are rare. The course of the disease is variable, but often it is protracted. Thus, cough, abnormal chest signs, and changes in the radiograph may persist for several weeks and relapse is a feature. The organisms also may persist in respiratory secretions despite antibiotic therapy, a feature occurring particularly in hypogammaglobulinaemic patients in whom excretion may continue for months or years rather than weeks [106]. Although a few very severe infections have been reported, usually in patients with immunodeficiency or sickle cell anaemia [35], death has occurred rarely; in one study [101], 3% of hospitalized patients with *M. pneumoniae* pneumonia died. In children, infection has been characterized occasionally by a prolonged illness with paroxysmal cough followed by vomiting, thus simulating the features of whooping cough.

Extrapulmonary manifestations

Disease caused by *M. pneumoniae* is limited usually to the respiratory tract, but a wide variety of extrapulmonary clinical conditions occurring during or as a sequel to the respiratory illness has been reported [35,82]. These complications and an estimation of their frequency are shown in Table 17.3. Whether any of them might be coincidental or due to co-infection with *M. genitalium* or to the latter entirely are moot points which may not be easy to resolve. Haemolytic

anaemia with crisis is brought about by the development and action of cold haemagglutinins (anti-I antibodies) [33]. There is dispute about the mechanism of their production, but *M. pneumoniae* organisms adhering to the I antigen receptor on erythrocytes may alter it sufficiently to stimulate an autoimmune response [34]; whether *M. genitalium* could do the same is unknown. It is likely that some of the other clinical conditions also have an autoimmune basis. In the case of the neurological complications, this may be possible too, but a direct effect on the central nervous system as a consequence of invasion cannot be discounted as there are several reports of the isolation of *M. pneumoniae* from cerebrospinal fluid, although not from the brain itself [59].

17.6.4 PATHOGENESIS

The following factors, most logically in the sequence presented, would seem to be important in pathogenesis.

Table 17.3 Extrapulmonary sequelae of *M. pneumoniae* infection

System	Manifestations	Estimated frequency
Cardiovascular	Myocarditis, pericarditis	<5%
Dermatological	Erythema multiforme; Stevens–Johnson syndrome; other rashes	Some skin involvement in about 25%
Gastrointestinal	Anorexia, nausea, vomiting and transient diarrhoea	Up to 45%
	Hepatitis	?
	Pancreatitis	?
Genitourinary	Tubo-ovarian abscess	Insignificant
Haematological	Cold haemagglutinin production	About 50%
	Haemolytic anaemia	?
	Thrombocytopenia	?
	Intravascular coagulation	Few cases reported
Musculoskeletal	Myalgia, arthralgia	Up to 45%
	Arthritis	?
Neurological	Meningitis, meningo-encephalitis, ascending paralysis, transient myelitis, cranial nerve palsy and poliomyelitis-like illness	6–7%
Renal	Acute glomerulonephritis	?

Aerosol particle size

M. pneumoniae organisms administered to hamsters in a large particle aerosol (8 μm) caused an infection which remained confined to the upper respiratory tract [51]. However, use of a small-particle aerosol (2.3 μm) resulted in infection of both the upper and lower respiratory tract. Thus, the size of particles containing the organisms in expelled respiratory secretions could be a factor in the development of pneumonia in humans.

Cytadherence

It is possible that the motility exhibited by *M. pneumoniae* organisms [89] assists them in passing through the mucus blanket covering the respiratory epithelium so that they are able to adhere to host cells (cytadherence). This would seem to be an important element in allowing the organisms to resist removal from the respiratory mucosa by mucociliary action. Examination of desquamated epithelial cells in sputum samples from patients with confirmed *M. pneumoniae* infections has revealed that the organisms are often orientated perpendicular to the host cell surface and anchored to it by the tip of the elongated neck of their bottle-shaped structure [25]. The same is true of experimentally infected trachea and lung tissue from hamsters [25] where the organisms are seen lying between cilia and attached to the epithelium in the same way (Figure 17.2). It is possible, however, that this orientation reflects, to some extent, the space constraints placed on the mycoplasmas by the closely arranged cilia, as opposed to the distribution of adhesins on the mycoplasmal surface. This view derives from the fact that the adherence of mycoplasmas incubated with non-ciliated cells or with erythrocytes, conditions in which *M. pneumoniae* need not contend with penetration between cilia to reach the cell surface, is not limited to the tip portion of the organisms [10,39]. Furthermore, immunoelectron microscopy using polyclonal antibodies to the adhesin protein P1 of *M. pneumoniae* has shown that this protein is not limited exclusively to the tip region but is scattered elsewhere on the mycoplasmal surface although, admittedly, less densely than at the tip. However, whatever the distribution of the adhesins, the critical role of cytadherence in virulence is apparent from the inability of non-cytadhering strains, which arise from multiple passage in broth medium, chemical mutagenesis, or spontaneous phase variation, to infect and cause disease in experimentally infected hamsters [60].

In addition to the P1 protein, a 32-kDa protein [5] is also involved in adherence. Furthermore, the results of other studies indi-

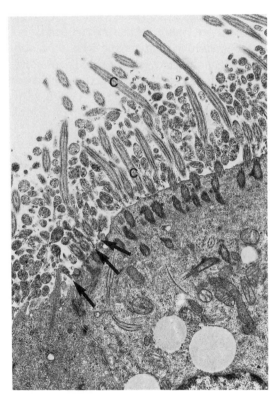

Figure 17.2 Electron micrograph of two ciliated epithelial cells in the tracheal mucosa of a hamster infected experimentally with *M. pneumoniae*. Cilia (c) and individual organisms (arrowed), some with the tip of the elongated structure pointing towards the eukaryotic cell membrane (original magnification× 13 000).

cate that *M. pneumoniae* adheres to host cells via specific long-chain sialo-oligosaccharides of the poly-*N*-acetyllactosamine type [70]. Such receptors are found in human and hamster bronchial epithelium on the microvilli and on the cilia [71]. *M. pneumoniae* has also been reported to adhere to sulphated glycolipids [61] so that it is clear that the receptor–adhesin mechanism is complex.

Toxic factors

Examination of sputum samples from patients infected with *M. pneumoniae* reveals organisms attached to degenerating epithelial cells [25], prompting speculation that the infection leads to host cell dysfunction and cell death. Furthermore, there is ciliary dysfunction and desquamation in hamster and fetal human tracheal rings infected *in vitro* [24]. At the ultrastructural level, freeze fracture/electron microscopy shows that, together with ciliary dysfunction, there are alterations in the distribution of host cell membrane particles [14], including a necklace structure encircling the cilia at the base of the shaft that is thought to be important in the regulation of ciliary motion [40]. These observations are consistent with clinical findings, as clearance from the respiratory tract can remain depressed for months following *M. pneumoniae* infections [12].

Like other mycoplasmas, *M. pneumoniae* produces hydrogen peroxide, but it differs from the other species that infect humans because it produces a larger quantity of peroxide. This accounts for the rapid β-haemolysis seen when colonies on agar are overlaid by guinea-pig erythrocytes compared with the less rapid α-haemolysis caused, for example, by *M. hominis* [98]. Superoxide as well as hydrogen peroxide are the normal by-products of *M. pneumoniae* metabolism [72,99], and result from the incomplete reduction of oxygen by a truncated electron transport system. It has been suggested that the accumulation of these oxygen metabolites in the extracellular milieu could overwhelm normal host cell mechanisms which defend against peroxidative injury [23]. This, of course, is speculation, as is the notion that adhesion and parasitism by the mycoplasmas could result in their acquisition of nutrients from the respiratory mucosa leading to host cell dysfunction.

Immunopathological factors

After cytadherence and epithelial cell damage, several observations suggest that immune mechanisms play an important role in the development of *M. pneumoniae* pneumonia in humans. Death attributable to the mycoplasma is a rare event (see above), so that a picture of histopathological changes has had to be built up by inference, mainly from experimental infection of hamsters and natural mycoplasmal disease in other animals. In these, the pneumonic infiltrate is predominantly a peribronchiolar and perivascular cuffing by lymphocytes, most of which are thymus dependent. It has been observed [103] that infected hamsters, depleted of T-cells in various ways, exhibit less lung histopathology, suggesting that the cellular immune response may be detrimental and important in the pathogenesis of *M. pneumoniae* pneumonia. The development of a cell-mediated immune response to *M. pneumoniae* in humans, initiated apparently by a T-cell epitope on the adhesin protein P1 [49], has been shown by positive lymphocyte transformation, macrophage migration inhibition and delayed hypersensitivity skin tests [102], delayed hypersensitivity correlating with disease severity [76]. The initial lymphocyte response is followed by a change in the character of the bronchiolar exudate with polymorphonuclear leucocytes and macrophages predominating. This sequence of events develops rather slowly on primary infection and contrasts with an accelerated and often more intense host response on reinfection. At least to some extent, therefore, the pneumonia caused by *M.*

pneumoniae would seem to be an immunopathological process. Young children, less than 5 years of age, often experience non-pneumonic infections and/or possess antibody [11], although it is not clear whether this is induced entirely by *M. pneumoniae* infection. Nevertheless, it is tempting to suggest that the pneumonia which occurs in older persons is an immunological over-response to re-infection, the lung being infiltrated by previously sensitized lymphocytes, despite the fact that, in adults, some protection is derived from a previous known episode of *M. pneumoniae* pneumonia in that second bouts are usually less and not more severe [35,37].

17.6.5 DIAGNOSIS

The diagnosis of *M. pneumoniae* infection has depended on cultural or non-cultural techniques to detect the organisms and/or on performing specific or non-specific (cold haemagglutinin) serological tests, overall the serological approach being favoured. The medium that has been employed most widely for the isolation of *M. pneumoniae* consists of beef-heart infusion broth, 20% (v/v) horse serum and 10% (v/v) fresh yeast extract (25 % w/v) [38]. However, attention should be paid to the use of SP4 medium which was developed originally to cultivate spiroplasmas [110]. This has improved the isolation rate not only of the more fastidious mycoplasmas, like *M. pneumoniae* [108], but also the more easily isolatable, such as *M. hominis*. SP4 medium comprises essentially a conventional mycoplasmal broth medium with fetal calf serum and a tissue-culture supplement. To either medium is added a broad-spectrum penicillin, and glucose with phenol red as a pH indicator. Such fluid medium, set initially at pH 7.5–7.8, is inoculated with sputum, throat washing, pharyngeal swab, or other specimen and incubated at 37°C. A colour change (red to yellow) due to a reduction in pH, occurs usually but not always within 4 to 21 days and signals the fermentation of glucose with production of acid due to multiplication of the organisms. This preliminary indication of the existence of a mycoplasma may be confirmed after subculturing to agar medium. Colonies of *M. pneumoniae* develop best in an atmosphere of air–5% CO_2 usually with a 'scrambled egg' rather than the classical 'fried egg' appearance, although such morphology may develop on subculture. Erythrocytes of various species adsorb to colonies of many different mycoplasmas (haemadsorption) [74]. Since erythrocytes, other than those of chicken origin, adsorb only to colonies of *M. pneumoniae* and *M. genitalium* of the mycoplasmas of human origin, a positive haemadsorption test provides a diagnostic clue, specific identification being made usually by demonstrating inhibition of colony development around discs impregnated with specific antiserum, the well-known agar growth inhibition test [21]. However, because of the close serological relationship between *M. pneumoniae* and *M. genitalium* [65], certainty in identifying either of them may require several techniques, more than one antiserum [64] and perhaps the use of monoclonal antibodies allied to Western blotting [78].

Antigen detection techniques have been developed for *M. pneumoniae*, antigen in respiratory exudates having been detected by direct immunofluorescence, counter immuno-electrophoresis, immunoblotting with monoclonal antibodies and several enzyme immunoassays [104]. However, while these methods have the virtue of speed, sensitivity is a problem with all of them; for example, 10^4 colony-forming units of *M. pneumoniae* per ml of sample were required to provide a positive result when tested with one of the immunoassays [58]. Inadequate sensitivity has also been an obstacle to the widespread use of DNA probes for *M. pneumoniae* [43]. However, the PCR assay should not have this drawback and DNA primers specific for *M. pneumoniae* have been developed and used for DNA amplification in this assay [7,53]. Such technology may be used by itself or in concert with culture, since it provides the opportunity for rapid determination of *M. pneumoniae* positivity and then, if isolates are required, continued culture only of those specimens that are PCR-positive. Currently, the PCR assay is used mainly as a research tool but commercial interest is almost certain to see it become a routine procedure in the future.

In view of the difficulties of isolating or detecting *M. pneumoniae* organisms, it has been inevitable that, in routine practice, reliance has been and still is placed almost wholly on serology for diagnosis. Antibody is detectable by a variety of procedures [104], but many are impractical. However, this is a criticism that cannot be levelled at the complement-fixation test which, with the chloroform–methanol-extractable lipid as antigen [56], has been the backbone of diagnosis. A four-fold or greater rise in antibody titre, with a peak at about 3–4 weeks after the onset of disease, has been estimated to occur in about 80% of cases and is usually taken as indicative of a recent infection (see below). A single antibody titre of 1:64–1:128 or greater in a suggestive clinical setting should at least be regarded as sufficient to institute therapy; a four-fold or greater fall in antibody titre, perhaps over 6 months, is of dubious value, particularly if there is difficulty in relating it to a particular prior illness. The problem created by the inability to distinguish between *M. pneumoniae* and *M. genitalium* by the complement-fixation test may be viewed in the following way. Current evidence suggests that *M. genitalium* does not occur often in the respiratory tract and the test, using *M. pneumoniae* antigen, is unlikely to be sufficiently sensitive to detect responses following genital tract infections, at least in humans. However, these issues need to be resolved before the diagnosis of *M. pneumoniae* infections can be based unconditionally on the complement-fixation test or others, such as the indirect haemagglutination test, which do not distinguish easily

between the two mycoplasmas. A sensible and realistic approach to serodiagnosis at the moment is to use complement-fixation with Western blotting to check on specificity in dubious cases [56], or to use the microimmunofluorescence test in which IgM antibody is sought [96]; this would seem sufficiently specific to provide some confidence in making an accurate diagnosis. In the future, more certain diagnosis may come from using an enzyme-linked immunosorbent assay with purified P1 adhesin protein [48] which should provide greater specificity.

Cold haemagglutinins, detected by agglutination of O Rh-negative erythrocytes at 4°C, develop in about half the patients, mostly those who are hospitalized with the more severe illnesses. The ease of undertaking the test and of reading it, a titre of 1:128 or greater being suggestive of a recent *M. pneumoniae* infection, make it worthwhile. However, it must be remembered that haemagglutinins are induced occasionally in a number of other conditions, and that the status of *M. genitalium* in being able to provoke a response is unknown.

17.6.6 TREATMENT

Mycoplasmas are indifferent to the penicillins, cephalosporins and other antibiotics that affect cell wall synthesis, but generally they are sensitive to antibiotics which inhibit protein synthesis. Thus, *M. pneumoniae*, like other mycoplasmas, is sensitive to the tetracyclines and apparently more sensitive to erythromycin than are the other mycoplasmas of human origin. In the case of pregnant women and children, it is advisable to use erythromycin rather than a tetracycline, and the former antibiotic has sometimes proved more effective than a tetracycline in adults. Overall, there should be no concern over therapeutic options because *M. pneumoniae* is also inhibited by the newer macrolides, such as clarithromycin and azithromycin, and the newer quinolones, such as sparfloxacin [6].

The value of antibiotic therapy in *M. pneumoniae*-induced disease was shown first in a controlled trial of dimethylchlortetracycline undertaken in US marine recruits, the duration of fever, pulmonary infiltration, and other signs and symptoms being reduced significantly [57]. Since then, other trials have continued to provide evidence for the effectiveness of various tetracyclines, as well as erythromycin and other macrolides [94]. It should be noted, however, that antibiotics tend to behave more effectively in planned trials than they do in routine clinical practice, probably because disease has become more established in practice before treatment is instituted. This should not be construed as meaning that antibiotic therapy is not worthwhile, although clinical improvement is not always accompanied by early eradication of the organisms from the respiratory tract [97]. The likely reason for this is that, apart from the quinolones, the earlier ones of which have only moderate activity against *M. pneumoniae* [6], the drugs inhibit multiplication of the organisms rather than killing them. This, in turn, is a possible explanation for clinical relapse in some patients and a plausible reason for recommending a 2 to 3-week course of antibiotic treatment rather than a shorter course. It is a moot point whether early treatment might prevent some of the complications but, nevertheless, it should commence as soon as possible. If facilities for rapid laboratory diagnosis, namely a PCR assay, are not available, confirmation of an *M. pneumoniae* infection inevitably will be slow. In this circumstance, it would seem wise to start antibiotic treatment on the basis of the clinical evidence alone, a cold haemagglutinin and/or suggestive single serum antibody titre perhaps providing some diagnostic assurance, despite the drawbacks, mentioned previously, of attempting to make the diagnosis in this way. Treatment of *M. pneumoniae* and also other mycoplasmal and ureaplasmal infections in patients who are immunodeficient, for example those with hypogammaglobulinaemia [106], may prove particularly challenging because the difficulty of eradicating the organisms from such patients is even greater than experienced with immunocompetent ones. This is an indication of the important contribution made by the immune system to successful treatment, a view which has been supported recently by the failure of antibiotic therapy to eliminate mycoplasmas from nude mice in contrast to elimination from their immunocompetent counterparts (D. Taylor-Robinson and P.M. Furr, unpublished data).

The treatment of infections by mycoplasmas which could potentially cause respiratory disease depends, of course, on their antibiotic susceptibility profiles. *M. hominis* is innately resistant to erythromycin and some of the other macrolides but susceptible to clindamycin and lincomycin, whereas the reverse is true for *U. urealyticum* [104]. *M. genitalium* has a profile similar to that for *M. pneumoniae*, being sensitive to the tetracyclines and a range of macrolides and streptogramins [90]. *M. fermentans* is resistant to erythromycin [44], having an antibiotic susceptibility profile similar to that for *M. hominis*. The development of antibiotic resistance has been seen particularly during the treatment of hypogammaglobulinaemic patients [105,106]. In the case of *M. pneumoniae* infections, clinical resistance to tetracycline with a response to erythromycin and the reverse have both been seen [26].

17.6.7 PREVENTION

Resistance to disease

It seems that infection early in life confers short-lived protection and there is speculation (see above) that such infection is then responsible for disease in teenagers and young adults. However, as the population grows older protection is maintained for longer periods. The attack rate diminishes with increasing age and although known cases of re-infection and disease have been reported

[37], this is rare in older adults. Overall, the suggestions are that infection with or without disease in early life is protective in later life and that immunization should provide a useful approach to the prevention of infection and disease.

Pulmonary disease induced in hamsters with virulent *M. pneumoniae* organisms given intratracheally protected them when challenged subsequently, whereas hamsters given the organisms parenterally were not protected, despite high levels of serum antibodies [2,32]. This suggests that local and/or cell-mediated immune mechanisms are involved in the protective response. These findings in the animal model are in concert with clinical observations. Thus, the presence of humoral antibody to *M. pneumoniae* correlates only partially with protection [73], since infection and the development of pneumonia may occur despite high levels of, for example, serum mycoplasmacidal antibody. Furthermore, resistance of adult volunteers to *M. pneumoniae*-induced disease has been related to the presence of IgA antibody in respiratory secretions [9]. This could act as a first line of defence by preventing attachment of the organisms to respiratory epithelial cells.

Vaccination against disease

The efficacy of formalin-inactivated *M. pneumoniae* vaccines in preventing pneumonia caused by this mycoplasma was assessed in 11 separate field trials involving more than 40 000 military recruits, University students and institutionalized children over a 15-year period in the USA [32,114]. Seroconversion rates ranged from 0% to 90%, and reduction in naturally occurring disease ranged from 28% to 67%. The failure of killed *M. pneumoniae* vaccines to protect fully may be explained by poor antigenicity of some preparations. However, poor protection associated with the induction of humoral antibody levels that in some cases were similar to those that develop after natural disease suggests that the failure of such vaccines may be due to their inability to stimulate cell-mediated immunity and/or local antibody production. With the latter in mind, Brunner and colleagues [8] developed live attenuated mutants of *M. pneumoniae*. Temperature-sensitive mutants were produced which multiplied at the temperature of the upper respiratory tract, but not at that of the lower tract. Some of these mutants produced pulmonary infection in hamsters without causing pathological changes and in so doing induced significant resistance to subsequent challenge with virulent wild-type *M. pneumoniae*. However, the same mutants caused moderately severe bronchitis or pneumonia in human volunteers [32] so that they were unacceptable for general human use and this approach to vaccination was abandoned. Currently, recombinant DNA vaccines involving the P1 adhesin and other proteins, and a live adenovirus recombinant vaccine developed by cloning a component of the *M. pneumoniae* P1 gene into an adenovirus vector, are being explored [32]. Whether they will ever come to fruition in a climate in which most fund-providing bodies probably regard *M. pneumoniae* vaccine development of low priority because the infection/disease is treatable, is open to question.

17.7 CHRONIC RESPIRATORY DISEASE

Since mycoplasmal infections of animals are often associated with or are the cause of chronic illnesses, the possible role of mycoplasmas in human chronic respiratory disease, particularly chronic bronchitis, is worth considering.

17.7.1 M. PNEUMONIAE INFECTIONS

The isolation of *M. pneumoniae* from some patients experiencing an acute exacerbation of chronic bronchitis, in addition to their having a serological response, suggests that this mycoplasma apart from viruses, is sometimes responsible for an exacerbation [13,20,45].

The occurrence of complement-fixing antibody to *M. pneumoniae* more frequently in the sera of patients suffering from chronic bronchitis than in those of normal subjects [63] is in keeping with this suggestion. However, the real contribution of *M. pneumoniae* in this situation is difficult to assess because it is also evident that patients with chronic bronchitis sometimes acquire a mycoplasmal infection without an apparent worsening of disease [81].

There is a tendency for *M. pneumoniae* organisms to persist in the respiratory tract after clinical recovery, as mentioned before, and occasionally the respiratory disease they cause has a protracted course. Furthermore, soon after a *M. pneumoniae* infection, tracheobronchial clearance is very much reduced and slow clearance may persist for many months [50]. Despite this, however, there is no evidence that *M. pneumoniae* is a primary cause of chronic bronchitis, or that it is responsible for maintaining chronic disease other than by causing, together with other microorganisms, some acute exacerbations.

17.7.2 OTHER MYCOPLASMAL INFECTIONS

M. salivarium, M. orale, and perhaps other mycoplasmas present in the oropharynx of healthy persons, spread to the lower respiratory tract of some patients suffering from chronic bronchitis [20]. There is no evidence that these mycoplasmas, which are regarded as commensals, are a cause of acute exacerbations, but antibody responses to them occur more often in association with such exacerbations than they do at other times [20]. This suggests that the organisms are more antigenic during exacerbations, probably due to increased multiplication and participation in tissue damage brought about primarily by viruses and bacteria. Although there can be no proof, it is tempting to conjecture that in this way the mycoplasmas play some part in perpetuating a chronic condition.

17.8 GENITAL MYCOPLASMAS ASSOCIATED WITH RESPIRATORY DISEASE

The term 'genital mycoplasmas' is used to denote those mycoplasmas that have the genitourinary tract as their primary site of colonization. The role of such mycoplasmas, for example *M. hominis* and *M. genitalium*, in causing respiratory disease in adults has been referred to earlier. In addition, mycoplasmas in the vagina of pregnant women may be transmitted to the infant and have the potential for causing respiratory disease. Such transmission occurs rarely *in utero*, but it often does so during birth and *U. urealyticum*, in particular, may be isolated from the throats and tracheal aspirates of new-borns. In a critical appraisal of four cohort studies [112], in which the relationship between ureaplasmas and chronic neonatal lung disease was analysed, it was concluded that there was strong but not definitive evidence that ureaplasmas are a cause of such disease and occasionally death in infants weighing less than 1250 g at birth. Several other observations have suggested ureaplasmal pathogenicity. For example, a greater risk of mortality was noted among infants with respiratory disease who developed elevated levels of ureaplasmal antibody [88] and the occurrence of ureaplasmas in the nasopharynx or trachea of very low-birth-weight infants was associated with an elevated blood leucocyte count [85], suggesting that these organisms are capable of eliciting an inflammatory response. In addition, bronchopulmonary dysplasia, a condition limited usually to pre-term infants, has been associated with ureaplasmal colonization [27].

Undertaking clinical treatment trials, targeting very low-birthweight infants, to determine whether eradication of ureaplasmas decreases the incidence of chronic lung disease would be logical and has been advocated [93]. Erythromycin has been indicated as the drug of choice [111]. This would not be effective against *M. hominis*, which has also been implicated in pneumonia soon after birth, albeit even more rarely than that associated with *U. urealyticum*. Whether *M. genitalium* might be involved is unknown, but the possibility exists because it has been detected in the vagina and cervix [86].

17.9 REFERENCES

1. Ainsworth, J.G., Hourshid, S., Clarke, J., Mitchell, D., Weber, J.N. and Taylor-Robinson, D. (1994) Detection of *Mycoplasma fermentans* in HIV-positive individuals undergoing bronchoscopy. IOM Letters, Vol. 3, Programme and Abstracts, 10th International Congress of the International Organisation for Mycoplasmology, pp. 319–20.

2. Barile, M.F., Chandler, D.K.F., Yoshida, H., Grabowski, M.W. and Razin, S. (1988) Hamster challenge potency assay for evaluation of *Mycoplasma pneumoniae* vaccines. *Infect. Immun.*, 56, 2450–7.

3. Barile, M.F. and Rottem, S. (1993) Mycoplasmas in cell culture, *in Rapid Diagnosis of Mycoplasmas*, (eds I. Kahane and A. Adoni), FEMS Symposium, no. 62, Plenum Press, New York, pp. 155–93.

4. Baseman, J.B., Dallo, S.F., Tully, J.G. and Rose, D.L. (1988) Isolation and characterization of *Mycoplasma genitalium* strains from the human respiratory tract. *J. Clin. Microbiol.*, 26, 2266–9.

5. Baseman, J.B., Morrison-Plummer, J., Drouillard, D., Puleo-Scheppke, B., Tryon, V.V. and Holt, S.C. (1987) Identification of a 32-kilodalton protein of *Mycoplasma pneumoniae* associated with hemadsorption. *Isr. J. Med. Sci.*, 23, 474–9.

6. Bebear, C., Dupon, M., Renaudin, H. and de Barbeyrac, B. (1993) Potential improvements in therapeutic options for mycoplasmal respiratory infections. *Clin. Infect. Dis.*, 17 (suppl. 1), 202–7.

7. Bernet, C., Garret, M., de Barbeyrac, B., Bebear, C. and Bonnet, J. (1989) Detection of *Mycoplasma pneumoniae* by using the polymerase chain reaction. *J. Clin. Microbiol.*, 27, 2492–6.

8. Brunner, H., Greenberg, H., James, W.D., Horswood, R.L. and Chanock, R.M. (1973) Decreased virulence and protective effect of genetically stable temperature-sensitive mutants of *Mycoplasma pneumoniae*. *Ann. N.Y. Acad. Sci.*, 225, 436–52.

9. Brunner, H., Greenberg, H.B., James, W.D., Horswood, R.L., Couch, R.B. and Chanock, R.M. (1973) Antibody to *Mycoplasma pneumoniae* in nasal secretions and sputa of experimentally infected human volunteers. *Infect. Immun.*, 8, 612–20.

10. Brunner, H., Krauss, H., Schaar, H. and Schiefer, H.-G. (1979) Electron microscopic studies on the attachment of *Mycoplasma pneumoniae* to guinea pig erythrocytes. *Infect. Immun.*, 24, 906–11.

11. Brunner, H., Prescott, B., Greenberg, H., James, W.D., Horswood, R.L. and Chanock, R.M. (1977) Unexpectedly high frequency of antibody to *Mycoplasma pneumoniae* in human sera as measured by sensitive techniques. *J. Infect. Dis.*, 135, 524–30.

12. Camner, P.C., Jarstrand, C. and Philipson, K. (1978) Tracheobronchial clearance 5–15 months after infection with *Mycoplasma pneumoniae*. *Scand. J. Infect. Dis.*, 10, 33–5.

13. Carilli, A.D., Gohd, R.S. and Gordon, W. (1964) A virologic study of chronic bronchitis. *N. Engl. J. Med.*, 270, 123–7.

14. Carson, J.L., Collier, A.M. and Hu, S.-C.S. (1980) Ultrastructural observations on cellular and subcellular aspects of experimental *Mycoplasma pneumoniae* disease. *Infect. Immun.*, 29, 1117–24.

15. Cassell, G.H., Yanez, A., Duffy, L.B. *et al.* (1994) Detection of *Mycoplasma fermentans* in the respiratory tract of children with pneumonia. IOM Letters, Vol. 3, Programme and Abstracts, 10th International Congress of the International Organisation for Mycoplasmology, p. 456.

16. Chanock, R.M. (1965) Mycoplasma infections of man. *N. Engl. J. Med.*, 273, 1199–206, 1257–64.

17. Chanock, R.M., Dienes, L., Eaton, M.D. *et al.* (1963) *Mycoplasma pneumoniae*: proposed nomenclature for atypical pneumonia organism (Eaton agent). *Science*, 140, 662.

18. Chanock, R.M., Hayflick, L. and Barile, M.F. (1962) Growth on artificial medium of an agent associated with atypical pneumonia and its identification as a PPLO. *Proc. Natl Acad. Sci. USA*, 48, 41–9.

19. Chanock, R.M., Steinberg, P. and Purcell, R.H. (1970) Mycoplasmas in human respiratory tract disease, *in The Role of Mycoplasmas and L forms of Bacteria in Disease*, (ed. J.T. Sharp), C.C. Thomas, Springfield, pp. 110–29.

20. Cherry, J.D., Taylor-Robinson, D., Willers, H. and Stenhouse, A.C. (1971) A search for mycoplasma infections in patients with chronic bronchitis. *Thorax*, **26**, 62–7.

21. Clyde, W.A. (1964) Mycoplasma species identification based upon growth inhibition by specific antisera. *J. Immunol.*, **92**, 958–65.

22. Clyde, W.A. Jr (1993) Clinical overview of typical *Mycoplasma pneumoniae* infections. *Clin. Infect. Dis.*, **17** (suppl. 1), 32–6.

23. Cohen, G. and Somerson, N.L. (1967) *Mycoplasma pneumoniae*: hydrogen peroxide secretion and its possible role in virulence. *Ann. N.Y. Acad. Sci.*, **143**, 85–7.

24. Collier, A.M. and Baseman, J.B. (1973) Organ culture techniques with mycoplasmas. *Ann. N.Y. Acad. Sci.*, **225**, 277–89.

25. Collier, A.M. and Clyde, W.A. Jr (1974) Appearance of *Mycoplasma pneumoniae* in lungs of experimentally infected hamsters and sputum from patients with natural disease. *Am. Rev. Respir. Dis.*, **110**, 765–73.

26. Couch, R.B. (1990) *Mycoplasma pneumoniae* (primary atypical pneumonia), in *Principles and Practice of Infectious Diseases*, 3rd edn, (eds G.L. Mandell, R.G. Douglas and J.E. Bennett), Churchill Livingstone, New York, pp. 1446–58.

27. Crouse, D.T., Odrezin, G.T., Cutter, G.R. *et al.* (1993) Radiographic changes associated with tracheal isolation of *Ureaplasma urealyticum* from neonates. *Clin. Infect. Dis.*, **17** (suppl. 1), 122–30.

28. DelGiudice, R.A., Tully, J.G., Rose, D.L. and Cole, R.M. (1985) *Mycoplasma pirum sp. nov.*, a terminal structured mollicute from cell cultures. *Int. J. Syst. Bacteriol.*, **35**, 285–91.

29. Dienes, L. and Edsall, G. (1937) Observations on the L-organism of Klieneberger. *Proc. Soc. Exp. Biol. Med.*, **36**, 740–4.

30. Eaton, M.D., Meiklejohn, G. and van Herick, W. (1944) Studies on the etiology of primary atypical pneumonia. A filterable agent transmissible to cotton rats, hamsters, and chicken embryos. *J. Exp. Med.*, **79**, 649–68.

31. Edward, D.G. and Freundt, E.A. (1956) The classification and nomenclature of organisms of the pleuropneumonia group. *J. Gen. Microbiol.*, **14**, 197–207.

32. Ellison, J.S., Olson, L.D. and Barile, M.F. (1992) Immunity and vaccine development, in *Mycoplasmas: Molecular Biology and Pathogenesis*, (eds. J. Maniloff, R.N. McElhaney, L.R. Finch and J.B. Baseman), American Society for Microbiology, Washington, DC, pp. 491–504.

33. Feizi, T. (1980) The monoclonal antibodies of cold agglutinin syndrome: immunochemistry and biological aspects of their target antigens with special reference to the Ii antigens. *Med. Biol.*, **58**, 123–7.

34. Feizi, T. (1987) Significance of carbohydrate components of cell surfaces, in *Autoimmunity and Autoimmune Disease*, (ed. D. Evered), Ciba Symposium No. 129, Wiley, Chichester, pp. 43–58.

35. Foy, H.M. (1993) Infections caused by *Mycoplasma pneumoniae* and possible carrier state in different populations of patients. *Clin. Infect. Dis.*, **17** (suppl. 1), 37–46.

36. Foy, H.M., Kenny, G.E., Cooney, M.K. and Allan, I.D. (1979) Long term epidemiology of infections with *Mycoplasma pneumoniae*. *J. Infect. Dis.*, **139**, 681–7.

37. Foy, H.M., Kenny, G.E., Sefi, R., Ochs, H.D. and Allan, I.D. (1977) Second attacks of pneumonia due to *Mycoplasma pneumoniae*. *J. Infect. Dis.*, **135**, 673–7.

38. Freundt, E.A. (1983) Culture media for classic mycoplasmas, in *Methods in Mycoplasmology*, Vol. 1, (eds S. Razin and J.G. Tully), Academic Press, New York, pp. 127–35.

39. Gabridge, M.G. and Taylor-Robinson, D. (1979) Interaction of *Mycoplasma pneumoniae* with human lung fibroblasts: role of receptor sites. *Infect. Immun.*, **25**, 455–9.

40. Gilula, N.B. and Satir, P. (1972) The ciliary necklace: a ciliary membrane specialization. *J. Cell Biol.*, **53**, 494–509.

41. Glezen, W.P., Clyde, W.A., Senior, R.J., Sheaffer, C.I. and Denny, F.W. (1967) Group A streptococci, mycoplasmas, and viruses associated with acute pharyngitis. *JAMA*, **202**, 455–60.

42. Grayston, J.T., Campbell, L.A., Kuo, C.-C. *et al.* (1990) A new respiratory tract pathogen: *Chlamydia pneumoniae* strain TWAR. *J. Infect. Dis.*, **161**, 618–25.

43. Harris, R., Marmion, B.P., Varkanis, G., Kok, T., Lunn, B. and Martin, J. (1988) Laboratory diagnosis of *Mycoplasma pneumoniae* infection: 2. Comparison of methods for direct detection of specific antigens or nucleic acid sequences in respiratory exudates. *Epidemiol. Infect.*, **101**, 685–94.

44. Hayes, M.M., Wear, D.J. and Lo, S.-C. (1991) *In vitro* antimicrobial susceptibility testing for the

newly identified AIDS-associated *Mycoplasma: Mycoplasma fermentans* (incognitus strain). *Arch. Pathol. Lab. Med.*, **115**, 464–6.

45. Hers, J.F.P. and Masurel, N. (1967) Infection with *Mycoplasma pneumoniae* in civilians in the Netherlands. *Ann. N.Y. Acad. Sci.*, **143**, 447–60.

46. Hill, A.C. (1991) *Mycoplasma spermatophilum*, a new species isolated from human spermatozoa and cervix. *Int. J. Syst. Bacteriol.*, **41**, 229–33.

47. Horner, P.J., Gilroy, C.B., Thomas, B.J., Naidoo, R.O.M. and Taylor-Robinson, D. (1993) Association of *Mycoplasma genitalium* with acute non-gonococcal urethritis. *Lancet*, **342**, 582–5.

48. Jacobs, E. (1993) Serological diagnosis of *Mycoplasma pneumoniae* infections: a critical review of current procedures. *Clin. Infect. Dis.*, 17 (suppl. 1), 79–82.

49. Jacobs, E., Rock, R. and Dalehite, L. (1990) A B cell-, T-cell-linked epitope located on the adhesin of *Mycoplasma pneumoniae*. *Infect. Immun.*, **58**, 2464–9.

50. Jarstrand, C., Camner, P. and Philipson, K. (1974) *Mycoplasma pneumoniae* and tracheobronchial clearance. *Am. Rev. Respir. Dis.*, **110**, 415–19.

51. Jemski, J.V., Hetsko, C.M., Helms, C.M., Grizzard, M.B., Walker, J.S. and Chanock, R.M. (1977) Immunoprophylaxis of experimental *Mycoplasma pneumoniae* disease: effect of aerosol particle size and site of deposition of *M. pneumoniae* on the pattern of respiratory infection, disease, and immunity in hamsters. *Infect. Immun.*, **16**, 93–8.

52. Jensen, J.S., Orsum, R., Dohn, B., Uldum, S., Worm, A.-M. and Lind, K. (1993) *Mycoplasma genitalium*: a cause of male urethritis? *Genitourin. Med.*, **69**, 265–9.

53. Jensen, J.S., Sondergard-Andersen, J., Uldum, S.A. and Lind, K. (1989) Detection of *Mycoplasma pneumoniae* in simulated clinical samples by polymerase chain reaction. *APMIS*, **97**, 1046–8.

54. Jensen, J.S., Uldum, S.A., Sondergard-Andersen, J., Vuust, J. and Lind, K. (1991) Polymerase chain reaction for detection of *Mycoplasma genitalium* in clinical samples. *J. Clin. Microbiol.*, **29**, 46–50.

55. Katseni, V.L., Gilroy, C.B., Ryait, B.K. *et al.* (1993) *Mycoplasma fermentans* in individuals seropositive and seronegative for HIV-1. *Lancet*, **341**, 271–3.

56. Kenny, G.E. (1992) Serodiagnosis, in *Mycoplasmas: Molecular Biology and Pathogenesis*, (eds J. Maniloff, R.N. McElhaney, L.R. Finch and J.B. Baseman), American Society for Microbiology, Washington, DC, pp. 505–12.

57. Kingston, J.R., Chanock, R.M., Mufson, M.A. *et al.* (1961) Eaton agent pneumonia. *JAMA*, **176**, 118–23.

58. Kok, T.-W., Varkanis, G., Marmion, B.P., Martin, J. and Esterman, A. (1988) Laboratory diagnosis of *Mycoplasma pneumoniae* infection. 1. Direct detection of antigen in respiratory exudates by enzyme immunoassay. *Epidemiol. Infect.*, **101**, 669–84.

59. Koskiniemi, M. (1993) CNS manifestations associated with *Mycoplasma pneumoniae* infections: summary of cases at the University of Helsinki and review. *Clin. Infect. Dis.*, **17** (suppl. 1), 52–7.

60. Krause, D.C. and Taylor-Robinson, D. (1992) Mycoplasmas which infect humans, in *Mycoplasmas: Molecular Biology and Pathogenesis*, (eds J. Maniloff, R.N. McElhaney, L.R. Finch and J.B Baseman), American Society for Microbiology, Washington, DC, pp. 417–44.

61. Krivan, H.C., Olson, L.D., Barile, M.F., Ginsburg, V. and Roberts, D.D. (1989) Adhesion of *Mycoplasma pneumoniae* to sulfated glycolipids and inhibition by dextran sulfate. *J. Biol. Chem.*, **264**, 9283–8.

62. Kundsin, R.B. and Praznik, J. (1967) Pharyngeal carriage of mycoplasma species in healthy young adults. *J. Epidemiol.*, **86**, 579–83.

63. Lambert, H.P. (1968) Antibody to *Mycoplasma pneumoniae* in normal subjects and in patients with chronic bronchitis. *J. Hyg. (Lond)*, **66**, 185–9.

64. Leach, R.H., Hales, A., Furr, P.M., Mitchelmore, D.L. and Taylor-Robinson, D. (1987) Problems in the identification of *Mycoplasma pirum* isolated from human lymphoblastoid cell cultures. *FEMS Microbiol. Lett.*, **44**, 293–7.

65. Lind, K. (1982) Serological cross-reactions between 'Mycoplasma genitalium' and *M. pneumoniae*. *Lancet*, ii, 1158–9.

66. Lind, K. and Bentzon, M.W. (1988) Changes in the epidemiological pattern of *Mycoplasma pneumoniae* infections in Denmark. *Epidemiol. Infect.*, **101**, 377–86.

67. Lo, S.-C., Dawson, M.S., Wong, D.M. *et al.* (1989) Identification of *Mycoplasma incognitus* infection in patients with AIDS: an immunohistochemical, *in situ* hybridization and ultra-

structural study. *Am. J. Trop. Med. Hyg.*, **41**, 601–16.

68. Lo, S.-C., Hayes, M.M., Wang, R.Y.-H., Pierce, P.F., Kotani, H. and Shih, J.W.-K. (1991) Newly discovered mycoplasma isolated from patients infected with HIV. *Lancet*, **338**, 1415–16.

69. Lo, S.-C., Wear, D.J., Green, S.L., Jones, P.G. and Legier, J.F. (1993) Adult respiratory distress syndrome with or without systemic disease associated with infections due to *Mycoplasma fermentans*. *Clin. Infect. Dis.*, **17** (suppl. 1), 259–63.

70. Loomes, L.M., Uemura, K.-i., Childs, R.A. *et al.* (1984) Erythrocyte receptors for *Mycoplasma pneumoniae* are sialylated oligosaccharides of Ii antigen type. *Nature*, **307**, 560–3.

71. Loveless, R.W., Griffiths, S., Fryer, P.R., Blauth, C. and Feizi, T. (1992) Immunoelectron microscopic studies reveal differences in distribution of sialo-oligosaccharide receptors for *Mycoplasma pneumoniae* on the epithelium of human and hamster bronchi. *Infect. Immun.*, **60**, 4015–23.

72. Lynch, R.E. and Cole, B.C. (1980) *Mycoplasma pneumoniae*: a pathogen which manufactures superoxide but lacks superoxide dismutase. *Proc. Fed. Eur. Biochem. Soc. Symp.*, **62**, 49–56.

73. McCormick, D.P., Wenzel, R.P., Senterfit, L.B. and Beam, W.E. (1974) Relationship of pre-existing antibody to subsequent infection by *Mycoplasma pneumoniae* in adults. *Infect. Immun.*, **9**, 53–9.

74. Manchee, R.J. and Taylor-Robinson, D. (1968) Haemadsorption and haemagglutination by mycoplasmas. *J. Gen. Microbiol.*, **50**, 465–78.

75. Marmion, B.P. and Goodburn, G.M. (1961) Effect of an inorganic gold salt on Eaton's primary atypical pneumonia agent and other observations. *Nature*, **189**, 24–8.

76. Mizutani, H., Mizutani, H., Kitayama, T. *et al.* (1971) Delayed hypersensitivity in *Mycoplasma pneumoniae* infections. *Lancet*, **i**, 186–7.

77. Montagnier, L. and Blanchard, A. (1993) Mycoplasmas as cofactors in infection due to the human immunodeficiency virus. *Clin. Infect. Dis.*, **17** (suppl. 1), 309–15.

78. Morrison-Plummer, J., Jones, D.H., Daly, K., Tully, J.G., Taylor-Robinson, D. and Baseman, J.B. (1987) Molecular characterization of *Mycoplasma genitalium* species-specific and cross-reactive determinants: identification of an immunodominant protein of *M. genitalium*. *Isr. J. Med. Sci.*, **23**, 453–7.

79. Mufson, M.A. (1983) *Mycoplasma hominis*: a review of its role as a respiratory tract pathogen of humans. *Sex. Transm. Dis.*, **10** (suppl.), 335–40.

80. Mufson, M.A., Ludwig, W.M., Purcell, R.H., Cate, T.R., Taylor-Robinson, D. and Chanock, R.M. (1965) Exudative pharyngitis following experimental *Mycoplasma hominis* type 1 infection. *JAMA*, **192**, 1146–52.

81. Mufson, M.A., Saxton, D., Schultz, P.S., Buscho, R.O. and Finch, E. (1974) Virus and mycoplasma infections in exacerbations of chronic bronchitis. *Clin. Res.*, **22**, 646A.

82. Murray, H.W., Masur, H., Senterfit, L.B. and Roberts, R.B. (1975) The protean manifestations of *Mycoplasma pneumoniae* infection in adults. *Am. J. Med.*, **58**, 229–42.

83. Nocard, E., Roux, E.R., Borrel, M.M., Salimbeni, Dujardin-Beaumetz (1898). Le microbe de la peripneumoniae. *Ann. Inst. Pasteur*, **12**, 240–62.

84. Nowak, J. (1929) Morphologie, nature et cycle evolutif du microbe de la peripneumoniae des bovides. *Ann. Inst. Pasteur*, **43**, 1330–52.

85. Ohlsson, A., Wang, E. and Vearncombe, M. (1993) Leukocyte counts and colonization with *Ureaplasma urealyticum* in preterm neonates. *Clin. Infect. Dis.*, **17** (suppl. 1), 144–7.

86. Palmer, H.M., Gilroy, C.B., Claydon, E.J. and Taylor-Robinson, D. (1991) Detection of *Mycoplasma genitalium* in the genitourinary tract of women by the polymerase chain reaction. *Int. J. STD and AIDS*, **2**, 261–3.

87. Palmer, H.M., Gilroy, C.B., Furr, P.M. and Taylor-Robinson, D. (1991) Development and evaluation of the polymerase chain reaction to detect *Mycoplasma genitalium*. *FEMS Microbiol. Lett.*, **77**, 199–204.

88. Quinn, P.A., Li, H.C.S., Th'ng, C., Dunn, M. and Butany, J. (1993) Serological response to *Ureaplasma urealyticum* in the neonate. *Clin. Infect. Dis.*, **17** (suppl. 1), 136–43.

89. Radestock, U. and Bredt, W. (1977) Motility of *Mycoplasma pneumoniae*. *J. Bacteriol.*, **129**, 1495–501.

90. Renaudin, H., Tully, J.G. and Bebear, C. (1992) *In vitro* susceptibilities of *Mycoplasma genitalium* to antibiotics. *Antimicrob. Agents Chemother.*, **36**, 870–2.

91. Ruiter, M. and Wentholt, H.M.M. (1952) The occurrence of pleuropneumonia-like organism

in fuso-spirillary infections of the human genital mucosa. *J. Invest. Dermatol.*, **18**, 313.

92. Saillard, C., Carle, P., Bove, J.M. *et al.* (1990) Genetic and serologic relatedness between *Mycoplasma fermentans* strains and a mycoplasma recently identified in tissues of AIDS and non-AIDS patients. *Res. Virol.*, **141**, 385–96.

93. Sanchez, P.J. (1993) Perinatal transmission of *Ureaplasma urealyticum*: current concepts based on review of the literature. *Clin. Infect. Dis.*, **17** (suppl. 1),107–11.

94. Shames, J.M., George, R.B., Holliday, W.B., Rasch, J.R. and Mogabgab, W.J. (1970) Comparison of antibiotics in the treatment of mycoplasmal pneumonia. *Arch. Intern. Med.*, **125**, 680–4.

95. Shepard, M.C. (1954) The recovery of pleuropneumonia-like organisms from negro men with and without non-gonococcal urethritis. *Am. J. Syph. Gonorrhea Vener. Dis.*, **38**, 113–24.

96. Sillis, M. (1990) The limitation of IgM assays in the serological diagnosis of *Mycoplasma pneumoniae* infection. *J. Med. Microbiol.*, **33**, 253–8.

97. Smith, C.B., Friedewald, W.T. and Chanock, R.M. (1967) Shedding of *Mycoplasma pneumoniae* after tetracycline and erythromycin therapy. *N. Engl. J. Med.*, **276**, 1172–5.

98. Somerson, N.L., Taylor-Robinson, D. and Chanock, R.M. (1963) Hemolysin production as an aid in the identification and quantification of Eaton agent *(Mycoplasma pneumoniae)*. *Am. J. Hyg.*, **77**, 122–8.

99. Somerson, N.L., Walls, B.E. and Chanock, R.M. (1965) Hemolysin of *Mycoplasma pneumoniae*: tentative identification as a peroxide. *Science*, **150**, 226–8.

100. Stadel, B.V., Foy, H.M., Nuckolls, J.W. and Kenny, G.E. (1975) *Mycoplasma pneumoniae* infection followed by *Haemophilus influenzae* pneumonia and bacteremia. *Am. Rev. Respir. Dis.*, **112**, 131–4.

101. Subcommittee of the Research Committee of the British Thoracic Society and the Public Health Laboratory Service (1987). Community-acquired pneumonia in adults in British hospitals in 1982–1983: a survey of aetiology, mortality, prognostic factors and outcome. *Q. J. Med.*, **62**, 195–220.

102. Taylor, G. and Taylor-Robinson, D. (1975) The part played by cell-mediated immunity in mycoplasma respiratory infections, in *International Symposium on Immunity to Infections of the Respiratory System in Man and Animals*, Develop Biol. Standard no. 28, Karger, Basel, pp. 195–210.

103. Taylor, G., Taylor-Robinson, D. and Fernald, G.W. (1974) Reduction in the severity of *Mycoplasma pneumoniae*-induced pneumonia in hamsters by immunosuppressive treatment with anti-thymocyte sera. *J. Med. Microbiol.*, **7**, 343–8.

104. Taylor-Robinson, D. (1995) Mycoplasmas and ureaplasmas, in *Manual of Clinical Microbiology*, 6th edn, (eds E.J. Baron and P.R. Murray), American Society for Microbiology, Washington, DC, pp. 652–62.

105. Taylor-Robinson, D. and Furr, P.M. (1986) Clinical antibiotic resistance of *Ureaplasma urealyticum*. *Pediatr. Infect. Dis.*, **5** (suppl.), 335–7.

106. Taylor-Robinson, D., Webster, A.D.B., Furr, P.M. and Asherson, G.L. (1980) Prolonged persistence of *Mycoplasma pneumoniae* in a patient with hypogammaglobulinaemia. *J. Infect.*, **2**, 171–5.

107. Tully, J.G. (1993) Current status of the mollicute flora of humans. *Clin. Infect. Dis.*, **17** (suppl. 1), 2–9.

108. Tully, J.G., Rose, D.L., Whitcomb, R.F. and Wenzel, R.P. (1979) Enhanced isolation of *Mycoplasma pneumoniae* from throat washings with a newly modified culture medium. *J. Infect. Dis.*, **139**, 478–82.

109. Tully, J.G., Taylor-Robinson, D., Cole, R.M. and Rose, D.L. (1981) A newly discovered mycoplasma in the human urogenital tract. *Lancet*, **i**, 1288–91.

110. Tully, J.G., Whitcomb, R.F., Clark, H.F. and Williamson, D.L. (1977) Pathogenic mycoplasmas: Cultivation and vertebrate pathogenicity of a new spiroplasma. *Science*, **195**, 892–4.

111. Waites, K.B., Crouse, D.T. and Cassell, G.H. (1993) Therapeutic considerations for *Ureaplasma urealyticum* infections in neonates. *Clin. Infect. Dis.*, **17** (suppl. 1), 208–14.

112. Wang, E.E.L., Cassell, G.H., Sanchez, P.J., Regan, J.A., Payne, N.R. and Liu, P.P. (1993) *Ureaplasma urealyticum* and chronic lung disease of prematurity: critical appraisal of the literature on causation. *Clin. Infect. Dis.*, **17** (suppl. 1), 112–16.

113. Wang, R.Y.-H., Shih, J.W.-K., Weiss, S.H. *et al.* (1993) *Mycoplasma penetrans* infection in male homosexuals with AIDS: high seroprevalence

and association with Kaposi's sarcoma. *Clin. Infect. Dis.*, **17**, 724–9.

114. Wenzel, R.P., Craven, R.B., Davies, J.A., Hendley, J.O., Hamory, B.H. and Gwaltney, J.M. (1976) Field trial of an inactivated *Mycoplasma pneumoniae* vaccine. I. Vaccine efficacy. *J. Infect. Dis.*, **134**, 571–6.

115. Williams, M.H. (1968) Recovery of mycoplasma from rheumatoid synovial fluid, in *Rheumatic Diseases (Pfizer Medical Monographs)*, (eds J.J.R. Duthie and W.R.M. Alexander), University Press, Edinburgh, pp. 171–81.

Brenda Thomas and David Taylor-Robinson

18.1 INTRODUCTION

Chlamydiae are bacteria since they: (i) are bounded by a cell wall which is similar to that possessed by Gram-negative bacteria; (ii) con-

Viral and Other Infections of the Human Respiratory Tract.
Edited by S. Myint and D. Taylor-Robinson. Published in 1996 by Chapman & Hall. ISBN 0 412 60070 6

tain both DNA and RNA; (iii) multiply by binary fission; and (iv) are susceptible to certain antibiotics. However, they become obligate intracellular parasites during part of their growth cycle and in this regard behave somewhat like viruses. Indeed, historically they have been dealt with mainly by virologists. The growth cycle of chlamydiae is one in which the infectious elementary body is taken into the cell by pinocytosis; it is reorganized into a reticulate body which, after about 12 hours, divides by binary fission so that a visible inclusion is produced in the cytoplasm of the cell close to the nucleus. The inclusion can be stained by the use of various vital dyes and by fluorescein-conjugated specific antiserum and its detection is an integral part of the cultural diagnosis of chlamydial infections. The reticulate bodies are reorganized into new elementary bodies which increase in number and after about 48–72 hours are released from the cell.

Of the chlamydial species known so far (see below), information on *Chlamydia pneumoniae* has accrued rapidly in the past few years and it is the subject of this review. Reference to the other species is made only where it is helpful in clarifying certain aspects, such as the serological cross-reactivity between the species.

18.2 CLASSIFICATION

There are four species of the genus *Chlamydia* within the class Chlamydiales. *Chlamydia psittaci* occurs in various animal species, particularly the avian, and may be transmitted to humans to cause pneumonia and very rarely abortion. *C. trachomatis*, as its name implies, causes the blinding disease, trachoma. The serovars responsible are A, B and C. Serovars D–K infect the human urogenital tract and are a cause of urethritis and cervicitis and consequent upper tract complications, such as epididymitis and pelvic inflammatory disease, as well as reactive arthritis. Other serovars (L1–L3) of this species are responsible for lymphogranuloma venereum. *C. pecorum* has been proposed as the fourth species of the genus *Chlamydia* on the basis of a genetic analysis of strains isolated from cattle and sheep with various diseases, including sporadic encephalitis, pneumonia, diarrhoea and polyarthritis [34].

C. pneumoniae causes a spectrum of upper and lower respiratory tract disease in humans. Strains which comprise this species were known in the past as the TWAR strains. The term TWAR is a composite of strain TW183, which was isolated in 1965 from the conjunctiva of a child in a trachoma vaccine study in Taiwan (TW for Taiwan), and strain AR39, which was isolated in 1983 from the throat of a Washington University student who was part of an acute respiratory disease study (AR for acute respiratory). The strain IOL207, which is the same serologically as the TWAR strains, was isolated at the Institute of Ophthalmology in London (IOL) in 1967, also from the conjunctiva of a child who had trachoma. It is interesting that these original strains of *C. pneumoniae* were recovered from the eye and yet, subsequently, there has been no substantive evidence to suggest that *C. pneumoniae* strains cause eye disease. Recently, a strain isolated from the respiratory tract of a horse has been found to be more closely related to *C. pneumoniae* than to the other three chlamydial species on the basis of analysis of the major outer membrane protein (MOMP) gene [80]. A strain responsible for blindness and infertility in koalas is also very closely related to *C. pneumoniae* [38].

Taxonomically, *C. pneumoniae* is distinguishable from the other three species in the following ways. First, little cross-reactivity is exhibited when specific antisera are used in the microimmunofluorescence (MIF) test. Second, the configuration of the elementary bodies of *C. pneumoniae* has been described as pear-shaped, in contrast to the round forms of the other species, although it has become evident that pear-shaped morphology is not a reliable taxonomic criterion [22,66]. Third, the restriction endonuclease pattern of each of the four chlamydial species is different. The sequence conservation of the variable domain IV region of the MOMP gene indicates that *C. pneumoniae* strains may be more genetically homogeneous than *C. psittaci* or *C. trachomatis* strains [36]. Fourth, the 98-kDa protein of *C. pneumoniae* may be species-specific [19] although the evidence is disputed [91]. Fifth, no extra-chromosomal DNA has been identified so far in the human strains of *C. pneumoniae*, whereas it has in the other species [18]. The sixth distinguishing feature is that the G+C content of the DNA of *C. pneumoniae* is intermediate between that of *C. psittaci* and of *C. trachomatis*, and there is less than 10% DNA homology between the three species [27].

18.2.1 SEROLOGICAL CROSS-REACTIVITY AND MISIDENTIFICATION OF CHLAMYDIAL SPECIES

The lack of specificity of the complement-fixation (CF) test, that is, its known inability to differentiate between the species, undoubtedly accounted for the misdiagnosis of early epidemics of *C. pneumoniae* infection. It is understandable how some outbreaks of infection were regarded as ornithosis due to strains of *C. psittaci*, when it is highly likely that they were due to *C. pneumoniae*. For example, epidemics in Norway, Sweden and Denmark in 1981–1983, considered to be ornithosis on the basis of CF serology, were identified subsequently as being due to *C. pneumoniae*, 55–70%

of the CF-antibody positive sera being *C. pneumoniae* antibody positive by the MIF test [45]. The same conclusion was drawn about an epidemic of ornithosis in Norway in 1987 [16]. As a further example, an outbreak of what was called psittacosis in a boys' boarding school in 1980 may be cited; avian contact was dubious and a subsequent specific serological evaluation indicated that the outbreak was probably due to *C. pneumoniae* [71].

18.3 INCIDENCE AND PREVALENCE OF *C. PNEUMONIAE*

In a few studies in which identification of *C. pneumoniae* has been attempted, up to 10% of patients with pneumonia and 5% of those with bronchitis have had their disease attributed to infection with *C. pneumoniae*. For example, in Seattle, in a 5-year study of students with acute respiratory disease, 20 (3%) of 647 were identified as *C. pneumoniae*-positive by isolation and serology, including 14 (10%) of 140 with pneumonia [44]. Subsequently, 4% of middle-aged and older patients with pneumonia, 5% with bronchitis, 5% with sinusitis and 2% with pharyngitis were found to have acute *C. pneumoniae* infections [85]. Similar rates have been reported in studies from the United Kingdom [79], Spain [3] and Italy [26], and rates as high as 19% in an epidemic among military recruits in Finland [58] and up to 16% and even 47%, respectively, in children and adults with pneumonia, in Brooklyn, New York [73].

In view of the difficulty of culturing *C. pneumoniae* and of detecting its antigens, it is not surprising that many of the earliest recognized outbreaks of *C. pneumoniae* infection were identified by CF and MIF tests alone. However, the reservations expressed previously and subsequently about the value of these serological methods means that the accuracy of incidence and prevalence rates based on early work must be viewed with suspicion. It would, therefore, seem pointless to

catalogue all the earliest studies in different parts of the world. Nevertheless, it is appropriate to mention those in Finland because, subsequently, they have been properly identified. Thus, in this country in 1978 there was an unusual epidemic of mild pneumonia in civilians in which 32 of 34 patients were considered to have had serological evidence of infection with an 'unusual *C. psittaci* strain' [77]. Indeed, this was the first evidence that the TW-183 strain of *C. pneumoniae* was associated with human disease. In similar epidemics in the military in 1985–1987, 70–75% of those susceptible were infected with *C. pneumoniae*, and 10% of those who developed *C. pneumoniae* antibody had pneumonia [58].

It is clear from the preceding text that *C. pneumoniae* infection has been found in different parts of the world. Furthermore, despite the reservation regarding serology, serological findings in a study undertaken in the UK [32] and in one undertaken in Seattle [88] also indicate that *C. pneumoniae* infection is worldwide. In the latter study, high rates of antibody were found in tropical countries and lower ones in more sparsely populated areas further north. Only in the Solomon Islands was there little evidence of infection. Infection seems to be endemic in most countries in which there may also be sporadic outbreaks and epidemics. Four-, 6- and 10-year cycles of epidemic infection have been reported [54,88], the longer cycles occurring in the north, where an epidemic itself may last 2–3 years, and the shorter ones in the more densely populated areas of the south, where the epidemics are short-lived.

At any time, up to 50% of the adult population has IgG antibody to *C. pneumoniae*, suggesting a cycle of infection and re-infection during a lifetime [41]. The antibody rate shows a distinctive rise with increasing age; antibody is almost non-existent in children below 5 years of age and is reported consistently to increase with age to the point where almost all 70-year-old individuals have evidence of a previous infection [74]. Acute infection with seroconversion occurs particularly between the ages of 8 and 16 years [48], whereas re-infection constitutes most acute infections among adults [2]. The finding of higher antibody rates among men than women [54,88] and the existence of only very few reports of intrafamilial infection [2], suggest that transmission of C. *pneumoniae*, unlike other respiratory pathogens, usually takes place outside the home.

18.4 TRANSMISSION

Knowledge about the infectivity of C. *pneumoniae* has been slow to accrue. Human-to-human transmission is assumed, particularly as there has been no evidence of an avian or mammalian vector. The recent finding of strains closely related to C. *pneumoniae* in horses and koalas would not seem to put the assumption in jeopardy. The organisms can survive in small-particle aerosols for a short time at high relative humidities to allow direct inhalation, and they may also remain viable in the environment for long enough to allow transfer from human respiratory secretions to fomites with subsequent auto-inoculation [31]. Asymptomatic infections play an important role in transmission, but transmission does not occur readily. This has been suggested by the fact that a case of infection occurs in a family often without involvement of other members. Details of incubation times and infectivity have been scanty, with failure to find contact cases and no evidence of a chain of transmission [42]. However, the results of recent studies in Japan [57] and Denmark [68], now indicate that the average incubation time for C. *pneumoniae* infection is 3–4 weeks.

18.5 CLINICAL MANIFESTATIONS

In children 1 to 5 years old, infection by C. *pneumoniae* is uncommon and in those 6 to 12 years old it is often asymptomatic or produces mild symptoms. In teenagers and young adults, again it is often asymptomatic but there may be mild pneumonia which is fre-

quently prolonged and also bronchitis with pharyngitis. Likewise, in adults younger than 40 years of age, infection is asymptomatic in many instances but mild to severe pneumonia may occur, often with prolonged bronchitis. Those who are over 40 years of age may experience pneumonia which is frequently complicated and severe, especially if they have an underlying chronic lung disease, and a few patients have died [43]. Despite this, most cases of pneumonia due to C. *pneumoniae* are mild and do not require admission to hospital.

Although asymptomatic C. *pneumoniae* infection is common and has been well documented [52], pneumonia is the most common syndrome associated with the infection. However, the clinical features are those of an atypical pneumonia and as such they are usually indistinguishable from those caused by *Mycoplasma pneumoniae* or some other infectious agents, such as *Legionella pneumophila*. Cough is a dominant feature, being seen in about 90% of cases, severe sore throat in 60% often with laryngitis, and fever in a similar proportion. Thus, a 'flu-like' illness is seen sometimes. Sinusitis, rare in M. *pneumoniae* infections, frequently accompanies the lower-tract disease. These features may have an acute onset. This is particularly characteristic of Legionnaires' disease which presents with a high fever, often accompanied by central nervous system and gastrointestinal abnormalities [53]. However, a more gradual onset is typical of C. *pneumoniae* infection. In this case, the disease sometimes appears biphasic, starting with a sore throat, often with hoarseness and usually with fever but without signs of pulmonary involvement. Symptomatic treatment leads to improvement but then there is relapse with cough and signs of lower respiratory tract involvement several days to weeks after the first symptoms. Rhonchi and rales are commonly heard on auscultation, even in patients with relatively mild symptoms. The leucocyte count is usually not elevated but the erythrocyte sedimentation rate is often increased. In 85% of the pneumonias, a single subsegmental lesion is seen radiographically. In the remainder, more extensive unilateral disease and bilateral lesions occur, and pleural effusion has been described [4]. Almost all cases that are diagnosed clinically as pneumonia are reported to have the disease radiographically. Infection may be prolonged (see below), with cough persisting for 1–2 months, and recurrent pneumonia may occur with a 2 to 3-month interval, the disease recurring sometimes in a different area of the lung.

There is evidence based on serological and limited antigen detection studies of a link between *C. pneumoniae* infection and acute exacerbations of chronic obstructive pulmonary disease [5,12]. An acute *C. pneumoniae* infection and repeated exposure to the organisms have been associated with asthma and/or an exacerbation of previously diagnosed asthma in adults and also in children [10,22,47]. In a recent collaborative study, involving ourselves, at St Mary's Hospital, London, serological evidence of acute infection by *C. pneumoniae* was found as frequently in patients with asthma as in those with acute non-asthmatic respiratory infections (B.S. Peters *et al.*, unpublished data). In addition, a sarcoid-like illness has been associated with *C. pneumoniae* infection [46] and also myocarditis [45]. Coronary artery involvement will be discussed later.

18.6 PERSISTENT INFECTION

There is speculation that *C. pneumoniae* infection may become chronic, especially in the lung, even though it may not be possible to detect the initiating organisms in the late phase [74]. However, there are also reports of the organisms persisting in a cultivable form. Thus, one group of investigators found that cultures for *C. pneumoniae* were positive on multiple occasions for periods of about 1 year in five patients originally presenting with pneumonia, despite the fact that three of them had received doxycycline or tetracycline therapy

[50]. In a later study by the same investigators, nine of 42 *C. pneumoniae*-positive children remained so during and after treatment with clarithromycin or erythromycin, even though the organisms recovered from them were susceptible to these antibiotics *in vitro* [73]. However, despite persistence of the organisms, the children appeared to improve clinically.

Persisting infection has been suggested to occur in patients with chronic obstructive pulmonary disease on the basis of serological tests; 81% of them were found to have high titres of IgG and IgA antibody in serum and sputum samples [51], but whether this is sufficient to indicate persisting infection without some direct evidence of persisting organisms is dubious.

18.7 RE-INFECTION AND IMMUNITY

Re-infection with *C. pneumoniae* is known to occur, but the effect that a previous infection has on a current one is not absolutely clear. Some investigators have suggested that more severe infections occur in older patients as a result of multiple exposure in earlier life [40]. This is supported by the occurrence in mice that had been infected previously of more severe inflammatory changes in the lungs after a second challenge, although the organisms were recovered from the lungs for a shorter period than after the primary infection [63].

On the other hand, others have indicated that milder disease occurs in middle-aged and older people [85]. Indeed, some evidence that a degree of immunity develops following exposure at an early age comes from outbreaks of *C. pneumoniae* infection among military recruits in Finland [29] and Sweden [7]. In both instances, patients with pre-existing antibody developed pneumonia less frequently, or required admission to hospital and multiple courses of antibiotics less frequently, than did those without previous exposure. In the mouse model too, pre-existing antibody has been shown to influence the course of a *C. pneumoniae* infection. Thus, passive immunization with high-titre antibody resulted in an inability to culture the organisms from the lungs after intranasal challenge, and no inflammatory changes occurred in the lungs [63].

18.8 DIAGNOSIS

Diagnosis, as for infection with other chlamydial species, may be made by detection of the organisms in cultured cells, by antigen detection using immunofluorescence and/or immunoperoxidase staining or enzyme immunoassays (EIAs), by the use of DNA probes and, most recently, by the polymerase chain reaction (PCR). Serological tests are widely used.

18.8.1 ISOLATION IN CELL CULTURES

C. pneumoniae seems to be more difficult to culture than *C. trachomatis*, although it may be that the optimal cell line and growth conditions remain to be discovered. Organisms in clinical samples produce significantly more inclusions in HL cells than in McCoy, HeLa or BHK-21 cells [25] and adaptation to continuous culture in HL cells is also successful [61]. Infectivity is reported to be enhanced by culturing the organisms in cells maintained in medium containing fetal calf serum [84] or, conversely, in serum-free medium [65], by centrifugation of the organisms onto the cell monolayers, and by using cells pre-treated with cycloheximide where, unlike *C. trachomatis*, *C. pneumoniae* can undergo a second cycle of replication. To date, however, there are only a few reports of the isolation of the organisms in cell culture from a sizeable proportion of subjects, whether they be patients with lower respiratory tract infections [42] or 'healthy' persons [39,49], and the sensitivity of current culture methods is estimated to be only about 50% [42]. In addition to the apparent inherent inadequacies of the culture technique, the small number of elementary bodies that may be available at some anatomical sites [74] and

the inclusion of cases of chronic disease may have contributed to the low isolation rates.

18.8.2 ANTIGEN AND DNA DETECTION

In the absence of a cultural 'gold standard' that could be used for comparison, accurate evaluation of the sensitivity and specificity of antigen detection methods is difficult. Furthermore, since there are no commercially available fluorescein-labelled antibodies to *C. pneumoniae* that have the quality of those used for *C. trachomatis*, using a direct fluorescent antibody (DFA) test for comparison would currently not seem feasible. In addition, the identification of small numbers of fluorescing elementary bodies in clinical samples is a problem. Indeed, it may be impossible if samples have been fixed in methanol, the fixative recommended for *C. trachomatis*, because it has been shown to destroy the antigenic activity of *C. pneumoniae* [89]. The sensitivity of DFA staining in relation to culture and seropositivity is estimated to be no higher than 60% [59].

EIAs may be more practical than DFA staining but positive results must be confirmed by a species specific assay to exclude *C. trachomatis* or *C. psittaci* infection [79]. If this is done, specificity is definitely improved. However, sensitivity is likely to be a problem. As an example, the limit of sensitivity of one EIA, the IDEIA (Dako), has been shown to be, at best, 10^4 elementary bodies [66]. The value of such an assay for throat swabs which contain only a few organisms must be limited.

18.8.3 PCR-BASED ASSAYS

PCR-based assays have not been widely applied to the diagnosis of *C. pneumoniae* infection, so their full potential may not yet have been realized. As an example, in some of the small number of studies reported so far, *C. pneumoniae* has been detected by this technique in a consistent proportion of patients with pneumonia or bronchitis who had serological evidence of acute *C. pneumoniae* infection.

Thus, the organism was found by a PCR assay in throat swabs from 12 (75%) of 16 such patients, only eight of whom were positive by culture [21], in nine (75%) of 12 students, seven of whom were culture positive [42] and in the throat of 15 (71%) of 21 middle-aged and older patients, only two of whom were culture-positive [85]. In none of these studies was the PCR assay positive in any patient in the seronegative control groups. These results indicate, not unexpectedly, that PCR-based assays for examining throat swabs are superior to culture, but that they are less sensitive than serology. The throat may be a suboptimal sampling site for the organism in patients with lower respiratory tract infection because only small numbers of inclusions have been found in cultures [21,42]. Sampling of sputum might improve the sensitivity of both antigen detection and cultural methods. The finding that 47% of nasal lavages from children with persistent cough and episodes of wheezing were PCR-positive [28] is interesting, but positive PCR results unsupported by any other method of detection raise questions about specificity. The sensitivity and specificity of PCR-based assays obviously need further evaluation, but there seems little doubt that accurate diagnosis of *C. pneumoniae* infection in the future will depend on their more widespread use.

18.8.4 SEROLOGY

In contrast to the inadequacy of serology for detecting *C. trachomatis* infection of the genital tract [82], serological methods have some value in identifying acute *C. pneumoniae* infections. Such value does not extend to the CF test, however, which is insensitive and, as mentioned before, non-specific as it is not possible to distinguish between the three chlamydial species. It may be useful for demonstrating antibody responses in young adults [29], but certainly not in the elderly. Thus, only six (10%) of 58 cases of chlamydial pneumonia were detected by its use in elderly

patients in an epidemic in Finland [30]. A recently developed EIA, in which lipopolysaccharide (LPS) is used as the antigen, was responsible for detecting 72% of the infections but, like the CF test, it lacks specificity [30]. However, antibody to LPS develops early in disease, so that an assessment based on an early single serum sample might be useful in diagnosis.

The bulk of evidence linking *C. pneumoniae* to respiratory disease has resulted from use of the MIF test. The indicators of acute infection based on examination of sera have been put forward as follows [43]: (i) the presence of IgM antibody, (ii) a four-fold or greater rise in the titre of IgG antibody; and (iii) a stable IgG antibody titre of ≥ 512. IgG antibody titres of >8 to <256 indicate past infection.

It is possible with experience to use the MIF test to distinguish between chlamydial species and even to detect differences between some strains [6,8]. There have been suggestions that, from a diagnostic point of view, antibody detection by the MIF test is at least as sensitive and perhaps even more sensitive than detection of the organisms [17]. As examples, in a study in the USA, 12 (22%) of 54 students with acute respiratory disease were seropositive but only nine of the 12 were also culture- and/or PCR-positive [42]. In another study, in Italy, in which the DFA test was used for detection of *C. pneumoniae*, 14 (13%) of 108 patients with pneumonia were seropositive and the organism was detected in the throat of 10 of them [9]. In contrast, in another study in the USA, the diagnostic sensitivity of the MIF test appeared to be much lower than indicated by the preceding observations. Thus, serological evidence of acute infection was found in only three of eight culture-positive patients with a lower respiratory tract infection, while a further 14 who were seropositive were culture- and PCR-negative [23]. It is clear, therefore, that despite the serological criteria for defining the existence of an acute infection, mentioned previously, the interpretation of MIF test results is not straightforward. It is complicated by the fact that: (i) both IgG and IgM antibodies to *C. pneumoniae* have been shown to persist for at least 2 years following an episode of *C. pneumoniae* pneumonia [42], so that the existence of IgM antibody is not necessarily pathognomonic of a current *C. pneumoniae* infection; (ii) IgM antibody is found infrequently, since it occurs almost exclusively in acute primary infections of the young – in elderly patients, non-specific IgM antibody due to the presence of rheumatoid factor may be a problem and should be removed by adsorption [30]; (iii) high IgG antibody titres may reflect not so much an acute infection by *C. pneumoniae*, but the development of chronic infection, although the need to establish the validity of this criterion has been commented on; (iv) immunocompromised and HIV-infected patients may not respond serologically to *C. pneumoniae* as often as others in the community [35], although this has not always been found to be the case [11], and (v) the chlamydial MIF test may be less specific than is generally appreciated [56] because infection with one species may invoke an anamnestic response to another species with which the subject has been infected previously [13], or there may be a broadening antibody response to chlamydial group antigen following repeated exposure, or antibody to one chlamydial species may cross-react [1,69]; of course, these events need not be mutually exclusive.

Our own observations throw some light on the specificity or otherwise of the MIF test and focus attention on areas where difficulties of interpretation remain. Sera from different groups of individuals have been examined for antibody to *C. pneumoniae* and *C. trachomatis* by this test (Table 18.1). Sera from 92 children were tested; 72 sera did not contain antibody to either species, while 20 contained antibody to *C. pneumoniae* alone. Since the children are unlikely to have had a *C. trachomatis* infection but quite likely to have had a *C. pneumoniae* respiratory infection, it may be assumed that

the antibody to *C. pneumoniae* was a consequence of infection by this species. Furthermore, such antibody did not cross-react with the *C. trachomatis* antigen in the test. While this indicates that the test itself is specific for *C. pneumoniae*, it does not exclude the possibility, in other situations, that the existence and detection of *C. pneumoniae* antibody is a consequence, at least partially, of a *C. trachomatis* infection. Relevant to this, we have found that mice given *C. trachomatis* intravenously produced antibody that reacts in the MIF test with the homologous antigen and to a lesser extent with *C. pneumoniae* antigen. On the other hand, mice given *C. pneumoniae* intravenously produced antibody that reacts only with the homologous antigen. Examination of sera collected from 69 men or women who were infected with *C. trachomatis* in the genital tract as indicated by a DFA test, showed that almost half ($n=31$) had titres of antibody to *C. pneumoniae* that were equal to or greater than those of antibody to *C. trachomatis*. If the test is specific, the assumption is that the antibody to *C. pneumoniae* had been stimulated by a previous *C. pneumoniae* infection, and likewise that the antibody to *C. trachomatis* had been stimulated by the *C. trachomatis* infection, but some heterotypic stimulation, particularly of *C. pneumoniae* antibody, cannot be excluded. In another study, 22 of 27 patients who had pneumonia of unknown aetiology at Northwick Park Hospital (northwest London) were found to have antibody to *C. pneumoniae* in titres greater than those of antibody to *C. trachomatis*. Since these patients were less likely to have been infected currently or in the past with *C. trachomatis* than subjects attending a sexually transmitted diseases clinic, it is likely that the antibody to *C. pneumoniae* was associated with the pneumonia. The results of examining sera from other patients with pneumonia, using the criteria for diagnosis mentioned earlier, indicated that four (13%) of 31 patients had a current *C. pneumoniae* infection, and 15 (48%) had had a previous *C.*

Table 18.1   Antibody to *C. pneumoniae* (C.p.) and *C. trachomatis* (C.t.)

Subjects studied	Antibody status	No. of subjects with indicated antibody status
Children (*n* = 92)	C.p. +, C.t.–	20
	C.p.–, C.t.+	0
	C.p.–, C.t.–	72
Men, women:	C.p.–, C.t.+	16
C.t. organism-positive* (*n* = 69)	C.p. < C.t.	8
	C.p. +, C.t.–	5
	C.p. > C.t.	5
	C.p. = C.t.	21
	C.p.–, C.t.–	14
Adult pneumonia (*n* = 27)	C.p.+, C.t.–	17
	C.p. > C.t.	5
	C.p.–, C.t.+	0
	C.p. < C.t.	1
	C.p. = C.t.	1
	C.p.–, C.t.–	3

* Detected in the genital tract by means of a direct fluorescent antibody (DFA) test

pneumoniae infection. However, the difficulty of making a diagnosis on serological grounds without attempting to detect the organisms is shown from the serological results obtained for one individual (Table 18.2). In this case, an antibody response was demonstrable to the three chlamydial species. It seems reasonable to suppose that the patient had not been infected simultaneously with all of the species, but whether infection was due to *C. pneumoniae* or *C. psittaci* is unclear.

18.9 TREATMENT

There are few *C. pneumoniae* strains available for antibiotic susceptibility testing *in vitro*, but those strains that have been tested are similar to each other, to the type strain TW183 and, apart from resistance to sulphonamides [24,60], to *C. trachomatis* in their susceptibilities. The tetracyclines, the quinolones and the macrolides have bactericidal activity against both of these chlamydial species, whereas the penicillins are bacteriostatic only.

The minimal inhibitory concentrations (MICs) of the tetracyclines range from 0.125 mg/l for demeclocycline (Deteclo) to 1.0 mg/l for tetracycline [72], the latter antibiotic being comparable in its activity to that of ofloxacin (MIC 1.0 mg/l), the most active of the quinolones. Of the macrolides, clarithromycin is the most effective *in vitro* (MIC 0.007 mg/l) followed by erythromycin (MIC 0.06 mg/l). Azithromycin is similar in activity to erythromycin, but its prolonged half-life and excellent tissue penetration enable lower doses to be given at longer intervals, thus improving patient compliance [33].

Studies of the efficacy of antibiotic treatment in *C. pneumoniae* infections are very limited and the results are confused by the diagnosis sometimes being unreliable. Although tetracyclines and erythromycin are established in the treatment of atypical pneumonias, several of the original cases which were thought to be chlamydial in aetiology did not respond to erythromycin, or the patients suffered relapses [44]. More recently, there has been a report of the continued recovery of *C. pneumoniae* organ-

Table 18.2 Antibody response of patient with pneumonia

Antigen used in tests	Antibody titres measured by MIF	
	May 19	*June 19*
CF (group-specific)	<4	64
C. pneumoniae	64 (32)*	2048
C. psittaci	<4	256
C. trachomatis (serovars D–K, LGV)	<4	256

* Result of repeat test. CF, complement-fixation; LGV, lymphogranuloma venereum; MIF, microimmunofluorescence

isms for a year following repeated courses of tetracycline [50], even though the isolates were susceptible *in vitro*. Clinical and microbiological resolution have been reported following treatment with azithromycin for 5 days, and with clarithromycin for 8 days, but episodes of both clinical and microbiological failure are no fewer with these than with the other antibiotics [14].

A mouse model of respiratory infection with *C. pneumoniae*, that has been developed recently, in which the animals have remained infected for up to 42 days, may help to resolve the problem of apparent resistance of the organism to therapy *in vivo* [92].

18.10 RELATIONSHIP OF *C. PNEUMONIAE* TO CORONARY ARTERY DISEASE

Evidence based on seroepidemiological, immunochemical and molecular biological studies suggests that there is an association between *C. pneumoniae* infection and coronary artery disease (CAD).

18.10.1 SEROEPIDEMIOLOGY

In 1988, Saikku and colleagues in Finland reported the first observations to link *C. pneumoniae* with CAD. They observed that high titres of *C. pneumoniae* IgG and IgA antibodies occurred in men with CAD and myocardial infarction significantly more often than in age-matched, randomly-selected controls [75]. A

later report from the USA was confirmatory in that the authors showed a similar association between IgG antibody and CAD in which the latter was defined as a 50% or greater narrowing of the coronary arteries [87]. However, two further studies followed in which other risk factors for heart disease, such as smoking and hypertension, were also included in the analyses. In one of these studies, in Finland, men receiving lipid-lowering medication who suffered a myocardial infarction or sudden cardiac death had significantly higher *C. pneumoniae* antibody levels, but not a greater antibody prevalence than controls in the same group [76]; conversely, in the other study, in the USA, patients with CAD had a significantly greater antibody prevalence but not a higher titre than controls without CAD [86]. Yet a further serological association has been noted, namely between circulating immune complexes and CAD detected by angiography [64], although the specificity of the serological test is crucial since non-specific immune complexes are commonly found following myocardial infarction.

Although there are some inconsistencies among these findings, which may be related to the choice of the control group or to the prevalence of *C. pneumoniae* infection in the population at the time of the studies, they all point to a positive association between antibody and CAD. In addition, the average of the odds ratio for all the results, that is, the relative risk of disease occurring in patients with antibody compared to those without, is about two. This two-fold increased risk seems to be associated consistently with previous or chronic infection with *C. pneumoniae*, since there is no reported association with the detection of IgM antibody.

18.10.2 DETECTION OF *C. PNEUMONIAE* IN ARTERIES

C. pneumoniae has been identified in arteriosclerotic lesions by electron microscopy, immunoperoxidase staining, and by a PCR assay, but not by isolation in cell culture. Pear-shaped *C. pneumoniae*-like organisms were observed first, by electron microscopy, in postmortem material from ten South African patients whose deaths were due to accidents and were not related to CAD. The organisms were confirmed as *C. pneumoniae* by immunoperoxidase staining with a specific monoclonal antibody and were found in coronary artery fatty streaks, in the lipid-rich core area of atheromatous plaques, and in intimal smooth muscle cells [78]. Application of the PCR technique and immunochemical staining to autopsy material from these and similar additional cases has resulted in the identification of *C. pneumoniae* in a total of 20 of 36 South African patients aged 20–83 years [62]. No organisms were identified in normal tissue adjacent to the sclerotic lesions or in normal coronary artery tissue from 11 control patients.

In a study in Seattle, *C. pneumoniae* was detected by immunoperoxidase staining and by a PCR assay in arterial tissue from six of 18 patients undergoing primary coronary atherectomy and from six of seven undergoing a secondary procedure to remove narrowing after a previous atherectomy [20]. Our own observations in a study at St Mary's Hospital, London, also provide some support for the existence of *C. pneumoniae* in the sclerotic arterial wall; *C. pneumoniae* has been found by means of a PCR-based assay on 11

occasions in aortic wall tissue from 25 patients undergoing repair of abdominal aortic aneurysms (G. Ong *et al.*, unpublished data). In contrast, the use of a PCR assay based on primers specific for the *C. pneumoniae* 16s rRNA gene provided no evidence for the existence of *C. pneumoniae* in arterial tissue from 33 patients undergoing atherectomy in Brooklyn [90], despite the organisms being endemic in the population and causing up to 20% of community-acquired pneumonia.

In the studies undertaken in both South Africa and the USA, no correlation was observed between the identification of *C. pneumoniae* organisms in coronary artery tissue and high titres of serum antibodies to *C. pneumoniae*. Indeed, somewhat to the contrary, in both geographical areas, the organisms could not be detected in a number of patients who had a high antibody titre, whereas they were found in others in whom antibody could not be found.

Although the presence of *C. pneumoniae* in atheromatous plaques has been documented, at least by some investigators, its role, if any, in the disease process is far from clear. *In vitro* studies and experiments in mice inoculated intranasally show that the organisms can replicate in macrophages [37,93] and thus may enter the circulation in such cells from the lungs and reach the coronary arteries. Indeed, *C. pneumoniae* was recovered from the heart of one of seven mice 3 days after intranasal inoculation. In humans, dissemination could happen at an early age, since most acute infections with *C. pneumoniae* occur between the ages of 5 and 15 years, and arteriosclerotic changes in young people are not uncommon; or it could happen later, during an episode of re-infection. In this way, the organisms could initiate infection within the endothelial cells and create a focus of intimal arterial damage upon which plaque could then develop. Conversely, macrophages containing *C. pneumoniae* organisms may simply be attracted to an already inflamed area of the arterial wall so that their occurrence at this site is secondary and has nothing to do with the initiation of the pathogenic process. Determining the role of *C. pneumoniae* in the aetiology of CAD will not be easy.

18.11 FUTURE CONSIDERATIONS

There is an undoubted need to improve the methods of detecting *C. pneumoniae* infections. If an analogy is drawn with the procedures used to diagnose *C. trachomatis* infections, EIAs and DNA probes will never be sufficiently sensitive. Better ways of culturing the organisms, improving the quality of fluorescent monoclonal antibodies and the commercial development of a sensitive PCR assay are likely to be the most rewarding ventures. In addition, further improvement in the specificity of serological tests, which in the past have led to so much misdiagnosis, will be helpful. For example, serodiagnosis of *C. pneumoniae* infections might be enhanced by the identification and synthesis of specific peptides, which could be used in an EIA. A measure of cell-mediated immunity might also be helpful, since specific lymphoproliferative responses to *C. pneumoniae* have been observed in patients with recent infections, although the response in healthy individuals is very varied [81]. In chronic infection, where diagnosis is especially problematic, it has been suggested that the detection of LPS- or protein-immune complexes might indicate the continuing presence of non-viable antigen [64]. The diagnostic value of antibody to the chlamydial heat-shock 60-kDa protein in *C. pneumoniae* infection remains to be elucidated [70].

It is interesting, from an immunological viewpoint, to reflect on whether a respiratory infection with *C. pneumoniae* could influence a urogenital infection with *C. trachomatis*, and vice versa, by completely or, more likely, partially protecting against disease or, alternatively, increasing its severity. Information derived from animal models and/or clinical studies is lacking but it is an issue that would

seem worth pursuing. Apart from this, there are various other clinical problems that need to be resolved or, at least, investigated. First, does *C. pneumoniae* account for any upper respiratory infections which might otherwise be considered as viral or for which an agent has not been identified? In this regard, there is some evidence that *C. pneumoniae* may cause common cold-like symptoms. It is unlikely that this occurs often, but the frequency with which it does happen would be useful to know since the prospect of rapid diagnosis as a result of improved methods followed by appropriate antibiotic treatment would be novel so far as colds are concerned. Second, to what extent does *C. pneumoniae* contribute to pneumonia occurring in patients with AIDS? On logical grounds, it would seem much more likely to do so than *C. trachomatis* which apparently makes a negligible contribution [67]. Third, in view of the fact that *C. trachomatis* is now recognized as one of the causes of sexually acquired reactive arthritis and Reiter's disease [55] and, also, one of the causes of seronegative arthritis in women [83], it is possible that *C. pneumoniae* is responsible in some patients for arthritis which appears not to be due to *C. trachomatis*. Indeed, it would seem reasonable to suppose that chlamydial organisms in the respiratory tract might be capable of initiating joint disease as readily as those in the genital tract. There is some support for this notion based on the occurrence of *C. pneumoniae*-specific synovial lymphocyte proliferation and high titres of specific antibody measured by the MIF test in a few patients who suffered reactive arthritis after infection with *C. pneumoniae* [15]. The existence of the organisms in affected joints has not been reported and this possibility needs to be investigated by the same methods that have been used to establish the involvement of *C. trachomatis* [55,83]. Fourth, although *C. trachomatis* causes neonatal conjunctivitis and pneumonia occasionally, as a result of acquisition of the organisms from the maternal genital tract, it is *C. pneumoniae* rather than *C. trachomatis* infections that occur in children. The question arises, therefore, of whether it is possible for some cases of juvenile chronic arthritis to be either initiated or exacerbated by *C. pneumoniae* infection. Our preliminary serological evidence does not indicate that this is so, but examination of synovial fluids and membranes for the organisms by the same techniques used successfully in sexually acquired reactive arthritis, namely MIF and the PCR, might be more revealing. Finally, the association between *C. pneumoniae* infection and CAD is sufficient to stimulate thoughts about the possibility of vaccination as a means of preventing such disease. The major requirement of a vaccine would be to inhibit haematogenous spread of the organisms from the respiratory tract to the heart and elsewhere, probably more easily achievable than preventing primary infection at a mucosal surface. While it might seem reasonable to curb such thoughts until the question of whether *C. pneumoniae* plays a primary or secondary role in CAD has been resolved, vaccination and/or prophylactic administration might, in fact, help in its resolution.

18.12 REFERENCES

1. Abeele van den, A.M., Renterghem van, L., Willems, K. and Plum, J. (1992) Prevalence of antibodies to *Chlamydia pneumoniae* in a Belgian population. *J. Infect.*, **25** (suppl. 1), 87–90.
2. Aldous, M.B., Grayston, J.T., Wang, S.P. and Foy, H.M. (1992) Seroepidemiology of *Chlamydia pneumoniae* TWAR infection in Seattle families, 1966–1979. *J. Infect. Dis.*, **166**, 646–9.
3. Almirall, J., Morato, I., Riera, F. *et al.* (1993) Incidence of community-acquired pneumonia and *Chlamydia pneumoniae* infection: a prospective multicentre study. *Eur. Resp. J.*, **6**, 14–18.
4. Augenbraun, M.H., Roblin, P.M., Mandel, L.J., Hammerschlag, M.R. and Schachter, J. (1991) *Chlamydia pneumoniae* pneumonia with pleural effusion: diagnosis by culture. *Am. J. Med.*, **91**, 437–8.
5. Beaty, C.D., Grayston, J.T., Wang, S.P., Kuo, C.C., Reto, C.S. and Martin, T.R. (1991) *Chlamydia pneumoniae*, strain TWAR, infection in patients with chronic obstructive pulmonary disease. *Am. Rev. Respir. Dis.*, **144**, 1408–10.
6. Berdal, B.P., Fields, P.I. and Melbye, H. (1991) *Chlamydia pneumoniae* respiratory tract infection: the interpretation of high titres in the complement fixation test. *Scand. J. Infect. Dis.*, **23**, 305–7.
7. Berdal, B.P., Scheel, O., Ogaard, A.R., Hoel, T., Gutteberg, T.J. and Anestad, G. (1992) Spread of subclinical *Chlamydia pneumoniae* infection in a closed community. *Scand. J. Infect. Dis.*, **24**, 431–6.
8. Black, C.M., Johnson, J.E., Farshy, C.E., Brown, T.M. and Berdal, B.P. (1991) Antigenic variation among strains of *Chlamydia pneumoniae*. *J. Clin. Microbiol.*, **29**, 1312–16.
9. Blasi, F., Cosentini, R., Legnani, D., Denti, F. and Allegra, L. (1993) Incidence of community-acquired pneumonia caused by *Chlamydia pneumoniae* in Italian patients. *Eur. J. Clin. Microbiol. Infect. Dis.*, **12**, 696–9.
10. Blasi, F., Cosentini, R., Raccanelli, R., Rinaldi, A., Tarsia, P. and Denti, F. (1994) *Chlamydia pneumoniae* causes acute exacerbations of asthma in adults, in *Chlamydial Infections, Proceedings of the 8th International Symposium on Human Chlamydial Infections*, Chantilly, (eds J. Orfila, G.I. Byrne, M.A. Chernesky *et al.*), pp. 477–9.
11. Blasi, F., Cosentini, R., Schoeller, M.C., Lupo, A. and Allegra, L. (1993) *Chlamydia pneumoniae* seroprevalence in immunocompetent and immunocompromised populations in Milan. *Thorax*, **48**, 1261–3.
12. Blasi, F., Legnani, D., Lombardo, V.M. *et al.* (1993) *Chlamydia pneumoniae* infection in acute exacerbations of COPD. *Eur. Respir. J.*, **6**, 19–22.
13. Bourke, S.J., Carrington, D., Frew, C.E., McSharry, C.P. and Boyd, G. (1992) A comparison of the seroepidemiology of chlamydial infection in pigeon fanciers and farmers in the UK. *J. Infect.*, **25** (suppl. 1), 91–8.
14. Bowie, W.R. (1994) Treatment of chlamydial infections, in *Chlamydial Infections, Proceedings of the 8th International Symposium on Human Chlamydial Infections*, Chantilly, (eds J. Orfila, G.I. Byrne, M.A. Chernesky *et al.*), pp. 621–30.
15. Braun, J., Laitko, S., Treharne, J., Eggens, U., Wu, P., Distler, A. and Sieper, J. (1994) *Chlamydia pneumoniae* – a new causative agent

of reactive arthritis and undifferentiated oligoarthritis. *Ann. Rheum. Dis.*, **53**, 100–5.

16. Bruu, A.L., Haukenes, G., Aasen, S. *et al.* (1991) *Chlamydia pneumoniae* infections in Norway 1981–1987 earlier diagnosed as ornithosis. *Scand. J. Infect. Dis.*, **23**, 299–304.

17. Campbell, J.F., Barnes, R.C., Kozarsky, P.E. and Spika, J.S. (1991) Culture-confirmed pneumonia due to *Chlamydia pneumoniae*. *J. Infect. Dis.*, **164**, 411–13.

18. Campbell, L.A., Kuo, C.C. and Grayston, J.T. (1987) Characterisation of a new Chlamydia agent, TWAR, as a unique organism by restriction endonuclease analysis and DNA:DNA hybridisation. *J. Clin. Microbiol.*, **25**, 1911–16.

19. Campbell, L.A., Kuo, C.C. and Grayston, J.T. (1990) Structural and antigenic analysis of *Chlamydia pneumoniae*. *Infect. Immun.*, **58**, 93–7.

20. Campbell, L.A., O'Brien, E.R., Cappuccio, A.L. *et al.* (1994) Detection of *Chlamydia pneumoniae* in atherectomy tissue from patients with symptomatic coronary artery disease, in *Chlamydial Infections, Proceedings of the 8th International Symposium on Human Chlamydial Infections*, Chantilly, (eds J. Orfila, G.I. Byrne, M.A. Chernesky *et al.*), pp. 212–15.

21. Campbell, L.A., Perez-Melgosa, M., Hamilton, D.J., Kuo, C.C. and Grayston, J.T. (1992) Detection of *Chlamydia pneumoniae* by polymerase chain reaction. *J. Clin. Microbiol.*, **30**, 434–9.

22. Carter, M.W., al-Mahdawi, S.A., Giles, I.G., Treharne, J.D., Ward, M.E. and Clarke, I.N. (1991) Nucleotide sequence and taxonomic value of the major outer membrane protein gene of *Chlamydia pneumoniae* IOL-207. *J. Gen. Microbiol.*, **137**, 465–75.

23. Chirgwin, K., Roblin, P.M., Gelling, M., Hammerschlag, M.R. and Schachter, J. (1991) Infection with *Chlamydia pneumoniae* in Brooklyn. *J. Infect. Dis.*, **163**, 757–61.

24. Chirgwin, K., Roblin, P.M. and Hammerschlag, M.R. (1989) *In vitro* susceptibilities of *Chlamydia pneumoniae* (*Chlamydia* sp. strain TWAR). *Antimicrob. Agents Chemother.*, **33**, 1634–5.

25. Cles, L.D. and Stamm, W.E. (1990) Use of HL cells for improved isolation and passage of *Chlamydia pneumoniae*. *J. Clin. Microbiol.*, **28**, 938–40.

26. Cosentini, R., Blasi, F., Rossi, S. *et al.* (1994) *Chlamydia pneumoniae* and severe community-acquired pneumonia, in *Chlamydial Infections,* *Proceedings of the 8th International Symposium on Human Chlamydial Infections*, Chantilly, (eds J. Orfila, G.I. Byrne, M.A. Chernesky *et al.*), pp. 453–6.

27. Cox, R.L., Kuo, C.C., Grayston, J.T. and Campbell, L.A. (1988) Deoxyribonucleic acid relatedness of *Chlamydia* sp. strain TWAR to *Chlamydia trachomatis* and *Chlamydia psittaci*. *Int. J. Syst. Bacteriol.*, **38**, 265–8.

28. Cunningham, A., Johnston, S., Julious, S., Sillis, M. and Ward, M.E. (1994) The role of *Chlamydia pneumoniae* and other pathogens in acute episodes of asthma in children, in *Chlamydial Infections, Proceedings of the 8th International Symposium on Human Chlamydial Infections*, Chantilly, (eds J. Orfila, G.I. Byrne, M.A. Chernesky *et al.*), pp. 480–3.

29. Ekman, M.R., Grayston, J.T., Visakorpi, R., Kleemola, M., Kuo, C.C. and Saikku, P. (1993) An epidemic of infections due to *Chlamydia pneumoniae* in military conscripts. *Clin. Infect. Dis.*, **17**, 420–5.

30. Ekman, M.R., Leinonen, M., Syrjala, H., Linnanmaki, E., Kujala, P. and Saikku, P. (1993) Evaluation of serological methods in the diagnosis *of Chlamydia pneumoniae* pneumonia during an epidemic in Finland. *Eur. J. Clin. Microbiol. Infect. Dis.*, **12**, 756–60.

31. Falsey, A.R. and Walsh, E.E. (1993) Transmission of *Chlamydia pneumoniae*. *J. Infect. Dis.*, **168**, 493–6.

32. Forsey, T., Darougar, S. and Treharne, J.D. (1986) Prevalence in human beings of antibodies to Chlamydia IOL-207, an atypical strain of Chlamydia. *J. Infect. Dis.*, **12**, 145–52.

33. Fraschini, F., Scaglione, F., Pintucci, G., Maccarinelli, G., Dugnani, S. and Demartini, G. (1991) The diffusion of clarithromycin and roxithromycin into nasal mucosa, tonsil and lung in humans. *J. Antimicrob. Chemother.*, **27** (suppl. A), 61–5.

34. Fukushi, H. and Hirai, K. (1992) Proposal of *Chlamydia pecorum* sp. nov. for Chlamydia strains derived from ruminants. *Int. J. Syst. Bacteriol.*, **42**, 306–8.

35. Gaydos, C.A., Fowler, C.L., Gill, V.J., Eiden, J.J. and Quinn, T.C. (1993) Detection of *Chlamydia pneumoniae* by polymerase chain reaction–enzyme immunoassay in an immunocompromised population. *Clin. Infect. Dis.*, **17**, 718–23.

36. Gaydos, C.A., Quinn, T.C., Bobo, L.D. and Eiden, J.J. (1992) Similarity of *Chlamydia pneumoniae* strains in the variable domain IV region of the major outer membrane protein gene. *Infect. Immun.*, **60**, 5319–23.

37. Gaydos, C.A., Summersgill, J.T., Sahney, N.N., Ramirez, J.A. and Quinn, T.C. (1994) Growth characteristics of *Chlamydia pneumoniae* in macrophages and endothelial cells, in *Chlamydial Infections, Proceedings of the 8th International Symposium on Human Chlamydial Infections*, Chantilly, (eds J. Orfila, G.I. Byrne, M.A. Chernesky *et al.*), pp. 216–19.

38. Girjes, A.A., Carrick, F.N. and Lavin, M.F. (1994) Remarkable sequence relatedness in the DNA encoding the major outer membrane protein of *Chlamydia psittaci* (koala type 1) and *Chlamydia pneumoniae*. *Gene (Amsterdam)*, **138**, 139–42.

39. Gnarpe, J., Gnarpe, H. and Sundelof, B. (1991) Endemic prevalence of *Chlamydia pneumoniae* in subjectively healthy persons. *Scand. J. Infect. Dis.*, **23**, 387–8.

40. Grayston, J.T. (1989) *Chlamydia pneumoniae*, strain TWAR. *Chest*, **95**, 664–9.

41. Grayston, J.T. (1990) *Chlamydia pneumoniae*, strain TWAR, in *Chlamydial Infections, Proceedings of the 7th International Symposium on Human Chlamydial Infections*, British Columbia, (eds W.R. Bowie, H.D. Caldwell, R.P. Jones *et al.*), pp. 389–401.

42. Grayston, J.T., Aldous, M.B., Easton, A. *et al.* (1993) Evidence that *Chlamydia pneumoniae* causes pneumonia and bronchitis. *J. Infect. Dis.*, **168**, 1231–5.

43. Grayston, J.T., Campbell, L.A., Kuo, C.C. *et al.* (1990) A new respiratory tract pathogen: *Chlamydia pneumoniae* strain TWAR. *J. Infect. Dis.*, **161**, 618–25.

44. Grayston, J.T., Kuo, C.C., Wang, S.P. and Altman, J. (1986) A new *Chlamydia psittaci* strain, TWAR, isolated in acute respiratory tract infections. *N. Engl. J. Med.*, **315**, 161–8.

45. Grayston, J.T., Mordhorst, C., Bruu, A.L., Vene, S. and Wang, S.P. (1989) Countrywide epidemics of *Chlamydia pneumoniae*, strain TWAR, in Scandinavia, 1981–1983. *J. Infect. Dis.*, **159**, 1111–14.

46. Grayston, J.T., Wang, S.P., Kuo, C.C. and Campbell, L.A. (1989) Current knowledge on *Chlamydia pneumoniae*, strain TWAR, an important cause of pneumonia and other acute respiratory diseases. *Eur. J. Clin. Microbiol. Infect. Dis.*, **8**, 191–202.

47. Hahn, D.L., Dodge, R.W. and Golubjatnikov, R. (1991) Association of *Chlamydia pneumoniae* (strain TWAR) infection with wheezing, asthmatic bronchitis and adult-onset asthma. *JAMA*, **266**, 225–30.

48. Haidl, S., Sveger, T. and Persson, K. (1994) Longitudinal pattern of antibodies to *Chlamydia pneumoniae* in children, in *Chlamydial Infections, Proceedings of the 8th International Symposium on Human Chlamydial Infections*, Chantilly, (eds J. Orfila, G.I. Byrne, M.A. Chernesky *et al.*), pp. 189–92.

49. Hammerschlag, M.R. (1993) *Chlamydia pneumoniae* infections. *Pediatr. Infect. Dis. J.*, **12**, 260–1.

50. Hammerschlag, M.R., Chirgwin, K., Roblin, P.M. *et al.* (1992) Persistent infection with *Chlamydia pneumoniae* following acute respiratory illness. *Clin. Infect. Dis.*, **14**, 178–82.

51. Hertzen, L. von, Leinonen, M., Koskinen, R., Liippo, K. and Saikku, P. (1994) Evidence of persistent *Chlamydia pneumoniae* infection in patients with chronic obstructive pulmonary disease, in *Chlamydial Infections, Proceedings of the 8th International Symposium on Human Chlamydial Infections*, Chantilly, (eds J. Orfila, G.I. Byrne, M.A. Chernesky *et al.*), pp. 473–6.

52. Hyman, C.L., Augenbraun, M.H., Roblin, P.M., Schachter, J. and Hammerschlag, M.R. (1991) Asymptomatic respiratory tract infection with *Chlamydia pneumoniae* TWAR. *J. Clin. Microbiol.*, **29**, 2082–3.

53. Johnson, D.H. and Cunha, B.A. (1993) Atypical pneumonias. Clinical and extrapulmonary features of *Chlamydia*, *Mycoplasma* and *Legionella* infections. *Postgrad. Med.*, **93**, 69–72, 75–6, 79–82.

54. Karvonen, M., Tuomilehto, J., Pitkaniemi, J. and Saikku, P. (1993) The epidemic cycle of *Chlamydia pneumoniae* infection in eastern Finland. *Epidemiol. Infect.*, **110**, 349–60.

55. Keat, A., Thomas, B., Dixey, J., Osborn, M., Sonnex, C. and Taylor-Robinson, D. (1987) *Chlamydia trachomatis* and reactive arthritis: the missing link. *Lancet*, **i**, 72–4.

56. Kern, D.G., Neill, M.A. and Schachter, J. (1993) A seroepidemiologic study of *Chlamydia pneumoniae* in Rhode Island. Evidence of serologic cross-reactivity. *Chest*, **104**, 208–13.

57. Kishimoto, T., Kimura, M., Kubota, Y., Miyashita, N., Niki, Y. and Soejima, R. (1994)

An outbreak of *Chlamydia pneumoniae* infection in households and schools, in *Chlamydial Infections, Proceedings of the 8th International Symposium on Human Chlamydial Infections*, Chantilly, (eds J. Orfila, G.I. Byrne, M.A. Chernesky *et al.*), pp. 465–8.

58. Kleemola, M., Saikku, P., Visakorpi, R., Wang, S.P. and Grayston, J.T. (1988) Pneumonia epidemics in military trainees in Finland caused by TWAR, a new Chlamydia organism. *J. Infect. Dis.*, **157**, 230–6.

59. Kuo, C.C., Chen, H.H., Wang, S.P. and Grayston, J.T. (1986) Identification of a new group of *Chlamydia psittaci* strains called TWAR. *J. Clin. Microbiol.*, **24**, 1034–7.

60. Kuo, C.C. and Grayston, J.T. (1988) *In vitro* drug susceptibility of *Chlamydia* sp. strain TWAR. *Antimicrob. Agents Chemother.*, **32**, 257–8.

61. Kuo, C.C. and Grayston, J.T. (1990) A sensitive cell line, HL cells, for isolation and propagation of *Chlamydia pneumoniae* strain TWAR. *J. Infect. Dis.*, **162**, 755–8.

62. Kuo, C.C., Shor, A., Campbell, L.A., Fukushi, H., Patton, D.L. and Grayston, J.T. (1993) Demonstration of *Chlamydia pneumoniae* in atherosclerotic lesions of coronary arteries. *J. Infect. Dis.*, **167**, 841–9.

63. Laitinen, K., Laurila, A., Leinonen, M. and Saikku, P. (1994). Experimental *Chlamydia pneumoniae* infection in mice; effect of reinfection and passive protection by immune serum, in *Chlamydial Infections, Proceedings of the 8th International Symposium on Human Chlamydial Infections*, Chantilly, (eds J. Orfila, G.I. Byrne, M.A. Chernesky *et al.*), pp. 545–8.

64. Linnanmaki, E., Leinonen, M., Mattila, K., Nieminen, M.S., Valtonen, V. and Saikku, P. (1993) *Chlamydia pneumoniae*-specific circulating immune complexes in patients with chronic coronary heart disease. *Circulation*, **87**, 1130–4.

65. Maass, M., Essig, A., Marre, R. and Henkel, W. (1993) Growth in serum-free medium improves isolation of *Chlamydia pneumoniae*. *J. Clin. Microbiol.*, **31**, 3050–2.

66. Miyashita, N., Kishimoto, T., Soejima, R. and Matsumoto, A. (1993) Reactivity of *Chlamydia pneumoniae* strains in the IDEIA CHLAMYDIA test kit designed for detection of *Chlamydia trachomatis*. *Kansenshogaku Zasshi*, **67**, 549–55.

67. Moncada, J.V., Schachter, J. and Wofsy, C. (1986) Prevalence of *Chlamydia trachomatis* lung infection in patients with acquired immune deficiency syndrome. *J. Clin. Microbiol.*, **23**, 986.

68. Mordhorst, C.H., Wang, S.P. and Grayston, J.T. (1994) Transmission of *Chlamydia pneumoniae* (TWAR), in *Chlamydial Infections, Proceedings of the 8th International Symposium on Human Chlamydial Infections*, Chantilly, (eds J. Orfila, G.I. Byrne, M.A. Chernesky *et al.*), pp. 488–91.

69. Moss, T.R., Darougar, S., Woodland, R.M., Nathan, M., Dines, R.J. and Cathrine, V. (1993) Antibodies to Chlamydia species in patients attending a genitourinary clinic and the impact of antibodies to *C. pneumoniae* and *C. psittaci* on the sensitivity and the specificity of *C. trachomatis* serology tests. *Sex. Transm. Dis.*, **20**, 61–5.

70. Peeling, R.W., Toye, B., Claman, P., Jessamine, P. and Laferriere, C. (1994) Seropositivity to *Chlamydia pneumoniae* and antibody response to the chlamydial heat shock protein, in *Chlamydial Infections, Proceedings of the 8th International Symposium on Human Chlamydial Infections*, Chantilly, (eds J. Orfila, G.I. Byrne, M.A. Chernesky *et al.*), pp. 502–5.

71. Pether, J.V., Wang, S.P. and Grayston, J.T. (1989) *Chlamydia pneumoniae*, strain TWAR, as the cause of an outbreak in a boys' school previously called psittacosis. *Epidemiol. Infect.*, **103**, 395–400.

72. Ridgway, G.L., Fenelon, L.E. and Mumtaz, G. (1990) The *in vitro* antibiotic susceptibility of *Chlamydia pneumoniae*, in *Chlamydial Infections, Proceedings of the 7th International Symposium on Human Chlamydial Infections*, British Columbia, (eds W.R. Bowie, H.D. Caldwell, R.P. Jones *et al.*), pp. 531–3.

73. Roblin, P.M., Montalban, G. and Hammerschlag, M.R. (1994) Susceptibilities of isolates of *Chlamydia pneumoniae* from children with pneumonia: relationship to clinical and microbiologic responses, in *Chlamydial Infections, Proceedings of the 8th International Symposium on Human Chlamydial Infections*, Chantilly, (eds J. Orfila, G.I. Byrne, M.A. Chernesky *et al.*), pp. 469–72.

74. Saikku, P. (1994) Diagnosis of acute and chronic *Chlamydia pneumoniae* infections, in *Chlamydial Infections, Proceedings of the 8th International Symposium on Human Chlamydial Infections*, Chantilly, (eds J. Orfila, G.I. Byrne, M.A. Chernesky *et al.*), pp. 163–72.

75. Saikku, P., Leinonen, M., Mattila, K. *et al.* (1988) Serological evidence of an association of novel

Chlamydia, TWAR, with chronic coronary heart disease and acute myocardial infarction. *Lancet*, **ii**, 983–6.

76. Saikku, P., Leinonen, M., Tenkanen, L. *et al.* (1992) Chronic *Chlamydia pneumoniae* infection as a risk factor for coronary heart disease in the Helsinki Heart Study. *Ann. Intern. Med.*, **116**, 273–8.

77. Saikku, P., Wang, S.P., Kleemola, M., Brander, E., Rusanen, E. and Grayston, J.T. (1985) An epidemic of mild pneumonia due to an unusual strain of *Chlamydia psittaci. J. Infect. Dis.*, **151**, 832–9.

78. Shor, A., Kuo, C.C. and Patton, D.L. (1992) Detection of *Chlamydia pneumoniae* in coronary arterial fatty streaks and atheromatous plaques. *S. Afr. Med. J.*, **82**, 158–61.

79. Sillis, M., White, P., Caul, E.O., Paul, I.D. and Treharne, J.D. (1992) The differentiation of *Chlamydia* species by antigen detection in sputum specimens from patients with community-acquired acute respiratory infections. *J. Infect.*, **25** (suppl. 1), 77–86.

80. Storey, C., Lusher, M., Yates, P. and Richmond, S. (1993) Evidence for *Chlamydia pneumoniae* of non-human origin. *J. Gen. Microbiol.*, **139**, 2621–6.

81. Surcel, H.M., Syrjala, H., Leinonen, M., Saikku, P. and Herva, E. (1993) Cell-mediated immunity to *Chlamydia pneumoniae* measured as lymphocyte blast transformation *in vitro*. *Infect. Immun.*, **61**, 2196–9.

82. Taylor-Robinson, D. and Thomas, B.J. (1991) Laboratory techniques for the diagnosis of chlamydial infections. *Genitourin. Med.*, **67**, 256–66.

83. Taylor-Robinson, D., Thomas, B.J., Dixey, J., Osborn, M.F., Furr, P.M. and Keat, A.C. (1988) Evidence that *Chlamydia trachomatis* causes seronegative arthritis in women. *Ann. Rheum. Dis.*, **47**, 295–9.

84. Theunissen, J.J., Heijst, B.Y. van, Wagenvoort, J.H., Stolz, E. and Michel, M.F. (1992) Factors influencing the infectivity of *Chlamydia pneumoniae* elementary bodies on HL cells. *J. Clin. Microbiol.*, **30**, 1388–91.

85. Thom, D.H., Grayston, J.T., Campbell, L.A., Kuo, C.C. and Wang, S.P. (1994) Respiratory infection with *Chlamydia pneumoniae* (TWAR) in middle-aged and older adults, in *Chlamydial Infections, Proceedings of the 8th International Symposium on Human Chlamydial Infections,* Chantilly, (eds J. Orfila, G.I. Byrne, M.A. Chernesky *et al.*), pp. 461–4.

86. Thom, D.H., Grayston, J.T., Siskovick, D.S., Wang, S.P., Weiss, N.S. and Daling, J.R. (1992) Association of prior infection with *Chlamydia pneumoniae* and angiographically demonstrated coronary artery disease. *JAMA*, **268**, 68–72.

87. Thom, D.H., Wang, S.P., Grayston, J.T. *et al.* (1991). *Chlamydia pneumoniae* strain TWAR antibody and angiographically demonstrated coronary artery disease. *Arterioscler. Thromb.*, **11**, 547–51.

88. Wang, S.P. and Grayston, J.T. (1990) Population prevalence antibody to *Chlamydia pneumoniae*, strain TWAR, in *Chlamydial Infections, Proceedings of the 7th International Symposium on Human Chlamydial Infections,* British Columbia, (eds W.R. Bowie, H.D. Caldwell, R.P. Jones *et al.*), pp. 402–5.

89. Wang, S.P. and Grayston, J.T. (1991) *Chlamydia pneumoniae* elementary body antigenic reactivity with fluorescent antibody is destroyed by methanol. *J. Clin. Microbiol.*, **29**, 1539–41.

90. Weiss, S., Roblin, P., Gaydos, C. *et al.* (1994) Failure to detect *Chlamydia pneumoniae* (Cp) in coronary atheromas of patients undergoing atherectomy, in *Chlamydial Infections, Proceedings of the 8th International Symposium on Human Chlamydial Infections,* Chantilly, (eds J. Orfila, G.I. Byrne, M.A. Chernesky *et al.*), pp. 220–3.

91. Wilson, P.A., Phipps, J. and Treharne, J.D. (1994) Antigenic specificity of the humoral immune response to infection by *C. pneumoniae,* in *Chlamydial Infections, Proceedings of the 8th International Symposium on Human Chlamydial Infections,* Chantilly, (eds J. Orfila, G.I. Byrne, M.A. Chernesky *et al.*), pp. 177–80.

92. Yang, Z.P., Kuo, C.C. and Grayston, J.T. (1993) A mouse model of *Chlamydia pneumoniae* strain TWAR pneumonitis. *Infect. Immun.*, **61**, 2037–40.

93. Yang, Z.P., Kuo, C.C. and Grayston, J.T. (1994) Systemic dissemination of *Chlamydia pneumoniae* following intranasal inoculation in mice, in *Chlamydial Infections, Proceedings of the 8th International Symposium on Human Chlamydial Infections,* Chantilly, (eds J. Orfila, G.I. Byrne, M.A. Chernesky *et al.*), pp. 549–52.

RESPIRATORY VIRUS AND MYCOPLASMAL INFECTIONS IN THE IMMUNOCOMPROMISED

Vincent Emery, Pat Furr, Eithne MacMahon, David Taylor-Robinson and David Webster

Viral and Other Infections of the Human Respiratory Tract.
Edited by S. Myint and D. Taylor-Robinson. Published in 1996 by Chapman & Hall. ISBN 0 412 60070 6

19.1 INTRODUCTION

Respiratory infections in the immunocompromised patient are an increasing clinical problem. This chapter examines the non-bacterial infections in three sections. The first of these deals with the more established viral pathogens, the second with enteroviruses and the third with mycoplasmal infections. The latter two sections are dealt with in, relatively, greater detail as these are areas that are not well covered elsewhere.

19.2 RESPIRATORY VIRAL INFECTIONS

The immunocompromised host is particularly susceptible to viral infections due to inadequacy of the T-cell-mediated arm of the immune system. Such inadequacy may result from immunosuppressive therapy, or arise in congenital or acquired immunodeficiency states. Severely immunocompromised patients are at risk of life-threatening pneumonic infections caused by DNA viruses of the human herpesvirus and adenovirus families. Respiratory disease may arise not only as a consequence of primary infection with these agents but also following reactivation of latent viral infection. Immunocompromised individuals are also prone to more frequent and severe viral pneumonias with the RNA viruses, which commonly cause lower respiratory tract disease in the general population (influenza, parainfluenza and respiratory syncytial (RS) viruses). These community or nosocomially acquired infections have been reviewed elsewhere [122], and will not be further discussed here. Instead, this section will concentrate on our current understanding of chest disease mediated by the DNA viruses, a problem largely confined to the immunocompromised host.

19.2.1 HERPESVIRUSES

The human herpesvirus family is a group of linear double-stranded DNA viruses that currently comprises herpes simplex virus (HSV) type-1 and type-2, varicella zoster virus (VZV), Epstein–Barr virus (EBV), cytomegalovirus (CMV) and human herpesviruses (HHV) type-6 and type-7. Infection is characterized by the failure of the normal host to eradicate virus following primary infection and the persistence of latent virus. Prior infection with each of the herpesviruses is associated with life-long virus-specific antibody production, enabling identification of individuals harbouring infection. The infection may be latent or reactivated with or without associated disease. Most herpesvirus infections are extremely common, with seroprevalences of between 50% and 100%, depending on the specific virus and the population under study. Life-threatening infection is rare in immunocompetent individuals. In seropositive immunocompromised individuals, reactivated virus may be associated with severe pulmonary disease. Immuno-compromised hosts are also at risk of primary infection or re-infection with herpesviruses. These may be transmitted in donor tissue, in transfused blood products, or by the usual routes in the community or in the hospital setting. In this section, those herpesviruses implicated as respiratory pathogens are discussed, with emphasis on current thinking and clinical management in the field. Strategies to control herpesvirus disease, including prevention of primary infection, pre-emptive therapy of productive (active) infection, and treatment of overt disease, will be discussed. Epstein–Barr virus has been associated primarily with extranodal lymphoproliferative disease and lymphomas, in the setting of immunodeficiency, and will not be considered further.

Herpes simplex virus

In the immunocompetent host, primary infection with or reactivation of HSV is not usually associated with respiratory disease. Localized mucocutaneous HSV disease is common in immunocompromised hosts and may progress to clinical pneumonitis in severely immunosuppressed or malnourished patients. Focal or multifocal infiltrates on chest X-ray are thought to arise by contiguous spread from the oropharynx via the trachea. More diffuse infiltrates are likely to arise by haematogenous spread in association with disseminated infection [81]. Transplant recipients are most vulnerable to HSV pneumonitis in the 4 weeks immediately following transplantation. The disease occurs with varying frequencies among the different transplant populations. Thus, whereas renal and liver recipients rarely show this complication, HSV pneumonitis may occur in 10% of heart–lung transplant patients, and even more frequently following bone marrow transplantation.

Higher than average pre-transplant antibody titres, possibly reflecting a propensity to frequent reactivation [8] can identify those at greatest risk. Detection of HSV in the cerebrospinal fluid (CSF) by the polymerase chain reaction (PCR) has largely replaced brain biopsy in the diagnosis of HSV encephalitis [3,52]. Such an approach has not been fully evaluated in pneumonia, where the use of a conventional cell culture technique usually provides a diagnosis within 5 days. However, rapid detection techniques utilizing specific anti-HSV antibodies can permit the detection of viral antigens in a matter of hours. At the time of writing, acyclovir is the drug of choice for treatment of HSV infections. The newer oral pro-drugs, valaciclovir and famciclovir, have yet to be evaluated formally in the suppression of HSV pneumonitis. Acyclovir has been used successfully in the prophylaxis of HSV disease, and is employed routinely for seropositive patients in different organ transplant groups [91]. Prophylactic acyclovir has also proved useful in controlling cytomegalovirus disease (see below). Recently, the possibility of reducing the severity of HSV disease through immunization has been studied [98]. However, whether this approach will have an impact on pneumonitis is debatable (see above) [8].

Varicella–zoster virus (VZV)

In immunocompetent adults, primary VZV infection is not infrequently complicated by pneumonitis [46]. For immunocompromised adults and children, primary infection is a life-threatening disorder, frequently associated with widespread organ dissemination [4]. Recurrent VZV infection or shingles tends to remain dermatomal, without pulmonary or other dissemination in immunocompromised hosts, with the exception of bone marrow recipients. In the latter group, reactivation may be associated with haematogenous spread to multiple organs, including the lungs, and has a high mortality rate.

Exposure of bone marrow transplant recipients to VZV warrants prompt passive immunization. Varicella zoster immune globulin (VZIG) should be administered, or an equivalent dose of intravenous immune globulin, where intramuscular injection is contraindicated by low platelet counts. In others patient groups, including organ transplant recipients, VZIG is usually reserved for seronegative subjects. Immune globulin prophylaxis does not afford 100% protection. Thus, while the skin rash may be attenuated, lethal VZV infection may still ensue. High-dose intravenous acyclovir therapy is indicated for treatment of VZV disease.

Cytomegalovirus (CMV)

Although common, CMV rarely causes significant disease in the normal host. In contrast, CMV infection is associated with severe illness in immunocompromised individuals. Although CMV may give rise to disease in many organs, pneumonitis is arguably the most significant manifestation in terms of morbidity and mortality. CMV-associated pulmonary disease differs among the various groups of immunocompromised patients. This is aptly demonstrated by comparing subjects with AIDS with allogeneic bone marrow recipients. In the former, patients do not appear to suffer from life-threatening CMV pneumonia, whereas in the latter CMV pneumonia occurs in 10–40%, with a mortality rate of up to 85% unless treated [79]. Pre-existing infection, as indicated by CMV seropositivity before transplantation, reduces – but does not eliminate – the risk of CMV disease. Indeed, the majority of cases of CMV disease in bone marrow transplant patients result from reactivation of virus in the recipient [125]. Solid organ recipients are also at risk of severe CMV disease, most commonly arising in the transplanted tissue. Thus, CMV pneumonitis is most frequently encountered in heart–lung transplant recipients.

Much of the current thinking on the pathogenesis of CMV pneumonitis centres on the hypothesis articulated by Grundy, Shanley and Griffiths [39]. Their thesis is based on data from both murine and human studies, which are outlined in Table 19.1. Note the association of CMV pneumonitis with allogeneic bone marrow transplantation (BMT), especially in the setting of graft-versus-host disease, and the lack of this complication following syngeneic or autologous transplantation. The control of CMV pneumonitis in the BMT recipient has proven difficult, even in the ganciclovir era. Despite substantial reductions in pulmonary viral load following therapy, the outcome was little changed [93]. However, with the addition of human immunoglobulin to the treatment regimen, significant improvement in survival rates was noted [25,82]. In the mouse, immunosuppression with a single low dose of cyclophosphamide induces CMV pneumonitis in infected animals. However, with repeated dosing regimens, CMV pneumonitis no longer occurs. The hypothesis thus states that CMV pneumonitis is not due to direct viral lysis *per se* but is an immunopathological condition. Such a situation is presumably mediated by T-cells and directed against target antigens (as yet unidentified) expressed on the infected cell. Therapeutic immunoglobulin preparations would then ameliorate disease by blocking such immunological reactions. Since this theory requires functional T-cell immunity for the development of pneumonitis, a logical extension would be that AIDS patients would not frequently suffer from pneumonitis attributable to CMV. This indeed is the case. Although CMV is frequently found in the lungs of HIV-infected subjects, there are relatively few, if any, cases of CMV pneumonitis in patients with AIDS

[10,87]. In contrast, patients undergoing acute HIV-seroconversion or in the early stages of HIV infection, with relatively well-preserved CD4 counts, have been shown to suffer from CMV pneumonitis [96].

The continued mortality from CMV pneumonitis, despite optimal available therapy, has resulted in increased emphasis on prevention. Serological testing before transplantation is of importance in establishing whether donor and recipient have been infected previously with CMV. Although matching of seronegative transplant recipients with seronegative donors is rarely feasible and may compromise HLA matching, steps can be taken to reduce transmission via transfusion. Thus, only CMV seronegative or leucocyte-depleted blood products should be administered to seronegative recipients [9]. Anti-viral prophylaxis with high-dose acyclovir has proved beneficial in reducing the mortality associated with CMV pneumonitis in seropositive BMT recipients [69]. Although no survival benefit was demonstrated in renal allograft recipients, diminished frequency of both CMV infection and disease was noted with oral acyclovir preventive therapy [5]. The relative safety of this agent makes it an attractive, if not fool-proof, prophylactic agent.

Table 19.1 Summary of data supporting the hypothesis that cytomegalovirus (CMV) pneumonitis is an immunopathological condition. (From [39].)

Animal species	Lung CMV	Pneumonitis
Mouse + MCMV	Yes	No
Mouse + MCMV + cyclophosphamide (1 dose)	Yes	Yes
Mouse + MCMV + cyclophosphamide (many doses)	Increased titre	No
Mouse + GVHD + MCMV	Yes	Increased
Man + syngeneic BMT	Yes	No
Man + allogeneic BMT	Yes	Yes
Man + allogeneic BMT + GVHD	Yes	Increased
Man + CMV pneumonitis + GCV	Reduced titre	Yes
Man + AIDS	Yes	No

MCMV, murine cytomegalovirus; BMT, bone marrow transplant; GVDH, graft-versus-host disease; GCV, ganciclovir

However, acyclovir is not sufficiently active against CMV for use as first-line treatment of pneumonitis. Although ganciclovir prophylaxis in CMV-seropositive bone marrow recipients reduces CMV infection and disease, the associated neutropenia limits its usefulness in this setting [35,125].

Routine surveillance is used to monitor bone marrow and organ transplant recipients for CMV infection or reactivation post-transplant. Serology is not useful, and may be misleading in the post-transplant setting. Conventional cell culture, detection of early antigen fluorescent foci (DEAFF), antigenaemia assays and PCR have been used to detect CMV in clinical specimens. The presence of CMV on culture of the blood of BMT patients is associated with a relative risk of greater than 7 of future disease attributable to CMV [115]. Similar data are available from PCR studies [51], where the relative risk is slightly lower than that for conventional cell culture, although the sensitivity is markedly improved. Thus, the value of surveillance is to identify a subgroup of patients with active CMV infection before the appearance of clinical CMV disease, permitting administration of pre-emptive therapy. A survival benefit has been demonstrated in the BMT group of patients using this approach [36].

In another placebo-controlled study of pre-emptive therapy, BMT recipients underwent routine bronchoscopy. Patients in whom routine broncheoalveolar lavage (BAL) fluid specimens collected at day 35 post-transplantation were CMV culture-positive were then randomized to receive ganciclovir or placebo. The results showed that ganciclovir was effective in preventing the development of CMV pneumonitis in patients with asymptomatic infection [89]. However, a significant proportion of patients without detectable virus in the BAL at this time of sampling went on to develop disease. The use of prophylactic acyclovir in the BMT group has been shown to delay the onset of CMV excretion and it is likely that routine lavage at day 35 post-transplant may not be the optimal time for sampling, and cell culture may not be the best method of detection.

Semi-quantitative PCR methods have been used to gain insight into CMV pathogenesis in recipients of bone marrow and heart–lung transplants. These data have shown the presence of increasing amounts of CMV DNA in sequential BAL samples taken before CMV interstitial pneumonitis [14]. The fact that CMV pneumonitis is probably an immunopathological condition, that increasing numbers of CMV genomes in BAL fluid precede pneumonitis and that pre-emptive therapy with ganciclovir can alleviate the effects of CMV pneumonitis, supports for the use of PCR monitoring for CMV in routine BAL samples in high-risk post-transplant groups. It should be noted that sequential bronchoscopy is not currently performed routinely in transplant recipients at most centres.

Human herpesvirus type 6 (HHV6)

HHV6 was identified only in 1986, and much less is known about the natural history and pathogenesis of infection by comparison with the other herpesviruses (excluding HHV7). HHV6 has nevertheless been associated with pneumonitis in BMT patients [15] and is often present in the lungs of AIDS patients [63]. The precise role that HHV6 plays in these situations remains to be elucidated, particularly, as HCMV is often found in the lungs at the same time. HHV6 infection in the lung appears to be restricted to macrophages.

19.2.2 ADENOVIRUSES

Adenoviruses are linear, double-stranded DNA viruses. They are non-enveloped and have icosahedral shells. The greater than 20 identified serotypes have been associated with respiratory, gastrointestinal and ophthalmic febrile illnesses. The majority of antigenic types have not been linked directly to specific clinical symptoms, possibly because they tend to cause only mild or subclinical infections in immuno-competent hosts. Furthermore, persistent ade-noviral carriage and viral shedding of these viruses limits the usefulness of viral isolation in making a diagnosis of acute adenovirus-medi-ated illness. For example, 10% of paediatric liver transplant recipients developed infection of the liver, lung or gastrointestinal tract at a median time of 26 days post-transplant [70]. Currently, there is no specific antiviral therapy available and hence treatment of adenoviral infection is limited to reduction or cessation of immunosuppressive therapy.

The adenoviruses types present in sub-groups B, C, D and E can all manifest as respi-ratory disease. Thus, types 3, 7, 11, 14, 16, 21, 34 and 35 (subgroup B) have been associated with acute fever and pneumonia; types 1, 2, 5 and 6 (subgroup C) have been associated with acute respiratory disease; types 8–10, 13, 15, 17, 19, 20, 22–30, 32, 33, 36, 37, 38 and 39 (sub-group D) have been associated with bronchi-tis, and type 4 has been associated with acute respiratory virus infection, with fever and atypical virus pneumonia in children [112]. In contrast, adenovirus types 38, 40 and 41 (sub-groups F and G) have only been associated with gastroenteritis in children. The patho-genesis of adenovirus lung disease in humans is relatively poorly understood. Direct tissue damage by adenovirus replication probably accounts for a large proportion of the patho-logical changes, although animal experiments have suggested that abortive replication may also participate in the pathogenesis [76]. The latter study showed that intranasal inocula-tion of adenovirus type 5 in cotton-tail rats produced pulmonary disease similar to that observed in human adenovirus infections. In a subsequent study, the early proteins E1b and E4 were shown to be responsible for the aden-ovirus pneumonia in this animal model [34]. Such a situation may parallel that observed in HCMV pneumonitis, with immunopathologi-cal mechanisms playing a major role in aden-ovirus lung pathology [55].

19.3 ENTEROVIRUSES IN ANTIBODY-DEFICIENT PATIENTS

The Medical Research Council in the United Kingdom set up a group to study patients with primary hypogammaglobulinaemia in 1970 at the Clinical Research Centre at Northwick Park Hospital; a few years later a 9-year-old boy with X-linked agammaglobulinaemia pre-sented with encephalitis and signs of dermato-myositis [122]. The facilities in Dr David Tyrrell's laboratories for culturing viruses enabled us to isolate an echovirus from this patient's CSF and, for the first time to propose a link between chronic echoviral infection and a dermatomyositis-like syndrome which had previously been described in similar patients in the USA [21,37]. Some 20 years later, chronic enteroviral infection remains an important and usually fatal complication in these patients.

19.3.1 CLINICAL FEATURES OF CHRONIC ECHOVIRUS DISEASE

Neurological manifestations

Our first patient had many of the typical fea-tures of echovirus disease which were subse-quently reviewed by Wilfert *et al.* [124]. The encephalitis was insidious with drowsiness, headache and hand tremor and was followed by status epilepticus after a cerebral angiogram. The CSF was under normal pres-sure, but contained moderately raised num-bers of white cells and protein, and echovirus 11 was isolated after culturing in human embryonic lung fibroblasts. The patient was treated with hyperimmune plasma obtained by screening donors for anti-echovirus 11 anti-body, and also plasma and white cells from his mother who was immunized with a formalin-inactivated echovirus 11 vaccine prepared at the MRC Common Cold Unit at Salisbury, England. A horse was immunized with this vaccine in Freund's adjuvant and the anti-serum (with a haemagglutination-inhibition titre of 1:1280) was given intravenously to the patient. No antibody to echovirus 11 was found in his CSF during this time, suggesting that very little of the infused antibody had crossed the blood brain barrier. Although there was some improvement in the short-term, the patient died 28 months after the initial presentation, apparently due to sudden respiratory centre failure [116].

A link between echovirus and chronic meningoencephalitis in patients with primary hypogammaglobulinaemia was clearly established as similar cases were reported in the USA and reviewed by Wilfert *et al.* [124], and later by McKinney, Ketz and Wilfert [65]. The fact that patients who are unable to make antibodies were particularly prone to this virus indicated that treatment should be directed towards passive immunotherapy, although giving antibodies systemically seemed to have little or no effect. Some clinicians then started injecting immunoglobulin containing anti-echovirus antibody directly into the CSF, usually surgically inserting a ventricular (Ommaya) reservoir through which regular injections could be given [24,66]. The procedure was to screen batches of commercially available intramuscular immunoglobulin for anti-echovirus antibody, and then to use batches with the highest titres. There is a strong impression that this benefited some patients, although there was some risk of bacterial superinfection of the reservoir. Some patients had prolonged remissions, and in a few this may have been permanent, but the majority eventually relapsed and died. The overall impression was that it was not possible to obtain immunoglobulin with a high enough titre of antibodies to be effective. A better approach might be to inject hyperimmune animal serum intrathecally, but this has not yet been attempted because of the high risk of severe inflammatory reactions; in our experience, goat or horse serum causes phlebitis when injected intravenously into antibody-deficient patients.

The neurological features associated with chronic echovirus disease are summarized in Table 19.2. They include signs of meningitis,

encephalitis and in some cases focal demyelination. We have recently reviewed 11 of our own patients in whom five have confirmed enteroviral infection, and have broadly classified the signs into three categories, the most common being focal encephalitis predominantly affecting the cortex [86]. Post-mortem studies are rare, but the brains from two of our patients have been studied in detail, and have shown extensive meningeal inflammation with fibrosis, together with pronounced cortical atrophy with foci of inflammation within the white matter. The meningitis often extends down the length of the spinal column, and it is easy to see how this might ultimately affect the blood supply to crucial centres in the brain stem. It is likely that all areas of the brain can be involved since there was loss of Purkinje cells in the cerebellum, and loss of cells in the anterior horns and posterior route ganglia in some areas of the spinal column in our first case [122]. More recently, we have seen cerebellar atrophy on computer-assisted tomography (CAT) scanning in another patient with advanced disease.

New techniques for diagnosis

In the 1980s we investigated three patients with primary hypogammaglobulinaemia and chronic central nervous system (CNS) disease, in whom attempts to culture viruses from their CSF failed. One of these, a patient with X-linked agammaglobulinaemia (XLA), had typical features of echovirus disease, his CSF having a moderately increased number of cells and raised protein level on a number of occasions. The other two patients were less typical, one having a progressive dementia over 12 years with cortical atrophy on CAT scan, and the other having signs of cerebellar involvement which remain stable 3 years later. Using a new PCR technique developed by Professor Harley Rotbart at Houston, Colorado, we were able to diagnose enteroviral infection [85,119]. In this technique, oligonucleotide primers are

Table 19.2   Clinical signs and symptoms of enteroviral disease in patients with primary immunodeficiency

---

Deterioration in writing and speech
Headache
Convulsions
8th nerve deafness
Sensory disturbances
Ataxia
Dementia
Rigidity of limb muscles
Flexion contractures of knees and elbows
Mild arthropathy
Erythematous skin rash

---

used to amplify a conserved segment at the 5′ end of all enteroviral genomes that have so far been sequenced. It is not possible to identify specific enteroviruses with this technique, so it is not clear whether these patients suffered from echovirus disease, or from some other enterovirus (e.g. Coxsackie A) which may be very difficult to culture *in vitro*. Furthermore, two infants with severe chronic encephalitis were PCR-negative, although in one the reaction generated a product of the right size which did not hybridize to the specific enteroviral probe. Another CSF from a patient with optic atrophy in The Netherlands was also negative in Professor Rotbart's laboratory, but on testing further samples with different primers a positive result was obtained. These findings suggest the current techniques are not yet optimal for recognizing all enteroviruses.

It could be argued that the presence of enteroviral RNA was unrelated to the patient's CNS disease, and that patients with primary immunodeficiency might frequently harbour small numbers of virions without overt disease. However, CSF specimens from four immunodeficient patients with minor self-limiting symptoms, such as headache and lethargy were negative.

The overall impression is that most of the CNS disease in patients with primary immunodeficiency is due to chronic enteroviral infec-tion, and with new PCR techniques it should be possible to diagnose the majority. Very few virions are likely to be present in the CSF of these patients and because their numbers are likely to vary from time to time it is sensible to search for evidence of infection in repeated samples. Furthermore, the introduction in the early 1980s of therapy with intravenous immunoglobulin, which contains small amounts of specific antibody to a range of enteroviruses, has probably modified the disease and been responsible for the recent failure to culture viruses from such patients. Brain disease appears to have become less common in these patients over the past 10 years, suggesting some protection; however, there are patients who have developed enteroviral disease after many years on intravenous immunoglobulin therapy [72].

Enteroviral infection in infants with primary immunodeficiency may explain the curious association between XLA and growth hormone deficiency [13]. Now that the XLA gene has been cloned and sequenced, it is clear that most of these patients have defects in this gene [23,112]. Thus, the growth hormone deficiency must be secondary to the primary immunodeficiency. Viruses have been implicated in the growth hormone deficiency associated with HIV infections in infants, and enteroviruses may play a role [58].

19.3.2 POLIOVIRUS

Cases of paralytic poliomyelitis following routine poliovirus vaccination were described in infants and children with both severe combined immunodeficiency and XLA in the 1970s [127,128]. Prolonged excretion of vaccine strains following immunization can occur in patients with primary hypogammaglobulinaemia, and patients with immunodeficiency are advised not to have live viral vaccines of any type [67,71]. Nevertheless, many patients with primary hypogammaglobulinaemia have been given attenuated oral poliovirus vaccine in childhood and have not suffered any complications, suggesting that infection with a revertant pathogenic virus is rare [62]. Furthermore, there have been no descriptions of immunodeficient patients developing paralytic poliomyelitis following routine immunization of an immunocompetent sibling, although most of these patients have been receiving intravenous immunoglobulin which contains IgG anti-poliovirus antibody; some of this probably reaches the saliva via the crevicular fluid where it neutralizes any virions acquired through the faecal–oral route. Thus, the available evidence indicates that the risk of acquiring pathogenic revertants from the community is remote, although poliovirus immunization is still contraindicated for the patients themselves.

19.3.3 THE DERMATOMYOSITIS-LIKE SYNDROME

Our first patient with echovirus encephalitis also developed signs of muscle, joint and skin involvement. The skin and joint signs were transient, with an arthropathy involving large joints and an erythematous rash on one of his arms, the latter associated with generalized thickening and oedema of the subcutaneous tissues of both arms. A muscle biopsy showed perivascular infiltration with mononuclear cells and some muscle fibre necrosis [117]. The most common presentation is of mild swelling of the lower extremities with gradual thickening of the muscles of the limbs, producing a stooped and flexed posture that is caused by gradual and sometimes extensive muscle fibrosis. Actual involvement of the joints is rare and significant chronic arthropathy has not been described.

Echovirus has been isolated from muscle biopsies, and in one of our cases the serotype was different from that of the virus found in the CSF. This presumably represents two different primary infections, since it is very unlikely that spontaneous mutations in the genes coding for the envelope proteins could be so extensive as to change the serotype. In our experience the dermatomyositic features of echovirus infection improve when patients are given intravenous immunoglobulin (IVIG) therapy, and significant muscle involvement has become a rare complication during the past decade. This can be explained by the presence of IgG antibody in IVIG preparations which should neutralize virions outside the CNS.

19.3.4 THE IMMUNOLOGICAL RESPONSE TO ENTEROVIRUSES

Very little work has been done in this area because there are no suitable animal models. The demonstration that antibody-deficient patients are particularly prone to chronic enteroviral infection clearly shows that specific antibodies play a central role in both preventing initial infection and eliminating established infection. This contrasts with infection with other common viruses such as varicella zoster, mumps and measles, from which hypogammaglobulinaemic patients usually recover uneventfully [2]. Thus, cellular immunity is much less important for eliminating enteroviral-infected cells than it is for most of the DNA viruses [107]. There is evidence in rats of immune lymphocytes from local lymph nodes gaining access to the CNS, after being primed by antigens originating in the CNS [19]. Immune B-lymphocytes may gain access to the CNS in this way and then differentiate into plasma cells and produce antibody, which in immunocompetent mice and humans can be found in the CSF following infection [49,110]. Patients with XLA have virtually no B-cells, and the B-cells in many patients with common variable immunodeficiency fail to differentiate *in vivo*, and would therefore be functionally inactive [11].

The mechanism whereby antibodies prevent progressive CNS infection is not known, but perhaps enteroviruses are shed from the infected cell into the extracellular environment before infecting a neighbouring cell. Antibody would therefore neutralize the virus in transit. It is unclear which cells are the targets for echoviruses and coxsackie viruses in the CNS, although poliovirus is well known to infect the anterior horn cells of the spinal column. In echovirus disease, inflammatory cells do accumulate around small blood vessels in the brain, suggesting that vascular endothelial cells might be the target, and this would explain the chronic meningitis that usually occurs. However, there is destruction of neurones without evidence of vasculitis, and it is likely that a variety of CNS cells are involved. *In situ* hybridization techniques need to be developed to shed light on this issue. Infection of lymphocytes and macrophages within the CNS seems less likely, and attempts to show echovirus antigen in CSF mononuclear cells by immunofluorescence in our first case were negative, suggesting that virions in the CSF are free of cells [122].

Another mechanism whereby antibody may eliminate viral infected cells is by antibody-dependent cytotoxicity. In this situation, specific antibody binds to viral encoded proteins on the cell surface, initiating a cytotoxic response from a macrophage or natural killer (NK) cell after binding to their Fc receptors [108]. This now needs to be explored *in vitro*, although it is not clear whether the CNS contains the appropriate cytotoxic effectors.

19.3.5 FUTURE STRATEGIES FOR TREATING ENTEROVIRAL DISEASE

There are no specific anti-enteroviral drugs available and it is doubtful whether it is commercially viable to develop these specifically to treat rare patients. We can only hope that drugs being developed against other RNA viruses (e.g. hepatitis C virus) may be useful. In the meantime, antibody therapy is the only practical way of influencing the course of the disease. It is sensible to give high dose (400 mg/kg at weekly intervals) intravenous immunoglobulin which contains some antibody to a wide range of enteroviruses because it is prepared from pooled plasma from about 20 000 healthy donors. However, although this might help eliminate virus outside the CNS, it will only have a limited effect within the CNS. Most clinicians are sceptical about the value of inserting a ventricular reservoir and injecting pooled immunoglobulin on a regular basis, mainly because the amount of antibody in these preparations is relatively small [71]. Assuming that a virus can be cultured from the CSF, vaccines could be prepared to immunize animals and, in principle, hyperimmune serum could be injected into these reservoirs with better effect, although the high risk of local reactions must be considered. Another approach would be to make monoclonal antibodies, or genetically engineered antibodies to a wide range of echovirus and coxsackie virus serotypes, and to inject these directly into a ventricular reservoir. This would be an expensive exercise but may be the best way forward. Meanwhile, infection in hypogammaglobulinaemic patients can be minimized by early diagnosis of the underlying immunodeficiency and insuring that the patients receive adequate immunoglobulin therapy so that their serum IgG levels are maintained within the normal range.

The current difficulty in culturing enteroviruses in patients already on IVIG is a major problem because specific immunotherapy is ruled out. Culture techniques will need to be improved, and in patients suspected of having infection suckling mice should be routinely injected with CSF in an attempt to propagate a virus. Repeated CSF samples may need to be taken to maximize the chance of culturing the agent.

19.4 MYCOPLASMA INFECTIONS

Members of the family Mycoplasmataceae, within the order Mycoplasmatales, are the smallest free-living organisms and are subdivided into two genera; the genus *Mycoplasma* and the genus *Ureaplasma* for organisms which are uniquely urea-hydrolysing. These are trivially called mycoplasmas and ureaplasmas respectively. Ureaplasmas of human origin belong to the species, *Ureaplasma urealyticum* [93]. In both human and veterinary medicine mycoplasmas and ureaplasmas have a predilection for mucous membranes so that they occur primarily in the oropharynx, respiratory and urogenital tracts. However some, which colonize mainly the oropharynx, are found also in the urogenital tract, presumably as a consequence of orogenital contact, and vice versa. In the case of some others, for example *Mycoplasma fermentans*, the primary site of colonization has been realized only recently. Because mycoplasmas and ureaplasmas are not confined entirely to one primary location, it is not always easy to determine the primary site of infection. Thus, when dissemination occurs in immunocompromised patients it is necessary to consider mycoplasmas and ureaplasmas not only in the respiratory tract, but also in the urogenital tract. The following discussion of our own experiences and those of others concerns the effect that immunodeficiency has on the proliferation, spread, and persistence of disease-producing capacity of these organisms.

19.4.1 PRIMARY HYPOGAMMAGLOBULINAEMIA

Over about 20 years, a large number of patients with primary hypogammaglobuli-

naemia have attended the Immunodeficiency Clinic at the MRC Clinical Research Centre in Harrow. About one-third of those attending were at one time or another investigated for possible mycoplasma or ureaplasma infection. An overview of these patients (53 males and 38 females) has recently been published [32]. Most were less than 50 years of age at first presentation, the range being 6–75 years (mean 32.5 years) for males and 15–80 years (mean 37 years) for females. The microbiological media and methods for the detection of mycoplasmas and ureaplasmas have already been described in detail [100,101].

19.4.2 MYCOPLASMAS AND UREAPLASMAS IN THE UROGENITAL TRACT

Arginine-hydrolysing mycoplasmas and ureaplasmas in urine specimens, and hence in the urogenital tract, were isolated from 31 patients with similar frequency in men and women (Table 19.3), but ureaplasmas were detected at least twice as often as the arginine-hydrolysing mycoplasmas. The dominance of ureaplasmas was seen also in swab specimens taken from the male urethra and in vaginal specimens, the latter being positive more often than urine specimens from women. Most of the males had symptoms and signs of urethritis, and in some this was associated with a chronic cystitis, and in two patients an epididymitis. Some of the women also had cystitis or symptoms and signs of vaginitis [120]. The link between mycoplasmas and chronic cystitis in hypogammaglobulinaemic patients was important because similar patients had been treated inappropriately before on the assumption that they had a bacterial cystitis, for instance with *E. coli*. In fact, our experience shows that bacterial urethritis and cystitis is extremely rare in hypogammaglobulinaemic patients, and that mycoplasma infection should be assumed, until proved otherwise, in patients presenting with urinary symptoms, including pyelonephritis [42]. These observa-

Table 19.3 The proportion of hypogammaglobulinaemic patients in whom mycoplasmas and ureaplasmas were found in urogenital tract specimens

Sex	Specimen	No. tested	No. of patients with indicated organism (%)		
			Arg*	Ureat	Both
Male	Urine	15	3 (20)	7 (47)	3 (20)
Female	Urine	16	4 (25)	9 (56)	4 (25)
Male	Urethral swab	16	2 (12.5)	7 (44)	2 (12.5)
Female	Vaginal swab	12	6 (50)	9 (75)	6 (50)

* Arginine-metabolizing mycoplasmas;
† Ureaplasmas

tions suggest that specific antibodies prevent mycoplasma colonization of the urethra and bladder, probably because antibody in the urine and urethral secretions may be sufficient to inhibit growth or attachment of these organisms to the mucosa [33]. In contrast, IgA antibodies to Gram-negative bacteria, such as *E. coli*, seem unable to prevent growth in the urine, and the infrequency of these infections in antibody-deficient patients leads one to speculate that IgA antibodies may be necessary for the development of cystitis, perhaps by preventing phagocytosis by 'blocking' Fc-mediated attachment of organisms to neutrophils and macrophages [38]

Comparison of urine in hypogammaglobulinaemic and immunocompetent patients

The sexual orientation, number of life-time sexual partners and other demographic details of the patients were unknown. Nevertheless, we compared the results of attempted mycoplasmal isolation from their urines with those from sex- and age-matched immunocompetent, non-venereal disease hospital patients and from healthy individuals for whom other demographic details were lacking [31]. Although the validity of such comparisons can be criticized, the arginine-hydrolysing mycoplasmas occurred two to six

times more frequently and the ureaplasmas two to three times more frequently in the hypogammaglobulinaemic patients than they did in the other two groups; differences which were statistically significant ($P<0.05$). This shows that hypogammaglobulinaemic patients are prone to mycoplasma colonization of the urinary tract, and supports the view that these organisms are directly responsible for the patients symptoms.

19.4.3 MYCOPLASMAS AND UREAPLASMAS IN THE SPUTUM AND THROAT

In an unselected subgroup of the 91 hypogammaglobulinaemic patients examined by us, arginine-hydrolysing mycoplasmas, which were not characterized further by speciation, occurred equally in men and women and were found almost twice as frequently in sputum as they were in throat swabs (Table 19.4). This discrepancy may have been due to inadequate swabbing. Ureaplasmas were detected with similar frequency in men and women, but were not detected in throat swabs from the women, although the number examined was small. Overall, the prevalence of arginine-hydrolysing mycoplasmas in the sputum and throat was certainly no greater than might be expected in immunocompetent patients [96] and the prevalence of ureaplasmas was in keeping with that found by others [84] who

Table 19.4 The proportion of hypogammaglobulinaemic patients in whom mycoplasmas and ureaplasmas were found in the sputum and throat

Sex	Specimen	Proportion (%) of patients from whom indicated organisms were isolated		
		Arg*	Urea†	Both
Male	Sputum	8/18 (44)	5/16 (31)	4/16 (25)
Female	Sputum	5/12 (42)	2/7 (29)	1/7 (14)
Male	Throat swab	5/24 (21)	7/22 (32)	3/22 (14)
Female	Throat swab	2/8 (25)	0/7 (0)	0/7 (0)

* Arginine-metabolizing mycoplasmas;
† Ureaplasmas

also have examined the respiratory tract of hypogammaglobulinaemic patients.

As part of the study mentioned above, *Mycoplasma pneumoniae* was sought in sputum and/or throat swab specimens from 41 male hypogammaglobulinaemic patients, but was found in only one and in none of 21 female patients. However, in the single case, the organisms persisted in respiratory secretions for 2 years [106] despite several courses of antibiotics. In addition, Foy *et al.* [29] reported *M. pneumoniae*-induced pneumonia in four patients with humoral immunodeficiency syndromes. All were severely ill and two, who had protracted cough, joint manifestations and skin rashes, required hospitalization. The lack of radiographic signs despite symptoms suggested that the pneumonic infiltrate in immunocompetent patients was, at least partially, an immune response of the host. This concept is supported by recent work in mice where pneumonitis to a related mycoplasma (*M. pulmonis*) depends on the presence of antibody [26]. One of the four patients had a second bout of pneumonia caused by *M. pneumoniae* a year later, but whether the organism had been carried subclinically during the intervening period was unknown [28]. Repeated episodes in patients with humoral deficiency syndrome (six cases) have been reported also by Roifman *et al.* [84].

19.4.4 MYCOPLASMAS AND UREAPLASMAS ISOLATED FROM JOINTS

In an earlier study [105] of 42 patients with inflammatory polyarthritis, the only patient from whom a mycoplasma, identified as *M. pneumoniae*, was isolated from the joint had primary hypogammaglobulinaemia. About the same time, ureaplasmas were identified in the joints of hypogammaglobulinaemic patients with septic arthritis [99,121]. Subsequently, there has been increasing evidence that ureaplasmas and mycoplasmas recovered from the joints of such patients with septic arthritis are the cause of the disease [27,44,45,53,54,56,68,95,111]. More frequent colonization at mucosal sites in hypogammaglobulinaemic patients, often with large numbers of organisms, increases their opportunity to disseminate to distant sites, such as joints and even bone [73,83]. In addition, the absence or reduced levels of antibody in these patients is likely to be an important factor in the enhanced ability of the organisms to spread. We have previously shown that mycoplasmas and ureaplasmas are phagocytosed by neutrophils in the absence of antibody, apparently by directly activating the first component of complement and generating activated complement which binds to complement receptors on the phagocyte. Furthermore, following phagocytosis *in vitro*, the organisms remain viable [118]. Thus, mycoplasmas may be disseminated within phagocytes which are likely to home to sites of inflammation, such as minor trauma within joints. This is supported by the fact that ureaplasma arthritis followed a traumatic sprain in one of our patients, and ureaplasma subcutaneous abscesses subsequently appeared at injection sites on his buttocks and thighs.

Of the 91 hypogammaglobulinaemic patients examined, which includes the ureaplasma-infected patient mentioned above [121], 21 (23%), comprising 12 men and 9 women, had arthritis of one or more joints in which there were polymorphonuclear leucocytes in the synovial fluid. Aerobic and anaerobic bacteria were not detected, but mycoplasmas and/or ureaplasmas were isolated from the joints of eight (38%) of the arthritic patients (see Table 19.5). Ureaplasmas were recovered most frequently, occurring in half of the patients who had organisms in the joints.

Whether any of the apparently mycoplasma- or ureaplasma-negative arthritides might be attributed to these microorganisms, culture attempts having failed perhaps due to previous antibiotic therapy, is not clear. Some of these patients may have had disease due to mycoplasmas which we did not seek. Those, such as *M. genitalium* and *M. fermentans*,

are difficult to isolate but may be detected by the much more sensitive PCR that has become available recently [47,77]. It is possible also that following an initial insult, mycoplasmas and/or ureaplasmas may remain in a non-viable form, as shown for *Chlamydia trachomatis* in sexually acquired reactive arthritis [48]. However, against this view is the finding that most of our patients with arthritis respond to tetracyclines, and patients exist with relapsing joint disease where it is clear that tetracyclines induce repeated remissions [30].

19.4.5 TREATMENT OF HYPOGAMMAGLOBULINAEMIC PATIENTS WITH ANTIBIOTICS

During the course of our observations, it became evident that most patients responded clinically to antibiotic therapy, in particular to doxycycline, and a 3-month course of oral doxycycline (100 mg/day) is now the standard first-line treatment. However, a few patients failed to respond, the organisms being repeatedly isolated from the joints despite their *in vitro* antibiotic sensitivity [104]. The ability of mycoplasmas, at least some species, to invade epithelial cells [60,103] may help explain their persistence. On the other hand, antibiotic resistance developed rapidly in some cases and

Table 19.5 Mycoplasmas and/or ureaplasmas isolated from the joints of hypogammaglobulinaemic patients with arthritis

Mycoplasma/Ureaplasma	*Case no.*	*Mycoplasma-positive joints*	
		No.	*Site*
M. pneumoniae	1	3	Thumb, ankles
	2	1	Knee
M. salivarium	3	1	Elbow
M. hominis	4	1	Knee
U. urealyticum	5*	4	Shoulder, elbow, wrist, knee
	6	3	Shoulder, wrist, knee
	7	1	Finger
	8	1	Knee

* Also *M. hominis* from shoulder and elbow

there may be value in always treating with more than one antibiotic in an attempt to overcome this problem. In recalcitrant cases, clinical improvement appeared to come from administration of antiserum prepared specifically in goats against the mycoplasma in question [104].

19.4.6 MYCOPLASMAS IN PATIENTS ON IMMUNOSUPPRESSIVE CHEMOTHERAPY

Severe manifestations of *M. pneumoniae* infection have been noted in a few patients undergoing immunosuppressive therapy for cancer or for other reasons [18,78]. Problems associated with ureaplasmas in this context appear to have been few [12,40], but there are many more reports of the proliferation and dissemination of *M. hominis* organisms in patients undergoing immunosuppressive chemotherapy [64]. Thus, invasion of the bloodstream or surgical wound infection has been seen in patients on cyclosporin and/or corticosteroids undergoing cardiac or renal transplantation.

19.4.7 PATIENTS WITH HUMAN IMMUNODEFICIENCY VIRUS INFECTION

The possible proliferation and role of mycoplasmas in the immunodeficiency associated with human immunodeficiency virus (HIV) disease warrants a brief discussion. In view of the known effect that immunodeficiency states have on the persistence and severity of *M. pneumoniae* infections, it is perhaps surprising that such infections have been described so infrequently in patients with the acquired immunodeficiency syndrome (AIDS). However, it is possible that some *M. pneumoniae* infections may go unnoticed if the diagnosis is based on a serological response, as it often is, rather than detection of the agent. In this regard, it is noteworthy that while *M. pneumoniae* was recovered from the bronchial washings [80] of two French children with AIDS complicated by pneumonia, both of whom were treated successfully, neither mounted a complement-fixing antibody response.

The notion that mycoplasmas might have some role in HIV-infected patients sprang from the observations of Lo and colleagues in the USA, in the mid-1980s, while searching for viruses in Kaposi's sarcoma (KS) [59]. They extracted DNA from KS tissue and thought they had succeeded in transfecting cell cultures because of the recovery of a virus-like infectious agent (VLIA). Antiserum prepared against it and used in an immunohistochemical procedure, together with electron microscopy, was instrumental in providing evidence of its existence in patients with AIDS and in a few non-AIDS patients with an acute fatal disease [59]. The electron microscopic appearance of the VLIA suggested that it might be a mycoplasma and subsequently it was cultured and termed *Mycoplasma incognitus* [61]. However, this name was unacceptable because further examination revealed that it was *M. fermentans* [88] of known human origin. Indeed, this mycoplasma had been isolated rarely from the human urogenital tract and was known also to contaminate cell cultures. Despite this, the findings of Lo and colleagues promoted the idea that mycoplasmas might have importance in the pathogenesis of AIDS. This idea was fuelled by the observations of Montagnier and colleagues [57] who were the first to note that tetracyclines inhibited the cytopathic effect induced by HIV-1 or HIV-2 in a human T-cell (CEM) line without suppressing virus growth. In addition, a fluoroquinolone was shown to protect an HIV-1-infected human cell line from death [75]. These observations suggested that an antibiotic-susceptible agent, possibly a mycoplasma, contributed to the cytopathic effect considered to be induced by HIV. Subsequently, several others have reported that mycoplasmas in cell cultures are able to influence HIV replication, usually in an enhancing way [17].

During attempts to recover HIV from the peripheral blood lymphocytes of AIDS patients, the French investigators detected *M. fermentans*, *M. pirum* and *M. genitalium* in a small proportion of cell cultures that had been inoculated with their lymphocytes [74]. *M. pirum* is a mycoplasma that, hitherto, had been recovered only from cell cultures [22]. *M. genitalium* has been associated with non-gonococcal urethritis in men [43] and was detected in the blood of male chimpanzees after intraurethral inoculation [109], so that invasion of the human bloodstream is possible.

Further observations

Whether mycoplasmas occur more frequently and are more invasive in HIV-positive individuals than in those without HIV cannot be deduced from any of the observations discussed above. However, Dawson *et al.* [20] reported that after PCR amplification they detected *M. fermentans* DNA in the urine of 10 (23%) of 43 HIV-positive patients being treated for AIDS but not in urine from 50 HIV-negative healthy military and civilian personnel; other investigators [16] came to similar conclusions using a culture method. Furthermore, Hawkins and colleagues [41] used the PCR and detected *M. fermentans* in the blood of about 9% of HIV-positive patients and, subsequent to their original observations, Lo [59] disclosed similar results. Interest in the mycoplasma–AIDS topic was stimulated further by Bauer *et al.* [7] who noted a strong association between *M. fermentans* and AIDS nephropathy. They studied 200 AIDS patients, 20 of whom had histopathological changes in the kidney characteristic of AIDS-associated nephropathy. Of these 20, 15 were examined and all had evidence of mycoplasmal antigen in the kidney, whereas antigen was not found in the kidneys of 15 AIDS patients who had normal renal histology, nor in those of five patients dying of non-AIDS related diseases.

Observations in London

Two observations relevant to the subject of mycoplasmas and HIV infection have been made. First, an asymptomatic HIV-antibody-positive patient was found to have focal glomerulosclerosis and *M. fermentans* DNA in a kidney biopsy, although he was not immunodeficient at the time. Some 15 months later, the mycoplasma was detected in three sites, namely peripheral blood lymphocytes, urine and the throat. Three months after this, the CD4 count had fallen and the patient developed pneumonia due to *Pneumocystis carinii*. These data strengthen the association between *M. fermentans* and HIV-induced nephropathy and illustrate that *M. fermentans* may precede rapid progression to AIDS [1].

Second, in view of the original concept of *M. fermentans* as a mycoplasma having a predilection for the genital tract, it was interesting to find it by a PCR technique in the throat of 23%, the peripheral blood lymphocytes of 10% and the urine of 7% of homosexual HIV-positive men attending the genitourinary medicine clinic at St Mary's Hospital. However, *M. fermentans* was found also in about the same proportions of similar specimens from HIV-negative subjects, predominantly homosexual, attending the same clinic [47]. Furthermore, there was no significant correlation between the occurrence of *M. fermentans* and the severity of HIV-related disease in the HIV-positive subjects, nor any correlation between the presence of the mycoplasma and the circulating viral load.

Mechanisms of possible mycoplasmal involvement in AIDS

While our observations may be somewhat negative in terms of defining a role for *M. fermentans* in AIDS, there are at least theoretical grounds for believing that some role might exist. Thus, it is possible that mycoplasmas flourish in the immunocompromised state caused by HIV, and so behave as an opportunistic invader. Furthermore, some mycoplasmas may cause immunosuppression in their own right, thus contributing to the immunodeficiency.

Dissemination and persistence of mycoplasmas in immunodeficient patients is mainly confined to those with severe primary antibody deficiency, and antibody production is generally preserved in HIV-infected individuals until the terminal stages of AIDS. In fact, there is good evidence of polyclonal B-lymphocyte activation in HIV-positive patients [90], and such unregulated antibody production might play some role in the associated nephropathy [91]. The finding that T-cell-deficient athymic nude mice have a more severe arthritis and an increased microbial load following parenteral injection of *M. pulmonis*, as compared with normal mice, suggests that cellular immunity is important in controlling infection from some mycoplasma strains [50]. However, specific antibody production was markedly delayed in the athymic mice. Nevertheless, antibody is clearly not sufficient in this murine model to eliminate *M. pulmonis*, and it is likely that T-cells are required to activate macrophages for killing of these organisms, in a similar way to that described for Listeria in another murine model [6]. *M. fermentans* in humans may have similar requirements for eradication, and be more resistant to the growth inhibitory effects of antibody as compared with *Ureaplasma urealyticum*, *M. hominis* and *M. pneumoniae*. It would also be useful to know whether *M. fermentans*, and the other mycoplasmas associated with HIV-positive patients will directly activate complement, as has been shown for *M. pneumoniae*, *M. hominis* and *M. salivarium* [118]. If they do, then one would expect spontaneous phagocytosis by neutrophils and monocytes in the circulation.

It is possible that by stimulating lymphocyte proliferation, mycoplasmas could enhance the replication of HIV and so speed the progression of AIDS. The way forward is to establish whether the existence of *M. fermentans* in the blood is transient, intermittent or persistent, as it is possible that the organism persists in AIDS patients but not in HIV-negative individuals. Further, our observations and those of others do not exclude the possibility that the possession of *M. fermentans* results in a more rapid progression to AIDS. To follow for years patients who do and do not possess mycoplasmas is a daunting endeavour, but the question may be answered sooner by determining the existence of *M. fermentans* in 'slow' and 'fast' progressors. Finally, whether other mycoplasmas, such as *M. genitalium*, *M. pirum* and *M. penetrans* flourish in HIV-positive patients and have any role in AIDS pathogenesis needs to be examined in the same way as for *M. fermentans*.

Observations on *M. penetrans*

This mycoplasma of human origin, the most recent to be identified, was isolated from the urine of about 5% of HIV-positive individuals, but not from that of HIV-negative, age-matched, healthy 'volunteers' [60]. Antibody to this mycoplasma, measured by an enzyme-linked immunosorbent assay, was found in 20–40% of HIV-positive individuals, but in only 0.3–1% of HIV-negative individuals [114], an observation which suggested that it was uniquely HIV-associated. However, this organism may only colonize sites such as the rectum, where it may not stimulate an antibody response, but then through sexual activity spread to the urethra and other sites to stimulate antibody [102]. Further work is now needed to clarify whether infection with these new mycoplasmas is related to homosexual activity rather than HIV infection, although the latter may compromise the ability of macrophages to control the microbial load.

19.5 REFERENCES

1. Ainsworth, J.G., Katseni, V., Hourshid, S. *et al.* (1995) Mycoplasma fermentans and HIV-associated nephropathy. *J. Infect.* (in press).
2. Asherson, G.L. and Webster, A.D.B. (1980) *Diagnosis and Treatment of Immunodeficiency Diseases*, Blackwell Scientific Publications, Oxford.

3. Aurelius, E., Johansson, B., Skoldenberg, B., Staland, A. and Forsgren, M. (1991) Rapid diagnosis of herpes simplex encephalitis by nested polymerase chain reaction assay of cerebrospinal fluid, *Lancet*, **337**, 189–92.

4. Balfour, H.H.Jr (1991) Varicella-zoster virus infections in the immuno-compromised host. Natural history and treatment. *Scand. J. Infect. Dis.*, **80**, 69–74.

5. Balfour, H.H.Jr, Chace, B.A., Stapleton, J.T., Simmons, R.L. and Fryd, D.S. (1989) A randomized, placebo-controlled trial of oral acyclovir for the prevention of cytomegalovirus disease in recipients of renal allografts. *N. Engl. J. Med.*, **320**, 1381–7.

6. Bancroft, G.J., Schreiber, R.D. and Unanue, E.R. (1991) Natural immunity: a T-cell independent pathway of macrophage activation, defined in the SCID mouse. *Immunol. Rev.*, **124**, 5–24.

7. Bauer, F.A., Wear, D.J., Angritt, P. and Lo, S.-C. (1991) *Mycoplasma fermentans* (incognitus strain) infection in the kidneys of patients with acquired immunodeficiency syndrome and associated nephropathy. *Hum. Pathol.*, **22**, 63–9.

8. Berry, N.J., Grundy, J.E. and Griffiths, P.D. (1987) Radioimmunoassay for the detection of IgG antibodies to herpes simplex virus and its use as a prognostic indicator of HSV excretion in transplant recipients. *J. Med. Virol.*, **21**, 147–54.

9. Bowden, R.A., Slichter, S.J., Sayers, M.H., Mori, M., Cays, M.J. and Meyers, J.D. (1991) Use of leukocyte-depleted platelets and cytomegalovirus-seronegative red blood cells for prevention of primary cytomegalovirus infection after marrow transplant, *Blood*, **78**, 246–50.

10. Bozzette, S.A., Arcia, J., Bartok, A.E. *et al.* (1992) Impact of *Pneumocystis carinii* and cytomegalovirus on the course and outcome of atypical pneumonia in advanced human immunodeficiency virus disease. *J. Infect. Dis.*, **165**, 93–8.

11. Bryant, A., Calver, N.C., Toubi, E., Webster, A.D.B. and Farrant, J. (1990) Classification of patients with common variable immunodeficiency by B cell secretion of IgM and IgG in response to anti-IgM and interleukin-2. *Clin. Immunol. Immunopathol.*, **56**, 239–48.

12. Burdge, D.R., Reid, G.D., Reeve, C.E., Robertson, J.A., Stemcke, G.W. and Bowie, W.R. (1988) Septic arthritis due to dual infection with *Mycoplasma hominis* and *Ureaplasma urealyticum*. *J. Rheumatol.*, **15**, 366–8.

13. Buzi, F., Notarangelo, L.D., Plebani, A. *et al.* (1994) X-linked agammaglobulinemia, growth hormone deficiency and delay of growth and puberty. *Acta Paediatr. Scand.*, **83**, 99–102.

14. Cagle, P.T., Buffone, G., Holland, V.A., Samo, T., Demmler, G.J. and Noon, G.P. (1992) Semiquantitative measurement of cytomegalovirus DNA in lung and heart–lung transplant patients by in vitro DNA amplification. *Chest*, **101**, 93–6.

15. Carrigan, D.R., Drobyski, W.R., Russler, S.K., Tapper, M.A., Knox, K.K. and Ash, R.C. (1991) Interstitial pneumonitis associated with human herpesvirus-6 infection after marrow transplantation, *Lancet*, **338**, 147–9.

16. Chirgwin, K.D., Cummings, M.C., DeMeo, L.R., Murphy, M. and McCormack, W.M. (1993) Identification of mycoplasmas in urine from persons infected with human immunodeficiency virus. *Clin. Infect. Dis.*, **17** (suppl. 1), 264–6.

17. Chowdhury, M.I.H., Munakata, T., Koyanaki, Y., Kobayashi, S., Arai, S. and Yamamoto, N. (1990) Mycoplasma can enhance HIV replication in vitro: a possible co-factor responsible for the progression of AIDS. *Biochem. Biophys. Res. Commun.*, **170**, 1365–70.

18. Cimolai, N. (1992) *Mycoplasma pneumoniae* in the immunocompromised host. *Chest*, **102**, 1303–4.

19. Cserr, H.F. and Knopf, P.M. (1992) Cervical lymphatics, the blood–brain barrier and the immunoreactivity of the brain: a new view. *Immunology Today*, **13**, 507–9.

20. Dawson, M.S., Hayes, M.M., Wang, R.Y.-H., Armstrong, D., Kundsin, R.B. and Lo, S.-C. (1993) Detection and isolation of *Mycoplasma fermentans* from urine of human immunodeficiency virus type 1-infected patients. *Arch. Pathol. Lab. Med.*, **117**, 511–14 .

21. Dayan, A.D. (1971) Chronic encephalitis in children with severe immuno-deficiency. *Acta Neuropathol.*, **19**, 234–41.

22. Del Giudice, R.A., Tully, J.G., Rose, D.L. and Cole, R.M. (1985) *Mycoplasma pirum* sp. nov., a terminal structured mollicute from cell cultures. *Int. J. Syst. Bacteriol.*, **35**, 285–91

23. Duriez, B., Duquesnoy, P., Dastot, F., Bougneres, P., Amselem, S. and Goossens, M. (1994) An exon skipping mutation in the btk gene of a patient with X-linked agammaglobulinemia and isolated growth hormone deficiency. *FEBS Lett.*, **346**, 165–70.

24. Dwyer, J.M. and Erlendsson, K. (1988) Intraventricular gamma-globulin for the management of enterovirus encephalitis. *Pediatr. Infect. Dis. J.*, **7**, S30–3.

25. Emanuel, D., Cunningham, I., Jules-Elysee, K. *et al.* (1988) Cytomegalovirus pneumonia after bone marrow transplantation successfully treated with the combination of ganciclovir and high-dose intravenous immune globulin. *Ann. Intern. Med.*, **109**, 777–82.

26. Evengard, B., Sandstedt, K., Bölske, G., Feinstein, R., Riesenfelt-Orn, I. and Smith, C.I.E. (1994) Intranasal inoculation of *Mycoplasma pulmonis* in mice with severe combined immunodeficiency *(scid)* causing a wasting disease with grave arthritis. *Clin. Exp. Clin. Immunol.*, **98**, 388–94.

27. Forgacs, P., Kundsin, R.B., Margles, S.W., Silverman, M.L. and Perkins, R.E. (1993) A case of *Ureaplasma urealyticum* septic arthritis in a patient with hypogammaglobulinaemia. *Clin. Infect. Dis.*, **16**, 293–4.

28. Foy, H.M., Kenny, G.E., Sefi, R., Ochs, H.D. and Allan, I.D. (1977) Second attacks of pneumonia due to *Mycoplasma pneumoniae*. *J. Infect. Dis.*, **135**, 673–7.

29. Foy, H.M., Ochs, H., Davis, S.D., Kenny, G.E. and Luce, R.R. (1973) *Mycoplasma pneumoniae* infections in patients with immunodeficiency syndromes: report of four cases. *J. Infect. Dis.*, **127**, 388–93 .

30. Franz, A., Furr, P.M., Taylor-Robinson, D. and Webster, A.D.B. (1995) Mycoplasma arthritis in primary immunodeficiency: diagnosis and management. (submitted) .

31. Furr, P.M. and Taylor-Robinson, D. (1987) Prevalence and significance of *Mycoplasma hominis* and *Ureaplasma urealyticum* in the urines of a non-venereal disease population. *Epidemiol. Infect.*, **98**, 353–9.

32. Furr, P.M., Taylor-Robinson, D. and Webster, A.D.B. (1994) Mycoplasmas and ureaplasmas in patients with hypogammaglobulinaemia and their role in arthritis: microbiological observations over twenty years. *Ann. Rheum. Dis.*, **53**, 183–7.

33. Gelfand, E.W. (1993) Unique susceptibility of patients with antibody deficiency to mycoplasma infection. *Clin. Infect. Dis.*, **17** (suppl. 1), 250–3.

34. Ginsberg, H.S., Valdesuso, J., Horswood, R., Chanock, R.M. and Prince, G. (1987) Adenovirus gene products affecting pathogenesis, in *Vaccines*, (eds R.M. Chanock, R.A. Lerner, F. Brown and H.S. Ginsberg), Cold Spring Harbor Laboratory, New York, pp. 322–6.

35. Goodrich, J.M., Bowden, R.A., Fisher, L., Keller, C., Schoch, G. and Meyers, J.D. (1993) Ganciclovir prophylaxis to prevent cytomegalovirus disease after allogeneic marrow transplant. *Ann. Intern. Med.*, **118**, 173–8.

36. Goodrich, J.M., Mori, M., Gleaves, C.A. *et al.* (1991) Early treatment with ganciclovir to prevent cytomegalovirus disease after allogeneic bone marrow transplantation. *N. Engl. J. Med.*, **325**, 1601–7.

37. Gotoff, S.P., Smith, R.D. and Sugar, O. (1972) Dermatomyositis with cerebral vasculitis in a patient with agammaglobulinemia. *Am. J. Dis. Child.*, **123**, 53–6.

38. Griffiss, J.M. (1983) Biologic function of the serum IgA system modulation of complement mediated effector mechanisms and conservation of antigenic mass. *Ann. N.Y. Acad. Sci.*, **409**, 697–707.

39. Grundy, J.E., Shanley, J.D. and Griffiths, P.D. (1987) Is cytomegalovirus interstitial pneumonitis in transplant recipients an immunopathological condition? *Lancet*, **ii**, 996–9.

40. Haller, M., Forst, H., Ruckdeschel, G., Denecke, H. and Peter, K. (1991) Peritonitis due to *Mycoplasma hominis* and *Ureaplasma urealyticum* in a liver transplant recipient. *Eur. J. Clin. Microbiol. Infect. Dis.*, **10**, 172.

41. Hawkins, R.E., Rickman, L.S., Vermund, S.H. and Carl, M. (1992) Association of mycoplasmas and human immunodeficiency virus infection: detection of amplified *M. fermentans* DNA in blood. *J. Infect. Dis.*, **165**, 581–5.

42. Hermaszewski, R.A. and Webster, A.D.B. (1993) Primary hypogammaglobulinaemia: a survey of clinical manifestations and complications. *Q. J. Med.*, **86**, 31–42.

43. Horner, P.J., Gilroy, C.B., Thomas, B.J., Naidoo, R.O.M. and Taylor-Robinson, D. (1993) Association of *Mycoplasma genitalium* with acute non-gonococcal urethritis. *Lancet*, **342**, 582–5.

44. Johnston, C.L.W., Webster, A.D.B., Taylor-Robinson, D., Rapaport, G. and Hughes, G.R.V. (1983) Primary late-onset common variable hypogammaglobulinaemia associated with

inflammatory polyarthritis and septic arthritis due to *Mycoplasma pneumoniae. Ann. Rheum. Dis.*, **42**, 108–10.

45. Jorup-Rönström, C., Ahl, T., Hammarström, L., Smith, C.I.E., Rylander, M. and Hallander, H. (1989) Septic osteomyelitis and polyarthritis with ureaplasma in hypogammaglobulinaemia. *Infection*, **17**, 301–3.

46. Joseph, S.G. and Oser, B. (1993) Complications of varicella pneumonia in adults. *J. Am. Osteopath. Assoc.*, **93**, 941–2.

47. Katseni, V.L., Gilroy, C.B., Ryait, B.K. *et al.* (1993) *Mycoplasma fermentans* in individuals seropositive and seronegative for HIV-1. *Lancet*, **341**, 271–3.

48. Keat, A., Thomas, B., Dixey, J., Osborn, M., Sonnex, C. and Taylor-Robinson, D. (1987) *Chlamydia trachomatis* and reactive arthritis: the missing link. *Lancet*, **i**, 72–4.

49. Kennedy, C.R., Robinson, R.O., Valman, H.B., Chrzanowska, K., Tyrrell, D.A.J. and Webster, A.D.B. (1986) A major role for viruses in acute childhood encephalopathy. *Lancet*, **i**, 989–91.

50. Keystone, E.C., Taylor-Robinson, D., Osborn, M.F., Ling, I., Pope, C. and Fornasier, V. (1980) Effect of T-cell deficiency on the chronicity of arthritis induced in mice by *Mycoplasma pulmonis. Infect. Immun.*, **27**, 192–6.

51. Kidd, I.M., Fox, J.C., Pillay, D., Charman, H., Griffiths, P.D. and Emery, V.C. (1993) Provision of prognostic information in immunocompromised patients by routine application of the polymerase chain reaction for cytomegalovirus. *Transplantation*, **56**, 867–71.

52. Klapper, P.E., Cleator, G.M., Dennett, C. and Lewis, A.G. (1990) Diagnosis of herpes encephalitis via Southern blotting of cerebrospinal fluid DNA amplified by polymerase chain reaction. *J. Med. Virol.*, **32**, 261–4.

53. Kraus, V.B., Baraniuk, J.N., Hill, G.B. and Allen, N.B. (1988) *Ureaplasma urealyticum* septic arthritis in hypogammaglobulinaemia. *J. Rheumatol.*, **15**, 369–71.

54. Lee, A.H., Ramanujam, T., Ware, P. *et al.* (1992) Molecular diagnosis of *Ureaplasma urealyticum* septic arthritis in a patient with hypogammaglobulinemia. *Arth. Rheum.*, **35**, 443–8.

55. Lee, S.-G. and Hung, P.P. (1993) Vaccines for control of respiratory disease caused by adenoviruses. *Rev. Med. Virol.*, **3**(4), 209–16.

56. Lehmer, R.R., Andrews, B.S., Robertson, J.A., Stanbridge, E.J., de la Maza, L. and Friou, G.J. (1991) Clinical and biological characteristics of *Ureaplasma urealyticum* induced polyarthritis in a patient with common variable hypogammaglobulinaemia. *Ann. Rheum. Dis.*, **50**, 574–6.

57. Lemaître, M., Guétard, D., Hénin, Y., Montagnier, L. and Zerial, A. (1990) Protective activity of tetracycline analogs against the cytopathic effect of the human immunodeficiency viruses in CEM cells. *Res. Virol.*, **141**, 5–16.

58. Lepage, P., van de Perre, P., van Vliet, G. *et al.* (1991) Clinical and endocrinologic manifestations in perinatally human immunodeficiency virus type 1 – infected children aged 5 years or older. *Am. J. Dis. Child.*, **145**, 1248–51.

59. Lo, S.-C. (1992) Mycoplasmas and AIDS, in *Mycoplasmas: Molecular Biology and Pathogenesis*, (eds J. Maniloff, R.N. McElhaney, L.R. Finch and J.B. Baseman), American Society for Microbiology, Washington, DC, pp. 525–45.

60. Lo, S.-C., Hayes, M.M., Wang, R.Y.-H., Pierce, P.F., Kotani, H. and Shih, J.W.-K. (1991) Newly discovered mycoplasma isolated from patients infected with HIV. *Lancet*, **338**, 1415–18.

61. Lo, S.-C., Shih, J.W.-K., Newton, P.B. *et al.* (1989) Virus-like infectious agent (VLIA) is a novel pathogenic mycoplasma: *Mycoplasma incognitus. Am. J. Trop. Med. Hyg.*, **41**, 586–600.

62. Lopez, C., Biggar, W.D., Park, B.H. and Good, R.A. (1974) Nonparalytic poliovirus infections in patients with severe combined immunodeficiency disease. *J. Pediatr.*, **84**, 497–502.

63. Lusso, P. and Gallo, R.C. (1994) Human herpesvirus 6 in AIDS. *Lancet*, **343**, 555–6.

64. Madoff, S. and Hooper, D.C. (1988) Non-genitourinary infections caused by *Mycoplasma hominis* in adults. *Rev. Infect. Dis.*, **10**, 602–13.

65. McKinney, R.E. Jr, Katz. S.L. and Wilfert, C.M. (1987) Chronic enteroviral meningoencephalitis in agammaglobulinemic patients. *Rev. Infect. Dis.*, **9**, 334–56.

66. Mease, P.J., Ochs, H.D. and Wedgwood, R.J. (1981) Successful treatment of echovirus meningoencephalitis and myositis-fasciitis with intravenous immune globulin therapy in a patient with X-linked agammaglobulinemia. *N. Engl. J. Med.*, **304**, 1278–81.

67. Medical Research Council Special Report Series (1971) *Hypogammaglobulinaemia in the United Kingdom*, 310 pp.

68. Meyer, R.D. and Clough, W. (1993) Extragenital *Mycoplasma hominis* infections in adults:

emphasis on immunosuppression. *Clin. Infect. Dis.*, **17** (suppl. 1), 243–9.

69. Meyers, J.D., Reed, E.C., Shepp, D.H. *et al.* (1988) Acyclovir for prevention of cytomegalovirus infection and disease after allogeneic marrow transplantation. *N. Engl. J. Med.*, **318**, 70–5.

70. Michaels, M.G., Green, M., Wald, E.R. and Starzl, T.E. (1992) Adenovirus infection in pediatric liver transplant recipients. *J. Infect. Dis.*, **165**, 170–4.

71. Misbah, S.A., Lawrence, P.A., Kurtz, J.B. and Chapel, H.M. (1991) Prolonged faecal excretion of poliovirus in a nurse with common variable hypogammaglobulinaemia. *Postgrad. Med. J.*, **67**, 301–3.

72. Misbah, S.A., Spickett, G.P., Ryba, P.C.J. *et al.* (1992) Chronic enteroviral meningoencephalitis in agammaglobulinemia: case report and literature review. *J. Clin. Immunol.*, **12**, 266–70

73. Mohiuddin, A.A., Corren, J., Harbeck, R.J., Teague, J.L., Volz, M. and Gelfand, E.W. (1991) *Ureaplasma urealyticum* chronic osteomyelitis in a patient with hypogammaglobulinemia. *J. Allergy Clin. Immunol.*, **87**, 104–7.

74. Montagnier, L. and Blanchard, A. (1993) Mycoplasmas as co-factors in infection due to the human immunodeficiency virus. *Clin. Infect. Dis.*, **17** (suppl. 1), 309–15.

75. Nozaki-Renard, J., Iino, T., Sato, Y., Marumoto, Y., Ohta, G. and Furusawa, M. (1990) A fluoroquinolone (DR-3355) protects human lymphocyte cell lines from HIV-l induced cytotoxicity. *AIDS*, **4**, 1283–6.

76. Pacini, D.L., Dubovi, E.J. and Clyde, W.A.Jr (1984) A new animal model for human respiratory tract disease due to adenovirus. *J. Infect. Dis.*, **150**, 92–7.

77. Palmer, H.M., Gilroy, C.B., Furr, P.M. and Taylor-Robinson, D. (1991) Development and evaluation of the polymerase chain reaction to detect *Mycoplasma genitalium*. *FEMS Microbiol. Lett.*, **77**, 199–204.

78. Parides, G.C., Bloom, J.W., Ampel, N.M. and Ray, C.G. (1988) Mycoplasmas and ureaplasmas in broncho-alveolar lavage fluids from immunocompromised hosts. *Diagn. Microbiol. Infect. Dis.*, **9**, 55–7.

79. Pecego, R., Hill, R., Appelbaum, F.R. *et al.* (1986) Interstitial pneumonitis following autologous bone marrow transplantation, *Transplantation*. **42**, 515–17.

80. Petitjean, J., Quibriac, M., Lechevalier, B., Brouard, J. and Freymuth, F. (1988) Infections respiratoires à *Mycoplasma pneumoniae* au cours du syndrome d'immuno-déficience acquise. *Presse Med.*, **17**, 1762.

81. Ramsey, P.G., Fife, K.H., Hackman, R.C., Meyers, J.D. and Corey, L. (1982) Herpes simplex virus pneumonia: clinical, virologic, and pathologic features in 20 patients. *Ann. Intern. Med.*, **97**, 813–20 .

82. Reed, E.C., Bowden, R.A., Dandliker, P.S., Lilleby, K.E. and Meyers, J.D. (1988) Treatment of cytomegalovirus pneumonia with ganciclovir and intravenous cytomegalovirus immunoglobulin in patients with bone marrow transplants. *Ann. Intern. Med.*, **109**, 783–8 .

83. Renton, P. and Webster, A.D.B. (1978) Mycoplasma osteomyelitis and pneumonia associated with primary hypogammaglobulinaemia. *Skeletal Radiol.*, **3**, 131–2.

84. Roifman, C.M., Rao, C.P., Lederman, H.M., Lavi, S., Quinn, P. and Gelfand, E.W. (1986) Increased susceptibility to *Mycoplasma* infection in patients with hypogammaglobulinemia. *Am. J. Med.*, **80**, 590–4.

85. Rotbart, H.A. (1990) Enzymatic RNA amplification of the enteroviruses. *J. Clin. Microbiol.*, **28**, 438–42.

86. Rudge, P., Webster, A.D.B., Warner, T., Espanol, T., Cunningham-Rundles, C. and Hyman, N. (1995) Encephalomyelitis in primary hypogammaglobulinaemia. (submitted)

87. Ruutu, P., Ruutu, T., Volin, L., Tukiainen, P., Ukkonen, P. and Hovi, T. (1990) Cytomegalovirus is frequently isolated in bronchoalveolar lavage fluid of bone marrow transplant recipients without pneumonia. *Ann. Intern. Med.*, **112**, 913–16 .

88. Saillard, C., Carle, P., Bové, J.M. *et al.* (1990) Genetic and serologic relatedness between *Mycoplasma fermentans* strains and a mycoplasma recently identified in tissues of AIDS and non-AIDS patients. *Res. Virol.*, **141**, 385–96.

89. Schmidt, G.M., Horak, D.A., Niland, J.C., Duncan, S.R., Forman, S.J. and Zaia, J.A. (1991) A randomized, controlled trial of prophylactic ganciclovir for cytomegalovirus pulmonary infection in recipients of allogeneic bone marrow transplants; The City of Hope–Stanford–Syntex CMV Study Group. *N. Engl. J. Med.*, **324**, 1005–11.

89. Schmittman, S.M., Lane, H.C., Higgins, S.E., Folks, T. and Fauci, A.S. (1986) Direct polyclonal activation of human B lymphocytes by the acquired immunodeficiency syndrome virus. *Science*, **233**, 1084–6.

90. Schoenfeld, P. and Feduska, N.J. (1990) Acquired immunodeficiency syndrome and renal disease: Report of the National Kidney Foundation – National Institutes of Health Task Force on AIDS and Kidney Disease. *Am. J. Kidney Dis.*, **16**, 14–25.

91. Selby, P.J., Powles, R.L., Easton, D. *et al.* (1989) The prophylactic role of intravenous and long-term oral acyclovir after allogeneic bone marrow transplantation. *Br. J. Cancer*, **59**, 434–8.

92. Shepard, M.C., Lunceford, C.D., Ford, D.K. *et al.* (1974) *Ureaplasma urealyticum* gen. nov. spp. nov. Proposed nomenclature for the human (T-strain) mycoplasma. *Int. J. Syst. Bacteriol.*, **24**, 160–71.

93. Shepp, D.H., Dandliker, P.S., de Miranda, P. *et al.* (1985) Activity of 9-[2-hydroxy-1-(hydroxymethyl) ethoxymethyl]guanine in the treatment of cytomegalovirus pneumonia. *Ann. Intern. Med.*, **103**, 368–73.

94. So, A.K.L., Furr, P.M., Taylor-Robinson, D. and Webster, A.D.B. (1983) Arthritis caused by *Mycoplasma salivarium* in hypogammaglobulinaemia. *Br. Med. J.*, **286**, 762–3.

95. Somerson, N.L. and Cole, B.C. (1979) The mycoplasma flora of human and nonhuman primates, in *The Mycoplasmas, Vol. II, Human and Animal Mycoplasmas*, (eds J.G. Tully and R.F. Whitcomb), Academic Press, New York , pp. 191–216.

96. Squire, S.B., Lipman, M.C., Bagdades, E.K. *et al.* (1992) Severe cytomegalovirus pneumonitis in HIV infected patients with higher than average CD4 counts. *Thorax*, **47**, 301–4.

97. Straus, S.E., Corey, L., Burke, R.L. *et al.* (1994) Placebo-controlled trial of vaccination with recombinant glycoprotein D of herpes simplex virus type 2 for immunotherapy of genital herpes. *Lancet* , **343**, 1460–3.

98. Stuckey, M., Quinn, P.A. and Gelfand, E.W. (1978) Identification of *Ureaplasma urealyticum* in a patient with polyarthritis. *Lancet*, **ii**, 917–20.

99. Taylor-Robinson, D. and Furr, P.M. (1981) Recovery and identification of human genital tract mycoplasmas. *Isr. J. Med. Sci.*, **17**, 648–53.

100. Taylor-Robinson, D. and Purcell, R.H. (1966) Mycoplasmas of the human urogenital tract and oropharynx and their possible role in disease: a review of some recent observations. *Proc. R. Soc. Med.*, **58**, 1112–16.

101. Taylor-Robinson, D. and Ainsworth, J. (1993) Antibodies *to Mycoplasma penetrans* in HIV-infected patients. *Lancet*, **341**, 557–8.

102. Taylor-Robinson, D., Davies, H.A., Sarathchandra, P. and Furr, P.M. (1991) Intracellular location of mycoplasmas in cultured cells demonstrated by immunocytochemistry and electron microscopy. *Int. J. Exp. Pathol.*, **72**, 705–14.

103. Taylor-Robinson, D., Furr, P.M. and Webster, A.D.B. (1986) *Ureaplasma urealyticum* infection in the immunocompromised host. *Pediatr. Infect. Dis.*, **5** (suppl.), 236–8.

104. Taylor-Robinson, D., Gumpel, J.M., Hill, A. and Swannell, A.J. (1978) Isolation of *Mycoplasma pneumoniae* from the synovial fluid of a hypogammaglobulinaemic patient in a survey of patients with inflammatory polyarthritis. *Ann. Rheum. Dis.*, **37**, 180–2.

105. Taylor-Robinson, D., Webster, A.D.B., Furr, P.M. and Asherson, G.L. (1980) Prolonged persistence of *Mycoplasma pneumoniae* in a patient with hypogammaglobulinaemia. *J. Infection*, **2**, 171–5.

106. Townsend, A.R., Rothbard, J., Gotch, F.M., Bahadur, G., Wraith, D. and McMichael, A.J. (1986) The epitopes of influenza nucleoprotein recognised by cytolytic lymphocytes can be defined with short synthetic peptides. *Cell*, **44**, 959–68.

107. Trinchieri, G. (1989) Biology of natural killer cells. *Adv. Immunol.*, **47**, 187–376.

108. Tully, J.G., Taylor-Robinson, D., Rose, D.L., Furr, P.M., Graham, C.E. and Barile, M.F. (1986) Urogenital challenge of primate species with *Mycoplasma genitalium* and characteristics of infection induced in chimpanzees. *J. Infect. Dis.*, **153**, 1046–54.

109. Tyor, W.R., Moench, T.R. and Griffin, D.E. (1989) Characterization of the local and systemic B cell response of normal and athymic nude mice with Sindbis virus encephalitis. *J. Neuroimmunol.*, **24**, 207–15.

110. Vogler, L.B., Waites, K.B., Wright, P.F., Perrin, J.M. and Cassell, G.H. (1985) *Ureaplasma urealyticum* polyarthritis in agammaglobulinemia. *Pediatr. Infect. Dis.*, **4**, 687–91.

111. Vorechovsky, I., Vetrie, D., Holland, J. *et al.* (1994) Isolation of cosmid and cDNA clones in

the region surrounding the BTK gene at Xq21.3-q22. *Genomics*, **21**, 517–24.

112. Wadell, G. (1984) Molecular epidemiology of human adenoviruses. *Curr. Top. Microbiol. Immunol.*, **110**, l91–220.

113. Wang, R.Y.-H., Shih, J.W.-K., Grandinetti, T. *et al.* (1992) High frequency of antibodies to *Mycoplasma penetrans* in HIV-infected patients. *Lancet*, **340**, 1312–16.

114. Webster, A., Blizzard, B., Pillay, D., Prentice, H.G., Pothecary, K. and Griffiths, P.D. (1993) Value of routine surveillance cultures for detection of CMV pneumonitis following bone marrow transplantation. *Bone Marrow Transpl.*, **12**, 477–81.

115. Webster, A.D.B. (1979) Infections in immunodeficient patients, *in Aspects of Slow and Persistent Virus Infections*, (ed. D.A.J. Tyrrell), Martinus Nijhoff for the Commission of the European Communities, pp. 255–266.

116. Webster, A.D.B. (1984) Echovirus disease in hypogammaglobulinaemic patients. *Baillières Clin. Rheumatol.*, **10**, 189–203.

117. Webster, A.D.B., Furr, P.M., Hughes-Jones, N.C., Gorick, B.D. and Taylor-Robinson, D. (1988) Critical dependence on antibody for defence against mycoplasmas. *Clin. Exp. Immunol.*, **71**, 383–7.

118. Webster, A.D.B., Rotbart, H.A., Warner, T., Rudge, P. and Hyman, N. (1993) Diagnosis of enterovirus brain disease in hypogammaglobulinemic patients by polymerase chain reaction. *Clin. Infect. Dis.*, **17**, 657–61.

119. Webster, A.D.B., Taylor-Robinson, D., Furr, P.M. and Asherson, G.L. (1978) Mycoplasma (ureaplasma) septic arthritis in hypogammaglobulinaemia. *Br. Med. J.*, **i**, 478–9.

120. Webster, A.D.B., Taylor-Robinson, D., Furr, P.M. and Asherson, G.L. (1982) Chronic cystitis and urethritis associated with ureaplasmal and mycoplasmal infection in primary hypogammaglobulinaemia. *Br. J. Urol.*, **54**, 287–91.

121. Webster, A.D.B., Tripp, J.H., Hayward, A.R. *et al.* (1978) Echovirus encephalitis and myositis in primary immunoglobulin deficiency. *Arch. Dis. Child.*, **53**, 33–7.

122. Whimbey, E. and Bodey, G.P. (1992) Viral pneumonia in the immunocompromised adult with neoplastic disease: the role of common community respiratory viruses. *Semin. Res. Infect.*, **7**, 122–31.

123. Wilfert, C.M., Buckley, R.H., Mohanakumar, T. *et al.* (1977) Persistent and fatal central nervous system echovirus infections in patients with agammaglobulinemia. *N. Engl. J. Med.*, **296**, 1485–9.

124. Winston, D.J., Ho, W.G., Bartoni, K., Du Mond, C., Ebeling, D.F. and Buhles, W.C. (1993) Ganciclovir prophylaxis of cytomegalovirus infection and disease in allogeneic bone marrow transplant recipients. Results of a placebo-controlled, double-blind trial. *Ann. Intern. Med.*, **118**, 179–84.

125. Winston, D.J., Ho, W.G. and Champlin, R.E. (1990) Cytomegalovirus infections after allogeneic bone marrow transplantation. *Rev. Infect. Dis.*, **12** (suppl. 7), S776–92.

126. Wright, P.F., Hatch, M.H., Kasselberg, A.G., Lowry, S.P., Wadlington, W.B. and Karzon, D.T. (1977) Vaccine-associated poliomyelitis in a child with sex-linked agammaglobulinaemia. *J. Pediatr.*, **91**, 408–12.

127. Wyatt, H.V. (1973) Poliomyelitis in hypogammaglobulinemics. *J. Infect. Dis.*, **128**, 802–6.

THE USE OF INTERFERONS IN RESPIRATORY VIRAL INFECTIONS

Geoff Scott

20.1 INTRODUCTION

It was at the National Institute of Medical Research, Mill Hill, in 1956–1957 when Alec Isaacs, with his postgraduate student Jean Lindenmann, made their profound discovery of interferon. Viral interference had been recognized for many years. Indeed, Edward Jenner observed that vaccination with cowpox sometimes did not 'take' when a subject had herpes infection. Definitive animal experiments illustrating the phenomenon were first reported in 1935 and by 1950, Henle [33] wrote an extensive review on the subject. While investigating viral interference with influenza viruses in 1957, Isaacs and Lindenmann [43,44] showed that fragments of chick chorioallantoic membrane secreted a substance into the tissue culture supernatant in response to killed influenza virus which was able to protect fresh pieces of membrane against live virus challenge. This was identified as a non-dialysable protein of molecular weight around 20 000 Da, resistant to pH 2. Recognizing that a substance produced by cells in response to virus could protect other cells was the genius of this experiment but the second brilliant move was to give the substance its name: 'interferon'. Legend has it that the name was suggested during afternoon tea while discussing the results with colleagues. The name was catchy, the first papers were clear and straight forward and the implications were exciting enough to catch the public's imagination and to be discussed in the lay press. Because many scientists were initially suspicious of the results, the substance was also named 'misinterpreton' for a time. However, over a number of years, as more and more groups found interferon activity in different cell–virus systems and then in animals in response to infection, the fact of the interferon response became inescapable.

The first demonstration that interferon was produced during a model influenza infection in mice was made by Isaacs and Hitchcock in 1960 [42] and in humans by Gresser and Dull in 1964 [21]. The peak of interferon coincided with decreased shedding of virus in the lungs

Viral and Other Infections of the Human Respiratory Tract.
Edited by S. Myint and D. Taylor-Robinson. Published in 1996 by Chapman & Hall. ISBN 0 412 60070 6

and as virus-specific antibody followed much later, interferon was thought to play a significant role in recovery of mice from this infection. A tantalizing observation by Baron and Isaacs in 1962 [6] was that extracts of lungs of patients who had died from fulminant influenzal pneumonia had no detectable interferon present (perhaps an observation anticipating the much later discovery of Mx influenza-susceptibility genes by Jean Lindenmann). It was well established that nasal washings from patients with experimental and natural colds and influenza contained interferon but secretion was rather variable in quantity and time. Some had interferon in the serum and this was more likely in those with a more severe influenzal illness [21,45,68]. The time of maximal interferon induction by live measles virus vaccine coincided with protection of volunteers against vaccinia virus [53].

In a mouse model with influenza A virus, Finter [15] showed that considerable amounts of mouse interferon given by injection had no effect on illness but there was a reduction of virus shedding if interferon was instilled intranasally. This led to the conclusion, never refuted, that interferon would best be given topically in the prevention or treatment of respiratory viral infection. Much work was performed through the 1960s and 1970s with interferon inducers but it has to be said that, although there were some tantalizing results from trials (e.g. with Poly 1: C against influenza A virus [39], or a substituted propanediamine against rhinovirus [52]), these did not yield a satisfactory product for clinical use.

The first clinical trial performed with interferon in human volunteers was against vaccinia virus. Intradermal injections of crude monkey kidney cell interferon protected skin sites against vaccinia virus challenge [58]. Because of enormous difficulties making sufficient interferon for clinical trials, it was almost 10 years before a trial could be done to show that interferon could protect volunteers against rhinovirus challenge.

Those buried in the cytokine network today might find it difficult to recognize that the unfolding of this network began with 25 years of hard labour with an intangible magical substance of very high specific activity which could not be properly purified. The most sophisticated techniques of protein separation in the early 1970s simply revealed all the interferon activity from a crude preparation to reside in a non-visible fraction of a gel (D.C. Burke, personal communication).

It was not until the late 1970s that non-chemical purification became possible by the manufacture of monoclonal antibodies and, perhaps more significantly, that individual human interferon genes were cloned. Supply problems were eased and it was then possible to repeat some of the earlier experiments in volunteers with more potent and purer preparations of interferon, in order to show that the properties of the earlier crude preparations were indeed due to interferon and not one of the other cytokines, interleukins and tumour necrosis factors with which we have now become familiar.

After 37 years of research, what do we now know about the interferon system? The key elements are those discovered in the 1950s – that interferons are proteins, some glycosylated, of molecular weight 10 000–100 000 Da, derived from cells in response to viral infection, which protect other cells against infection. There seems to be extraordinary re-duplication of the human alpha-interferon (IFN-α) gene system (on human chromosome 9) (the interferons derived from buffy coat leucocytes or lymphoblasts); there is one beta-interferon (IFN-β; from fibroblasts induced with polyI:polyC), a co-induced beta-2 interferon which has no antiviral activity but was found to be identical to hepatocyte stimulating factor and is now named IL-6; and gamma-interferon (initially distinguished by acid-lability), the gene of which is on Hu Chr 12, produced by T-lymphocytes in response to immune stimuli. Whereas IFN-α is a critical part of the innate immune response to acute virus infection, it is now thought that IFN-γ forms an important part of the learned immune response to infection with many different organisms, viral, bacterial and parasitic and this molecule has many immunological properties. All interferons are relatively species-specific: mouse interferon does not affect human cells and *vice versa*, but human interferon does have effects in monkey cells and human alpha interferons are surprisingly active in bovine cells. Their specific activity varies according to the cell–virus assay system used, and there is some dissociation between biological and immunoreactive activity. Something is known about the stimulation and control of interferon gene expression and about the response element: the induction of intracellular proteins and effects on virus replication. The body of knowledge has grown progressively since the discovery of interferon as a direct result of each new discovery in cellular biochemistry methods. Interferons modify the immune response and inhibit cell growth through a series of complex and interactive pathways and some of the paradoxes which will be seen in the effects of interferons on upper respiratory infections may reflect these processes. A useful recent review of current clinical applications is found in [7].

The discovery of viruses which cause common colds and influenza is described in other chapters. Much of this work was progressing as the interferon story began to unfold. It was hoped that because interferon was species- or cell-specific but not virus-specific, it would have a broad spectrum of activity against various viruses which were being discovered to cause the common cold, and, of course, against influenza viruses. However, it has proved less easy to predict *in vivo* activity of interferons against viruses from *in vitro* activity, than it has been for antibacterial agents. This is partly because of the unique effect of interferons on individual cell–virus systems which has led to a difficulty in establishing a

standard value for interferon activity. Activity tends to be measured in the most sensitive cell line using the most sensitive virus. When measuring interferon activity against a rhinovirus it has to be appreciated that this virus grows slowly and may take a week to cause a cytopathic effect; there is an analogy with the difficulty in establishing the activity of antimicrobials against slow-growing bacteria.

In the end, it was against natural rhinovirus infections that topical leucocyte interferon seemed to work best in the field, but initially rhinoviruses were not considered particularly susceptible to interferon *in vitro*. This was certainly so by comparison with influenza viruses. Similarly, although in volunteer trials, topical interferon could prevent colds caused by a number of agents including influenza virus and coronaviruses, natural colds with these agents were not apparently prevented in the definitive field trials, an observation which has not been satisfactorily explained.

The first attempts to prevent experimental colds were reported in 1965 by the UK Scientific Committee on Interferon [59]. In retrospect, very small doses of monkey kidney cell interferon were given only to a few volunteers, so it is not surprising that no effects were demonstrated. It was not until 1972 that the first successful experiment was done.

20.2 INTERFERON PRODUCTION FOR CLINICAL TRIALS

Early clinical work was done with leucocyte interferon derived from buffy coats taken from blood for transfusion stimulated with Sendai virus (n (natural) IFN-α Le (leucocyte)). In Finland, Kari Cantell pioneered the production of this interferon on a large scale and supplied material for virtually all the early controlled clinical trials. nIFN-α consists of many different subtypes which can be differentiated by antigenicity, cell species selectivity and nucleic acid chemistry. In the mid-1970s, two groups made fibroblast interferon (nIFN-β) from human embryo lung fibroblasts stimulated by PolyI:polyC. This material was found to be rather unstable. At the same time Finter, in the UK, started the development of large scale culture of human lymphoblasts from a Burkitt lymphoma, stimulated with Sendai virus (nIFN-αLy). These cells produce about 5% fibroblast interferon which is probably lost during purification.

In fact, most studies in the 1970s were done with very impure interferon preparations. It was never possible to make sufficiently large amounts of pure IFN-α to satisfy clinical trial demand until recombinant technology became available. However, towards the end of the decade, monoclonal antibody technology was discovered. A mouse monoclonal antibody (NK2) to components of nIFN-αLe was used to purify Cantell's material which was then tested in volunteers.

The most prominent species in a mixture of IFNs derived from leucocytes or lymphocytes is IFN-α_2 (Schering, or IFN-αA (Roche)). The gene for this protein was one of the first human genes cloned. It was inserted into *Escherichia coli* and the interferon produced in large quantities was then purified and finally administered to human volunteers – all within a period of some months. It was natural after the exciting volunteer studies with nIFN-αLe that these experiments should be repeated with the new material. Several species of IFN-α_2 were cloned by different workers. They have minor amino acid differences and are available as rIFN-α_{2a}, rIFN-α_{2b} and rIFN-α_{2c}. Recombinant gamma (rIFN-γ) and recombinant fibroblast interferons were also made, the latter containing an extra serine molecule (rIFN-βser) which conferred improved stability.

20.3 THE FIRST CLINICAL TRIALS AGAINST RHINOVIRUSES

The importance of the randomized double-blind trial design was clearly recognized during the evolution of the UK Medical Research Council Common Cold Unit (CCU) in Salisbury, UK. This design was used for all of the key interferon studies described below. The clinical observers were quite independent of those who gave the treatment. The only stratification necessary was according to pre-trial antibody titre which was measured during the volunteer quarantine period. Treatments were in general given by a different physician to those who arranged the random allocation and the code was not broken until all the data had been collected. When delivering interferon by spray, four spray guns often were used, two of each having active interferon and two placebo, so that the risk of volunteers and investigators discovering the code was reduced. When drops or self-administered nasal sprays were used, volunteers could have their own coded drug.

Leucocyte interferon nIFN-αLe was produced by Cantell for early trials of human interferon at the CCU. Tyrrell and Reed [67] first tested this preparation against rhinovirus 2. The total dose was about 2×10^5 tu in divided doses from 20 hours before until 43 hours after virus challenge, but this had no effect. Tom Merigan then came on sabbatical leave for a year from Stanford University. He found the *in vitro* sensitivities of rhinovirus 2 and influenza B virus to be about comparable with that of vesicular stomatitis virus in primary monkey kidney cells. Although influenza B virus was least sensitive in these cells, it seemed to be exquisitely sensitive in human fetal tracheal organ cultures.

In a trial in which volunteers were given a total of 8×10^5 tu interferon on the day before and the day after influenza B virus challenge, there was no effect on virus shedding, seroconversion or symptoms compared with those given placebo [49]. A slight delay in the onset of symptoms in the treated group, together with the results of previous *in vitro* studies on influenza virus, suggested that if treatment were continued for longer after virus challenge, it might be more effective. These were the facts known to the team as they moved towards the next trial.

After much discussion over the optimal protocol, IFN spray or placebo spray was started 1 day before virus challenge and given for a total of 4 days. The course of interferon was administered by one investigator who gave three doses of interferon or placebo within an hour, saturating the nose of each volunteer three times per day. The total dose of interferon given was 14 Mu and colds were almost completely prevented by this dose. Some question remained over efficacy because the number of volunteers developing definite colds in the placebo group was disappointingly small. However, it was not possible to repeat the experiment immediately, and certainly not to give a larger dose because there was not sufficient interferon available from the Finnish Red Cross Unit to pursue this approach. Rather, Cantell's interferon was used through the 1970s to treat herpes keratitis, reduce the risk of cytomegalovirus infection after renal transplantation, treat herpes zoster in lymphoma patients, treat patients with chronic hepatitis B virus infection and prevent herpes labialis after surgery to the trigeminal ganglion. Acyclovir was introduced and interferon became obsolete for the treatment of herpesvirus infections but it remains useful in the treatment of hepatitis. The other thrust of the research was cancer chemotherapy, and Strander performed a series of exciting but controversial trials in adjuvant management of osteosarcoma. Later, he was to suggest that systemic interferon treatment for osteosarcoma caused a reduction in trivial virus infections in his patients [65].

While this work was progressing worldwide with nIFN-αLe, two groups at Leuven and High Wycombe were making fibroblast (nIFN-β) interferon. The curious fact about this material was that it was very unstable against shear forces, particularly if syringed or pipetted. Substances which stabilize disulphide bonds were tried fairly successfully, but

the yields from the bulk culture of human lung fibroblasts were very disappointing. Sufficient material, however, was gathered for trials against rhinovirus 2. The interferon was given as nosedrops three times per day but otherwise according to the schedule established using nIFN-αLe [63]. The trial clearly failed and the interferon had indeed lost activity by the end. There was also concern about the increased tendency of nasal secretions to inactivate fibroblast interferon compared with leucocyte interferon shown by Harmon, Greenberg and Couch [23].

The disappointment of this experiment, however, turned to great excitement with two convincing trials using, first, highly purified nIFN-αLe, and then using rIFN-α₂. David Secher had worked with Milstein, improving methods of making monoclonal antibodies. He collaborated with Derek Burke to produce the first useful binding and neutralizing monoclonal antibody against leucocyte interferon, NK2, which was used in an affinity column to purify Cantell's nIFN-αLe to a very high degree. There was a little murine immunoglobulin detectable in the final preparation but it was considered to be >99% pure interferon. It is noteworthy that the earliest preparations were estimated as <0.1% pure and the material used in the Merigan trial as about 1% pure. The highly purified material was first given systemically to volunteers to establish that the febrile influenza-like reactions observed previously with crude interferon preparations were due to interferon and not to a contaminant, after which the next rhinovirus challenge experiment was carried out. By this time, rhinovirus 9 was popular in the CCU, and more material was available, so the dose was increased from 0.36 Mu to 2.5 Mu per dose (a total of 90 Mu over 4 days) but the same protocol (three doses in 1 hour, three times per day) was followed. The results were extraordinary, there being no definite colds in the treated group [61].

A trial using rIFN-α_2 was the next to be undertaken. Production of interferon from human cells would always tend to impose a limit to the amount of nIFN available. (Production problems of lymphoblastoid interferon were several years from being solved.) When it was realized that the technology was available to clone a pure protein, DNA clones from IFN mRNA extracted from induced cells were inserted into *E. coli*, which became the source of the first clinical product of the new age to be used in human volunteers. The development moved forward apace because of a widely held misconception that interferon was a potent, broad-spectrum anti-tumour agent and through competition. The cloning of the most prominent subtype (called IFN-α_2 by one group and IFN-αA by another) into *E. coli* was achieved almost simultaneously by two different groups.

The group producing IFN-α_2 (later termed IFN-α_{2b} (Schering)) was the first to make material suitable for clinical trial and without delay, systemic tolerance and common cold protection studies were performed. The same protocol as for NK2-purified nIFN-αLe was used and protection was complete [62]. At this stage, the results shown in Table 20.1 were sufficiently convincing to pursue further trials. It seems hardly worth performing a statistical test on these results, but around the time of these observations, Phillpotts was coming to the conclusion that, because of the small numbers of volunteers used at the CCU, the α and β errors of intervention trials were such that only large effects of treatment would be discovered. Luckily, interferon had such an effect. It is likely that some antivirals were discarded because their clinical effects were insufficiently dramatic.

These results were confirmed by many groups using different interferons and rhinoviruses. At the CCU, rIFN-βser [37] and nIFN-αLy [55] were both found to be effective in the same model.

20.4 PHARMACOLOGY OF INTRANASAL INTERFERON

While waiting for supplies of interferon to become available to perform volunteer studies, several groups studied the kinetics of interferon action in the nose and in nasal cells *in vitro*. Crude interferons protected human fetal lung cells against influenza virus and rhinovirus [49], and nasal biopsies against vesicular stomatitis virus and various respiratory viruses [18,19,24,25]. The length of time necessary for contact between the interferon and cells to ensure protection is, not surprisingly, dose-dependent and very short contact times were necessary if high concentrations of interferon were used. Greenberg and co-workers [19] used antihistamines to reduce the clearance of interferon, and continuous exposure to interferon in a cotton pledget. Clearly not convenient in clinical practice, this sort of strategy was developed simply because the supply of interferon was extremely restricted. The results of a recent study in sheep [5] have shown that a surfactant can enhance the effect of human (*sic*) interferon in inducing 2'5' oligoadenylate synthetase, an interferon response protein. Furthermore, when interferon was sprayed into one nostril, effects on the cells from the other nostril were observed and these changes persisted for as long as 3 days. This may be due to the direction of ciliary flow in the nose of sheep. Use of pharmacological agents to enhance the activity of interferon remains an interesting possibility for future research [22].

Aoki and Crawley [3] showed nosedrops of radioactive albumin to be superior to sprays in terms of the distribution and persistence in the upper nasal passages. Yet, although this is now recognized as important for common nasal preparations, most of the subsequent work with interferons was done with sprays. Perhaps this did not matter eventually: sprays were much more convenient to give and were, after all, effective in preventing rhinovirus colds.

Table 20.1 Early results at the CCU of the use of interferons to prevent common colds

	Interferon			Totals
	Crude Le	NK2 mca-purified Le	Cloned IFN-α₂	
Individual Dose*	~0.4	2.5	2.5	
Total dose	~14	90	90	
Cold o·s				
Placebo	5/16	8/11	8/22	21/49
Treatment	1/16	0/8	0/9	1/43

* Approximate doses in 10^6 units (Mu)

Johnson and co-workers [46] showed that nIFN-αLe was cleared rapidly from the noses of chimpanzees and volunteers, although relatively small doses of interferon (about 10 000 u) were applied.

At the CCU [10], the clearance of IFN-α₂ given by two different sorts of spray was compared; one was a coarse spray used by the physician investigator (the system used for the first three successful trials described above), and the other a hand-held pump spray operated by the volunteer. The amount of interferon recoverable in the nasal wash was similar for both methods. Although the decay in recoverable interferon was rapid in the first 2 hours, there was then a very prolonged decay and $>10^3$ u could still be recovered at 8 hours. This suggested that less frequent dosing might be effective. Efforts then went into finding out the minimal dosing schedule needed to prevent rhinovirus colds. In parallel, tests were done against other common viruses.

Little work was done in the treatment of established colds at this stage. In one study at the CCU using rhinovirus 9 and nIFN-αLe (unpublished), administration of interferon was delayed until after the virus had been given. There was a tantalizing, though non-significant, transient suppression or delay of symptoms, rather like the effect shown in the trial against influenza virus reported in 1973 [49]. As data emerged suggesting that interferon could have a direct unwanted effect on

nasal mucosa with symptoms, it was concluded that the symptoms of a cold might in part be due to a reaction to endogenous interferon and, if this was the case, treatment with exogenous interferon might not be expected to have any effect. This was a very simplistic view but it also was thought that very large numbers of volunteers would be needed to demonstrate convincingly the therapeutic activity of any antiviral once symptoms had begun. It was concluded that such studies could not economically be done at the CCU but would have to wait for larger field trials.

20.5 FURTHER ACTIVITY STUDIES: DOSE–RESPONSE

Successful experiments were done using smaller amounts and, more importantly, less frequent doses of interferon. For example, Greenberg *et al.* [20] looked at single or two-dose prophylaxis with interferon (nIFN-αLe) given by an intranasal pledget or by a nebulizer. Volunteers were also given antihistamines to reduce the clearance of interferon from the nose. The cumulative results showed a significant reduction in the number of colds in treated patients but the effects were less dramatic than those observed when larger numbers of doses had been given.

Samo and co-workers [56, 57] confirmed the activity of recombinant interferon using rIFN-αA (Roche) against rhinovirus 13. A high dose (10 Mu) was given twice a day. Lymphoblastoid interferon (nIFN-αLy), given at a dose of 2.7 Mu three times per day, suppressed colds due to rhinoviruses 9 and 14 [55]. Hayden and Gwaltney reported in 1983 [27] that they had compared a dose of about 10 Mu rIFN-α_2 four times per day with about 40 Mu given once per day and found that relatively infrequent dosing with high doses of interferon might be just as effective as the saturation regimen developed by the team at the CCU in 1972.

Investigators at the CCU attempted to determine the minimum effective dose of interferon against rhinovirus using the smallest number of volunteers [54]. A single dose of 2.3 Mu rIFN-α_2 per day for 4 days clearly prevented colds if the virus was given 2–3 hours after the first dose of interferon. However, when the virus was given 13-14 hours after the first dose, there was an impression that interferon was less effective, though this effect was not statistically significant, the rate of colds in the placebo group of this arm of the trial being rather low. This dose–response study complemented the important trials by Greenberg and colleagues [20].

20.6 PREVENTION OF OTHER EXPERIMENTAL RESPIRATORY VIRUS INFECTIONS

At the CCU, rIFN-αA at a dose of 2.7 Mu three times a day for 4 days suppressed coronavirus-induced colds [36]. Thus, 13 of 35 volunteers who were given placebo had colds compared with 2 of 35 treated with interferon. This result was confirmed by Turner and co-workers [66]. A similar dose of IFN-α_2 (5 Mu twice a day for 7 days) but started 2 days before virus caused some reduction in illness due to influenza A/Cal/78 H1N1 virus, though the effect was not so dramatic as for rhinovirus [11]. An inhibitory effect of rIFN-α_{2a} on experimental respiratory syncytial (RS) virus infection was also shown [38].

20.7 LARGE-SCALE SEASONAL PROPHYLAXIS STUDIES AND TOXICITY

Having demonstrated that intranasal interferon could suppress experimental common colds caused by rhinoviruses and coronaviruses and, to a lesser extent, influenza virus and RS virus, the next step was to decide how best to use interferon in the natural setting. The successful production of lymphoblastoid and recombinant interferons meant, at least, that there were no serious limitations on the supply of interferon for clinical trials and investigators were fairly certain of the minimum dose of intranasal interferon needed to prevent experimental rhinovirus colds. This dose was in the order of 1–2 Mu given one to three times per day. Furthermore, it had been shown that volunteers could self-administer interferon successfully using a simple spray and that antihistamines and pledgets were not necessary.

The first experiments were designed to give large doses of interferon or placebo to large groups of volunteers for a long period of time and to look for natural respiratory virus infections in the two groups ('seasonal prophylaxis') [8,12,14,29,51]. A representative trial was done by Farr and co-workers [14] using 10 Mu rIFN-α_2 per day in about 150 volunteers per group. Virologically proven colds were clearly suppressed (from 13 of 53 to none of 53) by treatment but many more subjects receiving interferon had definite nasal symptoms (41 of 153) compared with those given placebo (9 of 152). Very similar results were seen by Betts and colleagues [8] using a regimen of 5 Mu twice a day, by Monto and co-workers [51] using a dose of 1.5 Mu twice a day, and by Douglas *et al.* [12] in Adelaide using a dose of rIFN-α_2 1 Mu twice a day for 28 days. Altogether, very large numbers of volunteers were used in these studies. Protective efficacy against documented rhinoviruses was in the order of 70–100%. At the lower doses tested, the side effects were rather few but just as many volunteers developed nasal symptoms on interferon as on placebo. However, at the lower limit of the dose range needed to inhibit experimental colds, even the lower doses tested caused a significant reduction in the yield of rhinovirus in nasal washings.

To give some idea of the magnitude of the problem of local toxicity, about 25% of volunteers receiving 10 Mu/day of IFN-α had symptoms compared with 5% of those receiving 1 Mu/day [56,57]. Smaller groups of volunteers were given ascending doses of NK2-purified nIFN-αLe [60]. Side-effects were dose-related but it was clear that the symptoms due to interferon (nasal discomfort, dryness, blood-tinged nasal mucus and crusts) were quite different from those of genuine colds. In this protocol, the treatment was to be stopped when volunteers developed definite symptoms and 30% of volunteers had stopped taking the highest dose of interferon (4.4 Mu/day) by day 15.

Similar toxicity was seen with rIFN-βser at a dose of 12 Mu/day but substantially less at 3 Mu/day [30]. There was an impression that quantitatively less local toxicity occurred with fibroblast interferon than with leucocyte interferons but a direct comparison was not performed. Furthermore, it is impossible to accurately relate the dose of one interferon to another.

It was becoming clear that interferon given systemically caused a dose-related, febrile influenza-like reaction thought to be caused by a local effect in the thalamus or hypothalamus and probably mediated by inflammatory prostaglandin release. What of the pathological effects in the nose? Hayden and colleagues [30–32] took mucosal biopsies from the noses of volunteers and showed that within a few days of starting to spray high concentrations of interferon, the mucosa became suffused with lymphocytes and there was some superficial ulceration. These effects took longer with lower doses and all disappeared rapidly when treatment was stopped. The mechanism of this peculiar reaction has not been elucidated. No interleukin 2 is produced. There is no natural

parallel to the nasal mucosal changes due to interferon in the nose and this may be because of the short-term nature of the local production of endogenous interferon in the common cold. During colds, the mucosa tends to remain intact and is infiltrated by polymorphonuclear leucocytes rather than lymphocytes [69].

Many have observed that intranasal interferon treatment is associated with a definite trend towards the reduction of total white cell and neutrophil counts in the peripheral blood.

It was postulated that locally produced interferon might have immunological effects which might predispose to bronchospasm (e.g. enhancement of basophil migration [48]). Asthma was not a side effect noted of any interferon regimen in healthy volunteers but, by exclusion, none of these would have been prone to asthma. Injection of interferon in a dose sufficient to cause systemic symptoms did not cause increased airways sensitivity to histamine (G.M. Scott and T.A. Platts-Mills, unpublished). However, inhalation of a sufficiently large dose of NK2-purified nIFN-αLe to cause a systemic febrile reaction (some 40 Mu) did cause a significant reduction in the diffusion of lung carbon monoxide DL_{CO} and transfer of ^{99m}Tc across the alveolar–capillary membrane (G.M. Scott and C. Heneghan, unpublished). This experiment suggested that endogenous interferon may play a role in the systemic reactions and in the pulmonary dysfunction seen in influenza, and caused a wariness of using this method of delivering interferon for clinical purposes.

It was becoming increasingly apparent that the therapeutic ratio of intranasal interferon was close to or even above unity, but with one caveat. There had been trials reported which appeared to show that natural respiratory infections could be suppressed by much lower doses of intranasal interferon than were needed to inhibit experimental colds. For example, influenza-like illnesses in students and children [40,41] were apparently suppressed by 5000 u daily crude IFN-αLe given as nosedrops for a period of 2 months. Similar claims were made of less well-controlled studies from the Eastern Bloc using tiny amounts of crude interferon [4,9,64]. For a time, low-titre interferon could be purchased in Yugoslavia and the former USSR to treat common colds. A large glass ampoule was provided which contained about 1000 u IFN. If this treatment is still available, at least it is likely to be harmless.

Recently bovine alpha-interferon has been cloned and given systemically to cattle, at a dose of 5 mg (a very large dose by human standards) once a month for 6 months [2]. Weight gain was improved and respiratory infections were decreased in the active group of cattle compared with the placebo group.

If long-term intranasal interferon in doses sufficient to inhibit colds is not tolerable in humans, and systemic injections are unlikely to be tolerated or effective, then the obvious way to use interferon would be to give it in short courses at the time of greatest risk, that is, on known contact with someone with a cold. This was the next stage of the investigations.

20.8 CONTACT PROPHYLAXIS

A protocol was designed around the known epidemiology of colds within the family setting. When one member of the family came home with a cold, all the others would start interferon in an attempt to prevent themselves from catching the cold, but using it for a limited period only so as to reduce the risk of side effects. The groups of Hayden in Virginia [26] and Douglas in Adelaide [13] used the same protocol, in which families were allocated at random to placebo or interferon sprays. When one member of the family developed a cold, the treatment (5 Mu rIFN-α_2 or placebo once per day for 7 days) was started by all of the others. Respiratory symptoms were recorded in the conventional way on record cards and subjects with symptoms were seen and evaluated objectively and specimens taken for virus isolation. Both studies showed the same result,

namely an approximate 40% reduction in the risk of contacts acquiring a definite cold when interferon treatment was given. However, the major effect was against rhinoviruses only – infections by coronaviruses, parainfluenza and influenza viruses were not significantly inhibited. This was disappointing in view of the protection interferon afforded against these viruses in volunteer challenge experiments and this discrepancy has not been explained. Perhaps a larger dose is needed but the suspicion is that there is some idiosyncrasy about rhinovirus infections that makes them peculiarly sensitive to interferon in humans. The regime of treatment was well tolerated but some of the interferon recipients had nasal discomfort and bleeding.

Foy and co-workers [16] reported a similar study (though the dose and type of interferon were not stated), in which there was no significant reduction in respiratory infections among the contacts. One fundamental difference between this study and those mentioned above was that it was conducted between January and May when the predominant agent causing infection was influenza B virus. As expected, rhinoviruses were isolated relatively infrequently, and the influenza virus infections were not inhibited. This would explain the differences seen and also point to an important limitation of this strategy.

Herzog and colleagues [34,35] tested lower doses of rIFN-αA using a similar family contact prophylaxis protocol. They concluded that the optimal dose lay between 1.5 Mu and 5 Mu per day for 5–7 days. The effects of interferon on the risk of infection in contacts were negligible, but colds in those on active treatment were slightly less severe.

Wiselka *et al.* [70] tested rIFN-α_{2c}, 1.5 Mu taken twice per day for 5 days in patients with chronic respiratory disease who had been in contact with someone with a cold for more than 6 hours. There was no significant reduction in the risk of developing definite or mild, upper or lower respiratory infection or significant protection against pulmonary complications measured in terms of peak expiratory flow rates. There was a slight though not significant reduction in the length of respiratory illness in interferon compared with placebo recipients. Others have looked at the prevention of colds in asthmatics but in neither study, published in abstract form, was there any significant antiviral benefit of interferon [17,50].

20.9 TREATMENT OF ESTABLISHED COLDS

There is no doubt that intranasal interferon can prevent rhinovirus colds. In the experimental setting, interferon is active if given before the administration of virus. Rhinovirus colds in asymptomatic family contacts can also be inhibited. Surprisingly, however, after so much research it is still not clear how late in the course of an infection interferon can be given and have an effect. Mention has been made above of an unpublished study at the CCU in which the administration of interferon was delayed until the first symptom and the small but definite delay in the symptoms of the cold. For the reasons given, this observation was not pursued. Just *et al.* [47] reported that rIFN-αA given within 24 hours of the onset of a natural cold (3 Mu given intranasally four times per day for 5 days) could reduce subjective symptoms compared with the effect of a placebo spray. In contrast, Hayden and Gwaltney [28] showed that nasal spray or drops of rIFN-α$_{2b}$ at a dose of 9 Mu given three times per day starting 28 hours after rhinovirus inoculation had no effect on the infection at all.

Gwaltney [22] reasoned that if interferon failed to reduce nasal symptoms after a cold had begun, then inhibition of virus replication may be simply insufficient for a detectable therapeutic effect on symptoms. He therefore gave volunteers nosedrops of rIFN-α$_{2b}$, 3 Mu, three times per day, together with intranasal sprays of ipratropium (a parasympathetic inhibitor) and oral naproxen (a prostaglandin-synthetase inhibitor). The treatment started 24 hours after virus inoculation and continued for 4 days. Only small groups of volunteers were involved and the allocation was one-sided. Nevertheless, an effect on virus shedding was clearly demonstrated and this was associated with earlier termination of the symptoms, reductions in their severity with some subjective benefit in more of those on active treatment than on placebo.

The synergy between antiviral agents against rhinoviruses *in vitro* [1] seems not to have been pursued in clinical trials. Other agents which may enhance the activity of intranasal interferon include surfactant [5]. Clearly the need for controlled studies in volunteers has by no means come to an end.

20.10 CONCLUSIONS

This holy grail of interferon against the common cold has led to a vast amount of laboratory-based scientific work culminating in a series of well-conducted, double-blind, placebo-controlled trials in a very large number of healthy volunteers. There is no doubt that interferon can prevent rhinovirus colds, both experimentally produced and naturally occurring. However, many questions remain unanswered. How late can interferon be given and still have a useful clinical effect? Why does interferon not prevent natural colds due to viruses other than rhinoviruses in field trials? What is the mechanism for nasal epithelial toxicity? Can synergy with other antivirals, or with other pharmacological agents be refined so as to improve efficacy?

It is a disappointment to many that interferon is not available over the counter in a handy little nasal spray for the treatment or prevention of colds. No other 'pharmaceutical' product has gone through so many waves of unwarranted popularity and rejection since its discovery. However, interferon is now quietly established as an important therapy for several conditions, including several leukaemias, some tumours and chronic hepatitis.

20.11 REFERENCES

1. Ahmad, A.L.M. and Tyrrell, D.A.J. (1986) Synergism between anti-rhinovirals: various human interferons and a number of synthetic compounds. *Antiviral Res.*, **6**, 241–52.
2. Akiyama, K., Sugii, S. and Hirota, Y. (1983) A clinical trial of recombinant bovine interferon alpha 1 for the control of respiratory disease in calves. *J. Vet. Med. Sci.*, **55**, 449–52.
3. Aoki, F.Y. and Crawley, J.C.W. (1976) Distribution and removal of human serum albumin-technetium 99M instilled intranasally by drops and spray. *Br. J. Clin. Pharmacol.*, **3**, 869–78.
4. Arnaoudova, V. (1976) Treatment and prevention of acute respiratory virus infections in children with leukocytic interferon. *Rev. Roum. Med. Virol.*, **27**, 83–8.
5. Baglioni, C. and Phipps. R.J. (1990) Nasal absorption of interferon: enhancement by surfactant agents. *J. Interferon Res.*, **10**, 497.
6. Baron, S. and Isaacs. A. (1962) Absence of interferon in lungs from fatal cases of influenza. *Br. Med. J.*, **I**, 18–20.
7. Baron, S., Tyring. S.K., Fleischmann, R. *et al.* (1991) The Interferons: mechanisms of action and clinical applications. *JAMA*, **266**, 13pj75–83.
8. Betts, R.F., Erb, S., Roth, F. *et al.* (1983) A field trial of intranasal interferon. Proceedings of 13th International Congress of Chemotherapy, Vienna, SE 4.7/1–5, 60/13 (abstracts).
9. Busuek, G.P., Gailonskaia, I.N., Lozinskaya, T.M. *et al.* (1971) Prophylactic effect of human leukocytic interferon in pre-school children. *Vopr. Virusol.*, **16**, 226–9.
10. Davies, H.W. Scott, G.M., Robinson, J.A. *et al.* (1984) Comparative intranasal pharmacokinetics of interferon using two spray systems. *J. Interferon Res.*, **3**, 443–9.
11. Dolin, R., Betts, R.F., Treanor, J. *et al.* (1983) Intranasally administered interferon as prophylaxis against experimentally induced influenza A infection in humans. Proceedings of the 13th International Congress of Chemotherapy, Vienna, SE 4.7/1-7, 60/20–3.
12. Douglas, R.M., Albrecht, J.K., Miles, H.B. *et al.* (1985) Intranasal interferon-α_2 prophylaxis of natural respiratory virus infection. *J. Infect. Dis.*, **151**, 731–6.
13. Douglas, R.M., Moore, B.W., Miles, H.B. *et al.* (1986) Prophylactic efficacy of intranasal alpha interferon against rhinovirus infections in the family setting. *N. Engl. J. Med.* **314**, 65–70.
14. Farr, B.M., Gwaltney, J.M., Adams, K.F. and Hayden, F.G. (1984) Intranasal interferon-α_2: for prevention of natural rhinovirus colds. *Antimicrob. Agents Chemother.*, **26**, 31–4.
15. Finter, N.B. (1968) Of mice and men; studies with interferon, in *Interferon*, (eds G. Wolstenholme and M. O'Connor), Churchill, London, pp. 204–15.
16. Foy, H.M., Fox, J P. and Cooney, M.K. (1986) Efficacy of alpha$_2$-interferon against the common cold. *N. Engl. J. Med.*, **315**, 513–14.
17. Geha, R., Maguire, J., McIntosh, K. *et al.* (1986) A trial of intranasal alpha interferon in asthmatic children. *J. Allergy Clin. Immunol.*, **77** (suppl.), 160.
18. Greenberg, S.B. (1977) Activity of exogenous interferon in the human nasal mucosa. *Tex. Rep. Biol. Med.*, **35**, 491–5.
19. Greenberg, S.B., Harmon, M.W., Johnson, P.E. and Crouch, R.B. (1978) Antiviral activity of intranasally applied human leukocyte interferon. *Antimicrob. Agents Chemother.*, **14**, 596–600.
20. Greenberg, S.B., Harmon, M.W., Couch, R.B. *et al.* (1982) Prophylactic effect of low doses of leukocyte interferon against infection with rhinovirus. *J. Infect. Dis.*, **145**, 542–6.
21. Gresser, I. and Dull, H.B. (1964) A virus inhibitor in pharyngeal washings from patients with influenza. *Proc. Soc. Exp. Biol. Med.*, **115**, 192–6.
22. Gwaltney, J.M. Jr (1992) Combined antiviral and antimediator treatment for rhinovirus colds. *J. Infect. Dis.*, **166**, 776–82.
23. Harmon, M.W., Greenberg, S.B. and Couch, R.B. (1976) Effect of human nasal secretions on the antiviral activity of human fibroblast and leukocyte interferon. *Proc. Soc. Exp. Biol. Med.*, **152**, 598–602.
24. Harmon, M.W., Greenberg, S.B., Johnson, P.E. and Couch, R.B. (1977) A human nasal epithelial cell culture system: evaluation of the response to human interferons. *Infect. Immun.*, **16**, 480–5.
25. Harmon, M.W., Greenberg, S.B. and Johnson, P.E. (1980) Rapid onset of the interferon-induced antiviral state in human nasal epithelial and foreskin fibroblast cells. *Proc. Soc. Exp. Biol. Med.*, **164**, 146–8.
26. Hayden, F.G., Albrecht, J.K., Kaiser, D.L. and Gwaltney, J.M. Jr (1986) Prevention of natural

colds by contact prophylaxis with intranasal alpha-2 interferon. *N. Engl. J. Med.*, **314**, 71–5.

27. Hayden, F.G. and Gwaltney, J.M. Jr (1983) Intranasal interferon α_2 for prevention of rhinovirus infection and illness. *J. Infect. Dis.*, **148**, 543–55.

28. Hayden, F.G. and Gwaltney, J.M. Jr (1984) Intranasal interferon-α_2 treatment of experimental rhinoviral colds. *J. Infect. Dis.*, **150**, 174–88.

29. Hayden, F.G., Gwaltney, J.M. Jr and Johns, M.E. (1985) Prophylactic efficacy and tolerance of low-dose intranasal interferon-alpha, in natural respiratory viral infections. *Antiviral Res.*, **5**, 111–16.

30. Hayden, F.G., Innes, D.J., Mills, S.E. and Levine, P.A. (1986) Long term tolerance of intranasal recombinant interferon-beta-serine in man. *J. Interferon Res.*, **6** (suppl. 1), 31.

31. Hayden, F.G., Mills, S.E. and Johns, M.E. (1983) Human tolerance and histopathologic effects of long term interferon-α_2. *J. Infect. Dis.* **148**, 914–21.

32. Hayden, F.G., Winther, B., Donowitz, G.R., Mills, S.E. and Innes, D.J. (1987) Human nasal mucosal responses to topically applied recombinant leukocyte A interferon. *J. Infect. Dis.*, **156**, 64–72.

33. Henle, W. (1950) Interference phenomena between animal viruses: a review. *J. Immunol.*, **64**, 203–36.

34. Herzog, C., Berger, R., Fernex, M. *et al.* (1986) Intranasal interferon (IFN-αA, Ro 22-8181) for contact prophylaxis against common cold: a randomized, double-blind and placebo-controlled field study. *Antiviral Res.*, **6**, 171–6.

35. Herzog, C., Just, M., Berger, R., Havas, L. and Fernex, M. (1983) Intranasal interferon for contact prophylaxis against common cold in families. *Lancet*, **ii**, 962.

36. Higgins, P.G., Phillpotts, R.J., Scott, G.M., *et al.* (1983) Intranasal interferon as protection against experimental respiratory coronavirus infection in volunteers. *Antimicrob. Agents Chemother.*, **24**, 713–15.

37. Higgins, P.G., Al-Nakib, W., Willman, J. *et al.* (1986) Interferon β_{ser} as prophylaxis against experimental rhinovirus infection in volunteers. *J. Interferon Res.*, **6**, 153–9.

38. Higgins, P.G., Barrow, G.I., Tyrrell, D.A.J. *et al.* (1990) The efficacy of intranasal interferon-α_{2a}

in respiratory syncytial virus infection in volunteers. *Antiviral Res.*, **14**, 3–10.

39. Hill, D.A., Baron, S., Perkins, J.C. *et al.* (1972) Evaluation of an interferon inducer in viral respiratory disease. *JAMA.*, **219**, 1179–84.

40. Imanishi, J., Karaki, T., Sasaki, O. *et al.* (1980) The preventive effect of human interferon-alpha preparation on upper respiratory disease. *J. Interferon Res.*, **1**, 169–77.

41. Isomura, S., Ichikawa, T., Miyazu, M. *et al.* (1982) The preventive effect of human interferon-alpha on influenza infection: modification of clinical manifestations of influenza in children in a closed community. *Biken J.*, **25**, 131–7.

42. Isaacs, A. and Hitchcock, G. (1960) Role of interferon in recovery from virus infections. *Lancet*, **ii**, 69–71.

43. Isaacs, A. and Lindenmann, J. (1957) Virus interference. I. The interferon. *Proc. R. Soc. Lond. B.*, **147**, 258–67.

44. Isaacs, A., Lindenmann, J. and Valentine, R.C. (1957) Virus interference. II. Some properties of interferon. *Proc. R. Soc. Lond. B.*, **147**, 268–73.

45. Jao, R.L., Wheelock, E.F. and Jackson, G.G. (1970) Production of interferon in volunteers infected with Asian influenza. *J. Infect. Dis.*, **121**, 419–26.

46. Johnson, P.E., Greenberg, S.B., Harmon, M.W., Alford, B.R. and Couch, R.B. (1976) Recovery of applied human leukocyte interferon from the nasal mucosa of chimpanzees and humans. *J. Clin. Microbiol.*, **4**, 106–7.

47. Just, M., Berger, O., Ruuskanen, O., Ludin, M. and Linder, S. (1986) Intranasal interferon alpha-2A treatment of a common cold – a preliminary study. *J. Interferon Res.*, **6** (suppl. 1), 32.

48. Lett-Brown, M.A., Aelvoet, M., Thueson, D.O., Grant, J.A. and Deiss, W.P. (1980) Enhancement of *in vitro* basophil migration by viral-induced interferon. *Clin. Res.*, **28**, 505A.

49. Merigan, T.C., Reed, S.E., Hall, T.S. and Tyrrell, D.A.J. (1973) Inhibition of respiratory virus infection by locally applied interferon. *Lancet*, **i**, 563–7.

50. Michaels, D., Foy, H. and Krouse, H. *et al.* (1986) Intranasal interferon for prophylaxis of viral infections in adult and pediatric asthmatic patients. *J. Allergy Clin. Immunol.*, **77**, 160.

51. Monto, A.S., Shope, T.C., Schwartz, S.A. *et al.* (1986) Intranasal interferon-α_{2b} for seasonal prophylaxis of respiratory infection. *J. Infect.*

Dis., **154**, 128–33.

52. Panusarn, C., Stanley, E.D., Dirda, V., Rubenis, M. and Jackson, G.G. (1974) Prevention of illness from rhinovirus infection by a topical interferon inducer. *N. Engl. J. Med.*, **291**, 57–61.

53. Petralli, J.K., Merigan, T.C. and Wilbur, J.R. (1965) Circulating interferon after measles vaccination. *N. Engl. J. Med.*, **273**, 198–201.

54. Phillpotts, R.J., Scott, G.M., Higgins, P.G. *et al.* (1983) An effective dosage regimen for prophylaxis against rhinovirus infection by intranasal administration of HuIFN α₂. *Antiviral Res.*, **3**, 121–36.

55. Phillpotts, R.J., Higgins, P.G., Willman, J.S. *et al.* (1984) Intranasal lymphoblastoid interferon (Wellferon) prophylaxis against rhinovirus and influenza virus in volunteers. *J. Interferon Res.*, **4**, 535–41.

56. Samo, T.C. Greenberg, S.B., Couch, R.B. *et al.* (1983) Efficacy and tolerance of intranasally applied recombinant leukocyte A interferon in normal volunteers. *J. Infect. Dis.*, **148**, 535–42.

57. Samo, T.C., Greenberg, S.B., Palmer, J.M. *et al.* (1984) Intranasally applied recombinant leukocyte A interferon in normal volunteers. II. Determination of the minimal effective and tolerable dose. *J. Infect. Dis.*, **50**, 181–8.

58. Scientific Committee on Interferon (1962) Effect of interferon on vaccination in volunteers. *Lancet*, **i**, 873–5.

59. Scientific Committee on Interferon (1965) Experiments with interferon in man. *Lancet*, **i**, 505–6.

60. Scott, G.M., Onwubalili, J.K., Robinson, J.A. *et al.* (1985) Tolerance of one-month intranasal interferon. *J. Med. Virol.*, **17**, 99–106.

61. Scott, G.M., Phillpotts, R.J., Wallace, J. *et al.* (1982) Prevention of rhinovirus colds by human interferon alpha-2 from *Escherichia coli*. *Lancet*, **ii**, 186–8.

62. Scott, G.M., Phillpotts, R.J., Wallace, J. *et al.* (1982) Purified interferon as protection against rhinovirus infection. *Br. Med. J*,. **284**, 1822–5.

63. Scott, G.M., Reed, S.E., Cartwright, T.C. and Tyrrell, D.A.J. (1980) Failure of fibroblast interferon to protect against rhinovirus infection. *Arch. ges. Virus Forsch.*, **65**, 135–9.

64. Solov'ev, V.D. (1969) The results of controlled observations on the prophylaxis of influenza with interferon. *Bull. WHO*, **41**, 683–8.

65. Strander, H., Cantell, K., Carlström, G. *et al.* (1976) Acute infections in interferon-treated patients with osteosarcoma: preliminary report of a comparative study. *J. Infect. Dis.*, **133**, A245–8.

66. Turner, R.B., Felton, A., Kosak, K. *et al.* (1986) Prevention of experimental coronavirus colds with intranasal α-2b interferon. *J. Infect. Dis.*, **154**, 443–7.

67. Tyrrell, D.A.J. and Reed, S.E. (1973) Some possible practical implications of interferon and interference, in *Non-Specific Factors Influencing Host Resistance*, (eds W. Braun and J. Unger), Karger, Basel, pp. 438–42.

68. Wheelock, E.F. and Sibley, W.A. (1964) Interferon in human serum during clinical viral infections. *Lancet*, **ii**, 382–5.

69. Winther, B., Farr, B., Turner, R.B. *et al.* (1983) Histopathologic examination and enumeration of polymorphonuclear leukocyte in nasal mucosa in the nasal mucosa during experimental rhinovirus colds. *Acta Otolaryngol.*, Suppl. **413**, 19–24.

70. Wiselka, M.J., Nicholson, K.G., Kent, J., Cookson, J.B. and Tyrrell, D.A.J. (1991) Prophylactic intranasal alpha-2 interferon and viral exacerbations of chronic respiratory disease. *Thorax*, **46**, 706–11.

John Oxford and Ali Al-Jabri

Viral and Other Infections of the Human Respiratory Tract.
Edited by S. Myint and D. Taylor-Robinson. Published in 1996 by Chapman & Hall. ISBN 0 412 60070 6

21.1 INTRODUCTION

In this last decade of the twentieth century we inhabit a world of molecules which are being investigated for the first time at the atomic level, while biologists have begun to manipulate them to our advantage. The current scientific literature abounds with descriptions of transgenic animals, of chimaeric viruses, of viral proteins and enzymes analysed by X-ray crystallography, synthetic genes and, most recently, the ability of biologists to manipulate genes even to a point of fusion of the sperm and ovum.

The respiratory viruses co-inhabit this world of host DNA and RNA and indirectly manipulate host genetic machinery to produce new viral offspring. The entanglement of replication of viral nucleic acids, either RNA or DNA, with host cell nucleic acid and transcription and replication mechanisms has made the discovery of viral inhibitory molecules exceedingly difficult. In terms of specific inhibitors of influenza virus, only two compounds (namely amantadine and ribavirin) have been discovered to date. They have gained unequivocal acceptance as drugs for clinical use (reviewed in [69]) and no other drug against any other respiratory virus is used in the clinic. There remains, therefore, a unique opportunity for the development of a novel programme to discover new drugs. Indeed, the search for a small synthetic or a naturally occurring molecule, the virological counterpart of the antibacterial penicillin molecule, and which could target and inhibit the replication of a wide range of viral pathogens is perhaps more intense now than ever before.

21.2 THE MEDICAL RESEARCH COUNCIL COMMON COLD UNIT

The MRC Common Cold Unit in Salisbury, UK played a central role in early studies of antiviral chemotherapy, including the first controlled demonstration of the clinical activity of interferon against influenza and the clinical effectiveness of a virion-binding drug against the rhinovirus.

The research unit in Salisbury was established before the discovery of the first antiviral compounds in the mid-1950s. At this time two antivirals were being studied, namely methisazone against smallpox and idoxuridine against herpes infections of the eye (reviewed in [48]). However, the scientific attractions of specific molecules binding to virion proteins or inhibiting virus-specific enzymes were recognized as immense. It was also well recognized that the discovery of specific virus inhibitors could lead to fundamental discoveries about viruses themselves and this is particularly so when a drug-resistant virus can be selected.

Therefore, as the search for respiratory virus inhibitors quickened during the next decade, the Common Cold Unit (CCU) began to occupy a central stage position in this growing area of virology. By 1963 the first anti-influenza virus compound had been reported [16], while during the next two decades new anti-rhinovirus inhibitors came under scrutiny and investigations commenced of inhibitors to coronavirus [35,36] and respiratory syncytial (RS) virus. But during this period the most important contribution of the CCU was undoubtedly the clear demonstration that interferon could inhibit a virus in a clinical situation [42]. In later studies, the group investigated the effectiveness of purified interferon [60] and interferon alpha-2 produced by recombinant DNA [62] technology as well as other interferons produced in cell culture and administered by the intranasal route [54]. Not all interferons were found to be clinically

effective [61]. Undoubtedly it was a source of satisfaction to the staff of the CCU and to the scientific community worldwide when a compound which showed clear effects against the common cold virus was reported [1,4].

A wider and perhaps the major contribution to the developing science of antiviral chemotherapy to emerge from the Salisbury Unit was the establishment of a base for controlled clinical studies in a defined scientific environment [7]. The combination of clinical and laboratory science was a key element at the CCU. To my knowledge only two such clinical volunteer units now exist worldwide, one in the CIS (St Petersburg) and the other in the USA. Respiratory viruses are rather unique in the sense that true volunteers come forward for clinical experiments, in marked contrast to the other target viruses of antiviral chemotherapists: HIV, papillomavirus, hepatitis B virus and herpes simplex virus. Beare and Reed [7] described in a detailed chapter the day to day clinical and scientific work of the CCU.

In the last 18 months a new anti-influenza molecule has been discovered (Figure 21.2), or at least refined using the most modern methods of X-ray crystallography and computer graphics [72]. Normally the CCU was the first centre chosen for clinical experiments and it may be presumed that work would have been initiated there in volunteers with the new drug. The absence of the Unit means that there is now a serious gap for EC countries in the provision of clinical facilities to bring a new antiviral molecule to the registration stage. As the following discussion will illustrate, respiratory viruses are still very important pathogens and the need for a co-ordinated scientific and clinical approach for new inhibitors has not diminished since the inauguration of the CCU, rather it has increased. The chapter will highlight major discoveries in respiratory antiviral chemotherapy over the last three decades, with particular emphasis on influenza virus. The volunteer work carried out with this virus may be considered to be the most challenging for

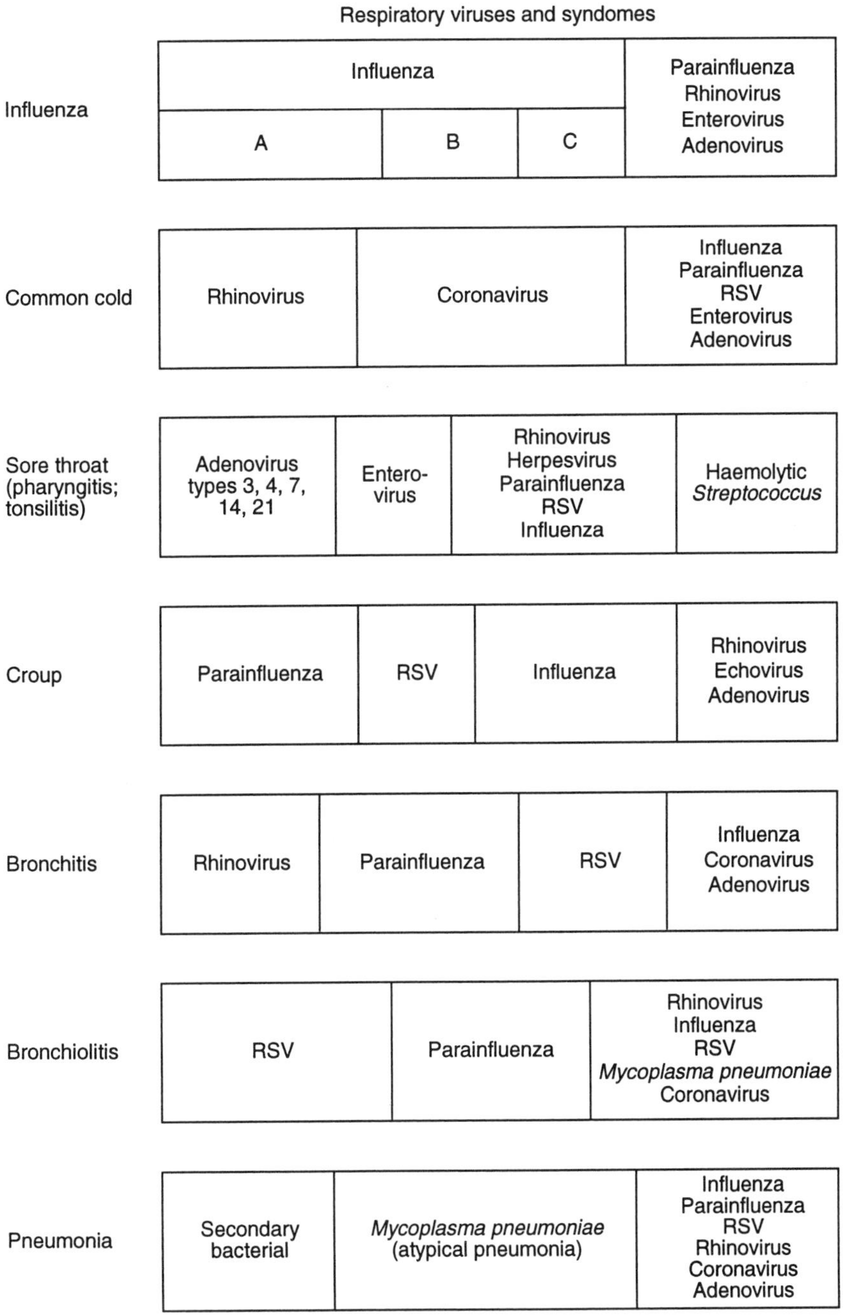

Respiratory viruses and syndomes

Figure 21.1 The range of respiratory viruses correlated with clinical disease. The relative size of the blocks is related to the proportion of viruses causing the disease. (From [66].)

the scientists and yet worrying for the clinicians. Unlike the other viruses utilized at the CCU, influenza virus has the ability to cause very serious infection in any person, volunteers included. It is a credit to the Salisbury Unit that volunteers came forward confidently for these experiments and the expertise at the Unit ensured that there was no clinical detriment to the volunteers. It should be clearly appreciated that basic studies into the genetic determinants of virulence of influenza viruses were also proceeding in parallel to studies of inhibitors (see Chapter 14) which may themselves lead in the near future to the discovery of new compounds.

21.3 CLINICAL NEED FOR ANTIVIRALS

A continued clinical need for the discovery and refinement of new antivirals against respiratory viruses is clearly apparent in the 1990s.

21.3.1 EPIDEMIC INFLUENZA

Influenza A and B viruses are notably the single major cause of several lower respiratory tract infections and death throughout the world and are seasonal in their impact (Table 21.1). For example, in the UK alone 26 000 excess deaths were caused by influenza A virus in the winter of 1975–1976 while in the winter of 1989–1990 there were in excess of 19 000 deaths. The viruses also cause severe economic disruption and in 1975 there were 900 000 excess new claims for sickness benefit in the UK alone. Not all sections of the community are at equal risk of mortality. Certain 'special risk' groups including older persons, diabetics of any age and persons of all ages with chronic heart, kidney or respiratory disease have been defined and are at a ten-fold greater risk of mortality following an infection with either influenza A or B virus. Not surprisingly, these groups of vulnerable persons in our community are the major target for vaccines and new chemotherapeutic agents.

21.3.2 SERIOUS LOWER RESPIRATORY INFECTIONS CAUSED BY RESPIRATORY SYNCYTIAL VIRUS

Another important virus causing severe lower respiratory tract infection and even death in

Table 21.1 The myriad of viruses causing pathology of the respiratory tract

Viruses causing respiratory tract infection	Target group	Antiviral agents which are clinically active
Upper respiratory tract Rhinoviruses	Young and old	Natural interferon, recombinant α-2a interferon*; virion binding molecules
Coronaviruses	Young and old	None (but see [35, 36])
Lower respiratory tract Influenza A, B and C	Young and old but mortality in 'at risk' group	Amantadine*, rimantadine* ribavirin*, neuraminidase inhibitor
Adenoviruses	Young and old – mortality rare	None
Parainfluenza types I, II, III and IV	Young and old – mortality rare	None
Respiratory syncytial virus	Young and old with mortality in both	Ribavirin*

* Licensed antiviral: clinically active.

the elderly and particularly in infants and young children, is respiratory syncytial (RS) virus. For example, in the USA alone RS virus causes 90 000 hospitalizations and 4500 deaths per year. RS virus is the most frequent cause of bronchiolitis and pneumonia in infants (Table 21.1). More recently [22], deaths have been reported in the elderly which may have been inadvertently attributed to influenza A virus. Like influenza virus, RS virus is seasonal and causes outbreaks worldwide but probably with greater regularity than influenza virus with, for example, yearly outbreaks in temperate climates for 3 months each winter. A unique feature of the epidemiology of RS virus is the ability of the virus to re-infect a patient. Since natural immunity seems short-lived it could be surmised that vaccination would be unsuccessful in controlling this disease. Therefore antiviral chemotherapy is expected to play an important role in prevention and cure of RS virus infections.

21.3.3 VIRUSES CAUSING UPPER RESPIRATORY TRACT INFECTION

A galaxy of viruses causes upper respiratory tract infection (Figure 21.1), including the 150 or so small RNA viruses, the rhinoviruses, while additional important viral pathogens of the upper respiratory tract are the coronaviruses and four types of parainfluenza virus (Figure 21.1). In the USA alone in 1985, colds accounted for 17% of all episodes of acute ill-ness requiring medical attention, while 23 million days of lost work were recorded. The rhinoviruses circulate all the year round and are not seasonal. The antigenic variability among the rhinoviruses would appear to argue against the usefulness of vaccines were it not for the presence of certain common amino acid sequences in the virus attachment site. Although the latter respiratory viruses are more commonly inhabitants of the upper respiratory tract, they can predispose a patient to serious secondary bacterial infections of the paranasal sinuses and middle ear and possibly contribute to increased mortality of children in third-world countries.

21.3.4 THE CHEMISTRY OF INHIBITORS OF RESPIRATORY VIRUSES

It would be difficult to give an accurate estimate of the number of antiviral molecules already screened *in vitro* against respiratory viruses. Over 100 000 molecules have been tested and the handful of molecules shown in Figure 21.2 are those with some clinical activity. It has been a gigantic task, and many first-rate chemists have devoted their life's work to this area of science. The objectives of this difficult but rewarding work are summarized in Table 21.2.

The molecules which have been discovered to date include natural products of plants, such as alkaloids, and chalcones and short peptides, benzimidazoles, nucleoside ana-

Table 21.2 The ideal requirements of an antiviral agent against respiratory viruses

Absence of toxicity and side effects

Broad antiviral spectrum to inhibit influenza A and B viruses, respiratory syncytial virus, rhinoviruses and coronaviruses*

Administered orally

Cheap to synthesize and heat stable for storage

Chemical variation possible to cope with subsequent emergence of drug-resistant viruses

* Ribavirin and interferon partially meet this requirement

logues, simple amines and thiosemicarbazones. It should be clearly recognized that none of these compounds has been 'designed' as an antiviral: all have been found as a result of random screening. Most large pharmaceutical companies will possess a 'compound library' of molecules synthesized over the years for many areas of medical science. Such libraries may contain 250 000 molecules, but for a screen only a portion of these would be tested in cell cultures against a target virus. Following the identification of a 'lead' compound, considerable chemical ingenuity has been and is still used to synthesize more effective derivatives. A particularly fruitful source of antiviral compounds, respiratory viruses not excluded, are the nucleoside analogues of purines and pyrimidines.

Thousands of analogues of the naturally occurring nucleosides have now been synthesized and tested in the laboratory and most pharmaceutical companies have large libraries of such compounds alone. These nucleoside analogues are particularly useful to chemists since they can be modified extensively. It will be appreciated that even a single atomic substitution may change an active to an inactive molecule or vice versa. Frustratingly, however, it has been a common experience that, *after* the initial discovery of an active molecule, subsequent chemical modification more often leads to compounds with less antiviral activity than the reverse. It is not often appreciated that many of these compounds are difficult to synthesize, even by experienced chemists, and may require many months of work. Often, therefore, only small quantities of drug are available for *in vitro* tests and in only a few cases is enough synthesized for animal experiments. In the particular case of drugs active against respiratory viruses which replicate at mucosal surfaces, the delivery vehicle can assume very great importance. In fact the use of nasal sprays and aerosols may avoid the ingestion orally of large quantities of the drug which would be required to achieve inhibiting drug levels in the respiratory tract.

21.4 INFLUENZA

21.4.1 AMANTADINE – THE FIRST ANTI-INFLUENZA A VIRUS DRUG

One of the first antivirals [16] against a respiratory virus to be tested extensively in humans was the primary acyclic amine, amantadine or Symmetrel (Figure 21.2), which inhibits influenza A virus but not influenza B virus replication. Addition of a methyl grouping to give the analogue rimantadine alters the pharmacological distribution of the drug and prevents its entry into the brain, thus reducing the particular side effect of amantadine described as 'jitteriness' but without affecting the antiviral activity or antiviral spectrum. Many hundreds of similar amantadine molecules have been synthesized (reviewed in [47]), but none apparently greatly exceeds the original compounds in its inhibitory effect against influenza virus.

Clinical studies

Worldwide clinical studies have demonstrated the prophylactic effectiveness of amantadine against influenza A viruses. From the time of the great influenza pandemic in 1918 chemotherapists have attempted to find and use selective inhibitors of the virus, but not until the discovery of the primary amine amantadine [16] did effective chemoprophylaxis and even chemotherapy of influenza A virus infections become a reality. It had been known for some time that simple amines and ammonium compounds inhibited the replication of influenza virus in cell cultures and in eggs but successful clinical application was only achieved with amantadine, presumably because its unusual 'birdcage' structure (Figure 21.2) confers pharmacological stability. Initially, Jackson and colleagues [65] showed that the drug could prevent the development of respiratory symptoms in volunteers. It

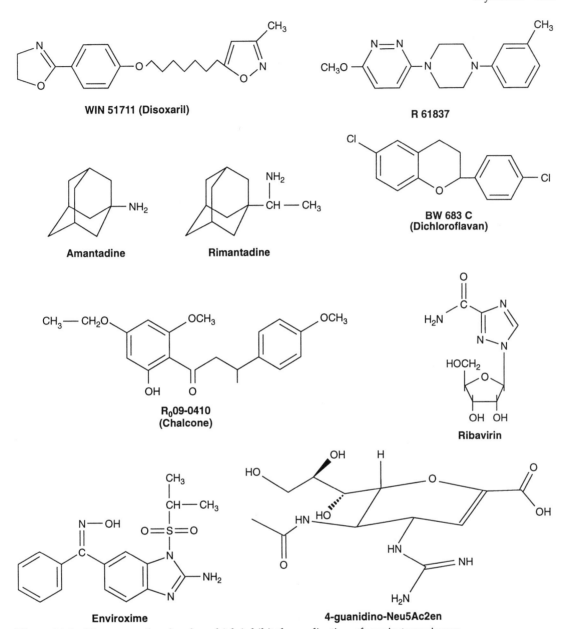

Figure 21.2 The range of molecules which inhibit the replication of respiratory viruses.

should be remembered that all these and subsequent clinical studies were pioneering in the sense that no drug had previously been shown to be active clinically against a respiratory virus, nor was it clear how such a compound could be used in the community to prevent disease. The properties of an ideal antiviral compound are summarized in Table 21.2 and it must be acknowledged that to date such a perfect molecule has not been found.

A study of amantadine in conjunction with the Royal College of General Practitioners

[24,25] is worthy of more detailed description because it was probably the first trial planned as a practical use of an antiviral, as well as a scientifically controlled study. Medical practitioners throughout the UK were asked to identify clinical cases of influenza and take blood (for serology) and nasal wash samples (for virus isolation) from this 'index case'. The index case was left untreated but immediate family contacts of the index case were given amantadine or placebo daily for 10 days in a double-blind trial. Then, the spread of virus from the untreated index case to the rest of the family was monitored. It was found that 14% of members of the families in the placebo group were subsequently infected with influenza A virus compared with 3.6% in the amantadine group. However, none of the latter persons had serologically confirmed influenza, whereas 10 (14.5%) of 69 persons in the placebo group had serologically confirmed influenza. Thus, amantadine used in a practical way successfully pre-

Table 21.3   Clinical trials with amantadine and rimantadine involving H3N2 and H1N1 viruses

Virus	Amantadine or rimantadine	Trial situation	Result	References
A/Shanghai/11/87/(H3N2)	Amantadine	Persons in nursing homes who had been vaccinated Prophylaxis	Viruses inhibited *in vitro* before prophylaxis. Drug-resistant viruses also isolated. Amantadine apparently reduced mortality	[41]
A/Bangkok/1/79 (H3N2)	Rimantadine	Double-blind, placebo-controlled therapeutic trial	Prompt reduction in virus titre in nasal secretion	[30]
A/USSR/77 (H1N1)	Rimantadine	Placebo-controlled trial in students		[70]
A/Leningrad/87 (H3N2)	Rimantadine	Placebo-controlled community prophylaxis study (low dosage 100 mg)	Low-dose rimantadine; 86% efficacy in illness reduction	[10]
A/Texas/1/85 (H1N1)	Amantadine	Prophylactic experimental challenge with wild type virus Adult volunteers (low dosage 100 mg)	78% protection against illness; 100-fold reduction in virus shedding	[63]
A/Bethesda/1/85 (H3N2)	Amantadine	Prophylactic experimental challenge with reduced dosage (100 mg)	Significant reduction in virus shedding Significant reduction in illness	[58]
Influenza A/H3N2 presumed 1985	Rimantadine	Therapeutic trial in children.	Viruses inhibited *in vitro* Significant reduction in fever and improvement in symptom scores. Resistant viruses also isolated	[12]
A/Philippines/2/82 (H3N2)	Amantadine	Prophylaxis in a partially vaccinated nursing home in reduced dosage (100 mg)	Beneficial effect on the decline of influenza	[5]
Typical influenza/A/H3N2 1986 virus	Amantadine	Uncontrolled trial in a boarding school (100 mg/day)	Significant reduction in illness compared with previous year's outbreak at the school	[15]
A/Philippines/82/(H3N2) A/Panama (H3N2) Influenza A H1N1	Rimantadine	Double-blind, placebo-controlled prophylactic study in children in families	73% reduction in rate of infection and 78% reduction in influenza illness	[13]
Typical influenza A/H3N2 1988 viruses	Rimantadine	Double-blind, placebo-controlled therapeutic study in families	Significant reduction in days of fever and symptoms Drug-resistant viruses isolated	[32]

vented spread of influenza A virus in persons in a family setting. With the advent of a new subtype of influenza virus the following year, namely A/Hong Kong/1/68 (H3N2) virus, a new trial was conducted which failed to show clinical efficacy [23].

However, numerous clinical trials, both in volunteers given attenuated influenza A viruses [67] and in the community where persons were infected with naturally virulent influenza A viruses, including influenza A (H3N2) viruses, confirmed and extended the initial studies of the prophylactic effect of amantadine (Table 21.3). Of particular significance has been the observation over the succeeding decades that all such subtypes of influenza A virus so far encountered in humans, namely H2N2, H1N1 and H3N2, have been inhibited both in the laboratory and also in clinical trials (see below).

The results of some studies recently have raised the question of whether amantadine has a mild suppressive effect on the development of immunity following infection. A lower dose of amantadine (50 mg) had no effect on local IgA production in the respiratory tract whereas a suppressive effect was noted in the systemic IgG response. For clinical use a recommended prophylactic and therapeutic dose of amantadine is 200 mg daily, reduced to 100 mg in those aged 10–15 years or over 65 years. The suggested dose in children aged 1–9 years is 2–4 mg/kg [26,45]. It is a truism that much of the pharmacology data for amantadine was collected using healthy volunteers whereas different problems occur in the use of any drug in the frail and elderly. In these, a 50 mg dosage of amantadine may lead to inadequate steady-state trough serum concentrations whereas 100 mg may lead to excessive levels and adverse effects, particularly in the presence of renal impairment. Some clinicians therefore recommend individualization of dosages in the elderly.

Therapeutic effect

Arguably the most significant discovery, at least for the future development of antivirals against respiratory viruses, was that amantadine had a *therapeutic* effect. In the initial studies, volunteers in prison in the USA who were infected with influenza A virus and were showing the first signs of clinical influenza were given amantadine or placebo (reviewed in [47]). The influenza illness, including objective parameters such as temperature subsided more rapidly in persons to whom amantadine was administered. Subsequently, Galbraith and his colleagues [47] planned a therapeutic trial in the UK on naturally infected patients in families and confirmed the therapeutic effect of oral amantadine administered not more than 24 hours after the first influenza symptoms appeared [47]. Perhaps, not surprisingly, the use of the antiviral later than this time failed to abrogate symptoms of the disease. It should be mentioned that before these trials most virologists considered that, concomitant with the peak of respiratory symptomatology, influenza virus replication would also reach a peak and therefore the therapeutic use of an inhibitor at this stage of the disease would be fruitless. Therefore, the demonstration of the therapeutic effect of amantadine, as well as being of considerable scientific interest, also emphasizes the need to perform experiments rather than placing too much emphasis on hypothesis. Few studies have actually been performed on the quantity of influenza virus excreted in the lower respiratory tract of infected humans. Postmortem examination of trachea specimens by immunofluorescence and cytology during the influenza A (H2N2) virus pandemic showed that virus replication was patchy and that many cells remained uninfected. This observation gives credence to the concept of the theoretical usefulness of therapeutic intervention since antiviral drugs would be expected to prevent spread of virus further down into the respiratory tract.

It must be acknowledged that the precise degree of clinical impact of amantadine or, indeed, any other antiviral drug against respiratory virus infection in the community has yet to be calculated. For example, it is quite possible that early use of the compound could prevent spread of the influenza A virus into the lower respiratory tract and lungs and so abrogate mortality [52]. Pharmacological experiments both in animals and humans have shown a somewhat preferential accumulation of amantadine in respiratory tissue including the lung (reviewed in [47]). To date, such clinical investigations of possible effects of amantadine on influenza virus-induced mortality in the elderly [14] have not been undertaken widely and in our opinion are urgently needed.

Amantadine is under-used as an antiviral

At present amantadine is not widely used in clinical medicine. For the future development of new antivirals it is important to understand how this situation has arisen. No pharmaceutical company would wish to invest scientific endeavour to discover a drug which perhaps for logistical reasons could not be used to its full potential. Three contributory reasons for the under-use of the drug are: concern about the mild side reactions of jitteriness noticed by 7% of patients taking the drug; the problem of self-administration (the ideal situation for

rapid use of the drug upon contact with an infected person) and the presumed absence of very dramatic clinical effects, such as prevention of mortality. In an event, which must be unique in the history of antimicrobials, two scientific meetings were convened to make suggestions about the clinical use of amantadine (Table 21.4). Both scientific panels recommended more extensive clinical use of amantadine and rimantadine. Amantadine, therefore, is a pioneering drug and the anticipation is that it and subsequent inhibitors might be more effectively used in near future.

The discovery and progression to clinical use of rimantadine, a drug closely related to amantadine, is not without relevance and interest to the future selection of new drugs. Firstly, rimantadine differs chemically from the parent molecule by the addition of a methyl group, but this change is sufficient to alter the pharmacological distribution of the drug without compromising effective antiviral action. Thus, rimantadine cannot penetrate the blood–brain barrier and so does not induce any neurological side effects sometimes noted with amantadine. Tissue and plasma levels are normally two-fold higher following oral

Table 21.4   Recommended usage of amantadine and rimantadine in prophylaxis and therapy*

1. There should be clear epidemiological and virological evidence of influenza A virus infection in the community

2. Amantadine or rimantadine should be used prophylactically each day for 6–8 weeks

3. Groups recommended for special consideration:†
   (a) High-risk persons, such as diabetics, the elderly and those with chronic heart, respiratory or circulatory disease, whether vaccinated or not
   (b) Special community groups such as public transport employees, doctors, nurses
   (c) Persons who may wish to be given prophylaxis or therapy

4. Drug prophylaxis should not replace vaccination because amantadine does not inhibit influenza B virus

5. Therapy should be initiated within 24 hours of first symptoms and given for 10 subsequent days

* These recommendations are an amalgam of those first published following the NIAID meeting (1980) and the Vienna WHO meeting [8]
† Amantadine or rimantadine cannot be used as a replacement for influenza vaccine in the elderly or those at special risk [14] because the drugs do not inhibit influenza B virus

administration of rimantadine (compared with amantadine) and thus dosage is reduced two-fold to 100 mg daily. The rimantadine compound was first discovered in the USA at the du Pont laboratories, but its apparent advantages in the clinic were first noted in the former USSR, where the drug has since been used extensively [38]. In the last few years scientific interest in rimantadine has been revived in the USA and the compound is now licensed there for clinical use. In spite of very large searches by different groups throughout the world no amantadine-like molecule with significantly improved antiviral efficacy has been discovered and, moreover, consensus opinion and also side-by-side comparison [20] would be that amantadine and rimantadine have a comparable clinical effect.

A potential advantage of rimantadine compared with amantadine is a lower risk of central nervous system effects such as light-headedness, difficulty in concentrating, nervousness and insomnia, which can be a problem in a minority of patients treated with amantadine. Dose-related, reversible gastrointestinal complaints such as nausea and vomiting are similar to those with amantadine. Since 200 mg/day doses can lead to high plasma levels of amantadine and an increased incidence of adverse effects in elderly nursing home residents, lowering the dosage to 100 mg/day may be advisable in such patients. Safety in pregnancy has not been established for either drug. There is increasing evidence that amantadine may be used as an adjunct to influenza vaccine in those persons at special risk.

Antiviral activity *in vitro*

In vitro sensitivity testing of influenza A virus subtypes demonstrates that most influenza A viruses are inhibited by amantadine or rimantadine. Earlier reported studies of the inhibitory activity of amantadine for influenza A viruses showed near universal inhibitory effects against all subtypes tested [59].

However, sensitivity testing has continued up to the present time and modern ELISA techniques have facilitated the testing of larger numbers of isolates. In a very extensive study, Belshe *et al.* [9] analysed the susceptibility of current influenza A viruses to amantadine and rimantadine. Some 65 influenza A (H1N1) viruses and 181 influenza A (H3N2) viruses isolated in rhesus monkey kidney cell culture from patients between 1978–1988 and 25 strains between 1984–1985 were tested. All were susceptible to inhibition (see Table 21.5). All of the 65 influenza A (H1N1) viruses isolated between 1980–1987 were susceptible to inhibition by amantadine and rimantadine, including A/England/80, A/Victoria 83 and A/Taiwan/85 like viruses. Similarly, all 96 influenza A (H3N2) viruses, namely A/Taiwan/79, A/Oregon/80, A/Philippines/82 and A/Mississippi/85-like viruses, tested between 1980–1986 were inhibited by the drugs. In addition, others have tested field isolates directly for resistance, including the following strains: A/Shanghai/1 1/87 (H3N2) [41] and influenza A (H3N2) viruses circulating in the USA in 1985 (Table 21.5).

Five drug-resistant viruses (A/Sichuan/87) were detected among 85 tested in 1987 and 1988 and these were from patients participating in a rimantadine trial. Therefore, sporadic and very occasional influenza A viruses when tested *in vitro* may show a lesser degree of inhibition to amantadine. The great majority of influenza A (H1N1, H2N2 and H3N2) viruses tested to date have been well inhibited by amantadine or rimantadine.

Emergence of drug-resistant viruses

Soon after the discovery of the anti-influenza activity of amantadine, emergence *in vivo* of drug-resistant influenza A (H2N2) viruses which were also resistant to other amantadine-like molecules was noted [49,51]. Early work indicated that gene 7 (matrix) was responsible for conferring sensitivity or resis-

Table 21.5 Susceptibility to amantadine and rimantadine of influenza A viruses from 1978–1988

Virus, epidemic year	No. susceptible to amantadine and rimantadine/No. tested	Antigenic characteristics of virus
H1N1		
1978–79	2/2	Not tested
1980–81	7/7	England/80-like
1982–83	11/11	England/80-like
1983–84	5/5	Victoria/83-like
1986–87	40/40	Taiwan/85-like
Total	65/65	
H3N2		
1980–81	32/32	Taiwan/79-like
1982–83	37/37	Oregon/80-like*
1984–85	25/25	Philippines/82-like
1985–86	2/2	Mississippi/85/-like
1987–88	80/85†	Sichuan/87-like
Total	176/181†	

* Some isolates were Texas/77-like
† Five resistant viruses were from patients participating in rimantadine versus placebo study (data from [9])

tance to amantadine [27,40]. In a wider virological context the selection of drug-resistant viruses is expected and, indeed, acknowledged to confirm the specific antiviral activity of a drug, thus indicating that the compound is not inhibiting viral replication indirectly via an effect on the host cell. In all other virus–drug combinations including HIV and zidovudine [39, 44], herpes simplex viruses and Zovirax [50] and rhinoviruses [73], drug-resistant viruses can be detected but, to date, they usually have had little clinical impact.

Influenza viruses resistant to amantadine or rimantadine have been detected at low frequency in the field, even in countries such as Germany, which have not utilized amantadine-like drugs [33] and also have been specifically isolated from ill patients undergoing prophylactic or therapeutic treatment with rimantadine [29, 31, 32] or amantadine [17]. Drug-resistant viruses are detected more frequently during treatment than prophylaxis [29]. However, even in such treated patients the drug appears to continue to exert beneficial clinical effects [32]. Drug-resistant variants occur with a frequency of one particle among 10 000 virions in cell cultures but do not appear to have any selective advantages in nature because their recovery is so rare [29]. It is quite possible that natural selection of antigenic variants and the disappearance of previous variants might prevent the emergence of amantadine-resistant viruses with changes in the M2 and HA genes.

Mechanism of action

The block to virus replication mediated by amantadine is thought to occur after the virus has bound to the cell but before uncoating occurs [11,19,21]. In the presence of the drug, intact complexes of the ribonucleoproteins (RNPs) and the viral membrane protein (M_1) can be isolated from cells, but these cannot be found in the absence of the drug [11]. It has been assumed that M_1 protein is selectively removed from the RNP structure at acidic pH (pH 5.5) in the endosomal compartment. However, as the M_1 protein resides inside the viral lipid bilayer, a mechanism, sensitive to amantadine, of making the interior of the virion accessible to a pH change is required. The M_1 protein is a peripheral membrane protein that associates with both the RNPs and the cytoplasmic face of the lipid bilayer. In contrast, the M_2 integral membrane protein (97 amino acids in length) is abundantly expressed at the plasma membrane of virus-infected cells but only a few (on average 23–60) molecules are incorporated into virus particles. The M_2 protein spans the lipid membrane once and is orientated such that it has a 23 amino acid N-terminal extracellular residue and a 54 residue C-terminal cytoplasmic domain; thus, M_2 is a model type III integral membrane protein. The native form of the M_2 protein is minimally a homotetramer, consisting of either a pair of disulphide-linked dimers or disulphide-linked tetramers, the disulphide bonds acting to stabilize the oligomer.

A most interesting study by Pinto, Holsinger and Lamb [57] provided direct evidence that the influenza virus M_2 protein forms an ion channel connecting the two sides of the lipid membrane. Experimental injection of wild-type M_2 mRNA into oocytes produced an inward current that could be blocked by addition of the anti-viral drug amantadine. However, amantadine did not block the conductance of oocytes expressing M_2 proteins that contained mutations that confer viral resistance to amantadine. In the virus particle, M_2 may be an ion channel permitting the flow of protons from endosomes into the virion interior to facilitate removal of M_1 protein from RNPs during virion uncoating in endosomes.

Rimantadine benefits against drug-resistant viruses

The recovery of amantadine or rimantadine-resistant influenza A virus from drug-treated children has been documented, and this finding has been confirmed in adults (reviewed in [29]). The transmission of drug-resistant influenza A virus from treated patients to contacts receiving drug prophylaxis has been described in the household setting during rimantadine use [31] and in nursing home outbreaks managed with amantadine.

The findings of a study by Hayden *et al.* [32] show that drug-resistant influenza A virus can be recovered from children or adults treated with rimantadine. Resistant virus was recovered from samples taken from one-third of rimantadine-treated patients collected on the fifth day of treatment. In the adult treatment study, which incorporated frequent samplings for virus isolation, resistant virus was detected in three of six rimantadine recipients and shedding began as early as the second day of treatment. Rimantadine recipients from whom virus was recovered on day four or five of treatment were likely to shed resistant virus in both the family-based (80%) and adult treatment (100%) studies. The results of the adult treatment study [32] confirmed earlier observations in children that once shedding of resistant virus develops, subsequent viral isolates from the same patient are also drug resistant. However, the clinical importance of recovering drug-resistant virus from treated patients is still unresolved. In the adult treatment study, rimantadine administration was associated with significant reductions in virus titres in the upper respiratory tract, despite the recovery of resistant virus from some patients. Thus, the drug still appeared to be exerting an antiviral effect. Patients who remain virus-positive after several days of drug treatment appeared to be extremely likely to shed drug-resistant virus, and one of the patients in the Hayden trial continued to shed high titres of drug-resistant virus during therapy (2.5 $\log_{10}$ $TCID_{50}/0.2$ ml) on treatment day five. Of course the clinical concern is that rimantadine-treated individuals may transmit resistant virus to close contacts, and this problem has been considered in households during rimantadine use and in a nursing home during amantadine use. In the household setting, avoiding treatment of ill index cases, specifically young children, would appear to reduce the likelihood of failure of drug prophylaxis as a result of the apparent transmission of resistant virus.

A sensible approach is to recommend that the presence or absence of drug-resistant viruses should continue to be monitored in clinical trials but to acknowledge that at present the problem at the clinical level is minimal and does not require changes in already published recommendations from the WHO consensus meeting [8] on the clinical use of amantadine or rimantadine. Essentially similar conclusions have been presented already in the UK [45], the USA [9, 43] and in the former USSR [38].

21.4.2 GENERAL RECOMMENDATIONS FOR THE CLINICAL USE OF AMANTADINE

A number of recommendations have been made from International and National Committees and Health Authorities regarding the clinical use of amantadine.

WHO consensus meeting (Vienna)

The committee recognized no significant difference in the degree of inhibition of human influenza A virus subtypes by amantadine [8]. It was concluded that the major deciding factor for the clinical use of amantadine was the prior isolation of influenza A virus (regardless of antigenic subtype) in the community. Under these circumstances prophylaxis of 6 weeks' duration or, alternatively, therapy of 10 days' duration, with amantadine was recommended.

Similarly, the NIH consensus meeting on the clinical use of amantadine recommended the drug's increased use (and now presumably also rimantadine) to combat influenza A infection in the community, not only particularly in the 'special risk' groups, but also in other groups in the community. No scientific discrimination was made between antigenic

subtypes of influenza A virus as regards sensitivity to inhibition by amantadine or rimantadine and all influenza A strains of different subtypes were considered sensitive. The consensus opinion is:

1. Amantadine and rimantadine are effective inhibitors of naturally occurring influenza A virus of all subtypes.
2. The efficacy of amantadine or rimantadine is equivalent or exceeds that of the influenza A virus component of currently licensed inactivated influenza virus vaccines. However, the antivirals do not inhibit influenza B virus which can cause mortality in the 'special risk' groups.
3. The compounds could with benefit be used more widely to provide extra protection in already immunized 'at risk' persons or in non-immunized 'at risk' persons or in special community groups.
4. Drug-resistant variants have emerged but are not at present a clinical problem and since the beneficial clinical effects of amantadine and rimantadine are apparent the compounds should continue to be used as recommended.
5. Monitoring should continue for drug-resistant variants.

Amantadine or rimantadine use in an influenza pandemic

In the event of a pandemic of influenza it is recognized that the Public Health Laboratory Service (PHLS) and other major laboratories will play a major role in isolating and identifying the new virus in the UK, monitoring its spread and initiating studies on vaccines and antivirals. The PHLS 'pandemic report' recognizes six phases in the emergence of a pandemic:

- phase 0   interpandemic period
- phase 1   emergence of a pandemic virus, presumably outside of the UK

- phase 2   epidemic or pandemic caused by the virus outside the UK
- phase 3   new virus isolated in UK – pandemic imminent
- phase 4   pandemic influenza in UK
- phase 5   return to 'background' influenza activity

The report specifically identifies amantadine, or rimantadine, as the compound to test in high-risk populations in phase 3. At present, amantadine is the only antiviral compound licensed for use against influenza in the UK. The report also recognizes the usefulness of assessing the degree of inhibition that amantadine exerts on a new influenza A virus during phase 2 before the new virus has arrived in the UK and caused clinical outbreaks. This is consistent with present policies of continued and voluntary monitoring of influenza A virus strains for sensitivity to inhibition by amantadine.

A minimal period of 36 weeks elapses from the isolation of a new influenza virus to the time of production of the first batches of inactivated vaccine. Therefore, it is apparent that a vaccine would not be available for high-risk individuals during the important first wave of the pandemic. Chemoprophylaxis with amantadine or rimantadine would have an important role to play both during the initial outbreak and in the succeeding waves.

Possible amantadine use for the first wave of an influenza pandemic

The EC licensing of amantadine and rimantadine for use against 'all influenza A viruses' as in the USA would permit rapid deployment of the drug in 'special risk' groups during the potentially dangerous first wave of the pandemic when no vaccine would be available. It is generally acknowledged that amantadine and rimantadine would have an important role to play during the evolution of a virus pandemic. It is also recognized [43] that the potential clinical benefits of the drug were compromised by previous restrictive licensing in the USA. Although *in vivo* studies had shown that all influenza A viruses of different subtypes were inhibited by amantadine the drug had, nevertheless, been licensed in 1966 in the USA for use against type A (H2N2) strains only. Re-licensing for use against the newly emerged pandemic influenza A subtype H3N2 in 1968 was not accomplished until 1976 when amantadine was approved in the USA for prophylactic and therapeutic use against *all* type A influenza viruses. These factors restricted the use of the drug during the important influenza A (H3N2) pandemic in 1968, but the change in licensing, was accomplished in time for the drug to be used effectively against the newly emerged influenza A (H1N1) virus in 1977.

Will amantadine and rimantadine inhibit new pandemic strains of influenza A virus?

There are two conflicting, and as yet unresolved, hypotheses of the origin of new pandemic influenza A strains. Firstly, a recycling theory whereby a virus re-emerges from elderly persons who many years previously have been infected with other subtypes [37] or, secondly, a theory of the emergence from 'mixing bowl' situations in pigs of new reassortant viruses with virulence for humans (reviewed in [66]). The latter hypothesis has most current scientific support, although the recycling theory is not discounted. In either case, however, it is apparent that a pandemic strain may not be 'new' as such but will have an HA or NA, either from an old human strain (e.g. H2) or from an existing avian or animal virus (H4–H13). In either case, evidence at present would suggest potential susceptibility of all subtypes of influenza A virus to inhibition by amantadine [59]. Recent studies in our laboratory have confirmed the *in vitro* inhibition of influenza A viruses of the majority of the avian subtypes (H1–H13) (J.S. Oxford and M. Wigg, unpublished data).

As noted above, the mode of action of amantadine would appear to be closely related to the M_2 protein of influenza A virus. Of the 120 isolates of amantadine or rimantadine-resistant influenza A viruses that have been characterized genetically so far, all have changes restricted to amino acids 27, 30, 31 or 34 in the M_2 protein [9,29]. With a change in antigenic subtype and the advent of a pandemic virus, the HA and NA proteins are completely different whereas other gene products including M_2 (coded by gene 7) may or may not remain unchanged.

The M_2 protein appears to be highly conserved among influenza A virus strains of all known subtypes (reviewed in [66]) and this would seem to provide the scientific basis of their universal susceptibility to amantadine. Therefore, it could be hypothesized that the emergence of a 'new' pandemic strain with completely novel HA and NA proteins for the human population would be expected to have few or no implications as regards susceptibility to amantadine, which at the low dosages used in human administration would be dependent on the amino acid sequence of the much less variable M_2 protein.

21.4.3 'DESIGNED' ANTI-NEURAMINIDASE DRUGS

Eleven years ago a group of scientists in Melbourne published the first data on the atomic X-ray structure of influenza A virus neuraminidase (reviewed in [66]). The latter enzyme is not virus-specific *per se* and similar enzymes are widespread in the animal kingdom. These enzymes are glycohydrolases which cleave terminal sialic acid from glycoproteins or glycolipids or oligosaccharides. In virological terms the enzyme functions at the stage of influenza virus release from an infected cell. The main study of anti-neuraminidase drugs commenced a quarter of a century ago with the synthesis of the Neu5Ac2en, a derivative of sialic acid which inhibited viral, bacterial and mammalian siali-

dases. Palese and Schulman (reviewed in [48]) tested a series of these compounds as anti-influenza virus drugs and although a number of analogues of Neu5Ac2en inhibited influenza virus replication in cell culture no antiviral activity was detected in animal models. In retrospect, this negative finding may be attributed to the rapid metabolism of the compounds tested in this early work, and now with newly synthesized, more effective analogues which can be applied topically, a specific anti-influenza virus effect can be measured *in vivo*. The new team of chemists, X-ray crystallographers and virologists have, in their own words, 'rationally designed a series of sialidase-based inhibitors of influenza virus' [72]. Firstly though it is advisable to review the basic structure of the neuraminidase molecule itself.

Influenza sialidase is a tetramer of identical subunits. There are six four-stranded, anti-parallel B-sheets arranged as if on the blades of a propeller (reviewed in [66]). The active enzyme site is a deep cavity on the neuraminidase protein surface and is lined entirely by amino acids that are invariant in sialidases of all strains of influenza A and B which have been characterized. In contrast, variable amino acids are found next to and encircling the active site and these are antigenic determinants reacting with post-infection anti-neuraminidase antibodies.

Sialic acid, the product of the enzyme-catalysed reaction, binds to the active site of the enzymes in a 'boat' configuration. The carboxylate interacts with three invariant arginine residues on the neuraminidase enzyme. In this conformation it resembles the geometry of the unsaturated sialic acid analogue (Neu5Ac2en), a potent sialidase inhibitor, as noted above. An important aspect of the active site is that strain invariance extends to a number of other amino acids which do not themselves make contact with sialic acid but which provide substructure or a scaffold on which amino acids contacting the bound sugar are supported.

The Melbourne group made predictions of energetically favourable substitutions to the unsaturated sialic acid analogue Neu5Ac2en [72]. Among the most interesting of these chemical modifications was the replacement of the hydroxyl group at the four position on Neu5Ac2en by an amino group. In theory, substitutions of the 4-hydroxyl group by an amino group should produce a significant increase in the overall binding interaction due to a salt bridge formation with the side chain carboxylic acid group of Glu119 in the enzyme. Similarly, the replacement of the same hydroxyl group by the significantly more basic guanidine group was predicted to produce an even tighter affinity of the substituted Neu5Ac2en for the active site as a result of lateral binding through the terminal nitrogens of the guanidino group with both Glu119 and Glu227. As predicted, direct measurements of viral neuraminidase enzyme inhibition showed that the 4-amino and 4-guanidino-substituted Neu5Ac2en (Figure 21.2) were high-affinity inhibitors for the influenza virus enzyme. Moreover, the compounds inhibited influenza virus plaque formation in cell culture. Most excitingly, the compounds had antiviral effects *in vivo* (Figure 21.3). Thus, inhibition of influenza virus replication was noted in both mice and ferrets, with both 4-amino and 4-guanidino-Neu5Ac2en inhibitor and, to a lesser extent, with Neu5Ac2en itself. The most comprehensive data have been obtained with 4-guanidino-Neu5Ac2en in mice infected with influenza A/Singapore/1/57 virus, comparison being made with amantadine and ribavirin used as positive control compounds .

Results obtained with 4-guanidino-Neu5Ac2en administered both before and during infection of ferrets at a dose of 50 µg/kg twice daily (b.d.) are shown in Figure 21.3. In this experimental system, virus shedding is observed typically in nasal washes from 3 to 5 days after inoculation, and an elevated body temperature recorded on days 2–3 after inocu-

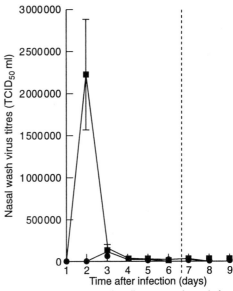

Figure 21.3 Treatment of influenza-infected ferrets with 4-guanidino-Neu5Ac2en. Influenza virus titres were measured in nasal washings from ferrets treated with 4-guanidino-Neu5Ac2en (treated, ●) or distilled water (controls, ■). Ferrets were infected intranasally with $10^{5.0}$ TCID$_{50}$ influenzaA/Mississippi/1/85. Treated and control groups ($n=5$) was treated twice daily (8 hours apart) from the day before infection to day 5 after infection. On the day of infection the first treatment dose was given 2 hours before infection (from [72]).

lation. At 50 µg/kg (b.d.) 4-guanidino-Neu5Ac2en effectively abolished virus shedding and reduced pyrexia fully in all of five animals. 4-Guanidino-Neu5Ac2en was approximately 1000 times more effective than amantadine when given intranasally to either species. The compound has now reached phase II and III clinical trials in various countries and has been shown to have prophylactic activity in students infected with an influenza A (H1N1) virus.

21.5 RESPIRATORY SYNCYTIAL (RS) VIRUS

21.5.1 INHIBITION BY RIBAVIRIN

Just as amantadine acted as a novel pioneer drug for a respiratory virus, namely influenza

A virus, ribavirin was also pioneered as a nucleoside analogue for possible long-term prophylaxis and therapy against RS virus and influenza virus (reviewed in [71]). When the compound was first discovered, only relatively toxic nucleoside analogues, such as iododeoxyuridine and adenine arabinoside, had been used for antiviral chemotherapy of serious herpesvirus infections. Now, 15 years later and with the additional extensive clinical experience of using the antiviral nucleoside analogue, acyclovir, it is generally acknowledged that not all analogues of naturally occurring purines or pyrimidines are inherently toxic for replicating mammalian cells.

Nevertheless, because of the relatively low therapeutic margin, ribavirin has been more extensively tested in aerosol form than other antivirals. As seen with the novel influenza anti-neuraminidase inhibitor above, aerosol administration has the theoretical advantage of direct deposition of the drug onto the tissues of the upper and lower respiratory tract (reviewed in [7]) without compromising the function of other organs where the presence of the drug is, in any case, not required for antiviral action. A potential disadvantage of aerosol use is the complicated equipment required, particularly for use of ribavirin in infants suffering from RS virus infection. Nevertheless, ribavirin is the only drug to have shown effectiveness in the clinic against potentially serious RS virus infections, although other compounds may be effective *in vitro* or in animal models. Pioneering studies were carried out by Hall *et al.* (reviewed in [48]) in controlled clinical trials, firstly in adults and then in children. The studies involved normal infants who were hospitalized with pneumonia or bronchiolitis but were otherwise normal, or infants with more serious underlying conditions, placing them at high risk from serious RS virus infection. Infants receiving ribavirin showed a significantly greater amelioration of their illness than those in the placebo group and improvement on the first day of therapy as well as on subsequent

dates. The results of these important studies show that antiviral therapy of RS virus is possible and gave encouragement and impetus in the search for other inhibitors of this otherwise uncontrolled virus infection, including interferons [34]. More recently, Fleming and Cross [22] have provided evidence of RS virus causing life-threatening infections in the elderly and ribavirin may have a clinical role to play in this group.

21.6 RHINOVIRUSES

21.6.1 ANTIVIRAL DRUGS INTERACTING WITH VIRUS STRUCTURAL PROTEINS

It might be expected that viruses are most vulnerable to interception when in the extracellular or transmission phase of their life cycle. Indeed, it is precisely at this point that many vaccines exert their effects via the induction of neutralizing antibody. However surprisingly, few antiviral compounds are known to inhibit or inactivate extracellular virus. Bile salts have been investigated recently [46] as a means of inactivating lipid-enveloped viruses. Some activity against influenza A virus infection was detected in an animal model system, but the main impetus of the research was towards direct inactivation of HIV. However, virologists interested in chemotherapy of rhinoviruses have discovered some remarkable molecules which bind to these viruses. A good example of molecules acting at this stage is the series of drugs which bind to the external VP1 coat protein of picornaviruses; typical examples are the arildone series of antiviral inhibitors dichloroflavan and chalcone (reviewed in [18,68]). These compounds are considered to *stabilize* the virus and so prevent the uncoating stage whereby the virus releases its nucleic acid intracellularly. In an extremely innovative study of the interaction of a virus with an antiviral (the first to be carried out at the atomic level using X-ray diffraction studies) Smith *et al.* [64] described the binding of WIN 51711 (Disoxaril) within a small canyon at the base of the receptor-binding pocket of rhinovirus type 14. It is hypothesized that the antiviral molecule blocks the entry, into the inner part of the virion, of ions which would normally participate in the process of uncoating of the rhinovirus RNA soon after the virion has entered the cell. Similarly, the cell receptor-binding site on influenza HA has been identified following X-ray crystallography studies of the protein; it is situated as a shallow saucer-like indentation at the tip of the molecule and in close proximity to the important antigenic site B. The precise configuration and spatial relationships of amino acids constituting the sides and floor of the site are known but to date this knowledge has not been exploited to develop antiviral inhibitors of influenza virus.

It is calculated that 72 molecules of WIN 51711 bind per rhinovirus virion. Fortunately, the electron density of WIN 51711 was unequivocal with the oxazoline and phenoxy groups appearing as two flattened bulges separated from the bulge of the iso-oxazole group. The oxazoline phenoxy (OP) groups were located in the hydrophilic region at an opening on the floor of the canyon of the receptor-binding site on the rhinovirus particle. The OP end may orientate by hydrogen bonding between the nitrogen in the oxazoline ring and a hydrogen of Asn219 amide group in VP1. The methyl iso-oxazole group inserts itself into the hydrophobic interior of the VP1 barrel. In this case, as mentioned above, the compound could act by blocking the entrance of ions into the channel which normally acts as a conduit to the inside of the virion. An interesting analysis was performed to determine the minimum inhibitory concentration (MIC) of WIN 51711 for two human rhinoviruses, namely HRV-2 and HRV-14. Also, the amino acid composition of their receptor-binding pockets was compared with those of other picornaviruses including meningovirus, and foot and mouth disease virus of cattle (FMDV) which are not inhibited

by WIN 51711. A possible explanation for this latter negative observation is the replacement of Asn219 in the WIN 51711-insensitive viruses by serine or glycine which does not allow correct orientation of the oxazoline group of the drug. In HRV-2, Val188 and Val191 have both been changed to leucine which would sterically hinder insertion of the isoxazole groups and aliphatic chain into the hydrophobic pocket. Concomitantly, the MIC value of WIN 51711 for HRV-2 increased compared with that for HRV-14. This classic study is now taken as an example of a future direction required for the selection of antiviral drugs. Nevertheless, the virological and pharmacological aspects and correlation are still very important since, for example, the X-ray study could not give an indication of whether a compound selected for very tight binding to the canyon would penetrate the cell plasma membrane to reach the intracellular virus or would be absorbed after oral administration and reach the tissues of the upper respiratory tract.

A compound known as SCH 38057 (I [6-(2-chloro-4-methoxy)-hexyl]imidazole hydrochloride) is a water-soluble antiviral agent that binds irreversibly to enterovirus and rhinovirus particles [74] and has also been subjected to X-ray crystallographic analysis. The structure of rhinovirus 14 complexed with SCH 38057 shows that this compound binds at the innermost end of the hydrophobic pocket of VP1, leaving the entrance unoccupied. This interaction induces conformational changes involving a stretch of at least 36 amino acids of VP1 and VP3.

A number of flavan derivatives have been synthesized and assayed against rhinovirus type 1B and 4,6-dichloroflavan blocked the replication of this rhinovirus in cultured cells. Clinical trials carried out at the CCU with the compound given orally or intranasally, failed to show antiviral effects. However, the compound could not be found in nasal secretions [56] which may provide a simple explanation for the failure.

The Roche Group have synthesized Ro 09-0410 (4'-ethoxy-2'-hydroxy-4,6'dimethoxy-chalcone), a compound active against rhinoviruses and which, like dichloroflavan, SCH 38057 and WIN 51711, binds to the viral capsid. Radioactively labelled compound binds specifically to rhinovirus particles and this binding is inhibited by dichloroflavan or RMI15,731, suggesting a similar mode of binding. Of 53 rhinovirus serotypes tested, 46 were inhibited by Ro 09-0410. Mutant rhinoviruses resistant to chalcone were also cross-resistant to dichloroflavan [73]. These results indicate that the three compounds had a similar mode of action based on their binding to virion particles. Several analogues of Ro-09-0410 have been synthesized, some of them being ten times more active than the parent compound. These analogues bind to the same site on the virus particles as Ro 09-0410 and have a similar mode of action [55]; however, the parent compound was not active clinically [2].

Finally, enviroxime (2-amino-1-(isopropyl-sulphonyl)-6-benzimidazole phenyl ketone oxime) is a benzimidazole derivative with high activity against rhinoviruses inhibiting RNA replication. Several human trials using enviroxime in an attempt to prevent colds induced by rhinoviruses have been conducted, but without any clear clinical benefit being observed [28,53].

21.6.2 VIRION-BINDING DRUGS ACTING AGAINST RHINOVIRUS INFECTIONS

Although the results of using dichloroflavan [3,56], Ro 09-0410 [2] and 44,081 RP [75] and other molecules in this group of virion-binding drugs clinically have been negative, recent trials with the drug R61837 have provided some striking evidence of efficacy. It is a truism that one of the major problems in the treatment of common colds is to attain drug levels in the nasal epithelium that are sufficiently high to block viral growth. When R61837 was given to volunteers as a nasal spray before they were inoculated with rhinovirus 9, the compound showed clinical efficacy, but failed to show effects when used therapeutically [1,6]. A significant factor may have been the use of a new vehicle which produced a soluble drug administered by intranasal spray. Indeed, the vital importance of the administration route of drugs with problems of solubility has been noted above with the new influenza anti-neuraminidase compound.

21.7 CONCLUSIONS

A series of crucial studies at the MRC Common Cold Unit signposted developments in virology and antiviral chemotherapy of the respiratory viruses over the past four decades. The discovery of methods for cultivating rhinoviruses and the use of organ cultures for growing coronaviruses have been key technical developments at the Unit and both discoveries resulted from keen attention to technical detail in the laboratory. The first demonstration of the clinical activity of interferon [42] and most recently the description of clear clinical effects of an anti-rhinovirus drug [I] have also shown the importance of extremely close monitoring of the clinical parameters of viral respiratory disease. At the heart of the Unit at Salisbury was the concept of the interaction of laboratory science and clinical medicine. Now, for the first time, the techniques of the molecular biologist (particularly X-ray crystallography) are being focused on the redesign or sensible alteration of existing antivirals and the last few years have witnessed the publication of some landmark papers [64,74]. The future will hold the design of an antiviral *de novo*. Concomitant with the latest discoveries in chemistry, there is now a requirement for the construction of a European facility for the clinical evaluation of these and hopefully the succession of new molecules over the next years. Antiviral chemotherapists certainly do not believe that these small molecules can be used to *eradicate* viral diseases. Compared with those of vaccinologists, the goals are smaller but nevertheless well defined, namely to search for and to discover small molecules that will interrupt the replication of respiratory and other viruses and hence alleviate human suffering. Viruses, and in particular the two great pandemic RNA genome viruses, influenza A and HIV, are continually evolving, and it must be appreciated that they pose a tremendous threat, ever capricious and unpredictable in nature to our well-being.

21.8 REFERENCES

1. Al-Nakib, W., Higgins, P.G., Barrow, G.I., Tyrrell, D.A.J., Andries, K., Vanden Bussche, G. Taylor, N. and Janssen, P.A.J. (1989) Suppression of colds in human volunteers challenged with rhinovirus by a new synthetic drug (R61837) *Antimicrob Agents Chemother.*, 33, 522–5.
2. Al-Nakib, W.. Higgins, P.G., Barrow, I., Tyrrell D.A.J., Lenox-Smith I. and Ishitsuka, H. (1987) Intranasal chalcone Ro 09-0410 as prophylaxis against rhinovirus infection in human volunteers. *J. Antimicrob. Chemother.*, 20, 887–92.
3. Al-Nakib, W., Willman, J., Higgins, P.G., Tyrrell, D.A.J., Shepherd, W.M. and Freestone, D.S. (1987) Failure of intranasally administered 4′, 6-dichloroflavan to protect against rhinovirus infection in man. *Arch. Virol.*, 92, 255–60.
4. Andries, K., Dewindt, B., Brabander, M., Stokbroekx, R. and Janssen, P.A.J. (1988) *In vitro* activity of R61837, a new antirhinovirus compound. *Arch. Virol.*, 101, 155–67.
5. Arden, N.H., Patriarca, P.A., Fasano, M.B., Lui, K.J., Harmon, M.W., Kendal. A.P. and Rimland, D. (1988) The roles of vaccination and amantadine prophylaxis in controlling a outbreak of influenza A (H3N2) in a nursing home. *Arch. Intern. Med.*, 148, 865–8.
6. Barrow, G.I., Higgins, P.G., Tyrrell, D.A.J. and Andries, K. (1990) An appraisal of the efficacy of the antiviral R61837 in rhinovirus infections in human volunteers. *Antiviral Chemistry and Chemotherapy*, 1, 279–83.
7. Beare, A.S. and Reed, S.E. (1977) The study of antiviral compounds in volunteers, in *Chemoprophylaxis and Virus Infections of the Respiratory Tract*, vol. 2, (ed. J.S. Oxford), CRC Press, Cleveland, pp. 27–55.
8. Bektimirov, F.A., Douglas, R.G., Dolin, R., Galasso, G.J., Krylov, V.F. and Oxford, J.S. (1985) Current status of amantadine and rimantadine as anti-influenza A agents: memorandum from a WHO meeting. *Bull. WHO*, 63, 51–6.
9. Belshe, R.B., Burk, B., Newman, F., Cerruti, R.L. and Sim, I.S. (1989) Resistance of influenza A virus to amantadine and rimantadine: results of one decade of surveillance. *J. Infect. Dis.*, 159, 430–5.
10. Brady, M.T., Sears, S.D., Pacini, D.L. *et al.* (1990) Safety and prophylactic efficacy of low dose rimantadine in adults during an influenza A epidemic. *Antimicrob. Agents Chemother.*, 34, 1633–6.
11. Bukrinskaya, A.G., Vorkunova, N.K., Kornilayeva, G.V., Narmanbetova, R.A. and Vorkunova G.K. (1982) Influenza virus uncoating in infected cells and effect of rimantadine. *J. Gen. Virol.*, 60, 49–59.
12. Breese-Hall, C., Dolin, R., Gala, C.L. *et al.* (1987) Children with influenza A infection: treatment with rimantadine. *Paediatrics*, 80, 275–82.
13. Crawford, S.A., Clover, R.D., Abell, T.D., Ramsey, C.N., Glezen, P. and Couch, R.B. (1988) Rimantadine prophylaxis in children: a follow up study. *Paediatr. Infect. Dis.*, 7, 379–83.
14. Cross, P.A. (1991) Current recommendations for the prevention and treatment of influenza in the older population. *Drugs & Ageing*, 1, 431–9.
15. Davies, J.R., Grilli, E.A., Smith, A.J. and Hoskins, T.W. (1988) Prophylactic use of amantadine in a boarding school outbreak of influenza A. *J. R. Coll. Gen. Pract.*, 38, 346–8.
16. Davies, W.L., Grunert, R.R., Haff, R.F. *et al.* (1964) Antiviral activity of 1-adamantanamine (amantadine) *Science*, 144, 862–3.
17. Degelau, J., Somani, S.K., Cooper, S.L., Guay, D.R.P. and Crossley, K.B. (1992) Amantadine resistant influenza A in a nursing facility. *Arch. Intern. Med.*, 152, 390–3.
18. Diana, G.D., Nitz, T.J., Mallamo, J.P. and Treasurywala, A. (1993) Antipicornavirus compounds: use of rational drug design and molecular modelling. *Antiviral Chem. Chemother.*, 4, 1–10.
19. Dimmock, N.J. (1982) Initial stages in infection with animal viruses. *J. Gen. Virol.*, 59, 1–22.
20. Dolin, R., Reichman, R.C., Madore, H.P., Maynard, R., Linton, P.N., and Webber-Jones, J. (1982) A controlled trial of amantadine and

rimantadine in the prophylaxis of influenza A infection. *N. Engl. J. Med.*, **307**, 580–4.

21. Dourmashkin, R.R. and Tyrrell, D.A.J. (1974) Electron microscopic observations on the entry of influenza virus into susceptible cells. *J. Gen. Virol.*, **24**, 129–41.

22. Fleming, D.M. and Cross, K.W. (1993) Respiratory syncytial virus or influenza? *Lancet*, **342**, 1507–10.

23. Galbraith, A.W., Oxford, J.S., Schild, G.C. and Watson, G.I. (1969) Study of l-adamantanamine hydrochloride used prophylactically during the Hong Kong influenza epidemic in the family environment. *Bull. WHO*, **41**, 677–82.

24. Galbraith, A.W., Oxford, J.S., Schild, G.C. and Watson, G.I. (1969) Protective effect of 1-adamantanamine hydrochloride against influenza A2 in the family environment. *Lancet*, ii, 1026.

25. Galbraith, A.W., Oxford, J.S., Schild, G.C. and Watson G.I. (1970) Protective effect of aminoadamantane on influenza A2 infections in the family environment. *Ann. N.Y. Acad. Sci.*, **173**, 29–43.

26. Hall, C.B., Dolin, R., Gala, C.L. *et al.* (1987) Children with influenza A infection: treatment with rimantadine. *Pediatrics*, **80**, 275–82.

27. Hay, A.J., Kennedy, N.C., Skehel, J.J. and Appleyard, G. (1979) The matrix protein gene determines amantadine-sensitivity of influenza viruses. *J. Gen. Virol.*, **42**, 189–91.

28. Hayden, F.G. and Gwaltney, J.M. Jr (1982) Prophylactic activity of intranasal enviroxime against experimentally induced rhinovirus type 39 infections. *Antimicrob. Agents Chemother.*, **21**, 892–7.

29. Hayden, F.G. and Hay, A.J. (1992) Emergence and transmission of influenza A viruses resistant to amantadine and rimantadine. *Curr. Topics Microbiol. Immunol.*, **176**, 119–30.

30. Hayden, F.G. and Monto, A.S. (1986) Oral rimantadine hydrochloride therapy of influenza A virus H3N2 subtype infection in adults. *Antimicrob. Agents Chemother.*, **29**, 339–41.

31. Hayden, F.G., Belshe, R.B., Clover, R.D., Hay, A.J., Oakes, M.G. and Soo, W. (1989) Emergence and apparent transmission of rimantadine resistant influenza A virus in families. *N. Engl. J. Med.*, **321**, 1696–702.

32. Hayden, F.G., Sperber, S.J., Belshe, R.B., Clover, R.D., Hay, A.J. and Pyke, S. (1991)

Recovery of drug resistant influenza A virus during therapeutic use of rimantadine. *Antimicrob. Agents Chemother*, **35**, 1741–7.

33. Heider, H., Adamczyk, B., Presber, H.W., Schroeder, C., Feldblum, R. and Indulen, M.K. (1981) Occurrence of amantadine and rimantadine resistant influenza A virus strains during the 1980 epidemic. *Acta Virologica*, **25**, 395–400.

34. Higgins, P.G., Barrow, G.I., Tyrrell, D.A.J., Isaacs, D. and Gauci, C.L. (1990) The efficacy of intranasal interferon α-2A in respiratory syncytial virus infection in volunteers. *Antiviral Res.*. **14**, 3–10.

35. Higgins, P.G., Barrow, G.I., Tyrrell, D.A.J., Snell, N.J.C., Jones, K. and Jolley, W.B. (1991) A study of the efficacy of the immunomodulatory compound 7-thia-8-oxoguanosine in coronavirus 229E infections in human volunteers. *Antiviral Chem. Chemother.*, **2**, 61–3.

36. Higgins, P.G., Phillpotts, R.J., Scott, G.M., Wallace, J., Bernhardt, L.L. and Tyrrell, D.A.J. (1983) Intranasal interferon as protection against experimental respiratory coronavirus infection in volunteers. *Antimicrob. Agents Chemother.*, **24**, 713–15.

37. Hope-Simpson, E. (1993) *The Origin of Epidemic Influenza*, Plenum Press, New York.

38. Kubar, O.I., Brjantseva, E.A., Nikitina, L.E. and Zlydnikov, D.M. (1989) The importance of virus drug resistance in the treatment of influenza with rimantadine. *Antiviral Res.*, **11**, 313–16.

39. Levantis, P. and Oxford, J.S. (1992) Molecular aspects of AZT resistance in HIV-I. *Res. Virol*, **143**, 136–42.

40. Lubeck, M.D., Schulman, J.L. and Palese, P. (1978) Susceptibility of influenza A virus to amantadine is influenced by the gene coding for M protein. *J. Virol.*, **28**, 710–16.

41. Mast, E.E., Harmon, M.W., Gravenstein, I. *et al.* (1991) Emergence and possible transmission of amantadine resistant viruses during nursing home outbreaks of influenza A (H3N2). *Am. J. Epidemiol.*, **134**, 988–97.

42. Merrigan, T.C., Reed, S.E., Hall, T.S. and Tyrrell, D.A.J. (1973) Inhibition of respiratory virus infection by locally applied interferon. *Lancet*, **i**, 563–7.

43. Monto, A.S. and Arden, N.H. (1992) Implications of viral resistance to amantadine in control of influenza A. *Clin. Infect. Dis.* **15**, 362–7.

44. Muckenthaler, M., Gunkel. N., Levantis, P. *et al.* (1992) Sequence analysis of an HIV-I isolate which displays unusually high-level AZT resistance *in vitro. J. Med. Virol.*, **36**, 79–83.

45. Nicholson, K.G. and Wiselka, M.J. (1991) Amantadine for influenza A. *Br. Med. J.*, **302**, 425–6.

46. Oxford, J.S., Zuckerman, M.A., Race, E., Dourmashkin, R., Broadhurst, K. and Sutton, P.M. (1994) Sodium deoxycholate exerts a direct destructive effect on HIV and influenza viruses in vitro and inhibits retrovirus induced pathology in an animal model. *Antiviral. Chem. Chemother.*, **5**, 176–181.

47. Oxford, J.S. and Galbraith, A. (1980) Antiviral activity of amantadine: a review of laboratory and clinical data. *Pharmacol. Ther.*, **11**, 181–262.

48. Oxford, J.S. and Oberg, B. (1985) *Conquest of Viral Diseases*, Elsevier Biomedical Press, Amsterdam.

49. Oxford, J.S. and Potter, C.W. (1973) Amino adamantine resistance strains of influenza A2 virus. *J. Hygiene.*, **71**, 227–35.

50. Oxford, J.S., Field, H.J. and Reeves, D.S. (eds) (1986) *Drug Resistance in Viruses, other Microbes and Eukaryotes*, Academic Press, London.

51. Oxford, J.S., Logan, I.S. and Potter, C.W. (1970) *In vivo* selection of an influenza A2 strain resistant to amantadine. *Nature*, **226**, 82–3.

52. Patriarca, P.A., Kater, N.A., Kendal, A.P., Bregman, D.J., Smith, J.D. and Sikes, R.K. (1984) Safety of prolonged administration of rimantadine hydrochloride in the prophylaxis of influenza A virus infections in nursing homes. *Antimicrob. Agents Chemother.*, **26**, 101–3.

53. Phillpotts, R.J., DeLong, D.C., Wallace, J., Jones, R.W., Reed, S.E. and Tyrrell, D.A.J. (1981) The activity of enviroxime against rhinovirus infection in man. *Lancet*, **i**, 1342–4.

54. Phillpotts, R.J., Higgins, P.G., Willman, J.S. *et al.* (1984) Intranasal lymphoblastoid interferon ('Wellferon') prophylaxis against rhinovirus and influenza virus in volunteers. *J. Interferon Res.*, **4**, 535–41.

55. Phillpotts, R.J., Higgins, P.G., Willman, J.S., Tyrrell, D.A.J. and Lenox-Smith, I. (1984) Evaluation of the antirhinovirus chalcone Ro 09-0415 given orally to volunteers. *J. Antimicrob. Chemother.*, **14**, 403–9.

56. Phillpotts, R.J., Wallace, J., Tyrrell, D.A.J., Freestone, D.S. and Shepherd, W.M. (1983) Failure of oral 4', 6-dichloroflavan to protect against rhinovirus infection in man. *Arch. Virol.* **75**, 115–21.

57. Pinto, L.H., Holsinger, L.J. and Lamb, R.A. (1992) Influenza virus M2 protein has ion channel activity. *Cell*, **69**, 517–28.

58. Reuman, P.D., Bernstein, D.I., Keefer, M.C., Young, E.C., Sherwood, J.R. and Schiff, G.M. (1989) Efficacy and safety of low dosage amantadine hydrochloride as prophylaxis for influenza. *Antiviral Res.*, **11**, 27–40.

59. Schild, G.C. and Sutton, R.N.P. (1965) Inhibition of influenza viruses in vitro and in vivo by l-adamantanamine hydrochloride. *Br. J. Exp. Pathol.*, **46**, 263–73.

60. Scott, G.M., Phillpotts, R.J., Wallace, J., Secher, D.S., Cantell, K. and Tyrrell, D.A.J. (1982) Purified interferon as protection against rhinovirus infection. *Br. Med. J.*, **284**, 1822–5.

61. Scott, G.M., Reed, S., Cartwright, T. and Tyrrell, D.A.J. (1980) Failure of human fibroblast interferon to protect against rhinovirus infection. *Arch. Virol.*, **65**, 135–9.

62. Scott, G.M., Wallace, J., Greiner, J., Phillpotts, R.J., Gauci, C.L. and Tyrrell, D.A.J. (1982) Prevention of rhinovirus colds by human interferon alpha-2 from *Escherichia coli. Lancet*, **i**, 186–8.

63. Sears, S.D. and Clements, M.L. (1987) Protective efficacy of low dose amantadine in adults challenged with wild type influenza A virus. *Antimicrob. Agents Chemother.*, **31**, 1470–3.

64. Smith, T.J., Kremer, H.J., Luo, M. *et al.* (1986) The site of attachment in human rhinovirus 14 for antiviral agents that inhibit uncoating. *Science*, **233**, 1286–93.

65. Stanley, E.D., Muldoon, R.E., Akers, L.W. and Jackson, G.G. (1965) Evaluation of antiviral drugs: the effect of amantadine on influenza volunteers. *Ann. N.Y. Acad. Sci.*, **130**, 44–51.

66. Stuart-Harris, C.H., Schild, G.C. and Oxford, J.S. (1985) *Influenza: the Viruses and the Disease*, Edward Arnold, London.

67. Togo, Y., Hornick, R.B. and Dawkins, A.T. (1968) Studies on induced influenza in man. Double-blind studies designed to assess prophylactic efficacy of amantadine hydrochloride against A2/Rockville/1/65 strain. *JAMA*, **203**, 1089–91.

68. Tyrrell, D.A.J. and Al-Nakib, W. (1988) Prophylaxis and treatment of rhinovirus infections, in *Clinical Use of Antiviral Drugs*, (ed. E.

De Clercq), Martinus Nijhoff Publishing, The Hague, pp. 241–76.

69. Tyrrell, D.A.J. and Oxford, J.S. (1985) Antiviral chemotherapy and interferon. *Br. Med. Bull.*, **41**, 307–405.

70. Van Voris, L.P., Betts, R.F., Hayden, F.G., Christmas, W.A. and Douglas, R.G. (1981) Successful treatment of naturally occurring influenza A (USSR/77 [H1N1]) *JAMA*, **245**, 1128–31.

71. Van Voris, L.P. and Newell, P.M. (1992) Antivirals for the chemoprophylaxis and treatment of influenza. *Semin. Respir. Infect.*, **7**, 61–70.

72. Von Itzstein, M., Wu, W.Y., Kok, G.B. *et al.* (1993) *Nature*, **363**, 418–23.

73. Yasin, S.R., Al-Nakib, W. and Tyrrell, D.A.J. (1990) Isolation and preliminary characterisation of chalcone Ro 09-041 0-resistant human rhinovirus type 2. *Antiviral Chem. Chemother.*, **1**, 149–54.

74. Zhang, A., Nanni, R.G., Li, T. *et al.* (1993) Structure determination of antiviral compound SCH 38057 complexed with human rhinovirus 14. *J. Mol. Biol.*, **230**, 857–67.

75. Zerial, A., Werner. G.H., Phillpotts, R.J., Willman, J.S., Higgins, P.G. and Tyrrell, D.A.J. (1985) Studies on 44 081 R.P., a new antirhinovirus compound, in cell cultures and in volunteers. *Antimicrob. Agents Chemother.*, **27**, 846–50.

Nigel J.Dimmock

22.1 INTRODUCTION

Defective interfering (DI) viruses have a long history; they were discovered as auto-interfering elements in influenza A virus preparations by von Magnus [74–76,78] and for many years were named after him. The term DI virus was used when it became clear that these mutants

Viral and Other Infections of the Human Respiratory Tract.
Edited by S. Myint and D. Taylor-Robinson. Published in 1996 by Chapman & Hall. ISBN 0 412 60070 6

were not confined to influenza virus but comprised a unique class of defective virus with an almost universal distribution, and were redefined by Huang and Baltimore [59]. Interest in DI viruses reached a peak in the 1970s but then waned due to an over-extravagant expectation of their *in vivo* antiviral activity which was not forthcoming at that time. Now understanding has improved and it is possible to take a more optimistic view of the clinical potential of DI genomes, particularly in a prophylactic context. Recently, Baltimore [4] has advocated the use of 'intracellular immunization' as an antiviral approach applied to whole organisms, meaning the introduction into the cell of molecules, such as anti-sense RNAs or antibodies, with antiviral activity. Some, but not all, DI genomes fall into this category and these molecules are being analysed to determine the relationship between sequence and activity. Hopefully these can be engineered to improve that activity. This chapter is concerned with the transient resistance to experimental influenza in mice which is invoked by treatment with DI influenza viruses. It is thought that this has clinical potential in preventing influenza virus infection of humans, horses and poultry, all of which are naturally susceptible. Beyond this we look to the constitutive expression of DI RNA as a transgene in order to confer resistance to specific virus diseases.

The DI genome is a deleted form of the genome of the infectious virus which gave rise to it and has several unique properties which distinguish it from other types of defective viral nucleic acid molecules:

1. It is non-infectious, and is replicated only when its genome enters a cell which is also infected by the virus from which it arose (sometimes also termed 'helper' virus). Heterotypic interactions are rare except with type A influenza viruses.
2. It is encapsidated into virus particles which are usually indistinguishable in size and protein composition from infectious virus particles.
3. After arising *de novo*, the DI genome is rapidly amplified in concentration relative to that of the genome of infectious virus, so that within a few passages there is more DI virus in a population than infectious virus.
4. It has the ability to interfere intracellularly with, that is specifically inhibit, the multiplication of infectious virus.

A more detailed account of the properties of DI viruses in general [11,28,49,58,110] and of DI influenza viruses in particular [93,94] can be found in the recent literature.

DI genomes have been much used to study the control sequences required for replication and encapsidation since, by definition, these must be retained if the DI genome is to be replicated and encapsidated by proteins expressed by the genome of the infectious virus. Thus, the DI genome is seen as a smaller version of the infectious virus genome which provides less complex, yet authentic, sequences to study. The other interesting aspect of DI viruses is their unique ability to interfere with the multiplication of infectious virus in cell culture systems, and the possibility of using DI viruses to protect animals, including humans, from virus diseases.

DI viruses are natural products which are antiviral, and occur almost universally. Thus, it is relevant that they should be studied with a view to employing them clinically. However, this aspect has made slow progress due largely

to the failure to realize that, although all DI virus genomes interfere *in vitro*, many do not have antiviral activity *in vivo*. The reason for this appears to be that the replication of a DI genome and its interfering properties can be strongly cell-dependent (see below), and in some cells DI genomes are not replicated at all. However, more commonly the cell influences the expression of the DI genome quantitatively. It seems likely that, in the past, *in vivo* studies have been abandoned because the DI virus chosen for study had an inappropriate genome which was not active in that particular system.

There is little information about the presence of DI influenza virus during natural, or even experimental, infections. The exception is the association of subgenomic RNAs with avirulence in A/chicken/Pennsylvania/1/83 virus, whereas the virulent form lacked these extra RNAs [19]. Chickens infected experimentally with a mixture of avirulent and virulent strains enjoyed greatly reduced mortality, but it is not clear that this was due to the DI RNAs or to an early stimulation of host defence mechanisms. Further work indicated that there were also other factors influencing the attenuation of the avirulent strain, so that the role of the DI RNAs could not be apportioned [122].

The study of DI viruses *in vivo* is thus still at an elementary stage. Three systems have been studied in some detail: vesicular stomatitis virus (Rhabdoviridae), Semliki Forest virus (SFV: Togaviridae) [6], and type A influenza viruses (Orthomyxoviridae). This chapter will concentrate on the latter.

22.1 GENERAL PROPERTIES OF TYPE A INFLUENZA VIRUSES

Details of the type A influenza viruses can be found in a number of excellent reviews [66,121]. In terms of their overall natural history, type A influenza viruses infect birds, primarily terns, waterfowl and shore birds, but are also found in humans. They are divided into subtypes based on the antigenicity of the major surface proteins, the haemagglutinin (HA) and neuraminidase (NA). In all there are 14 HAs, numbered H1–H14 and nine NAs, numbered N1–N9; a subtype would, for example, be H6N8. All possible subtypes are found in birds, but only H1N1, H2N2, and H3N2 have been isolated from humans.

In wild birds most influenza A viruses cause little or no disease but an H5N3 strain caused a fatal infection in terns [12]. In domestic poultry there are occasional outbreaks of serious disease caused by H5 and H7 subtypes, probably as a result of contact with virus carried by migrating feral birds. These outbreaks can have severe economic consequences as they are usually contained by slaughter. The cost of a major outbreak in chickens in the USA in 1983–1984 was put at $60 million to government, $15 million to the farmers, and $349 million to the consumer in increased costs [2]. Inactivated avian influenza virus vaccines are used mainly in the USA with free-range poultry. Because of the cost and because they are effective only against homologous or closely related strains, appropriate subtype vaccines are stock-piled and used to control outbreaks as they occur, rather than as a routine preventive measure [2].

Influenza A viruses also infect pigs, horses and some marine mammals. In recent years horses have been infected with H3N8 strains, although before 1977 H7N7 strains were a problem. Both cause respiratory disease with fever, loss of appetite and muscle soreness and there is a high incidence of interstitial myocarditis with H3N8 viruses. Secondary bacterial infection almost always follows. Recently, a new strain of H3N8 has been causing 20% fatalities and 70% morbidity in horses in China [123]. Economic losses to the horse racing industry are a major problem. Currently an inactivated vaccine is used which provides low level, strain-specific protection of only short duration and thus requires repeated administrations [90,91,129].

In humans, influenza is a continuing problem: the pandemics of 1918 (H1N1), 1957 (H2N2) and 1968 (H3N2) were caused by the sudden introduction of a new subtype which globally displaced its predecessor within the season. The mechanism by which this rapid changeover (antigenic shift) occurs is not understood (see [55] and [66] for different views on this subject). Pandemics cause widespread morbidity and significant mortality; high mortality was an unusual feature of the 1918 pandemic with over 20 million deaths worldwide, particularly in young adults. However, death is normally confined to a high-risk group which includes the over 65s, those with chronic cardiac, pulmonary, renal or endocrine problems, and the immunosuppressed. Overall, a far greater toll is taken in the winter epidemics which occur about every 4 years. Apart from the morbidity, there is an annual mortality of around 15 000 in the UK alone. The problem is the antigenic evolution of the HA and the NA (antigenic drift) whereby immunity acquired one year is rendered ineffective to the strain circulating some 4 years later. Drift is seen only in humans, presumably because the high immune selection afforded by this long-lived species. An inactivated 'split' vaccine which contains purified HA and NA of the currently circulating H1N1, H3N2 and influenza type B virus strains is available. The policy is to provide this to the high-risk groups and to their carers. It has to be administered every year and gives around 70% protection.

Figure 22.1 shows a schematic version of an influenza virion. It acquires its membrane as it buds from the apical surface of the infected cell. Into the membrane are inserted the virion proteins, the most numerous type being the HA, which has attachment and fusion functions, and which carries the most important neutralization epitopes. Cellular plasma membrane proteins are excluded from the virion envelope by a mechanism which is not understood. Inside the virion is the genome which comprises eight unique segments of single-stranded, negative-sense RNA (i.e. complementary to mRNA). It is not clear how the virion organizes the packaging of the eight segments which comprise the genome, and it is variously suggested that there is a mechanism which specifically selects the eight segments [115] (J. McCauley and S. Duhaut, personal communication) or that there is no selection process [34]. In the latter case it may be that each virion packages more than eight segments per virion so that by chance sufficient virions will have at least eight different segments. There are about 10 physical particles per infectious unit and it has been calculated that statistically this ratio could be obtained if each virion packaged more RNA segments than the minimum required to form a complete genome [21].

The terminally linked sugar molecule, *N*-acetylneuraminic acid, is the specific cell recep-

tor which binds to an attachment site on the HA. Virus is endocytosed and in the slightly acidic environment of the endosome the viral and vesicle membranes fuse to release the viral core into the cytoplasm. Further uncoating takes place to remove the M1 of the core and the viral ribonucleoprotein (RNP) enters the nucleus. A minor component of each RNP is a RNA-dependent RNA transcriptase which transcribes the virion RNA segment into a shorter mRNA and a full-length complementary molecule which serves as a template for the synthesis of new virion RNAs. All RNA synthesis occurs in the cell nucleus. New virions are assembled in the cytoplasm and eventually bud from the apical surface of the infected cell. Action by the neuraminidase removes *N*-acetylneuraminic acid from the cell surface and from the viral glycoproteins, and hence aids dispersal of new virions.

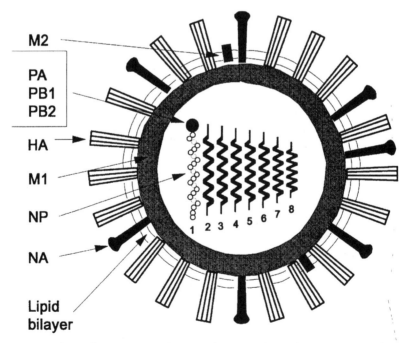

Figure 22.1 Structure of an influenza virion showing the integral membrane proteins: the major proteins, haemagglutinin (HA) and neuraminidase (NA), and the minor M2 ion channel; the capsid formed by the matrix protein (M1); and the eight nucleoprotein complexes containing an RNA segment, the major NP protein and the minor proteins of the transcriptase complex (PA, PB1, PB2). The virion diameter is about 120 nm. (Modified from [100].)

22.3 STRUCTURE AND PACKAGING OF DI INFLUENZA VIRUS GENOMES

Most viruses have a genome consisting of a single molecule of nucleic acid. Typically, DI genomes are deleted forms of this and they lack most, if not all, of the potential for coding functional proteins. One of the most common types of DI RNA genome is the centrally deleted form in which the DI genome retains the 5' and 3' terminal sequences (a few hundred nucleotides) plus some internal sequences. In this way the infectious genome of SFV (11 442 nucleotides) is reduced to a DI genome of 1244 nucleotides [119]. The internal region may be made up of one or a few discrete regions of coding sequence. These may, but usually do not, contain an open reading frame (ORF) and express a truncated version of one of the virion polypeptides. Sometimes there are rearrangements of the virion sequences, and/or duplications, and/or further deletions. Two of the extremes are represented by DI genomes of Semliki Forest virus (SFV), one of which has no rearrangements after the initial formation of the DI genome, while the other has four sequence blocks of about 273 nucleotides repeated and other modifications [69,119]. It is difficult to determine the number of DI genomes per virion, but the kinetics of ultraviolet (UV)-inactivation of interference are single-hit, suggesting that that there is one genome per virion.

Jennings *et al.* [62] sequenced 35 subgenomic RNAs isolated from a preparation of the PR8 strain of influenza A virus (H1N1) prepared at high multiplicity of infection (m.o.i.). These and other sequenced subgenomic RNAs from DI PR8 [128], A/NT/60/68 (H3N2) [37,88] and DI WSN (H1N1) [95,114], including four from RNAs 1, 2 and 3 isolated from the lungs of infected mice [98a]; and see below, DI RNAs which mediate protection *in vivo* are only putative DI RNAs as it has not been directly shown that they have interfering activity. Taken together, most such 'DI' RNAs (32 of 42, or 76%) are derived from the largest virion RNA segments PB1, PB2 and PA. Of these,

57% (24 of 42) originate from the virion RNA encoding the PB2 polypeptide, while there are five originating from the PA RNA, and three from the PB1 RNA. Deleted versions of all the other segments, except from RNA 7, were found. Cloning technicalities made isolation of the latter very unlikely [62], but putative RNA 7 'DI' RNAs have since been isolated (see Table 22.1). The best indication that these subgenomic RNAs are the mediators of interference is the activity of smaller than normal RNPs which have been isolated from DI virus. These isolated RNPs contain small RNAs and are interfering [61]. The formal proof of what constitutes a DI influenza RNA must await the encapsidation of a cloned DI RNA into a virion which as a result acquires interfering activity.

All 'DI' influenza RNAs retain the 5' and 3' termini of the virion RNA. This contrasts with the most common form of DI VSV RNA, the '5' or 'panhandle', which lacks the 3' terminus of the virion RNA and in its place has a complementary copy of the 5' terminus [101,106]. Both viruses have single-stranded, negative-sense genomes, and the reason for the different types of DI RNA is not understood. Four types of 'DI' influenza RNA have been described (Table 22.1). How these might be generated is discussed below. The majority (88%) of 'DI' RNAs have a single deletion of up to 84% of the full-length virion RNA segment from which they arose. The deletion appears unbiased as the number of nucleotides derived from either the 5' and 3' ends is roughly 100–400. Double deletion 'DI' RNAs have one major deletion and another minor deletion of about 50 nucleotides. There seems no pressure for either of these categories of 'DI' RNAs to encode any (truncated) polypeptide as only 28% retain an ORF. This proportion suggests that the occurrence of the ORF is a random process. The one complex 'DI' RNA identified has a double deletion and an insertion which results in a transposition. The one mosaic 'DI' RNA is a true inter-segmental recombinant between virion RNAs 3 and 1 with multiple insertions and deletions.

In contrast to the single RNA molecule packaged by viruses with non-segmented genomes, an influenza virion must package the eight unique segments of single-stranded RNA which constitute its genome. DI viruses arising from the former package a single DI molecule, but how many RNA molecules does a DI influenza virus particle contain? Extraction and analysis of RNA from such preparations which contain only 0.1% of the infectivity expected of a non-DI preparation,

shows that the majority of full-length virion segments and at least one DI RNA are present in approximately equimolar proportions, suggesting that DI virions package both DI and full-length RNA. However it appears that if the DI RNA is derived from, say the PB2 RNA, the proportion of full-length PB2 RNAs in the population is decreased [1,92,98]. This suggests that in the DI virion, a DI RNA is packaged in place of the full-length virion RNA from which it arose, and that all other RNAs in that virion are full-length virion RNAs (Figure 22.2).

Table 22.1 Classification of putative 'DI' influenza RNAs*

Type	Frequency†
Single deletion	37/42
Double deletion	3/42
Complex	1/42
Mosaic	1/42

* Adapted from [94]
† Number of DI RNAs in the category/total number of sequenced putative 'DI' RNAs.

Table 22.2 Change in size and distribution of 'DI' RNAs when egg-grown DI WSN together with 10 LD$_{50}$ WSN is inoculated intranasally into C3H/He-mg mice*

		'DI' RNAs isolated by RT–PCR from			
		embryonated eggs inoculated with DI WSN + WSN		lung tissue from mice inoculated with DI WSN + WSN	
RNA		~1 kb	~0.5 kb	~1 kb	~0.5 kb
1	(PB2)	+	+	−	+†
2	(PB1)	+	+	−	+†
3	(PA)	+	+,+	−	+
4	(HA)	−	+	−	−
5	(NP)	+,+,+	+,+	−	−
6	(NA)	−	±	−	−
7	(M1 + M2)	−	+	−	+†,+
8	(NS1 + NS2)	+	±,±	−	−
Total number of 'DI' RNAs		18		5	

+, denotes presence of a strong band of 'DI' DNA; ±, presence of a weaker band of 'DI' DNA; −, absence of 'DI' DNA.
* 'DI' RNAs were identified using reverse transcriptase followed by the polymerase chain reaction (RT–PCR) with segment-specific primers. Resulting DNAs were analysed by gel electrophoresis.
† not the same size as the major ~0.5 kb band from egg-grown DI WSN
Data from [97a]

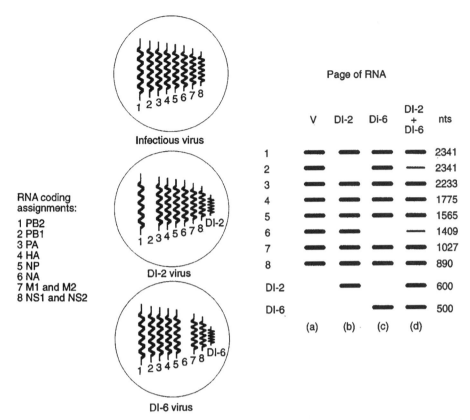

Figure 22.2 RNA content of DI influenza virus particles based on the hypothetical selective packaging of RNA segments. Numbers 1–8 represent full-length virion RNAs. The left panel shows the RNAs of infectious virus and of two DI particles, one containing a DI RNA derived from segment 2 and the other containing a DI RNA derived from segment 6. Each lacks the respective full-length RNA segment. The right panel shows an analysis of RNA by polyacrylamide gel electrophoresis (PAGE) of (a) infectious virions (V) containing the full-length virion RNAs 1–8, (b) a pure population of DI-2, (c) a pure population of DI-6, and (d) a population consisting of a mixture of DI-2 and DI-6. In the latter, the amounts of full-length RNAs 2 and 6 are reduced in proportion to the relative amounts of each DI virus in the mixture. If one DI virion contained two DI RNAs (say DI RNA 2 and DI RNA 6) it would have the same RNA profile as the mixture of DI-2 virions and DI-6 virions shown.

The natural situation is frequently more complex as a population of DI influenza viruses usually contains several different-sized DI RNA molecules [23,96]. Theoretically, these can all be derived from the same virion RNA, or can be DI RNAs derived from different virion RNA segments. In the latter case the population could comprise (i) DI virions each of which contains several DI RNAs (and possibly each replacing the corresponding full-length RNA – see above), or (ii) a mixture of DI virions each of which contains a single but different DI RNA (Figure 22.2d). In practice it is difficult to distinguish between these possibilities and it is not known if a DI particle can contain more than one DI RNA segment.

22.4 GENERATION, AMPLIFICATION AND PROPAGATION OF DI GENOMES

Generation is the initial event by which the truncated DI genome is created by imperfect transcription of the full-length RNA. This is a spontaneous process and two mechanisms have been put forward to explain it (Figure 22.3) [94]. The first 'jumping polymerase' mechanism involves the transcribing enzyme detaching from the original template and re-attaching downstream to the same template molecule or an adjacent template molecule. In the second 'looping-out' mechanism, the replicating enzyme misses out a loop of the template RNA formed by its secondary structure, and proceeds across the gap without detaching. With either process the end result is identical and the resulting viral genome contains a large deletion. Most (41 of 42) DI influenza RNAs arise from just one of the RNA segments (usually PA, PB1 or PB2) of the infectious parent. However, the mosaic DI RNA contains sequences derived from RNA 3 and RNA 1 of the virion (see above) [88]. This apparently arises by the transcriptase switching templates but, since the RNA is not sequenced directly, a cloning artefact cannot be excluded. A similar event commonly occurs in the creation of DI Sindbis virus RNAs which have most of the cellular tRNA[asp] sequence attached to their 5' terminus, presumably as a result of the transcriptase switching between virion RNA and cellular RNA templates [111]. The selective advantage of this type of DI RNA is not known.

The efficiency with which DI RNAs are generated depends on the processivity of the replicating enzyme and/or its ability to continue chain elongation on a new template. If the enzyme was infallible, no DI genomes would be generated and, indeed, this occurs in a small number of cell types (see below: this section). Influenza virus strains differ enormously in their ability to produce DI progeny [85]. Since the polymerase is presumed to be a complex of virus- and cell-coded polypeptides, variation in either can affect the generation of DI RNA.

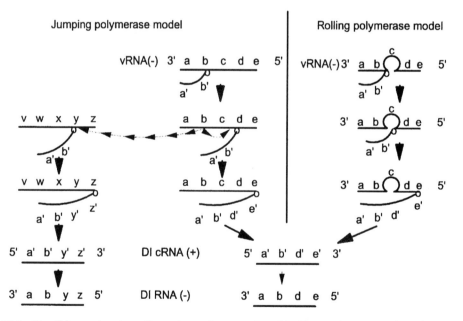

Figure 22.3   Possible mechanisms ('jumping polymerase' and 'rolling polymerase') by which single and double deletion influenza DI RNAs are generated. A complex DI RNA would arise by subsequent mutations (not shown). A mosaic DI RNA would arise by the jumping polymerase detaching from its original template and reattaching to a second RNA template (see left of figure). Generation of DI RNAs could occur during transcription of RNA–, as shown, or of RNA+. (Modified from [94].)

An increased ability to generate DI virus was acquired by the re-assortant influenza virus Wa-182, which has the NS gene of A/Aichi (H3N2) in a WSN (H1N1) background [98]. Mutations in the NS gene resulted in two amino acid substitutions in the NS2 protein, and it is therefore likely that this is a component of the polymerase. The cellular component of the polymerase can also affect amplification/propagation of a DI genome in a cell type other than that in which it was created. This is of especial relevance to the *in vivo* situation where a virus may infect several different types of cell, and greatly complicates the interpretation of *in vivo* experiments.

DI genomes are far from immutable and continue to evolve on passage. The reason is that the majority of DI genomes do not encode essential information and the majority of the sequence, apart from the polymerase recognition and encapsidation sites, is dispensable. However, this conclusion is based on *in vitro* studies, and there is *in vivo* evidence from work with cloned DI Semliki Forest Virus that the sequence outside the control signals is also essential for interfering activity (M. Thomson and N.J. Dimmock, unpublished data). Repetition of the generation process could be responsible for further deletions and a smaller DI genome, while duplication of sequence blocks produces an overall increase in the size of the genome. Thus, creation of DI genomes primarily resides in recombination and indicates, incidentally, just how commonly this process occurs between RNA molecules. Generation is independent of multiplicity of m.o.i., but for the next stage, amplification, to occur the new DI particle must enter an infected cell, since it has an absolute requirement for the replicative and encapsidation equipment provided by the infectious genome. Hence, amplification is favoured by a m.o.i. sufficient to ensure that all cells contain one or more infectious virus genomes.

Amplification is probably a property of the small size of the DI genome since in theory more copies of the DI genome can be made in unit time than can be made of the larger infectious genome (assuming that in both cases addition of a nucleotide occurs with the same efficiency). Some support for this is lent by DI poliovirus genomes which have relatively small deletions (about 20%) which would not be expected to exert much of a replicative advantage and indeed, it routinely takes many (15–20) passages to establish a DI poliovirus preparation [20]. Without amplification DI viruses would be an irrelevance, since at the moment of generation the new DI genome represents only one molecule in a huge excess (perhaps 10^9) of infectious genomes. The power of amplification is such that within as few as two passages, the concentration of DI virions can exceed that of infectious virus. However the rate of amplification depends on the virus–cell system and the sequence of the DI genome (see below). At this stage there is no interference, since this depends on the attainment of a certain high ratio of DI: infectious virus genomes (see below).

Both generation and amplification are affected by the properties of the host cell, and DI viruses are not generated by certain cell–virus combinations but are in most others. For example, DI viruses are not generated in chick embryo lung cells (Sendai virus) [67], in chicken's eggs (Newcastle disease virus) [108], in primary human fibroblasts (vesicular stomatitis virus (VSV) [65], and in certain sublines of HeLa cells (VSV) [51,54], poliovirus [84,102] and SFV [117] (also N.J. Dimmock, unpublished data). However, in some cases cells which cannot generate DI virus can propagate DI virus with which they have been inoculated (e.g. [65]). Generation appears to involve a host-coded function as treatment of DI VSV-generating cells with actinomycin D, an inhibitor of DNA-dependent RNA synthesis, abrogates the generation of DI virus [64], and in human cells this has been mapped to chromosome 16 [65]. This function may relate to cellular proteins which interact with the viral polymerase to form the holoenzyme. In the past it has been difficult to distinguish specific from adventitious binding, but the new two fusion protein system has recently been used to isolate cell proteins interacting with a protein of the polymerase complex of influenza virus [98a]. Thus, a polymerase complex formed in a particular cell may replicate the infectious virus genome perfectly well and may or may not generate DI genomes. Again, propagation of DI genomes is a separate issue and once a DI genome is made, the polymerase may replicate it as efficiently as the infectious genome, less efficiently, with intermediate efficiency, or not at all (Figure 22.4). The cell can also determine the spectrum of 'DI' RNAs present in a DI virus preparation [22] or have no effect on it [24].

22.5 INTERFERENCE

By definition, interference by DI viruses is a cell culture phenomenon which is related to, although not identical with, protection of the whole animal from disease (see below). Interference operates intracellularly which distinguishes it from extracellular interference phenomena like the blockade of receptors. Interference requires co-infection of a cell by DI and infectious virus but there is also a quantitative dimension since although interference requires at least one DI genome per cell, it generally needs considerably more than one DI genome per cell. Interference in different cells requires different ratios of DI: infectious virus (SFV [6]; influenza A virus) [80]. Using one of the most sensitive assays recorded, it was found that a m.o.i. of one physical DI particle per cell protected Madin–Darby canine kidney (MDCK) cells from infection with influenza A virus [80]. In some cell lines there is no interference. Obviously, this results if the cell does not replicate the DI genome, but DI VSV is replicated in HeLa cells by VSV without any interference being manifested; there is no explanation [54].

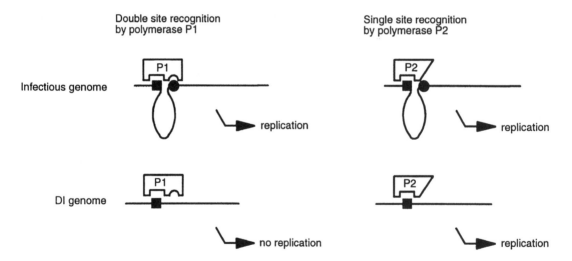

Figure 22.4 Hypothetical scheme to explain how the genome of infectious virus is replicated in two cell types, whereas the DI genome is replicated in only one: the polymerase holoenzymes, P1 and P2, are both complexes of virus-coded polypeptide(s) and cell-coded polypeptide(s). The latter are cell type-specific, and result in P1 requiring two recognition sites on the genome, and P2 requiring only one recognition site. Since the second recognition site has been deleted in the formation of the DI genome, it is replicated only in the cell which synthesize the P2 polymerase. (Modified from [18].)

The mechanism(s) of interference are in general incompletely understood. *A priori*, an explanation based on an extension of amplification is attractive but there is little factual evidence to support this or other suggestions. The argument proposes that some component of replication such as the polymerase itself (which is synthesized only by the infectious genome) is limiting. Thus, once the DI genome reaches a critical concentration in the cell it effectively sequesters the component with the result that replication of the infectious genome is inhibited. Indeed, competition between DI VSV and infectious genomes for the L/NS protein complex has been demonstrated in a cell-free system [44]. This results in a decline in the production of the limiting component, and as a result replication of the DI genome itself is also inhibited. It is likely that there are a number of different mechanisms by which DI genomes from viruses with very different replication strategies exert their interference. For example, some DI genomes have portions of their genomes repeated up to four times (SFV DI 309) [69] and it is tempting to speculate that this is of biological significance. Deletion analysis has shown that the repeat contains a sequence which is essential for the propagation of DI virions (M. Thomson and N.J. Dimmock, unpublished data) and may be involved in encapsidation or afford a more efficient way of sequestering the limiting component referred to above. There is no information about the mechanism of interference by DI influenza virus. A general discussion of possible interference mechanisms can be found elsewhere [28,110].

Classically the consequence of interference is a cycle of events:

decline of production of infectious virions

which results in:

decline in production of DI virions

which results in:

increase in production of infectious virions

which results in:

increase in the production of DI virions, etc., etc.

The first description of cycling, which was with DI influenza A virus, was by von Magnus [78] and more examples are listed by Roux, Simon and Holland [110]. Theoretically, the notion of cycling is sound but in practice cycling is often irregular or absent. One obvious proviso is the need for the continued existence of both DI and infectious virus throughout the cycle, and this may not always occur. For example, cells inoculated with 50 plaque-forming units (p.f.u.) SFV per cell and 'an excess' of DI SFV produce no virus at all, and show no sign of even being infected as regards production of a cytopathic effect or of virion polypeptides [10]. Evidently an infection can be cured completely if sufficient DI virus is present and cycling cannot take place under such conditions.

22.5.1 HOMOTYPIC VERSUS HETEROTYPIC INTERFERENCE

In general DI viruses only interfere homotypically. This is because they depend absolutely on the functions of the infectious virus for their continued existence; the polymerase must bind to recognition sequences to permit replication to take place, and virion structural proteins must recognize packaging signals to allow encapsidation into DI particles. This has been verified for the VSV system by studying the evolution of a mutant infectious virus which had become resistant to interference from its DI virus. Sequencing showed that there was a mutation in the polymerase recognition site [43]. Further study showed that the mutant virus was not intrinsically refractory to the generation or action of DI virus and generated its own DI virus which interfered with its multiplication in the normal way. Occasional heterotypic interactions have been reported between different strains of VSV [105,112], some but not all strains of SFV and between SFV and DI Sindbis virus or vice versa [9,125], and between different strains of West Nile virus [25]. Presumably there is sufficient sequence similarity in the appropriate regions to permit the required recognition and interaction.

The type A influenza viruses appear be the exception as there is reciprocal cross-interference *in vitro* between viruses of the human H1N1, H2N2, and H3N2 subtypes [24]. Further, all these viruses were able to support the replication of each others DI RNAs. More limited interference was found between the DI virus of a human H1N1 strain and an avian H7N1 virus [83]. Heterologous interference is in fact to be expected, as there is virtually free exchange of full-length virion RNA segments (genetic re-assortment) between all influenza A viruses, and representative viruses carrying all permutations of the 14 HA genes and nine NA genes are known. Obviously, for this to happen there has to be efficient cross-recognition of all polymerase and packaging signals on every RNA segment. Indeed all type A influenza virion RNAs bear putative recognition signals in their conserved terminal sequences [79]:

5' AGUAGAAACAAGG-
-CCUGCUUUCGCU 3' for RNAs 1–3, and RNA 7 of human strains
-CCUGCUUUUGCU 3' for RNAs 4–8

22.6 PERSISTENT INFECTIONS AND DI VIRUS

Persistent infections can be initiated *in vitro* by many different viruses in a number of different ways. One of these involves the homologous DI virus in the establishment, and sometimes maintenance, of persistence (e.g. Rhabdoviridae [27,52,68], Paramyxoviridae [109], Togaviridae [124], Arenaviridae [38], Flaviviridae [103] and Herpesviridae [46]). A mathematical model for persistence has been proposed [5]. However, persistent infections are not necessarily easy to establish and may need special requirements such as *ts* infectious virus, or added interferon. Frequently cell

populations go into crisis with an extensive cytopathology and require resuscitation by the addition of fresh non-infected cells [100] or by assiduous cultivation of a few surviving cells.

With influenza virus, persistence could not be established in several cell types with *ts*+ virus, with or without added DI virus, but was achieved when MDCK or HeLa cells were inoculated with a *ts* mutant together with DI virus [24]. In these persistent infections most cells expressed viral antigen. The authors suggested that the frequent crises that the cells endured assisted the evolution of an infectious virus variant which was capable of less cytopathic multiplication. Persistence was also established in BHK cells by WSN of presumably *ts*+ status [39]. In both experimental systems, the virus which emerged was mutant in a number of characteristics including reduced yield and plaque phenotype. The persistent virus studied by Frielle, Huang and Younger [39] had decreased neuraminidase activity, decreased HA activity and little infectivity. In neither system did DI virus appear to persist. Persistently infected cells were resistant to infection with homologous virus [24,39], but not to VSV or a paramyxovirus [24].

Persistent infections *in vivo* have been established in the main by co-inoculation of very young animals with mixtures of infectious and DI virus (reovirus in rats [116], lymphocytic choriomeningitis virus in mice [104]). Both were achieved by intracerebral inoculation. However, when Welsh, Lampert and Oldstone [126] prevented lymphocytic choriomeningitis virus (LCMV)-mediated disease with DI LCMV, also in mice inoculated intracerebrally, no persistent infection resulted. Persistent VSV-infection of hamsters was established by intraperitoneal inoculation of VSV + DI VSV but not of VSV + UV-inactivated DI VSV. Obviously, an active DI genome was required but a similar result was achieved by stimulating interferon instead of administering DI VSV, and it was thought that protection and persistence were brought about by DI virus in conjunction with interferon [40,41]. In contrast, various studies in mice failed to establish persistence of VSV [18,53] but as the sequence of the DI VSV genomes employed as well as the host species differed from those of the previous studies, no conclusion can be drawn. Mice which survived an otherwise lethal SFV infection as a result of treatment with DI SFV appeared clinically normal from the day of infection, but infectious virus could be isolated reproducibly directly from the brains of a small proportion of the animals. This ranged from 0–27% up to 6.5 months post-infection; none was isolated after that time [3]. All mice, but one, were clinically normal throughout this period. No persistently infected mouse has been detected after intranasal infection with influenza A virus + DI influenza virus even though lung homogenates were serially passaged in embryonated eggs to detect trace amounts of infectious virus [29].

22.7 STABILITY OF DI RNA IN CELLS *IN VITRO*

MDCK cells inoculated with a high multiplicity of completely non-infectious DI influenza A/WSN are rendered refractory to subsequent infection by WSN, and this proved to be a relatively long-lived phenomenon. Cells divided at the normal rate and daughter cultures prepared by trypsinization in the usual way were also resistant to WSN. This suggested that the DI RNA segregated to both progeny cells, otherwise the cell which received no DI genome would have produced virus normally. Eventually the cells regained their original sensitivity to infection and a half-life of interfering activity of 25 days was calculated [16]. This was in good agreement with the half-life of the DI RNA itself. Such stability was not a unique property of DI RNA as WSN RNA 6, which encodes the HA, was similarly long-lived in MDCK cells inoculated with non-infectious UV-irradiated virus (half-life of 13 days) [15]. (The dose of UV was calculated to

inactivate only virion RNA segments larger than RNA 6.) However, influenza RNA stability appears to be at the end of a spectrum, since the half-life of the RNA of DI measles virus, DI Sendai virus, DI VSV inoculated onto cells in culture was 3 days, 1 day, and up to 18 hours, respectively [26,89]; the RNA of DI SFV was similarly unstable [30]. The reason why influenza virus RNA is more stable than that of other negative strand viruses is not known, but it may be that its location in the nucleus [48,60,113,118] protects it from degradation by nucleases. Attempts to find long-lived influenza virus RNA *in vivo* following infection of mice or ducks have so far proved negative [31,120], but the technique used – reverse transcription followed by polymerase chain reaction (PCR) – may not be sufficiently sensitive in the presence of excess cellular RNA, or the correct tissues may not have been examined. The half-life of DI RNA *in vitro* varied according to the type of cell inoculated and was lower in BHK cells [16].

22.8 USE OF DI VIRUS TO PREVENT EXPERIMENTAL INFLUENZA IN MICE

22.8.1 TRANSIENT INTRACELLULAR IMMUNIZATION OF MICE WITH DI INFLUENZA VIRUS

Protection is due to the activity of the DI genome

There is a long history of experiments which demonstrate that DI influenza virus interferes with infectious virus multiplication *in vivo* (chicken embryos [14,35,36,74–78,127]) or affords protection against disease induced by infection (encephalitis in mice [14,42]; pneumonia in mice [14,45,56,57,76,107]). Holland and Doyle [50] failed to demonstrate protection by either route but demonstrated that the virus load in the lung was diminished. However, in general these experiments were not sufficiently well controlled to be able to unequivocally attribute the beneficial effects to the activity of the DI genome rather than to the stimulation of host defence responses by protein and other constituents of the DI particle. Indeed, Rabinowitz and Huprikar [107] concluded that in their system protection was due to an augmented humoral response.

More recently, there has been a series of papers in which protection by DI virus of mice from an otherwise lethal respiratory infection with its homologous strain, the human influenza A virus, WSN (H1N1) has been reported. These studies have all been controlled by inoculating mice with the same preparation of DI WSN after its interfering and mouse-protecting activity had been inactivated by incubation with β-propiolactone (BPL) [29] or prolonged UV-irradiation [98]. BPL is used in the manufacture of inactivated human and veterinary vaccines, and acts by alkylating and acylating nucleic acids. UV irradiation induces chain breaks in RNA. Both have little effect on protein at the doses used, and leave the HA and NA activities unaffected. The results of many experiments show that treatment of infected mice with active DI virus allows 80% of otherwise lethally infected mice to survive, whereas all mice given the control inactivated DI virus succumb to a lethal pneumonia [29,81–83,86,87]. DI-mediated protection has been demonstrated in three different strains of mouse [87].

The UV target size of the protective activity of DI WSN virus in mice is also consistent with the DI RNA being responsible for protection. This value is calculated from the ratio of the rates of inactivation of infectious and DI influenza virus. Since the rate is proportional to the size of the genome, it is possible to extrapolate to the size of the DI genome responsible for protection. This was equivalent to 350 nucleotides, a value in good agreement with the UV target size of the interfering activity of the same preparation in MDCK cells (405 nucleotides) [87]. Both estimates are smaller than the smallest full-length virion RNA 8 (890 nucleotides) and around the size of previously isolated putative DI RNAs [93,94] and of the

major putative DI RNAs isolated from the lungs of mice infected with WSN + DI WSN, and now cloned and sequenced [97a].

Different DI virus preparations have genomes of different sequence and hence have potentially different biological properties, as discussed above, so that in the initial exploration of DI WSN-mediated protection care was taken to use a DI preparation which varied as little as possible. This was achieved by starting with biologically cloned virus, making a stock of second-passage DI virus at high multiplicity which was stored in liquid nitrogen, and then making a single, final high multiplicity passage for use in mice. This material has behaved with exemplary reproducibility, unlike DI SFV where a protecting DI preparation can in one passage lose its *in vivo* activity while retaining its *in vitro* activity [7,8]. More recently protection of mice has been achieved with DI virus generated from an entirely different source, A/equine/Newmarket/7339/79 (H3N8: EQV). This DI virus is equally active against the homologous parent virus and heterologous (H1N1) human influenza viruses [97] (also see below).

Eventually it is hoped that reverse genetics [13,32,33] will permit the construction of molecularly defined DI influenza virions which contain DI RNA of known sequence, so that the sequence–function relationships can be explored, as has been done with the positive-strand RNA alphaviruses [111,119].

Isolation of 'DI' RNAs which may mediate protection *in vivo*

S. Noble and N.J. Dimmock [97a] have used reverse transcription followed by PCR with segment-specific 3' primers and a segment-common 5' primer to identify putative 'DI' RNAs present in mouse-protecting preparations of DI WSN. The original infectious virus was plaque-picked three times and passaged three times in eggs at high m.o.i. Analysis of the PCR products obtained from the egg-grown DI virus showed that there was at least one 'DI' RNA derived from each RNA segment and four 'DI' RNAs derived from the NP-encoding RNA 5 (Table 22.2). Although the technique is not quantitative, all 'DI' DNA bands were equally strong except for fainter bands derived from RNA 6 (NA) and RNA 8 (NS). The preparation contained a total of 18 'DI' RNAs.

Putative DI RNAs were next isolated from the lungs of mice 5 days after inoculation with 10 LD_{50} WSN together with a protecting dose of the egg-grown DI WSN. The major and reproducible finding was that the pattern of 'DI' RNAs was altered considerably: first, only 'DI' RNAs originating from the PB2, PB1, PA and M RNAs were now present, giving a total of 5 'DI' RNAs; second, the distribution of these 'DI' RNAs differed from the 'DI' RNAs present in the egg-grown DI virus; and third, only 'DI' RNAs of around 0.5 kb were represented (Table 22.3). Of four P gene 'DI' RNAs which have now been cloned and sequenced, three are *identical* (including nucleotide deletions and substitutions) with 'DI' RNAs from the inoculum. All have a single major internal deletion and range in size from 377–425 nucleotides [97a]. Clearly there is preferential replication (selection) of certain 'DI' RNAs in the lung. In addition, it is possible that there is concomitant evolution of 'DI' RNAs in the lung. This study demonstrates some of the difficulties and pitfalls of attempting to analyse DI influenza viruses, and suggests that further investigation of the properties of the 'DI' RNAs isolated from the lung would be worthwhile.

Timing of treatment relative to infection

Two doses of DI virus, the first given 2 hours before infection and the second co-inoculated, are optimal. The reason for this is not clear. Mice treated with the same two doses of DI EQV or DI WSN up to 5 days before infection show significant resistance to infection (Table 22.3) [97] (also L. McLain and N. J. Dimmock, unpublished data). However, protection declines as time elapses between treatment and infection. Such prophylaxis required an active DI genome, but inactivated DI virus mediated a degree of protection, presumably by stimulating host responses, which became significant after a few days and protected about 50% of infected animals.

This protection was short-lived. The idea that DI virus can provide immediate protection through the activity of its genome and at the same time act as a conventional 'live' vaccine will be explored later. DI EQV or DI WSN afford no protection when inoculated one day

Table 22.3 Duration of effectiveness of DI A/equine/Newmarket/7339/79 (H3N8:EQV) prophylaxis against 10 LD_{50} A/WSN (H1N1)

Day of treatment relative to infection*	Protection (%) mediated by	
	DI EQV	Inactivated DI EQV
−7	44	56
−5	69	44
−3	94	44
−1	88	31
0	100	19
+1	27	27

* Five-week-old mice were treated with two intranasal doses of the material indicated and then infected on the day shown with 10 LD50 WSN by the same route. Mice inoculated on day 0 were given one dose of DI EQV or inactivated DI EQV followed 2 hours later by co-inoculation of the same material containing WSN. Each group comprised 15 or 16 mice; 19% (3 of 16) of infected mice which received no DI EQV survived.

after infection (Table 22.3; L. McLain and N.J. Dimmock, unpublished data).

How many DI particles are needed for protection?

The total number of particles in a DI virus preparation is conveniently indicated by its HA titre, although this says nothing about its other properties, such as its interfering efficiency. Over many experiments it has been found that 400 HAU per mouse (about 1.2 mg) gives maximum protection against a lethal dose of virus, and 40 HAU gives around 33% protection. This dose appears to be independent of mouse strain [87] and a number of different virus–DI virus permutations [29,97].

This situation changed when a lower dose of infectious virus, chosen as better simulating natural influenza, was used [87]. By itself this inoculum caused respiratory disease in 70% of mice, from which most recovered. Under these conditions, 4 HAU (which is equivalent to 4×10^7 particles or 12 ng) of DI virus per mouse afford 100% protection. This finding has two implications: first, it indicates that low doses of DI virus can be protective; and second, it says something about the number of cells in the respiratory tract which support virus multiplica-

tion, since *in vitro* over 50% cells are required to contain at least one DI particle per cell for there to be significant interference. Less than 4×10^7 DI particles are required for protection from morbidity, which, applying the Poisson distribution, means that 2.5×10^7 cells in the respiratory tract of the mouse need to contain a DI genome. If protection requires more than one DI particle per cell then even fewer cells are involved.

A review of the number of DI particles per animal required for protection by different DI viruses in a variety of experimental situations is shown in Table 22.4. The fact that there is protection at all is significant as all these infections, with the exception of the influenza mouse morbidity model, are lethal and of short (2–7 days) duration. More favourable results would probably be achieved with milder infections which better reflect the human situation.

DI EQV affords heterotypic protection to mice against lethal infection with WSN and PR8

Normally DI viruses only interfere with the strain of virus from which they were derived. However, as shown in Table 22.5, DI EQV

Table 22.4 Titration of DI particles required for protection *in vivo*

Virus	DI particles* (particles/mouse)	Infectivity (pfu/mouse)	Conditions Animal	Route†	Reference
VSV	5×10^9	10^2	Adult mouse	i.c.	[63]
	3×10^8	10^2	Adult mouse	i.c.	[63]
	10^5	10^4	Adult hamster	i.p.	[41]
	10^4				
SFV	$\sim 10^7$	6×10^3	Adult mouse	i.n.	[30]
WSN	4×10^8	10^3	Adult mouse	i.n.	‡
	$4 \times 10^{8*}$	2×10^3	Adult mouse	i.n.	[98]
	$\leqslant 10^{7*}$	2×10^2	Adult mouse	i.n.	[98]

SFV, Semliki Forest virus; VSV, vesicular stomatitis virus; WSN, influenza virus A/WSN
* The homologous DI virus was used except by Noble and Dimmock [98] who used the heterologous DI equine influenza virus (H3N8) to protect against WSN (H1N1).
† Inoculation route: i.c., intracerebral; i.p. intraperitoneal; i.n., intranasal.
‡ L. McLain and N.J. Dimmock, unpublished data.

(H3N8) completely protected mice, inoculated as described above, from either of the human type A influenza viruses WSN or PR8 (H1N1). There was no mortality; only a few mice showed any sign of morbidity and this was always of a mild nature. In comparison, there was about 80% mortality of WSN-infected mice and 60–70% mortality of PR8-infected mice which were treated with inactivated DI EQV or were mock-treated with phosphate-buffered saline (PBS). Inactivated DI EQV gave no significant protection against WSN or PR8.

It was interesting that DI EQV (H3N8) protected as efficiently against WSN and PR8 (both H1N1) as against the homologous H3N8 strain since, in general, DI viruses interact only with the virus strain from which they arose (see above). This would seem to reflect the commonality of control (particularly replication and encapsidation) sequences in the DI genome as indicated by the substantial cross-interference and replication found *in vitro* between DI viruses of H1N1, H2N2 and H3N2 subtypes [24] and more limited interference *in vitro* between DI H1N1 virus and a H7N1 subtype virus [80]. However, this is not universal as one strain of DI WSN (H1N1) did not protect mice against PR8 (H1N1) [83], although

mice were completely protected in the same system by DI EQV [97].

DI EQV and DI WSN exert different mechanisms of protection

Essential to the understanding of the action of DI influenza virus in modulating infection in mice is the fact that the disease is primarily immune-mediated, with T-cells playing the major role in the immunopathogenesis [17,47,70–73,86]. The protection mediated by DI EQV against challenge by WSN differed substantially from protection, in exactly the same experimental system, mediated by DI WSN (Table 22.6) [29]. While both DI EQV and DI WSN were equally effective at preventing death (in terms of the amount of DI virus inoculated per mouse), a body of evidence suggests that DI EQV protected mice by inhibiting virus multiplication in the lung. This resulted in reduction of the characteristic lung immunopathology and consolidation, and almost complete abrogation of morbidity (Table 22.6). In contrast, DI WSN does not inhibit virus multiplication in mouse lung *in vivo*, although it inhibits multiplication in cell culture and embryonated eggs. All mice suffer substantial clinical disease before recovering, even though consolidation is reduced relative

Table 22.5 DI A/equine/Newmarket/7339/79 (H3N8: EQV) gives heterotypic protection against 10 LD$_{50}$ A/WSN (H1N1) and A/PR8 (H1N1)

Treatment*	Challenge virus	Protection†(%)
DI EQV	WSN	28/28 (100)
Inactivated DI EQV	WSN	5/23 (22)
PBS	WSN	5/25 (20)
DI EQV	PR8	10/10 (100)
Inactivated DI EQV	PR8	3/10 (30)
PBS	PR8	4/10 (40)

* Five-week-old mice were treated with an intranasal dose of the material indicated, and then 2 hours later were given a second dose of the same material containing 10 LD^{50} of WSN or PR8[97].
† Number of mice surviving at 12 days after infection/number of mice inoculated.
PBS, phosphate-buffered saline.

to that occurring in lethally infected mice [29,86,87]. To explain this paradox it was postulated that immunopathogenesis during infection with WSN results from the infection of T-cells and suppression of T-cell functions as a consequence of the expression of viral gene products. There is only limited viral gene expression in lymphocytes as the infection is non-permissive.

Experimental support for the hypothesis came from the finding that lung T-cell mitogenesis in response to lectin stimulation is inhibited during infection with WSN [86]. It was further postulated that the immunosuppression inhibits the production of cellular signals which normally control the accumulation of T-cells in the lung. It is this uncontrolled accumulation which results in consolidation and is the ultimate cause of death. If DI WSN also entered T-cells it could interfere with infectious virus gene expression, in the same manner as in cell culture, with the result that the T-cells function normally. Indeed, it was found that T-cells from the lungs of mice inoculated with WSN + DI WSN were not suppressed and gave the highest mitogenic response to lectin stimulation [86]. Thus, it appears that interference mediated by DI WSN in the mouse is cell-specific, and prevents infectious virus gene expression in T- (and other) cells, but not in the cells of the lung which are permissive for virus multiplication [86]. DI EQV on the other hand inhibits virus multiplication in the lung, thus decreasing the antigen available for stimulat-

ing T- and B-cells and implicitly reducing immune damage by this route. We have not investigated whether or not DI EQV is interfering in immune cells. The difference between DI WSN and DI EQV probably represents sequence differences between DI RNAs rather than any generic difference between the DI particles of WSN and EQV.

22.8.2 A NEW FORM OF LIVE VACCINE

A new form of live vaccine is now available by attenuation of virulent influenza virus in *trans* by co-inoculation of defective interfering virus. The recent experimental work described above indicates that mice suffer no clinical disease when co-inoculated with a mixture of DI EQV and virulent virus. However, these mice were immune to challenge some weeks later with a lethal dose of infectious virus homologous with the DI component of the vaccine or a lethal dose of virus homologous with the infectious component of the vaccine (S. Noble and N.J. Dimmock, unpublished data). Thus, we have, in effect, a live vaccine in which virulent virus is attenuated in *trans* by the activity of the DI virus, rather than by the conventional route of mutating the genome of the virulent virus itself. For example, DI EQV abrogates the disease caused by co-inoculation of a lethal dose of the heterosubtypic WSN (H1N1) strain. Under these conditions infectious virus still multiplies in the lung but the titre is diminished by 40- to 100-fold. When mice were chal-

Table 22.6 Comparison of protection by DI A/equine/Newmarket/7339/79 (H3N8: EQV) and DI A/WSN (H1N1) against a lethal dose of A/WSN

Protection	DI EQV	DI WSN*
Prevention of death	+	+
Prevention of disease	+	−
Inhibition of virus multiplication in the lungs	+	−
Inhibition of lung consolidation (%)	80	50

* Data from [29, 87]

lenged 3 weeks later with a very large dose (about 1000 LD$_{50}$) of infectious virus (which can be EQV, that is homologous with the DI virus, or WSN, homologous with the infectious component, or PR8, that is homologous with neither), they were completely resistant. It seems likely that this resistance is due to conventional acquired immunity rather than persistence of the DI genome, as the same DI genome had an effective half-life of about 5 days when inoculated without infectious virus (see above). However, this has yet to be formally demonstrated. There is self-evidently cross-immunity to PR8, but the basis of the immunity to EQV is not known. In theory, it could be heterotypic and mediated by cross-reactive antibody and/or T-cells which are stimulated by the infectious WSN component of the vaccine, or it could be homotypic and due to immunity generated to DI EQV. The latter would depend upon DI EQV carrying at least some full-length RNAs encoding EQV proteins, and their expression in a WSN-infected cell. There is no evidence of persistence of infectious virus in the lungs of these mice.

These data suggest the possibility of developing a novel live influenza vaccine. However, there is much to be discovered. For example, although the infectious load in the mouse lung is reduced, the possible spread of virulent virus to susceptible individuals without the accompanying attenuating DI virus requires investigation. Nevertheless, the potential of a live, one-shot, intranasal, cross-reactive vaccine which avoids the many-fold difficulties of attenuation, is an attractive proposition.

22.9 CONCLUSIONS

The potential of DI influenza virus as an agent of transient intracellular immunization which could provide anti-influenza prophylactic treatment of susceptible animal species including man has been discussed. Nanogram amounts are active and protection can be mediated by a single intranasal dose.

Protection lasts for 5 days and extends to heterologous subtypes of type A influenza virus. In addition, a new format of live vaccine is described in which co-administration of DI virus attenuates the virulence of the infectious strain. Inoculation of mice with DI virus + virulent infectious virus produces a subclinical infection which, nevertheless, establishes a solid heterotypic immunity to a later lethal dose of challenge virus.

ACKNOWLEDGEMENTS

Lesley McLain, David Morgan and Simon Noble have all made major contributions in the author's laboratory to the understanding of DI influenza viruses. This work has been variously supported by the SERC, the AFRC, the CRC and the MRC and the Research and Innovations Fund of the University of Warwick. The author warmly acknowledges the devotion to this project over many years of Carol Hill, Barry Gardner and Jean Westerman, and thanks Lesley McLain for preparing the figures.

22.10 REFERENCES

1. Akkina, R.K., Chambers, T.M. and Nayak, D.P. (1984) Mechanism of interference by defective-interfering particles of influenza virus: differential reduction of intracellular synthesis of specific viral polymerase proteins. *Virus Res.*, 1, 687–702.
2. Alexander, D.J. (1993) Orthomyxovirus infection, in *Virus Infections of Birds*, (eds J.B. McFerran and M.S. McNulty), Elsevier, Amsterdam, pp. 287–316.
3. Atkinson, T., Barrett, A.D.T., Mackenzie, A. and Dimmock, N.J. (1986) Persistence of virulent Semliki Forest virus in mouse brain following co-inoculation with defective interfering particles. *J. Gen. Virol.*, 67, 1189–94.
4. Baltimore, D. (1988) Gene therapy: intracellular immunization. *Nature*, 335, 395–6.
5. Bangham, C.R.M. and Kirkwood, T.B.L. (1990) Defective interfering particles: effects on modulating virus growth and persistence. *Virology*, 179, 821–6.

6. Barrett, A.D.T., Crouch, C.F. and Dimmock, N.J. (1981) Assay of defective-interfering Semliki Forest virus by the inhibition of synthesis of virus-specified RNAs. *J. Gen. Virol.*, **54**, 273–80.

7. Barrett, A.D.T., Crouch, C.F. and Dimmock, N.J. (1984) Defective interfering Semliki Forest virus populations are biologically and physically heterogeneous. *J. Gen. Virol.*, **65**, 1273–83.

8. Barrett, A.D.T. and Dimmock, N.J. (1984) Modulation of Semliki Forest virus-induced infection of mice by defective interfering virus. *J. Infect. Dis.*, **150**, 98–104.

9. Barrett, A.D.T. and Dimmock, N.J. (1984) Variation in homotypic and heterotypic interference by defective interfering viruses derived from different strains of Semliki Forest virus and Sindbis virus. *J. Gen. Virol.*, **65**, 1119–22.

10. Barrett, A.D.T. and Dimmock, N.J. (1985) Differential effects of defective interfering Semliki Forest virus on cellular and virus polypeptide synthesis. *Virology*, **142**, 59–67.

11. Barrett, A.D.T. and Dimmock, N.J. (1986) Defective interfering viruses and infections of animals. *Curr. Topics Microbiol. Immunol.*, **128**, 55–84.

12. Becker, W.B. (1966) The isolation and classification of tern virus: influenza virus A/tern/South Africa/61. *J. Hyg. (Camb)*, **64**, 309–20.

13. Bergmann, M., Garcia-Sastre, A. and Palese, P. (1992) Transfection-mediated recombination of influenza A virus. *J. Virol.*, **66**, 7576–80.

14. Bernkopf, H. (1950) Study of infectivity and haemagglutination of influenza virus in de-embryonated eggs. *J. Immunol.*, **65**, 571–83.

15. Cane, C. and Dimmock, N.J. (1990) Intracellular stability of the gene encoding influenza virus haemagglutinin. *Virology*, **175**, 385–90.

16. Cane, C., McLain, L. and Dimmock, N.J. (1987) Intracellular stability of the interfering activity of a defective interfering influenza virus in the absence of virus multiplication. *Virology*, **159**, 259–64.

17. Cate, T.R. and Mold, N.G. (1975) Increased influenza mortality of mice adoptively immunized with node and spleen cells sensitized by inactivated but not live virus. *Infect. Immun.*, **11**, 908–14.

18. Cave, D.R., Hendrickson, F.M. and Huang, A.S. (1985) Defective interfering virus particles modulate virulence. *J. Virol.*, **55**, 366–73.

19. Chambers, T.M. and Webster, R.G. (1987) Defective interfering virus associated with A/Chicken/Pennsylvania/83 influenza virus. *J. Virol.*, **61**, 1517–23.

20. Cole, C.N. (1975) Defective interfering (DI) particles of poliovirus. *Prog. Med. Virology*, **20**, 180–207.

21. Compans, R.W., Dimmock, N.J. and Meier-Ewert, H. (1970) An electron microscopic study of the influenza virus-infected cell, in *The Biology of Large RNA Viruses*, (eds R.D. Barry and B.W.J. Mahy), Academic Press, New York, pp. 87–108.

22. Crumpton, W.C., Avery, R.J. and Dimmock, N.J. (1981) Influence on the host cell on the genomic and subgenomic RNA content of defective interfering particles. *J. Gen. Virol.*, **53**, 173–7.

23. Crumpton, W.C., Dimmock, N.J., Minor, P.D. and Avery, R.J. (1978) The RNAs of defective-interfering influenza virus. *Virology*, **90**, 370–8.

24. De, B.K. and Nayak, D.P. (1980) Defective-interfering influenza viruses: establishment and maintenance of persistent influenza virus infection in MDBK and HeLa cells. *J. Virol.*, **36**, 847–59.

25. Debnath, N.C., Tiernery, R., Sil, B.K., Wills, M.R. and Barrett, A.D.T. (1991) *In vitro* homotypic and heterotypic interference by defective interfering particles of West Nile virus. *J. Gen. Virol.*, **72**, 2705–11.

26. DePolo, N.J. and Holland, J.J. (1986) The intracellular half-lives of non-replicating nucleocapsids of DI particles of wild type and mutant strains of vesicular stomatitis virus. *Virology*, **151**, 371–8.

27. DePolo, N.J. and Holland, J.J. (1986) Very rapid generation/amplification of defective interfering particles by vesicular stomatitis virus variants isolated from persistent infections. *J. Gen. Virol.*, **67**, 1195–8.

28. Dimmock, N.J. (1991) The biological significance of defective interfering viruses. *Rev. Med. Virol.*, **1**, 165–76.

29. Dimmock, N.J., Beck, S. and McLain, L. (1986) Protection of mice from lethal influenza: evidence that defective interfering virus modulates the immune response and not virus multiplication. *J. Gen. Virol.*, **67**, 839–50.

30. Dimmock, N.J. and Kennedy, S.I.T. (1978) Prevention of death in Semliki Forest virus-infected mice by administration of defective

interfering Semliki Forest virus. *J. Gen. Virol.*, **39**, 231–42.

31. Eichelberger, M.C., Wang, M., Allan, W., Webster, R.G. and Doherty, P.C. (1991) Influenza virus RNA in the lung and lymphoid tissue of immunologically intact and CD4-depleted mice. *J. Gen. Virol.*, **72**, 1695–8.

32. Enami, M., Luytjes, W., Krystal, M. and Palese, P. (1990) Introduction of site-specific mutations into the genome of influenza virus. *Proc. Natl Acad. Sci. USA*, **87**, 3802–5.

33. Enami, M. and Palese, P. (1991) High efficiency formation of influenza virus transfectants. *J. Virol.*, **65**, 2711–13.

34. Enami, M., Sharma, G., Benham, C. and Palese, P. (1991) An influenza virus containing nine different RNA segments. *Virology*, **185**, 291–8.

35. Fazekas, de St., Groth, S. and Graham, D.M. (1954) Artificial production of incomplete influenza virus. *Nature*, **173**, 637–8.

36. Fazekas, de St., Groth, S. and Graham, D.M. (1954) The production of incomplete influenza virus particles among influenza strains. Experiments in eggs. *Br. J. Exp. Pathol.*, **35**, 60–74.

37. Fields, S. and Winter, G. (1982) Nucleotide sequences of influenza virus segments 1 and 3 reveal mosaic structure of a small viral RNA segment. *Cell*, **28**, 303–13.

38. Francis, S.J. and Southern, P.J. (1988) Deleted viral RNAs and lymphocytic choriomeningitis virus persistence *in vitro*. *J. Gen. Virol.*, **69**, 1893–902.

39. Frielle, D.W., Huang, D.D. and Youngner, J.S. (1984) Persistent infection with influenza A virus: evolution in virus mutants. *Virology*, **138**, 103–7.

40. Fultz, P., Shadduck, J.A., Kang, C.Y. and Streilein, J.W. (1981) On the mechanism of DI particle protection against lethal VSV infection in hamsters, in *The Replication of Negative Strand Viruses*, (eds. D.H.L. Bishop and R.W. Compans), Elsevier-North Holland, New York, pp. 893–9.

41. Fultz, P.N., Shadduck, J.A., Kang, C.Y. and Streilein, J.W. (1982) Mediators of protection against lethal systemic vesicular stomatitis virus infection in hamsters: defective interfering particles, polyinosinate-polycytidylate, and interferon. *Infect. Immun.*, **37**, 679–86.

42. Gamboa, E.T., Harter, D.H., Duffy, P.E. and Hsu, K.C. (1976) Murine influenza virus encephalomyelitis. III Effect of defective interfering virus particles. *Acta Neuropathol.*, **34**, 157–69.

43. Giachetti, C. and Holland, J.J. (1988) Altered replicase specificity is responsible for resistance to defective interfering particle interference of an Sdi- mutant of vesicular stomatitis virus. *J. Virol.*, **62**, 3614–21.

44. Giachetti, C. and Holland, J.J. (1989) Vesicular stomatitis virus and its defective interfering particles exhibit in vitro transcriptional and replicative competition for purified L-NS polymerase molecules. *Virology*, **170**, 264–7.

45. Ginsberg, H.S. (1954) Formation of non-infectious influenza virus in mouse lungs: its dependence upon extensive pulmonary consolidation initiated by the viral inoculum. *J. Exp. Med.*, **100**, 581–603.

46. Harty, R.N., Holden, R. and O'Callaghan, D.J. (1993) Transcriptional and translational analysis of the UL2 gene of equine herpesvirus 1: a homolog of UL55 of herpes simplex virus type 1 that is maintained in the genome of defective interfering particles. *J. Virol.*, **67**, 2255–65.

47. Hers, J.F.P., Mulder, J., Masurel, N., van der Kuip, L. and Tyrrell, D.A.J. (1962) Studies on the pathogenesis of influenza virus pneumonia in mice. *J. Pathol. Bacteriol.*, **83**, 207–17.

48. Herz, C., Stavnezer, E., Krug, R.M. and Gurney, T. (1981) Influenza virus, an RNA virus, synthesizes its messenger RNA in the nucleus. *Cell*, **26**, 254–6.

49. Holland, J.J. (1990) Defective viral genomes, in *Virology*, 2nd edn, Vol. 1, (eds. B.N. Fields and D.M. Knipe), Raven Press, New York, pp. 151–65.

50. Holland, J.J. and Doyle, M. (1973) Attempts to detect homologous autointerference in vivo with influenza virus and vesicular stomatitis virus. *Infect. Immun.*, **7**, 526–31.

51. Holland, J.J., Semler, B.L., Jones, C., Perrault, J., Reid, L. and Roux, L. (1978) Role of DI, virus mutation, and host response in persistent infections by enveloped RNA viruses, in *Persistent Viruses: ICN–UCLA Symposium on Molecular and Cellular Biology*, Vol. 11, (eds J.G. Stevens, G.J. Todaro and C.F. Fox), Academic Press, New York, pp. 57–73.

52. Holland, J.J. and Villarreal, L.P. (1974) Persistent noncytocidal vesicular stomatitis virus infections mediated by defective interfer-

ing T particles that suppress virion transcriptase. *Proc. Natl Acad. Sci. USA*, **71**, 2956–60.

53. Holland, J.J. and Villarreal, L.P. (1975) Purification of defective interfering T particles of vesicular stomatitis and rabies viruses generated in vivo in brains of newborn mice. *Virology*, **67**, 438–49.

54. Holland, J.J., Villarreal, L.P. and Breindel, M. (1976) Factors involved in the generation and replication of rhabdovirus defective T particles. *J. Virol.*, **17**, 805–15.

55. Hope-Simpson, R.E. (1992) *The Transmission of Epidemic Influenza Virus*, Plenum Press, New York.

56. Horsfall, F.L. (1954) On the reproduction of influenza virus. Quantitative studies with procedures which enumerate infective and hemagglutinating virus particles. *J. Exp. Med.*, **100**, 135–61.

57. Horsfall, F.L. (1955) Reproduction of influenza viruses. Quantitative investigations with particle enumeration procedures on the dynamics of influenza A and B virus reproduction. *J. Exp. Med.*, **102**, 441–73.

58. Huang, A.S. (1988) Modulation of viral disease processes by defective interfering particles, in *RNA Genetics*, Vol. 3, (eds E. Domingo, J.J. Holland and P. Ahlquist), CRC Press, Boca Raton, Florida, pp. 195–208.

59. Huang, A.S. and Baltimore, D. (1970) Defective viral particles and viral disease processes. *Nature*, **226**, 325–7.

60. Jackson, D.A., Caton, A.J., McCready, S.J. and Cook, P.R. (1982) Influenza virus RNA is synthesized at fixed sites in the nucleus. *Nature*, **296**, 366–8.

61. Janda, J.M. and Nayak, D.P. (1979) Defective influenza viral ribonucleoproteins cause interference. *J. Virol.*, **32**, 679–702.

62. Jennings, P.A., Finch, J.T., Winter, G. and Robertson, J.S. (1983) Does the higher order of the influenza virus ribonucleoprotein guide sequence rearrangements in influenza viral RNA. *Cell*, **34**, 619–27.

63. Jones, C.L. and Holland, J.J. (1980) Requirements for DI particle prophylaxis against vesicular stomatitis virus infection *in vivo. J. Gen. Virol.*, **49**, 215–20.

64. Kang, C.Y. and Allen, R. (1978) Host function-dependent induction of defective-interfering particles of vesicular stomatitis virus. *J. Virol.*, **24**, 202–6.

65. Kang, C.Y., Weide, L.G. and Tischfield, J.A. (1981) Suppression of vesicular stomatitis virus defective interfering particle generation by a function(s) associated with human chromosome 16. *J. Virol.*, **40**, 946–52.

66. Kilbourne, E.D. (1987) *Influenza*. Plenum Medical Book Co., New York.

67. Kingsbury, D.W. and Portner, A. (1970) On the genesis of incomplete Sendai virions. *Virology*, **42**, 872–9.

68. von Laer, D.M., Mack, D. and Kruppa, J. (1988) Delayed formation of detective interfering particles in vesicular stomatitis virus-infected cells: kinetic studies of viral protein and RNA synthesis during autointerference. *J. Virol.*, **62**, 1323–9.

69. Lehtovaara, P., Söderlund, H., Keränen, S., Pettersson, R.F. and Kääriäinen, L. (1982) Extreme ends of the genome are conserved and rearranged in the defective interfering RNAs of Semliki Forest virus. *J. Mol. Biol.*, **156**, 731–48.

70. Leung, K.N. and Ada, G.L. (1980) Cells mediating delayed-type hypersensitivity in the lungs of mice infected with an influenza A virus. *Scand. J. Immunol.*, **12**, 393–400.

71. Leung, K.N. and Ada, G.L. (1982) Different functions of helper T cells on the primary *in vitro* production of delayed hypersensitivity to influenza virus. *Cell. Immunol.*, **67**, 312–24.

72. Liew, F.Y. and Russell, S.M. (1980) Delayed-type hypersensitivity to influenza virus: induction of antigen-specific suppressor T cells for delayed hypersensitivity to haemagglutinin during influenza virus infection in mice. *J. Exp. Med.*, **151**, 799–814.

73. Liew, F.Y. and Russell, S.M. (1983) Inhibition of pathogenic effect of effector T cells by specific suppressor T cells during influenza infection of mice. *Nature*, **304**, 541–3.

74. von Magnus, P. (1947) Studies on interference in experimental influenza. I. Biological observations. *Ark. Kemi. Mineral. Geol.*, **24b**, 1.

75. von Magnus, P. (1951) Propagation of the PR/8 strain of influenza virus in chick embryos. III Properties of the incomplete virus produced in serial passages of undiluted virus. *Acta Pathol. Microbiol. Scand.*, **29**, 157–81.

76. von Magnus, P. (1951) Propagation of the PR8 strain of influenza virus in chick embryos. II. The formation of 'incomplete' virus following

the inoculation of large doses of seed virus. *Acta Path. Microbiol. Scand.*, **28**, 278–93.

77. von Magnus, P. (1952) Propagation of the PR8 strain of influenza virus in chick embryos. IV Studies on the factors involved in the formation of incomplete virus upon serial passage of undiluted virus. *Acta Pathol. Microbiol. Scand.*, **30**, 311–35.

78. von Magnus, P. (1954) Incomplete forms of influenza virus. Adv. Virus Res., 21, 59–79.

79. McCauley, J.W. and Mahy, B.W.J. (1983) Structure and function of the influenza virus genome. *Biochem. J.*, **211**, 281–94.

80. McLain, L., Armstrong, S.J. and Dimmock, N.J. (1988) One defective interfering particle per cell prevents influenza virus-mediated cytopathology: an efficient assay method. *J. Gen. Virol.*, **69**, 1415–19.

81. McLain, L. and Dimmock, N.J. (1989) Protection of mice from lethal influenza by adoptive transfer of non-neutralizing haemagglutination-inhibiting IgG obtained from the lungs of infected animals treated with defective-interfering virus. *J. Gen. Virol.*, **70**, 2615–24.

82. McLain, L. and Dimmock, N.J. (1991) An influenza haemagglutinin-specific IgG enhances class I MHC-restricted CTL killing *in vitro*. *Immunology*, **73**, 12–18.

83. McLain, L., Morgan, D.J. and Dimmock, N.J. (1992) Protection of mice from lethal influenza by defective interfering virus: T cell responses. *J. Gen. Virol.*, **73**, 375–81.

84. McLaren, L.C. and Holland, J.J. (1974) Defective interfering particles of poliovirus vaccine and vaccine reference strains. *Virology*, **60**, 579–83.

85. Meier-Ewert, H. and Dimmock, N.J. (1970) The role of the neuraminidase of the infecting virus in the generation of noninfectious (von Magnus) interfering virus. *Virology*, **42**, 794–8.

86. Morgan, D.J. and Dimmock, N.J. (1992) Defective interfering virus inhibits immunopathological effects of infectious virus in the mouse. *J. Virol.*, **66**, 1188–92.

87. Morgan, D.J., McLain, L. and Dimmock, N.J. (1993) Protection of three strains of mice against lethal influenza in vivo by defective interfering virus. *Virus Res.*, **29**, 179–93.

88. Moss, B.A. and Brownlee, G.G. (1981) Sequence of DNA complementary to a small segment of influenza virus A/NT/60/68. *Nucleic Acids Res.*, **9**, 1941–7.

89. Mottet, G., Curran, J. and Roux, L. (1990) Intracellular stability of non-replicating paramyxovirus nucleocapsids. *Virology*, **176**, 1–7.

90. Mumford, J.A., Wood, J.M., Folkers, C. and Schild, G.C. (1988) Protection against experimental infection with influenza virus A/equine/Miami/63 (H3N8) provided by inactivated whole virus vaccines containing homologous virus. *Epidemiol. Infect.*, **100**, 501–10.

91. Mumford, J.A., Wood, J.M., Scott, A.M., Folkers, C. and Schild, G.C. (1983) Studies with inactivated equine influenza vaccine. 2 Protection against experimental infection with A/equine/Newmarket/79 (H3N8). *J. Hygiene (Camb)*, **90**, 385–95.

92. Nakajima, K., Ueda, M. and Suguira, A. (1979) Origin of small RNA in von Magnus virus particles of influenza virus. *J. Virol.*, **29**, 1142–8.

93. Nayak, D.P., Chambers, T.M. and Akkina, R.K. (1985) Defective-interfering (DI) RNAs of influenza viruses: origin, structure, expression and interference. *Curr. Topics Microbiol. Immunol.*, **114**, 103–51.

94. Nayak, D.P., Chambers, T.M. and Akkina, R.M. (1989) Structure of defective-interfering RNAs of influenza virus and their role in interference, in *The Influenza Viruses*, (ed. R.M. Krug), Plenum Press, New York, pp. 269–317.

95. Nayak, D.P., Sivasubramanian, N., Davis, A.R., Cortini, R. and Sung, J. (1982) Complete sequence analyses show that two defective interfering influenza viral RNAs contain a single internal deletion of a polymerase gene. *Proc. Natl Acad. Sci. USA*, **79**, 2216–20.

96. Nayak, D.P., Tobita, K., Janda, J.M., Davis, A.R. and De, B.K. (1978) Homologous interference mediated by defective interfering influenza virus derived from a temperature-sensitive mutant of influenza virus. *J. Virol.*, **28**, 375–86.

97. Noble, S. and Dimmock, N.J. (1994) Defective interfering type A equine influenza virus (H3N8) protects mice from morbidity and mortality caused by homologous and heterologous subtypes of type A influenza virus. *J. Gen. Virol.*, **75**, 3485–91.

98. Odagiri, T. and Tobita. K. (1990) Mutation in NS2, a non-structural protein of influenza A virus, extragenically causes aberrant replication and expression of the PA gene, and leads to the generation of defective interfering particles. *Proc. Natl Acad. Sci. USA*. **87**, 5988–92.

98a. O'Neill, R.E. and Palese, P. (1995) NPI-1, the human homologue of SRP-1, interacts with influenza virus nucleoprotein. *Virology,* **206,** 116–25.

99. Outlaw, M.C. and Dimmock, N.J. (1991) Insights into neutralization of animal viruses gained from the study of influenza viruses. *Epidemiol. Infect.,* **106,** 205–20.

100. Palma, E.L. and Huang, A.S. (1974) Cyclic production of vesicular stomatitis virus caused by defective interfering particles. *J. Infect. Dis.,* **129,** 402–10.

101. Perrault, J. (1981) Origin and replication of defective interfering particles. *Curr. Topics Microbiol. Immunol.,* **93,** 151–207.

102. Philips, B.A., Lundquist, R.E. and Maizel, J.V. (1980) Absence of viral particles and assembly activity in HeLa cells infected with defective-interfering (DI) particles of poliovirus. *Virology,* **100,** 116–24.

103. Poidinger, M., Coelen, R. and MacKenzie, J.S. (1991) Persistent infection of Vero cells by the flavivirus Murray Valley encephalitis virus. *J. Gen. Virol.,* **72,** 573–8.

104. Popescu, M. and Lehmann-Grube, F. (1977) Defective interfering particles in mice infected with lymphocytic choriomeningitis virus. *Virology,* **77,** 78–83.

105. Prevec, L. and Kang, C.Y. (1970) Homotypic and heterotypic interference by defective interfering particles of vesicular stomatitis virus. *Nature,* **288,** 25–7.

106. Pringle, C.R. (1987) Rhabdovirus genetics, in *The Rhabdoviruses,* (ed. R.R. Wagner), Plenum Press, New York, pp. 176–243.

107. Rabinowitz, S.G. and Huprikar, J. (1979) The influence of defective-interfering particles of the PR-8 strain of influenza A virus on the pathogenesis of pulmonary infection in mice. *J. Infect. Dis.,* **140,** 305–15.

108. Roman, J.M. and Simon, E.H. (1976) Defective interfering particles in monolayer-propagated Newcastle disease virus. *Virology,* **69,** 298–303.

109. Roux, L. and Holland, J.J. (1979) Role of defective interfering particles of Sendai virus in persistent infections. *Virology,* **93,** 91–103.

110. Roux, L., Simon, A.E. and Holland, J. J. (1991) Effects of defective interfering viruses on viral replication and pathogenesis *in vitro* and *in vivo. Adv. Virus Res.,* **40,** 181–211.

111. Schlesinger, S. and Weiss, B.G. (1986) Defective RNAs of alphaviruses, in *The Togaviridae and Flaviviridae,* (eds S. Schlesinger and M.J. Schlesinger), Plenum Press, New York, London, pp. 149–69.

112. Schnitzlein, W.M. and Reichman, M.E. (1976) The size and cistronic origin of defective interfering vesicular stomatitis virus particle RNAs in relation to homotypic and heterotypic interference. *J. Mol. Biol.,* **101,** 307–25.

113. Shapiro, G.I., Gurney, T. and Krug, R.M. (1987) Influenza virus genome expression: control mechanisms at early and late times of infection and nuclear-cytoplasmic transport of virus-specific RNAs. *J. Virol.,* **61,** 764–73.

114. Sivasubramanian, N. and Nayak, D.P. (1983) Defective interfering influenza RNAs of polymerase 3 gene contain single as well as multiple internal deletions. *Virology,* **124,** 232–7.

115. Smith, G.L. and Hay, A.J. (1982) Replication of the influenza virus genome. *Virology,* **118,** 96–108.

116. Spandidos, D.A. and Graham, A.F. (1976) Generation of defective virus after infection of newborn rats with reovirus. *J. Virol.,* **20,** 234–47.

117. Stark, C. and Kennedy, S.I.T. (1978) The generation and propagation of defective-interfering particles of Semliki Forest virus in different cell types. *Virology,* **89,** 285–99.

118. Stephenson, J.R. and Dimmock, N.J. (1975) Early events in influenza virus multiplication. I Location and fate of the input RNA. *Virology,* **65,** 77–86.

119. Thomson, M. and Dimmock, N.J. (1994) Sequences of defective interfering Semliki Forest virus necessary for anti-viral activity in vivo and interference and propagation *in vitro. Virology,* **199,** 354–65.

120. Wang, M. and Webster, R.G. (1990) Lack of persistence of influenza virus genetic information in ducks. *Arch. Virol.,* **111,** 263–7.

121. Webster, R.G., Bean, W.J., Gorman, O.T., Chambers, T.M. and Kawaoka, Y. (1992) Evolution and ecology of influenza A viruses. *Microbiol. Rev.,* **56,** 152–79.

122. Webster, R.G., Kawaoka, Y. and Bean, W.J. (1986) Molecular changes in A/chicken/Pennsylvania/83 influenza virus associated with acquisition of virulence. *Virology,* **149,** 165–73.

123. Webster, R.G. and Yuanji, G. (1991) New influenza virus in horses. *Nature,* **351,** 527.

124. Weiss, B., Rosenthal, R. and Schlesinger, S. (1980) Establishment and maintenance of per-

sistent infection by Sindbis virus in BHK cells. *J. Virol.*, **33**, 463–74.

125. Weiss, B. and Schlesinger, S. (1981) Defective interfering particles of Sindbis virus do not interfere with the homologous virus obtained from persistently infected BHK cells but do interfere with Semliki Forest virus. *J. Virol.*, **37**, 840–4.

126. Welsh, R.M., Lampert, P.W. and Oldstone, M.B.A. (1977) Prevention of virus-induced cerebellar disease by defective interfering lymphocytic choriomeningitis virus. *J. Infect. Dis.*, **136**, 391–9.

127. Werner, G.H. (1956) Quantitative studies on influenza virus infection of the chick embryo by the amniotic route. *J. Bacteriol.*, **71**, 505–15.

128. Winter, G., Fields, S. and Ratti, G. (1981) The structure of two sub-genomic RNAs from human influenza virus A/PR/8/34. *Nucleic Acids Res.*, **9**, 6907–15.

129. Wood, J.M., Mumford, J.A., Folkers, C., Scott, A.M. and Schild, G.C. (1983) Studies with inactivated equine influenza vaccine. I . Serological responses of ponies to graded doses of vaccine. *J. Hygiene (Camb)*, **90**, 371–84.

PSYCHOLOGY OF COMMON COLDS AND OTHER INFECTIONS

23

Sheldon Cohen and Andrew Smith

23.1 INTRODUCTION

There is increasing evidence that the central nervous system (CNS) and immune system interact with one another. Hormones released from the brain circulate through the bloodstream and alter the functional activity of immune cells and immune cells produce chemical messengers that alter brain function. During the last several years of operation of the Medical Research Council's Common Cold Unit (CCU), we together with David Tyrrell and his staff embarked on a series of studies addressing the possible interrelations between the CNS and immune system. In particular, we studied how psychological states might influence pathogenesis of upper respiratory infections and how infections might influence the processing of information. In this chapter, we review the literature relevant to each of these projects and summarise the work done at the CCU.

23.1 PSYCHOLOGICAL STRESS AND SUSCEPTIBILITY TO INFECTIOUS DISEASE

On exposure to an infectious agent, only a proportion of people develop clinical disease [14,17]. Moreover, severity and duration of symptomatology vary widely among those who do become ill. Reasons for variability in response are not well understood and the possibility that psychological stress plays a role has received increased attention [5,11,58]. This section of the chapter addresses the possible role of stress in the aetiology and progression of infectious diseases.

Viral and Other Infections of the Human Respiratory Tract. Edited by S. Myint and D. Taylor-Robinson. Published in 1996 by Chapman & Hall. ISBN 0 412 60070 6

23.2.1 PATHWAYS LINKING STRESS TO INFECTIOUS DISEASE SUSCEPTIBILITY

When demands imposed by life events exceed a person's ability to cope, a psychological stress response composed of negative cognitive and emotional states is elicited [41]. These responses trigger behavioural and biological changes that often suppress immune function and as a consequence are thought to put persons at higher risk for infectious disease. Psychological stress may influence immunity through direct innervation of the CNS and immune system or through neuroendocrine–immune pathways. Direct neural pathways linking the CNS to the immune system have been identified [15,16]. In the case of hormonal pathways, catecholamines secreted by the adrenal medulla in response to stress, and stress-triggered, pituitary-mediated hormones such as cortisol and prolactin have been associated with modulation of immune function [27,28]. Moreover, receptors for adrenocorticotrophic hormone (ACTH), thyroid-stimulating hormone (TSH), growth hormone, prolactin and catecholamines have been found on lymphocytes.

Behavioural changes that occur as adaptations or coping responses to psychological stress may also influence immunity. For example, persons experiencing stress often engage in poor health practices, for example smoking, poor diets and poor sleeping habits [11,13] that may have immunosuppressive effects [36]. Aggressive or affiliative behaviours triggered by prolonged psychological stress may also influence immunity. Thus, it may be that it is these behaviours (and not the stressors themselves) that trigger sympathetic or endocrine response [9,34,43].

Stress may also play a role in reactivating latent pathogens such as herpesviruses. Reactivation could occur through hormonal or neural stimulation of pathogen reproduction or through suppression of cellular immune processes that might otherwise hold the pathogen in check [20,35].

23.2.2 STRESS, IMMUNITY AND SUSCEPTIBILITY TO INFECTION

Human and infra-human studies indicate that various stressors modulate both cellular and humoral measures of immune function. This includes human research on immunomodulating effects of *acute* laboratory stressors such as difficult mental tasks [43] as well as studies of naturalistic *chronic* stressors such as separation and divorce [33,36], care-giving for Alzheimer patients [33], and bereavement [4]. It also includes experimental studies of social stressors on immunity in non-human primates. For example, the separation of offspring from their mothers results in suppression of both mitogen-stimulated lymphocyte proliferation and antibody production in response to an antigenic challenge in the young animals [8,39,40]. Similarly, our own work indicates that adult cynomolgus monkeys randomly assigned to 2 years of exposure to an unstable social environment demonstrate relatively suppressed mitogen-stimulated lymphocyte proliferation when compared with animals assigned to a stable social environment.

Although the effects of stressors on immune response are often described as immunosuppressive, the implications of stressor-induced immune changes for disease susceptibility are not clear [7,30]. First, in studies of stressor effects on immunity, the immune responses of stressed persons fall within normal ranges [37,38]. Second, there are few data on immune status in healthy persons as a predictor of susceptibility to disease. Finally, the immune system is complex. One or even several measures of immune function may not provide an adequate representation of host resistance.

23.2.3 STRESS AND SUSCEPTIBILITY TO INFECTION

Although human research in this area is sparse, existing epidemiological evidence on stress influences on infectious disease is provocative [12,38]. These studies do not manipulate exposure to stressors. Instead, stressor exposure and/or emotional response to stressors are assessed among healthy persons and disease prevalence (retrospective studies) or later disease incidence (prospective studies) is predicted.

First, there is evidence suggesting that stress increases risk for verified upper respiratory infections. In two prospective epidemiological studies it was found that family conflict and disorder predicted serologically verified infectious illness [24,45]. Converging evidence comes from viral-challenge studies, where volunteers who fill out stressful life event or emotional distress scales are subsequently challenged with a cold or influenza virus. The results of early work with this paradigm provided mixed support for a relation between stress and susceptibility to upper respiratory infections. Thus, the work of Broadbent *et al.* [6] and Totman *et al.* [60] was supportive but that of Greene *et al.* [26] and Locke and Heisel [42] was not. Our own work (described later) employing a large sample and more sophisticated methodology provides strong evidence for a dose–response relation between self-reported psychological stress and risk of illness [10].

Second, there is growing evidence that stress may trigger *reactivation* of herpesviruses and hence, recurrence of disease among those with previous exposure to herpes. Indirect support for stress-triggered reactivation comes from a series of studies indicating increased antibodies to three herpes viruses (herpes simplex virus, HSV-1; cytomegalovirus, CMV; Epstein–Barr virus, EBV) under stress [21,22], while direct support derives from prospective studies of unpleasant moods on oral [18,32] and genital herpes [23,44]. The results of a single prospective study [31] also indicate the possibility of stress-triggered primary EBV infection and clinical disease (mononucleosis). Although these data provide provocative evidence for a relation between stress and infectious disease, none of these studies has demonstrated either biological or behavioural pathways through which stress influences disease susceptibility [12].

23.3 THE COMMON COLD STUDIES

In work carried out in collaboration with David Tyrrell and the staff at the MRC Common Cold Unit (CCU), the question of whether psychological stress places people at greater risk for infectious disease was pursued and at the same time an attempt was made to identify the behavioural and biological pathways through which such relations operate. For stressful events to influence susceptibility, they are presumed to be appraised as stressful (as exceeding the ability to cope) and to consequently elicit an emotional response. This emotional response is thought to trigger either behavioural (e.g. increased smoking) or neuroendocrine (e.g. increases in adrenaline, noradrenaline or cortisol) responses which are considered to influence the ability of the immune system to respond to a challenge. The work described was designed to examine the psychological, behavioural and biological pathways thought to link stressful events to illness susceptibility, while carefully controlling for a variety of other factors that might influence the risk for infectious disease.

The data described here are from a trial conducted at the CCU between 1986 and 1989 [10]. In this study, stressful life events, perceived stress, and negative effect was assessed *before* experimentally exposing subjects to a common cold virus. The subjects were monitored carefully for the development of infection and clinical illness. By intentionally exposing people to an upper respiratory virus, the possible effects of stressful events on exposure to infectious agents (as opposed to their effects on host resistance) could be eliminated. In the remainder of this section, the relation between the measures of stress and risk for clinical colds is examined and potential pathways through which stress might influence susceptibility to infectious disease are evaluated.

23.3.1 METHODS

The subjects were 154 men and 266 women volunteers 18 to 54 years old. All reported no chronic or acute illness or regular medication regimen and were judged in good health following examination. During their first 2 days in the CCU, they were given a thorough medical examination, completed psychological stress, personality, and health practice questionnaires and had blood drawn for immune and cotinine assessments. Subsequently, the volunteers were exposed via nasal drops to a low infectious dose of one of five respiratory viruses: rhinovirus types 2 ($n=86$), 9 ($n=122$), 14 ($n=92$), respiratory syncytial (RS) virus ($n=40$), and coronavirus type 229E ($n=54$). An additional 26 volunteers received saline. For 2 days before and 7 days after viral-challenge, the volunteers were quarantined in large flats (alone or with one or two others). Starting 2 days before viral challenge and continuing through 6 days post-challenge, each volunteer was examined daily by a clinician using a standard respiratory sign–symptom protocol. Examples of items on the protocol include sneezing, watering of eyes, nasal stuffiness, sore throat, hoarseness and cough. The protocol also included an objective count of the number of tissues used daily by a volunteer and body temperature (oral) assessed twice each day. Samples of nasal secretions were also collected daily to assess whether volunteers were infected by the virus. Approximately 28 days after challenge a second serum sample was collected to assess changes in viral-specific antibody. All investigators were blind to the volunteers' psychological status and to whether they received virus or saline.

Infections and clinical colds

Infection was detected directly by culturing nasal secretion samples (viral isolation) or indirectly through establishing significant increases in viral-specific antibody. Nasal wash samples for viral isolation were collected before inoculation and on days 2 to 6 six after viral inoculation. They were mixed with broth and stored in aliquots at –70°C. Rhinoviruses were detected in O-HeLa cells, RS virus in Hep2 cells and coronavirus in C-16 strain of continuous human fibroblast cells. When a characteristic cytopathic effect was observed the tissue culture fluids were passed into further cultures; rhinoviruses and coronaviruses were identified by neutralization tests with specific rabbit immune sera, and RS virus by immunofluorescence staining of infected cells. Titres of antibodies were determined before and at 28 days after challenge, those for rhinoviruses by neutralization tests with homologous virus [19], a four-fold rise being regarded as significant. Viral-specific IgA and IgG levels for rhinoviruses, coronavirus and RS virus were determined by enzyme-linked immunosorbent assays.

A person can be infected without developing clinical illness. The criteria for clinical illness were both infection and a positive clinical diagnosis. At the end of the trial, the clinician judged the severity of each volunteer's cold on a scale ranging from nil (O) to severe (4). Ratings of mild cold (2) or greater were considered positive clinical diagnoses. Some 82% (325) of the 394 volunteers receiving virus were infected and 38% (148) developed clinical colds. None of the 26 saline controls developed colds. The subjects also rated the severity of their colds on the same scale. The clinical diagnosis was in agreement with the subject's rating in 94% of the cases.

Psychological stress

It was mentioned earlier that when demands imposed by events exceed ability to cope, a psychological stress response is elicited and that this response is composed of negative cognitive and emotional states. In order to assess the various components of this process, three kinds of measures of psychological stress were administered before viral challenge: (i) number of major stressful life events judged by the respondent as having a negative impact; (ii) perception that current demands exceed capabilities to cope; and (iii) current negative affect. The major stressful life events scale consisted of events that might happen in the life of the respondent (41 items) or close others (26 items). The scale score was the number of negative events reported as occurring during the last year. The Perceived Stress Scale was used to assess the degree to which situations in life are perceived as stressful. Items in the scale were designed to tap how unpredictable, uncontrollable and overloading respondents find their lives. Finally, the negative affect scale included 15 items from Zevon and Tellegen's list of negative emotions. Examples of emotions on the list include sad, angry, nervous, distressed and scared. Because these three scales were correlated with one another, the data presented are based on analyses using a psychological stress index that combines the three scales. The index was created by quartiling each scale and summing quartile ranks for each subject, resulting in a scale with scores ranging from 3 to 12. Although not reported in this chapter, analysis of data from each of the individual stress scales suggests similar results.

Standard control variables

Each analysis statistically controls (co-varies) for the possible effects of a series of variables that might provide alternative explanations for a relation between stress and illness. These include pre-challenge serostatus for the experimental virus, age, gender, education, allergic status, bodyweight, season, number of other subjects with whom the volunteer was housed, whether an apartment mate was infected, and the challenge virus.

Health practice measures

Health practices were assessed as possible pathways linking stress and susceptibility.

Measures including smoking (serum cotinine), drinking alcohol, exercise, quality of sleep and diet were administered before viral challenge.

White cell counts and total immunoglobulin levels

White cell counts and total immunoglobulin levels were also assessed in blood samples collected before viral challenge as possible factors linking psychological stress and susceptibility to illness. White cells were counted with an automatic cell counter, and differential counts (lymphocytes, monocytes and neutrophils) were calculated from 200 cells in a stained film. Total serum and nasal-wash IgA and IgE levels and total nasal-wash protein levels were assessed by using an enzyme-linked immunosorbent assay.

Personality measures

Because psychological stress could reflect stable personality styles rather than responses to environmental stressors, self-esteem and personal control (two personality characteristics closely associated with stress) were assessed before viral challenge. A third personality characteristic, introversion–extroversion was also assessed. These measures were also administered before viral challenge.

23.3.2 RESULTS

Stress and susceptibility to clinical illness

As shown in Figure 23.1, rates of clinical illness increased in a dose–response manner with scores on the psychological stress index. To determine whether any of these effects might be attributable to relations between stress and health practices or stress and white cell counts or total immunoglobulin levels, additional conservative analyses were carried out, including the five health practices, three white cell differentials, and total immunoglobulin measures, in the equations along with the ten standard control variables. This procedure tests whether stress is associated with greater susceptibility after the possible effects of these variables are subtracted. The addition of health practices, white cell counts and total immunoglobulins did not alter the results. To determine whether these relations might be attributable to the stress scales actually reflecting personality characteristics, an additional analysis was undertaken in which the three personality factors were added to the equation. Again, the relations between stress and illness were independent of these personality characteristics.

Are stress effects consistent across the five viruses?

The analyses described so far have collapsed across viruses (including statistical controls for virus in the regression equation). However, a test of whether the effects of stress were consistent across the viruses (interaction of stress and virus type) indicated that they were. The influence of stress on each virus is shown in Figure 23.2. This suggests the possibility that the relation between psychological stress and upper respiratory illness is non-specific, that

is, independent of the pathogenesis of the specific virus. Figure 23.2 also suggests that the dose–response type relationship in Figure 23.1 occurred in all cases, with each increase in stress associated with an increase in colds. (A detailed analysis of the dose–response issue is reported in [10].)

Is stress associated with increased infection or increased illness?

Stress-associated increases in clinical illness could be attributable to an association between

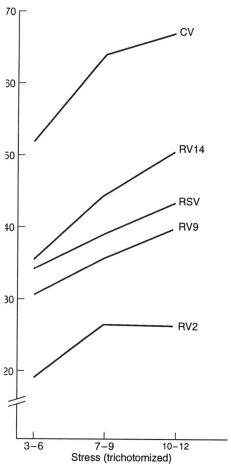

Figure 23.2 Association between the Psychological Stress Index and the rate of clinical colds for each of the five viruses. CV, coronavirus; RSV, respiratory syncytial virus; RV, rhinovirus.

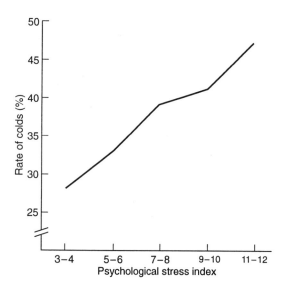

Figure 23.1 Association between the Psychological Stress Index and the rate of clinical colds collapsing across viruses. (Adapted from [10].)

stress and increased probability of infection or to an association between stress and increased probability of infected persons developing clinical symptoms. Additional analyses addressed this issue. The first analysis assessed whether the reported relation between the stress index and clinical colds was partly or wholly attributable to an association between these scales and increased infection. As apparent from Figure 23.3, the probability of becoming infected (independent of symptoms) increased with increases in the stress index. The second analysis assessed whether the reported relations between the various stress measures and clinical colds were partly or wholly attributable to associations between stress and becoming sick (developing clinical symptoms) *following* infection. Because this analysis included only persons who were infected, the results are independent of earlier analyses predicting infection. There was no association between stress and the development of illness among infected persons.

23.3.3 DISCUSSION

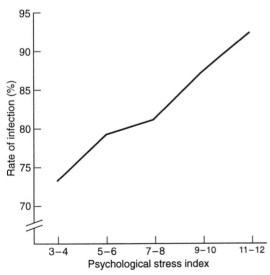

Figure 23.3 Association between the psychological stress index and rate of infection collapsing across viruses. (Adapted from [10].)

It was found that increases in stress were associated in a dose–response manner with increases in rates of clinical illness. This relation could not be explained by factors thought to be associated with stress including age, gender, education, bodyweight and allergic status, or design factors such as the virus the subject was exposed to and environmental characteristics associated with the design of the study. The relations were also not explicable in terms of either stress-induced differences in health practices, white cell counts, immunoglobulin levels or associations between stress and the three personality characteristics that were measured, namely self-esteem, personal control, and introversion–extroversion.

The consistency of the stress–illness relation among three very different viruses – rhinovirus, coronavirus, and RS virus (as well as among rhinovirus types) – was impressive. This observation suggests that stress is associated with the suppression of a general resistance process in the host, leaving persons susceptible to multiple infectious agents (or at least agents attacking the upper respiratory tract), or that stress is associated with the suppression of many different immune processes, with similar results. It is also possible that stress is associated with some general change in the host such as the ability to produce mucus or the quality of mucus produced.

Interestingly, stress was associated with the development of infection rather than the development of disease symptoms among infected persons. This suggests that stress might influence some non-specific aspects of the primary response of the host to infection rather than the production of symptom mediators.

In short, it was found that psychological stress is associated with increased risk for acute respiratory infectious illness in a dose–response manner. This increased risk is attributable to increased rates of infection. It is known that these effects are not attributable to differential viral exposure or health practices. It is assumed, however, that they are mediated by primary (as opposed to memory) functional immune response and this issue is being pursued currently.

23.4 INFECTION AND PERFORMANCE OF COGNITIVE TASKS

The study of the effects of upper respiratory tract infections and illnesses on mental performance is important for two reasons. First, it is of theoretical importance in that it provides additional evidence on interactions between the immune system and the CNS. Secondly, it is of great practical relevance in that many people continue their normal range of activities when they have minor illnesses such as the common cold and it is essential to determine whether their performance efficiency is reduced and whether there are potential risks to safety. The following section shows that there has been little previous interest in these topics and the main aim of our research programme was to provide initial data on the above issues.

23.4.1 ANECDOTAL EVIDENCE OF EFFECT OF INFLUENZA ON PERFORMANCE

Tye [61] reviewed a series of anecdotal reports which suggested that influenza was associated with an increased incidence of many different types of accidents. The impetus for the review was the following account given by a man who had trapped the fingers of his small son in his car door: 'I wasn't thinking', he said. 'Normally I always watch for that sort of thing, but my head was a bit muzzy at the time. I had a touch of flu coming on ...'. On the basis of a series of case histories Tye concluded that 'influenza *is* an invisible factor in many accidents; it *does* cost the nation millions of pounds when the judgement of individuals is 'off-peak' due to an approaching influenza attack; and it *can* wipe out in one instant the safety sense in individuals which has taken years to develop.'

It is interesting that Tye's report cites cases where performance was impaired just before

the illness started. When an individual has influenza he or she will often retire to bed and the question of impaired efficiency will not arise. However, performance efficiency may be reduced in the incubation period of the illness and could also extend into the period after the symptoms have gone. Grant [25] argues that post-influenzal effects may occur and that these can influence the judgements of highly skilled professional staff. The outstanding features of the case histories reported by Grant were that individuals who had been ill with influenza but no longer had the primary symptoms, frequently made technical errors which they failed to notice when they returned to work. There was firm rejection of advisory comments from colleagues, yet the mistakes could not be attributed to poor motivation or general lack of ability. A typical case history is summarized briefly below:

'The individual concerned was responsible for calibration of a spectrophotometer before commencing a day's work ... He had previously been off work for 2 days with influenza and returned alleging health ... During the first part of the morning he made 11 attempts to correctly prepare the instrument. On each occasion elementary faults were observed. Despite the incorrectness of the last calibration, the individual commenced work, compiling results which were finally discarded by himself 3 weeks later.'

The above reports suggested that experimental studies of the effects of infectious diseases on performance needed to be carried out, and the next section reviews early studies of this topic.

23.4.2 EXPERIMENTAL STUDIES OF INFECTION

While there has been considerable interest in the structural damage caused to the brain by certain viruses, there has been little research carried out on infection and human performance [62]. In the early 1970s Alluisi and his colleagues examined the effects of severe infections (for example rabbit fever – a febrile disease characterized by headache, photophobia, nausea and myalgia) on performance [1,2,59]. In one study those who became ill showed an average drop in performance of about 25% and when tested a few days after recovery their performance was still 15% below that of the control group. Furthermore, the effects of the illness were selective, with some tasks being more impaired than others.

The illnesses studied in the above experiments were very severe and analogous effects rarely occur in everyday life. In contrast to this, colds and influenza are widespread and frequent and it is of great importance to know more about their effects on performance. In one of the few reports of a controlled experiment on naturally occurring colds, Heazlett and Whaley [29], who had examined the effects of having a cold on childrens' perception and reading comprehension, showed that the latter was unimpaired whereas auditory and visual perception were worse when the children had colds. The tasks used in this experiment were crude and yet they were able to detect selective impairments associated with having a cold.

Why has there been little research on the effects of upper respiratory virus infections on performance? There are probably two main reasons. The first is that people feel they already know about behavioural effects of these illnesses (from personal experience) and therefore it is a waste of time carrying out such research. A second reason is that it is difficult to study naturally occurring infections and illnesses because they are hard to predict and it is unclear which virus (if any) led to the symptoms. Objective measures of symptoms are difficult to obtain, and even when this is possible the results will only portray the effects of the clinical illness. It is also possible that subclinical infections may influence behaviour and these need to be identified using the appropriate virological techniques. Problems such as these have been overcome by examining the effects of experimentally induced influenza and colds at the CCU, Salisbury.

The next sections review our research on the effects of experimentally induced influenza and colds.

23.4.3 INFLUENZA AND PERFORMANCE

Smith *et al.* [50] examined the effects of experimentally induced influenza B virus illnesses on two tasks requiring subjects to detect targets appearing at irregular intervals, and another measuring hand–eye coordination (a tracking task). Subjects performed the tasks before challenge with an influenza B virus. They were tested again a week later when some of the volunteers had developed influenza (as defined by the clinician's rating). The subjects with influenza were compared with those who remained uninfected (as defined by the failure to isolate the virus or detect a rise in antibody titre). Subjects with influenza responded more slowly in both detection tasks (Figure 23.4) but were not impaired on the tracking task. It should be pointed out that the effects of influenza on the detection tasks were very large (in the simple reaction time task there was a 57% impairment) and the magnitude of the effect can be illustrated by comparing it with that produced by a moderate dose of alcohol, or by having to perform in the middle of the night, both of which typically produce an impairment of 5–10%.

Smith *et al.* [51] confirmed that influenza B virus illnesses impair performance of detection tasks. Again, the illnesses did not reduce the speed of movements or the accuracy of hand–eye coordination. Furthermore, a working memory task (involving logical reason) was not impaired in the subjects with influenza. The major finding of this study, however, was that subclinical influenza virus infections also impaired performance of the selective attention task.

A study of the effects of influenza A virus is reported by Smith *et al.* [53]. Subjects with influenza

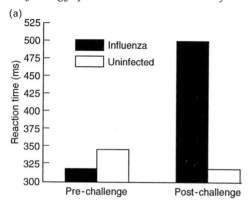

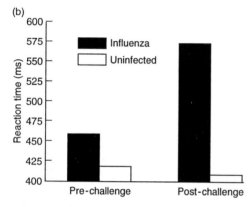

Figure 23.4   (a) Mean reaction time in the simple reaction time task and (b) fives detection task for influenza B virus-challenged and uninfected groups.

illnesses were slower at a search task where they were uncertain in which location a target would appear. In contrast to this, they were unimpaired when they knew where the target was going to be presented. In summary, the results from studies of experimentally induced influenza virus infections and illnesses have shown that selective performance impairments are observed. The tasks most sensitive to the effects of influenza were those where the person did not know exactly when to respond or where the target stimulus was going to be presented. In other words, it appeared that influenza impaired attentional mechanisms rather than affecting motor functions or aspects of cognition such as working or seman-

tic memory. Before considering a possible mechanism underlying these effects it is necessary to examine whether comparable results are obtained with naturally occurring illnesses.

Recent studies of naturally occurring illnesses

Unfortunately, no further influenza virus trials were carried out at the CCU. However, we have recently replicated the results obtained with experimentally induced illnesses in a study involving naturally occurring influenza B virus infections [49]. In addition, it was demonstrated that the effects of influenza A virus on performance can last for several weeks after the primary symptoms have gone [47]. Another extension of the research has demonstrated that identical performance impairments to those seen with influenza are apparent in the acute stage of glandular fever [28].

23.4.4 INTERFERON-ALPHA AND PERFORMANCE

The most important study carried out in this area was an attempt to examine a possible mechanism underlying the effects of influenza-like illnesses on performance. Interferon-alpha can be found in circulation during viral illnesses such as influenza, and it is now clear that such peptide mediators have an effect on the CNS. It was postulated that the performance deficits observed in influenza may be due to interferon or some similar molecule. This was investigated by injecting volunteers with different doses of interferon-alpha or saline placebo. Those subjects who received the largest dose (1.5 Mu) showed symptoms similar to those produced by influenza (increased temperature, myalgia, etc.) and it was predicted that this group, but not the others receiving smaller doses or placebo, would show performance impairments comparable with those observed in our earlier studies of experimentally induced influenza.

The data from the simple reaction time task [52,55] showed that an injection of 1.5 Mu produced an identical change to that seen in subjects with influenza (Figure 23.5). However, other detection tasks which were impaired by influenza were unaffected by interferon-alpha, whereas performance of the peg-board tasks was impaired following interferon challenge, even though earlier studies had failed to demonstrate an effect of influenza on this task. These discrepant effects could reflect differences between virally induced interferon production and direct challenge, or the fact that other peptide mediators (e.g. interleukin-1) are involved in the behavioural effects of influenza. While the interferon-alpha explanation for the effects of influenza on mental performance is clearly too simplistic, the general view that a virally induced immune response influences the brain and behaviour is very important. However, in the case of the common cold there are few systemic symptoms and one might, therefore, expect to find a different profile of performance effects. Our

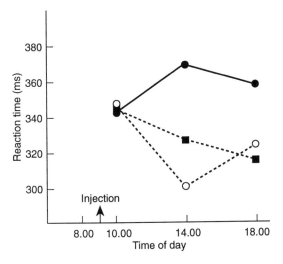

Figure 23.5 Effects of three doses of interferon-alpha on performance on the variable foreperiod simple reaction time task. Interferon-alpha dose levels: ●, 1.5 Mu; o, 0.5 Mu; ■, 0.1 Mu.

studies on this topic are reviewed in the next section.

23.4.5 EXPERIMENTALLY-INDUCED COLDS AND PERFORMANCE

Smith *et al.* [50] examined the effects of colds following rhinovirus or coronavirus challenge on the same three tests used in the influenza trials. The results showed that subjects with colds did not have an impaired performance when they undertook the two detection tasks but they were worse at the tracking task than those who remained well (Figure 23.6). A similar impairment in hand–eye coordination was found in subjects with colds by Smith *et al.* [51]. In this experiment subjects were challenged with a RS virus and hand–eye coordination was tested using a peg-board task, in which subjects had to transfer pegs from a full solitaire set to the same position in an empty one as quickly as possible. In contrast to the influenza studies, no effects of cold-producing viruses were found in the incubation period, nor were there any effects of subclinical infections. In another study, subclinical infections were associated with deficits on a measure of hand–eye coordination and self-paced responding. In this task, subjects were presented with a series of problems. In each, one of five possible stimuli was presented and required an appropriate response from one of five keys on the computer keyboard. Although there are some inconsistencies in our data, there is increasing evidence of the performance effects of infection with cold-producing viruses which are present before or in the absence of clinical symptoms.

Anecdotal evidence suggests that the effects of upper respiratory tract illnesses may persist after the primary symptoms have gone. This was investigated by Smith *et al.* [53] in a trial in which volunteers stayed at the CCU for 3 weeks during which time it was possible to test them not only when they were symptomatic but also when symptoms were no longer observable. The results showed that the per-

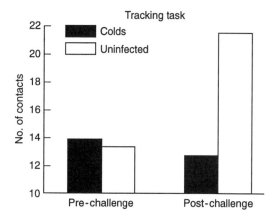

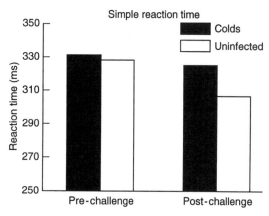

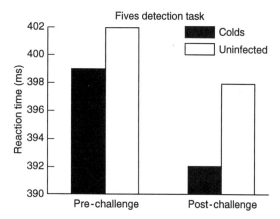

Figure 23.6 Mean scores for volunteers with colds and uninfected subjects on three performance tasks.

formance of subjects who remained well improved over the course of the trial, whereas those who developed colds were slower when they were symptomatic and still worse than baseline 1 week after the symptoms had gone. At the moment it is unclear why such 'after-effects' of viral illnesses occur. One possibility is that the performance tests are sensitive to the immunological changes that occur after symptoms have disappeared. Another possibility is that subjects continue at a lower level after their cold because they have 'learned' the task when ill.

Before considering some of the possible mechanisms which could underlie the effects of the common cold on psychomotor performance, it is necessary to describe briefly some largely negative results from experiments investigating colds and memory. Smith *et al.* [54] found that many aspects of memory, such as the ability to recall a string of digits in order or to recall a list of words, appeared to be unaffected by having a cold. Similarly, there was little evidence that colds impair retrieval of information from semantic memory. However, having a cold did produce difficulties in learning and recall of information in a story. Subjects with a cold had more difficulty following the theme of a story and instead focused on less relevant information. There was no sign of impaired retrieval of material learnt before the cold.

Mechanisms underlying the effects of the common cold on performance

The initial studies we carried out showed that having a cold impaired tasks involving hand–eye coordination. In studies designed to determine the effects of colds on aspects of vision [49], few impairments were found in visual functioning. Infections with certain viruses (e.g. enteroviruses) may produce muscle damage and recent evidence [46] confirms that even upper respiratory tract infections can influence muscle function, so providing a plausible explanation for certain performance effects.

Another possibility is that some other cytokine, for example interleukin-2 (IL-2) is involved. Again, there is evidence that IL-2 has an effect on muscles, which could plausibly account for the psychomotor impairments. Alternatively, the impairments could reflect reduced arousal due to reduced stimulation via the trigeminal nerves in the nose. Evidence for this view comes from two studies in which it was shown that drugs which probably influence the sensory afferents, namely sodium nedocromil [3] and zinc gluconate [56], removed the cold-induced performance impairments. Other explanations are clearly possible but have received little support in our studies. For example, it is possible that subjects with colds differ in task-related motivation, but Smith *et al.* [57] found no evidence for this view. Similarly, the effects could be due to increased distraction (from sneezing or other nasal irritations). The main difficulty for such explanations lies in accounting for the selective impairments which are observed when a person has a cold.

Recent studies of naturally occurring colds

As in the case of influenza, it has been possible recently to duplicate many of the effects observed at the CCU by studying naturally occurring colds [28,48]. In addition, we have examined whether having a cold makes a person more sensitive to other stressors, such as noise, and to drugs such as alcohol, and whether the impairments can be removed by mild stimulants such as caffeine. The results have demonstrated that individuals with colds are more susceptible to other forms of stress or pharmacological challenge. Again, such effects are both of theoretical importance and practical relevance. They are of theoretical importance in that they help our understanding of the mechanisms (both cognitive and physiological) underlying such changes in state. From a practical point of view, these results suggest that levels of exposure thought to be safe on the basis of studies carried out with healthy volunteers may still lead to impair-

ments in people suffering from upper respiratory tract illnesses. Indeed, it is now essential to continue the study of the combined effects of factors which influence performance, rather than examining them in isolation.

23.4.6 DISCUSSION

The research carried out at the CCU with David Tyrrell and his staff showed that upper respiratory virus infections and illnesses can reduce performance efficiency. The effects were selective in that they depended on the nature of the virus and the type of activity being carried out. The impairments were not restricted to times when the person was symptomatic, and there were often low correlations between the magnitude of the performance impairments and symptom severity. Several possible mechanisms underlying these effects have been put forward and many of these seem plausible but need to be examined in more detail. This will be difficult now that the CCU has closed and more recent research has addressed more applied questions and considered the implications of upper respiratory virus illnesses for safety and efficiency. Overall, therefore, the studies have had an impact because they have been both theoretically interesting and of potential practical importance.

23.5 CONCLUSIONS

The evidence we have presented provides additional support for the interactive nature of the central nervous and immune systems. We found that psychological states are associated with susceptibility to infectious agents, and that infection itself can alter cognitive processes. Our ongoing research continues to pursue the behavioural and biological mechanisms that explain the relations we report. However, without the collaboration and assistance of David Tyrrell and the staff of the MRC Common Cold Unit, this work would have not been possible.

23.6 REFERENCES

1. Alluisi, E.A., Beisel, W.R., Bartelloni, P.J. and Coates, G.D. (1973) Behavioral effects of tularensis and sandfly fever in man. *J. Infect. Dis.*, **128**, 710–7.
2. Alluisi, E.A., Thurmond, J.B. and Coates, G.D. (1971) Behavioral effects of infectious disease: respiratory *Pasteurella tularensis* in man. *Percept. Mot. Skills*, **32**, 647–88.
3. Barrow, G.I., Higgins, P.G., Al-Nakib, W., Smith, A.P., Wenham, R.B.M. and Tyrrell, D.A.J. (1990) The effect of intranasal nedocromil sodium on viral upper respiratory tract infections in human volunteers. *Clin. Allergy*, **20**, 45–51.
4. Bartrop, R.W., Lazarus, L., Luckhurst, E., Kiloh, L.G. and Penny, R. (1977) Depressed lymphocyte function after bereavement. *Lancet*, **i**, 834–6.
5. Bierman, S.M. (1983) A possible psychoneuroimmunological basis for recurrent genital herpes simplex. *West. J. Med.*, **139**, 547–52.
6. Broadbent, D.E., Broadbent, M.H.P., Phillpotts, R.J. and Wallace, J. (1984) Some further studies on the prediction of experimental colds in volunteers by psychological factors. *J. Psychosom. Res.*, **28**, 511–23.
7. Calabrese, J.R., Kling, M.A. and Gold, P.W. (1987) Alterations in immunocompetence during stress, bereavement, and depression: focus on neuroendocrine regulation. *Am. J. Psychiatr.*, **144**(9), 1123–34.
8. Coe, C.L., Rosenberg, L.T. and Levine, S. (1988) Effect of maternal separation on the complement system and antibody responses in infant primates. *Int. J. Neurosci.*, **40**, 289–302.
9. Cohen, S., Evans, G.W., Stokols, D. and Krantz, D.S. (1986) *Behavior, Health and Environmental Stress*, New York, Plenum Press.
10. Cohen, S., Tyrrell, D.A.J. and Smith, A.P. (1991) Psychological stress and susceptibility to the common cold. *N. Engl. J. Med.*, **325**, 606–12.
11. Cohen, S. and Williamson, G. (1988) Perceived stress in a probability sample of the United States, in *The Social Psychology of Health*, (eds S. Spacapan and S. Oskamp), Sage, Newbury Park, CA, pp. 31–67.
12. Cohen, S. and Williamson, G. (1991) Stress and infectious disease in humans. *Psychol. Bull.*, **109**, 5–24.
13. [illegible] Vard, H.W. and stress and vari-

14. Cornfeld, D. and Hubbard, J.P. (1964) A four-year study of the occurrence of betahemolytic streptococci in 64 school children. *N. Engl. J. Med.*, **264**, 211–1 5.
15. Felten, D.L., Felten, S.Y., Carlson, S.L., Olschowka, J.A. and Livnat, S. (1985) Noradrenergic sympathetic innervation of lymphoid tissue. *J. Immunol.*, **135**, 755–65.
16. Felten, S.Y. and Olschowka, J.A. (1987) Noradrenergic sympathetic innervation of the spleen: II. Tyrosine hydroxylase (TH)-positive nerve terminals from synaptic-like contacts on lymphocytes in the splenic white pulp. *J. Neurosci. Res.*, **18**, 37.
17. Fernald, G.W., Collier, A.M. and Clyde, W.A. (1975) Respiratory infections due to Mycoplasma pneumonia in infants and children. *Pediatrics*, **355**, 327–31.
18. Friedmann, E., Katcher, A.H. and Brightman, V.J. (1977) Incidence of recurrent herpes labialis and upper respiratory infection: a prospective study of the influence of biologic, social and psychologic predictors. *Oral Surg. Oral Med. Oral Pathol.*, **43**, 873–8.
19. Friedman, S.B., Ader, R. and Glasgow, L.A. (1965) Effects of psychological stress in adult mice inoculated with Coxsackie B viruses. *Psychosom. Med.*, **27**, 361.
20. Glaser, R. and Gotlieb-Stematsky, T.E. (1982) *Human Herpesvirus Infections: Clinical Aspects*, New York, Marcel Dekker.
21. Glaser, R., Kiecolt-Glaser, J.K., Speicher, C.E. and Holliday, J.E. (1985) Stress, loneliness, and changes in herpesvirus latency. *J. Behav. Med.*, **8**, 249–60.
22. Glaser, R., Rice, J., Sheridan, J. *et al.* (1987) Stress-related immune suppression: health implications. *Brain Behav. Immun.*, **1**, 7–20.
23. Goldmeier, D. and Johnson, A. (1982) Does psychiatric illness affect the recurrence rate of genital herpes? *Br. J. Ven. Dis.*, **54**, 40–3.
24. Graham, N.M.H., Douglas, R.B. and Ryan, P. (1986) Stress and acute respiratory infection. *Am. J. Epidemiol.*, **124**, 389–401.
25. Grant, J. (1972) Post-influenzal judgement deflection among scientific personnel. *Asian J. Med.*, **8**, 535–9.
26. Greene, W.A., Betts, R.F., Ochitill, H.N., Iker, H.P. and Douglas, R.G. (1978) Psychosocial fac-

tors and immunity: Preliminary report. *Psychosom. Med.*, **40**, 87.

27. Hall, N.R. and Goldstein, A.L. (1981) Neurotransmitters and the immune system, in *Psychoneuroimmunology*, (ed. R. Ader), Academic Press, New York, pp. 521–43.

28. Hall, S.R. (1993) Behavioral effects of acute and chronic viral illnesses. PhD Thesis, University of Wales.

29. Heazlett, M. and Whaley, R.F. (1976) The common cold: its effect on perceptual ability and reading comprehension among pupils of a seventh grade class. *J. Sch. Health*, **46**,145–7.

30. Jemmott, J.B. III and Locke, S.E. (1984) Psychosocial factors, immunologic mediation, and human susceptibility to infectious diseases: how much do we know? *Psychol. Bull.*, **95**, 78–108.

31. Kasl, S.V., Evans, A.S. and Niederman, J.G. (1979) Psychosocial risk factors in the development of infectious mononucleosis. *Psychosom. Med.*, **41**, 445–66.

32. Katcher, A.H., Brightman, V.J., Luborsky, L. and Ship, I. (1973) Prediction of the incidence of recurrent herpes labialis and systemic illness from psychological measures. *J Dent. Res.*, **52**, 49–58.

33. Kiecolt-Glaser, J.K., Glaser, R., Shuttleworth, E.C., Dyer, C.S., Ogrocki, P. and Speicher, C.E. (1987) Chronic stress and immunity in family caregivers of Alzheimer's disease victims. *Psychosom. Med.*, **49**, 523–35.

34. Kiecolt-Glaser, J.K. and Glaser, R. (1988) Methodological issues in behavioral immunology research with humans. *Brain Behav. Immun.*, **2**, 67–78.

35. Kiecolt-Glaser, J.K. and Glaser, R. (1987) Psychosocial influences on Herpesvirus latency, in *Viruses, Immunity and Mental Disorders*, (eds E. Kurstak, Z.J. Lipowski and P.V. Morozov), Plenum Press, New York, pp. 403–11.

36. Kiecolt-Glaser, J.K., Kennedy, S., Malkoff, S., Fisher, L., Speicher, C.E. and Glaser, R. (1988) Marital discord and immunity in males. *Psychosom. Med.*, **50**, 213–29.

37. Laudenslager, M.L. (1988) The psychobiology of loss: lessons from humans and non-human primates. *J. Soc. Issues*, **44**, 19–36.

38. Laudenslager, M.L. (1987) Psychosocial stress and susceptibility to infectious disease, in *Viruses, Immunity and Mental Disorders*, (eds E.

Kurstak, Z.J. Lipowski and P.V. Morozov), Plenum Press, New York, pp. 391–402.

39. Laudenslager, M.L., Reite, M.L. and Held, P.E. (1986) Early mother/infant separation experiences impair the primary but not the secondary antibody response to a novel antigen in young pigtail monkeys. *Psychosom. Med.*, **48**(3/4), 304.

40. Laudenslager, M., Reite, M. and Harbeck, R. (1982) Immune status during mother–infant separation. *Psychosom. Med.*, **44**, 303.

41. Lazarus, R.S. and Folkman, S. (1984) *Stress, Appraisal, and Coping*, Springer Publishing, new York.

42. Locke, S.E. and Heisel, J.S. (1977) The influence of stress and emotions on the human immune response. *Biofeedback Self Regul.*, **2**, 320.

43. Manuck, S.B., Harvey, A., Lechleiter, S. and Neal, K. (1978) Effects of coping on blood responses to threat of aversive stimulation. *Psychophysiology*, **15**, 544–9.

44. McLarnon, L.D. and Kaloupek, D.G. (1988) Psychological investigation of genital herpes recurrence: prospective assessment and cognitive-behavioral intervention for a chronic physical disorder. *Health Psychol.*, **7**, 231–49.

45. Meyer, R.J. and Haggerty, R.J. (1962) Streptococcal infections in families. *Pediatrics*, **29**, 539–49.

46. Mier-Jedzrejowicz, A., Brophy, C. and Green, M. (1988) Respiratory muscle weakness during upper respiratory tract infections. *Am. Rev. Respir. Dis.*, **138**, 5–7.

47. Smith, A.P. (1992) Chronic fatigue syndrome and performance, in *Handbook of Human Performance, Vol. 2: Health and Performance*, (eds. A.P. Smith and D.H. Jones), Academic Press, London, pp. 261–78.

48. Smith, A.P., Thomas, M., Brockman, P., Kent, J. and Nicholson, K.G. (1993) Effect of influenza B virus infection on human performance. *Br. Med. J.*, **306**, 760–1 .

49. Smith, A.P., Tyrrell, D.A.J., Barrow, C.I. *et al.* (1992) The common cold, pattern sensitivity and contrast sensitivity. *Psychol. Med.*, **22**, 487–94.

50. Smith, A.P., Tyrrell, D.A.J., Al-Nakib, W. *et al.* (1987) Effects of experimentally induced respiratory virus infections and illness on psychomotor performance. *Neuropsychobiology*, **18**, 144–8.

51. Smith, A.P., Tyrrell, D.A.J., Al-Nakib, W. *et al.* (1988) The effects of experimentally induced respiratory virus infections on performance. *Psychol. Med.*, **18**, 65–71.

52. Smith, A.P., Tyrrell, D.A.J., Coyle, K.B. and Higgins, P.G. (1988) Effects of interferon alpha on performance in man: a preliminary report. *Psychopharmacology*, **96**, 414–16.

53. Smith, A.P., Tyrrell, D.A.J., Al-Nakib, W. *et al.* (1989) Effects and after-effects of the common cold and influenza on human performance. *Neuropsychobiology*, **21**, 90–3.

54. Smith, A.P., Tyrrell, D.A.J., Barrow, G.I. *et al.* (1990) The effects of experimentally induced colds on aspects of memory. *Percept. Motor Skills*, **71**, 1207–15.

55. Smith, A.P., Tyrrell, D.A.J., Coyle, K.B. and Higgins, P.G. (1991) Effects and after-effects of interferon alpha on human performance, mood and physiological functions. *J. Psychopharmacol.*, **5**, 243–50.

56. Smith, A.P., Tyrrell, D.A.J., Al-Nakib, W., Barrow, I., Higgins, P. and Wenham, R. (1991) The effects of zinc gluconate and nedocromil sodium on performance deficits produced by the common cold. *J. Psychopharmacol.*, **5**, 251–4.

57. Smith, A.P., Tyrrell, D.A.J., Coyle, K. and Willman, J.S. (1987) Selective effects of minor illnesses on human performance. *Br. J. Psychol.*, **78**, 183–8.

58. Stein, M. (1981) A biopsychosocial approach to immune function and medical disorders. *Pediatr. Clin. North Am.*, **4**, 203–21.

59. Thurmond, J.B., Alluisi, E.A. and Coates, G.D. (1971) An extended study of the behavioral effects of respiratory *Pasteurella tularensis* in man. *Percept. Mot. Skills*, **33**, 439–54.

60. Totman, R., Kiff, J., Reed, S.E. and Craig, J.W. (1980) Predicting experimental colds in volunteers from different measures of recent life stress. *J. Psychosom. Res.*, **24**, 155–63.

61. Tye, J. (1960) *The invisible factor – an inquiry into the relationship between influenza and accidents.* British Safety Council, London.

62. Warm, J.S. and Alluisi, E.A. (1967) Behavioral reactions to infections: review of the psychological literature. *Percept. Mot. Skills*, **24**, 755–83.

THE ETHICAL USE OF HUMAN VOLUNTEERS IN CLINICAL RESEARCH

24

Norma Watson and Walter Nimmo
Special contribution by Sylvia Reed

24.1 INTRODUCTION

Continuous assessment of standards should accompany a search for quality of practice in most aspects of daily living. In particular, doctors have a responsibility to ensure that the therapies they advocate have been evaluated properly with regard to efficacy, safety and quality. It follows that they should be prepared to contribute to this process of evaluation [17]. Clinical trials and research in general

Viral and Other Infections of the Human Respiratory Tract.
Edited by S. Myint and D. Taylor-Robinson. Published in 1996 by Chapman & Hall. ISBN 0 412 60070 6

become more (rather than less) important with time and, when resources are scarce, a greater proportion of them must be channelled into evaluation [16].

Medical research always raises unique moral questions about the use of animals, volunteers or patients because the research is often directed towards the furtherance of medical knowledge rather than the individual's personal and particular gain. Until the 1960s there was little public interest in decision making about medical research, but the mood has shifted. Ethics committees have now been established all over the civilized world to review the ethics of medical research on a case-by-case basis [21] and this reflects the high public interest in the safety and morality of medical research.

In the aftermath of the thalidomide tragedy, government agencies were established to assess the quality, safety and efficacy of a drug before its widespread use. No other therapy such as surgery or physical treatment has ever come under such scrutiny. The need for regulatory approval has undoubtedly increased the amount of research conducted, particularly in animals and human volunteers. It may be that the extensive toxicity testing of new drugs in animals cannot be wholly justified in terms of the information gained, but early studies in humans are usually of great value. In a climate in which testing in human volunteers is not mandatory, these studies continue. The ethics and proper conduct of these studies is a matter of great interest to researchers, drug developers and the general public who are asked to volunteer.

24.2 HISTORY

Historical examples of the use of human volunteers in medical experimentation include Edward Jenner's use of cowpox vaccine in the treatment of a young boy suffering from smallpox. This was regarded as the forerunner of so many successful immunization programmes. In the late nineteenth century, German physicians inoculated healthy human subjects with the organism discovered by Neisser and thought to be responsible for gonorrhoea in order to prove its pathogenic role.

Self-experimentation has been reported also. Werner Forssman passed a catheter into the right side of his own heart and thus began the development of a technique which has saved many thousands of lives.

It is likely that a modern ethics committee would not approve these studies despite their importance. Experiments carried out in concentration camps by Nazi doctors produced only revulsion in the civilized world despite the importance of the objectives they encompassed – prevention of malaria, typhus, deaths at high altitude or from exposure to cold.

For the past 35 years, medical freedom to conduct research on human subjects has been limited by voluntary codes with additional legal restrictions in some countries [17].

24.3 REGULATIONS AND GUIDELINES

Much has been written about legal issues in clinical research and the relationship between the law and ethics but specific legal rules are virtually non-existent [18]. This is in contrast to research on animals which has been restricted legally since the passage of the law relating to Cruelty to Animals in 1876.

In 1962, the Medical Research Council (MRC) in the United Kingdom laid down guidelines for research in humans carried out

under its sponsorship. This preceded the World Medical Association Declaration of Helsinki (1964) which set out the basic ethical principles of clinical research.

Neither the Declaration of Helsinki nor the MRC statements provided guidance about the setting up or monitoring of systems to protect the subjects of research. In 1967, the Royal College of Physicians of London (RCP) recommended that all clinical research should be subject to ethical review.

In 1970, the Association of the British Pharmaceutical Industry (ABPI) issued a report setting out standards of practice for member companies to provide safeguards for staff volunteers in drug studies. These guidelines also acted as a basis for volunteer studies conducted outside the pharmaceutical industry. An update on this report was issued in 1984 to reflect the changes in research practice and opinion.

In 1986 the Royal College of Physicians produced a report entitled 'Research on Healthy Volunteers' [24] and subsequently the ABPI set up a working party to review and re-draft the guidelines relating to volunteer studies, taking into account the conclusions reached by the Royal College of Physicians [2].

The most definitive document with respect to volunteer studies of drugs is that of the EC/CPMP working party entitled 'Good Clinical Practice for trials on medicinal products in the European Community' approved by the CPMP in 1990 for implementation in 1991. This stated that 'Pre-established, systematic written procedures for the organization, conduct, data collection, documentation and verification of clinical trials are necessary to ensure that the rights and integrity of the trial subjects are thoroughly protected and to establish the credibility of data and to improve the ethical, scientific and technical quality of trials'.

The guidelines cover all aspects of clinical trials and provide recommendations for all areas including ethics committee approval, liability for injury to subjects and design of protocols, recognizing that: 'It is unethical to establish the co-operation of human subjects in trials which are not adequately designed'.

All of the guidelines omit advice about the precision and reproducibility of measurement [31]. It is obviously unacceptable to conduct studies in which the instrument of measurement is inaccurate or of unknown accuracy. A recent example involves the Hawksley random sphygmomanometer [12] which has been known to have a systematic error for over 20 years but has been used in many clinical trials. On average, it reads systolic pressure 3.5 mmHg lower and diastolic pressure 7.5 mmHg lower than the standard mercury sphygmomanometer.

24.3.1 ETHICS

The accepted basis for clinical trial ethics is the current revision of the Declaration of Helsinki which must be familiar to and followed by all persons engaged in research on human beings. Since 1967, it has been accepted that all clinical research conducted in institutions within the National Health Service should be approved by an independent ethics committee. It is interesting that ethics committees have not reviewed informal and uncontrolled research such as a new surgical technique or drug therapy outwith its licensed indications.

It is now agreed that all volunteer study protocols should obtain approval by an independent and properly constituted Ethics Committee. The 1990 document, *'Guidelines on the Practice of Ethics Committees in Medical Research Involving Human Subjects'* published by the Royal College of Physicians of London and supported by the ABPI is the definitive report on Ethics Committees. These guidelines cover all aspects of ethics committee function from membership of the committee and terms of reference to payment of subjects and volunteer consent [25].

Informed consent is central to the ethical conduct of clinical investigation. Subjects must have enough information in a form that is comprehensible to them to enable them to make a voluntary, deliberated judgement whether or not to participate. One of the functions of an Ethics Committee is to ensure that sufficient information, especially about potential risks and benefits, has been provided and therefore the volunteer information sheet is reviewed as well as the study protocol.

The ABPI guidelines for the conduct of non-patient volunteer studies also recommended that a register should be established of units conducting such studies. In order to ensure that such units provided suitable facilities for the safe conduct of clinical studies, further guidelines on the standards which should be achieved by such units were issued. These guidelines cover staffing, emergency procedures and equipment as well as support services and archiving of records [3].

24.4 DEFINITION OF A VOLUNTEER

There appears to be no official definition of the term 'non-patient volunteer'. Accepted key elements include:

1. The individual cannot be expected to derive therapeutic benefits from the proposed study.
2. He or she is not known to suffer any significant illness relevant to the proposed study.
3. His or her mental state is such that he is able to understand the procedures involved.
4. He or she is willing and able to give valid consent for the study.

By definition, subjects participating in a trial of a new vaccine for example cannot be 'non-patient' as they will derive therapeutic benefit. Also, any 'healthy' volunteer group will contain the normal percentage of any population who suffer allergies, other minor illnesses and abnormalities.

Any investigator must ensure that a volunteer is healthy and suitable before inclusion in the proposed clinical trial [7]. Thus all potential volunteers should be screened in the period 4 weeks before the study begins. A history and medical examination appropriate to the study should be carried out and this will normally include blood and urine tests. In a hospital screening healthy subjects in France, 33 of 465 subjects were excluded for clinically significant laboratory findings. A further 11 of 339 subjects developed abnormal clinical pathology results between screening and the start of the study. Most of these abnormalities were in liver function tests [27]. Guidelines for screening laboratory tests have been published [26].

Contact should be made with the volunteer's general practitioner to inform him or her of the volunteer's participation but also to obtain any further information on the medical history and current medication.

In a recent study, of all male subjects applying to participate in clinical trials, 25% were not suitable because of pre-existing disease – 14.1% were excluded on the basis of information given on the application form, 5.4% as result of findings of the medical examination, including blood and urine tests, and a further 5.3% on the basis of information obtained from the potential volunteer's general practitioner [35].

At present, volunteer screening includes tests for HIV, hepatitis and drugs of abuse. It is important that confidentiality is maintained, that informed consent is obtained for this screening and that appropriate advice with counselling is available in the event of a positive result.

24.5 DEFINITION OF NORMALITY

It is often difficult to define normality – particularly in relation to clinical pathology in which many variables are measured and some may be outwith what is described routinely as normal. The research programme on a new drug may require 3000 volunteers and patients each studied on three occasions (pre-study, mid-study, end of study). The standard number of

clinical pathology investigations is 25, therefore the total number of laboratory investigations will be 225 000. The laboratory normal range is based on the mean ± 2 standard deviations. One in 20 will be abnormal as a result of chance alone, i.e. false positives. If we are only interested in an abnormally high or an abnormally low value, this figure is 1 in 40.

Therefore, in the type of programme described, approximately 5625 laboratory abnormal values will have arisen by chance. These tests will have cost approximately £250 000 and it will take considerable time to establish which results are truly abnormal. Thus one should use laboratory tests intelligently and selectively [28].

Screening of liver function tests in the population will reveal that up to 4% have an isolated raised bilirubin concentration (Gilbert's disease). Bilirubin concentrations may be also elevated by fasting. This may have no association with abnormalities of hepatic enzyme function and these subjects may safely be included in clinical trials.

The electrocardiograph (ECG) is used frequently before volunteer studies to exclude significant heart disease. Once again it may be difficult to decide what is abnormal. For example, the PR interval of the ECG is usually described as being <0.20 s duration. A duration >0.20 s indicates first-degree heart block. However, longer duration may be observed in healthy individuals and 0.65–1.1% of asymptomatic healthy subjects have a PR interval >0.20 s [6]. Long-term follow-up studies have indicated that, although the risk of coronary artery disease might be slightly increased, the risk of sudden death, syncope or advanced atrioventricular block is not. No detailed investigation is necessary [6]. The PR interval may be variable in a healthy subject on repeated measurements (Table 24.1).

Dynamic cardiography (DCG) for a period of 24 hours may demonstrate drug efficacy in clinical trials of anti-arrhythmic drugs and may be useful in diagnosing the cause of

Table 24.1 The PR interval in one healthy subject receiving no drug therapy recorded on repeated visits over a period of 1 month

	PR interval (s)
Visit 1	0.189
Visit 2	0.221
Visit 3	0.199
Visit 4	0.199
Visit 5	0.249
Visit 5 repeated	0.218

All electrocardiagrams were recorded on the same Hewlett Packard cardiograph. (Normal range is up to 0.20 s)

symptoms such as palpitations or fainting [10]. When used in symptom-free volunteers during drug safety or tolerability studies, it is difficult to identify what is abnormal. For example, ventricular ectopic beats (VEs) may appear alarming although a frequency of up to 200/h is unlikely to have predictive value in the absence of symptoms [11]. In healthy asymptomatic subjects very frequent ventricular extrasystoles (VEs) may be seen (Table 24.2). Other variations may also be seen in normal subjects (Table 24.3). The value of DCG in volunteer studies may be limited by this difficulty in identifying what is abnormal in the absence of symptoms.

24.5.1 SPECIAL GROUPS OF VOLUNTEERS

In the past, healthy volunteers have been mostly male aged between 20 and 45 years. However, it may be appropriate to include females and other groups of volunteers. The National Institute of Health of USA now requires the inclusion of women and minorities in research study populations so that 'research findings can be of benefit to all persons at risk of the disease, disorder or condition under study' [20].

Women of childbearing potential have not normally been accepted as volunteers in early studies of new chemical entities but may be the group of choice for studies of oral contraceptives. In this situation, satisfactory reproductive toxicology studies should have been performed and adequate safeguards taken to ensure absence of a pre-existing pregnancy. In the evaluation of oestrogen preparations, the volunteer group of choice is post-menopausal females. Their baseline oestrogenic hormone concentrations give a constant background against which to evaluate the pharmacokinetics of the administered oestrogen preparations.

Elderly volunteers may be studied in an attempt to identify any altered metabolism or excretion [13]. This should allow appropriate dosage selection for studies in patients. The definition of elderly and normality may be particularly difficult. It is likely that any changes in elderly patients, apart from the predictable reduction in renal function, results from the increased prevalence of disease and other drugs in this group.

Drugs which are metabolized extensively by a specific pathway which exhibits phenotypic variation, e.g. acetylation should be studied in the appropriate phenotypic subgroup. Thus identification of acetylator status of the volunteers may be undertaken at screening.

24.6 TYPES OF STUDIES

Most volunteer studies occur in new drug or formulation development. Types are listed in Table 24.4.

Table 24.2   Dynamic cardiography of a healthy subject measured over a 24-hour period using the Compass monitor

Time	Heart rate (b.p.m)	SVE/h	VE/h
11:00	48–97	7	161
12:00	67–106	2	10
13:00	57–101	4	130
14:00	57–105	2	41
15:00	60–101	1	12
16:00	51–94	0	114
17:00	55–105	1	2
18:00	61–106	0	18
19:00	65–101	0	0
20:00	57–102	3	34
21:00	53–101	0	187
22:00	52–96	5	376
23:00	48–90	0	484
24:00	47–91	0	681
01:00	47–87	1	408
02:00	47–84	3	998
03:00	47–121	0	962
04:00	47–91	0	865
05:00	48–91	0	339
06:00	47–99	0	437
07:00	63–105	0	88
08:00	65–111	0	33
09:00	70–121	0	6
10:00	57–105	1	6

* SVE, supraventricular extrasystole; VE, ventricular extrasystole

Table 24.3   Prevalence of selected arrhythmias detected by DCG in asymptomatic subjects ([30])

Arrhythmia	Prevalence (%)
Bradycardia	60
Atrial fibrillation	20
Supraventricular extrasystole	35
Ventricular extrasystole	50–100
Multiform ventricular extrasystoles	3
Ventricular tachycardia	2
Bigeminy	0.2
R on T ventricular extrasystoles	0.01

Table 24.4   Volunteer studies during new drug development

- Single-dose tolerability of a new clinical entity
- Multiple-dose tolerability of a new clinical entity
- Radiolabelled study of metabolism
- Pharmacokinetic including bioavailability
- Clinical pharmacology

24.6.1 TOLERABILITY STUDY

The most important step in the development of a new drug is the progression from animal studies to human studies – single dose and then multiple dose tolerability. There are published guidelines describing the pre-clinical toxicology required before administration to humans. These should be regarded as the minimum requirements and additional investigations should be considered depending on the pharmacology of the drugs and the types of clinical trials to be undertaken [9].

Properly designed volunteer tolerability studies allow the safe escalation of the dose of a new drug with the provision of early pharmacokinetic data. They also provide information on the nature of side effects or toxic effects and the dose at which they occur.

The initial dose given in first administration studies is estimated from animal studies. It may be a fraction (e.g. 1/200) of the dose producing the desired pharmacological effect or no adverse effect in the most sensitive animal species.

It is possible to use pharmacokinetic principles to estimate the initial dose to be given to humans. The volume of distribution of the drug may be estimated from the lipid/water partition coefficient (PC) and the extent of protein binding (p) to 4% human albumin [23].

For a one-compartment model
$$V = (0.955. PC + 1.2232).(1 - p).BW \quad (ml)$$
For a two-compartment model
$$V_c = (0.0397. PC + 0.0273).(1 - p).BW \quad (ml)$$
$$V_{ss} = (0.1141. PC + 0.6611).(1 - p).BW \quad (ml)$$

where V is the volume of distribution, V_c is the central volume, V_{ss} is the volume of distribution at steady state, and BW is bodyweight in kg

Thus, if PC is 5.4 and p is 85%, V = 18.2 litres, V_c = 2.54 litres and V_{ss} = 13.41 litres. The predicted drug concentration in blood may be calculated from the dose/V (oral administration) or dose/V_c (intravenous administration).

There is a variety of designs for such tolerability studies but all involve gradual increases in dose size with exposure in a few subjects at each dose. In such single-dose studies, the dose is increased until signs of possible toxicity occur indicating side effects unacceptable for patient studies. There should then follow a multiple dose study to investigate the tolerance of the drug at the dose and duration of therapy proposed for the initial patient studies.

24.6.2 METABOLIC STUDIES INCLUDING RADIOLABELLED DRUGS

Healthy volunteer studies of radiolabelled drugs with measurement of radioactivity in biological samples may be the most precise way of determining the patterns of metabolites in plasma, urine and faeces. In addition, this type of study may be the only practicable way of comparing administered dose with amount excreted to determine if there is residual drug which may accumulate with multiple dosing.

Most of these studies involve beta-emitting isotopes – either ^{14}C or ^{3}H, and before progressing to dosing in humans, synthesis of the radiolabelled drug and a series of investigations using it in at least two species of laboratory animals must be undertaken. These investigations allow the development of appropriate separation and identification procedures for the metabolites, ensure adequate recovery of the drug radioactivity and allow measurement of the time-course of tissue distribution of radioactivity so that whole body and tissue radioactivity exposure in humans can be estimated.

In the UK, all experiments involving radioactivity must be approved by the Administration of Radioactive Substances Advisory Committee (ARSAC) who have published guidelines on the investigative use of radioactivity in humans [15].

The selection of volunteers for such studies is similar to that described for other clinical trials but females should not participate and males should be 30 years old or over. The exclusion of females avoids radioactivity in undiagnosed early pregnancy but also this type of study may be carried out at a very early stage of drug development before full teratogenicity studies have been completed. The preference for older male volunteers is to reduce any risk to the individual of cumulative radiation exposure.

24.6.3 OTHER STUDIES

Volunteer studies may also be required to look in more detail at the clinical pharmacology of a drug using special investigations. The nature of these investigations will be determined by the animal studies and the human tolerance studies. For example, it may be necessary to look at the effect of a new drug on various hormone production pathways within the body or to study a drug's cardiovascular effects more closely.

Predictions of human metabolism of pesticides are needed for the interpretation of biological monitoring data. Extrapolation from animal studies can be very misleading and the benefits of using human volunteers for metabolism studies at low doses outweigh the minimal risks involved [37].

24.7 RISKS OF VOLUNTEER STUDIES (ADVERSE EVENTS)

The Declaration of Helsinki states that 'Biomedical research involving human subjects cannot legitimately be carried out unless the importance of the object is in proportion to the inherent risk to the subject'. Therefore, the assessment of the risk involved in any experimentation is an important part of the consideration of an ethics committee.

Clinical studies involving healthy volunteers have been shown to be extremely safe when subjects are screened carefully and the research conducted in a safe environment by qualified personnel. Exclusion of volunteers at risk by careful screening avoids serious adverse events [27,35]. Of the four deaths of

volunteers reported, two were related to inappropriate screening [14,19,27].

In a review of adverse events occurring in studies conducted during a 1-year period, 39% of subjects receiving active drug developed at least one adverse event. This was almost identical to the frequency of adverse events experienced by subjects receiving placebo (39%). None of the adverse events was serious and the pattern of adverse events was similar in the two groups. Adverse drug reactions (ADRs) 'definitely related' to drug administration represented 4% of the reported adverse events after active drug. There was none after placebo [36].

24.8 THE FUTURE

It seems inevitable that volunteer studies will continue in the future. With appropriate screening, qualified staff and recognized centres they are widely regarded as safe. Animal studies may decrease in number in new drug evaluation because early volunteer studies with pharmacokinetic analysis will allow the most appropriate animal species for chronic toxicology and toxicokinetic studies to be identified [22]. *In vitro* metabolic studies will identify the specific type of cytochrome P450 enzyme involved in metabolizing the drug so that the most important drug interactions can be predicted, thus reducing the number of human studies required [32].

Better, more reliable and precise clinical measurement will allow better clinical pharmacology and pharmacodynamic studies. This may allow surrogate end points of drug action to be identified in volunteer studies at an early stage of drug development [29].

24.9 THE USE OF HUMAN VOLUNTEERS AT THE COMMON COLD UNIT

The use of human volunteers in medical research has a long history, particularly in respect of infectious diseases. The Common Cold Unit (CCU) existed from 1946 to 1989 at the Harvard Hospital, Salisbury under the aegis of the Medical Research Council, and successfully undertook experiments using upwards of 20 000 volunteers. The CCU was unique in Britain and indeed, worldwide. It existed to allow experiments involving inoculation of volunteers with the viruses of common colds and related infections. It operated principally under two directors, C.H. (later Sir Christopher) Andrewes (1946–1961) and D.A.J. Tyrrell (1962–1989).

The CCU was housed in 'temporary' buildings constructed in 1941 by Harvard University Medical School as an isolation hospital for research on the potential epidemics of war-time. The buildings were well suited to the initial purpose of the new unit as proposed by C.H. Andrewes. This was to identify one or more viruses causing common colds, which could not then be cultivated *in vitro* or in laboratory animals. Later, with this primary object largely achieved, these same viruses were used in many other studies of respiratory infections. Remarkably, the procedures for recruitment of volunteers and organization of the experiments or 'trials' remained in essence unchanged throughout the 43 years of the CCU's existence. That this happened was a tribute to the effectiveness of the procedures in providing not only a more-or-less continuous supply of volunteers but also a logistical framework for the production of significant experimental results. The procedures used were outlined in numerous scientific papers, which have been reviewed [1,5,33] and also described in a less scientific vein [30].

The CCU functioned through years of changing perspectives on the ethics of experimentation in humans and in animals. Throughout its work the ethical principle remained valid, that young (18 to 50-year-old), healthy volunteers, both men and women, could legitimately be asked to accept an inoculation of an infectious agent such as they might commonly encounter in the course of daily life. During the CCU's existence, guidelines on research in humans and official recognition of ethical standards in such work came into existence, and the CCU came to operate with the supervision of the Ethics Committee of the Clinical Research Centre at Northwick Park Hospital. Studies using antiviral drugs took place within the framework of the proper studies of toxicology, human tolerance and pharmacology carried out by the developers of each potentially beneficial compound.

Volunteers were recruited from all walks of life, many being students, Publicity leaflets outlined what was required for 'a free holiday' and the opportunity to help in medical research. The volunteers were required to live at the CCU for 10 days, isolated in groups of two or three, in spacious and comfortable if somewhat basic accommodation. They gave informed consent to cooperate in an experiment in which they would receive an intranasal inoculum containing either a common respiratory virus or a simple saline solution. This would allow the recording of their clinical response and its evaluation under double-blind conditions, the taking of specimens for laboratory analysis, and perhaps other procedures as required by the nature of that particular experiment.

In the early years when the cultivation of common cold viruses in vitro was not possible, volunteers were used simply as a test system for the presence or absence of a virus. With their aid, techniques for cultivation of rhinoviruses [34] and later some coronaviruses [8] were developed. These viruses and suitable strains of influenza viruses could then be used in a wide range of volunteer studies relating to immunity to respiratory viruses, antiviral substances and live, attenuated influenza virus vaccines. Some studies of viruses causing 'winter vomiting disease' were also carried out. Trials could be carefully planned and performed using relatively few subjects under strictly controlled conditions. It was clear that studies like these, done directly in the human host, could provide quick and accurate answers to questions about which information could be gleaned only slowly, inaccurately, or not at all using experimental animals or field trials. The experiments in human volunteers proved particularly useful for the evaluation of antiviral compounds, including interferon and its inducers [5], and for assessment of attenuation and antigenicity of influenza for use as live vaccines [4].

What can be learned from the experience of the CCU about the use of such a uniquely valuable resource as the human volunteer? The nature of the motivation of volunteers taking part was believed to be of primary importance in the success of the CCU. It was considered essential that no pressure to enrol should be exerted and that each volunteer should make the first move by sending for information and an enrolment form. Volunteers were fully informed about the nature of the experiments and what would be required of them, and were medically examined on arrival at the Unit. A fundamental precept was that they should be paid no more than notional pocket money. In return for their help they received free travel to the CCU and free board and accommodation for 10 days in a very pleasant rural environment on the edge of Salisbury, Wiltshire, with telephone, television and facilities for country walks and some sports.

The CCU was run by a small staff of about 20, including scientific, domestic and administrative personnel, some of whom were dedicated and long-serving. The relatively small size of the unit, taking in groups of no more than 30, and often 15–25 volunteers at a time, helped to produce the friendly ambience which the volunteers appreciated. Mutual trust between volunteers and staff was essential, ensuring that each would fulfil their side of the agreement. This trust was seldom abused. Very many individual volunteers were sufficiently satisfied with the arrangements to return for repeated visits. For the laboratory staff, the system of continuous recruitment of volunteers allowed over 20 trials per year, necessitating a similar flow of suitable experimental protocols and associated laboratory work, and producing the potential for abundant scientific results. The disadvantage of working in a small unit physically somewhat isolated from mainstream academic activity was countered by the availability of this unique scientific tool to scientists and by the wide spectrum of national and international visiting scientists and collaborators attracted to work at the Unit.

The CCU's output of experimental results, evaluated over the years, was repeatedly judged to justify continued funding. Perhaps the circumstances were unique. Indeed, when the possibility of re-building the Unit or transferring it elsewhere was considered, it was recognized that the CCU's spacious buildings and pleasant environment were not insignificant factors in its success. It may be concluded that the methods and experiences of this successful research facility are surely worthy of examination by anyone considering embarking on studies involving the use of volunteers.

24.10 REFERENCES

1. Andrewes, C.H. (1973) *In Pursuit of the Common Cold*, William Heinemann Medical Books, London.

2. Association of the British Pharmaceutical Industry (1988) *Guidelines for Medical Experiments in Non-Patient Human Volunteers*. The Association of the British Pharmaceutical Industry, London.

3. Association of the British Pharmaceutical Industry (1989) *Facilities for Non-Patient Volunteer Studies*. The Association of the British Pharmaceutical Industry, London.

4. Beare, A.S. (1975) Live viruses for immunisation against influenza. *Prog. Med. Virol.*, **20**, 49–83.

5. Beare, A.S. and Reed, S.E. (1977) The study of antiviral compounds in volunteers, *in Chemoprophylaxis and Virus Infections of the Respiratory Tract*, vol. 2, (ed. J.S. Oxford), CRC Press, Cleveland, Ohio, pp. 27–55.

6. Beaton, R.S. and Camm, A.J. (1984) First degree atrioventricular block. *Eur. Heart J.*, **5**, 107–9.

7. Bechtel, P.R. and Alvan, G. (1989) Criteria for the choice and definition of healthy volunteers and or patients for Phase I and Phase II studies in drug development. *Eur. J. Clin. Pharmacol.*, **36**, 549–50.

8. Bradburne, A.F. and Tyrrell, D.A.J. (1971) Coronaviruses of man. *Prog. Med. Virol.*, **13**, 373–403.

9. Breckenridge, A.M. (1991) A clinical scientist's view of preclinical drug testing. *Hum. Exp. Toxicol.*, **10**, 395–7.

10. Campbell, R.W.F. (1993) Dynamic electrocardiography. *Medicine International*, **21**, 338–45

11. Campbell, R.W.F. (1993) Ventricular ectopic beats and non-sustained ventricular tachycardia. *Lancet*, **341**, 1454–8.

12. Conroy, R.M., O'Brien, E.O., O'Malley, K. and Atkins, N. (1993) Measurement error in the Hawksley random zero sphygmomanometer: what damage has been done and what can we learn? *Br. Med. J.*, **306**, 1319–22

13. Crome, P. and Flanagan, R. J. (1994) Pharmacokinetic studies in elderly people. Are they necessary? *Clin. Pharmacokinet.*, **26**, 243–7.

14. Darragh, A., Lambe, R., Kenny, M. *et al.* (1985) Sudden death of a volunteer. *Lancet*, **i**, 193–4.

15. Department of Health and Social Security (1988) *Notes for Guidance on the Administration of Radioactive Substances to Persons for Purposes of Diagnosis, Treatment or Research*. Department of Health and Social Security, London.

16. Hampton, J.R. (1983) The end of clinical freedom. *Br. Med. J.*, **287**, 1237–8.

17. Hoffenberg, R. (1986) Ethical constraints – Freedom to perform research, in *Clinical Freedom*, Nuffield Provincial Hospitals Trust, London, pp. 49–62.

18. Katz, J. (1972) *Experimentation with Human Beings*. Russell Sage Foundation, New York

19. Kolata, G.B. (1980) The death of a research subject. *The Hastings Center*, **10**, 5–6.

20. National Institutes of Health (1993) Protecting human research subjects; Institutional Review Board Guidebook, National Institutes of Health, Office of Extramural Research, Office for Protection from Research Risks, Bethesda.

21. Neuberger, J. (1992) *Ethics and Health Care*. Kings Fund Institute, 125 Albert Street, London NW1 7NF.

22. Nimmo, W.S. and Watson, N. (1994) What a clinical pharmacologist requires from toxicokinetic studies. *Drug Information J.*, **28**, 185–6.

23. Ritschel, W.A. (1992) First dose size in man, in *Handbook of Basic Pharmacokinetics*, 4th edn, Drug Intelligence Publications, Hamilton, pp. 452–5.

24. Royal College of Physicians (1986) Research on healthy volunteers. *J. R. Coll. Physicians*, **20**, 1–17.

25. Royal College of Physicians (1990) *Guidelines on the Practice of Ethics Committees in Medical Research Involving Human Subjects*. Royal College of Physicians of London, London.

26. Sibille, M. and Vital Durand, D. (1990) Laboratory screening method for selection of healthy volunteers. *Eur. J. Clin. Pharmacol.*, **39**, 475–9.

27. Sibille, M., Vital Durand, D. and Levrat, R. (1992) Intercurrent disease in healthy volunteers in pharmacological research. *Lancet*, **339**, 1058.

28. Stephens, M.D.B. (1988) Laboratory investigations, in *The Detection of New Adverse Drug Reactions*, 2nd edn, Stockton Press, Basingstoke, pp. 51–72.

29. Temple, R.J. (1994) A regulatory agency view of surrogate end points, in *Clinical Measurement in New Drug Development*, 2nd edn, (eds W.S. Nimmo and G.T. Tucker), John Wiley, London (in press).

30. Thompson, K.R. (1991) *Harvard Hospital and its Volunteers*, Danny Howell Books, Warminster, Wiltshire.

31. Tucker, G.T. (1991) Clinical measurement and its interpretation in drug evaluation, in *Clinical Measurement in New Drug Evaluation*, (eds W.S. Nimmo and G.T. Tucker), Wolfe, London, pp. 1–4.

32. Tucker, G.T. (1992) Drug–drug interactions – inhibition of metabolism, *in Drug–drug and Drug–food Interactions*, (eds D.D. Breimer and F.W.H. Merkus), Center for Bio-Pharmaceutical Sciences, Leiden University, pp. 57–71.

33. Tyrrell, D.A.J. (1965) *Common Colds and Related Diseases*, Edward Arnold, London.

34. Tyrrell, D.A.J., Bynoe, M.L., Buckland, F.E. and Hayflick, L. (1962) The cultivation in human embryo cells of a virus (D.C.) causing colds in man. *Lancet*, **ii**, 320–2.

35. Watson, N. and Wyld, P.J. (1992) The importance of general practitioner information in the selection of volunteers for Phase I clinical trials. *Br. J. Clin. Pharmacol.*, **33**, 197–9.

36. Watson, N., Wyld, P. and Nimmo, W.S. (1994) Adverse events in healthy volunteer studies (in press).

37. Wilkes, M.F., Woollen, B.H., Marsh, J.R. et al. (1993) Biological monitoring for pesticide exposure – the role of human volunteer studies. *Int. Arch. Occup. Environ. Health*, **65**, S189–92.

INDEX

Note: Page numbers in *italic* refer to figures and/or tables